An Introduction to
Chemical Engineering
Kinetics & Reactor Design

AN INTRODUCTION TO CHEMICAL ENGINEERING KINETICS & REACTOR DESIGN

CHARLES G. HILL, JR.
The University of Wisconsin

JOHN WILEY & SONS
New York   Chichester
Brisbane   Toronto
Singapore

To my family:
Parents, Wife, and Daughters

Copyright © 1977, by John Wiley & Sons, Inc.

Library of Congress Cataloging in Publication Data:

Hill, Charles G   1937–
    An introduction to chemical engineering kinetics
and reactor design.

    Bibliography: p.
    Includes indexes.
    1.   Chemical reaction, Rate of.   2.   Chemical
reactors—Design and construction.   I.   Title.
QD502.H54   660.2′83   77–8280
ISBN 0-471-39609-5

Printed in the United States of America

Preface

One feature that distinguishes the education of the chemical engineer from that of other engineers is an exposure to the basic concepts of chemical reaction kinetics and chemical reactor design. This textbook provides a judicious introductory level overview of these subjects. Emphasis is placed on the aspects of chemical kinetics and material and energy balances that form the foundation for the practice of reactor design.

The text is designed as a teaching instrument. It can be used to introduce the novice to chemical kinetics and reactor design and to guide him until he understands the fundamentals well enough to read both articles in the literature and more advanced texts with understanding. Because the chemical engineer who practices reactor design must have more than a nodding acquaintance with the chemical aspects of reaction kinetics, a significant portion of this textbook is devoted to this subject. The modern chemical process industry, which has played a significant role in the development of our technology-based society, has evolved because the engineer has been able to commercialize the laboratory discoveries of the scientist. To carry out the necessary scale-up procedures safely and economically, the reactor designer must have a sound knowledge of the chemistry involved. Modern introductory courses in physical chemistry usually do not provide the breadth or the in-depth treatment of reaction kinetics that is required by the chemical engineer who is faced with a reactor design problem. More advanced courses in kinetics that are taught by physical chemists naturally reflect the research interests of the individuals involved; they do not stress the transmittal of that information which is most useful to individuals engaged in the practice of reactor design. Seldom is significant attention paid to the subject of heterogeneous catalysis and to the key role that catalytic processes play in the industrial world.

Chapters 3 to 7 treat the aspects of chemical kinetics that are important to the education of a well-read chemical engineer. To stress further the chemical problems involved and to provide links to the real world, I have attempted where possible to use actual chemical reactions and kinetic parameters in the many illustrative examples and problems. However, to retain as much generality as possible, the presentations of basic concepts and the derivations of fundamental equations are couched in terms of the anonymous chemical species A, B, C, U, V, etc. Where it is appropriate, the specific chemical reactions used in the illustrations are reformulated in these terms to indicate the manner in which the generalized relations are employed.

Chapters 8 to 13 provide an introduction to chemical reactor design. We start with the concept of idealized reactors with specified mixing characteristics operating isothermally and then introduce complications such as the use of combinations of reactors, implications of multiple reactions, temperature and energy effects, residence time effects, and heat and mass transfer limitations that are often involved when heterogeneous catalysts are employed. Emphasis is placed on the fact that chemical reactor design represents a straightforward application of the bread and butter tools of the chemical engineer the material balance and the energy balance. The fundamental design equations in the second half of the text are algebraic descendents of the generalized material balance equation

$$\frac{\text{Rate of}}{\text{input}} = \frac{\text{Rate of}}{\text{output}} + \frac{\text{Rate of}}{\text{accumulation}} + \frac{\text{Rate of disappearance}}{\text{by reaction}} \qquad \text{(P.1)}$$

In the case of nonisothermal systems one must write equations of this form both for energy and for the chemical species of interest, and then solve the resultant equations simultaneously to characterize the effluent composition and the thermal effects associated with operation of the reactor. Although the material and energy balance equations are not coupled when no temperature changes occur in the reactor, the design engineer still must solve the energy balance equation to ensure that sufficient capacity for energy transfer is provided so that the reactor will indeed operate isothermally. The text stresses that the design process merely involves an extension of concepts learned previously. The application of these concepts in the design process involves equations that differ somewhat in mathematical form from the algebraic equations normally encountered in the introductory material and energy balance course, but the underlying principles are unchanged. The illustrations involved in the reactor design portion of the text are again based where possible on real chemical examples and actual kinetic data. The illustrative problems in Chapter 13 indicate the facility with which the basic concepts may be rephrased or applied in computer language, but this material is presented only after the student has been thoroughly exposed to the concepts involved and has learned to use them in attacking reactor design problems. I believe that the subject of computer-aided design should be deferred to graduate courses in reactor design and to more advanced texts.

The notes that form the basis for the bulk of this textbook have been used for several years in the undergraduate course in chemical kinetics and reactor design at the University of Wisconsin. In this course, emphasis is placed on Chapters 3 to 6 and 8 to 12, omitting detailed class discussions of many of the mathematical derivations. My colleagues and I stress the necessity for developing a "seat of the pants" feeling for the phenomena involved as well as an ability to analyze quantitative problems in terms of design framework developed in the text.

The material on catalysis and heterogeneous reactions in Chapters 6, 12, and 13 is a useful framework for an intermediate level graduate course in catalysis and chemical reactor design. In the latter course emphasis is placed on developing the student's ability to analyze critically actual kinetic data obtained from the literature in order to acquaint him with many of the traps into which the unwary may fall. Some of the problems in Chapter 12 and the illustrative case studies in Chapter 13 have evolved from this course.

Most of the illustrative examples and problems in the text are based on actual data from the kinetics literature. However, in many cases, rate constants, heats of reaction, activation energies, and other parameters have been converted to SI units from various other systems. To be able to utilize the vast literature of kinetics for reactor design purposes, one must develop a facility for making appropriate transformations of parameters from one system of units to another. Consequently, I have chosen not to employ SI units exclusively in this text.

Like other authors of textbooks for undergraduates, I owe major debts to the instructors who first introduced me to this subject matter and to the authors and researchers whose publications have contributed to my understanding of the subject. As a student, I benefited from instruction by R. C. Reid, C. N. Satterfield, and I. Amdur and from exposure to the texts of Walas, Frost and Pearson, and Benson. Some of the material in Chapter 6 has been adapted with permission from the course notes of Professor C. N. Satterfield of MIT, whose direct and indirect influence on my thinking is further evident in some of the data interpretation problems in Chapters 6 and 12. As an instructor I have found the texts by Levenspiel and Smith to be particularly useful at the undergraduate level; the books by Denbigh, Laidler, Hinshelwood, Aris, and Kramers and Westerterp have also helped to shape my views of chemical kinetics and reactor design. I have tried to use the best ideas of these individuals and the approaches that I have found particularly useful in the classroom in the synthesis of this textbook. A major attraction of this subject is that there are many alternative ways of viewing the subject. Without an exposure to several viewpoints, one cannot begin to grasp the subject in its entirety. Only after such exposure, bombardment by the probing questions of one's students, and much contemplation can one begin to synthesize an individual philosophy of kinetics. To the humanist it may seem a misnomer to talk in terms of a philosophical approach to kinetics, but to the individuals who have taken kinetics courses at different schools or even in different departments and to the individuals who have read widely in the kinetics literature, it is evident that several such approaches do exist and that specialists in the area do have individual philosophies that characterize their approach to the subject.

The stimulating environment provided by the students and staff of the Chemical Engineering Department at the University of Wisconsin has provided much of the necessary encouragement and motivation for writing this textbook. The Department has long been a fertile environment for research and textbook writing in the area of chemical kinetics and reactor design. The text by O. A. Hougen and K. M. Watson represents a classic pioneering effort to establish a rational approach to the subject from the viewpoint of the chemical engineer. Through the years these individuals and several members of our current staff have contributed significantly to the evolution of the subject. I am indebted to my colleagues, W. E. Stewart, S. H. Langer, C. C. Watson, R. A. Grieger, S. L. Cooper, and T. W. Chapman, who have used earlier versions of this textbook as class notes or commented thereon, to my benefit. All errors are, of course, my own responsibility.

I am grateful to the graduate students who have served as my teaching assistants and who have brought to my attention various ambiguities in the text or problem statements. These include J. F. Welch, A. Yu, R. Krug, E. Guertin, A. Kozinski, G. Estes, J. Coca, R. Safford, R. Harrison, J. Yurchak, G. Schrader, A. Parker, T. Kumar, and A. Spence. I also thank the students on whom I have tried out my ideas. Their response to the subject matter has provided much of the motivation for this textbook.

Since drafts of this text were used as course notes, the secretarial staff of the department, which includes D. Peterson, C. Sherven, M. Sullivan, and M. Carr, deserves my warmest thanks for typing this material. I am also very appreciative of my wife's efforts in typing the final draft of this manuscript and in correcting the galley proofs. Vivian Kehane, Jacqueline Lachmann, and Peter Klein of Wiley were particularly helpful in transforming my manuscript into this text.

   My wife and children have at times been neglected during the preparation of this textbook; for their cooperation and inspiration I am particularly grateful.

Madison, Wisconsin   CHARLES G. HILL, Jr.

Supplementary References

Since this is an introductory text, all topics of potential interest cannot be treated to the depth that the reader may require. Consequently, a number of useful supplementary references are listed below.

A. References Pertinent to the Chemical Aspects of Kinetics

1. I. Amdur and G. G. Hammes, *Chemical Kinetics: Principles and Selected Topics*, McGraw-Hill, New York, 1966.

2. S. W. Benson, *The Foundations of Chemical Kinetics*, McGraw-Hill, New York, 1960.

3. M. Boudart, *Kinetics of Chemical Processes*, Prentice-Hall, Englewood Cliffs, N.J., 1968.

4. A. A. Frost and R. G. Pearson, *Kinetics and Mechanism*, Wiley, New York, 1961.

5. W. C. Gardiner, Jr., *Rates and Mechanisms of Chemical Reactions*, Benjamin, New York, 1969.

6. K. J. Laidler, *Chemical Kinetics*, McGraw-Hill, New York, 1965.

B. References Pertinent to the Engineering or Reactor Design Aspects of Kinetics

1. R. Aris, *Introduction to the Analysis of Chemical Reactors*, Prentice-Hall, Englewood Cliffs, N.J., 1965.

2. J. J. Carberry, *Chemical and Catalytic Reaction Engineering*, McGraw-Hill, New York, 1976.

3. A. R. Cooper and G. V. Jeffreys, *Chemical Kinetics and Reactor Design*, Oliver and Boyd, Edinburgh, 1971.

4. H. W. Cremer (Editor), *Chemical Engineering Practice*, Volume 8, *Chemical Kinetics*, Butterworths, London, 1965.

5. K. G. Denbigh and J. C. R. Turner, *Chemical Reactor Theory*, Second Edition, Cambridge University Press, London, 1971.

6. H. S. Fogler, *The Elements of Chemical Kinetics and Reactor Calculations*, Prentice-Hall, Englewood Cliffs, N.J., 1974.

7. H. Kramers and K. R. Westerterp, *Elements of Chemical Reactor Design and Operation*, Academic Press, New York, 1963.

8. O. Levenspiel, *Chemical Reaction Engineering*, Second Edition, Wiley, New York, 1972.

9. E. E. Petersen, *Chemical Reaction Analysis*, Prentice-Hall, Englewood Cliffs, N.J., 1965.

10. C. N. Satterfield, *Mass Transfer in Heterogeneous Catalysis*," MIT Press, Cambridge, Mass., 1970.

11. J. M. Smith, *Chemical Engineering Kinetics*, Second Edition, McGraw-Hill, New York, 1970.

C. G. H., Jr.

Contents

1 Stoichiometric Coefficients and Reaction Progress Variables

1.0 INTRODUCTION

Without chemical reaction our world would be a barren planet. No life of any sort would exist. Even if we exempt the fundamental reactions involved in life processes from our proscription on chemical reactions, our lives would be extremely different from what they are today. There would be no fire for warmth and cooking, no iron and steel with which to fashion even the crudest implements, no synthetic fibers for clothing, and no engines to power our vehicles.

One feature that distinguishes the chemical engineer from other types of engineers is the ability to analyze systems in which chemical reactions are occurring and to apply the results of his analysis in a manner that benefits society. Consequently, chemical engineers must be well acquainted with the fundamentals of chemical kinetics and the manner in which they are applied in chemical reactor design. This textbook provides a systematic introduction to these subjects.

Chemical kinetics deals with quantitative studies of the rates at which chemical processes occur, the factors on which these rates depend, and the molecular acts involved in reaction processes. A description of a reaction in terms of its constituent molecular acts is known as the *mechanism* of the reaction. Physical and organic chemists are primarily interested in chemical kinetics for the light that it sheds on molecular properties. From interpretations of macroscopic kinetic data in terms of molecular mechanisms, they can gain insight into the nature of reacting systems, the processes by which chemical bonds are made and broken, and the structure of the resultant product. Although chemical engineers find the concept of a reaction mechanism useful in the correlation, interpolation, and extrapolation of rate data, they are more concerned with applications of chemical kinetics in the development of profitable manufacturing processes.

Chemical engineers have traditionally approached kinetics studies with the goal of describing the behavior of reacting systems in terms of macroscopically observable quantities such as temperature, pressure, composition, and Reynolds number. This empirical approach has been very fruitful in that it has permitted chemical reactor technology to develop to a point that far surpasses the development of theoretical work in chemical kinetics.

The dynamic viewpoint of chemical kinetics may be contrasted with the essentially static viewpoint of thermodynamics. A kinetic system is a system in unidirectional movement toward a condition of thermodynamic equilibrium. The chemical composition of the system changes continuously with time. A system that is in thermodynamic equilibrium, on the other hand, undergoes no net change with time. The thermodynamicist is interested only in the initial and final states of the system and is not concerned with the time required for the transition or the molecular processes involved therein; the chemical kineticist is concerned primarily with these issues.

In principle one can treat the thermodynamics of chemical reactions on a kinetic basis by recognizing that the equilibrium condition corresponds to the case where the rates of the forward and reverse reactions are identical. In this sense kinetics is the more fundamental science. Nonetheless, thermodynamics provides much vital information to the kineticist and to the reactor designer. In particular, the first step in determining the economic feasibility of producing a given material from a given reactant feed stock should be the determination of the product yield at equilibrium at the conditions of the reactor outlet. Since this composition represents the goal toward which the kinetic process is moving, it places a maximum limit on the product yield that may be obtained. Chemical engineers must also use thermodynamics to determine heat transfer requirements for proposed reactor configurations.

1.1 BASIC STOICHIOMETRIC CONCEPTS

1.1.1 Stoichiometric Coefficients

Consider the following general reaction.

$$bB + cC + \cdots \rightleftharpoons sS + tT + \cdots \quad (1.1.1)$$

where b, c, s, and t are the stoichiometric coefficients of the species B, C, S, and T, respectively. We define generalized stoichiometric coefficients v_i for the above reaction by rewriting it in the following manner.

$$0 = v_B B + v_C C + \cdots + v_S S + v_T T + \cdots \quad (1.1.2)$$

where

$$v_B = -b \qquad v_S = s$$
$$v_C = -c \qquad v_T = t$$

The generalized stoichiometric coefficients are defined as positive quantities for the products of the reaction and as negative quantities for the reactants. The coefficients of species that are neither produced nor consumed by the indicated reaction are taken to be zero. Equation 1.1.2 has been written in inverted form with the zero first to emphasize the use of this sign convention, even though this inversion is rarely used in practice.

One may further generalize equation 1.1.2 by rewriting it as

$$0 = \sum_i v_i A_i \qquad (1.1.3)$$

where the sum is taken over all components A_i present in the system.

There are, of course, many equivalent ways of writing the stoichiometric equation for a reaction. For example, one could write the carbon monoxide oxidation reaction in our notation as

$$0 = 2CO_2 - 2CO - O_2$$

instead of in the more conventional form, which has the reactants on the left side and the products on the right side.

$$2CO + O_2 = 2CO_2$$

This second form is preferred, provided that one keeps the proper sign convention for the stoichiometric coefficients in mind.

$$v_{CO} = -2 \qquad v_{O_2} = -1 \qquad v_{CO_2} = 2$$

Alternatively, the reaction could be written as

$$0 = CO_2 - CO - \tfrac{1}{2}O_2$$

with

$$v_{CO} = -1 \qquad v_{O_2} = -\tfrac{1}{2} \qquad v_{CO_2} = 1$$

The choice is a matter of personal convenience. The essential point is that the ratios of the stoichiometric coefficients are unique for a given reaction {i.e., $v_{CO}/v_{O_2} = (-2/-1) = [-1/(-1/2)] = 2$}. Since the reaction stoichiometry can be expressed in various ways, one must always write down a stoichiometric equation for the reaction under study during the initial stages of the analysis and base subsequent calculations on this reference equation. If a consistent set of stoichiometric coefficients is used throughout the calculations, the results can be readily understood and utilized by other workers in the field.

1.1.2 Reaction Progress Variables

In order to measure the progress of a reaction it is necessary to define a parameter, which is a measure of the degree of conversion of the reactants. We will find it convenient to use the concept of the *extent* or degree of *advancement* of reaction. This concept has its origins in the thermodynamic literature, dating back to the work of de Donder (1).

Consider a closed system (i.e., one in which there is no exchange of matter between the system and its surroundings) where a single chemical reaction may occur according to equation 1.1.3. Initially there are n_{i0} moles of constituent A_i present in the system. At some later time there are n_i moles of species A_i present. At this time the molar extent of reaction is defined as

$$\xi = \frac{n_i - n_{i0}}{\nu_i} \quad (1.1.4)$$

This equation is valid for all species A_i, a fact that is a consequence of the law of definite proportions. The molar extent of reaction ξ is a time-dependent extensive variable that is measured in moles. It is a useful measure of the progress of the reaction because it is not tied to any particular species A_i. Changes in the mole numbers of two species j and k can be related to one another by eliminating ξ between two expressions that may be derived from equation 1.1.4.

$$n_k = n_{k0} + \left(\frac{\nu_k}{\nu_j}\right)(n_j - n_{j0}) \quad (1.1.5)$$

If more than one chemical reaction is possible, an extent may be defined for each reaction. If ξ_k is the extent of the kth reaction, and ν_{ki} the stoichiometric coefficient of species i in reaction k, the total change in the number of moles of species A_i because of R reactions is given by

$$n_i - n_{i0} = \sum_{k=1}^{k=R} \nu_{ki}\xi_k \quad (1.1.6)$$

Another advantage of using the concept of extent is that it permits one to specify uniquely the rate of a given reaction. This point is discussed in Section 3.0. The major drawback of the concept is that the extent is an extensive variable and consequently is proportional to the mass of the system being investigated.

The fraction conversion f is an intensive measure of the progress of a reaction, and it is a variable that is simply related to the extent of reaction. The fraction conversion of a reactant A_i in a closed system in which only a single reaction is occurring is given by

$$f = \frac{n_{i0} - n_i}{n_{i0}} = 1 - \frac{n_i}{n_{i0}} \quad (1.1.7)$$

The variable f depends on the particular species chosen as a reference substance. In general, the initial mole numbers of the reactants do not constitute simple stoichiometric ratios, and the number of moles of product that may be formed is limited by the amount of one of the reactants present in the system. If the extent of reaction is not limited by thermodynamic equilibrium constraints, this limiting reagent is the one that determines the maximum possible value of the extent of reaction (ξ_{max}). We should refer our fractional conversions to this stoichiometrically limiting reactant if f is to lie between zero and unity. Consequently, the treatment used in subsequent chapters *will define fractional conversions in terms of the limiting reactant.*

One can relate the extent of reaction to the fraction conversion by solving equations 1.1.4 and 1.1.7 for the number of moles of the limiting reagent n_{lim} and equating the resultant expressions.

$$n_{lim} = n_{lim,0} + \nu_{lim}\xi = n_{lim,0}(1 - f) \quad (1.1.8)$$

or

$$\xi = -\frac{fn_{lim,0}}{\nu_{lim}} \quad \text{and} \quad \xi_{max} = -\frac{n_{lim,0}}{\nu_{lim}} \quad (1.1.9)$$

In some cases the extent of reaction is limited by the position of chemical equilibrium, and this extent (ξ_e) will be less than ξ_{max}. However, in many cases ξ_e is approximately equal to ξ_{max}. In these cases the equilibrium for the reaction highly favors formation of the products, and only an *extremely small* quantity of the limiting reagent remains in the system at equilibrium. We will classify these reactions as *irreversible.* When the extent of reaction at equilibrium differs measurably from ξ_{max}, we will classify the reaction involved as *reversible*. From a thermodynamic point of view, all reactions are reversible. However, when one is analyzing a reacting system, it is often convenient to neglect the reverse reaction in order to simplify the analysis. For "irreversible" reac-

tions, one then arrives at a result that is an extremely good approximation to the correct answer.

LITERATURE CITATION

1. De Donder, Th., *Leçons de Thermodynamique et de Chemie-Physique*, Paris, Gauthier-Villus, 1920.

2 Thermodynamics of Chemical Reactions

2.0 INTRODUCTION

The science of chemical kinetics is concerned primarily with chemical changes and the energy and mass fluxes associated therewith. Thermodynamics, on the other hand, is concerned with equilibrium systems . . . systems that are undergoing *no net change* with time. This chapter will remind the student of the key thermodynamic principles with which he should be familiar. Emphasis is placed on calculations of equilibrium extents of reaction and enthalpy changes accompanying chemical reactions.

Of primary consideration in any discussion of chemical reaction equilibria is the constraints on the system in question. If calculations of equilibrium compositions are to be in accord with experimental observations, one must include in his or her analysis all reactions that occur at appreciable rates relative to the time frame involved. Such calculations are useful in that the equilibrium conversion provides a standard against which the actual performance of a reactor may be compared. For example, if the equilibrium yield of a given reactant system is 75%, and the observed yield from a given reactor is only 30%, it is obviously possible to obtain major improvements in the process yield. On the other hand, if the process yield were close to 75%, the potential improvement in the yield is minimal and additional efforts aimed at improving the yield may not be warranted. Without a knowledge of the equilibrium yield, one might be tempted to look for catalysts giving higher yields when, in fact, the present catalyst provides a sufficiently rapid approach to equilibrium.

The basic criterion for the establishment of chemical reaction equilibrium is that

$$\sum_i v_i \mu_i = 0 \qquad (2.0.1)$$

where the μ_i are the chemical potentials of the various species in the reaction mixture. If r reactions may occur in the system and equilibrium is established with respect to each of these reactions, it is required that

$$\sum_i v_{ki}\mu_i = 0 \qquad k = 1, 2, \ldots, r \quad (2.0.2)$$

These equations are equivalent to a requirement that the Gibbs free energy change for each reaction (ΔG) be zero at equilibrium.

$$\Delta G = \sum v_i \mu_i = 0 \text{ at equilibrium} \quad (2.0.3)$$

2.1 CHEMICAL POTENTIALS AND STANDARD STATES

The activity a_i of species i is related to its chemical potential by

$$\mu_i = \mu_i^0 + RT \ln a_i \qquad (2.1.1)$$

where

R is the gas constant

T is the absolute temperature

μ_i^0 is the standard chemical potential of species i in a reference state where its activity is taken as unity

The choice of the standard state is largely arbitrary and is based primarily on experimental convenience and reproducibility. The temperature of the standard state is the same as that of the system under investigation. In some cases, the standard state may represent a hypothetical condition that cannot be achieved experimentally, but that is susceptible to calculations giving reproducible results. Although different standard states may be chosen for various species, *throughout any set of calculations it is important that the standard state of a component be kept the same so as to minimize possibilities for error.*

Certain choices of standard states have found such widespread use that they have achieved

5

the status of recognized conventions. In particular, those listed in Table 2.1 are used in calculations dealing with *chemical reaction equilibria*. In all cases the temperature is the same as that of the reaction mixture.

Table 2.1
Standard States for Chemical Potential Calculations (for Use in Studies of Chemical Reaction Equilibria)

State of aggregation	Standard state
Gas	Pure gas at unit fugacity (for an ideal gas the fugacity is unity at 1 atm pressure; this is a valid approximation for most real gases).
Liquid	Pure liquid in the most stable form at 1 atm
Solid	Pure solid in the most stable form at 1 atm.

Once the standard states for the various species have been established, one can proceed to calculate a number of standard energy changes for processes involving a change from reactants, all in their respective standard states, to products, all in their respective standard states. For example, the Gibbs free energy change for this process is

$$\Delta G^0 = \sum v_i \mu_i^0 \qquad (2.1.2)$$

where the superscript zero on ΔG emphasizes the fact that this is a process involving standard states for both the final and initial conditions of the system. In a similar manner one can determine standard enthalpy (ΔH^0) and standard entropy changes (ΔS^0) for this process.

2.2 ENERGY EFFECTS ASSOCIATED WITH CHEMICAL REACTIONS

Since chemical reactions involve the formation, destruction, or rearrangement of chemical bonds, they are invariably accompanied by changes in the enthalpy and Gibbs free energy of the system. The enthalpy change on reaction provides information that is necessary for any engineering analysis of the system in terms of the first law of thermodynamics. It is also useful in determining the effect of temperature on the equilibrium constant of the reaction and thus on the reaction yield. The Gibbs free energy is useful in determining whether or not chemical equilibrium exists in the system being studied and in determining how changes in process variables can influence the yield of the reaction.

In chemical kinetics there are two types of processes for which one calculates changes in these energy functions.

1. A chemical process whereby reactants, each in its standard state, are converted into products, each in its standard state, under conditions such that the initial temperature of the reactants is equal to the final temperature of the products.
2. An actual chemical process as it might occur under either equilibrium or nonequilibrium conditions in a chemical reactor.

One must be very careful not to confuse actual energy effects with those that are associated with the process whose initial and final states are the standard states of the reactants and products respectively.

In order to have a consistent basis for comparing different reactions and to permit the tabulation of thermochemical data for various reaction systems, it is convenient to define enthalpy and Gibbs free energy changes for standard reaction conditions. These conditions involve the use of stoichiometric amounts of the various reactants (each in its standard state at some temperature T). The reaction proceeds by some unspecified path to end up with complete conversion of reactants to the various products (each in its standard state at the same temperature T).

The enthalpy and Gibbs free energy changes for a standard reaction are denoted by the symbols ΔH^0 and ΔG^0, where the superscript zero is used to signify that a "standard" reaction is involved. Use of these symbols is restricted to the case where the extent of reaction is 1 mole for the reaction as written. The remaining discussion in this chapter refers to this basis.

Because G and H are state functions, changes in these quantities are independent of whether the reaction takes place in one or in several steps. Consequently, it is possible to tabulate data for relatively few reactions and use this data in the calculation of ΔG^0 and ΔH^0 for other reactions. In particular, one tabulates data for the standard reactions that involve the formation of a compound from its elements. One may then consider a reaction involving several compounds as being an appropriate algebraic sum of a number of elementary reactions, each of which involves the formation of one compound. The dehydration of n-propanol

$$CH_3CH_2CH_2OH(\ell) \rightarrow$$
$$H_2O(\ell) + CH_3CH{=}CH_2(g)$$

may be considered as the algebraic sum of the following series of reactions.

called the enthalpy (or heat) of formation of the compound and is denoted by the symbol ΔH_f^0.

Thus,

$$\Delta H_{overall}^0 = \Delta H_{f\,water(\ell)}^0 + \Delta H_{f\,propylene}^0 - \Delta H_{f\,propanol(\ell)}^0 \tag{2.2.3}$$

and

$$\Delta G_{overall}^0 = \Delta G_{f\,water(\ell)}^0 + \Delta G_{f\,propylene}^0 - \Delta G_{f\,propanol(\ell)}^0 \tag{2.2.4}$$

where ΔG_f^0 refers to the standard Gibbs free energy of formation.

This example illustrates the principle that values of ΔG^0 and ΔH^0 may be calculated from values of the enthalpies and Gibbs free energies of formation of the products and reactants. In more general form,

$$\Delta H^0 = \sum_i \nu_i \, \Delta H_{f,i}^0 \tag{2.2.5}$$

$$\Delta G^0 = \sum_i \nu_i \, \Delta G_{f,i}^0 \tag{2.2.6}$$

When an element enters into a reaction, its standard Gibbs free energy and standard enthalpy of formation are taken as zero if its state of aggregation is that selected as the basis for

$CH_3CH_2CH_2OH(\ell) \rightarrow 3C(\beta\ graphite) + 4H_2(g) + \frac{1}{2}O_2(g)$	ΔH_1^0	ΔG_1^0
$3C(\beta\ graphite) + 3H_2(g) \rightarrow CH_3CH{=}CH_2(g)$	ΔH_2^0	ΔG_2^0
$H_2(g) + \frac{1}{2}O_2(g) \rightarrow H_2O(\ell)$	ΔH_3^0	ΔG_3^0
$CH_3CH_2CHOH(\ell) \rightarrow H_2O(\ell) + CH_3CH{=}CH_2(g)$	ΔH^0	ΔG^0

For the overall reaction,

$$\Delta H^0 = \Delta H_1^0 + \Delta H_2^0 + \Delta H_3^0 \tag{2.2.1}$$

$$\Delta G^0 = \Delta G_1^0 + \Delta G_2^0 + \Delta G_3^0 \tag{2.2.2}$$

However, each of the individual reactions involves the formation of a compound from its elements or the decomposition of a compound into those elements. The standard enthalpy change of a reaction that involves the formation of a compound from its elements is

the determination of the standard Gibbs free energy and enthalpy of formation of its compounds. If ΔH^0 is negative, the reaction is said to be *exothermic*; if ΔH^0 is positive, the reaction is said to be *endothermic*.

It is not necessary to tabulate values of ΔG^0 or ΔH^0 for all conceivable reactions. It is sufficient to tabulate values of these parameters only for the reactions that involve the formation of a compound from its elements. The problem of data compilation is further simplified by the fact that it is unnecessary to record ΔG_f^0 and ΔH_f^0 at all temperatures, because of the relations that exist between these quantities and other thermodynamic properties of the reactants and products. The convention that is commonly accepted in engineering practice today is to report values of standard enthalpies of formation and Gibbs free energies of formation at 25 °C (298.16 °K) or at 0 °K. The problem of calculating a value for ΔG^0 or ΔH^0 at temperature T thus reduces to one of determining values of ΔG_f^0 and ΔH_f^0 at 25 °C or 0 °K and then adjusting the value obtained to take into account the effects of temperature on the property in question. The appropriate techniques for carrying out these adjustments are indicated below.

The effect of temperature on ΔH^0 is given by

$$\Delta H_T^0 = \Delta H_{298.16}^0 + \int_{298.16\,°K}^{T} \left(\sum_i v_i C_{p,i}^0 \right) dT$$

$$(2.2.7)$$

where $C_{p,i}^0$ is the constant pressure heat capacity of species i in its standard state.

In many cases the magnitude of the last term on the right side of equation 2.2.7 is very small compared to $\Delta H_{298.16}^0$. However, if one is to be able to evaluate properly the standard heat of reaction at some temperature other than 298.16 °K, one must know the constant pressure heat capacities of the reactants and the products as functions of temperature as well as the heat of reaction at 298.16 °K. Data of this type and techniques for estimating these properties are contained in the references in Section 2.3.

The most useful expression for describing the variation of standard Gibbs free energy changes with temperature is:

$$\left[\frac{\partial \left(\dfrac{\Delta G^0}{T} \right)}{\partial T} \right]_P = -\frac{\Delta H^0}{T^2} \qquad (2.2.8)$$

In Section 2.5 we will see that the equilibrium constant for a chemical reaction is simply related to $\Delta G^0/T$ and that equation 2.2.8 is useful in determining how this parameter varies with temperature. If one desires to obtain an expression for ΔG^0 itself as a function of temperature, equation 2.2.7 may be integrated to give ΔH^0 as a function of temperature. This relation may then be used with equation 2.2.8 to arrive at the desired relation.

The effect of pressure on ΔG^0 and ΔH^0 depends on the choice of standard states employed. When the standard state of each component of the reaction system is taken at 1 atm pressure, whether the species in question is a gas, liquid, or solid, the values of ΔG^0 and ΔH^0 refer to a process that starts and ends at 1 atm. For this choice of standard states, the *values of* ΔG^0 *and* ΔH^0 *are independent of the system pressure at which the reaction is actually carried out.* It is important to note in this connection that we are calculating the enthalpy change for a hypothetical process, not for the actual process as it occurs in nature. This choice of standard states at 1 atm pressure is the convention that is customarily adopted in the analysis of chemical reaction equilibria.

For cases where the standard state pressure for the various species is chosen as that of the system under investigation, changes in this variable will alter the values of ΔG^0 and ΔH^0. In such cases thermodynamic analysis indicates that

$$\Delta H_P^0 = \Delta H_{1\,\text{atm}}^0 + \sum_i v_i \int_1^P \left[V_i - T \left(\frac{\partial V_i}{\partial T} \right)_P \right] dP$$

$$(2.2.9)$$

where V_i is the molal volume of component i in its standard state and where each integral is evaluated for the species in question along an isothermal path. The term in brackets represents the variation of the enthalpy of a component with pressure at constant temperature $(\partial H/\partial P)_T$.

It should be emphasized that the choice of standard states implied by equation 2.2.9 is *not* that which is conventionally used in the analysis of chemically reacting systems. Furthermore, *in the vast majority of cases the summation term on the right side of the equation is very small compared to the magnitude of $\Delta H^0_{1\,atm}$ and, indeed, is usually considerably smaller than the uncertainty in this term.*

The Gibbs free energy analog of equation 2.2.9 is

$$\Delta G^0_P = \Delta G^0_{1\,atm} + \sum_i v_i \int_1^P V_i\, dP \quad (2.2.10)$$

where the integral is again evaluated along an isothermal path. For cases where the species involved is a condensed phase, V_i will be a very small quantity and the contribution of this species to the summation will be quite small unless the system pressure is extremely high. For ideal gases, the integral may be evaluated directly as $RT \ell n\, P$. For nonideal gases the integral is equal to $RT \ell n\, f_i^0$, where f_i^0 is the fugacity of pure species i at pressure P.

2.3 SOURCES OF THERMOCHEMICAL DATA

Thermochemical data for several common species are contained in Appendix A. Other useful standard references are listed below.

1. F. D. Rossini, et al., *Selected Values of Physical and Thermodynamic Properties of Hydrocarbons and Related Compounds*, Carnegie Press, Pittsburgh, 1953; also loose-leaf supplements. Data compiled by Research Project 44 of the American Petroleum Institute.

2. F. D. Rossini, et al., "Selected Values of Chemical Thermodynamic Properties," National Bureau of Standards, Circular 500 and Supplements, 1952.

3. E. W. Washburn (Editor), *International Critical Tables*, McGraw-Hill, New York, 1926.

4. T. Hilsenrath, et al., "Thermal Properties of Gases," National Bureau of Standards Circular 564, 1955.

5. D. R. Stull and G. C. Sinke, "Thermodynamic Properties of the Elements," *Adv. Chem. Ser.*, 18, 1956.

6. *Landolt-Börnstein Tabellen*, Sechste Auflage, Band II, Teil 4, Springer-Verlag, Berlin, 1961.

7. *Janaf Thermochemical Tables*, D. R. Stull, Project Director, PB 168370, Clearinghouse for Federal Scientific and Technical Information, 1965.

The following references contain techniques for estimating thermochemical data.

1. R. C. Reid and T. K. Sherwood, *The Properties of Gases and Liquids*, Second Edition, McGraw-Hill, New York, 1966.

2. S. W. Benson, *Thermochemical Kinetics*, Wiley, New York, 1968.

3. G. J. Janz, *Estimation of Thermodynamic Properties of Organic Compounds*, Academic Press, New York, 1958.

2.4 THE EQUILIBRIUM CONSTANT AND ITS RELATION TO ΔG^0

The basic criterion for equilibrium with respect to a given chemical reaction is that the Gibbs free energy change associated with the progress of the reaction be zero.

$$\Delta G = \sum_i v_i \mu_i = 0 \quad (2.4.1)$$

The standard Gibbs free energy change for a reaction refers to the process wherein the reaction proceeds isothermally, starting with stoichiometric quantities of reactants each in its standard state of unit activity and ending with products each at unit activity. In general it is nonzero and given by

$$\Delta G^0 = \sum_i v_i \mu_i^0 \quad (2.4.2)$$

Subtraction of equation 2.4.2 from equation 2.4.1 gives

$$\Delta G - \Delta G^0 = \sum_i v_i(\mu_i - \mu_i^0) \quad (2.4.3)$$

This equation may be rewritten in terms of the activities of the various species by making use of equation 2.1.1.

$$\Delta G - \Delta G^0 = RT \sum_i v_i \ell n\, a_i = RT \ell n \left(\prod_i a_i^{v_i} \right)$$
$$(2.4.4)$$

where the $\prod_i$ symbol denotes a product over i species of the term that follows.

For a general reaction of the form

$$bB + cC + \cdots \rightleftharpoons sS + tT + \cdots \quad (2.4.5)$$

the above equations become:

$$\Delta G - \Delta G^0 = RT \ln \left(\frac{a_S^s a_T^t \cdots}{a_B^b a_C^c \cdots} \right) \quad (2.4.6)$$

For a system at equilibrium, $\Delta G = 0$, and

$$\Delta G^0 = -RT \ln \left[\frac{a_S^s a_T^t \cdots}{a_B^b a_C^c \cdots} \right] = -RT \ln K_a$$

$$(2.4.7)$$

where the equilibrium constant for the reaction (K_a) at temperature T is defined as the term in brackets. The subscript a has been used to emphasize that an equilibrium constant is properly written as a product of the activities raised to appropriate powers. Thus, in general,

$$K_a = \prod_i a_i^{\nu_i} = e^{-\Delta G^0/RT} \quad (2.4.8)$$

As equation 2.4.8 indicates, the equilibrium constant for a reaction is determined by the temperature and the standard Gibbs free energy change (ΔG^0) for the process. The latter quantity in turn depends on temperature, the definitions of the standard states of the various components, and the stoichiometric coefficients of these species. Consequently, in assigning a numerical value to an equilibrium constant, one must be careful to specify the three parameters mentioned above in order to give meaning to this value. Once one has thus specified the point of reference, this value may be used to calculate the equilibrium composition of the mixture in the manner described in Sections 2.6 to 2.9.

2.5 EFFECTS OF TEMPERATURE AND PRESSURE CHANGES ON THE EQUILIBRIUM CONSTANT FOR A REACTION

Equilibrium constants are quite sensitive to temperature changes. A quantitative description

of the influence of temperature changes is readily obtained by combining equations 2.2.8 and 2.4.7.

$$\left[\frac{\partial \left(-\frac{\Delta G^0}{T} \right)}{\partial T} \right]_P = \left(\frac{R \partial \ln K_a}{\partial T} \right)_P = \frac{\Delta H^0}{T^2} \quad (2.5.1)$$

or

$$\left(\frac{\partial \ln K_a}{\partial T} \right)_P = \frac{\Delta H^0}{RT^2} \quad (2.5.2)$$

and

$$\left(\frac{\partial \ln K_a}{\partial \left(\frac{1}{T} \right)} \right)_P = -\frac{\Delta H^0}{R} \quad (2.5.3)$$

For cases where ΔH^0 is essentially independent of temperature, plots of $\ln K_a$ versus $1/T$ are linear with slope $-(\Delta H^0/R)$. For cases where the heat capacity term in equation 2.2.7 is appreciable, this equation must be substituted in either equation 2.5.2 or equation 2.5.3 in order to determine the temperature dependence of the equilibrium constant. For exothermic reactions $(\Delta H^0$ negative) the equilibrium constant decreases with increasing temperature, while for endothermic reactions the equilibrium constant increases with increasing temperature.

For cases where the standard states of the reactants and products are chosen as 1 atm, the value of ΔG^0 is pressure independent. Consequently, equation 2.4.7 indicates that K_a is also pressure independent for this choice of standard states. For the unconventional choice of standard states discussed in Section 2.2, equations 2.4.7 and 2.2.10 may be combined to give the effect of pressure on K_a.

$$\left(\frac{\partial \ln K_a}{\partial P} \right)_T = -\frac{\sum_i \nu_i V_i}{RT} \quad (2.5.4)$$

where the V_i are the standard state molal

volumes of the reactants and products. However, this choice of standard states is extremely rare in engineering practice.

2.6 DETERMINATION OF EQUILIBRIUM COMPOSITIONS

The basic equation from which one calculates the composition of an equilibrium mixture is equation 2.4.7.

$$\Delta G^0 = -RT \ln K_a$$

$$= -RT \ln \left(\frac{a_S^s a_T^t}{a_B^b a_C^c} \right) \qquad (2.6.1)$$

In a system that involves gaseous components, one normally chooses as the standard state the pure component gases, each at unit fugacity (essentially 1 atm). The activity of a gaseous species B is then given by

$$a_B = \frac{\hat{f}_B}{f_{B,SS}} = \frac{\hat{f}_B}{1} = \hat{f}_B \qquad (2.6.2)$$

where $\hat{f}_B$ is the fugacity of species B as it exists in the reaction mixture and $f_{B,ss}$ is the fugacity of species B in its standard state.

The fugacity of species B in an ideal solution of gases is given by the Lewis and Randall rule

$$\hat{f}_B = y_B f_B^0 \qquad (2.6.3)$$

where y_B is the mole fraction B in the gaseous phase and f_B^0 is the fugacity of pure component B evaluated at the temperature and total pressure (P) of the reaction mixture. Alternatively,

$$\hat{f}_B = y_B \left(\frac{f}{P} \right)_B P \qquad (2.6.4)$$

where $(f/P)_B$ is the fugacity coefficient for pure component B at the temperature and total pressure of the system.

If all of the species are gases, combination of equations 2.6.1, 2.6.2, and 2.6.4 gives

$$K_a = \left(\frac{y_S^s y_T^t}{y_B^b y_C^c} \right) \left[\frac{\left(\frac{f}{P} \right)_S^s \left(\frac{f}{P} \right)_T^t}{\left(\frac{f}{P} \right)_B^b \left(\frac{f}{P} \right)_C^c} \right] P^{s+t-b-c} \qquad (2.6.5)$$

The first term in parentheses is assigned the symbol K_y, while the term in brackets is assigned the symbol $K_{f/P}$.

The quantity $K_{f/P}$ is constant for a given temperature and pressure. However, unlike the equilibrium constant K_a, the term $K_{f/P}$ is affected by changes in the system pressure as well as by changes in temperature.

The product of K_y and $P^{s+t-b-c}$ is assigned the symbol K_P.

$$K_P \equiv K_y P^{s+t-b-c} = \frac{(y_S P)^s (y_T P)^t}{(y_B P)^b (y_C P)^c} = \frac{P_S^s P_T^t}{P_B^b P_C^c} \qquad (2.6.6)$$

since each term in parentheses is a component partial pressure. Thus

$$K_a = K_{f/P} K_P \qquad (2.6.7)$$

For cases where the gases behave ideally, the fugacity coefficients may be taken as unity and the term K_P equated to K_a. At higher pressures where the gases are no longer ideal, the $K_{f/P}$ term may differ appreciably from unity and have a significant effect on the equilibrium composition. The corresponding states plot of fugacity coefficients contained in Appendix B may be used to calculate $K_{f/P}$.

In a system containing an inert gas I in the amount of n_I moles, the mole fraction of reactant gas B is given by

$$y_B = \frac{n_B}{n_B + n_C + \cdots + n_S + n_T + \cdots + n_I} \qquad (2.6.8)$$

Combination of equations 2.6.5 to 2.6.7 and defining equations similar to equation 2.6.8 for the various mole fractions gives:

$$K_a = K_{f/P} \left(\frac{n_S^s n_T^t}{n_B^b n_C^c}\right) \left(\frac{P}{n_B + n_C + \cdots + n_S + n_T + \cdots n_I}\right)^{s+t-b-c}$$

(2.6.9)

This equation is extremely useful for calculating the equilibrium composition of the reaction mixture. The mole numbers of the various species at equilibrium may be related to their values at time zero using the extent of reaction. When these values are substituted into equation 2.6.9, one has a single equation in a single unknown, the equilibrium extent of reaction. This technique is utilized in Illustration 2.1. If more than one independent reaction is occurring in a given system, one needs as many equations of the form of equation 2.6.9 as there are independent reactions. These equations are then written in terms of the various extents of reaction to obtain a set of independent equations equal to the number of unknowns. Such a system is considered in Illustration 2.2.

ILLUSTRATION 2.1 CALCULATION OF EQUILIBRIUM YIELD FOR A CHEMICAL REACTION

Problem

Calculate the equilibrium composition of a mixture of the following species.

N_2 15.0 mole percent
H_2O 60.0 mole percent
C_2H_4 25.0 mole percent

The mixture is maintained at a constant temperature of 527 °K and a constant pressure of 264.2 atm. Assume that the only significant chemical reaction is

$$H_2O(g) + C_2H_4(g) \rightleftharpoons C_2H_5OH(g)$$

Use only the following data and the fugacity coefficient chart.

Compound	T_C (°K)	P_C (atm)
$H_2O(g)$	647.3	218.2
$C_2H_4(g)$	283.1	50.5
$C_2H_5OH(g)$	516.3	63.0

Compound	$\Delta G^0_{f298.16}$ (kcal)	$\Delta H^0_{f298.16}$ (kcal)
$H_2O(g)$	−54.6357	−57.7979
$C_2H_4(g)$	16.282	12.496
$C_2H_5OH(g)$	−40.30	−56.24

The standard state of each species is taken as the pure material at unit fugacity.

Solution

Basis: 100 moles of initial gas
In order to calculate the equilibrium composition one must know the equilibrium constant for the reaction at 527 °K.
From the values of ΔG^0_f and ΔH^0_f at 298.16 °K and equations 2.2.5 and 2.2.6:

$$\Delta G^0_{298} = (1)(-40.30) + (-1)(16.282) + (-1)(-54.6357) = -1.946 \text{ kcal/mole}$$
$$\Delta H^0_{298} = (1)(-56.24) + (-1)(12.496) + (-1)(-57.7979) = -10.938 \text{ kcal/mole}$$

The equilibrium constant at 298.16 °K may be determined from equation 2.4.7.

$$\Delta G^0 = -RT \ln K_a$$

$$\ln K_a = -\frac{(-1946)}{(1.987)(298.16)} = 3.28$$

The equilibrium constant at 527 °K may be found using equation 2.5.3.

$$\left[\frac{\partial \ln K_a}{\partial \left(\frac{1}{T}\right)}\right]_P = -\frac{\Delta H^0}{R}$$

rewritten as

$$K_a = \left(\frac{y_{C_2H_5OH}}{y_{H_2O} y_{C_2H_4}}\right) \left[\frac{\left(\frac{f}{P}\right)_{C_2H_5OH}}{\left(\frac{f}{P}\right)_{H_2O}\left(\frac{f}{P}\right)_{C_2H_4}}\right] \frac{1}{P} \quad (A)$$

The fugacity coefficients (f/P) for the various species may be determined from the generalized chart in Appendix B if one knows the reduced temperature and pressure corresponding to the species in question. Therefore,

Species	Reduced temperature at 527 °K	Reduced pressure at 264.2 atm	f/P
H_2O	$527/647.3 = 0.815$	$264.2/218.2 = 1.202$	0.190
C_2H_4	$527/283.1 = 1.862$	$264.2/50.5 = 5.232$	0.885
C_2H_5OH	$527/516.3 = 1.021$	$264.2/63.0 = 4.194$	0.280

If one assumes that ΔH^0 is independent of temperature, this equation may be integrated to give

$$\ln K_{a,2} - \ln K_{a,1} = \frac{\Delta H^0}{R}\left(\frac{1}{T_1} - \frac{1}{T_2}\right)$$

For our case,

$$\ln K_{a,2} - 3.28 = -\frac{10,938}{1.987}\left(\frac{1}{298.16} - \frac{1}{,527}\right)$$

$$= -8.02$$

or

$$K_{a,2} = 8.74 \times 10^{-3} \text{ at } 527 °K$$

Since the standard states are the pure materials at unit fugacity, equation 2.6.5 may be

From the stoichiometry of the reaction it is possible to determine the mole numbers of the various species in terms of the extent of reaction and their initial mole numbers.

$$n_i = n_{i0} + v_i \xi$$

	Initial moles	Moles at extent ξ
N_2	15.0	15.0
H_2O	60.0	$60.0 - \xi$
C_2H_4	25.0	$25.0 - \xi$
C_2H_5OH	0.0	ξ
Total	100.0	$100.0 - \xi$

The various mole fractions are readily determined from this table. Note that the upper limit on ξ is 25.0.

Substitution of numerical values and expressions for the various mole fractions in equation A gives:

$$8.74 \times 10^{-3} = \frac{\left(\dfrac{\xi}{100.0 - \xi}\right)}{\left(\dfrac{60.0 - \xi}{100.0 - \xi}\right)\left(\dfrac{25.0 - \xi}{100.0 - \xi}\right)} \frac{(0.280)}{(0.190)(0.885)} \frac{1}{(264.2)}$$

or

$$\frac{\xi(100.0 - \xi)}{(60.0 - \xi)(25.0 - \xi)} = 8.74 \times 10^{-3}(264.2)\frac{(0.190)(0.885)}{0.280} = 1.39$$

This equation is quadratic in ξ. The solution is $\xi = 10.8$. On the basis of 100 moles of starting material, the equilibrium composition is as follows.

	Mole numbers	Mole percentages
N_2	15.0	16.8
H_2O	49.2	55.2
C_2H_4	14.2	15.9
C_2H_5OH	10.8	12.1
Total	89.2	100.0

2.7 THE EFFECT OF REACTION CONDITIONS ON EQUILIBRIUM YIELDS

Equation 2.6.9 is an extremely useful relation for determining the effects of changes in process parameters on the equilibrium yield of a given product in a system in which only a single gas phase reaction is important. It may be rewritten as

$$\frac{n_S^s n_T^t}{n_B^b n_C^c} = \frac{K_a}{K_{f/P}}\left(\frac{n_B + n_C + \cdots + n_S + n_T + \cdots + n_I}{P}\right)^{s+t-b-c} \tag{2.7.1}$$

Any change that increases the right side of equation 2.7.1 will increase the ratio of products to reactants in the equilibrium mixture and thus correspond to increased conversions.

2.7.1 Effect of Temperature Changes

The temperature affects the equilibrium yield primarily through its influence on the equilibrium constant K_a. From equation 2.5.2 it follows that the equilibrium conversion is decreased as the temperature increases for exothermic reactions. The equilibrium yield increases with increasing temperature for endothermic reactions. Temperature changes also affect the value of $K_{f/P}$. The changes in this term, however, are generally very small compared to those in K_a.

2.7.2 Effect of Total Pressure

The equilibrium constant K_a is independent of pressure for those cases where the standard states are taken as the pure components at 1 atm. This case is the one used as the basis for deriving equation 2.6.9. The effect of pressure changes then appears in the terms $K_{f/P}$ and $P^{s+t+\cdots-b-c\cdots}$. The influence of pressure on $K_{f/P}$ is quite small. However, for cases where there is no change in the total number of gaseous moles during the reaction, this is the only term by which pressure changes affect the equilibrium yield. For these cases the value of $K_{f/P}$ should be calculated from the fugacity coefficient charts for the system and conditions of interest in order to determine the effect of pressure on the equilibrium yield. For those cases where the reaction produces a change in the total number of gaseous species in the system, the term that has the largest effect on the equilibrium yield of products is $P^{s+t+\cdots-b-c\cdots}$. Thus, if a reaction produces a decrease in the total number of gaseous components, the equilibrium yield is increased by an increase in pressure. If the total number of gaseous moles is increased by reaction, the equilibrium yield decreases as the pressure increases.

2.7.3 Effect of Addition of Inert Gases

The only term in equation 2.7.1 that is influenced by the addition of inert gases is n_I. Thus, for reactions in which there is no change in the total number of gaseous moles, addition of inerts has no effect on the equilibrium yield. For cases where there is a change, the effect produced by addition of inert gases is in the same direction as that which would be produced by a pressure decrease.

2.7.4 Effect of Addition of Catalysts

The equilibrium constant and equilibrium yield are independent of whether or not a catalyst is

present. If the catalyst does not remove any of the passive restraints that have been placed on the system by opening up the possibility of additional reactions, the equilibrium yield will not be affected by the presence of this material.

2.7.5 Effect of Excess Reactants

If nonstoichiometric amounts of reactants are present in the initial system, the presence of excess reactants tends to increase the equilibrium fractional conversion of the limiting re-

agent above that which would be obtained with stoichiometric ratios of the reactants.

2.8 HETEROGENEOUS REACTIONS

The fundamental fact on which the analysis of heterogeneous reactions is based is that when a component is present as a pure liquid or as a pure solid, its activity may be taken as unity, provided the pressure on the system does not differ very much from the chosen standard state pressure. At very high pressures, the effect of pressure on solid or liquid activity may be determined using the Poynting correction factor.

$$\ln\left(\frac{a_{P=P}}{a_{P=1\ \text{atm}}}\right) = \int_1^P \frac{V\ dP}{RT} \qquad (2.8.1)$$

where V is the molal volume of the condensed phase. The activity ratio is essentially unity at moderate pressures.

If we now return to our generalized reaction 2.4.5 and add to our gaseous components B, C, S, and T a pure liquid or solid reactant D and a pure liquid or solid product U with stoichiometric coefficients d and u, respectively, the reaction may be written as

$$bB(g) + cC(g) + dD\ (\ell\ \text{or}\ s) + \cdots \rightleftharpoons sS(g) + tT(g) + uU(\ell\ \text{or}\ s) + \cdots \qquad (2.8.2)$$

The equilibrium constant for this reaction is

$$K_a = \frac{a_S^s a_T^t a_U^u}{a_B^b a_C^c a_D^d} \qquad (2.8.3)$$

When the standard states for the solid and liquid species correspond to the pure species at 1 atm pressure or at a low equilibrium vapor pressure of the condensed phase, the activities of the pure species at equilibrium are taken as unity at all moderate pressures. Consequently, the gas phase composition at equilibrium will not be affected by the amount of solid or liquid present. At very high pressures equation 2.8.1 must be used to calculate these activities. When solid or liquid solutions are present, the activities of the components of these solutions are no longer unity even at moderate pressures. In this case one needs data on the activity coefficients of the various species and the solution composition in order to determine the equilibrium composition of the system.

2.9 EQUILIBRIUM TREATMENT OF SIMULTANEOUS REACTIONS

The treatment of chemical reaction equilibria outlined above can be generalized to cover the situation where multiple reactions occur simultaneously. In theory one can take all conceivable reactions into account in computing the composition of a gas mixture at equilibrium. However, because of kinetic limitations on the rate of approach to equilibrium of certain reactions, one can treat many systems as if equilibrium is achieved in some reactions, but not in others. In many cases reactions that are thermodynamically possible do not, in fact, occur at appreciable rates.

In practice, additional simplifications occur because at equilibrium many of the possible reactions occur either to a negligible extent, or else proceed substantially to completion. One criterion for determining if either of these conditions prevails is to examine the magnitude of the equilibrium constant in question. If it is many orders of magnitude greater than unity, the reaction may be said to go to completion. If it is orders of magnitude less than unity, the reaction may be assumed to go to a negligible extent. In either event, the number of chemical species that must be considered is reduced and the analysis is thereby simplified. After the simplifications are made, there may still remain a group of reactions whose equilibrium constants are neither very small nor very large, indicating that appreciable amounts of both products and reactants are

present at equilibrium. All of these reactions must be considered in calculating the equilibrium composition.

In order to arrive at a consistent set of relationships from which complex reaction equilibria may be determined, one must develop the same number of independent equations as there are unknowns. The following treatment indicates one method of arriving at a set of chemical reactions that are independent. It has been adopted from the text by Aris (1).*

If R reactions occur simultaneously within a system composed of S species, then one has R stoichiometric equations of the form

$$\sum_{i=1}^{S} v_{ki}A_i = 0 \qquad k = 1, 2, \ldots, R \quad (2.9.1)$$

where v_{ki} is the stoichiometric coefficient of species i in reaction k.

Since the same reaction may be written with different stoichiometric coefficients, the importance of the coefficients lies in the fact that the ratios of the coefficients of two species must be identical no matter how the reaction is written. Thus the stoichiometric coefficients of a reaction are given up to a constant multiplier λ. The equation

$$\sum_{i} \lambda v_{ki}A_i = 0 \qquad (2.9.2)$$

has the same meaning as equation 2.9.1, provided that λ is nonzero. If three or more reactions can be written for a given system, one must test to see if any is a multiple of one of the others and if any is a linear combination of two or more others. We will use a set of elementary reactions representing a mechanism for the $H_2 - Br_2$ reaction as a vehicle for indicating how one may determine which of a set of reactions are independent.

_____
* Rutherford Aris, _Introduction to the Analysis of Chemical Reactors_, copyright 1965, pp. 10–13. Adapted by permission of Prentice-Hall, Inc., Englewood Cliffs, NJ.

$$Br_2 \rightarrow 2Br$$
$$Br + H_2 \rightarrow HBr + H$$
$$H + Br_2 \rightarrow HBr + Br \quad (2.9.3)$$
$$H + HBr \rightarrow H_2 + Br$$
$$2Br \rightarrow Br_2$$

If we define

$$A_1 = Br_2 \quad A_2 = Br \quad A_3 = H_2$$
$$A_4 = H \quad A_5 = HBr \quad (2.9.4)$$

then the above reactions may be written as

$$-A_1 + 2A_2 = 0$$
$$-A_2 - A_3 + A_4 + A_5 = 0$$
$$-A_1 + A_2 - A_4 + A_5 = 0 \quad (2.9.5)$$
$$A_2 + A_3 - A_4 - A_5 = 0$$
$$A_1 - 2A_2 = 0$$

To test for independence, form the matrix of the stoichiometric coefficients of the above reactions with v_{ki} in the kth row and the ith column.

$$\begin{matrix} -1 & 2 & 0 & 0 & 0 \\ 0 & -1 & -1 & 1 & 1 \\ -1 & 1 & 0 & -1 & 1 \\ 0 & 1 & 1 & -1 & -1 \\ 1 & -2 & 0 & 0 & 0 \end{matrix} \quad (2.9.6)$$

Next, take the first row with a nonzero element in the first column and divide through by the leading element. If $v_{11} \neq 0$, this will give a new first row of

$$1 \frac{v_{12}}{v_{11}} \frac{v_{13}}{v_{11}} \cdots \frac{v_{1S}}{v_{11}} \quad (2.9.7)$$

This new row may now be used to make all other elements of the first column zero by subtracting v_{k1} times the new first row from the corresponding element in the kth row. This row then becomes

$$0 \left(v_{k2} - v_{k1} \frac{v_{12}}{v_{11}} \right) \left(v_{k3} - v_{k1} \frac{v_{13}}{v_{11}} \right) \cdots \left(v_{kS} - v_{k1} \frac{v_{1S}}{v_{11}} \right) \quad (2.9.8)$$

In the present example the revised coefficient matrix becomes

$$\begin{matrix} 1 & -2 & 0 & 0 & 0 \\ 0 & -1 & -1 & 1 & 1 \\ 0 & -1 & 0 & -1 & 1 \\ 0 & 1 & 1 & -1 & -1 \\ 0 & 0 & 0 & 0 & 0 \end{matrix} \quad (2.9.9)$$

The next step is to ignore the first row and the first column of this matrix and repeat the above reduction on the resultant reduced matrix containing $R - 1$ rows. Thus, equation 2.9.9 becomes

$$\begin{matrix} 1 & -2 & 0 & 0 & 0 \\ 0 & 1 & 1 & -1 & -1 \\ 0 & 0 & 1 & -2 & 0 \\ 0 & 0 & 0 & 0 & 0 \\ 0 & 0 & 0 & 0 & 0 \end{matrix} \quad (2.9.10)$$

This procedure may be repeated as often as necessary until one has 1's down the diagonal as far as possible and zeros beneath them. In the present case we have reached this point. If this had not been the case, the next step would have been to ignore the first two rows and columns and to repeat the above operations on the resultant array. The number of independent reactions is then equal to the number of 1's on the diagonal.

Once the number of independent reactions has been determined, an independent subset can be chosen for subsequent calculations.

ILLUSTRATION 2.2 DETERMINATION OF EQUILIBRIUM COMPOSITIONS IN THE PRESENCE OF SIMULTANEOUS REACTIONS [ADAPTED FROM STRICKLAND-CONSTABLE (2)]

Consider a system that initially consists of 1 mole of CO and 3 moles of H_2 at 1000 °K. The system pressure is 25 atm. The following reactions are to be considered.

$$2CO + 2H_2 \rightleftharpoons CH_4 + CO_2 \qquad (A)$$
$$CO + 3H_2 \rightleftharpoons CH_4 + H_2O \qquad (B)$$
$$CO_2 + H_2 \rightleftharpoons H_2O + CO \qquad (C)$$

When the equilibrium constants for reactions A and B are expressed in terms of the partial pressures of the various species (in atmospheres), the equilibrium constants for these reactions have the following values.

$$K_{P,A} = 0.046 \qquad K_{P,B} = 0.034$$

Determine the equilibrium composition of the mixture.

Solution

The first step in the analysis is to determine if the chemical equations A to C are independent by applying the test described above. When one does this one finds that only two of the reactions are independent. We will choose the first two for use in subsequent calculations. Let the variables ξ_A and ξ_B represent the equilibrium degrees of advancement of reactions A and B, respectively. A mole table indicating the mole numbers of the various species present at equilibrium may be prepared using the following form of equation 1.1.6.

$$n_i = n_{i0} + \nu_{Ai}\xi_A + \nu_{Bi}\xi_B$$

Species	Initial mole number	Equilibrium mole number
CO	1	$1 - 2\xi_A - \xi_B$
H$_2$	3	$3 - 2\xi_A - 3\xi_B$
CH$_4$	0	$\xi_A + \xi_B$
CO$_2$	0	ξ_A
H$_2$O	0	ξ_B
Total	4	$4 - 2\xi_A - 2\xi_B$

The fact that none of the mole numbers can ever be negative places maximum values of 1/2 on ξ_A and 1 on ξ_B.

The values of K_P for reactions A and B are given by

$$K_{P,A} = \frac{P_{CH_4}P_{CO_2}}{P_{CO}^2 P_{H_2}^2} = \frac{y_{CH_4}y_{CO_2}}{y_{CO}^2 y_{H_2}^2 P^2}$$

$$K_{P,B} = \frac{P_{CH_4}P_{H_2O}}{P_{CO}P_{H_2}^3} = \frac{y_{CH_4}y_{H_2O}}{y_{CO}y_{H_2}^3 P^2}$$

The mole fractions of the various species may be expressed in terms of ξ_A and ξ_B so that the above expressions for K_P become

$$K_{P,A} = \frac{\left(\dfrac{\xi_A + \xi_B}{4 - 2\xi_A - 2\xi_B}\right)\left(\dfrac{\xi_A}{4 - 2\xi_A - 2\xi_B}\right)}{\left(\dfrac{1 - 2\xi_A - \xi_B}{4 - 2\xi_A - 2\xi_B}\right)^2\left(\dfrac{3 - 2\xi_A - 3\xi_B}{4 - 2\xi_A - 2\xi_B}\right)^2 P^2}$$

$$= \frac{(\xi_A + \xi_B)(\xi_A)(4 - 2\xi_A - 2\xi_B)^2}{(1 - 2\xi_A - \xi_B)^2(3 - 2\xi_A - 3\xi_B)^2 P^2}$$

$$K_{P,B} = \frac{\left(\dfrac{\xi_A + \xi_B}{4 - 2\xi_A - 2\xi_B}\right)\left(\dfrac{\xi_B}{4 - 2\xi_A - 2\xi_B}\right)}{\left(\dfrac{1 - 2\xi_A - \xi_B}{4 - 2\xi_A - 2\xi_B}\right)\left(\dfrac{3 - 2\xi_A - 3\xi_B}{4 - 2\xi_A - 2\xi_B}\right)^3 P^2}$$

$$= \frac{(\xi_A + \xi_B)(\xi_B)(4 - 2\xi_A - 2\xi_B)^2}{(1 - 2\xi_A - \xi_B)(3 - 2\xi_A - 3\xi_B)^3 P^2}$$

Substitution of numerical values for P, $K_{P,A}$ and $K_{P,B}$ gives two equations in two unknowns. The resultant equations can be solved only by trial and error techniques. In this case a simple graphical approach can be employed in which one plots ξ_A versus ξ_B for each equation and notes the point of intersection. Values of $\xi_A = 0.128$ and $\xi_B = 0.593$ are consistent with the equations. Thus, at equilibrium,

Species	Mole number	Mole fraction
CO	0.151	0.059
H$_2$	0.965	0.377
CH$_4$	0.721	0.282
CO$_2$	0.128	0.050
H$_2$O	0.593	0.232
Total	2.558	1.000

2.10 SUPPLEMENTARY READING REFERENCES

The following texts contain adequate discussions of the thermodynamics of chemical reactions; they can be recommended without implying judgment on others that are not mentioned.

1. K. G. Denbigh, *The Principles of Chemical Equilibrium*, Third Edition, Cambridge University Press, Cambridge, 1971.

2. O. A. Hougen, K. M. Watson, and R. A. Ragatz, *Chemical Process Principles, Part I, Material and Energy Balances and Part II, Thermodynamics*, Second Edition, Wiley, New York, 1954, 1959.

3. I. Prigogine and R. Defay, *Chemical Thermodynamics*, translated by D. H. Everett, Wiley, New York, 1954.

4. F. T. Wall, *Chemical Thermodynamics*, Second Edition, W. H. Freeman, San Francisco, 1965.

5. K. S. Pitzer and L. Brewer, revision of *Thermodynamics*, by G. N. Lewis and M. Randall, McGraw-Hill, New York, 1961.

6. M. Modell and R. C. Reid, *Thermodynamics and Its Applications*, Prentice-Hall, Englewood Cliffs, N. J., 1974.

LITERATURE CITATIONS

1. Aris, R., *Introduction to the Analysis of Chemical Reactors*, Prentice-Hall, Englewood Cliffs, N.J., copyright © 1965. Used with permission.

2. Strickland-Constable, R. F., "Chemical Thermodynamics," in Volume 8 of *Chemical Engineering Practice*, edited by H. W. Cremer and S. B. Watkins, Butterworths, London, 1965. Used with permission.

PROBLEMS

1. In the presence of an appropriate catalyst, acetone reacts with hydrogen to give isopropanol.

$$CH_3COCH_3 + H_2 \rightleftharpoons CH_3CHOHCH_3$$

Standard enthalpies and Gibbs free energies of formation at 25 °C and 101.3 kPa are as follows.

	ΔH_f^0 (kJ/mole)	ΔG_f^0 (kJ/mole)
Acetone (*g*)	−216.83	−152.61
Isopropanol (*g*)	−261.30	−159.94

Tabulate the effects of the changes below on:
(a) Reaction velocity.
(b) Equilibrium constant.
(c) Equilibrium degree of conversion.
(d) Actual degree of conversion obtained in a very short time interval at the start of the reaction.

Variables:
(1) Increase in temperature.
(2) Increase of pressure.
(3) Substitution of a less active catalyst.
(4) Dilution with argon at constant total pressure.
(5) Increase in the Reynolds number of the fluid in the reactor.

Appropriate tabular entries are: (increases), (decreases), (no effect), (indeterminate).

2. Consider the ammonia synthesis reaction being carried out at 450 °C and 101.3 MPa.

$$\tfrac{1}{2}N_2 + \tfrac{3}{2}H_2 \rightleftharpoons NH_3$$

The feed stream consists of 60 mole percent hydrogen, 20% nitrogen, and 20% argon. Calculate the composition of the exit gases, assuming equilibrium is achieved in the reactor. Make sure that you take deviations from the ideal gas law into account. The equilibrium constant expressed in terms of activities relative to standard states at 1 atm may be assumed to be equal to 8.75×10^{-3}. The fugacity of pure H_2 at 450 °C and 101.3 MPa may be assumed to be equal to 136.8 MPa.

3. A new process for the manufacture of acetylene has been proposed. The process will involve the dehydrogenation of ethane over a suitable catalyst (yet to be found). Pure ethane will be fed to a reactor and a mixture of acetylene, hydrogen, and unreacted ethane will be withdrawn. The reactor will operate at 101.3 kPa total pressure and at some as yet unspecified temperature T.

The reaction

$$C_2H_6 \rightarrow C_2H_2 + 2H_2$$

may be assumed to be the only reaction occurring in the proposed reactor. The following data on the heats of formation of ethane and acetylene are available.

	ΔG^0_{f298} (kJ/mole)	ΔH^0_{f298} (kJ/mole)
Ethane	− 32.908	− 84.724
Acetylene	209.340	226.899

The standard states of these materials are taken as the pure components at 298 °K and a pressure of 101.3 kPa. The following data on the absolute entropies of the hydrocarbons at 298 °K are available.

	Entropy (J/mole · K)
Ethane	229.65
Acetylene	200.95

(a) What is the standard Gibbs free energy change at 25 °C for the reaction as written?

(b) What is the standard enthalpy change at 25 °C for the reaction as written?

(c) What is the absolute entropy of gaseous hydrogen at 25 °C and a pressure of 101.3 kPa? Use only the above data in evaluating this quantity.

(d) What is the equilibrium constant for the reaction at 25 °C?

(e) If the standard enthalpy change may be assumed to be essentially independent of temperature, what is the equilibrium constant for the reaction at 827 °C?

(f) If we supply pure ethane at 827 °C and a pressure of 101.3 kPa to the reactor described above and if equilibrium with respect to the reaction

$$C_2H_6 \rightleftharpoons C_2H_2 + 2H_2$$

is obtained, what is the composition of the effluent mixture? The reactor may be assumed to operate isothermally at 101.3 kPa total pressure. Neglect any other reactions that may occur.

4. As a thermodynamicist working at the Lower Slobbovian Research Institute, you have been asked to determine the standard Gibbs free energy of formation and the standard enthalpy of formation of the compounds *cis*-butene-2 and *trans*-butene-2. Your boss has informed you that the standard enthalpy of formation of butene-1 is 1.172 kJ/mole while the standard Gibbs free energy of formation is 72.10 kJ/mole where the standard state is taken as the pure component at 25 °C and 101.3 kPa.

Your associate, Kem Injuneer, has been testing a new catalyst for selective butene isomerization reactions. He says that the only reactions that occur to any appreciable extent over this material are:

$$\text{butene-1} \rightleftharpoons \textit{cis}\text{-butene-2}$$

$$\textit{cis}\text{-butene-2} \rightleftharpoons \textit{trans}\text{-butene-2}$$

He has reported the following sets of data from his system as being appropriate for equilibrium conditions.

Run I	
Reactor pressure	53.33 kPa
Reactor temperature	25 °C
Gas composition (mole percent):	
butene-1	3.0
cis-butene-2	21.9
trans-butene-2	75.1

Run II	
Reactor pressure	101.3 kPa
Reactor temperature	127 °C
Gas composition (mole percent):	
butene-1	8.1
cis-butene-2	28.8
trans-butene-2	63.1

Kem maintains that you now have enough data to determine the values of $\Delta G_f{}^0$ and $\Delta H_f{}^0$ for the two isomers of butene-2 at 25 °C.

Proceed to evaluate these quantities. State specifically what assumptions you must make in your analysis and comment on their validity. Use *only* the above data.

5. Hydrogen can be manufactured from carbon monoxide by the water gas shift reaction

$$CO + H_2O \rightleftharpoons CO_2 + H_2$$

At 900 °F the equilibrium constant for this reaction is 5.62 when the standard states for all species are taken as unit fugacity. If the reaction is carried out at 75 atm, what molal ratio of steam to carbon monoxide is required to produce a product mixture in which 90% of the inlet CO is converted to CO_2?

6. (a) What is the composition of an equilibrium mixture of NO_2 and N_2O_4 at 25 °C and 101.3 kPa? It may be assumed that the only chemical reaction involved is:

$$N_2O_4(g) \rightleftharpoons 2NO_2(g)$$

(b) Calculate values of the Gibbs free energy of mixtures of these two substances at 25 °C and 101.3 kPa for several different compositions from 0 to 1.0 mole fraction N_2O_4. Base your calculations on a mixture containing 2 *gram atoms of nitrogen*. Plot the results versus composition. Compare the composition at the minimum with that determined in part (a).

7. At 25 °C the standard Gibbs free energy change for the reaction

$$SO_2 + \tfrac{1}{2}O_2 \rightleftharpoons SO_3$$

is -70.04 kJ/mole where the standard states are taken as the pure components at 101.3 kPa and 25 °C. At 227 °C and a total system pressure of 1.013 kPa, the following equilibrium composition was determined experimentally.

Component	Mole percent
O_2	0.10
SO_2	0.12
SO_3	78.18
Helium	21.60

(a) What is the equilibrium constant for the reaction at 25 °C and 1.013 kPa?
(b) What is the equilibrium constant for the reaction at 227 °C and 1.013 kPa?
(c) What is the standard enthalpy change for the reaction if it is assumed to be temperature independent?
(d) Will the equilibrium constant for the reaction at 25 °C and 101.3 kPa be greater than, equal to, or less than that calculated in part (a)? Explain your reasoning.

8. A company has a large ethane (C_2H_6) stream available and has demands for both ethylene (C_2H_4) and acetylene (C_2H_2). The relative amounts of these two chemicals required varies from time to time, and the company proposes to build a single plant operating at atmospheric pressure to produce either material.
(a) Using the data below, calculate the maximum temperature at which the reactor may operate and still produce essentially ethylene (not more than 0.1% acetylene).
(b) Calculate the minimum temperature at which the reactor can operate and produce essentially acetylene (not more than 0.1% ethylene).
(c) At what temperature will one produce equimolal quantities of acetylene and ethylene?
Data and Notes: Assume that only the following reactions occur.

$$C_2H_6 \rightleftharpoons C_2H_4 + H_2$$
$$C_2H_6 \rightleftharpoons C_2H_2 + 2H_2$$

Neglect the effect of temperature on ΔH^0.

9. A gas mixture containing only equimolal quantities of CO_2 and H_2 is to be "reformed" by

flowing it over a powerful catalyst at 1000 °K and at various pressures. Over what range of pressures may carbon deposit on the catalyst? Over what range of pressures will it not deposit? *Note.* Assume that only the reactions given in the table occur, and that equilibrium is attained.

	K
$C + H_2O \rightleftharpoons CO + H_2$	3.16
$C + 2H_2O \rightleftharpoons CO_2 + 2H_2$	5.01
$CO_2 + C \rightleftharpoons 2CO$	2.00
$CO + H_2O \rightleftharpoons CO_2 + H_2$	1.58

For an initial ratio of CO_2 to H_2 of unity, what must the pressure be if exactly 30% of the carbon present in the feed can precipitate as solid carbon?

The equilibrium constants are based on a standard state of unit fugacity for the gaseous species and on a standard state corresponding to the pure solid for carbon.

10. Butadiene can be produced by the dehydrogenation of butene over an appropriate catalyst.

$$C_4H_8 \rightleftharpoons H_2 + C_4H_6$$

In order to moderate the thermal effects associated with the reaction, large quantities of steam are usually injected with the butene feed. Steam/butene ratios of 10 to 20 are typical of the conditions employed in many industrial reactors. If equilibrium is achieved within the reactor, determine the effluent composition corresponding to the conditions enumerated below.

11. Methanol may be synthesized from hydrogen and carbon monoxide in the presence of an appropriate catalyst.

$$CO + 2H_2 \rightleftharpoons CH_3OH$$

If one has a feed stream containing these gases in stoichiometric proportions ($H_2/CO = 2$) at 200 atm and 275 °C, determine the effluent composition from the reactor:
(a) If it operates isothermally and equilibrium is achieved.
(b) If it operates adiabatically and equilibrium is achieved. (Also determine the effluent temperature.)

Data:
(1) $\Delta H^0 = -17{,}530 - 18.19T + 0.0141T^2$ for ΔH^0 in calories per gram mole and T in degrees Kelvin.
(2) Molal heat capacities at constant pressure

$$H_2 \qquad C_p = 6.65 + 0.00070T$$
$$CO \qquad C_p = 6.89 + 0.00038T$$
$$CH_3OH \quad C_p = 2.0 + 0.03T$$

for C_p in calories per gram mole-degree Kelvin.
(3) Equilibrium constant expressed in terms of fugacities

$$\log_{10} K_f = \frac{3835}{T} - 9.150 \log_{10} T$$
$$+ 3.08 \times 10^{-3}T + 13.20$$

(4) Note that in part (b) a trial and error solution is required. (*Hint.* The effluent temperature will be close to 700 °K.)

Effluent temperature	820 °K	
Effluent pressure	101.3 kPa	
Feed composition (mole percent)		
Steam (H_2O)	92%	
Butene	8%	
Thermodynamic data at 800 °K	ΔG_f^0 (kJ/mole)	ΔH_f^0 (kJ/mole)
H_2O (g)	-203.66	-246.53
Butene (g)	207.04	-21.56
Butadiene (g)	228.10	97.34
H_2 (g)	0	0

12.* In a laboratory investigation a high-pressure gas reaction $A \rightleftharpoons 2B$ is being studied in a flow reactor at 200 °C and 10.13 MPa. At the end of the reactor the gases are in chemical equilibrium, and *their composition is desired.*

Unfortunately, to make any analytical measurements on this system, it is necessary to bleed off a small side stream through a low-pressure conductivity cell operating at 1 atm. It is found that when the side stream passes through the sampling valve, the temperature drops to 100 °C and the conductivity cell gives compositions of $y_A = 0.5$ and $y_B = 0.5$ (mole fractions).

* Adapted from Michael Modell and Robert C. Reid, *Thermodynamics and Its Applications*, copyright © 1974. Reprinted by permission of Prentice-Hall, Inc. Englewood Cliffs, N. J.

From these experimental data and the physical properties listed below:
(a) Calculate the composition of the gas stream before the sampling valve.
(b) Are the gases at, near, or far from chemical equilibrium after the sampling valve? (Show definite proof for your answer.)
Data and Allowable Assumptions:
(1) Heat of the reaction, $\Delta H = 29.31$ kJ/mole of A reacting, independent' of temperature.
(2) Heat capacities: 58.62 J/mole·°K for A and 29.31 J/mole·°K for B, independent of temperature.
(3) The gas mixture is ideal at all pressures, temperatures, and compositions.
(4) The flow through the sampling valve is a true Joule-Thompson expansion.
(5) Assume no heat loss in the sampling line or across sampling valve.

3 Basic Concepts in Chemical Kinetics—Determination of the Reaction Rate Expression

3.0 INTRODUCTION

This chapter defines a number of terms that are used by the chemical kineticist and treats some of the methods employed in the analysis of laboratory data to determine empirical rate expressions for the systems under investigation.

It is convenient to approach the concept of reaction rate by considering a closed, isothermal, constant pressure homogeneous system of uniform composition in which a single chemical reaction is taking place. In such a system the rate of the chemical reaction (r) is defined as:

$$r = \frac{1}{V}\frac{d\xi}{dt} \qquad (3.0.1)$$

where

$$V = \text{system volume}$$
$$\xi = \text{extent of reaction}$$
$$t = \text{time}$$

Several facts about this definition should be noted.

1. The rate is defined as an intensive variable. Note that the reciprocal of system volume is outside the derivative term. This consideration is important in treating variable volume systems.
2. The definition is independent of any particular reactant or product species.
3. Since the reaction rate almost invariably changes with time, it is necessary to use the time derivative to express the instantaneous rate of reaction.

Many sets of units may be used to measure reaction rates. Since the extent of reaction is expressed in terms of moles, the reaction rate has the units of moles transformed per unit time per unit volume. The majority of the data reported in the literature is expressed in some form of the metric system of units, (e.g., mole/liter · sec or molecules/cm^3·sec).

Changes in the mole numbers n_i of the various species involved in a reaction are related to the extent of reaction by equation 1.1.4

$$\xi = \frac{n_i - n_{i0}}{\nu_i} \qquad (3.0.2)$$

Differentiation of this equation with respect to time and substitution in equation 3.0.1 gives

$$r = \frac{1}{\nu_i}\frac{1}{V}\frac{dn_i}{dt} \qquad (3.0.3)$$

If one defines the rate of increase of the moles of species i as

$$r_i \equiv \frac{1}{V}\frac{dn_i}{dt} \qquad (3.0.4)$$

then

$$r_i = \nu_i r \qquad (3.0.5)$$

Since the ν_i are positive for products and negative for reactants, and since the reaction rate r is intrinsically positive, the various r_i will have the same sign as the corresponding ν_i and dn_i/dt will have the appropriate sign (i.e., positive for products and negative for reactants).

In the analysis of engineering systems, one frequently encounters systems whose properties vary from point to point within the system. Just as it is possible to define local temperatures, pressures, concentrations, etc., it is possible to generalize equations 3.0.1 and 3.0.4 to define local reaction rates.

In constant volume systems it is convenient to employ the extent per unit volume ξ^*

$$\xi^* = \frac{\xi}{V} \qquad (3.0.6)$$

and to denote the rate in such cases as

$$r_V = \frac{1}{V}\frac{d\xi}{dt} = \frac{d\xi^*}{dt} \qquad (3.0.7)$$

In terms of molar concentrations, $C_i = n_i/V$, and equation 3.0.3 becomes

$$r_V = \frac{1}{v_i}\frac{dC_i}{dt} = \frac{r_{i,V}}{v_i} \qquad (3.0.8)$$

The rate of reaction at constant volume is thus proportional to the time derivative of the molar concentration. However, *it should be emphasized that in general the rate of reaction is not equal to the time derivative of a concentration.* Moreover, omission of the $1/v_i$ term frequently leads to errors in the analysis and use of kinetic data. When one substitutes the product of concentration and volume for n_i in equation 3.0.3, the essential difference between equations 3.0.3 and 3.0.8 becomes obvious.

$$r = \frac{1}{v_i}\frac{1}{V}\frac{d}{dt}(C_iV) = \frac{1}{v_i}\frac{dC_i}{dt} + \frac{C_i}{v_iV}\frac{dV}{dt} \qquad (3.0.9)$$

In variable volume systems the dV/dt term is significant. Although equation 3.0.9 is a valid one arrived at by legitimate mathematical operations, its use in the analysis of rate data is extremely limited because of the awkward nature of the equations to which it leads. Equation 3.0.1 is preferred.

Many reactions take place in heterogeneous systems rather than in a single homogeneous phase. These reactions often occur at the interface between the two phases. In such cases it is appropriate to define the reaction rate in terms of the interfacial area (S) available for reaction.

$$r'' = \frac{1}{S}\frac{d\xi}{dt} \qquad (3.0.10)$$

The double prime superscript is used to emphasize the basis of unit surface area.

In many cases, however, the interfacial area is not known, particularly when one is dealing with a heterogeneous catalytic reaction involving a liquid phase and a solid catalyst. Consequently, the following definitions of the reaction rate are sometimes useful.

$$r_m = \frac{1}{W}\frac{d\xi}{dt} \qquad (3.0.11)$$

$$r''' = \frac{1}{V'}\frac{d\xi}{dt} \qquad (3.0.12)$$

where W and V' are the weight and volume of the solid particles dispersed in the fluid phase. The subscript and superscript emphasize the definition employed.

The choice of the definition of the rate to be used in any given situation is governed by convenience in use. The various forms of the definition are interrelated, and kineticists should be capable of switching from one form to another without excessive difficulty.

Many process variables can affect the rate at which reactants are converted into products. The conversion rate should be considered as a phenomenological property of the reaction system under the given operating conditions. The nature of the dependence of the conversion rate on macroscopic or laboratory variables cannot be completely determined on an *a priori* basis. On the contrary, recourse to experimental data on the reaction involved and on the relative rates of the physical and chemical processes involved is almost always necessary. Among the variables that can influence the rate of conversion are the system temperature, pressure and composition, the properties of a catalyst that may be present, and the system parameters that govern the various physical transport processes (i.e., the flow conditions, degree of mixing, and the heat and mass transfer parameters of the system). Since several of these variables may change from location to location within the reactor under consideration, a knowledge of the relationship between these variables and the conversion rate is needed if one is to be able to integrate the appropriate material balance equations over the reactor volume. It is important to note that in many situations of practical engineering importance, *the conversion rate is not identical with the intrinsic chemical reaction rate evaluated using the bulk fluid properties.* The conversion rate takes into account the effects of both chemical and physical rate processes. The intrinsic rate may be thought of as the conversion rate that would exist if all physical rate processes occurred at infinitely fast rates.

Chapter 12 treats situations where both physical and chemical rate processes influence the conversion rate; the present chapter is concerned only with those situations where physical rate processes are unimportant. This approach permits us to focus our concern on the variables that influence intrinsic chemical reaction rates (i.e., temperature, pressure, composition, and the presence or absence of catalysts in the system).

In reaction rate studies one's goal is a phenomenological description of a system in terms of a limited number of empirical constants. Such descriptions permit one to predict the time-dependent behavior of similar systems. In these studies the usual procedure is to try to isolate the effects of the different variables and to investigate each independently. For example, one encloses the reacting system in a thermostat in order to maintain it at a constant temperature.

Several generalizations can be made about the variables that influence reaction rates. Those that follow are in large measure adapted from Boudart's text (1).

1. The rate of a chemical reaction depends on the temperature, pressure, and composition of the system under investigation.
2. Certain species that do not appear in the stoichiometric equation for the reaction under study can markedly affect the reaction rate, even when they are present in only trace amounts. These materials are known as catalysts or inhibitors, depending on whether they increase or decrease the reaction rate.
3. At a constant temperature, the rate of reaction generally decreases monotonically with time or extent of reaction.
4. If one considers reactions that occur in systems that are far removed from equilib-

rium, the rate expressions can generally be written in the form

$$r = k\phi(C_i) \qquad (3.0.13)$$

where $\phi(C_i)$ is a function that depends on the concentrations (C_i) of the various species present in the system (reactants, products, catalysts, and inhibitors). This function $\phi(C_i)$ may also depend on the temperature. The coefficient k is called the *reaction rate constant.* It does not depend on the composition of the system and is consequently independent of time in an isothermal system.

5. The rate constant k generally varies with the absolute temperature T of the system according to the law proposed by Arrhenius.

$$k = Ae^{-E/RT} \qquad (3.0.14)$$

where
E is the apparent activation energy of the reaction
R is the gas constant
A is the preexponential factor, sometimes called the frequency factor, which is assumed to be a temperature independent quantity

6. Very often the function $\phi(C_i)$ in equation 3.0.13 is temperature independent and, to a high degree of approximation, can be written as

$$\phi(C_i) = \prod_i C_i^{\beta_i} \qquad (3.0.15)$$

where the product $\prod$ is taken over all components of the system. The exponents β_i are the *orders of the reaction with respect to each of the i species* present in the system. The algebraic sum of the exponents is called the *total order* or *overall order* of the reaction.

7. If one considers a system in which both forward and reverse reactions are important, the net rate of reaction can generally be expressed as the difference between the rate in the forward direction $\vec{r}$ and that in the opposite direction $\overleftarrow{r}$.

$$r = \vec{r} - \overleftarrow{r} \qquad (3.0.16)$$

3.0.1 Reaction Orders

The manner in which the reaction rate varies with the concentrations of the reactants and products is indicated by stating the order of the reaction. If equation 3.0.15 is written in more explicit form as

$$r = kC_A^{\beta_A}C_B^{\beta_B}C_C^{\beta_C} \cdots \qquad (3.0.17)$$

the reaction is said to be of the β_Ath order with respect to A, β_Bth order with respect to B, etc. The overall order of the reaction (m) is simply

$$m = \bar{\beta}_A + \beta_B + \beta_C + \cdots \qquad (3.0.18)$$

These exponents β_i may be small integers or fractions, and they may take on both positive and negative values as well as the value zero. In many cases these exponents are independent of temperature. In other cases where the experimental data have been forced to fit expressions of the form of equation 3.0.17, the exponents will vary slightly with temperature. In these cases the observed correlation should be applied only in a restricted temperature interval.

It must be emphasized that, in general, the individual orders of the reaction (β_i) are *not related* to the corresponding stoichiometric coefficients ν_i. *The individual β_i's are quantities that must be determined experimentally.*

It is important to recognize that by no means can all reactions be said to have an order. For example, the gas phase reaction of H_2 and Br_2 to form HBr has a rate expression of the following form:

$$r = \frac{k(H_2)(Br_2)^{1/2}}{1 + \dfrac{k'(HBr)}{(Br_2)}} \qquad (3.0.19)$$

where k and k' are constants at a given temperature and where the molecular species contained in brackets refer to the concentrations of these species. This rate expression is discussed in more detail in Section 4.2.1.

When one reactant is present in very large excess, the amount of this material that can be consumed by reaction is negligible compared to the total amount present. Under these circumstances, its concentration may be regarded as remaining essentially constant throughout the course of the reaction, and the product of the reaction rate constant and the concentration of this species raised to the appropriate order will also be constant. This product is then an apparent or empirical pseudo rate constant, and a corresponding pseudo reaction order can be determined from the new form of the rate expression.

3.0.2 The Reaction Rate Constant

The term reaction rate constant is actually a misnomer, since k may vary with temperature, the solvent for the reaction, and the concentrations of any catalysts that may be present in the reaction system. The term is in universal use, however, because it implies that the parameter k is independent of the concentrations of reactant and product species.

The reaction rate is properly defined in terms of the time derivative of the extent of reaction. It is necessary to define k in a similar fashion in order to ensure uniqueness. Definitions in terms of the various r_i would lead to rate constants that would differ by ratios of their stoichiometric coefficients.

The units of the rate constant will vary depending on the overall order of the reaction. These units are those of a rate divided by the mth power of concentration, as is evident from equations 3.0.17 and 3.0.18.

$$(k) = \frac{(r)}{(C)^m} = \frac{(\text{moles/volume-time})}{(\text{moles/volume})^m} \qquad (3.0.20)$$

or

$$(k) = \text{time}^{-1}(\text{moles/volume})^{-m+1} \qquad (3.0.21)$$

For a first-order reaction, the units of k are time^{-1}; for the second-order case, typical units are m^3/mole·sec.

3.1 MATHEMATICAL CHARACTERIZATION OF SIMPLE REACTION SYSTEMS

Although the reaction rate function can take on a variety of mathematical forms and the reaction orders that one observes in the laboratory are not necessarily positive integers, a surprisingly large number of reactions have an overall order that is an integer. This section treats the mathematical forms that the integrated rate expression will take for several simple cases. The discussion is restricted to *irreversible* reactions carried out *isothermally*. It provides a framework for subsequent treatment of the results that are observed in the laboratory. We start by treating constant volume systems that lead to closed form solutions and then proceed to the complications present in variable volume systems. We have chosen to place a "V" to the right of certain equation numbers in this section to emphasize to the reader that these equations are not general, but are restricted to constant volume systems. The use of ξ^*, the extent of reaction per unit volume in a constant volume system, will also emphasize this restriction.

3.1.1 Mathematical Characterization of Simple Constant Volume Reaction Systems

3.1.1.1 First-Order Reactions in Constant Volume Systems.
In a first-order reaction the reaction rate is proportional to the first power of the concentration of one of the reacting substances.

$$r = kC_A \qquad (3.1.1)$$

For a constant volume system

$$C_A = C_{A0} + v_A \xi^* \qquad (3.1.2)\text{V}$$

Combination of this equation with equations 3.0.7 and 3.1.1 gives

$$\frac{d\xi^*}{dt} = k(C_{A0} + v_A \xi^*) \qquad (3.1.3)\text{V}$$

Separation of variables and integration subject

to the condition that $\xi^* = 0$ at $t = 0$ gives

$$\frac{1}{v_A} \ell n \left(\frac{C_{A0} + v_A \xi^*}{C_{A0}} \right) = kt \qquad (3.1.4)\text{V}$$

Thus

$$\xi^* = \frac{C_{A0}}{v_A} (e^{v_A kt} - 1) \qquad (3.1.5)\text{V}$$

The species concentrations at various times can now be determined from the reaction stoichiometry.

$$C_i = C_{i0} + v_i \xi^* = C_{i0} + \frac{v_i}{v_A} C_{A0}(e^{v_A kt} - 1)$$
$$(3.1.6)\text{V}$$

If one is interested in the time dependence of the concentration of species A and if the stoichiometric coefficient of A is equal to -1, this relation becomes

$$C_A = C_{A0}e^{-kt} \qquad (3.1.7)\text{V}$$

or

$$\ell n \left(\frac{C_A}{C_{A0}} \right) = -kt \qquad (3.1.8)\text{V}$$

In graphical form, these two relations imply that for first-order reactions, plots of $\ell n\ C_A$ versus time will be linear with a slope equal to $(-k)$ and an intercept equal to $\ell n\ C_{A0}$. Since this type of plot is linear, it is frequently used in testing experimental data to see if a reaction is first order.

A great many reactions follow first-order kinetics or pseudo first-order kinetics over certain ranges of experimental conditions. Among these are many pyrolysis reactions, the cracking of butane, the decomposition of nitrogen pentoxide (N_2O_5), and the radioactive disintegration of unstable nuclei.

3.1.1.2 Second-Order Reactions in Constant Volume Systems.
There are two primary types of second-order reactions: for the first the rate is proportional to the square of the concentration of a single reacting species; for the second the

rate is proportional to the product of the concentrations of two different species.

Class I: $\qquad r = kC_A^2 \qquad$ (3.1.9)

Class II: $\qquad r = kC_A C_B \qquad$ (3.1.10)

For Class I second-order rate expressions, combining equations 3.0.7, 3.1.2, and 3.1.9 gives

$$\frac{d\xi^*}{dt} = k(C_{A0} + v_A\xi^*)^2 \qquad (3.1.11)\text{V}$$

Integration of this equation subject to the initial condition that $\xi^* = 0$ at $t = 0$ gives

$$-\frac{1}{v_A}\left(\frac{1}{C_{A0} + v_A\xi^*} - \frac{1}{C_{A0}}\right) = kt \qquad (3.1.12)\text{V}$$

or

$$\frac{1}{C_A} - \frac{1}{C_{A0}} = -v_A kt \qquad (3.1.13)\text{V}$$

The concentrations of the various species can then be determined by solving equation 3.1.12 for ξ^* and employing basic stoichiometric relations.

$$C_i = C_{i0} + v_i\xi^* = C_{i0} + \frac{v_i k C_{A0}^2 t}{1 - v_A k C_{A0} t}$$

$$(3.1.14)\text{V}$$

If v_A is equal to -1, equation 3.1.13 becomes

$$\frac{1}{C_A} - \frac{1}{C_{A0}} = kt \qquad (3.1.15)\text{V}$$

In testing experimental data to see if it fits this type of rate expression, one plots $1/C_A$ versus t. If the data fall on a straight line, the rate expression is of the form of equation 3.1.9, and the slope and intercept of the line are equal to $-v_A k$ and $1/C_{A0}$, respectively.

Many second-order reactions follow Class I rate expressions. Among these are the gas-phase thermal decomposition of hydrogen iodide ($2HI \rightarrow H_2 + I_2$), dimerization of cyclopentadiene ($2C_5H_6 \rightarrow C_{10}H_{12}$), and the gas phase thermal decomposition of nitrogen dioxide ($2NO_2 \rightarrow 2NO + O_2$).

For Class II second-order rate expressions of the form of equation 3.1.10, the rate can be expressed in terms of the extent of reaction per unit volume as

$$r_V = \frac{d\xi^*}{dt} = k(C_{A0} + v_A\xi^*)(C_{B0} + v_B\xi^*)$$

$$(3.1.16)\text{V}$$

When the stoichiometric coefficients of species A and B are identical and when one starts with equal concentrations of these species, the Class II rate expression will collapse to the Class I form because, under these conditions, one can always say that $C_A = C_B$.

Separation of variables and integration of equation 3.1.16 leads to the following relation.

$$\ln\left[\frac{(C_{B0} + v_B\xi^*)(C_{A0})}{(C_{A0} + v_A\xi^*)(C_{B0})}\right] = (C_{A0}v_B - C_{B0}v_A)kt$$

$$(3.1.17)\text{V}$$

or

$$\xi^* = \frac{C_{A0}\{1 - \exp[(C_{A0}v_B - C_{B0}v_A)kt]\}}{v_A \exp[(C_{A0}v_B - C_{B0}v_A)kt] - v_B\dfrac{C_{A0}}{C_{B0}}}$$

$$(3.1.18)\text{V}$$

Since two of the terms within the brackets in equation 3.1.17 are the residual reactant concentrations at time t or extent ξ^*, it is often useful to rewrite this equation as

$$\ln\left[\left(\frac{C_B}{C_A}\right)\left(\frac{C_{A0}}{C_{B0}}\right)\right] = (C_{A0}v_B - C_{B0}v_A)kt$$

$$(3.1.19)\text{V}$$

or

$$\ln\left[\frac{(C_B)}{(C_A)}\right] + \ln\left[\frac{(C_{A0})}{(C_{B0})}\right] = (C_{A0}v_B - C_{B0}v_A)kt$$

$$(3.1.20)\text{V}$$

These equations are convenient for use in determining if experimental rate data follow Class II second-order kinetics in that they predict a linear relationship between $\ln(C_B/C_A)$ and time. The y intercept is $\ln(C_{B0}/C_{A0})$ and the slope is $(C_{A0}v_B - C_{B0}v_A)k$.

Class II second-order rate expressions are one of the most common forms one encounters in the laboratory. They include the gas phase reaction of molecular hydrogen and iodine ($H_2 + I_2 \rightarrow 2HI$), the reactions of free radicals with molecules (e.g., $H + Br_2 \rightarrow HBr + Br$), and the hydrolysis of organic esters in nonaqueous media.

3.1.1.3 Third-Order Reactions in Constant Volume Systems.

Third-order reactions can be classified into three primary types, according to the general definition.

Class I: $\quad r = kC_A^3$ (3.1.21)

Class II: $\quad r = kC_A^2 C_B$ (3.1.22)

Class III: $\quad r = kC_A C_B C_Q$ (3.1.23)

If one uses reactants in precisely stoichiometric concentrations, the Class II and Class III rate expressions will reduce to the mathematical form of the Class I rate function. Since the mathematical principles employed in deriving the relation between the extent of reaction or the

concentrations of the various species and time are similar to those used in Sections 3.1.1.1 and 3.1.1.2, we will list only the most useful results.

Class I Third-Order Rate Expression:

$$\frac{1}{C_A^2} - \frac{1}{C_{A0}^2} = -2v_A kt \quad (3.1.24)V$$

and

$$\xi^* = \frac{C_{A0}}{v_A}\left(\frac{1}{\sqrt{1 - 2v_A k C_{A0}^2 t}} - 1\right) \quad (3.1.25)V$$

Class II Third-Order Rate Expression:

$$\frac{1}{C_A} - \frac{1}{C_{A0}} + \frac{1}{C_{A0} - C_{B0}\dfrac{v_A}{v_B}} \ln\left(\frac{C_B C_{A0}}{C_A C_{B0}}\right)$$

$$= (C_{A0}v_B - C_{B0}v_A)kt \quad (3.1.26)V$$

In this case one obtains an expression for ξ^* that cannot be manipulated to yield a simple algebraic form. However, if the concentration of one species is known as a function of time, the concentrations of all other species may be determined from the definition of the extent of reaction per unit volume; that is,

$$\xi^* = \frac{C_i - C_{i0}}{v_i} = \frac{C_j - C_{j0}}{v_j} \quad (3.1.27)V$$

Hence,

$$C_j = C_{j0} + \frac{v_j}{v_i}(C_i - C_{i0}) \quad (3.1.28)V$$

Class III Third-Order Rate Expression:

$$\left.\begin{array}{l} v_A(v_Q C_{B0} - v_B C_{Q0})\ln\left(\dfrac{C_A}{C_{A0}}\right) \\[2mm] + v_B(v_A C_{Q0} - v_Q C_{A0})\ln\left(\dfrac{C_B}{C_{B0}}\right) \\[2mm] + v_Q(v_B C_{A0} - v_A C_{B0})\ln\left(\dfrac{C_Q}{C_{Q0}}\right) \end{array}\right\} = kt(v_A C_{B0} - v_B C_{A0})(v_A C_{Q0} - v_Q C_{A0})(v_Q C_{B0} - v_B C_{Q0}) \quad (3.1.29)V$$

The last equation has extremely limited utility and is presented more as a subject of academic interest than of practicality.

Gas phase third-order reactions are rarely encountered in engineering practice. Perhaps the best-known examples of third-order reactions are atomic recombination reactions in the presence of a third body in the gas phase and the reactions of nitric oxide with chlorine and oxygen: ($2NO + Cl_2 \rightarrow 2NOCl$; $2NO + O_2 \rightarrow 2NO_2$).

3.1.1.4 Fractional and Other Order Reactions in Constant Volume Systems.

In chemical kinetics, one frequently encounters reactions whose orders are not integers. Consider a reaction involving only a single reactant A whose rate expression is of the form

$$\frac{d\xi^*}{dt} = kC_A^n = k(C_{A0} + \nu_A \xi^*)^n \quad (3.1.30)\text{V}$$

Systems composed of stoichiometric proportions of reactants also have rate expressions that will often degenerate to the above form.

Except for the case where n is unity, equation 3.1.30 can be integrated to give

$$\frac{1}{(C_{A0} + \nu_A \xi^*)^{n-1}} - \frac{1}{C_{A0}^{n-1}} = (1 - n)\nu_A kt$$

$$(3.1.31)\text{V}$$

or

$$\frac{1}{C_A^{n-1}} - \frac{1}{C_{A0}^{n-1}} = (1 - n)\nu_A kt$$

$$(3.1.32)\text{V}$$

Among the reactions whose orders are not integers are the pyrolysis of acetaldehyde ($n = 3/2$), and the formation of phosgene from CO and Cl_2 [$r = k(Cl_2)^{3/2}(CO)$].

3.1.2 Mathematical Characterization of Simple Variable Volume Reaction Systems

From the viewpoint of an engineer who must design commercial reactors to carry out gaseous reactions involving changes in the total number of moles present in the system, it is important to recognize that such reactions are usually accompanied by changes in the specific volume of the system under study. These considerations are particularly important in the design of continuous flow reactors. For these systems one must employ the basic definition of the reaction rate given by equation 3.0.1.

$$r = \frac{1}{V}\frac{d\xi}{dt} \quad (3.0.1)$$

Unfortunately, when one combines this relation with the rate functions for various reaction orders, the situation is entirely different from that which prevails in the constant volume case. One cannot develop explicit closed form expressions for the extent of reaction as a function of time for all the cases treated in Section 3.1.1. Since the only common case for which one can develop such a solution is the first-order reaction, we will start by considering this case. Since

$$C_A = \frac{n_A}{V} = \frac{n_{A0} + \nu_A \xi}{V} \quad (3.1.33)$$

the first-order rate expression becomes

$$\frac{1}{V}\frac{d\xi}{dt} = k\left(\frac{n_{A0} + \nu_A \xi}{V}\right) \quad (3.1.34)$$

Solution of this equation subject to the condition that $\xi = 0$ at $t = 0$ gives

$$\nu_A kt = \ln\left(\frac{n_{A0} + \nu_A \xi}{n_{A0}}\right) = \ln\left(\frac{n_A}{n_{A0}}\right) \quad (3.1.35)$$

or

$$\xi = \frac{n_{A0}}{\nu_A}(e^{\nu_A kt} - 1) \quad (3.1.36)$$

For reaction orders other than unity, one must treat the various reactions on an individual basis and be able to express the total volume of the mixture as a function of the composition, temperature, and pressure of the system. In many cases of interest in kinetics, one is interested in reactions that occur at constant pressure and temperature. In such situations one can say that

$$V = V(n_1, n_2, \cdots) = V(n_j) \quad (3.1.37)$$

If the functional form of this relation is known (e.g., if one is dealing with a gaseous system that behaves ideally), this relationship can be combined with equation 3.0.1 and the appropriate rate function to obtain a differential equation, which can then be integrated numerically or in explicit form. If we consider a generalized rate expression of the form

$$r = k \prod_i C_i^{\beta_i} \qquad (3.1.38)$$

one can combine this relation with equation 3.0.1 to give

$$\frac{1}{V(n_j)} \frac{d\xi}{dt} = k \prod_i \left[\frac{n_i}{V(n_j)} \right]^{\beta_i} \qquad (3.1.39)$$

The various mole numbers can be expressed in terms of the extent of reaction and the initial mole numbers. If the functional form of V is known (e.g., ideal gas behavior), one obtains a differential equation that can be integrated even if it is necessary to resort to numerical methods to do so.

In the analysis of experimental rate data taken in variable volume systems it is possible to develop another expression for the rate of formation of species i that is more convenient to use than equation 3.1.39. This alternative approach has been popularized by Levenspiel (2). It involves the use of fraction conversion rather than concentration as a primary variable and is applicable only to systems in which the volume varies linearly with the fraction conversion.

$$V = V_0(1 + \delta_A f_A) \qquad (3.1.40)$$

The fraction conversion is based on the limiting reagent, in this case assumed to be species A. The parameter δ_A is the fraction change in the volume of a closed reacting system between zero conversion and complete conversion. As such it may take on both positive and negative values. Hence

$$\delta_A = \frac{V \,(\text{at } f_A = 1) - V \,(\text{at } f_A = 0)}{V \,(\text{at } f_A = 0)} \qquad (3.1.41)$$

This parameter takes into account the presence of inerts, the use of nonstoichiometric quantities of reactants, and the presence of one or more of the reaction products in the original system. To illustrate this point, let us consider as an example the isothermal gas phase hydrogenation of ethylene ($C_2H_4 + H_2 \rightarrow C_2H_6$) taking place at constant pressure under conditions such that deviations from the ideal gas law are negligible. The total volume of the system is then given by

$$V = n_{\text{total}} \frac{RT}{P} \qquad (3.1.42)$$

Thus

$$\delta_A = \frac{[n_{\text{total}}(\text{at } f_A = 1) - n_{\text{total}}(\text{at } f_A = 0)] \dfrac{RT}{P}}{[n_{\text{total}}(\text{at } f_A = 0)] \dfrac{RT}{P}}$$

$$= \frac{n_{\text{total}}(\text{at } f_A = 1) - n_{\text{total}}(\text{at } f_A = 0)}{n_{\text{total}}(\text{at } f_A = 0)} \qquad (3.1.43)$$

If one starts with a mixture consisting of 2 moles of hydrogen, 4 moles of ethylene, and 3 moles of inert gases, the following mole table may be established.

Moles	$f_A = 0$	$f_A = 1$
H_2	2	0
C_2H_4	4	2
C_2H_6	0	2
Inerts	3	3
Total	9	7

In this case the hydrogen is the limiting reagent and equation 3.1.43 gives

$$\delta_A = \frac{7 - 9}{9} = -\frac{2}{9}$$

If 2 moles of ethane were present in the original mixture used in the last example, the value of δ_A becomes $-(2/11)$. The reader should verify this point.

If one is dealing with a gaseous system in which deviations from ideality are negligible, one may take variations in the absolute temperature and the absolute pressure into account by a slight modification of equations 3.1.40 and

3.1.41. In this case,

$$V = V_0(1 + \delta_A f_A) \frac{T}{T_0} \frac{P_0}{P} \quad (3.1.44)$$

with

$$\delta_A = \frac{V_{f_A=1,T=T_0,P=P_0} - V_{f_A=0,T=T_0,P=P_0}}{V_{f_A=0,T=T_0,P=P_0}}$$

$$(3.1.45)$$

where the temperature T and the pressure P correspond to a given fraction conversion f_A.

To determine the concentration of the limiting reagent at a given fraction conversion one need only note that

$$n_A = n_{A0}(1 - f_A) \quad (3.1.46)$$

and that at constant temperature and pressure,

$$C_A = \frac{n_A}{V} = \frac{n_{A0}(1 - f_A)}{V_0(1 + \delta_A f_A)} = C_{A0} \frac{(1 - f_A)}{(1 + \delta_A f_A)}$$

$$(3.1.47)$$

The concentrations of the other species present in the reaction mixture may be found by using the extent concept. From equation 1.1.9,

$$\xi = \frac{-n_{A0} f_A}{\nu_A} \quad (3.1.48)$$

Thus

$$n_i = n_{i0} + \nu_i \xi = n_{i0} - \frac{\nu_i n_{A0} f_A}{\nu_A} \quad (3.1.49)$$

The concentration of species i is then given by

$$C_i = \frac{n_i}{V} = \frac{n_{i0} - \frac{\nu_i}{\nu_A} n_{A0} f_A}{V_0(1 + \delta_A f_A)} \quad (3.1.50)$$

Equations 3.1.47 and 3.1.50 express the relation between system concentrations and the fraction conversion for variable-volume systems that satisfy the linearity assumption of equation 3.1.40. This assumption is a reasonably unrestrictive one that is valid for all practical purposes in isothermal constant pressure systems in

which one need not be concerned with consecutive reactions. The assumption is also valid for many nonisothermal condensed phase systems. For nonisothermal or variable pressure *gaseous* systems a modification of the form of equation 3.1.44 is more appropriate for use.

To develop expressions for the reaction rate in variable volume systems, one need only return to the fundamental definition of the reaction rate (3.0.1) and combine this relation with equations 3.1.40 and 3.1.48.

$$r = \frac{1}{V} \frac{d\xi}{dt} = \frac{1}{V_0(1 + \delta_A f_A)} \left(-\frac{n_{A0}}{\nu_A} \frac{df_A}{dt} \right)$$

$$= -\frac{C_{A0}}{(1 + \delta_A f_A)} \cdot \frac{1}{\nu_A} \frac{df_A}{dt} \quad (3.1.51)$$

For variable volume systems equation 3.1.51 is much simpler in form and easier to use than equation 3.0.9.

$$r = \frac{1}{\nu_i} \frac{dC_i}{dt} + \frac{C_i}{\nu_i V} \frac{dV}{dt} \quad (3.0.9)$$

The δ_A concept will prove to be particularly useful in the design of tubular reactors for gas phase reactions.

3.2 EXPERIMENTAL ASPECTS OF KINETIC STUDIES

The chief significance of reaction rate functions is that they provide a satisfactory framework for the interpretation and evaluation of experimental kinetic data. This section indicates how a chemical engineer can interpret laboratory scale kinetic data in terms of such functions. Emphasis is placed on the problems involved in the evaluation and interpretation of kinetic data.

One should be very careful to distinguish between the problem of determining the reaction rate function and the problem of determining the mechanism of the reaction. The latter involves a determination of the exact series of molecular processes involved in the reaction. It is by far the more difficult problem. From the viewpoint of the chemical engineer, who is interested in the design of commercial scale reactors, a knowledge of the reaction mechanism is useful, but not essential. Such individuals are more concerned with the problem of determining reaction rate expressions for use in design calculations. Since this function must be determined experimentally, we now turn our attention to some of the points to be considered in obtaining kinetic data.

3.2.1 Preliminary Questions to be Answered in Experimental Kinetics Studies

Since the proper design of chemical reactors requires a knowledge of the reaction rate expression for the system under consideration, it is essential that the chemical engineer thoroughly understand the methods by which these empirical functions are determined. One should be extremely careful in experimental work to obtain a valid rate expression by taking appropriate precautions in the laboratory and in the interpretation of the data. Since many kinetic data in the literature are in error in one important respect or another and since many experienced chemists and chemical engineers have erred in carrying out kinetics experiments, it is crucial for the beginning student to be aware of some of the pitfalls. The sections that follow treat several aspects of kinetic experimentation that must be considered if one is to obtain meaningful data. The discussion is patterned after Bunnett's (3) treatment in *Technique of Organic Chemistry*.

3.2.1.1 Is the Reaction Under Study Properly Identified? It may seem ridiculous to ask the experimenter whether the reaction whose kinetics are being studied is actually the reaction he or she thinks it is. Nonetheless, the literature contains many examples of the incorrect identification of reactions. One cannot always reason that the products of a reaction are the corresponding analogs of a well-characterized

reaction. Even though the reaction under investigation may be reported in textbooks, isolation and identification of the products should be carried out *with the products of the reaction run under the conditions of the rate measurement*. Errors exist in the literature, and one may have components present in his or her system that may promote side reactions and/or inhibit the reaction that one thinks is being studied.

3.2.1.2 Are Side Reactions Important? What is the Stoichiometry of the Reaction? When a mixture of various species is present in a reaction vessel, one often has to worry about the possibility that several reactions, and not just a single reaction, may occur. If one is trying to study one particular reaction, side reactions complicate chemical analysis of the reaction mixture and mathematical analysis of the raw data. The stoichiometry of the reaction involved and the relative importance of the side reactions must be determined by qualitative and quantitative analysis of the products of the reaction at various times. If one is to observe the growth and decay of intermediate products in series reactions, measurements must be made on the reaction system before the reaction goes to completion.

3.2.1.3 Are the Conditions of the Experiment Properly Specified? All kineticists should recognize the importance in experimental work of using pure materials, of controlling and measuring temperatures accurately, and of recording times properly. Nonetheless, it is very common for individuals to "save time" by using reactants or solvents that are not properly purified.

The first tendency of the chemist involved in experimental kinetics work might well be to say that one should purify the reagents and solvents as much as possible. On the other hand, the first tendency of a practicing engineer would be to use a minimum of purification procedures, particularly if the ultimate goal is to design a commercial scale reactor that will use a relatively impure feedstock. In practice, what one does is to accept reagents and solvents at a certain degree of purification as being conditionally satisfactory subject to the reservation that preliminary experiments might indicate the need for further purification. These preliminary experiments may also indicate that the chemical engineer will have to allow for purification of the feedstock in the design of the commercial scale facility.

The reason for stressing the importance of working with relatively pure reagents and solvents is that the rates of many reactions are extremely sensitive to the presence of trace impurities in the reaction system. If there is reason to suspect the presence of these effects, a series of systematic experiments may be carried out to explore the question by seeing how the reaction rate is affected by the intentional addition of impurities. In many cases, lack of reproducibility between experiments may be an indication that trace impurity effects are present.

Temperature control for most reactions in solution can be provided by immersion of the reactor in a liquid constant temperature bath that can be controlled to ± 0.1 °C or better, depending on the desired degree of control. For gas phase reactions, particularly those that occur at high temperatures, the reactor can be immersed in a fluidized sand bath. The fact that these thermostatted baths can be held at a constant temperature for prolonged periods of time may lead to such a sense of complacency on the part of the experimenter that he or she fails to record the temperature at periodic intervals. One of the laws of experimental work ("If anything can go wrong, it will") dictates that the investigator adhere to a strict timetable for recording the system temperature. When the kinetic data of different laboratories disagree, the failure to calibrate temperatures properly is often at fault. Consequently, it is wise to calibrate periodically the device used to record the bath temperature.

Although some precautions must be observed in timing a reaction, the direct measurement of time seldom affects the accuracy of a rate determination. The time required for sampling, for initiating, or for quenching a reaction is likely to introduce larger uncertainties in the rate measurement than is inherent in the performance of a good timer. One generally determines the value of the reaction rate constant by plotting some function of the instantaneous concentrations versus time and determining the slope of the straight line that results. Since this slope is determined by differences between the time points and not by their absolute values, it is not essential in most cases to know these absolute values of time. However, a word of caution is appropriate when one is concerned with the analysis of consecutive reactions. In such cases a knowledge of the true zero of time is required. (See Section 5.3.)

3.2.1.4 Does the Analytical Method Properly Represent the Extent of Reaction?   One should test the analytical procedure by analyzing mixtures of known compositions that are similar to those expected for samples from the reaction vessel. Even if this test gives an excellent check, one must still face the possibility that the primary reaction under investigation may be accompanied by unexpected side reactions that consume reactants or produce products in such a way as to distort the meaning of the analytical results. One general approach that provides a measure of protection against being misled by analytical information is to follow the progress of the reaction by two or more different and independent analytical methods. Where possible, each of these methods should involve analysis for a different species.

3.2.1.5 Is the Experiment Properly Planned to Provide Significant Data?   One cannot directly measure a reaction rate. One is restricted to measurements of the concentrations of various species or of a physical property of the system as a function of time. Thus one must plan the experiments so as to obtain significant differences in the quantities that are observed in the laboratory. In a properly planned experiment the

kineticist chooses experimental techniques that provide differences between successive samples that are many times greater than the expected error in the measurement. This procedure permits the kineticist to derive a meaningful rate constant from the original measurements.

One decision that the kineticist must make in planning experiments is the number of observations to be made during a run and the times at which these observations are to be made. One usually tries to work up the data in a form such that a plot of some function of the concentrations in the reaction vessel versus time gives a straight line. The reaction rate constant is then derived from the slope of this line. The question then becomes, "How many points does it take to establish the linearity and slope of the line?" Two points are adequate to determine a straight line, but they are not sufficient to establish linearity of data. Any number of curved lines could be drawn through two points. Various functional forms $\phi(C_i)$ corresponding to different reaction orders could be used as the ordinate in these plots, and one could not distinguish between them on the basis of two experimental points. The straight line joining these points will not give any indication of whether or not other experimental points would lie on the same straight line. Moreover, this approach to determining a rate constant cannot reveal any of the experimental defects or complications that are often called to one's attention by curved or scattered plots.

If one has three or four completely accurate points, an adequate test of the linearity of the assumed concentration function could be made. However, one cannot be sure whether departures from linearity are the result of scatter of the data, of a slowly curving function, or of some combination of these circumstances. When one further recognizes that one or two points in a set of data may be in error as a result of an experimental mishap and that these points may deviate substantially from the line established by the rest of the data, one comes to the conclusion that it is desirable to have eight to ten data points per run.

Such numbers are usually adequate to allow curvature to be distinguished from linearity, even when one needs to reject one or two points because of scatter of the data.

3.2.1.6 Are the Data Reproducible?

If experimental results are to be accepted as meaningful by the scientific community, they must be capable of being reproduced by investigators in other laboratories as well as by the original investigator. An individual should assure this reproduction by replicating the experiment.

There are several sources of irreproducibility in kinetics experimentation, but two of the most common are individual error and unsuspected contamination of the materials or reaction vessel used in the experiments. An individual may use the wrong reagent, record an instrument reading improperly, make a manipulative error in the use of the apparatus, or plot a point incorrectly on a graph. Any of these mistakes can lead to an erroneous rate constant. The probability of an individual's repeating the same error in two successive *independent* experiments is small. Consequently, every effort should be made to make sure that the runs are truly independent, by starting with fresh samples, weighing these out individually, etc. Since trace impurity effects also have a tendency to be time-variable, it is wise to check for reproducibility, not only between runs over short time spans, but also between runs performed weeks or months apart.

It is commendable experimental procedure to repeat each run in duplicate and to be satisfied if the two results agree, but this is expensive in terms of the labor costs involved. Moreover, repetition of each run is not always necessary. For example, if one is studying the effect on the reaction rate of a variable such as temperature or reactant concentration, a series of experiments in which the parameter under investigation is systematically varied may be planned. If a plot of the results versus this parameter yields a smooth curve, one generally assumes that the reproducibility of the data is satisfactory.

In addition to worrying about the reproducibility of the data, one should also take care to insure that the calculations are reproducible by an independent calculation by a colleague. The literature contains numerous values of miscalculated rate constants. One has the usual problem of conversion of units and the problem of conversion from natural logarithms to common logarithms. Thus, even though one's raw data might be highly accurate, the calculated results may be off by a large multiple. If a colleague uses the raw data, determines the rate constant for a particular experimental run with reference only to first principles, and obtains the same rate constant as the original investigator, one has a good indication that a systematic error in calculations has not been made.

3.2.2 Experimental Techniques and Apparatus

In order to determine reaction rate constants and reaction orders, it is necessary to determine reactant or product concentrations at known times and to control the environmental conditions (temperature, homogeneity, pH, etc.) during the course of the reaction. The experimental techniques that have been used in kinetics studies to accomplish these measurements are many and varied, and an extensive treatment of these techniques is far beyond the intended scope of this textbook. It is nonetheless instructive to consider some experimental techniques that are in general use. More detailed treatments of the subject are found in the following books.

1. R. B. Anderson, *Experimental Methods in Catalytic Research*, Academic Press, New York, 1968.
2. H. W. Melville, and B. G. Gowenlock, *Experimental Methods in Gas Reactions*, Macmillan, New York, 1964.
3. S. L. Freiss, E. S. Lewis, and A. Weissberger, *Technique of Organic Chemistry*, Volume VIII, "Investigation of Rates and Mechanisms of Reactions," Parts One and Two, Second Edition, Wiley-Interscience, New York, 1961.
4. A. Weissberger, *Techniques of Chemistry*, Volume VI, "Investigation of Rates and Mechanisms of Reactions," Parts I and II, Third Edition, Wiley-Interscience, New York, 1974.

3.2.2.1 General Reactor Types. In planning experiments the chemist is most apt to think in terms of closed systems and has traditionally used some form of batch reactor. The chemical engineer, on the other hand, is accustomed to dealing with open systems and the eventual design of large-scale continuous processes. Consequently, the chemical engineer will often choose some form of flow reactor for laboratory scale investigations. The discussion in this chapter is restricted to batch reactors, but the relations developed in Chapter 8 may be used to analyze kinetic data from flow reactors.

3.2.2.2 Methods of Following the Course of a Reaction. A general direct method of measuring the *rate* of a reaction does not exist. One can only determine the amount of one or more product or reactant species present at a certain time in the system under observation. If the composition of the system is known at any one time, then it is sufficient to know the amount of any one species involved in the reaction as a function of time in order to be able to establish the complete system composition at any other time. This statement is true of any system whose reaction can be characterized by a single reaction progress variable (ξ or f_A). In practice it is always wise where possible to analyze occasionally for one or more other species in order to provide a check for unexpected errors, losses of material, or the presence of side reactions.

In more complex cases when several reactions are occurring simultaneously in the system under observation, calculations of the composition of the system as a function of time will require the knowledge of a number of independent composition variables equal to the number of independent chemical equations used to characterize the reactions involved.

In principle one can use any of the many tools and methods of the analytical chemist in carrying out these determinations. In practice analytical methods are chosen on the basis of their specificity, accuracy, ease of use, and rapidity of measurement. For our discussion these methods may be classified into two groups: physical and chemical. Regardless of the method chosen, however, it must meet the following criteria.

1. It must not disturb the system under investigation by affecting the kinetic processes occurring therein.
2. The measurement should be representative of the system at the time it is made, or at the time the sample analyzed was taken.
3. The method must provide a true measure of the extent of reaction.

In the choice of an analytical technique for following the course of a reaction, it is important to recognize that no aspect of the measurement should affect the kinetic processes occurring in the system. For example, a solution conductivity method that uses platinized electrodes should not be used in the study of a reaction that is catalyzed by platinum black.

Perhaps the most obvious method of studying kinetic systems is to periodically withdraw samples from the system and to subject them to chemical analysis. When the sample is withdrawn, however, one is immediately faced with a problem. The reaction will proceed just as well in the test sample as it will in the original reaction medium. Since the analysis will require a certain amount of time, regardless of the technique used, it is evident that if one is to obtain a true measurement of the system composition at the time the sample was taken, the reaction must somehow be quenched or inhibited at the moment the sample is taken. The quenching process may involve sudden cooling to stop the reaction, or it may consist of elimination of one of the reactants. In the latter case, the concentration of a reactant may be reduced rapidly by precipitation or by fast quantitative reaction with another material that is added to the sample mixture. This material may then be back-titrated. For example, reactions between iodine and various reducing agents can be quenched by addition of a suitably buffered arsenite solution.

It is useful at this point to consider briefly the relative merits of chemical and physical methods for monitoring the course of a reaction. A *chemical* method involves removal of a portion of the reacting system, inhibition of the reaction occurring within the sample, and subsequent analysis of the system composition. This analysis may be carried out using conventional wet chemistry techniques, a spectroscopic method, or any one of a variety of other analytical techniques. The essence of the chemical method, however, lies in the fact that it requires removal of a sample from the reacting system. Chemical methods give absolute values of the concentrations of the various species present in the reaction mixture. However, it should be stressed that even the most *precise* of these methods has an *accuracy* that is no better than the *accuracy* of the standard analytical procedure that was used to calibrate the method. Another disadvantage inherent in the use of chemical methods is that the sampling procedure involved does not provide a *continuous* record of the reaction progress. It is impossible to study very fast reactions using these methods.

Physical methods involve the measurement of a physical property of the system as a whole while the reaction proceeds. The measurements are usually made in the reaction vessel so that the necessity for sampling with the possibility of attendant errors is eliminated. With physical methods it is usually possible to obtain an essentially continuous record of the values of the property being measured. This can then be transformed into a continuous record of reactant and product concentrations. It is usually easier to accumulate much more data on a given reaction system with such methods than is possible with chemical methods. There are certain limitations on physical methods, however. There must be substantial differences in the contributions of the reactants and products to the value of the particular physical property used as a measure of the reaction progress. Thus one would not use pressure measurements to follow the course of a gaseous reaction that does not involve a change in the total number of moles in the gaseous phase.

It is always wise to calibrate physical methods of analysis using mixtures of known composition under conditions that approximate as closely as practicable those prevailing in the reaction system. This procedure is recommended because side reactions can introduce large errors and because some unforeseen complication may invalidate the results obtained with the technique. For example, in spectrophotometric studies of reaction kinetics, the absorbance that one measures can be grossly distorted by the presence of small amounts of highly colored absorbing impurities or by-products. For this reason, when one uses indirect physical methods in kinetic studies, it is essential to verify the stoichiometry of the reaction to ensure that the products of the reaction and their relative mole numbers are known with certainty. For the same reason it is recommended that more than one physical method of analysis be used in detailed kinetic studies.

In principle, any physical property that varies during the course of the reaction can be used to follow the course of the reaction. In practice one chooses methods that use physical properties that are simple exact functions of the system composition. The most useful relationship is that the property is an additive function of the contributions of the different species and that each of these contributions is a linear function of the concentration of the species involved. This physical situation implies that there will be a linear dependence of the property on the extent of reaction. As examples of physical properties that obey this relationship, one may cite electrical conductivity of dilute solutions, optical density, the total pressure of gaseous systems under nearly ideal conditions, and rotation of polarized light. In sufficiently dilute solutions, other physical properties behave in this manner to a fairly good degree of approximation. More complex relationships than the linear one can be utilized but, in such cases, it is all the more imperative that the experimentalist prepare care-

ful calibration curves relating the property being measured to the extent of reaction or species concentrations.

3.2.2.3 Physical Methods that have been Used to Monitor Reaction Kinetics.

In this section some physical property measurements of general utility are discussed. One of the oldest and most useful techniques used in kinetics studies involves the measurement of the total pressure in an *isothermal constant volume reactor*. This technique is primarily used to follow the course of homogeneous gas phase reactions that involve a change in the total number of gaseous molecules present in the reaction vessel (e.g., the hydrogenation of propylene).

$$H_2 + C_3H_6 \rightarrow C_3H_8 \qquad (3.2.1)$$

Pressure measurements can be accomplished by any one of a number of different types of manometric devices without disturbing the system being observed. Another type of reaction system that can be monitored by pressure measurements is one in which one of the products can be quantitatively removed by a solid or liquid reagent that does not otherwise affect the reaction. For example, acids formed by reactions in the gas phase can be removed by absorption in hydroxide solutions.

From a knowledge of the reaction stoichiometry and measurements of the total pressure as a function of time, it is possible to determine the extent of reaction and the partial pressures or concentrations of the various reactant and product species at the time at which the measurement is made. Illustration 3.3 indicates how pressure measurements can be used to determine a reaction rate function.

The various forms of spectroscopy find widespread application in kinetic studies. They are usually well suited for application to *in situ* studies of the characteristics of the reaction mixture. The absorption by a reacting system of electromagnetic radiation (light, microwaves, radio-frequency waves, etc.) is a highly specific property of the system composition and the dimensions of the reaction vessel. Both the magnitude of the absorption coefficient and the wavelengths at which absorption maxima occur are characteristic of the absorbing compound and, to a lesser extent, its physical state. By appropriate choice of the wavelength of the incident radiation, one is able to take advantage of the remarkable specificity of the various forms of spectroscopy to monitor the progress of the reaction. Among the various forms of spectroscopy that can be used in *in situ* kinetic studies are visible, ultraviolet, infrared, microwave, nuclear magnetic resonance, and electron spin resonance spectroscopy. For treatments of the limitations and uses of spectroscopic techniques, consult treatises on analytical chemistry.

In addition to spectrophotometric or spectroscopic measurements, there are a number of other optical measurements that can be used to monitor the course of various reactions. Among the optical properties that can be used for these studies are optical rotation, refractive indices, fluorescence, and colorimetry.

There are several electrical measurements that may be used for analysis of solutions under *in situ* conditions. Among the properties that may be measured are dielectric constants, electrical conductivity or resistivity, and the redox potential of solutions. These properties are easily measured with instrumentation that is readily adapted to automatic recording operation. However, most of these techniques should be used only after careful calibration and do not give better than 1% accuracy without unusual care in the experimental work.

3.3 TECHNIQUES FOR THE INTERPRETATION OF REACTION RATE DATA

In Section 3.1 the mathematical expressions that result from integration of various reaction rate functions were discussed in some detail. Our present problem is the converse of that considered earlier (i.e., given data on the concentra-

tion of a reactant or product as a function of time, how does one proceed to determine the reaction rate expression?).

The determination of the reaction rate expression involves a two-step procedure. First, the concentration dependence is determined at a fixed temperature. Then the temperature dependence of the reaction rate constant is evaluated to give a complete reaction rate expression. The form of this temperature dependence is given by equation 3.0.14, so our present problem reduces to that of determining the form of the concentration dependence and the value of the rate constant at the temperature of the experiment.

Unfortunately, there is no completely general method of determining the reaction rate expression or even of determining the order of a reaction. Usually one employs a trial-and-error procedure based on intelligent guesses and past experience with similar systems. Very often the stoichiometry of the reaction and a knowledge of whether the reaction is "reversible" or "irreversible" will suggest the form of the rate equation to try first. If this guess is incorrect, the investigator may then try certain forms that are suggested by assumptions about the mechanism of the reaction. Each reaction presents a unique problem, and success in fitting a reaction rate expression to the experimental data depends on the ingenuity of the individual investigator.

The discussion below is largely confined to irreversible reactions with simple rate expressions of the form of equation 3.0.17.

$$r = kC_A^{\beta_A}C_B^{\beta_B}C_C^{\beta_C} \cdots \qquad (3.0.17)$$

but the methods developed are more generally applicable. We have chosen this course to keep the discussion as simple as possible and to present the material in a manner that avoids the introduction of more complex rate expressions. Most of the methods presented below are applicable, regardless of the mathematical form of the rate expression, and they may be readily extended to cover the rate expressions that will be encountered in subsequent chapters.

The techniques used to determine reaction rate functions may be broken down into three general categories.

1. Integral methods based on integration of the reaction rate expression. In these approaches one analyzes the data by plotting some function of the reactant concentrations versus time.
2. Differential methods based on differentiation of experimental concentration versus time data in order to obtain the actual rate of reaction. In these approaches one analyzes the data by postulating various functional relations between the rate of reaction and the concentrations of the various species in the reaction mixture and tests these hypotheses using appropriate plots.
3. Methods based on simplification of the reaction rate expression. In these approaches one uses a vast excess of one or more of the reactants or stoichiometric ratios of the reactants in order to permit a partial evaluation of the form of the rate expression. They may be used in conjunction with either a differential or integral analysis of the experimental data.

Each of these general categories is discussed in the subsections that follow.

3.3.1 Differential Methods for the Treatment of Reaction Rate Data

In experimental kinetics studies one measures (directly, or indirectly) the concentration of one or more of the reactant and/or product species as a function of time. If these concentrations are plotted against time, smooth curves should be obtained. For constant volume systems the reaction rate may be obtained by graphical or numerical differentiation of the data. For variable volume systems, additional numerical manipulations are necessary, but the process of determining the reaction rate still involves differentiation of some form of the data. For example,

from a knowledge of the concentration of one of the reactants or products at some time t, the initial composition of the system, and the reaction stoichiometry (from which δ_A may be determined), it is possible to use equation 3.1.47 or equation 3.1.50 to determine the fraction conversion of the limiting reagent at this time. Equation 3.1.51 may then be used to determine the reaction rate at this conversion level.

$$r = \left(\frac{-C_{A0}}{1 + \delta_A f_A} \right) \frac{1}{v_A} \frac{df_A}{dt} \qquad (3.1.51)$$

Thus, for both variable and constant volume systems, one can manipulate concentration versus time data to obtain values of the reaction rate as a function of time or as a function of the concentrations of the various species present in the reaction mixture. The task then becomes one of fitting this data to a reaction rate expression of the form of equation 3.0.13.

$$r = k\phi(C_i) \qquad (3.0.13)$$

Since data are almost invariably taken under isothermal conditions to eliminate the temperature dependence of reaction rate constants, one is primarily concerned with determining the concentration dependence of the rate expression $[\phi(C_i)]$ and the rate constant at the temperature in question. We will now consider two differential methods that can be used in data analysis.

3.3.1.1 Differentiation of Data Obtained in the Course of a Single Experimental Run.
If one has experimental results in the form of concentration versus time data, the following general differential procedure may be used to determine $\phi(C_i)$ and k at the temperature in question.

1. Set forth a hypothesis as to the form of the concentration dependent portion of the rate function, $\phi(C_i)$.
2. From the experimental concentration-time data, determine the reaction rate at various times.

3. At the selected times, prepare a table listing the reaction rate and the concentrations of the various species present in the reaction mixture. Calculate $\phi(C_i)$ at each of these points.
4. Prepare a plot of the reaction rate versus $\phi(C_i)$. If the plot is linear and passes through the origin, the form of ϕ is consistent with the experimental data. Consequently, insofar as the accuracy of the data is concerned, the form of the rate expression is satisfactory for subsequent use in design calculations. The slope of this straight line is the reaction rate constant. If the plot is not linear or does not pass through the origin, one must return to step 1 and hypothesize a new form for the concentration dependence of the reaction rate.

In its application to specific kinetics studies this general procedure may take on a variety of forms that are minor modifications of that outlined above. One modification does not require an explicit assumption of the form of $\phi(C_i)$ including numerical values of the orders of the reaction with respect to the various species, but merely an assumption that the rate expression is of the following form.

$$r = k\phi(C_i) = kC_A^{\beta_A}C_B^{\beta_B}C_C^{\beta_C} \cdots \quad (3.3.1)$$

The reaction rate r and the various C_i may be determined from the experimental data. Taking the logarithms of both sides of this equation:

$$\log r = \log k + \beta_A \log C_A + \beta_B \log C_B$$

$$+ \beta_C \log C_C + \cdots \quad (3.3.2)$$

Data may be fitted to this equation by the *method of least squares* in order to determine values of the constants $\log k$, β_A, β_B, etc. The goodness of fit may be shown in graphical form by using the values of β_i determined in this manner to calculate $\phi(C_i)$. A plot of the reaction rate versus this function should then meet the criteria of the general method outlined above.

A second common modification of the general method is that which ensues when one uses stoichiometric ratios of the reactants. One must also assume that the rate expression is of the form of equation 3.3.1 with the β_i *for all species other than reactants taken as zero.*

From the use of a stoichiometric feed ratio,

$$\frac{n_{A0}}{\nu_A} = \frac{n_{B0}}{\nu_B} = \frac{n_{C0}}{\nu_C} \quad (3.3.3)$$

From the definition of extent of reaction,

$$\xi = \frac{n_A}{\nu_A} - \frac{n_{A0}}{\nu_A}; \quad \xi = \frac{n_B}{\nu_B} - \frac{n_{B0}}{\nu_B}; \quad \text{and} \quad \xi = \frac{n_C}{\nu_C} - \frac{n_{C0}}{\nu_C} \quad (3.3.4)$$

From equations 3.3.3 and 3.3.4, one can see that at all times, with a stoichiometric feed ratio,

$$\frac{n_A}{\nu_A} = \frac{n_B}{\nu_B} = \frac{n_C}{\nu_C} \quad (3.3.5)$$

or

$$\frac{C_A}{\nu_A} = \frac{C_B}{\nu_B} = \frac{C_C}{\nu_C} = \frac{C_i}{\nu_i} \quad (3.3.6)$$

since all species occupy the same volume. Consequently, equation 3.3.1 may be written in the form

$$r = k \prod_{\text{reactants}} C_i^{\beta_i} = k\left(\frac{C_A}{\nu_A}\right)^{\sum \beta_i} \prod_{\text{reactants}} (\nu_i)^{\beta_i} = k'C_A^{m} \quad (3.3.7)$$

where we have defined a new rate constant k' to include the effect of the stoichiometric coefficients and have replaced the overall order of the reaction $(\sum_{\text{reactants}} \beta_i)$ by m.

If one now takes the logarithm of both sides of this expression, the following expression results.

$$\log r = \log k' + m \log C_A \quad (3.3.8)$$

The constants k' and m may be determined from a log-log plot of the rate versus C_A. This procedure leads to a value for the overall order of the reaction. Experiments with nonstoichiometric ratios of reactants can then be used to determine the orders of the reaction with respect to each of the individual species.

Differential procedures are illustrated schematically in Figure 3.1. The first diagram indicates how the rate may be determined from concentration versus time data in a constant volume system; the second schematic illustrates the method just described. The third diagram indicates the application of our general differential method to this system.

It is always preferable to use as much of the data as possible to determine the reaction rate function. This often implies that one should use some sort of graphical procedure to analyze the data. From such plots one may see that certain points are seriously in error and should not be weighted heavily in the determination of the reaction rate function. The consistency and precision of the data can also be evaluated by eye by observing its deviation from a smooth curve (ideally its deviation from a straight line).

The following example illustrates the use of the differential method for the analysis of kinetic data. It also exemplifies some of the problems

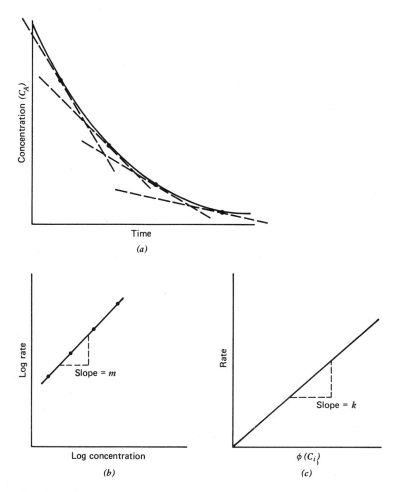

Figure 3.1
Schematic representation of the application of the differential method for data analysis.

one has in attempting to utilize a differential approach in his analysis.

ILLUSTRATION 3.1 USE OF A DIFFERENTIAL METHOD TO DETERMINE A PSEUDO REACTION RATE EXPRESSION FOR THE IODINE CATALYZED BROMINATION OF m-XYLENE

Neyens (4) has studied the bromination of meta-xylene at 17 °C. The reaction is carried out by introducing small quantities of iodine and bromine into pure liquid xylene and following the rate of disappearance of bromine by titrating samples removed from the liquid to determine their bromine content. The iodine serves as a catalyst for the reaction. Since the concentrations of xylene and catalyst remain essentially unchanged during the course of the reaction, it may be assumed that the rate expression is of the form

$$r = kC_{Br_2}^m = -\frac{dC_{Br_2}}{dt} \quad \text{(A)}$$

where k is a pseudo rate constant that will depend on the iodine and xylene concentrations. Use a differential method to determine the order of the reaction and the reaction rate constant.

Data

Time, t (min)	Bromine concentration (moles/liter)	Time, t (min)	Bromine concentration (moles/liter)
0	0.3335	19.60	0.1429
2.25	0.2965	27.00	0.1160
4.50	0.2660	30.00	0.1053
6.33	0.2450	38.00	0.0830
8.00	0.2255	41.00	0.0767
10.25	0.2050	45.00	0.0705
12.00	0.1910	47.00	0.0678
13.50	0.1794	57.00	0.0553
15.60	0.1632	63.00	0.0482
17.85	0.1500		

Solution

Two differential approaches to the analysis of the above data will be presented. The first of these is based on the similarity of equation A to equation 3.3.7. The second is the general approach outlined earlier.

In logarithmic form equation A becomes

$$\log\left(-\frac{dC_{Br_2}}{dt}\right) = \log k + m \log C_{Br}, \quad \text{(B)}$$

The term in parentheses on the left side of equation B may be determined from the data in several ways. The bromine concentration may be plotted as a function of time and the slope of the curve at various times determined graphically. Alternatively, any of several methods of numerical differentiation may be employed. The simplest of these is used in Table 3.I.1 where dC/dt is approximated by $\Delta C/\Delta t$. Mean bromine concentrations corresponding to each derivative are also tabulated.

Figure 3.2 is a plot of reaction rate versus mean bromine concentration using logarithmic coordinates. The slope of this plot (m) is 1.54, which is approximately 1.5. Neyens concluded that the reaction was 3/2 order in bromine. The rate constant can be determined from the value of the rate corresponding to a bromine concentration of 1 mole/liter. Thus $k = 1.0 \times 10^{-1}$ (liters/mole)$^{1/2}$ min^{-1}. The value reported by Neyens was 0.91×10^{-1} based on the use of an integral method for the analysis of the data. However, slight shifts in the slope of the straight line in Figure 3.2 could bring about rather large changes in the intercept. It should also be obvious from this figure that the fit of the data to the straight line is not nearly as good as one would like. One frequently encounters problems of this type in attempting to use differential methods for the analysis of data taken in a batch reactor. Some of the scatter could probably be removed, however, if one prepared a concentration versus time plot and determined the slope graphically.

Table 3.I.1
Data Workup for Illustration 3.1

Time, t (min)	Bromine concentration (moles/liter)	$-\dfrac{\Delta C}{\Delta t} \times 10^3$	$\bar{C}$ (moles/liter)	$\bar{C}^{1.5}$ (moles/liter)$^{1.5}$
0	0.3335			
2.25	0.2965	16.44	0.3150	0.1762
4.50	0.2660	13.56	0.2812	0.1491
6.33	0.2450	11.48	0.2555	0.1291
8.00	0.2255	11.68	0.2353	0.1141
10.25	0.2050	9.11	0.2153	0.0999
12.00	0.1910	8.00	0.1980	0.0881
13.50	0.1794	7.73	0.1852	0.0797
15.60	0.1632	7.71	0.1713	0.0709
17.85	0.1500	5.87	0.1566	0.0620
19.60	0.1429	4.06	0.1465	0.0561
27.00	0.1160	3.64	0.1295	0.0466
30.00	0.1053	3.23	0.1107	0.0369
38.00	0.0830	2.79	0.0942	0.0289
41.00	0.0767	2.10	0.0799	0.0226
45.00	0.0705	1.55	0.0736	0.0200
47.00	0.0678	1.35	0.0692	0.0182
57.00	0.0553	1.25	0.0615	0.0153
63.00	0.0482	1.18	0.0518	0.0118

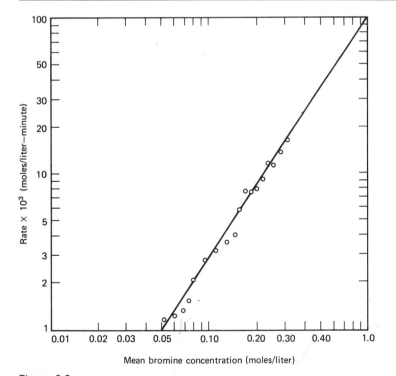

Figure 3.2
Log-log plot of reaction rate versus bromine concentration for illustration 3.1.

The second method that will be used for the solution of the problem posed in this illustration is the general procedure outlined at the beginning of this section.

1. Assume that the concentration dependent portion of the rate function $\phi(C_i)$ is equal to $C_{Br_2}^{1.5}$.
2. From the experimental concentration-time data, determine the reaction rate at various times. (See Table 3.I.1.)
3. Calculate values of $C_{Br_2}^{1.5}$ at those times for which the reaction rate has been determined. (See Table 3.I.1.)
4. The reaction rate is plotted versus $C_{Br_2}^{1.5}$, as shown in Figure 3.3. The plot is reasonably linear and passes through the origin. I have also plotted the rate versus C_{Br_2} and $C_{Br_2}^2$. Both of these plots show marked curvature, particularly at the points corresponding to high bromine conversions or low bromine concentrations. One plot is concave and the other convex. From the linear relationship that exists for the 1.5 order case, one may conclude that this order is approximately correct. Again, however, one should note the fact that there is appreciable scatter in the data when one attempts to apply this method.

The value of the pseudo reaction rate constant may be determined from the slope of the straight line in Figure 3.3. This value is 9.2×10^{-2} (liters/mole)$^{1/2}$ min^{-1}. This number is consistent with the value determined by Neyens (9.1×10^{-2}).

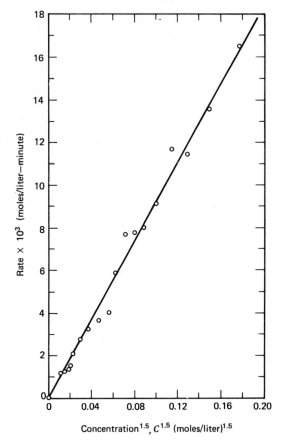

Figure 3.3
Plot of reaction rate versus $C_{Br_2}^{1.5}$ to determine reaction order in illustration 3.1.

3.3.1.2 Initial Rate Measurements. Another differential method useful in the determination of reaction rate expressions is the initial rate approach. It involves a series of rate measurements at different initial reactant concentrations but restricted to very small conversions of the limiting reagent (5 to 10% or less). This technique differs from those discussed previ-

ously in that lower conversions are used and each rate measurement involves a new experiment.

In an initial rate study, one focuses on the conditions that prevail at the start of the reaction. Since the concentrations of the various species do not undergo large changes during this period (varying by 10% at the most), one may characterize these concentrations by initial or average values that can then be substituted directly into the trial rate expression. One may determine the values of the reaction rate corresponding to zero time by measuring the initial slopes of concentration versus time curves in constant volume systems or by measuring the initial slope of a fraction conversion versus time curve and using equation 3.1.51 to calculate the initial reaction rate. Numerical calculations could also be used. In either event, for each run one obtains a value of the initial reaction rate r_0 at a given mixture composition. If the reaction rate expression is of the form

$$r_0 = kC_{A0}^{\beta_A} C_{B0}^{\beta_B} C_{C0}^{\beta_C} = k \prod_i C_{i0}^{\beta_i} \quad (3.3.9)$$

one may determine the order of the reaction with respect to an individual component (e.g., A) by making a rate measurement at two different initial concentrations of this species while holding all other concentrations constant between the two runs. If we denote the two observed rates by r_0 and r'_0 and the corresponding initial concentrations by C_{A0} and C'_{A0}, equation 3.3.9 may be written for these runs as

$$r_0 = (kC_{B0}^{\beta_B} C_{C0}^{\beta_C} \cdots)C_{A0}^{\beta_A} \quad (3.3.10)$$

and

$$r'_0 = (kC_{B0}^{\beta_B} C_{C0}^{\beta_C} \cdots)C_{A0}'^{\beta_A} \quad (3.3.11)$$

Division of equation 3.3.10 by equation 3.3.11 and solution of the resulting expression for β_A gives

$$\beta_A = \frac{\log\left(\dfrac{r_0}{r'_0}\right)}{\log\left(\dfrac{C_{A0}}{C'_{A0}}\right)} \quad (3.3.12)$$

By varying the initial concentration of each component of the reaction mixture in turn, it is possible to determine the order of the reaction with respect to each species. After this has been established, equation 3.3.9 may be used to determine the reaction rate constant.

If one has initial rate data available at several different concentrations of species A and at the same initial concentrations of all other species, a more accurate value of β_A may be determined from a log-log plot of equation 3.3.10. In many cases orders determined in this fashion will not

be simple integers or half-integers. When this is the case, one should note that the values of β that fit the data may not have mechanistic significance. Instead, they merely provide exponents for a reasonable mathematical approximation to the true rate expression, which itself may be a much more complex mathematical function.

One advantage of the initial rate method is that complex rate functions that may be extremely difficult to integrate can be handled in a convenient manner. Moreover, if one uses initial reaction rates, the reverse reactions can be neglected and attention can be focused solely on the reaction rate function for the forward reaction. More complex rate functions may be tested by the choice of appropriate coordinates for plotting the initial rate data. For example, a reaction rate function of the form

$$r_0 = \frac{kC_{A0}}{1 + k'C_{A0}} \quad (3.3.13)$$

may be tested by plotting the reciprocal of the initial rate versus the reciprocal of the initial concentration of species A. Since

$$\frac{1}{r_0} = \frac{1}{kC_{A0}} + \frac{k'}{k} \quad (3.3.14)$$

this plot should be linear with an intercept equal to k'/k and a slope equal to $1/k$.

3.3.2 Integral Methods for the Treatment of Reaction Rate Data

When integral methods are used in data analysis, measured concentrations are used in tests of proposed mathematical formulations of the reaction rate function. One guesses a form of the reaction rate expression on the basis of the reaction stoichiometry and assumptions concerning its mechanism. The assumed expression is then integrated to give a relation between the composition of the reaction mixture and time. A number of such relations were developed in

Section 3.1. The present section indicates how experimental data are tested for consistency with these relations. Several methods based on integration of the reaction rate expression are considered.

3.3.2.1 A General Integral Method for the Analysis of Kinetic Data—Graphical Procedure.

The general integral technique for the determination of reaction rate functions consists of the following trial-and-error procedure.

1. Set forth a hypothesis as to the mathematical form of the reaction rate function.

$$\frac{1}{V}\frac{d\xi}{dt} = k\phi(C_i) \qquad (3.3.15)$$

2. Separate variables and integrate.

$$t = \int_0^\xi \frac{d\xi}{kV\phi(C_i)} \qquad (3.3.16)$$

In order to evaluate this integral it is necessary to express all terms on the right in terms of a single variable. For isothermal systems k is a constant, and this equation can be written as

$$kt = \int_0^\xi \frac{d\xi}{V\phi(C_i)} \qquad (3.3.17)$$

Since our present goal is the determination of $\phi(C_i)$, we will restrict the subsequent discussion to isothermal systems. If the integral is represented by $\psi(C_i)$,

$$kt = \psi(C_i) \qquad (3.3.18)$$

3. From experimentally determined values of the various concentrations, or from the value of a single measured concentration and the reaction stoichiometry, calculate the value of $\psi(C_i)$ at the times corresponding to these measurements. In some cases it may be necessary to resort to graphical integration to determine $\psi(C_i)$.

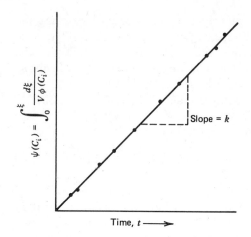

Figure 3.4
Test of reaction rate data by the integral method of analysis.

4. Plot the calculated values of $\psi(C_i)$ versus t as shown in Figure 3.4.
5. If the data yield a satisfactory straight line passing through the origin, the reaction rate expression assumed in step 1 is said to be consistent with the data and may be accepted as a basis for subsequent work in the same temperature-concentration regime of operation. The slope of this line is equal to the reaction rate constant k. If the data do not fall on a satisfactory straight line, one must return to step 1 and assume a new mathematical form of the reaction rate expression.

Slight modifications of the above procedure are often employed to reduce the numerical calculations required. For example, the value of $\psi(C_i)$ corresponding to a Class II second-order rate expression is

$$kt = \psi(C_i) = \frac{1}{C_{A0}\nu_B - C_{B0}\nu_A} \ln\left(\frac{C_{A0}C_B}{C_{B0}C_A}\right) \qquad (3.3.19)$$

A plot of this function versus time should be

linear in time with a slope equal to the rate constant k. However, it is more convenient to rearrange this equation as

$$\ln\left(\frac{C_B}{C_A}\right) = \ln\left(\frac{C_{B0}}{C_{A0}}\right) + (C_{A0}v_B - C_{B0}v_A)kt$$

$$(3.3.20)$$

and plot the left side versus time. One still requires the data plot to be linear, but the restriction that the line pass through the origin is modified so that the Y intercept becomes $\ln(C_{B0}/C_{A0})$.

The graphical approach to the analysis of kinetic data has several advantages. The best straight line through the experimental points can often be found by inspection (i.e., by moving a transparent straight edge until it appears to fit the data with a minimum of deviation). Since the errors in the concentration measurements are usually much greater than those in the time measurements, in estimating the best straight line through a series of points one should think of the residuals as parallel to the concentration axis (i.e., vertical), not as perpendicular to the line. The graphical method readily shows trends and deviations from linear behavior. Scattered points that are obviously in error can be recognized easily and eliminated from further consideration. In those rare cases where the accuracy of the data exceeds that obtained in plotting on a reasonable scale, one should use the numerical methods described in Section 3.3.2.2.

Since the integral method described above is based on the premise that some rate function exists that will lead to a value of $\psi(C_i)$ that is linear in time, deviations from linearity (or curvature) indicate that further evaluation or interpretation of the rate data is necessary. Many mathematical functions are roughly linear over sufficiently small ranges of variables. In order to provide a challenging test of the linearity of the data, one should perform at least one experimental run in which data are taken at 80%, 90%, or higher conversions of the limiting reagent.

Perhaps the most discouraging type of deviation from linearity is random scatter of the data points. Such results indicate that something is seriously wrong with the experiment. The method of analysis may be at fault or the reaction may not be following the expected stoichiometry. Side reactions may be interfering with the analytical procedures used to follow the progress of the reaction, or they may render the mathematical analysis employed invalid. When such plots are obtained, it is wise to reevaluate the entire experimental procedure and the method used to evaluate the data before carrying out additional experiments in the laboratory.

Another form of deviation from linearity that is often encountered in plots of $\psi(C_i)$ versus time is curvature of the data, as shown in Figure 3.5. If the assumed order is greater than the true order of the reaction, upward curvature will be observed. If the assumed order is less than the true order, downward curvature will result.

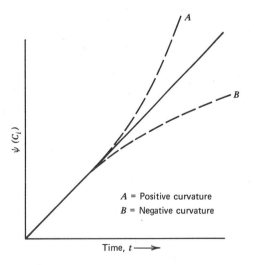

Figure 3.5
Observation of curvature in tests of reaction rate data.

In some cases one attempts to cause a simplification of the rate expression by using stoichiometric ratios of reactants. For example, a mixed second-order rate expression ($r = kC_AC_B$) becomes a Class I second-order rate expression ($r = kC_A^2$) if a stoichiometric mixture of A and B is used. If, by mistake, a non-stoichiometric mixture is used, positive or negative deviations can be observed, depending on which species is present in excess.

If a reaction is reversible and if one has assumed a rate function that does not take the reverse reaction into account, one observes a downward curvature. As equilibrium is approached, the slope of this curve approaches zero. Another cause of curvature is a change in temperature during the course of the experiment. An increase in temperature causes an increase in the reaction rate, leading to an upward curvature. Bunnett (3) has discussed a number of other sources of curvature, including changes in pH and ionic strength, impurity effects, autocatalysis, and side reactions.

Illustrations 3.2 and 3.3 are examples of the use of the graphical integral method for the analysis of kinetic data.

ILLUSTRATION 3.2 USE OF A GRAPHICAL INTEGRAL METHOD FOR DETERMINING THE RATE CONSTANT FOR A CLASS II SECOND-ORDER REACTION

R. T. Dillon (5) has studied the reaction between ethylene bromide and potassium iodide in 99% methanol.

$$C_2H_4Br_2 + 3KI \xrightarrow{k} C_2H_4 + 2KBr + KI_3$$

Given the data in the table below, determine the second-order reaction rate constant.

Data

Temperature: 59.72 °C
Initial KI concentration: 0.1531 kmole/m³

Initial $C_2H_4Br_2$ concentration:
0.02864 kmole/m³

Time, t (ksec)	Fraction dibromide reacted
29.7	0.2863
40.5	0.3630
47.7	0.4099
55.8	0.4572
62.1	0.4890
72.9	0.5396
83.7	0.5795

Solution

The stoichiometry of the reaction is of the form

$$A + 3B \rightarrow \text{Products}$$

Assume that the rate expression is of the form

$$\frac{d\xi^*}{dt} = kC_AC_B \tag{A}$$

The instantaneous concentrations C_A and C_B can be expressed in terms of the initial values and the extent of reaction per unit volume.

$$C_A = C_{A0} - \xi^* \qquad C_B = C_{B0} - 3\xi^*$$

Thus

$$\frac{d\xi^*}{dt} = k(C_{A0} - \xi^*)(C_{B0} - 3\xi^*)$$

Separation of variables and integration gives

$$kt = \int_0^{\xi^*} \frac{d\xi^*}{(C_{A0} - \xi^*)(C_{B0} - 3\xi^*)}$$

$$= \frac{1}{C_{B0} - 3C_{A0}} \ell n \left[\frac{(C_{B0} - 3\xi^*)}{(C_{A0} - \xi^*)} \frac{C_{A0}}{C_{B0}} \right] \tag{B}$$

The fraction dibromide (A) reacted is given by

$$f_A = \frac{\xi^*}{C_{A0}} \tag{C}$$

Elimination of ξ^* between equations B and C gives

$$kt = \frac{1}{C_{B0} - 3C_{A0}} \ln \left[\frac{\left(\dfrac{C_{B0}}{C_{A0}} - 3f \right)}{(1 - f)} \frac{C_{A0}}{C_{B0}} \right]$$

(D)

If the reaction is second order, it is evident from equation D that a plot of

$$\ln\{[(C_{B0}/C_{A0}) - 3f]/(1 - f)\}$$

versus time should be linear with a slope equal to $[(C_{B0} - 3C_{A0})k]$. The results of the calculations necessary to determine the logarithmic term are given below.

$$\frac{C_{B0}}{C_{A0}} = \frac{0.1531}{0.02864} = 5.3457$$

Time, t (ksec)	f	$1 - f$	$\dfrac{C_{B0}}{C_{A0}} - 3f$	$\dfrac{\dfrac{C_{B0}}{C_{A0}} - 3f}{1 - f}$
0	0	1	5.3457	5.346
29.7	0.2863	0.7137	4.4868	6.260
40.5	0.3630	0.6370	4.2567	6.682
47.7	0.4099	0.5901	4.1160	6.975
55.8	0.4572	0.5428	3.9741	7.321
62.1	0.4890	0.5110	3.8787	7.590
72.9	0.5396	0.4604	3.7269	8.095
83.7	0.5795	0.4205	3.6072	8.578

From a plot on semilogarithmic coordinates of $\{[(C_{B0}/C_{A0}) - 3f]/(1 - f)\}$ versus time (Figure 3.6) it may be determined that

$$(C_{B0} - 3C_{A0})k = 5.69 \times 10^{-3} \text{ ksec}^{-1}$$

Thus

$$k = 5.69 \times 10^{-3}/0.0672$$
$$= 8.47 \times 10^{-2} \text{ m}^3/\text{kmol·ksec}$$

In more conventional units this is 0.305 liters/mole·hour which is comparable to the value of 0.300 liters/mole·hour reported by Dillon.

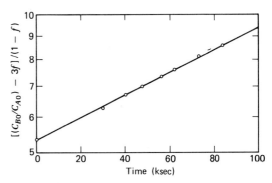

Figure 3.6
Plot for determination of rate constant in Illustration 3.2

ILLUSTRATION 3.3 USE OF THE GRAPHICAL INTEGRAL METHOD TO DETERMINE THE RATE EXPRESSION FOR A GAS PHASE CHEMICAL REACTION MONITORED BY RECORDING THE TOTAL PRESSURE OF THE SYSTEM

Hinshelwood and Askey (6) have investigated the gas phase decomposition of dimethyl ether according to the following reaction.

$$CH_3OCH_3 \rightarrow CH_4 + CO + H_2$$
$$E \rightarrow B + C + D$$

The following data were recorded in an isothermal (552 °C) constant volume reactor.

Time, t (sec)	Pressure (mm Hg)
0	420
57	584
85	662
114	743
145	815
182	891
219	954
261	1013
299	1054

Use a graphical integral method to determine the order of the reaction and the reaction rate constant.

Solution

Assume that the reaction follows first-order kinetics. The integral form of the reaction rate expression is given by equation 3.1.8.

$$\ln\left(\frac{C_E}{C_{E0}}\right) = -kt \qquad (A)$$

At the temperature and pressures of this experiment, ideal gas behavior may be assumed. Thus

$$C_E = \frac{P_E}{RT} = Y_E \frac{\pi}{RT} \qquad (B)$$

where π is the total pressure in the system. Combination of equations A and B gives

$$\ln\left(\frac{Y_E \pi}{\pi_0}\right) = -kt \qquad (C)$$

If the mole fraction ether (E) can be expressed in terms of the total pressure of the system, it can be used in conjunction with equation C to give an integral equation against which the data can be tested. In order to develop this relation, it is helpful to prepare a mole table.

Time, t	0	t	∞
Mole number			
CH_3OCH_3	n_0	$n_0 - \xi$	0
CH_4	0	ξ	$\xi_\infty = n_0$
H_2	0	ξ	$\xi_\infty = n_0$
CO	0	ξ	$\xi_\infty = n_0$
Total moles	n_0	$n_0 + 2\xi$	$3n_0$
Pressure	π_0	π	$\pi_\infty = 3\pi_0$

Now, from this table and the ideal gas law, it is evident that

$$\frac{\pi - \pi_0}{\pi_\infty - \pi_0} = \frac{[(n_0 + 2\xi) - n_0]\dfrac{RT}{V}}{(3n_0 - n_0)\dfrac{RT}{V}} = \frac{2\xi}{2n_0}$$

or

$$\frac{\xi}{n_0} = \frac{\pi - \pi_0}{3\pi_0 - \pi_0} = \frac{1}{2}\left(\frac{\pi}{\pi_0} - 1\right) \qquad (D)$$

The mole fraction ether at time t is given by

$$Y_E = \frac{n_0 - \xi}{n_0 + 2\xi} = \frac{1 - \dfrac{\xi}{n_0}}{1 + \dfrac{2\xi}{n_0}} \qquad (E)$$

Combination of equations D and E gives

$$Y_E = \frac{1 - \dfrac{1}{2}\left(\dfrac{\pi}{\pi_0} - 1\right)}{1 + \left(\dfrac{\pi}{\pi_0} - 1\right)} = \frac{\dfrac{3}{2} - \dfrac{1}{2}\dfrac{\pi}{\pi_0}}{\dfrac{\pi}{\pi_0}} = \frac{3\pi_0 - \pi}{2\pi}$$

Substitution of this relation in equation C gives

$$\ln\left(\frac{3\pi_0 - \pi}{2\pi_0}\right) = -kt$$

This last equation is one against which the data can be tested to see if the assumption of first-order kinetics is appropriate. The data are worked up below.

Time, t (sec)	Pressure (mm Hg)	$3\pi_0 - \pi$
0	420	840
57	584	676
85	662	598
114	743	517
145	815	445
182	891	369
219	954	306
261	1013	247
299	1054	206

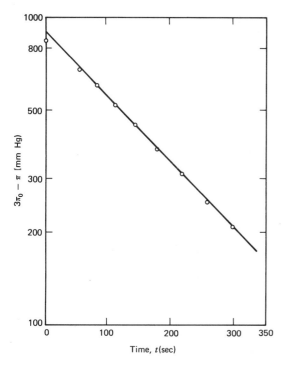

Figure 3.7
Graphical test of data for illustration 3.3.

The last column is plotted versus time using semilogarithmic coordinates in Figure 3.7. The fact that the data fits a straight line is indicative of first-order kinetics. The reaction rate constant may be determined from the slope of this plot.

$$k = -\left[\frac{d \ln(3\pi_0 - \pi)}{dt}\right] = -\left(\frac{\ln 907 - \ln 264}{0 - 250}\right)$$

$$= 4.94 \times 10^{-3} \sec^{-1}$$

3.3.2.2 Integral Methods for the Analysis of Kinetic Data—Numerical Procedures. While the graphical procedures discussed in the previous section are perhaps the most practical and useful of the simple methods for determining rate constants, a number of simple numerical procedures exist for accomplishing this task.

Some of these procedures are of doubtful utility, while others are widely used because they are embodiments of techniques for the statistical analysis of experimental data. All of these procedures are based on the calculation of k from the integrated form of a reaction rate expression such as equations 3.1.8, 3.1.15, 3.1.19, etc. For a detailed evaluation of the various averaging procedures that can be employed in the calculation of k, see the works of Livingston (7) and Margerison (8). They have been used as the basis for the present discussion. Regardless of the averaging procedure employed, it is good practice to arrange the computed values of k in order of increasing time or to plot k versus time so that systematic trends may be observed. Such trends indicate significant departures from the assumed rate law.

The *long-interval* and *short-interval* methods are simple computational procedures that the budding kineticist might be tempted to use. They avoid the subjective weighting of the various experimental points that is inherent in any graphical method, but they have the common disadvantage of weighting the points in arbitrary, illogical fashion.

The long-interval method involves the calculation of k using the *initial* values of reactant concentrations successively with each of the other values of the measured concentrations and times. If there are $(n + 1)$ measurements of the concentrations of interest (including the initial value), the procedure yields n values of k. The "average" value of k is then taken to be the arithmetic average of these computed values.

$$\bar{k} = \left(\frac{1}{n}\right) \sum_{i=1}^{i=n} k_{0,i} \qquad (3.3.21)$$

where $k_{0,i}$ is the value of the rate constant computed using the concentration value determined at time t_i. For a Class I, second-order reaction $(v_A = -1)$ taking place in a constant volume system, the several values of the rate constant may be determined from equation 3.1.15.

$$\left.\begin{array}{l} k_{0,1} = \dfrac{1}{t_1}\left(\dfrac{1}{C_{A1}} - \dfrac{1}{C_{A0}}\right) \\[2ex] k_{0,2} = \dfrac{1}{t_2}\left(\dfrac{1}{C_{A2}} - \dfrac{1}{C_{A0}}\right) \\[2ex] k_{0,i} = \dfrac{1}{t_i}\left(\dfrac{1}{C_{Ai}} - \dfrac{1}{C_{A0}}\right) \end{array}\right\} \quad (3.3.22)$$

The arithmetic mean of these values is:

$$\bar{k} = \frac{1}{n}\sum_{i=1}^{i=n}\frac{1}{t_i C_{Ai}} - \frac{1}{nC_{A0}}\sum_{i=1}^{i=n}\frac{1}{t_i} \quad (3.3.23)$$

This method assigns a much greater weight to the first point than to any succeeding one. Moreover, as the time interval increases, the contribution of the corresponding concentration term to the average rate constant becomes less and less, so that $\bar{k}$ is essentially determined by the early concentration observations and the heavily overweighted initial value. It is reasonable to employ this method of computation only in cases where the initial concentration is known much more accurately than any of the succeeding values. This condition is achieved in practice when the initial reaction mixture can be made up exactly, the time of mixing is short compared to the interval between experimental data points, and the analytical determinations of C_{Ai} are relatively inaccurate. More often the standard deviation of C_{A0} is approximately equal to those of the C_{Ai}.

In the short-interval method, one computes a value of the rate constant $(k_{i-1,i})$ for each successive pair of data points. The arithmetic average of the rate constants computed in this manner is assumed to be a representative value of the rate constant. However, it can be shown that *when the time interval between experimental observations is constant, the short-interval method for computing $\bar{k}$ is equivalent to rejecting all but the first and the last measurements!* The intermediate observations might just as well have not been made.

When the time intervals are approximately, but not exactly, equal, the result is not much

different. The greatest weights are placed on the inital and final measurements (which in practice are often the least accurate), and relatively small, varying weights are placed on the others.

If one does not desire to employ elaborate statistical methods for determining a rate constant, there is one good, simple method for determining $\bar{k}$ that depends on careful planning of the experimental work. The essence of the experimental method is as follows.

One makes a series of $(n + 1)$ measurements of the concentrations of interest at times 0, t_1, $2t_1$, $3t_1$, nt_1. The total time interval (nt_1) is taken to be less than the time for 50% conversion of the limiting reagent $(\tau_{1/2})$. After a known period of time Δ that is as large as or larger than $\tau_{1/2}$, one makes a second series of $(n + 1)$ measurements of the concentrations at times that differ by the constant increment t_1.

One then proceeds to calculate a value of the rate constant for each pair of points separated by a time Δ [i.e., a value is calculated from the points corresponding to (0 and Δ), t_1 and ($\Delta + t_1$), $2t_1$ and ($\Delta + 2t_1$), etc.]. The arithmetic mean of these values is a good representative value of the rate constant. In this technique each data point is used once and only once, and the probable errors of the quantities that are averaged are all of the same order of magnitude. For the first-order case it is apparent from equation 3.1.8 that the average value of the rate constant is given by

$$\bar{k} = \frac{1}{(n+1)\Delta}\sum_{i=0}^{i=n}\ln\left(\frac{C_{Ai}}{C'_{Ai}}\right) \quad (3.3.24)$$

where C_{Ai} and C'_{Ai} are the concentrations of the species of interest at times it_1 and $\Delta + it_1$, respectively.

This approach can be used with other simple rate expressions in order to determine a representative value of the reaction rate constant. Moreover, the experimental plan on which this technique is based will provide data over such a range of fraction conversions that it is readily adapted to various graphical techniques. Illustration 3.4 indicates how this approach is used to determine a reaction rate constant.

One may also use the methods of the statistician to determine average rate constants (e.g., the standard unweighted least squares procedure).

The unweighted least squares analysis is based on the assumption that the best value of the rate constant k is the one that minimizes the sum of the squares of the residuals. In the general case one should regard the zero time point as an adjustable constant in order to avoid undue weighting of the initial point. An analysis of this type gives the following expressions for first- and second-order rate constants

First order:

$$\bar{k} = \frac{\sum(t_i \ln C_{Ai}) - [(\sum t_i)(\sum \ln C_{Ai})/n]}{[(\sum t_i)^2/n] - \sum t_i^2} \quad (3.3.25)$$

Second order:

$$\bar{k} = \frac{\sum\left(\dfrac{t_i}{C_{Ai}}\right) - \dfrac{1}{n}\left(\sum\dfrac{1}{C_{Ai}}\right)\left(\sum t_i\right)}{\sum(t_i^2) - \dfrac{1}{n}(\sum t_i)^2} \quad (3.3.26)$$

These expressions apply to constant volume systems in which $v_A = -1$. The sums are taken from $i = 1$ to $i = n$. Although the use of these equations is somewhat laborious for hand calculations, they are easily handled by even the simplest types of computers.

While the unweighted least squares method of data analysis is commonly used for the determination of reaction rate constants, it *does not* yield the *best possible* value for $\bar{k}$. There are two principal reasons for this failure.

1. All points are assigned equal weights.
2. Functions of the measured concentrations must be used in the equations defining the residuals.

Statisticians have developed general calculation procedures that avoid these difficulties.

Many computer libraries contain programs that perform the necessary statistical calculations and relieve the engineer of this burden. For discussions of the use of weighted least squares methods for the analysis of kinetic data, see Margerison's review (8) on the treatment of experimental data and the treatments of Kittrell et al. (9), and Peterson (10).

3.3.2.3 Integral Methods for the Analysis of Kinetic Data—Fractional Life Methods.

The time necessary for a given fraction of a limiting reagent to react will depend on the initial concentrations of the reactants in a manner that is determined by the rate expression for the reaction. This fact is the basis for the development of the fractional life method (in particular the half-life method) for the analysis of kinetic data. The half-life, or half-period, of a reaction is the time necessary for one half of the original reactant to disappear. In constant volume systems it is also the time necessary for the concentration of the limiting reagent to decline to one half of its original value.

The fractional life approach is most useful as a means of obtaining a preliminary estimate of the reaction order. It is not recommended for the accurate determination of rate constants. Moreover, it cannot be used for systems that do not obey nth order rate expressions.

If one combines the definition of the reaction rate in variable volume systems with a general nth-order rate expression, he finds that the time necessary to achieve a specified fraction conversion is given by

$$t_f = \int_0^{\xi} \frac{d\xi}{V k C_A^n} = \int_0^{f_A} \frac{n_{A0}\, df_A}{-v_A V k C_A^n} \quad (3.3.27)$$

or, using equations 3.1.40 and 3.1.47,

$$t_f = \frac{1}{C_{A0}^{n-1}} \int_0^{f_A} \frac{(1 + \delta_A f_A)^{n-1}\, df_A}{-k v_A (1 - f_A)^n} \quad (3.3.28)$$

This relation indicates that for all values of n, k, v_A, and δ_A, the fractional life is inversely proportional to the initial concentration raised to the $(n-1)$ power. A given fractional life is independent of initial concentration for first-order reactions. It increases with C_{A0} for n less than 1, and it decreases with C_{A0} for n greater than 1.

For constant volume conditions $\delta_A = 0$ and

$$t_f = \frac{1}{v_A(1-n)kC_{A0}^{n-1}}\left[\frac{1}{(1-f)^{n-1}} - 1\right]$$

(3.3.29)V

with

$$t_{1/2} = \frac{2^{n-1}-1}{v_A k C_{A0}^{n-1}(1-n)}$$

(3.3.29a)V

when $n \neq 1$.

For first-order reactions,

$$t_f = \frac{\ell n(1-f)}{kv_A}$$

(3.3.30)V

with

$$t_{1/2} = \frac{\ell n\,2}{k} \quad \text{for} \quad v_A = -1$$

(3.3.30a)V

Equations 3.3.28 and 3.3.29 indicate that plots of $\log t_f$ versus $\log C_{A0}$ are linear, and reaction orders may be determined from the slopes of such plots (slope $= 1 - n$). For constant volume systems there is another method based on equations 3.3.29 and 3.3.30 by which one can obtain preliminary estimates of the reaction

Table 3.1
Fractional Life Relations for Constant Volume Systems

Reaction order n:	0	1	2	3	n
Partial reaction time			Useful relationships		
$t_{1/4}$	$-\dfrac{1}{4}\dfrac{C_{A0}}{v_A k}$	$\dfrac{\ell n(4/3)}{k}$	$\dfrac{1}{3}\dfrac{1}{v_A k C_{A0}}$	$\dfrac{7}{18}\dfrac{1}{v_A k C_{A0}^2}$	$\dfrac{(4/3)^{n-1}-1}{(1-n)v_A k C_{A0}^{n-1}}$
$t_{1/3}$	$-\dfrac{1}{3}\dfrac{C_{A0}}{v_A k}$	$\dfrac{\ell n(3/2)}{k}$	$\dfrac{1}{2}\dfrac{1}{v_A k C_{A0}}$	$\dfrac{5}{8}\dfrac{1}{v_A k C_{A0}^2}$	$\dfrac{(3/2)^{n-1}-1}{(1-n)v_A k C_{A0}^{n-1}}$
$t_{1/2}$	$-\dfrac{1}{2}\dfrac{C_{A0}}{v_A k}$	$\dfrac{\ell n\,2}{k}$	$-\dfrac{1}{v_A k C_{A0}}$	$-\dfrac{3}{2}\dfrac{1}{v_A k C_{A0}^2}$	$\dfrac{2^{n-1}-1}{(1-n)v_A k C_{A0}^{n-1}}$
t_f	$-f\dfrac{C_{A0}}{v_A k}$	$-\dfrac{\ell n(1-f)}{k}$	$-\dfrac{f/(1-f)}{v_A k C_{A0}}$	$-\dfrac{(2f-f^2)}{2(1-f)^2 v_A k C_{A0}^2}$	$\dfrac{\left(\dfrac{1}{1-f}\right)^{n-1}-1}{(1-n)v_A k C_{A0}^{n-1}}$
$t_{1/2}/t_{1/4}$	2.000	2.409	3.000	3.857	$\dfrac{2^{n-1}-1}{(4/3)^{n-1}-1}$
$t_{1/2}/t_{1/3}$	1.500	1.709	2.000	2.400	$\dfrac{2^{n-1}-1}{(3/2)^{n-1}-1}$
$t_{f'}/t_{f''}$	f'/f''	$\dfrac{\ell n(1-f')}{\ell n(1-f'')}$	$\left(\dfrac{f'}{1-f'}\right)\left(\dfrac{1-f''}{f''}\right)$	$\dfrac{f'(2-f')}{(1-f')^2}\dfrac{(1-f'')^2}{f''(2-f'')}$	$\dfrac{\left(\dfrac{1}{1-f'}\right)^{n-1}-1}{\left(\dfrac{1}{1-f''}\right)^{n-1}-1}$

order. From the data of one experimental run or from different runs using the same initial composition, one may determine the times necessary to achieve different fraction conversions. The ratio of these times (t_{f1} and t_{f2}) is given by

$$\frac{t_{f1}}{t_{f2}} = \frac{\dfrac{1}{(1 - f_1)^{n-1}} - 1}{\dfrac{1}{(1 - f_2)^{n-1}} - 1} \quad \text{for} \quad n \neq 1 \tag{3.3.31}V$$

and

$$\frac{t_{f1}}{t_{f2}} = \frac{\ell n(1 - f_1)}{\ell n(1 - f_2)} \quad \text{for} \quad n = 1 \tag{3.3.31a}V$$

The value of this ratio is characteristic of the reaction order. Table 3.1 contains a tabulation of partial reaction times for various rate expressions of the form $r = kC_A{}^n$ as well as a tabulation of some useful ratios of reaction times. By using ratios of the partial reaction times based on experimental data, one is able to obtain a quick estimate of the reaction order with minimum effort. Once this estimate is in hand one may proceed to use a more exact method of determining the reaction rate parameters.

3.3.2.4 Integral Methods for the Analysis of Kinetic Data—Guggenheim's Method for First-Order Reactions.

Guggenheim (11) has developed a special method that is useful in obtaining the rate constant *for a first-order reaction* when an accurate value of the initial reactant concentration is not available. It requires a series of readings of the parameter being used to follow the progress of the reaction at times t_1, t_2, t_3, etc. and at times $t_1 + \Delta$, $t_2 + \Delta$, $t_3 + \Delta$, etc. The time increment Δ should be two or three times the half life of the reaction. If we denote the extent of reaction at times t_1 and $t_1 + \Delta$ by ξ_1 and ξ_1', respectively, equation 3.1.36 indicates

that

$$\xi_1 = \frac{n_{A0}}{\nu_A}(e^{\nu_A k t_1} - 1) \tag{3.3.32}$$

and

$$\xi_1' = \frac{n_{A0}}{\nu_A}[e^{\nu_A k(t_1 + \Delta)} - 1] \tag{3.3.33}$$

where n_{A0} need not be known. Subtraction of equation 3.3.33 from equation 3.3.32 gives

$$\xi_1 - \xi_1' = \frac{n_{A0}}{\nu_A} e^{\nu_A k t_1}(1 - e^{\nu_A k \Delta}) \tag{3.3.34}$$

Rearrangement of this equation yields

$$\ell n(\xi_1 - \xi_1') - \nu_A k t_1 = \ell n\left[\frac{n_{A0}}{\nu_A}(1 - e^{\nu_A k \Delta})\right] \tag{3.3.35}$$

Similar equations are valid at times t_2, t_3, etc. In all cases, however, the right side of these equations will be a constant, since the time increment Δ is a constant. Thus, at time t_i,

$$\ell n(\xi_i - \xi_i') = \nu_A k t_i + \text{constant} \tag{3.3.36}$$

A plot of $\ell n(\xi - \xi')$ versus t will therefore give a straight line with a slope equal to $\nu_A k$ for first-order kinetics.

This technique is readily adaptable for use with the generalized additive physical approach discussed in Section 3.3.3.2. It is applicable to systems that give apparent first-order rate constants. These include not only simple first-order irreversible reactions but also irreversible first-order reactions in parallel and reversible reactions that are first-order in both the forward and reverse directions. The technique provides an example of the advantages that can be obtained by careful planning of kinetics experiments instead of allowing the experimental design to be dictated entirely by laboratory convention and experimental convenience.

Guggenheim's technique has been extended to other order reactions (12, 13), but the final expressions are somewhat cumbersome.

ILLUSTRATION 3.4 USE OF GUGGENHEIM'S METHOD AND A NUMERICAL INTEGRAL PROCEDURE TO DETERMINE THE RATE CONSTANT FOR THE HYDRATION OF ISOBUTENE IN HYDROCHLORIC ACID SOLUTION

Ciapetta and Kilpatrick (14) have used a dilatometric technique to investigate the kinetics of the hydration of isobutene in perchloric acid solution at 25 °C.

$$CH_3-\underset{\underset{CH_3}{|}}{C}=CH_2 + H_2O \rightarrow CH_3-\underset{\underset{CH_3}{|}}{\overset{\overset{CH_3}{|}}{C}}-OH$$

The dilatometer readings (h) listed below are given in arbitrary units and are arranged in pairs taken at a fixed interval of 2 hr. They are related to the extent of reaction by

$$\frac{\xi}{\xi_\infty} = \frac{h - h_0}{h_\infty - h_0}$$

1. Use Guggenheim's method to determine the pseudo first-order rate constant.
2. Given the additional fact that the reading at "infinite" time is 12.16, use a numerical averaging procedure to determine the reaction rate constant.

Data

Initial concentrations (moles/liter):

$$HClO_4 = 0.3974; C_4H_8 = 0.00483.$$

Time, t (min)	Dilatometer reading at time t	Dilatometer reading at time $t + 2$ hr
0	18.84	13.50
10	17.91	13.35
20	17.19	13.19
30	16.56	13.05
40	16.00	12.94
50	15.53	12.84
60	15.13	12.75
70	14.76	12.69

Solution

Equation 3.3.36 indicates that the natural logarithm of the difference in extents of reactions at times t and $(t + \Delta)$ should be linear in t.

Taking the difference between the extent of reaction at times t and $(t + \Delta)$ gives

$$\xi - \xi' = \left(\frac{h - h'}{h_\infty - h_0}\right)\xi_\infty \qquad (A)$$

Combination of equations A and 3.3.36 gives

$$\ell n(h - h') = v_A kt + \ell n\left(\frac{h_\infty - h_0}{\xi_\infty}\right) + \text{constant}$$

If one notes that $v_A = -1$ and that the quantity $\ell n[(h_\infty - h_0)/\xi_\infty]$ is itself a constant

$$\ell n(h - h') = -kt + \text{a new constant} \qquad (B)$$

Note that k can be determined from the slope of a plot of the left side of equation B versus time and that this plot can be prepared without a knowledge of the dilatometer readings at times zero and infinity. The data are worked up below and plotted in Figure 3.8.

Time, t (min)	$h - h'$
0	5.34
10	4.56
20	4.00
30	3.51
40	3.06
50	2.69
60	2.38
70	2.07

From the slope of the plot,

$$k = \frac{\ell n\dfrac{5.26}{2.06}}{70} = 1.34 \times 10^{-2} \text{ min}^{-1}$$

This value compares favorably with the literature value of 1.32×10^{-2} min^{-1}, which is based on much more data.

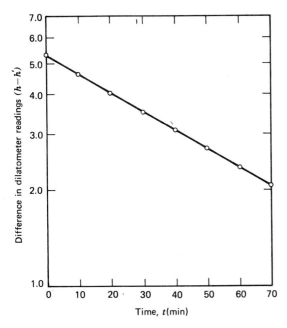

Figure 3.8
Use of a semilogarithmic Guggenheim plot to determine first-order rate constant.

The half-life of this reaction is given by

$$t_{1/2} = \frac{0.693}{k} = \frac{0.693}{1.34 \times 10^{-2}} = 51.7 \text{ min}$$

The time Δ between readings is thus greater than the reaction half-life, and the numerical averaging procedure leading to equation 3.3.24 could be used for the analysis of these data if an estimate of the reading at infinite time can be obtained. In order to manipulate this equation to a form that can make use of the available data, several points must be noted.

1. $\dfrac{C_{Ai}}{C'_{Ai}} = \dfrac{C_{A0} - \xi^*}{C_{A0} - \xi^{*'}} = \dfrac{1 - (\xi^*/C_{A0})}{1 - (\xi^{*'}/C_{A0})}$

2. $\dfrac{h - h_0}{h_\infty - h_0} = \dfrac{\xi}{\xi_\infty} = \dfrac{\xi^*}{\xi^*_\infty} = \dfrac{\xi^*}{C_{A0}}$

3. $1 - \dfrac{\xi^*}{C_{A0}} = \dfrac{h_\infty - h}{h_\infty - h_0}$

4. $\dfrac{C_{Ai}}{C'_{Ai}} = \dfrac{h_\infty - h}{h_\infty - h'}$ (from 1 and 3)

Thus the average value of the reaction rate constant can be calculated numerically from

$$\bar{k} = \frac{1}{(n+1)\Delta} \sum_{i=0}^{i=n} \ln\left(\frac{h_\infty - h_i}{h_\infty - h'_i}\right)$$

where n is the number of pairs of data points. Numerical calculations indicate that

$$\bar{k} = \frac{1}{(7+1)(120)}(12.768) = 0.0133 \text{ min}^{-1}$$

This value is consistent with that obtained above.

3.3.3 Techniques for the Analysis of Reaction Rate Data that are Suitable for Use with Either Integral or Differential Methods

3.3.3.1 Use of Excess Concentrations—The Isolation Method. The method of isolation for elucidating the form of the reaction rate expression is based on the simplifications that occur when the concentration of one or more of the reactants appearing in the rate expression is much greater than that called for by the stoichiometry of the reaction.

Consider a reaction rate expression of the form

$$r = kC_A^{\beta_A} C_B^{\beta_B} C_C^{\beta_C} \qquad (3.3.37)$$

If one now plans an experiment such that $C_{B0} \gg C_{A0}$ and $C_{C0} \gg C_{A0}$, the only concentration that will change appreciably during the course of the experiment is that of species A. This situation implies that the above rate expression will simplify to the following form.

$$r = k'C_A^{\beta_A} \qquad (3.3.38)$$

where

$$k' = kC_{B0}^{\beta_B} C_{C0}^{\beta_C} \qquad (3.3.39)$$

We thus have a pseudo β_A th-order reaction and an effective rate constant k'. The methods discussed in Sections 3.3.1 and 3.3.2 may now be used to determine k' and β_A. It should be noted that this technique should always be used in conjunction with other methods of determining the rate expression, because it forces the data to fit equation 3.3.37 when, in fact, the rate expression may be more complex. If one observes that the exponents β_i are nonintegers that vary continuously with temperature or pressure, this is an indication that the rate expression is more complex than equation 3.3.37.

3.3.3.2 Use of Physical Property Measurements as a Measure of the Extent of Reaction.

The most useful physical properties for use in kinetics studies are those that are an additive function of the contributions of the various constituents, the contribution of each species being a linear function of its concentration. Total pressure, absorbance, optical rotation, and the electrical conductivity of dilute solutions are all properties of this type. This section indicates how such physical property measurements may be simply related to the extent of reaction per unit volume.

Consider some physical property λ that results from the contributions of the various species present in the reaction mixture

$$\lambda = \sum_i \lambda_i \qquad (3.3.40)$$

where λ is the experimentally measured quantity and λ_i is the contribution of the ith species. Furthermore, for the derivation presented below, it is necessary that the contribution of a species be a linear function of the concentration of that species

$$\lambda_i = g_i + h_i C_i \qquad (3.3.41)$$

where g_i and h_i are constants characteristic of species i. The constants g_i are normally zero.

If one now considers the following general reaction,

$$v_A A + v_B B + v_Z Z = 0 \qquad (3.3.42)$$

the concentrations of the various species at time

t may be expressed in terms of the extent of reaction per unit volume ξ^*.

$$C_i = C_{io} + v_i \xi^* \qquad (3.3.43)$$

The experimental variable λ may now be written in the following manner.

$$\lambda = \lambda_M + \lambda_A + \lambda_B + \lambda_Z \qquad (3.3.44)$$

where λ_M includes the contribution of the solvent or medium in which the reaction is carried out as well as the contributions of any inert species that are present. It also includes any contributions arising from the reaction vessel itself. Combination of equations 3.3.41 and 3.3.44 gives

$$\lambda = \lambda_M + g_A + h_A C_A + g_B + h_B C_B + g_Z + h_Z C_Z \qquad (3.3.45)$$

If the initial property value is denoted by λ_0,

$$\lambda_0 = \lambda_M + g_A + h_A C_{A0} + g_B + h_B C_{B0} + g_Z + h_Z C_{Z0} \qquad (3.3.46)$$

The change in the value of the property λ between time zero and time t is given by:

$$\lambda - \lambda_0 = h_A(C_A - C_{A0}) + h_B(C_B - C_{B0}) + h_Z(C_Z - C_{Z0}) \qquad (3.3.47)$$

or

$$\lambda - \lambda_0 = h_A v_A \xi^* + h_B v_B \xi^* + h_Z v_Z \xi^* = \xi^* \sum_i h_i v_i \qquad (3.3.48)$$

As equation 3.3.48 indicates, the change in λ is directly proportional to the extent of reaction per unit volume. Similarly, the change in λ between times zero and infinity is given by

$$\lambda_\infty - \lambda_0 = \xi_\infty^* \sum_i h_i v_i \qquad (3.3.49)$$

The ratio of the extent of reaction at time t to that at equilibrium is then

$$\frac{\xi^*}{\xi_\infty^*} = \frac{\lambda - \lambda_0}{\lambda_\infty - \lambda_0} \qquad (3.3.50)$$

Equation 3.3.50 is an extremely useful relation

that is applicable to a very large number of physical properties. The value of ξ_∞^* can easily be determined from the equilibrium constant for the reaction if the reaction is reversible. If the reaction is irreversible. it may be determined from the initial concentration of the limiting reagent.

$$\xi_\infty^* = -\frac{C_{\text{lim},0}}{v_{\text{lim}}} \qquad (3.3.51)$$

The concentrations of the various species present in the reaction mixture can be determined by combining equations 3.3.50 and 3.3.43.

$$C_i = C_{i0} + v_i \xi_\infty^* \left(\frac{\lambda - \lambda_0}{\lambda_\infty - \lambda_0}\right) \qquad (3.3.52)$$

These concentrations may be used in the various integral and differential methods for the analysis of kinetic data that have been described in previous sections. An example of the use of this approach is given in Illustration 3.5.

ILLUSTRATION 3.5 USE OF CONDUCTIVITY MEASUREMENTS IN CONJUNCTION WITH THE GRAPHICAL INTEGRAL METHOD FOR THE ANALYSIS OF REACTION RATE DATA

Biordi (15) has studied the methanolysis of benzoyl chloride at 25 °C in methanol solution.

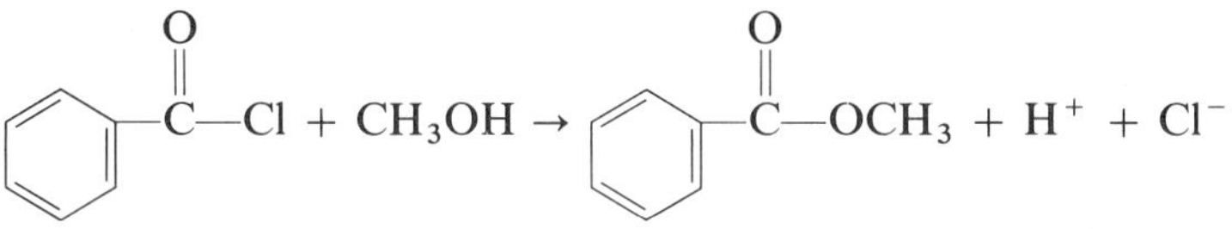

Data

Time, t (sec)	Solution conductivity $\times 10^4$ (Ω^{-1} cm^{-1})
0	negligible
27	0.352
48	0.646
55	0.732
62	0.813
70	0.900
79	0.969
86	1.07
93	1.12
100	1.21
105	1.26
114	1.33
120	1.40
$\vdots$	$\vdots$
10,800	3.50

Solution

Since a vast excess of methanol is used, the concentration of methanol is essentially invariant, and it is appropriate to assume a rate expression of the form $r = k' C_A^\beta{}_A$.

Since the conductivity of the solution is a property that is an additive function of contributions that are linear in concentration, the

$$A + B \rightarrow C + D + E$$

Conductivity measurements were used to monitor the progress of the reaction.

Under the conditions of the experiment, it is known that the reaction may be considered as irreversible. From the data below determine the order of the reaction with respect to benzoyl chloride and the pseudo reaction rate constant under these conditions.

generalized physical property approach may be used:

$$\frac{\kappa - \kappa_0}{\kappa_\infty - \kappa_0} = \frac{\xi^*}{\xi_\infty^*} = \frac{\xi^*}{C_{A0}} \qquad (A)$$

where κ_∞ is taken to be the reading at 10,800 sec.

The integral form of the expression for a first-order reaction is given by equation 3.1.4.

$$-kt = \ell n \left(\frac{C_{A0} - \xi^*}{C_{A0}} \right) = \ell n \left(1 - \frac{\xi^*}{C_{A0}} \right) \quad \text{(B)}$$

Combination of equations A and B gives

$$-kt = \ell n \left[1 - \left(\frac{\kappa - \kappa_0}{\kappa_\infty - \kappa_0} \right) \right] = \ell n \left(\frac{\kappa_\infty - \kappa}{\kappa_\infty - \kappa_0} \right)$$

or

$$\ell n(\kappa_\infty - \kappa) = \ell n(\kappa_\infty - \kappa_0) - kt \quad \text{(C)}$$

If a plot of the left side of equation C versus time is linear, the reaction is first order, and the slope of this plot is equal to $-k$. A plot of the data indicates that the reaction is indeed first order with respect to benzoyl chloride. From the slope of the line the pseudo first-order rate constant is found to be $4.3 \times 10^{-3} \ \text{sec}^{-1}$.

3.3.4 Determination of the Activation Energy

This section focuses on the problem of determining the temperature dependence of the reaction rate expression (i.e., the activation energy of the reaction). Virtually all rate constants may be written in the Arrhenius form:

$$k = Ae^{-E/RT} \quad \text{(3.3.53)}$$

where

    E is the activation energy
    A is the frequency factor for the reaction
    R is the gas constant
    T is the absolute temperature

The variation of k with temperature may be determined by differentiating the logarithmic form of this equation.

$$\frac{d \ell n \ k}{dT} = \frac{E}{RT^2} \quad \text{(3.3.54)}$$

Since virtually all known activation energies are positive quantities, this relation indicates that the reaction rate constant will almost always increase with temperature.

A useful alternative form of this relation is:

$$\frac{d \ell n \ k}{d \left(\frac{1}{T} \right)} = -\frac{E}{R} \quad \text{(3.3.55)}$$

If one has data on the reaction rate constant at several temperatures, this equation provides the basis for the most commonly used method for determining the activation energy of a reaction. If E is temperature invariant, a plot of $\ell n \ k$ versus the reciprocal of the absolute temperature should be linear with slope $-(E/R)$. A typical plot is shown in Figure 3.9 for the reaction $H_2 + I_2 \rightarrow 2HI$. The slope corresponds to an activation energy of 44.3 kcal/mole.

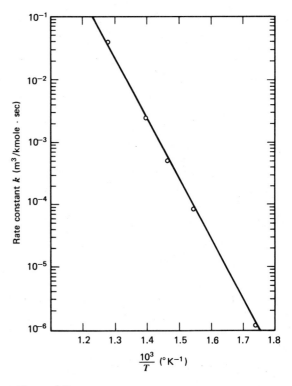

Figure 3.9
Arrhenius plot of the rate constant for the reaction $H_2 + I_2 \rightarrow 2HI$ (16).

A somewhat less accurate method for determining the activation energy involves integration of equation 3.3.55 over the interval between two data points, assuming that E is constant.

$$\ell n \left(\frac{k_2}{k_1} \right) = -\frac{E}{R} \left(\frac{1}{T_2} - \frac{1}{T_1} \right) \quad \text{(3.3.56)}$$

When more than two data points are available, the graphical method is much better to use than common averaging techniques. It gives one a visual picture of the fit of the data to 3.3.55. If one has several data points and estimates of the uncertainty in each point, a weighted least squares fit of the data would be appropriate.

3.3.5 Precision of Rate Measurements for Simple Irreversible Reactions

In order to obtain a feeling for the major sources of uncertainty and error in the calculation of reaction rate constants, it is useful to consider the nature of the errors inherent in the measurement of these parameters.

3.3.5.1 Precision of Reaction Rate Constants.
A reaction rate constant can be calculated from the integrated form of a kinetic expression if one has data on the state of the system at two or more different times. This statement assumes that sufficient measurements have been made to establish the functional form of the reaction rate expression. Once the equation for the reaction rate constant has been determined, standard techniques for error analysis may be used to evaluate the expected error in the reaction rate constant.

In the general case of a dependent variable $y = f(x_1, x_2, x_3, \ldots, x_m)$, which is a known function of the m independent variables $x_1, x_2, x_3, \ldots, x_m$, the expected relative error in y due to the relative errors in $x_1, x_2, x_3, \ldots, x_m$ is given by

$$\left(\frac{\Delta y}{y}\right)^2 = \sum_{i=1}^{m} \left(\frac{\partial \ln f}{\partial \ln x_i}\right)^2 \left(\frac{\Delta x_i}{x_i}\right)^2 \quad (3.3.57)$$

Equation 3.1.32 applies to a constant volume system that follows nth-order kinetics. If we take $v_A = -1$ it can be rewritten as

$$k = \frac{C_{A1}^{n-1} - C_{A2}^{n-1}}{(n-1)(t_2 - t_1)C_{A2}^{n-1}C_{A1}^{n-1}} \quad (3.3.58)$$

where C_{A2} and C_{A1} are the concentrations of A present at times t_2 and t_1, respectively, and where $n \neq 1$.

If one assumes that the errors in the four quantities t_1, t_2, C_{A1}, and C_{A2} are independent and that n is a known constant, the expected random error in the rate constant Δk may be expressed as follows.

$$\left(\frac{\Delta k}{k}\right)^2 = \left(\frac{\partial \ln k}{\partial \ln t_1}\right)^2 \left(\frac{\Delta t_1}{t_1}\right)^2$$
$$+ \left(\frac{\partial \ln k}{\partial \ln t_2}\right)^2 \left(\frac{\Delta t_2}{t_2}\right)^2$$
$$+ \left(\frac{\partial \ln k}{\partial \ln C_{A1}}\right)^2 \left(\frac{\Delta C_{A1}}{C_{A1}}\right)^2$$
$$+ \left(\frac{\partial \ln k}{\partial \ln C_{A2}}\right)^2 \left(\frac{\Delta C_{A2}}{C_{A2}}\right)^2$$
$$(3.3.59)$$

or

$$\left(\frac{\Delta k}{k}\right)^2 = \left(\frac{\Delta t_1}{t_2 - t_1}\right)^2 + \left(\frac{\Delta t_2}{t_2 - t_1}\right)^2$$
$$+ \left[\frac{(n-1)C_{A2}^{n-1}}{C_{A1}^{n-1} - C_{A2}^{n-1}}\right]^2 \left(\frac{\Delta C_{A1}}{C_{A1}}\right)^2$$
$$+ \left[\frac{(1-n)C_{A1}^{n-1}}{C_{A1}^{n-1} - C_{A2}^{n-1}}\right]^2 \left(\frac{\Delta C_{A2}}{C_{A2}}\right)^2$$
$$(3.3.60)$$

For the case of $n = 1$, the corresponding equation is

$$\left(\frac{\Delta k}{k}\right)^2 = \left(\frac{\Delta t_1}{t_2 - t_1}\right)^2 + \left(\frac{\Delta t_2}{t_2 - t_1}\right)^2$$
$$+ \left[\frac{1}{\ln\left(\frac{C_{A1}}{C_{A2}}\right)}\right]^2 \left(\frac{\Delta C_{A1}}{C_{A1}}\right)^2$$
$$+ \left[\frac{1}{\ln\left(\frac{C_{A1}}{C_{A2}}\right)}\right]^2 \left(\frac{\Delta C_{A2}}{C_{A2}}\right)^2$$
$$(3.3.61)$$

To illustrate the use of equation 3.3.60 for estimating the relative uncertainty in the reaction rate constant, consider an example where $n = 2$, $(t_2 - t_1) = 100$ sec, the uncertainty in each measurement (Δt_1 and Δt_2) is 1 sec, $C_{A2} = 0.9\, C_{A1}$ (10% reaction), and the relative uncertainty in each concentration measurement is 0.01 (1%). Substitution of these values into equation 3.3.60 gives

$$\left(\frac{\Delta k}{k}\right)^2 = \left(\frac{1}{100}\right)^2 + \left(\frac{1}{100}\right)^2$$

$$+ \left(\frac{0.9}{1 - 0.9}\right)^2 (0.01)^2$$

$$+ \left(\frac{1}{1 - 0.9}\right)^2 (0.01)^2$$

$$= 10^{-4} + 10^{-4}$$

$$+ 81 \times 10^{-4} + 100 \times 10^{-4}$$

$$= 183 \times 10^{-4}$$

Thus, the relative uncertainty in the rate constant is

$$\frac{\Delta k}{k} = (183 \times 10^{-4})^{1/2} = \pm 13.7 \times 10^{-2}$$

$$= \pm 0.137$$

The major sources of error are the concentration measurements and attempts to improve the accuracy of the rate constant should focus on trying to improve the accuracy of the concentration measurements. Benson (17) has presented a useful table that summarizes the converse of the problem we have just considered (i.e., what precision in analytical capability is required to measure k to a given degree of accuracy). It is reproduced here as Table 3.2. Benson states that the analytical error is about the same for orders of reaction from zero to four and is primarily dependent on the extent of reaction occurring between the two points chosen. Inspection of this table illustrates the dilemma the kineticist faces in planning experiments. In order to measure k to within 1% accuracy he or she must be able to analyze for his or her reactants with a precision of approximately 0.1% at a conversion level of approximately 15%. If higher precision in k is desired, much higher analytical precision must be obtained and higher degrees of conversion must be used.

The situation is significantly changed if it is possible to monitor the appearance of products instead of the disappearance of reactants. The mathematics of the situation lead one to the general principle that whenever possible it is advantageous to monitor the products during the initial stages and the reactants during the final stages of the reaction.

Table 3.2
Errors in Calculated Rate Constants Caused by Analytical Errors

	Percent change in reactant species monitored						
	1	5	10	20	30	40	50
Analytical precision (%)	Error in k (%)						
± 0.1	14	2.8	1.4	0.7	0.5	0.4	0.3
± 0.5	70	14	7	3.5	2.5	2	1.5
± 1.0	> 100	28	14	7	5	4	3
± 2.0	> 100	56	28	15	10	8	6

Valid for a rate expression of the form $\dfrac{dC_A}{dt} = -kC_A^n$ ($n = 0, \frac{1}{2}, 1, \ldots, 4$)

From *The Foundations of Chemical Kinetics* by S. W. Benson. Copyright © 1960. Used with permission of McGraw-Hill Book Company.

3.3.5.2 Precision of Activation Energy Measurements.

The activation energy of a reaction can be determined from a knowledge of the reaction rate constants at two different temperatures. The Arrhenius relation may be written in the following form.

$$E = \frac{RT_1 T_2}{T_2 - T_1} \ln\left(\frac{k_2}{k_1}\right) \qquad (3.3.62)$$

If errors in each of the experimental quantities k_1, k_2, T_1 and T_2 are random, the relative error in the Arrhenius activation energy is given by

$$
\begin{aligned}
\left(\frac{\Delta E}{E}\right)^2 =& \left(\frac{T_2}{T_2 - T_1}\right)^2 \left(\frac{\Delta T_1}{T_1}\right)^2 \\
&+ \left(\frac{T_1}{T_2 - T_1}\right)^2 \left(\frac{\Delta T_2}{T_2}\right)^2 \\
&+ \left[\frac{1}{\ln\left(\frac{k_2}{k_1}\right)}\right]^2 \left[\left(\frac{\Delta k_1}{k_1}\right)^2 + \left(\frac{\Delta k_2}{k_2}\right)^2\right]
\end{aligned}
$$

$$(3.3.63)$$

From this equation it can be seen that the relative error in the activation energy is strongly dependent on the size of the temperature interval chosen. Decreases in the size of the temperature interval greatly increase the uncertainty in E because they not only increase the contributions of the first two terms but also simultaneously increase the contribution of the last term, since the term $\ln(k_2/k_1)$ in the denominator approaches zero as k_2 approaches k_1.

In order to measure E over a 10 °C temperature interval to within $\pm 0.5\%$ one generally requires temperature uncertainties of less than ± 0.03 °C and rate constant uncertainties of 0.3%. As Table 3.2 indicates, the latter requirement will in turn necessitate analytical precision of $\pm 0.1\%$ over an extended range of concentration changes. These numbers indicate the difficulty one must face in attempting precise measurements of the activation energy and why it is so difficult to observe the variation of E with temperature.

At room temperature the relative uncertainty in measuring E over 10 °C temperature intervals is generally about $\pm 5\%$ for gas phase reactions and about $\pm 3\%$ for liquid phase reactions.

LITERATURE CITATIONS

1. Boudart, M., *Kinetics of Chemical Processes*, Chapter 1, Prentice-Hall, Englewood Cliffs, N.J., copyright © 1968. Used with permission.

2. Levenspiel, O., *Chemical Reaction Engineering*, Second Edition, Wiley, New York, 1972.

3. Bunnet, J. F., "The Interpretation of Rate Data" in *Investigation of Rates and Mechanisms of Reactions*, Volume VIII, Part I of *Technique of Organic Chemistry*, edited by S. L. Freiss, E. S. Lewis, and A. Weissberger, Interscience, New York, copyright © 1961. Used with permission.

4. A. Neyens, Unpublished results cited on page 88 of "Cinétique Chimique Appliquée" by J. C. Jungers, J. C. Balacéanu, F. Coussemant, F. Eschard, A. Giraud, M. Hellin, P. Le Prince, and G. E. Limido, Société des Éditions Technip, Paris, 1958. Table VII, p. 88, from the book *Cinetique Chimique Appliquée* by J. C. Jungers et al., published in 1958 by Editions Technip, 27, rue Ginoux, 75737 Paris Cedex 15, France.

5. Dillon, R. T., *J. Am. Chem. Soc.*, *54* (952), 1932.

6. Hinshelwood, C. N., and Askey, P. J., *Proc. Roy. Soc. (London)*, *A115* (215), 1927.

7. Livingston, R., "Evaluation and Interpretation of Rate Data," in *Investigation of Rates and Mechanisms of Reactions*, Volume VIII, Part I of *Technique of Organic Chemistry*, edited by S. L. Freiss, E. S. Lewis, and A. Weissberger, Interscience, New York, 1961.

8. Margerison, D., "The Treatment of Experimental Data" in *The Practice of Kinetics*, Volume I of *Comprehensive Chemical Kinetics*, edited by C. H. Bamford and C. F. H. Tipper, Elsevier, New York, 1969.

9. Kittrell, J. R., Mezaki, R., and Watson, C. C., *Ind. Eng. Chem.*, *58* (50), May 1966.

10. Peterson, T. I., *Chem. Eng. Sci.*, *17*, (203), 1962.

11. Guggenheim, E. A., *Phil. Mag.*, *1* (538), 1926.

12. Roseveare, W. E., *J. Am. Chem. Soc.*, *53* (1951), 1931.

13. Sturtevant, J. M., *J. Am. Chem. Soc.*, *59* (699), 1937.

14. Ciapetta, F. G., and Kilpatrick, M., *J. Am. Chem. Soc.*, *70* (641), 1948.

15. Biordi, J. C., *J. Chem. Eng. Data*, *15* (166), 1970.

16. Moelwyn-Hughes, E. A., *Physical Chemistry*, p. 1109, Pergamon Press, New York, 1957.

17. Benson, S. W., *The Foundations of Chemical Kinetics*, McGraw-Hill, New York, copyright © 1960. Used with permission.

PROBLEMS

1. Moelwyn-Hughes (*Physical Chemistry*, page 1109, Pergamon Press, New York, 1957) has tabulated the following values of the rate constant for the reaction

$$N_2O_5 \rightarrow N_2O_4 + \tfrac{1}{2}O_2$$

Temperature T (°K)	k (sec^{-1})
288.1	1.04×10^{-5}
298.1	3.38×10^{-5}
313.1	2.47×10^{-4}
323.1	7.59×10^{-4}
338.1	4.87×10^{-3}

If the rate constant is of the form of equation 3.0.14, determine the parameters A and E.

2. The decomposition of hexaphenylethane to triphenylmethyl radicals in liquid chloroform has been studied at 0 °C.

$$(C_6H_5)_3CC(C_6H_5)_3 \rightarrow 2(C_6H_5)_3C$$

The following results were obtained.

Time, t (sec)	C/C_0 for hexaphenylethane
0	1.000
17.4	0.941
35.4	0.883
54.0	0.824
174.0	0.530
209	0.471
313	0.324
367	0.265
434	0.206
584	0.118
759	0.059

Determine the reaction order and rate constant for the reaction by both differential and integral methods of analysis. For orders other than one, C_0 will be needed. If so, incorporate this term into the rate constant.

3. The data given below are typical of the polymerization of vinyl phenylbutyrate in dioxane solution in a batch reactor using benzoyl peroxide as an initiator. The reaction was carried out isothermally at 60 °C using an initial monomer concentration of 73 kg/m^3. From the following data determine the order of the reaction and the reaction rate constant. Note that there is an induction period at the start of the reaction so that you may find it useful to use a lower limit other than zero in your integration over time. The reaction order may be assumed to be an integer.

Time, t (ksec)	Monomer concentration (kg/m^3)
7.2	40.6
10.8	23.2
14.4	13.3
18.0	7.4
21.6	4.16
25.2	2.32
28.8	1.30
32.4	0.74

4. Huang and Dauerman [*Ind. Eng. Chem. Product Research and Development, 8* (227), 1969] have studied the acetylation of benzyl chloride in dilute solution at 102 °C.

$$NaAc + C_6H_5CH_2Cl \rightarrow$$

$$C_6H_5CH_2Ac + Na^+ + Cl^-$$

Using equimolal concentrations of sodium acetate and benzyl chloride (0.757 kmole/m^3), they reported the following data on the fraction benzyl chloride remaining unconverted versus time.

Time, t (ksec)	$(C_6H_5CH_2Cl)/(C_6H_5CH_2Cl)_0$
10.80	0.945
24.48	0.912
46.08	0.846
54.72	0.809
69.48	0.779
88.56	0.730
109.44	0.678
126.72	0.638
133.74	0.619
140.76	0.590

Determine the order of the reaction and the reaction rate constant at this temperature.

5. Dvorko and Shilov [*Kinetics and Catalysis*, 4 (212), 1964] have studied the iodine catalyzed addition of HI to cyclohexene in benzene solution.

The reaction is believed to be first order in each reactant and second order overall. The following data were reported for their experiments at 20 °C using an iodine concentration of 0.422×10^{-3} kmoles/m^3.

Time, t (sec)	HI concentration (kmoles/m^3)
0	0.106
150	0.099
480	0.087
870	0.076
1500	0.062
2280	0.050

The initial cyclohexene concentration was 0.123 kmole/m^3.

Are these data consistent with the proposed rate expression? If so, determine the reaction rate constant.

6. Dyashkovskii and Shilov [*Kinetics and Catalysis*, 4 (808), 1963] have studied the kinetics of the reaction between ethyl lithium and ethyl iodide in decalin solution.

$$C_2H_5Li + C_2H_5I \rightarrow 2C_2H_5\cdot + LiI$$

$$2C_2H_5\cdot \rightarrow \text{butane, ethane, and ethylene}$$

The following data are typical of those observed by these authors at 20 °C. They correspond to initial ethyl lithium and ethyl iodide concentrations of 2.0 and 1.0 kmoles/m^3, respectively.

Time, t (sec)	C_{LiI} (kmoles/m^3)
600	0.167
1200	0.306
1800	0.412
2400	0.498
3000	0.569

(a) What rate expression is consistent with these data? What is the reaction rate constant at 20 °C?

(b) Using a 10:1 ratio of ethyl iodide to ethyl lithium, the data below were reported at 22 and 60 °C. What is the activation energy of the reaction? The initial concentration of ethyl lithium was 0.5 kmole/m^3 in both instances.

Time, t (sec)	C_{LiI} (kmoles/m^3)	Temperature
1200	0.0169	
2400	0.0327	22 °C
3600	0.0495	
3900	0.0536	
600	0.162	
900	0.219	60 °C
1200	0.269	
1500	0.31	

7. The following initial rate data $[-d(B_2H_6)/dt]$ were reported for the gas-phase reaction of diborane and acetone at 114 °C.

$$B_2H_6 + 4Me_2CO \rightarrow 2(Me_2CHO)_2BH$$

Run	Initial pressure (torr)		Initial rate $\times 10^3$ (torr/sec)
	B_2H_6	Me_2CO	
1	6.0	20.0	0.5
2	8.0	20.0	0.63
3	10.0	20.0	0.83
4	12.0	20.0	1.00
5	16.0	20.0	1.28
6	10.0	10.0	0.33
7	10.0	20.0	0.80
8	10.0	40.0	1.50
9	10.0	60.0	2.21
10	10.0	100.0	3.33

If one postulates a rate expression of the form:

$$Rate = kP^n_{B_2H_6}P^m_{Me_2CO}$$

determine n, m, and k. Be sure to express k in appropriate units.

8. The following data are typical of the pyrolysis of dimethylether at 504 °C.

$$CH_3OCH_3 \rightarrow CH_4 + H_2 + CO$$

The reaction takes place in the gas phase in an isothermal constant volume reactor. Determine the order of the reaction and the reaction rate constant. The order may be assumed to be an integer.

Time, t (sec)	P (kPa)
0	41.6
390	54.4
777	65.1
1195	74.9
3155	103.9
∞	124.1

What does the fact that the final observed pressure is 124.1 kPa imply?

9. The reaction rate constant for the gas phase decomposition of ethylene oxide is 0.0212 min^{-1} at 450 °C.

$$CH_2 - CH_2(g) \rightarrow CH_4(g) + CO(g)$$
$$\diagdown O \diagup$$

At time zero, pure ethylene oxide is admitted to a constant temperature, constant volume reaction vessel at a pressure of 2.0 atm and a temperature of 450 °C. After 50 min, what is the total pressure in the reaction vessel? Ideal gas behavior may be assumed.

10. Chawla et al. [*J. Am. Chem. Soc.*, *89* (557), 1967] have studied the bromination of a hydrated amine complex in aqueous solution at 25 °C.

$$(Complex) + Br_2 \rightarrow (Complex\ Br) + Br^- + H^+$$

The initial bromine concentration was 72.6 mmoles/m^3, while the initial complex concentration was 1.49 moles/m^3. The reaction may be considered to be essentially irreversible. The following data were reported.

Time, t (sec)	Br_2 concentration (mmoles/m^3)
0	72.6
432	63.6
684	58.9
936	55.3
1188	51.6
1422	48.1
1632	45.2
2058	39.8
2442	35.1

(a) If the stoichiometry of the reaction is as indicated above, what is the percent change in the complex concentration during the time interval studied?

(b) What is the order of the reaction with respect to Br_2 concentration? What is the product of the true rate constant and the term corresponding to the dependence of the rate on complex concentration (pseudo rate constant)?

11. Svirbley and Roth [*J. Am. Chem. Soc., 75* (3106), 1953] have reported the following data for the reaction between HCN and C_3H_7CHO in aqueous solution at 25 °C.

Time, t (sec)	HCN (moles/m³)	C_3H_7CHO (moles/m³)
166.8	99.0	56.6
319.8	90.6	48.2
490.2	83.0	40.6
913.8	70.6	28.2
1188.	65.3	22.9
∞	42.4	0

Determine the order of the reaction and the reaction rate constant.

12. Baciocchi, et al. [*J. Am. Chem. Soc., 87* (3953), 1965] have investigated the kinetics of the chlorination of several hexa-substituted benzenes. They have reported the following data for the chlorination of dichlorotetramethylbenzene in acetic acid solution at 30 °C.

Initial concentrations:

$$C_6(CH_3)_4Cl_2 = 34.7 \text{ moles/m}^3$$
$$Cl_2 = 19.17 \text{ moles/m}^3$$

Time, t (ksec)	Fraction conversion
0	0
48.42	0.2133
85.14	0.3225
135.3	0.4426
171.3	0.5195
222.9	0.5955
257.4	0.6365

Determine the order of the reaction and the reaction rate constant. The order is an integer.

13. When one makes use of the assumption that the volume of a system is linear in the fraction conversion

$$V = V_0(1 + \delta f)$$

the parameter δ is based on the limiting reagent. Calculate δ for the following gaseous reactions and the feed concentrations given. I refers to the concentration of inerts in the feed stream.

(a) $C_2H_5OH \rightarrow H_2O + C_2H_4$
Initial concentration ratios
$C_2H_5OH:H_2O:C_2H_4:I \approx 1:1:1:2$

(b) $CH_3CHO + H_2 \rightarrow CH_3CH_2OH$
Initial concentration ratios
$CH_3CHO:H_2:CH_3CH_2OH:I \approx 1:2:0:1$

(c) $CH_3CHO + H_2 \rightarrow CH_3CH_2OH$
Initial concentration ratios
$CH_3CHO:H_2:CH_3CH_2OH:I \approx 2:1:0:0$

(d) $C_2H_6 + Cl_2 \rightarrow HCl + C_2H_5Cl$
Initial concentration ratios
$C_2H_6:Cl_2:HCl:C_2H_5Cl:I \approx 1:2:1:0:3$

(e) $CH_2{=}CH{-}CH{=}CH_2 + 2H_2 \rightarrow C_4H_{10}$
Initial concentration ratios
$CH_2{=}CH{-}CH{=}CH_2:H_2:C_4H_{10}:I \approx 2:1:1:0$

(f) $2N_2O_5 \rightarrow 2N_2O_4 + O_2$
Initial concentration ratios
$N_2O_5:I:N_2O_4:O_2 \approx 1:0:0:0$

14. The reaction of cyclohexanol and acetic acid in dioxane solution as catalyzed by sulfuric acid was studied by McCracken and Dickson [*Ind. Eng. Chem. Proc. Des. and Dev., 6* (286), 1967]. The esterification reaction can be represented by the following stoichiometric equation.

$$
\begin{array}{ccccccc}
A & + & B & \underset{k_b}{\overset{k_f}{\rightleftharpoons}} & C & + & W \\
\text{acetic} & & \text{cyclo-} & & \text{cyclo-} & & \text{water} \\
\text{acid} & & \text{hexanol} & & \text{hexyl} & & \\
& & & & \text{acetate} & &
\end{array}
$$

The reaction was carried out in a well-stirred batch reactor at 40 °C. Under these conditions, the esterification reaction can be considered as irreversible at conversions less than 70%.

$$A + B \xrightarrow{k_f} C + W$$

The following data were obtained using identical sulfuric acid concentrations in both runs.

Run 1: $C_{A0} = C_{B0} = 2.5$ kmoles/m³

C_B (kmoles/m³)	Time, t (ksec)
2.070	7.2
1.980	9.0
1.915	10.8
1.860	12.6
1.800	14.4
1.736	16.2
1.692	18.0
1.635	19.8
1.593	21.6
1.520	25.2
1.460	28.8

Run 2: $C_{A0} = 1$ kmole/m³
            $C_{B0} = 8$ kmoles/m³

C_A (kmoles/m³)	Time, t (ksec)
0.885	1.8
0.847	2.7
0.769	4.5
0.671	7.2
0.625	9.0
0.544	12.6
0.500	15.3
0.463	18.0

(a) Determine the order of the reaction with respect to each reactant and the rate constant for the forward reaction under the conditions of the two runs. The rate constant should differ between runs, but the conditions are such that the order will not differ. The individual and overall orders are integers.

(b) If acetic acid exists primarily as a dimer $(C_2H_4O_2)_2$ in solution at 40 °C, postulate a set of mechanistic equations that will be consistent with this rate expression.

15. The saponification of the ester phenylacetate by sodium phenolate has the following stoichiometry.

$$H_2O + CH_3COOC_6H_5 + Na^{+-}OC_6H_5 \rightarrow$$
$$CH_3COO^-Na^+ + 2C_6H_5OH$$

The reaction takes place in aqueous solution. Equimolal concentrations of the ester and the phenolate are used. These concentrations are equal to 30 moles/m³. By the time the samples are brought to thermal equilibrium in the reactor and efforts made to obtain data on ester concentrations as a function of time, some saponification has occurred. At this time the concentration of ester remaining is 26.29 moles/m³, and the concentration of phenol present in the reactant mixture is 7.42 moles/m³. The rate expression for the reaction is believed to be of the form

$$\text{Rate} = \frac{kC_{ester}C_{phenolate}}{C_{phenol}}$$

Determine the reaction rate constant from the following data.

Time, t (ksec)	Ester concentration (moles/m³)
0	26.29
0.72	22.00
2.16	18.30
4.32	15.29
8.64	12.15
14.34	9.99
20.64	8.50
27.24	7.30

Be very careful to note the appropriate stoichiometric coefficients for use in your analysis.

16. Consider a gas phase dehydrogenation reaction that occurs at a constant temperature of 1000 °K.

$$RCH_2OH(g) \rightarrow RCHO(g) + H_2(g)$$

At this temperature the reaction rate constant is equal to 0.082 liters/mole-min. If the reaction takes place in a *constant pressure* reactor, starting with pure gaseous alcohol at a pressure of 2 atm, find the time necessary to reach 20% decomposition:

(a) In the sense that the number of moles of alcohol in the reactor decreases by 20%.
(b) In the sense that the concentration of alcohol in the reactor decreases by 20%.

Ideal gas laws may be assumed in both cases.

17. Hinshelwood and Burk [*Proc. Roy. Soc.*, *106A* (284), 1924] have studied the thermal decomposition of nitrous oxide. Consider the following "adjusted" data at 1030 °K.

Initial pressure of N_2O (mm Hg)	Half-life (sec)
82.5	860
139	470
296	255
360	212

(a) Determine the order of the reaction and the reaction rate constant.
(b) The following additional data were reported at the temperatures indicated.

Temperature, T (°K)	Initial pressure (mm Hg)	Half-life (sec)
1085	345	53
1030	360	212
967	294	1520

What is the activation energy for the reaction?

18. Daniels (*Chemical Kinetics*, page 19, Cornell University Press, 1938) has carried out a series of investigations of the N_2O_5 decomposition reaction. The following data are typical of the time for 50% decomposition of a given initial charge of N_2O_5 in a constant volume reactor at various temperatures. It may be assumed that the initial concentration of N_2O_5 is the same in all cases and that each run is carried out isothermally.

Temperature, T (°C)	Half-life (sec)
300	3.9×10^{-5}
200	3.9×10^{-3}
150	8.8×10^{-2}
100	4.6
50	780

What is the activation energy of the reaction?

19. The gas phase dimerization of trifluorochloroethylene may be represented by

$$2CF_2{=}CFCl \rightarrow \begin{matrix} CF_2-CFCl \\ | \qquad | \\ CF_2-CFCl \end{matrix}$$

The following data are typical of this reaction at 440 °C as it occurs in a constant volume reactor.

Time, t (sec)	Total pressure (kPa)
0	82.7
100	71.1
200	64.0
300	60.4
400	56.7
500	54.8

Determine the order of the reaction and the reaction rate constant under these conditions. The order is an integer.

20. Daniels (*Chemical Kinetics*, page 9, Cornell University Press, 1938) has reported the following data on the rate of decomposition of N_2O_5

in the gas phase at 45 °C. The reaction takes place in a constant volume system.

Time, t (min)	$P_{N_2O_5}$ (mm of Hg)
0	?
10	247
20	185
30	140
40	105
50	78
60	58
70	44
80	33
90	24
100	18

The pertinent stoichiometry is:

$$2N_2O_5 \rightarrow 2N_2O_4 + O_2$$

Use both integral and differential approaches to determine the order of the reaction and the reaction rate constant. The order is either 0, 1, or 2.

21. Frost and Pearson have indicated that the following data are typical of the reaction.

$$C_2H_5NO_2 + C_5H_5N + I_2 \rightarrow$$
$$C_2H_4INO_2 + C_5H_5NH^+ + I^-$$

Original concentrations	Time, t (sec)	Resistance (Ω)
Nitroethane 100 moles/m^3	0	2503
Pyridine 100 moles/m^3	300	2295
Iodine 4.5 moles/m^3	600	2125
	900	1980
Temperature 25°C	1200	1850
Water-alcohol-solvent	1500	1738
	1800	1639
	∞	1470

Adapted from *Kinetics and Mechanism*, 2nd edition, by A. A. Frost and R. G. Pearson, copyright © 1961. Reprinted by permission of John Wiley and Sons, Inc.

Determine the apparent order of reaction and the apparent rate constant in suitable units.
Hints
(1) The resistance is inversely proportional to the conductance of the solution. The latter is an additive property of the contributions of the various species in the solution. The reaction order is either 0, 1, or 2. Note that the concentrations of nitroethane and pyridine are much greater than that of the iodine.
(2) *Show* that the extent of reaction at time t relative to the extent at time infinity is given by

$$\frac{\xi}{\xi_\infty} = \left[\frac{R - R_0}{R_\infty - R_0} \right] \frac{R_\infty}{R}$$

and make use of this fact in your analysis.

22. NMR is an analytical technique that has been applied to the studies of chemical reactions with promising results. The chemical shift parameter may be used as a measure of the relative proportions of different species that are present in solution. These shifts are measured relative to a standard reference sample.

Benzhydryl bromide (BHB) undergoes solvolysis in aqueous dioxane to produce benzhydryl alcohol (BHA) and hydrogen bromide.

$$BHB + H_2O \rightarrow BHA + H^+ + Br^- \quad (A)$$

The rate of formation of hydrogen ions can be determined by observing the chemical shift for the water molecules in the solution. Since the rate of protonation is very rapid compared to a typical NMR time scale, the chemical shift may be said to be a linear combination of the contributions of the protonated water molecules and the unprotonated water molecules. That is,

$$\delta = \delta_{H_2O} X_{H_2O} + \delta_{H_3^+O} X_{H_3^+O}$$

where

δ is the observed chemical shift

δ_{H_2O} is the chemical shift that would be observed in the absence of unprotonated species

$\delta_{H_3^+O}$ is the chemical shift that would be observed if all species were protonated

X_{H_2O} is the fraction of the water molecules that are not protonated

$X_{H_3^+O}$ is the fraction of the water molecules that are protonated

It may be assumed that all the protons formed by reaction immediately react to form protonated water molecules. The following chemical shifts were observed as a function of time.

Time, t (sec)	δ
240	40.1
480	27.0
720	19.3
960	14.8
1200	12.3
1440	10.9
1680	10.1
∞	9.0

[Data taken from article by Horman and Strauss, *J. Chem. Education*, 46 (114), 1969.] If reaction (A) may be considered as an irreversible reaction, what is the order of this reaction with respect to BHB? The initial concentration of BHB is 360 moles/m³. The water is present in sufficient excess that its concentration may be considered to be essentially constant. What is the reaction rate constant?

23. Tuulmets [*Kinetics and Catalysis*, 5 (59), 1964] has studied the kinetics of the reaction of ethyl magnesium bromide with pinacolin. He used a calorimetric technique to monitor the progress of the reaction. The overall temperature increase of the reaction mixture was less than a degree. Mixture temperatures were determined with a sensitive potentiometer. The data below

have been corrected to allow for heat losses from the system (i.e., adiabatic operation of the batch reactor may be assumed). The reaction was carried out in diethyl ether solution using initial concentrations of C_2H_5MgBr and pinacolin of 403 and 10 moles/m³, respectively. The stoichiometry of the reaction may be represented as

$$A + B \rightarrow C$$

Changes in the potentiometer reading may be assumed to be proportional to the temperature change of the reaction mixture. Determine the order of the reaction with respect to pinacolin and the apparent reaction rate constant.

Time, t (sec)	Potentiometer reading
10	65.0
20	96.5
30	107.6
40	114.2
50	116.2
60	117.5
70	118.2
280	118.5

Unfortunately, the potentiometer reading at time zero is uncertain.

24. Pannetier and Souchay* have reported the data below as an example of the application of dilatometry to kinetics studies.

The hydration of ethylene oxide

$$CH_2\!\!-\!\!CH_2 + H_2O \Rightarrow \begin{array}{c} CH_2OH \\ | \\ CH_2OH \end{array}$$

was studied in aqueous solution at 20 °C using an initial oxide concentration of 120 moles/m³ and a catalyst concentration of 7.5 moles/m³.

---

* Adapted from *Chemical Kinetics* by G. Pannetier and P. Souchay, p. 68, copyright ©1964. Used with permission of Elsevier Scientific Publishing Company.

Time, t (ksec)	Height of capillary level (cm)
0	18.48
1.80	18.05
3.60	17.62
5.40	17.25
7.20	16.89
14.40	15.70
23.40	14.62
∞	12.30

What is the order of the reaction and the reaction rate constant? The reverse reaction may be neglected. The volume of the solution as determined by the height of the meniscus in the capillary may be assumed to be a measure of the fraction conversion (i.e., the volume change is proportional to the extent of reaction).

25. Burkhart and Newton [*J. Phys. Chem., 73* (1741), 1969] have studied the kinetics of the reaction between vanadium (II) and neptunium (IV) in aqueous perchlorate solutions

$$Np(IV) + V(II) \rightarrow Np(III) + V(III)$$

The reaction may be considered as *irreversible*. The progress of the reaction was monitored by observing the absorbance of the solution at 723 nm where Np(IV) is the principal absorbing species. From the data below determine the reaction rate constant for this reaction if the reaction is first order in both Np(IV) and V(II).

Temperature = 25.2 °C

Initial Np(IV) concentration = 0.358 mole/m^3

Initial V(II) concentration = 1.30 moles/m^3

Time, t (sec)	Absorbance	Time, t (sec)	Absorbance
0	0.600	520	0.150
80	0.472	600	0.128
160	0.376	680	0.109
240	0.301	760	0.094
320	0.244	840	0.081
440	0.180	∞	0.000

26. Marvel, Dec, and Cooke [*J. Am. Chem. Soc.*, 62 (3499), 1940] have used optical rotation measurements to study the kinetics of the polymerization of certain optically active vinyl esters. The change in rotation during the polymerization may be used to determine the reaction order and reaction rate constant. The specific rotation angle in dioxane solution is a linear combination of the contributions of the monomer and of the polymerized mer units. The optical rotation due to each mer unit in the polymer chain is independent of the chain length. The following values of the optical rotation were recorded as a function of time for the polymerization of d-s-butyl α-chloroacrylate

Time, t (ksec)	Rotation
0	2.79
3.6	2.20
7.2	1.84
10.8	1.59
14.4	1.45
18.0	1.38
∞	1.27

Determine the order of the reaction and the reaction rate constant if the initial monomer concentration is M_0.

27. The data below have been reported for the decomposition of diazobenzene chloride.

$$C_6H_5N_2Cl \rightarrow C_6H_5Cl + N_2$$

The reaction was followed by measuring the nitrogen evolved from the liquid mixture.

The initial concentration of diazobenzene chloride was 5.0 kg/m^3. The reaction is believed to be first order and irreversible. Determine if the reaction is indeed first order and evaluate the reaction rate constant.

Time, t (sec)	N_2 evolved (cm^3 at STP)
360	19.3
540	26.0
720	32.6
840	36.0
1080	41.3
1320	45.0
1440	46.5
1560	48.4
1800	50.4
∞	58.3

28. The following addition reaction takes place in the gas phase at 400 °C.

$$A + B \xrightarrow{k_1} C$$

The initial composition is as follows (mole percents).

A	0.40
B	0.40
C	0.11
Inerts	0.09

The reaction takes place in an isothermal reactor that is maintained at a *constant pressure* of 1 atm. If the reaction is first-order in A and first-order in B, and if the value of k_1 is 2 liters/mole-min, how long will it be before the mole fraction C reaches 0.15?

4 Basic Concepts in Chemical Kinetics—Molecular Interpretations of Kinetic Phenomena

4.0 INTRODUCTION

This chapter treats the descriptions of the molecular events that lead to the kinetic phenomena that one observes in the laboratory. These events are referred to as the *mechanism* of the reaction. The chapter begins with definitions of the various terms that are basic to the concept of reaction mechanisms, indicates how elementary events may be combined to yield a description that is consistent with observed macroscopic phenomena, and discusses some of the techniques that may be used to elucidate the mechanism of a reaction. Finally, two basic molecular theories of chemical kinetics are discussed—the kinetic theory of gases and the transition state theory. The determination of a reaction mechanism is a much more complex problem than that of obtaining an accurate rate expression, and the well-educated chemical engineer should have a knowledge of and an appreciation for some of the techniques used in such studies.

There are at least two levels of sophistication at which one may approach the problem of providing a molecular description of the phenomena that occur during the course of a chemical reaction. At the first level the sequence of molecular events is described in terms of the number and type of molecules and molecular fragments that come together and react in the various steps. The second level of description contains all of the elements of the first but goes beyond it to treat the geometric and electronic configurations of the various species during the different stages of the reaction sequence. For this textbook, the first level of description is adequate. The second level is more appropriate for study in courses of physical organic chemistry or advanced physical chemistry.

The mechanism of a reaction is a hypothetical construct. It is a provisional statement based on available experimental data, representing a suggestion as to the sequence of molecular events that occur in proceeding from reactants to products. It does not necessarily represent the actual events that occur during the reaction process, but *it must be consistent with the available experimental facts*. Often there will be more than one mechanism that is consistent with these facts. The problem of designing an experiment that will eliminate one or more of the competing mechanisms is a challenging problem for the kineticist. In some cases it is an impossible task.

Implicit in the use of the term "sequence of molecular events" is the idea that the chemical transformation that one observes in the laboratory is not the result of a single molecular process but is the end result of a number of such processes. If one considers reactions such as

$$C_7H_{16} + 11O_2 \rightarrow 7CO_2 + 8H_2O$$
$$(4.0.1)$$

$$5SO_3^= + 2IO_3^- + 2H^+ \rightarrow 5SO_4^= + I_2 + H_2O$$
$$(4.0.2)$$

it should be evident from a purely statistical viewpoint that there is virtually zero probability that the number of molecular species involved in these reactions would simultaneously be in spatial and electronic configurations such that more than a few chemical bonds could be broken and/or made in a single step. These equations merely describe the stoichiometry that is observed in the laboratory. Nonetheless, they do reflect the conversion of reactant molecules into product molecules. It is quite plausible to assume that the events that occur on a molecular level are encounters at which atomic rearrangements occur and that the observed reaction may be interpreted as the sum total of the changes that occur during a number of such encounters. This assumption is the foundation on which all studies of reaction mechanisms are based. The major justifications for its use lie in the tremendous success it has had in providing a molecular interpretation of kinetic phenomena and the fact that it has been possible to observe experimentally some of the intermediates postulated in sequences of elementary reactions.

Each elemental process contributing to the overall mechanism is itself an irreducible chemical reaction. The elementary reactions may also be referred to as simple reactions, mechanistic reactions, or reaction steps. The superposition in time of elementary reactions leads to the experimentally observed reaction. Since each step in the mechanism is itself a chemical reaction and is written as such, the equation representing an elemental process looks the same as an equation that represents the stoichiometry of a chemical reaction. The mechanistic equation, however, has an entirely different meaning than the stoichiometric equation. It is only from the context of their use that the reader knows which type of equation is being used, and the beginning kineticist must be sure to distinguish between their meanings in his or her thinking.

The number of chemical species involved in a single elementary reaction is referred to as the *molecularity* of that reaction. Molecularity is a theoretical concept, whereas stoichiometry and order are empirical concepts. A simple reaction is referred to as uni-, bi-, or termolecular if one, two, or three species, respectively, participate as reactants. The majority of known elementary steps are bimolecular, with the balance being unimolecular and termolecular.

Since an elementary reaction occurs on a molecular level exactly as it is written, its rate expression can be determined by inspection. A unimolecular reaction is a first-order process, bimolecular reactions are second-order, and termolecular processes are third-order. However, the converse statement is not true. Second-order rate expressions are not necessarily the result of an elementary bimolecular reaction. While a stoichiometric chemical equation remains valid when multiplied by an arbitrary factor, a mechanistic equation loses its meaning when multiplied by an arbitrary factor. Whereas stoichiometric coefficients and reaction orders may be integers or nonintegers, the molecularity of a reaction is always an integer. The following examples indicate the types of rate expression associated with various molecularities.

Unimolecular: $\quad A \xrightarrow{k_1} B + C \quad r = k_1 C_A$
$$(4.0.3)$$

Bimolecular: $\quad M + N \xrightarrow{k_2} P \quad\quad r = k_2 C_M C_N$
$$(4.0.4)$$

Termolecular: $R + 2S \xrightarrow{k_3} T + U \quad r = k_3 C_R C_S^2$
$$(4.0.5)$$

Note that for the last reaction the stoichiometric coefficients require that

$$r_S = -2k_3 C_R C_S^2 \qquad (4.0.6)$$

Each elementary reaction must fulfill the following criteria.

1. On a molecular scale the reaction occurs exactly as written.
2. The simple reaction should break or form as few bonds as possible. Normally only one bond is made and/or broken. Occasionally two bonds are broken and two new ones made in a four-center reaction. Only very rarely do more complex processes occur.
3. The simple reaction must be feasible with respect to bond energies, atomic geometry, and allowed electron shifts.

In all but the simplest cases, the mechanism of a reaction consists of a number of steps, some of which involve reacting species that do not appear in the overall stoichiometric equation for the reaction. Some of these *intermediate* species are stable molecules that can be isolated in the laboratory. Others are highly reactive species that can be observed only by using sophisticated experimental techniques. These reaction intermediates are sometimes referred to as *active centers.*

Among the various intermediate species that may participate in a reaction sequence are stable molecules, ions, free atoms, free radicals, carbanions, carbonium ions, molecular and ionic complexes, and tautomeric or excited forms of stable molecules. If the intermediate is, indeed, a stable substance, then its presence can be detected by any of the standard techniques of chemical analysis, provided that the intermediate can be isolated (i.e., prevented from participation in the processes that would normally follow its formation). If isolation is impossible, then the techniques available for the study of stable intermediates are the same as those for the study of highly reactive species. For a detailed discussion of appropriate experimental techniques, consult the references listed in Section 3.2.2 or the review by Wayne (1).

4.1 REACTION MECHANISMS

At first glance the problem of "finding the mechanism" of a reaction may appear to the beginning student as an exercise in modern alchemy. Nonetheless, there are certain basic principles underlying the reasoning that intervenes between experimental work in the laboratory and the postulation of a set of molecular events that give rise to the chemical reaction being investigated. This section is a discussion of the nature of the problem, the means by which one derives a rate expression from a proposed sequence of reactions, and some techniques that are useful in the elucidation of reaction mechanisms.

4.1.1 The Nature of the Problem

The postulation of a reaction mechanism is the result of inductive rather than deductive thinking. Even though the kinetics researcher may present the ideas and experiments that lead to a proposed mechanism in a logical

orderly manner, the thought processes leading to these results involve elements of experience, intuition, luck, knowledge, and guesswork.

An individual who has started work on a kinetics research project will usually have some preconceived ideas about the mechanism of the reaction being investigated and will usually know the major products of the reaction before starting any kinetics experiments. As a result of a literature search, the kineticist often has a knowledge of experimental rate expressions and proposed mechanisms for similar reactions. As the kineticist proceeds with the experimental work and determines the complete product distribution, the order of the reaction, and the effects of temperature, pressure, solvent, etc., on the reaction rate, his or her ideas relative to the mechanism will evolve into a coherent and logical picture. Along the path, however, the preliminary ideas may be shuffled, added to, discarded, broadened, or refurbished before they can be manipulated into a logical sequence of events that is consistent with all of the available facts. Each mechanism study evolves in its own fashion as the investigator designs experiments to test provisional mechanisms that have been postulated on the basis of earlier experiments. These experiments constitute a more or less systematic attempt to make mechanistic order out of what may appear to the uninitiated observer to be a random collection of experimental facts. By the time the experimental research is completed and the results are ready for publication, the investigator will have resolved his or her ideas about the mechanism into a fully coherent picture that can be presented in a systematic and rational manner. Since this end result is all that is published for the judgment of the scientific community, the uninitiated observer obtains the false impression that the research work itself was actually carried out in this systematic and rational fashion. In the actual program there may have been many false starts and erroneous ideas that had to be overcome before arriving at the final result. To a fellow kineticist, however, the investigator's systematic and logical rendering of the experimental facts is an efficient means of communicating the results of the research.

The problem of determining the mechanism of a chemical reaction is one of the most interesting and challenging intellectual and experimental problems that a chemical engineer or chemist is likely to encounter. Even for relatively simple mechanisms, it is occasionally difficult to determine from the assumptions about the elementary reactions just what the exact form of the predicted dependence of the rate on reactant concentrations should be. In many cases the algebra becomes intractable and, without an *a priori* knowledge of the magnitudes of various terms, it may be difficult or even impossible for the theoretician to make simplifications that will permit these predictions to be made. Many chemical reactions are believed to have mechanisms involving a large number of elementary steps [e.g., the mechanism of reaction between H_2 and O_2 contains at least 26 steps (2)], and it becomes an extremely formidable task to manipulate the corresponding mathematical relations to predict the dependence of the rate on the concentrations of stable molecules. How much more formidable is the problem that our investigator must tackle—that of deriving such a mechanism consistent with all available experimental data!

Since the problem of deriving a rate expression from a postulated set of elementary reactions is simpler than that of determining the mechanism of a reaction, and since experimental rate expressions provide one of the most useful tests of reaction mechanisms, we will now consider this problem.

4.1.2 Basic Assumptions Involved in the Derivation of a Rate Expression from a Proposed Reaction Mechanism

The mechanism of a chemical reaction is a microscopic description of the reaction in terms of its constituent elementary reactions. The fundamental principle from which one starts is that the rate of an elementary reaction is proportional to the frequency of collisions indicated by the stoichiometric-mechanistic equation for the reaction (i.e., to the product of the concentrations indicated by the molecularity of the reaction). In addition, one usually bases the analysis on one or more of the following simplifications in order to make the mathematics amenable to closed form solution.

First, one often finds it convenient to *neglect reverse reactions* (i.e., the reaction is considered to be "irreversible"). This assumption will be valid during the initial stages of any reaction, since the number of product molecules available to serve as reactants for the reverse reaction will be small during this period. If the equilibrium constant for the reaction is very large, this assumption will also be true for intermediate and later stages of the reaction.

Second, it may be convenient to assume that one elementary reaction in the sequence occurs at a much slower rate than any of the others. The overall rate of conversion of reactants to products may be correctly calculated on the assumption that this step governs the entire process. The concept of a *rate limiting step* is discussed in more detail below.

Third, it is often useful to assume that the concentration of one or more of the intermediate species is not changing very rapidly with time (i.e., that one has a *quasistationary* state situation). This approximation is also known as the *Bodenstein steady-state* approximation for intermediates. It implies that the rates of production and consumption of intermediate species are nearly equal. This approximation is particularly good when the intermediates are highly reactive.

Fourth, one often finds it convenient to assume that one or more reactions is in a *quasi-equilibrium condition* (i.e., the forward and reverse rates of this reaction are much greater than those of the other reactions in the sequence). The net effect is that these other reactions do not produce large perturbations of the first reaction from equilibrium.

In the sequence of elementary reactions making up the overall reaction, there often is one step that is very much slower than all the subsequent steps leading to reaction products. In these cases the rate of product formation may depend on the rates of all the steps preceding the last slow step, but will not depend on the rates of any of the subsequent more rapid steps. This last slow step has been termed the *rate controlling, rate limiting,* or *rate determining* step by various authors.

In a mechanism consisting of several consecutive elementary reaction steps, fragments of the initial reactant pass through a number of intermediate stages, finally ending up as the products of the reaction. The total time necessary to produce a molecule of product is simply the sum of the discrete times necessary to pass through each individual stage of the overall reaction. Where a reversible reaction step is involved, the net time for that step is necessarily increased by the "feedback" of intermediate accompanying that step. The mean reaction time (t_{mean}) is thus

$$t_{mean} = t_1 + t_2 + t_3 + \cdots + t_n \quad (4.1.1)$$

where the t_i represent *effective times necessary on the average* to accomplish each step.

The overall reaction rate r may be defined as the reciprocal of this mean reaction time.

$$r = \frac{1}{t_{mean}} \quad (4.1.2)$$

Thus

$$r = \frac{1}{t_1 + t_2 + t_3 + \cdots + t_n} \quad (4.1.3)$$

The rates of each individual step are given by analogs of equation 4.1.2. Thus

$$r = \frac{1}{\dfrac{1}{r_1} + \dfrac{1}{r_2} + \dfrac{1}{r_3} + \cdots + \dfrac{1}{r_n}} \quad (4.1.4)$$

If one of the times (say t_{RLS}) in the denominator of equation 4.1.3 is very much larger than any of the others, then this equation becomes

$$r \cong \frac{1}{t_{RLS}} = r_{RLS} \quad (4.1.5)$$

where the subscript *RLS* refers to the rate limiting step.

If some of the processes prior to the rate limiting step are characterized by times comparable to that of the rate determining step, equation 4.1.4 may be rewritten as

$$r = \frac{1}{\dfrac{1}{r_1} + \dfrac{1}{r_2} + \dfrac{1}{r_3} + \cdots + \dfrac{1}{r_{RLS}}} \quad (4.1.6)$$

where terms corresponding to events beyond the last slow step in the process have been dropped. Kinetic events preceding the *last* slow step can influence the overall reaction rate, but subsequent steps cannot.

4.1.3 Preliminary Criteria for Testing a Proposed Reaction Mechanism— Stoichiometry and Derivation of a Rate Expression for the Mechanism

There are two crucial criteria with which a proposed mechanism must be consistent.

1. The net effect of the elementary reactions must be the overall reaction and must explain the formation of all observed products.
2. It must yield a rate expression that is consistent with the kinetics observed experimentally.

In the case of relatively simple reaction mechanisms, the net or overall effect of the elementary reactions can be determined by adding them together. For example, the stoichiometric equation for the decomposition of nitrogen pentoxide is

$$2N_2O_5 \rightarrow 4NO_2 + O_2 \quad (4.1.7)$$

A mechanism that might be proposed to explain this reaction is

$$N_2O_5 \underset{k_2}{\overset{k_1}{\rightleftharpoons}} NO_2 + NO_3 \qquad (4.1.8)$$

$$NO_2 + NO_3 \overset{k_3}{\rightarrow} NO + O_2 + NO_2 \qquad (4.1.9)$$

$$NO + NO_3 \overset{k_4}{\rightarrow} 2NO_2 \qquad (4.1.10)$$

If one multiplies equation 4.1.8 by two and adds the result to equations 4.1.9 and 4.1.10, equation 4.1.7 is obtained.

For mechanisms that are more complex than the above, the task of showing that the net effect of the elementary reactions is the stoichiometric equation may be a difficult problem in algebra whose solution will not contribute to an understanding of the reaction mechanism. Even though it may be a fruitless task to find the exact linear combination of elementary reactions that gives quantitative agreement with the observed product distribution, it is nonetheless imperative that the mechanism qualitatively imply the reaction stoichiometry. Let us now consider a number of examples that illustrate the techniques used in deriving an overall rate expression from a set of mechanistic equations.

ILLUSTRATION 4.1 DEMONSTRATION OF THE FACT THAT TWO PROPOSED MECHANISMS CAN GIVE RISE TO THE SAME RATE EXPRESSION

The stoichiometric equation for the reaction between nitric oxide and hydrogen is:

$$2NO + 2H_2 \rightarrow N_2 + 2H_2O \qquad (A)$$

The overall rate expression is third order.

$$r_{N_2} = k[NO]^2[H_2] \qquad (B)$$

Two different mechanisms have been proposed for this reaction. They are described below.

Mechanism A

$$2NO + H_2 \overset{k_1}{\rightarrow} N_2 + H_2O_2 \qquad \text{(slow)} \quad (C)$$

$$H_2O_2 + H_2 \overset{k_2}{\rightarrow} 2H_2O \qquad \text{(fast)} \quad (D)$$

Mechanism B

$$2NO \underset{k_4}{\overset{k_3}{\rightleftharpoons}} N_2O_2 \qquad \text{(fast)} \quad (E)$$

$$N_2O_2 + H_2 \overset{k_5}{\rightarrow} N_2 + H_2O_2 \qquad \text{(slow)} \quad (F)$$

$$H_2O_2 + H_2 \overset{k_2}{\rightarrow} 2H_2O \qquad \text{(fast)} \quad (G)$$

Show that both of these mechanisms are consistent with the observed rate expression and the stoichiometry of the reaction.

Solution

First consider mechanism **A**. Addition of reactions C and D gives reaction A, so the stoichiometry of the mechanism is consistent. Reaction C is the rate limiting step, so the overall reaction rate is given by

$$r_{N_2} = k_1[NO]^2[H_2] \qquad (H)$$

This rate expression is consistent with the observed kinetics, so this combination of a slow termolecular step with a rapid bimolecular step is a plausible mechanism based on the information we have been given.

Now consider mechanism **B**. Addition of reactions E, F, and G again yields equation A. Thus the stoichiometry is consistent. The second step in the mechanism is the rate controlling step. Thus

$$r_{N_2} = k_5[N_2O_2][H_2] \qquad (I)$$

This expression contains the concentration of an intermediate species N_2O_2. It may be eliminated by recognizing that reaction E is essentially at equilibrium. Hence

$$k_3[NO]^2 = k_4[N_2O_2] \qquad (J)$$

The concentration of the intermediate is given by

$$[N_2O_2] = \frac{k_3}{k_4}[NO]^2 = K_e[NO]^2 \qquad (K)$$

where K_e is an equilibrium constant for reaction E.

Substitution of this expression for the N_2O_2 concentration in equation I gives

$$r_{N_2} = \frac{k_3 k_5}{k_4}[NO]^2[H_2] = k_5 K_e[NO]^2[H_2]$$

(L)

These equations are consistent with the observed kinetics if we identify k as

$$k = \frac{k_3 k_5}{k_4} = k_5 K_e$$

(M)

Thus mechanism **B**, which consists solely of bimolecular and unimolecular steps, is also consistent with the information that we have been given. This mechanism is somewhat simpler than the first in that it does not requite a termolecular step. This illustration points out that the fact that a mechanism gives rise to the experimentally observed rate expression is by no means an indication that the mechanism is a unique solution to the problem being studied. We may disqualify a mechanism from further consideration on the grounds that it is inconsistent with the observed kinetics, but consistency merely implies that we continue our search for other mechanisms that are consistent and attempt to use some of the techniques discussed in Section 4.1.5 to discriminate between the consistent mechanisms. It is also entirely possible that more than one mechanism may be applicable to a single overall reaction and that parallel paths for the reaction exist. Indeed, many catalysts are believed to function by opening up alternative routes for a reaction. In the case of parallel reaction paths each mechanism proceeds independently, but the vast majority of the reaction will occur via the fastest path.

ILLUSTRATION 4.2 USE OF STEADY-STATE AND PSEUDO EQUILIBRIUM APPROXIMATIONS FOR INTERMEDIATE CONCENTRATIONS

The thermal decomposition of nitrogen pentoxide to oxygen and nitrogen dioxide

$$2N_2O_5 \rightarrow O_2 + 4NO_2$$

(A)

is a classic reaction in kinetics because it was the first gas phase, first-order reaction to be reported (3).

The reaction has been studied by a number of investigators, and the following mechanism is the one that appears to give the best agreement with the available experimental facts.

$$N_2O_5 \underset{k_2}{\overset{k_1}{\rightleftharpoons}} NO_2 + NO_3$$

(B)

$$NO_3 + NO_2 \overset{k_3}{\rightarrow} NO + O_2 + NO_2 \quad (slow) \quad (C)$$

$$NO + NO_3 \overset{k_4}{\rightarrow} 2NO_2$$

(D)

In reaction C the NO_2 itself does not react but plays the role of a collision partner that may effect the decomposition of the NO_3 molecule. The NO_2 and NO_3 molecules may react via the two paths indicated by the rate constants k_2 and k_3. The first of these reactions is believed to have a very small activation energy; the second reaction is endothermic and consequently will have an appreciable activation energy. On the basis of this reasoning, Ogg (4) postulated that k_3 is much less than k_2 and that reaction C is the rate controlling step in the decomposition. Reaction D, which we have included, differs from the final step postulated by Ogg.

Derive an expression for the overall reaction rate based on the above mechanism.

Solution

The rate of the overall reaction is given by either of the following expressions.

$$r = \frac{-\frac{1}{2}d[N_2O_5]}{dt} = \frac{d[O_2]}{dt}$$

(E)

The second is more useful for our purposes, since oxygen is produced in only one step of the mechanism and that step is the rate controlling step. Thus

$$r = r_{RLS} = k_3[NO_3][NO_2]$$

(F)

Equation F contains the concentration of an intermediate (NO_3) that does not appear in the overall stoichiometric equation. This term may be eliminated by means of the Bodenstein steady-state approximation.

$$\frac{d[NO_3]}{dt} = k_1[N_2O_5] - k_2[NO_2][NO_3] - k_3[NO_3][NO_2] - k_4[NO][NO_3] = 0 \qquad (G)$$

Thus

$$[NO_3]_{ss} = \frac{k_1[N_2O_5]}{k_2[NO_2] + k_3[NO_2] + k_4[NO]} \qquad (H)$$

The subscript ss refers to a steady-state value.

Equation H also involves the concentration of a reaction intermediate $[NO]$. If we make the steady-state approximation for this species,

$$\frac{d[NO]}{dt} = k_3[NO_2][NO_3] - k_4[NO][NO_3] = 0 \qquad (I)$$

or

$$[NO]_{ss} = \frac{k_3}{k_4}[NO_2] \qquad (J)$$

Combination of equations H and J gives

$$[NO_3]_{ss} = \frac{k_1[N_2O_5]}{k_2[NO_2] + 2k_3[NO_2]} \qquad (K)$$

This relation may now be substituted in equation F to obtain an expression for the overall rate of reaction that contains only the concentrations of species that appear in the stoichiometric equation

$$r = \frac{k_1 k_3 [N_2O_5]}{k_2 + 2k_3} \qquad (L)$$

This expression is consistent with the first-order kinetics that are observed experimentally.

Instead of using the steady-state approximation in the manipulation of the individual rate expressions, the same result may be reached by assuming that a pseudo equilibrium condition is established with respect to reaction B and that reaction C continues to be the rate limiting step. From the equilibrium condition,

$$[NO_3]_{PE} = \frac{k_1}{k_2} \frac{[N_2O_5]}{[NO_2]} \qquad (M)$$

where the subscript PE is used to indicate the pseudo equilibrium concentration. As above,

$$r = r_{RLS} = k_3[NO_3][NO_2] \qquad (N)$$

By combination of equations M and N,

$$r = \frac{k_1 k_3}{k_2}[N_2O_5] \qquad (O)$$

This expression is the same as that obtained by the steady-state approach if one makes the assumption that $k_2 \gg k_3$.

The previous examples indicate how one employs the assumptions that are frequently made in the derivation of a reaction rate expression from a proposed mechanism. Additional examples involving the use of these assumptions are presented in conjunction with the discussion of chain reactions in Section 4.2.

4.1.4 From Stoichiometry and Rate Expression to Reaction Mechanism

To the uninitiated student, the task of postulating a suitable mechanism for a complex chemical reaction often seems to be an exercise in extrasensory perception. Even students who have had some exposure to kinetics often cannot understand how the kineticist can write down a series of elementary reactions and avow that the mechanism is reasonable. Nonetheless, there is a set of guidelines within which the kineticist works in postulating a mechanism. Since these working principles are known to experienced kineticists, they are able to make implicit chemical judgments so automatically that they often cannot see the problem from the student's viewpoint. These principles are seldom stated explicitly in the literature. Edwards et al. (5) have assembled perhaps the most extensive collection of guidelines. Those that are presented below permit a kineticist to judge the reasonableness of a proposed mechanism. Often they will permit the elimination of a proposed sequence of reactions as being unreasonable in the light of available experimental data.

Guideline 1. *The most fundamental basis for mechanistic speculation is a complete analysis of the reaction products.* It is important to obtain a *complete* quantitative and qualitative analysis for *all* products of the reaction. Inasmuch as many chemical reactions give a complex array of products, the relative proportions of which change as the time of reaction increases, it is very useful to carry out a complete analysis for the shortest possible reaction time and for successively longer periods. In this manner one can differentiate between primary products formed directly from the reactants and secondary products formed by subsequent reaction of the primary products. Such analyses can give valuable clues as to the identity of reaction intermediates.

Guideline 2. *The atomic and electronic structure of the reactants and products may provide important clues as to the nature of possible intermediate species.* The degree of atomic and electronic rearrangement that takes place will often indicate which portions of the reactant molecules participate in the reaction act and which would be involved in elementary reactions leading to the formation of reaction intermediates. The structural arrangement of atoms in the molecules that react must correspond at the instant of reaction to interatomic distances appropriate for the formation of new species.

Guideline 3. *All of the elementary reactions involved in a mechanistic sequence must be feasible with respect to bond energies.* Compared to the average density of molecular energies in a reacting mixture, the energies required to break interatomic bonds in a molecule are quite large. Because of this, elementary reactions will normally involve relatively simple acts in which few bonds are involved. Complicated rearrangements consisting of a series of concerted motions must be viewed with skepticism in the absence of strong supporting evidence.

Bond energy considerations also lead to the conclusion that highly endothermic elementary reactions will be slow processes because of the large activation energies normally associated with these reactions.

Guideline 4. *A number of elementary reactions sufficient to provide a complete path for the formation of all observed products must be employed.*

Guideline 5. *All of the intermediates produced by the elementary reactions must be consumed by other elementary reactions so that there will be no net production of intermediate species.*

Guideline 6. *The great majority of known elementary steps are bimolecular, the remainder being unimolecular or termolecular.* Any reaction where the stoichiometric coefficients of the reactants add up to four or more must involve a multiplicity of steps. The ammonia synthesis reaction is known to occur by a number of steps rather than as

$$N_2 + 3H_2 \rightarrow 2NH_3 \qquad (4.1.11)$$

Guideline 7. *A postulated mechanism for a reaction in the forward direction must also hold for the reverse reaction.* This guideline is a consequence of the principle of microscopic reversibility. (See Section 4.1.5.4.) Three corollaries of this guideline should also be kept in mind when postulating a reaction mechanism. First, the rate limiting step for the reverse reaction must be the same as that for the forward reaction. Second, the reverse reaction cannot have a molecularity greater than three, just as the forward reaction is so limited. Consequently, the ammonia decomposition reaction

$$2NH_3 \rightarrow N_2 + 3H_2 \qquad (4.1.12)$$

cannot occur as a simple bimolecular process. Third, if the reaction rate expression for the forward reaction consists of two or more independent terms corresponding to parallel reaction paths, there will be the same number of independent terms in the rate expression for the reverse reaction. At equilibrium not only is the total rate of the forward reaction equal to the total rate of the reverse reaction, but the forward rate by each path is equal to the reverse rate for that particular path.

Guideline 8. *Transitory intermediates (highly reactive species) do not react preferentially with*

Guideline 9. *When the overall order of a reaction is greater than three, the mechanism probably has one or more equilibria and intermediates prior to the rate determining step.*

Guideline 10. *Inverse orders arise from rapid equilibria prior to the rate determining step.* In order to illustrate this point, consider the oxidation of arsenious acid in an aqueous solution.

$$H_3AsO_3 + I_3^- + H_2O = H_3AsO_4 + 3I^- + 2H^+$$
$$(4.1.13)$$

The experimentally observed rate expression is

$$r = \frac{k[H_3AsO_3][I_3^-]}{[I^-]^2[H^+]} \qquad (4.1.14)$$

and two proposed mechanisms are based on the following sequence of elementary reactions.

$$H_3AsO_3 \rightleftharpoons H_2AsO_3^- + H^+ \qquad \text{(Equilibrium constant } K_1) \qquad (4.1.15)$$

$$I_3^- + H_2O \rightleftharpoons 2I^- + H_2OI^+ \qquad \text{(Equilibrium constant } K_2) \qquad (4.1.16)$$

$$H_2OI^+ + H_2AsO_3^- \xrightarrow{k'} H_2AsO_3I + H_2O \qquad \text{(Slow step for mechanism I)}$$
$$\underset{(\rightleftharpoons)}{} \qquad \text{(Equilibrium constant } K_3 \text{ for mechanism II)} \quad (4.1.17)$$

$$H_2AsO_3I \rightleftharpoons I^- + H_2AsO_3^+ \qquad \text{(Fast step for mechanism I)}$$
$$\underset{(k'')}{\rightarrow} \qquad \text{(Slow step for mechanism II)} \qquad (4.1.18)$$

$$H_2AsO_3^+ + H_2O \rightleftharpoons H_3AsO_4 + H^+ \qquad (4.1.19)$$

---

one another to the exclusion of their reaction with stable species. The concentration of these intermediates is usually so low that the number of encounters an intermediate may undergo with other intermediates will be very small compared to the number of encounters it will undergo with stable molecules.

For the first mechanism the overall reaction rate may be taken to be that of the rate limiting step.

$$r = k'[H_2OI^+][H_2AsO_3^-] \qquad (4.1.20)$$

With the use of the equilibrium steps 4.1.15 and 4.1.16, this expression may be converted to a form consistent with the experimental rate expression

$$r = k'K_1K_2 \frac{[I_3^-][H_2O]}{[I^-]^2} \frac{[H_3AsO_3]}{[H^+]} \quad (4.1.21)$$

since the solvent concentration is just another constant.

For the second mechanism, equation 4.1.18 is the rate controlling step, and the arrows in parentheses indicate the assumptions made regarding the reversibility of reactions 4.1.17 and 4.1.18. The overall reaction rate is now

$$r = k''[H_2AsO_3I] \quad (4.1.22)$$

With the use of the equilibrium expressions for the first three reactions, equation 4.1.22 may be converted to the following form,

$$r = k''K_1K_2K_3 \frac{[H_3AsO_3][I_3^-]}{[H^+][I^-]^2} \quad (4.1.23)$$

which is also consistent with the experimental rate expression.

In both cases the negative reaction orders arise from equilibria that are established prior to the rate controlling step. A final rate expression depends only on equilibria that are established by elementary reactions *prior* to the rate determining step. Subsequent equilibria (e.g. 4.1.19) do not influence its form.

Guideline 11. *Whenever a rate law contains non-integer orders, there are intermediates present in the reaction sequence.* When one observes a fractional order in an empirical rate expression for a homogeneous reaction, it is often an indication that an important part of the mechanism is the splitting of a molecule into free radicals or ions.

Guideline 12. *If the magnitude of the stoichiometric coefficient of a reactant exceeds the order of the reaction with respect to that species, there are one or more intermediates and reactions after the rate determining step.* Before applying this rule, one must write the stoichiometric equation for the reaction in a form such that all coefficients are integers. (See Illustrations 4.1 and 4.2.)

Guideline 13. *If the order of a reaction with respect to one or more species increases as the concentration of that species increases, it is an indication that the reaction may be proceeding by two or more parallel paths.* Liebhafsky and Mohammed (6) have reported that for the reaction between hydrogen peroxide and iodide ion, the order with respect to hydrogen ion increases from zero to one as the pH is decreased.

$$r = k_1[H_2O_2][I^-] + k_2[H_2O_2][H^+][I^-] \quad (4.1.24)$$

Such expressions are typical of many acid and base catalyzed reactions. (See Section 7.3.1.)

Guideline 14. *If there is a decrease in the order of a reaction with respect to a particular substance as the concentration of that species increases, the dominant form of that species in solution may be undergoing a change brought about by the change in concentration.* A decrease in reaction order with respect to hydrogen ion concentration with increasing acidity has frequently been observed for reactions involving weak acids.

The guidelines enumerated above are by no means complete, but they do provide a starting point for the beginning student. Further details are available in publications by Edwards et al. (5) and King (7).

4.1.5 Additional Methods and Principles used in Investigations of Reaction Mechanisms

Although reaction rate expressions and reaction stoichiometry are the experimental data most often used as a basis for the postulation of reaction mechanisms, there are many other experimental techniques that can contribute to the elucidation of these molecular processes. The conscientious investigator of reaction mechanisms will draw on a wide variety of experimental and theoretical methods in his or her research program in an attempt to obtain information about the elementary reactions taking place in the system of interest. Some useful techniques are described below. The information that they supply may provide a basis for choice between two or more alternative mechanisms, or it may provide additional circumstantial evidence that a proposed mechanism is, indeed, correct.

4.1.5.1 Studies of Reaction Intermediates.

In the past two decades the most fruitful progress in experimental studies of reaction mechanisms has come about through the development of instrumental techniques for detecting and identifying the trace amounts of active intermediates produced in complex reaction systems. Such studies place a mechanism on a much firmer basis than exists in those cases where the identity of intermediates can only be surmised.

A stable species that is suspected to act as an intermediate in a complex chemical reaction can often be added to a reaction mixture and its effects observed. If the appropriate products are formed *at a rate no less than that of the "uninterrupted" or "original" reaction*, this is strong evidence that the reaction proceeds through the intermediate that has been isolated. This evidence is, however, not unequivocal. Similarly, the failure of other presumed intermediates to give the correct products under appropriate reaction conditions will cause a kineticist to revise his or her ideas concerning the mechanism of the reaction being investigated. For example, the Clemmensen reduction of carbonyl ($C=O$) groups to methylene (CH_2) groups, using zinc amalgam and hydrochloric acid, cannot proceed through the corresponding carbinol ($HCOH$), since the carbinols themselves are not generally reduced by the same combination of reactants (8).

For detailed treatments of the experimental methods used in studies of reaction intermediates, consult the works of Wayne (1), Melville and Gowenlock (9), Anderson (10), Friess, Lewis, and Weissberger (11), Lewis (12), and Hammes (13).

4.1.5.2 Isotopic Substitution Techniques.

If one can discover where the various portions of reactant molecules end up in the product molecules, a great deal of insight into the mechanism of the reaction can be gained. By isotopically labeling a functional group or atom in a molecule and examining the distribution (or lack thereof) of the tagged species after reaction, it is possible to determine which bonds are broken and formed. In studies with radioactive and stable isotopes it is generally assumed that the kinetic behavior of a labeled atom is essentially the same as that of an unlabeled atom. The important exceptions are studies involving deuterium and tritium.

There are three basic types of mechanistic information that are derived from isotopic experiments.

1. A knowledge of which bonds are broken and thus give rise to intramolecular or intermolecular rearrangements.
2. A knowledge of whether the reaction proceeds in an isolated molecule or requires the participation of more than one molecule.
3. A knowledge of which products are the precursors of others.

As an example, consider the thermal and photochemical decomposition of acetaldehyde

$$CH_3CHO \rightarrow CH_4 + CO \quad (4.1.25)$$

One proposed mechanism involved an intramolecular rearrangement, while a second involved a free radical chain mechanism composed of the following sequence of elementary reactions:

$$CH_3CHO \rightarrow CH_3\cdot + CHO\cdot \quad (4.1.26)$$
$$CH_3\cdot + CH_3CHO \rightarrow CH_4 + CH_3CO\cdot \quad (4.1.27)$$
$$CH_3CO\cdot \rightarrow CH_3\cdot + CO \quad (4.1.28)$$

where reactions 4.1.27 and 4.1.28 occur many, many times for each time that reaction 4.1.26 occurs. If one uses a mixture of CH_3CHO and CD_3CDO to study this reaction, the first mechanism predicts that only CD_4 and CH_4 will be formed, while the second predicts that a mixture of CD_4, CH_4, CH_3D, and CD_3H will be formed. The fact that the statistically expected distribution of the latter methane species was observed experimentally (14–16) is taken as evidence that the major path for this reaction is the chain reaction sequence.

4.1.5.3 Stereochemical Methods.

Many organic reactions involve reactants that can exist in stereoisomeric forms. If an optically active reactant yields a product that is capable of exhibiting optical activity, but that does not exhibit such activity, it is usually assumed that the reaction must involve an intermediate that is not optically active (e.g., a carbonium ion intermediate). If it can be determined that the configuration of an optically active product differs from or is similar to that of the optically active reactant, one gains information about whether there is an *inversion of configuration*. Retention of configuration might imply a two-step process, the first step turning the molecule "inside out," and the second step turning it "outside in" again. It could also mean that none of the bonds of the optically active carbon atom are broken in the reaction.

4.1.5.4 The Principle of Microscopic Reversibility.

The principle of microscopic reversibility is based on statistical mechanical arguments and was first formulated by Tolman (17) in 1924.

In a system at equilibrium, any molecular process and the reverse of that process occur on the average at the same rate.

The most significant consequence of the principle for kineticists is that if in a system at equilibrium there is a flow of reacting molecules along a particular reaction path, there must be

an equal flow in the opposite direction. This implies that the reaction path established as most probable for the forward direction must also be the most probable path for the reverse reaction. This consequence is also known as the principle of *detailed balancing* of chemical reactions. Its relationship to the principle of microscopic reversibility has been discussed by Denbigh (18). If we consider a substance that can exist in three intraconvertible isomeric forms, *A*, *B*, and *C* (e.g., *trans*-butene-2, *cis*-butene-2, and 1-butene), there is more than one independent reaction that occurs at equilibrium. The conditions for thermodynamic equilibrium would be satisfied if there were a steady unidirectional flow at the molecular level around the cycle

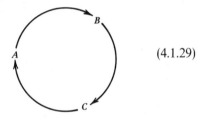

$$(4.1.29)$$

such that the concentration of each species remains constant. However, this flow would not be in accord with the principle of microscopic reversibility. If this principle were not applicable, then the concentrations of the various species would show oscillations if one started with a nonequilibrium system and allowed it to approach equilibrium. Several attempts have been made to observe oscillatory phenomena of this type, but they have not led to definitive results. One concludes that each of the reactions should be balanced individually.

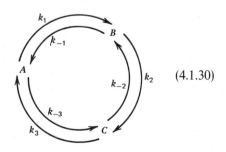

$$(4.1.30)$$

The requirement of detailed balancing has implications with regard to the relationships that must exist between reaction rate constants and equilibrium constants. It requires that at equilibrium each reaction must be balanced in the forward and reverse directions; that is,

$$k_1[A] = k_{-1}[B] \tag{4.1.31}$$
$$k_2[B] = k_{-2}[C] \tag{4.1.32}$$
$$k_3[C] = k_{-3}[A] \tag{4.1.33}$$

or

$$\frac{B}{A} = \frac{k_1}{k_{-1}} = K_1 \tag{4.1.34}$$

$$\frac{C}{B} = \frac{k_2}{k_{-2}} = K_2 \tag{4.1.35}$$

$$\frac{A}{C} = \frac{k_3}{k_{-3}} = K_3 \tag{4.1.36}$$

The principle of detailed balancing provides an automatic check on the self-consistency of postulated reaction mechanisms when equilibrium can be approached from both sides.

4.1.5.5 Activation Energy Considerations.
Activation energy considerations can provide a basis for eliminating certain elementary reactions from a sequence of reactions. Unfortunately, the necessary activation energy data is seldom available, and one must estimate these parameters by empirical rules and generalizations that *are of doubtful reliability*.

Activation energies for bimolecular reactions have been correlated with bond energy data by the use of the Hirschfelder rules (19).

1. For a simple displacement reaction involving atoms or radicals, such as $A + BC \rightarrow AB + C$, where the reaction is written in the exothermic direction, the activation energy is 5.5% of the dissociation energy of the bond which is broken. For the reverse endothermic reaction, the standard energy change of reaction must be added to this quantity in order to obtain the activation energy.
2. For an exchange reaction such as

$$AB + CD \rightarrow AC + BD$$

where the reaction is again written in the exothermic direction, the activation energy is 28% of the sum of the AB and CD bond strengths. For the reverse endothermic reaction, the activation energy is the sum of the standard energy change for the reaction and the activation energy for the exothermic reaction.

These rules provide only very crude estimates of activation energies and are not sufficiently sensitive to show differences in a series of related reactions.

Semenov (20) has suggested that for *exothermic* abstraction and addition reactions of atoms and small radicals, the following relation is useful.

$$E = 11.5 - 0.25q \tag{4.1.37}$$

where q is the heat *evolved* in the reaction expressed in kilocalories and E is also expressed in kilocalories. For the endothermic case,

$$E = 11.5 + 0.75q \tag{4.1.38}$$

Other theoretical approaches to the problem of predicting reaction activation energies exist (21–23). For our purposes, however, it is sufficient to recognize that "ball-park" estimates are the best one can expect. Such estimates are often adequate for purposes of differentiating between alternative mechanisms on the basis of a comparison of predicted and actual activation energies.

4.1.5.6 Families of Reactions.
Research in chemical kinetics can be influenced in several ways by similarities in reactivity among compounds having similar chemical structures. Consequently, one of the first steps in a kinetics investigation should be a careful study of the published literature on related reactions. While one may occasionally be led astray by tentative conclusions drawn from a survey of the literature, the possibility of achieving large savings in time and laboratory effort is so much more likely that it should be considered a must. Such a survey furnishes a large bank of tested ideas and hypotheses about elementary reactions on which

the investigator may draw for inspiration and stimulation. They provide a framework against which one can orient and test ideas about the mechanism of a specific reaction.

The fact that compounds having similar chemical structures often react in similar ways implies that they follow corresponding mechanisms in proceeding from reactants to products. One must have a very sound basis in experimental fact to be able to defend successfully a proposed reaction mechanism that contains elementary reactions that represent a major departure from those accepted by the scientific community as being a proper mechanistic interpretation of related reactions. On occasions such departures are necessary, and they themselves may provide the key to an improved understanding of the common family of related reactions.

4.1.5.7 Occam's Razor—A Rule of Simplicity.
Another principle occasionally used as a basis for choice between two alternative mechanisms is the rule of simplicity that is derived from a more general philosophical principle known as Occam's Razor. The kineticist resorts to this rule when all of the experimental and theoretical information that has been brought to bear on the problem of determining the mechanism of a reaction leaves two (or more) mechanisms, both consistent with the facts. In this case one assumes that nature prefers the simple to the complex and that the reaction follows the simplest path. In many cases a serious question may exist as to which mechanism is simpler but, in others, the investigator may decide that one mechanism is more likely because it involves only bimolecular steps and no termolecular processes, or because it involves a smaller change in chemical structure than an alternative mechanism. Complicated atomic rearrangements that occur in a single step are regarded with suspicion, and simpler stepwise processes are generally considered more probable.

Occam's Razor states that "multiplicity ought not to be posited without necessity." This principle is named for the fourteenth-century philosopher William Ockham (William of Occam), who was famous for his hardheaded approach to problem solving. He believed in shaving away all extraneous details—hence the term Occam's Razor. In essence his axiom states that when there are several possible solutions to a problem, the right one is probably the most obvious. Therefore one should not postulate or accept mechanisms that are more complex than necessary to explain all observed experimental facts.

4.2 CHAIN REACTIONS

The reaction mechanisms treated thus far have involved the conversion of reactants to products by a sequence of elementary reactions proceeding in simple stepwise fashion. Once these sequences have been completed, all of the reaction intermediates have disappeared and only stable product molecules remain. These types of reaction are classified as *open sequence* reactions because they always proceed in the stagewise fashion with no closed reaction cycles wherein a product of one elementary reaction is fed back to react with another species in an earlier elementary reaction. (Reversible reactions are, however, permitted in open sequence mechanisms.) This section deals with *closed sequence* or *chain reaction* mechanisms in which one of the reaction intermediates is regenerated during one stage of the reaction and is fed back to an earlier stage to react with other species so that a closed loop or cycle results. The intermediate species are thus periodically renewed by reaction, and the final products are the result of a cyclic repetition of the intervening processes. Because chain reactions play such an important role in systems of practical industrial significance, it is important that the chemical engineer understand their basic nature. Failure to comprehend the implications of chain reaction processes can lead to hazardous explosions.

To illustrate the nature of chain reactions, we will consider a classic example—the formation of hydrogen bromide from molecular hydrogen and bromine in a homogeneous gas phase reaction.

$$H_2 + Br_2 \rightleftharpoons 2HBr \qquad (4.2.1)$$

This stoichiometric equation does not give any indication of the complex nature of the mechanism by which this reaction proceeds.

In certain highly energetic collisions with any molecule M in the system, a bromine molecule may be dissociated in a homolytic split of the bond joining two bromine atoms.

$$Br_2 + M \rightarrow Br\cdot + Br\cdot + M \qquad (4.2.2)$$

The collision must be sufficiently energetic that enough energy is available to break the chemical bond linking the two bromine atoms. This type of reaction is called an *initiation* reaction because it generates a species that can serve as a *chain carrier* or *active center* in the following sequence of elementary reactions.

$$Br\cdot + H_2 \rightarrow HBr + H\cdot \qquad (4.2.3)$$

$$H\cdot + Br_2 \rightarrow HBr + Br\cdot \qquad (4.2.4)$$

$$Br\cdot + H_2 \rightarrow HBr + H\cdot \qquad (4.2.5)$$

$$H\cdot + Br_2 \rightarrow HBr + Br\cdot \qquad (4.2.6)$$

etc.

Equations 4.2.3 and 4.2.4 are the elementary reactions responsible for product formation. Each involves the formation of a chain carrying species (H· for 4.2.3 and Br· for 4.2.4) that propagates the reaction. Addition of these two relations gives the stoichiometric equation for the reaction. These two relations constitute a single closed sequence in the cycle of events making up the chain reaction. They are referred to as *propagation* reactions because they generate product species that maintain the continuity of the chain.

The chain carriers need not all react with reactant molecules in chain propagation reactions. Some will disappear in *termination re-actions* that do not involve the formation of species capable of maintaining the chain; for example,

$$Br\cdot + Br\cdot + M \rightarrow Br_2 + M \qquad (4.2.7)$$

where M represents a third body or molecule capable of soaking up and carrying away the energy released by the formation of a Br-Br bond.

Bond energy considerations indicate that the initiation reaction (4.2.2) should be quite slow because its activation energy must be quite high (at least equal to the bond dissociation energy). If one were dealing with an open sequence reaction mechanism, such a step would imply that the overall reaction rate would also be low because in these cases the overall reaction becomes approximately equal to that of the rate limiting step. In the case of a chain reaction, on the other hand, the overall reaction rate is usually much faster because the propagation steps occur *many* times for each time that an initiation step occurs.

The *chain length* of a reaction is defined as the average number of cycles in which a chain carrying species participates from the time that it is first formed until the time that it is destroyed in a termination reaction. Depending on the relative rates of the propagation and termination steps, a chain may involve only a few links or it may be extremely long. For example, a chain length of 10^6 has been determined for the gas phase chain reaction between hydrogen and chlorine. Chain length represents a statistical concept; some cycles will terminate before and others after the average number of propagation steps. This fact leads to a distribution of molecular weights in the polymers formed by chain reactions. In terms of quantities that can be determined in the laboratory, we find it convenient to define the chain length by the following ratio.

$$\mathscr{L} = \frac{\text{rate of disappearance of reactants}}{\text{rate of chain carrier formation by initiation}}$$

$$(4.2.8)$$

In the case of the hydrogen-bromine reaction, each of the elementary propagation reactions led to the formation of a single chain carrier. This type of reaction is said to be a *straight* or *linear chain* reaction. Some mechanisms involve elementary propagation reactions in which more than a single chain carrier is formed by the reaction. This type of reaction is known as a *branching* reaction. Examples of such reactions are

$$O + H_2 \rightarrow OH + H \quad \text{and} \quad H + O_2 \rightarrow OH + O$$
$$(4.2.9)$$

Both of these reactions involve the production of two active centers where there was only one before. When reactions of this type occur to a significant extent, the total number of active centers present in the system can increase very rapidly, since a multiplication effect sets in as the chains propagate. The growth of chain carriers in a branched chain reaction is pictured below.

When chain termination processes are operative such that they destroy chain carriers at a rate that makes the average net production of chain carriers unity, one has a situation corresponding to a nuclear power reactor generating energy at steady-state conditions. The effect of the destruction processes is to cause the chain branching process to degenerate to a straight chain process. On the other hand, as the sketch indicates, there can be a multiplication of chain carriers. In physical terms this can lead to the chemical analog of a nuclear explosion. (See Section 4.2.5.)

4.2.1 The Reaction between Hydrogen and Bromine $H_2 + Br_2 \rightleftharpoons 2HBr$

In 1906 Bodenstein and Lind (24) investigated the gas phase homogeneous reaction between molecular bromine and molecular hydrogen at pressures in the neighborhood of 1 atm. They fitted their experimental data with a rate expression of the form

$$\frac{d[HBr]}{dt} = \frac{k[H_2][Br_2]^{1/2}}{1 + k'\left(\dfrac{[HBr]}{[Br_2]}\right)} \quad (4.2.10)$$

where k' is a constant independent of temperature and k follows the normal Arrhenius form for a reaction rate constant with an activation energy of 40.2 kilocalories per mole.

In 1919 Christiansen (25), Herzfeld (26), and Polanyi (27) all suggested the same mechanism for this reaction. The key factor leading to their success was recognition that hydrogen atoms and bromine atoms could alternately serve as chain carriers and thus propagate the reaction. By using a steady-state approximation for the concentrations of these species, these individuals were able to derive rate expressions that were consistent with that observed experimentally.

Their original mechanism consists of the following elementary reactions.
Initiation:
$$Br_2 \overset{k_1}{\rightarrow} 2Br\cdot \qquad (4.2.11)$$

Propagation:
$$Br\cdot + H_2 \overset{k_2}{\rightarrow} H\cdot + HBr \qquad (4.2.12)$$
$$H\cdot + Br_2 \overset{k_3}{\rightarrow} Br\cdot + HBr \qquad (4.2.13)$$
$$H\cdot + HBr \overset{k_4}{\rightarrow} H_2 + Br\cdot \qquad (4.2.14)$$

Termination:
$$2Br\cdot \overset{k_5}{\rightarrow} Br_2 \qquad (4.2.15)$$

Other reactions need not be considered on the basis of the arguments presented below. Reaction 4.2.14 is the reverse of reaction 4.2.12 and is responsible for the inhibition of the reaction by HBr. Steps 4.2.12 to 4.2.14 all produce one chain carrier to replace the chain carrier destroyed by the reaction. The sharp-eyed observer may have noted that the initiation and termination reactions proposed by these individuals differ slightly from those discussed previously. The reasons for this are dealt with later.

From these equations, the overall rate of formation of HBr is given by

$$\frac{d[\text{HBr}]}{dt} = k_2[\text{Br}][\text{H}_2] + k_3[\text{H}][\text{Br}_2]$$
$$- k_4[\text{H}][\text{HBr}] \qquad (4.2.16)$$

This relation expresses the fact that under steady-state conditions, the rate of the initiation reaction is equal to the rate of the termination reaction, and the steady-state bromine atom concentration is equal to that which would arise from the equilibrium $\text{Br}_2 \rightleftharpoons 2\text{Br}$; that is,

$$[\text{Br}]_{ss} = \left(\frac{k_1}{k_5}[\text{Br}_2]\right)^{1/2} \qquad (4.2.20)$$

Equation 4.2.20 may be combined with the steady-state form of either equation 4.2.17 or 4.2.18 to give

$$k_2\left(\frac{k_1}{k_5}[\text{Br}_2]\right)^{1/2}[\text{H}_2] - (k_3[\text{Br}_2] + k_4[\text{HBr}])[\text{H}] = 0 \quad (4.2.21)$$

As it stands, this expression is awkward and inconvenient to test because it contains the concentrations of bromine and hydrogen atoms, parameters that are not easily measured. In principle these quantities could be eliminated by solving this equation simultaneously with the differential equations for each of these species.

$$\frac{d[\text{H}]}{dt} = k_2[\text{Br}][\text{H}_2] - k_3[\text{H}][\text{Br}_2]$$
$$- k_4[\text{H}][\text{HBr}] \qquad (4.2.17)$$

$$\frac{d[\text{Br}]}{dt} = 2k_1[\text{Br}_2] - k_2[\text{Br}][\text{H}_2] + k_3[\text{H}][\text{Br}_2]$$
$$+ k_4[\text{H}][\text{HBr}] - 2k_5[\text{Br}]^2 \qquad (4.2.18)$$

However, the complete solution of these three simultaneous differential equations is difficult to obtain and is no more instructive than the approximate solution that can be obtained by means of the steady-state approximation for intermediates. If one sets the time derivatives in equations 4.2.17 and 4.2.18 equal to zero and adds these equations, their sum is found to be

$$2k_1[\text{Br}_2] - 2k_5[\text{Br}]^2 = 0 \qquad (4.2.19)$$

which, in turn, may be solved for the steady-state concentration of atomic hydrogen.

$$[\text{H}]_{ss} = \frac{k_2\left(\frac{k_1}{k_5}[\text{Br}_2]\right)^{1/2}[\text{H}_2]}{k_3[\text{Br}_2] + k_4[\text{HBr}]} \qquad (4.2.22)$$

Equations 4.2.20 and 4.2.22 may be combined with equation 4.2.16 in straightforward algebraic manipulation to yield an expression for the overall rate of formation of HBr.

$$\frac{d[\text{HBr}]}{dt} = \frac{2k_2\left(\frac{k_1}{k_5}\right)^{1/2}[\text{H}_2][\text{Br}_2]^{1/2}}{1 + \frac{k_4}{k_3}\frac{[\text{HBr}]}{[\text{Br}_2]}} \qquad (4.2.23)$$

Equation 4.2.23 has the same form as the empirical rate expression and will agree with it quantitatively if

$$k = 2k_2\left(\frac{k_1}{k_5}\right)^{1/2} \quad \text{and} \quad k' = \frac{k_4}{k_3} = \frac{1}{10} \qquad (4.2.24)$$

Recognition that the kinetics of this reaction could be explained on the basis of the chain reaction mechanism presented above was one of the major breakthroughs in the evolution of the theory of chemical reaction mechanisms. Since the mechanism was first published, modifications in the initiation and termination steps have been required as additional experimental facts have been established by subsequent investigators. It is enlightening to consider these modifications and the rationale underlying the omission of certain elementary reactions from the mechanism.

Because the steady-state assumption leads to the equilibrium relation for the bromine atom concentration (4.2.20), it does not matter what mechanism is assumed to be responsible for establishing this equilibrium. Alternative elementary reactions for the initiation and termination processes, which give rise to the same equilibrium relationship, would also be consistent with the observed rate expression for HBr formation. For example, the following reactions give rise to the same equilibrium:

Initiation: $Br_2 + M \xrightarrow{k_1'} 2Br + M$ (4.2.25)

Termination: $M + 2Br \xrightarrow{k_5'} Br_2 + M$ (4.2.26)

where M is a third body. It represents any available molecule that is capable of supplying to bromine molecules the energy necessary for dissociation of the Br–Br bond and of carrying away enough energy in the termolecular reaction so that the Br–Br bond does not immediately dissociate.

If equations 4.2.25 and 4.2.26 are substituted for equations 4.2.11 and 4.2.15, respectively, in the mechanism described above, the effect is to replace k_1 by $k_1'[M]$ and k_5 by $k_5'[M]$ everywhere that they appear. Since these quantities appear as a ratio in the final rate expression, the third body concentration will drop out and (k_1/k_5) becomes identical with (k_1'/k_5'). The necessity for the use of the third body concentration thus is not obvious in kinetic studies of the thermal reaction. However, from studies of photochemical reaction between hydrogen and bromine, there is strong evidence that the termination reaction is termolecular. This fact and

others based on studies of various atomic recombination reactions imply that the "correct" initiation and termination reactions are not reactions 4.2.11 and 4.2.15 but reactions 4.2.25 and 4.2.26 (28).

The next step in analysis of the mechanism is to indicate why we have limited the number of elementary reactions in the mechanism to five. To one uninitiated in the task of dealing with reaction mechanisms, it is difficult to see why elementary reactions such as

$$H_2 + Br_2 \rightarrow 2HBr \qquad (4.2.27)$$

$$H_2 + M \rightarrow 2H + M \qquad (4.2.28)$$

$$HBr + M \rightarrow H + Br + M \qquad (4.2.29)$$

$$Br + HBr \rightarrow Br_2 + H \qquad (4.2.30)$$

$$H + Br + M \rightarrow HBr + M \qquad (4.2.31)$$

$$H + H + M \rightarrow H_2 + M \qquad (4.2.32)$$

were rejected on an *a priori* basis and what considerations led to their rejection. We will see that they may be eliminated from consideration through the use of some of the concepts of probability and the bond energy requirements outlined previously. If any of these reactions were important, the rate expression for the revised mechanism would have a different mathematical form than the empirical rate expression.

Equation 4.2.27 may be rejected even if it provides a parallel path competitive with the mechanism described above because its presence would require a second-order term in the experimental reaction rate expression.

The initiation reaction is the dissociation of molecular bromine rather than the dissociation of molecular hydrogen because the energy necessary to dissociate the latter is much greater than that required to dissociate the former. This fact is evident from a consideration of the standard enthalpies of reaction.

$$Br_2 + M \rightarrow 2Br + M \qquad \Delta H_I = 46.1 \text{ kcal/g-mole}$$
$$(4.2.33)$$

$$H_2 + M \rightarrow 2H + M \qquad \Delta H_{II} = 103.4 \text{ kcal/g-mole}$$
$$(4.2.34)$$

Since the activation energy of an endothermic elementary reaction cannot be less than ΔH, the reaction with the significantly lower activation energy will occur much more frequently, other factors being equal. Similar arguments permit one to eliminate the HBr dissociation reaction from consideration.

The Br + HBr reaction will be unimportant in competition with the Br + H$_2$ reaction because energy considerations again dictate that the former will have a much higher activation energy than the latter.

$$Br + HBr \rightarrow Br_2 + H \qquad \Delta H_{IV} = 41.2 \text{ kcal/g-mole}$$
$$(4.2.35)$$

$$Br + H_2 \rightarrow HBr + H \qquad E_A = 17.6 \text{ kcal/g-mole}$$
$$(4.2.36)$$

Both reactions are endothermic, but the interaction of bromine atoms with HBr is much more so than the interaction with molecular hydrogen. Consequently, the former reaction will occur much less frequently than the latter.

Equations 4.2.31 and 4.2.32 represent radical recombination reactions that might be considered as alternative termination reactions. Since these reactions and reaction 4.2.26 are radical recombination reactions, they will all have very low activation energies. The relative rates of these processes will then be governed by the collision frequencies and thus the concentrations of the reacting species. The relative concentrations of hydrogen and bromine atoms can be determined from the steady-state form of equation 4.2.17.

$$\frac{[H]}{[Br]} = \frac{k_2[H_2]}{k_3[Br_2] + k_4[HBr]} \qquad (4.2.37)$$

If we consider this ratio during the initial stages of the reaction when the HBr concentration is extremely low and when the concentrations of hydrogen and bromine are nearly equal, this expression reduces to

$$\frac{[H]}{[Br]} \approx \frac{k_2}{k_3} \qquad (4.2.38)$$

If the preexponential factors of these rate constants are comparable in magnitude, further simplification is possible.

$$\frac{[H]}{[Br]} \approx e^{-(E_2 - E_3)/RT} \qquad (4.2.39)$$

If the estimation techniques discussed in Section 4.1.5.5 are employed, one finds that E_3 is equal to 2.5 kcal. From 4.2.36, E_2 is equal to 17.6 kcal. At 540 °K the ratio of atomic concentrations is thus of the order of 10^{-6}. Consequently, the reaction between atomic hydrogen and atomic bromine is only one millionth as likely as a reaction between two bromine atoms and that between two hydrogen atoms is only $1/10^{12}$ as likely. Although the atomic hydrogen concentration is extremely low under reaction conditions, it has been estimated to be four orders of magnitude greater than that which would be in equilibrium with molecular hydrogen. The excess hydrogen atom concentration over that which would exist at equilibrium is "paid for" by the energy liberated by the reacting mixture. This feature is a general characteristic of chain reactions. Superequilibrium steady-state concentrations of highly reactive intermediates can be produced as a result of the energy released by reaction.

The reaction between hydrogen and bromine illustrates many of the general features of straight chain reactions. This reaction system has been studied by many investigators, and the body of accumulated experimental data represents as nearly a consistent and complete collection of data as is available for any reaction in the literature.

4.2.2 Chain Reaction Mechanisms— General Comments

The essential characteristic of a chain reaction mechanism is the existence of a closed cycle of reactions in which unstable or highly reactive intermediates react in propagation steps with stable reactant molecules or other intermediates and are regenerated by the sequence of reactions that follows. The mathematical methods and approximations used to express the overall rate of reaction in terms of the individual rate constants and the concentrations of stable species are merely extensions of those discussed earlier.

The elementary reactions comprising the chain reaction mechanism are generally classified as initiation, propagation, or termination reactions. In the initiation reaction an active center or chain carrier is formed. Often these are atoms or free radicals, but ionic species or other intermediates can also serve as chain carriers. In the propagation steps the chain carriers interact with the reactant molecules to form product molecules and regenerate themselves so that the chain may continue. The termination steps consist of the various methods by which the chain can be broken.

The key assumption that permits one to proceed from mechanistic equations to an overall rate expression is that the steady-state approximation for intermediates is valid. One is also usually required to make the *long-chain* approximation (i.e., the by-products resulting from the initiation and termination reactions are assumed to represent only a very small fraction of the total products). The reaction then will remain stoichiometrically simple to within an approximation that improves with increasing chain length. It is also sometimes necessary to make the additional assumption that at steady state the rate of the initiation process must be equal to the rate at which the various chain breaking processes are occurring. Otherwise, the concentration of chain carriers would depend explicitly on time. Often this equality is obvious from the algebraic equations resulting from the use of the steady-state approximation. When there are two or more alternative termination reactions, the algebraic relations may be so complex that the equality is not apparent.

In any attempt to elucidate the mechanism of a chemical reaction, it is very important to determine at an early stage whether or not a chain reaction is occurring. Consequently, we now note some of the characteristics by which the presence of such mechanisms may be recognized. Not all chain reactions will exhibit all of the characteristics enumerated below, but the presence of several of them should be a strong indication to the investigator that a chain reaction is probably taking place. The absence of any of these criteria should not be regarded as particularly significant.

Some pertinent criteria for recognizing the presence of a chain reaction are:

1. An induction period is present.
2. Extremely large increases or decreases in the reaction rate occur when relatively small amounts of other substances are added to the reaction mixture. These materials may act either as initiators or inhibitors of the reaction, depending on whether they participate in the reaction as initiators or terminators of the chain.
3. Small changes in pressure, temperature, or composition can markedly affect the overall reaction rate or cause an explosion.
4. In gas phase reactions an increase in the surface to volume ratio of the reaction vessel may reduce the reaction rate, while addition of "inert" gases may increase the reaction rate. (See Section 4.2.5.)
5. Abnormally high quantum yields may occur in photochemical reactions. Einstein's law of photochemical equivalence is the principle that light is absorbed by molecules in discrete amounts as an individual molecular process (i.e., one molecule absorbs one photon at a time). From optical measurements it is possible to determine quantitatively the number of photons absorbed in the course of a reaction and, from analyses of the product mixture, it is possible to determine the number of molecules that have reacted. The quantum yield is defined as the ratio of the number of molecules reacting to the number of photons absorbed. If this quantity exceeds unity, it provides unambiguous evidence for the existence of secondary processes and thus indicates the presence of unstable intermediates.

While yields greater than unity provide evidence for chain reactions, yields less than unity do *not* indicate the absence of a chain reaction. Quantum yields as high as 10^6 have been observed in the photochemical reaction between H_2 and Cl_2.

6. Complex rate expressions or fractional reaction orders with respect to individual reactants are often indicative of chain reaction mechanisms. However, other mechanisms composed of many elementary reactions may also give rise to these types of rate expressions, so this criterion should be applied with caution.

Because of the variety of chain reaction mechanisms that exist, it is difficult to develop completely general rate equations for these processes. However, some generalizations with regard to the overall rate expression can be made for certain classes of reactions.

Some of these are considered in Section 4.2.3. Illustration 4.3 indicates the techniques used in the derivation of a reaction rate expression from a chain reaction mechanism.

ILLUSTRATION 4.3 USE OF THE BODENSTEIN STEADY-STATE APPROXIMATION TO DERIVE A RATE EXPRESSION FROM A CHAIN REACTION MECHANISM

The following rate expression has been determined for the low temperature chlorine catalyzed decomposition of ozone.

$$\frac{d[O_3]}{dt} = -k[Cl_2]^{1/2}[O_3]^{3/2}$$

The following chain reaction mechanism has been proposed for this reaction.

Initiation: $\qquad Cl_2 + O_3 \overset{k_1}{\to} ClO\cdot + ClO_2\cdot$

Propagation: $\begin{cases} ClO_2\cdot + O_3 \overset{k_2}{\to} ClO_3\cdot + O_2 \\ ClO_3\cdot + O_3 \overset{k_3}{\to} ClO_2\cdot + 2O_2 \end{cases}$

Termination: $\begin{cases} ClO_3\cdot + ClO_3\cdot \overset{k_4}{\to} Cl_2 + 3O_2 \\ ClO\cdot + ClO\cdot \overset{k_5}{\to} Cl_2 + O_2 \end{cases}$

What rate expression results from this mechanism? Is this expression consistent with the experimentally determined rate expression?

Solution

The processes by which ozone will disappear are reactions 1, 2, and 3. We will subsequently make use of the fact that the amount of ozone disappearing via the initiation step will be negligible compared to the amount that is consumed by the propagation steps. The overall rate of disappearance of ozone is given by:

$$\frac{d[O_3]}{dt} = -k_1[Cl_2][O_3] - k_2[ClO_2\cdot][O_3]$$

$$- k_3[ClO_3\cdot][O_3] \qquad (A)$$

This expression contains the concentrations of two intermediate free radicals, $ClO_2\cdot$ and $ClO_3\cdot$. These terms may be eliminated by using the Bodenstein steady-state approximation.

$$\frac{d[ClO_2\cdot]}{dt} = k_1[Cl_2][O_3] - k_2[ClO_2\cdot][O_3] + k_3[ClO_3\cdot][O_3] \approx 0 \quad (B)$$

$$\frac{d[ClO_3\cdot]}{dt} = k_2[ClO_2\cdot][O_3] - k_3[ClO_3\cdot][O_3] - 2k_4[ClO_3\cdot]^2 \approx 0 \quad (C)$$

Adding equations B and C gives

$$k_1[Cl_2][O_3] = 2k_4[ClO_3\cdot]^2 \qquad (D)$$

Thus

$$[ClO_3\cdot]_{ss} = \sqrt{\frac{k_1[Cl_2][O_3]}{2k_4}} \tag{E}$$

Substitution of this expression into equation B gives

$$k_1[Cl_2][O_3] - k_2[ClO_2\cdot][O_3] + k_3\sqrt{\frac{k_1[Cl_2][O_3]}{2k_4}}[O_3] = 0 \tag{F}$$

or

$$[ClO_2\cdot]_{ss} = \frac{k_1[Cl_2] + k_3\sqrt{\dfrac{k_1[Cl_2][O_3]}{2k_4}}}{k_2} \tag{G}$$

Substitution of the steady-state concentrations of the two intermediates into the equation for the rate of disappearance of ozone gives

$$\frac{d[O_3]}{dt} = -k_1[Cl_2][O_3] - k_2\left[\frac{k_1[Cl_2] + k_3\sqrt{\dfrac{k_1[Cl_2][O_3]}{2k_4}}}{k_2}\right][O_3] - k_3\sqrt{\frac{k_1[Cl_2][O_3]}{2k_4}}[O_3] \tag{H}$$

or

$$\frac{d[O_3]}{dt} = -2k_1[Cl_2][O_3] - 2k_3\sqrt{\frac{k_1[Cl_2][O_3]}{2k_4}}[O_3] \tag{I}$$

The first term in equation I refers to ozone consumption via the initiation step. This term is negligible compared to the second. Thus,

$$\frac{d[O_3]}{dt} = -k_3\sqrt{\frac{2k_1}{k_4}}[Cl_2]^{1/2}[O_3]^{3/2}$$

This form is consistent with the experimentally determined rate expression.

Even though this reaction mechanism meets the requirements imposed by experimental rate data and the stoichiometry of the reaction, it is not consistent with all of our principles because the reverse of termination reaction 4 would require a molecularity of four. Consequently this reaction may be regarded as suspect. An alternate termination reaction might be

$$ClO_3\cdot + ClO_3\cdot \rightarrow Cl_2 + 2O_3$$

Since the propagation reactions are the only ones that significantly affect the concentration

of ozone, a mechanism using this equation as a termination step would give a rate expression of the same form as that described above.

4.2.3 Rice-Herzfeld Mechanisms

The thermal decomposition reactions of many organic compounds obey relatively simple rate laws. Consequently, it was assumed for many years that they are simple elementary' processes. It was not until the middle of the 1930s that it was recognized that free radicals play an essential role in these reactions. Rice and Herzfeld (29) have postulated some general principles that are applicable to/ pyrolysis reactions and have proposed detailed mechanisms for several reactions. They recognized that free radical chain reaction mechanisms could give rise to simple rate expressions that would be integer or half-integer order with respect to the material being pyrolyzed. This fact had not been recognized previously because earlier chain reaction mechanisms had led to rate expressions that were more complicated than those normally associated with organic pyrolysis reactions.

The basic premises on which Rice-Herzfeld mechanisms are based are as follows.

Initiation:
1. Free radicals are formed by scission of the weakest bond in the molecule.

Propagation:
2. One or both of the radicals formed in the initiation step abstracts a hydrogen atom from the parent compound to form a small saturated molecule and a new free radical.
3. The new free radical stabilizes itself by splitting out a simple molecule such as an olefin or CO.

$$RCH_2 - CH_2 \cdot \rightarrow R \cdot + CH_2 = CH_2 \quad (4.2.40)$$

Termination:
4. The chain is broken by a combination or disproportionation reaction between two radicals.

These basic premises go a long way in correlating and tying together the extraordinary complexity of many pyrolysis reactions. In terms of mechanistic equations, they may be written as

Initiation:

$$(1) \ M \overset{k_1}{\rightarrow} R_1 + R_1' \qquad (4.2.41)$$

Propagation:

$$(2) \ R_1 + M \overset{k_2}{\rightarrow} R_2 + R_1 H \qquad (4.2.42)$$

$$(3) \ R_2 \overset{k_3}{\rightarrow} R_1 + P_1 \qquad (4.2.43)$$

Termination:

$$(4a) \ R_1 + R_1 \overset{k_{4a}}{\rightarrow} P_2 \qquad (4.2.44)$$

$$(4b) \ R_1 + R_2 \overset{k_{4b}}{\rightarrow} P_3 \qquad (4.2.45)$$

$$(4c) \ R_2 + R_2 \overset{k_{4c}}{\rightarrow} P_4 \qquad (4.2.46)$$

The initiation step has been written as a unimolecular reaction. In terms of the theory presented in Section 4.3.1.3, this initiation reaction will shift to a bimolecular process at low pressures.

This particular mechanism assumes that R_1 and R_1' are different radicals and that the latter do not participate in the propagation reactions. In the more general case the radical R_1' can participate in propagation reactions analogous to reactions (2) and (3). These propagation steps consist of a bimolecular hydrogen abstraction reaction followed by a unimolecular decomposition reaction.

The three possible termination reactions have been written as combination reactions when, in fact, disproportionation reactions may also occur. Under a given set of experimental conditions, only one of the three chain breaking steps (4a), (4b), or (4c) can be expected to be important.

With this general sequence of reactions as our proposed mechanism, we are now prepared to use the corresponding rate expressions and our standard assumptions to show that rate expressions with reaction orders of 1/2, 1, and 3/2 can be derived. The form of the rate expression that results is an indication of the nature of the chain breaking process.

A first-order rate expression results when the termination process involves the dissimilar radicals, R_1 and R_2. That is, reaction (4b) is the chain breaking step.

Based on the set of mechanistic equations set forth above, the rate of disappearance of reactant M is given by:

$$\frac{d[M]}{dt} = -k_1[M] - k_2[R_1][M] \quad (4.2.47)$$

For long chain reactions the amount of reactant undergoing decomposition via the initiation reaction is small compared to that decomposed by the propagation reactions (i.e., $k_2[R_1] \gg k_1$. Thus, to a good approximation,

$$\frac{d[M]}{dt} = -k_2[R_1][M] \quad (4.2.48)$$

The concentrations of the chain carriers R_1 and R_2 may be determined by the use of the stationary state assumption for each species.

$$\frac{d[R_1]}{dt} = k_1[M] - k_2[R_1][M] + k_3[R_2] - k_{4b}[R_1][R_2] \approx 0 \tag{4.2.49}$$

$$\frac{d[R_2]}{dt} = k_2[R_1][M] - k_3[R_2] - k_{4b}[R_1][R_2] \approx 0 \tag{4.2.50}$$

Adding these two equations and rearranging, we find that

$$[R_2] = \frac{k_1[M]}{2k_{4b}[R_1]} \tag{4.2.51}$$

Substitution of this relation into equation 4.2.50 gives

$$k_2[R_1][M] - \frac{k_1 k_3[M]}{2k_{4b}[R_1]} - \frac{k_1[M]}{2} = 0 \tag{4.2.52}$$

On the basis of our earlier statement that $k_2[R_1] \gg k_1$, the last term is negligible. Therefore

$$[R_1] = \left(\frac{k_1 k_3}{2k_{4b}k_2}\right)^{1/2} \tag{4.2.53}$$

Substitution of this relation into equation 4.2.48 gives

$$\frac{d[M]}{dt} = -\left(\frac{k_1 k_2 k_3}{2k_{4b}}\right)^{1/2}[M] \tag{4.2.54}$$

which is a first-order rate expression. If each of the individual rate constants is written in the

Arrhenius form,

$$k = \left(\frac{A_1 A_2 A_3}{2A_{4b}}\right)^{1/2} e^{-(E_1 + E_2 + E_3 - E_{4b})/2RT} \tag{4.2.55}$$

and it is apparent that the overall activation energy of the reaction is given by

$$E = \frac{E_1 + E_2 + E_3 - E_{4b}}{2} \tag{4.2.56}$$

The average chain length $\mathscr{L}$ is given by the following ratio.

$$\mathscr{L} = \frac{\text{rate of disappearance of reactants}}{\text{rate of initiation process}}$$

$$= \frac{\dfrac{d[M]}{dt}}{k_1[M]} \tag{4.2.57}$$

Thus the chain length $\mathscr{L}$ is given by

$$\mathscr{L} = \left(\frac{k_2 k_3}{2k_1 k_{4b}}\right)^{1/2} \tag{4.2.58}$$

an expression that is independent of the initial reactant concentration.

The following mechanism has been proposed for the decomposition of ethane into ethylene and hydrogen. The overall rate expression is first order in ethane.

Initiation: $\qquad C_2H_6 \xrightarrow{k_1} 2CH_3 \tag{4.2.59}$

Chain transfer: $\quad CH_3 + C_2H_6 \xrightarrow{k_2} CH_4 + C_2H_5 \tag{4.2.60}$

Propagation: $\qquad C_2H_5 \xrightarrow{k_3} C_2H_4 + H \tag{4.2.61}$

$\qquad\qquad\qquad H + C_2H_6 \xrightarrow{k_4} H_2 + C_2H_5 \tag{4.2.62}$

Termination: $\qquad H + C_2H_5 \xrightarrow{k_5} C_2H_6 \tag{4.2.63}$

Since these equations are of the same form as those discussed above, you should be able to show that

$$\frac{d[C_2H_6]}{dt} = -\frac{k_4 k_1}{k_5} \frac{1}{\left[\frac{k_1}{2k_3} \pm \sqrt{\left(\frac{k_1}{2k_3}\right)^2 + \frac{k_1 k_4}{k_3 k_5}}\right]} [C_2H_6] \qquad (4.2.64)$$

which is a first-order rate expression. Physical arguments indicate that k_1/k_3 must be very small compared to unity ($k_1/k_3 \ll 1$). Thus equation 4.2.64 becomes:

$$\frac{d[C_2H_6]}{dt} = -\left(\frac{k_1 k_3 k_4}{k_5}\right)^{1/2} [C_2H_6] \quad (4.2.65)$$

A reaction rate expression that is proportional to the square root of the reactant concentration results when the dominant termination step is reaction (4c), that is, the termination reaction occurs between two of the radicals that are involved in the unimolecular propagation step. The generalized Rice-Herzfeld mechanism contained in equations 4.2.41 to 4.2.46 may be employed to derive an overall rate expression for this case.

As before, the time rate of change of reactant concentration is given by

$$\frac{d[M]}{dt} = -k_1[M] - k_2[R_1][M] \quad (4.2.66)$$

The steady-state approximation may be used for each of the chain carrying species.

$$\frac{d[R_1]}{dt} = k_1[M] - k_2[R_1][M] + k_3[R_2] \approx 0 \qquad (4.2.67)$$

$$\frac{d[R_2]}{dt} = k_2[R_1][M] - k_3[R_2] - 2k_{4c}[R_2]^2 \approx 0 \qquad (4.2.68)$$

Adding these two equations and rearranging, one finds that:

$$R_2 = \left(\frac{k_1[M]}{2k_{4c}}\right)^{1/2} \qquad (4.2.69)$$

Substitution of this result into equation 4.2.67 leads to a relation for $[R_1]$.

$$[R_1] = \frac{k_1}{k_2} + \frac{k_3}{k_2}\left(\frac{k_1}{2k_{4c}}\right)^{1/2} \frac{1}{[M]^{1/2}} \quad (4.2.70)$$

Combination of equations 4.2.66 and 4.2.70 gives

$$\frac{d[M]}{dt} = -2k_1[M] - k_3\left(\frac{k_1}{2k_{4c}}\right)^{1/2}[M]^{1/2} \qquad (4.2.71)$$

Since the first term on the right side of this equation merely represents twice the rate of initiation, it is negligible compared to the second term. Hence

$$\frac{d[M]}{dt} = -k_3\left(\frac{k_1}{2k_{4c}}\right)^{1/2}[M]^{1/2} \quad (4.2.72)$$

and one has half-order kinetics.

In order for the overall rate expression to be 3/2 order in reactant for a first-order initiation process, the chain terminating step must involve a second-order reaction between two of the radicals responsible for the second-order propagation reactions. In terms of our generalized Rice-Herzfeld mechanistic equations, this means that reaction (4a) is the dominant chain breaking process. One may proceed as above to show that the mechanism leads to a 3/2 order rate expression.

For Rice-Herzfeld mechanisms the mathematical form of the overall rate expression is strongly influenced by the manner in which the chains are broken. It can also be shown that changing the initiation step from first order to second order also increases the overall order of the reaction by 1/2 (30). These results are easily obtained from the equations derived previously by substituting k'_1 [M] for k_1 everywhere that the latter term appears. Since k_1 always appears to the 1/2 power in the final rate expressions, the exponent on M in the existing rate expression must be added to 0.5 in order to obtain the overall order of the reaction corresponding to the bimolecular initiation step. In like manner, shifts from bimolecular to termolecular termination reactions will decrease the overall order of the reaction by 1/2.

Although the above Rice-Herzfeld mechanisms lead to simple overall rate expressions, do *not* get the impression that this is always the case. More detailed discussions of these types of reactions may be found in textbooks (31–34) and in the original literature.

4.2.4 Inhibitors, Initiators, and Induction Periods

Since the complex decomposition reactions considered in the previous section are propagated by a series of elementary steps, it might be expected that they and other chain reaction processes will show an unusual sensitivity to any substance or physical condition that interferes with the propagation of the chain. It is therefore not surprising to find that the rates of such reactions can be markedly reduced either by the presence of trace quantities of certain chemical substances or by changes in the physical condition of the surface of the vessel in which the reaction is being studied. The latter changes include both variations in the surface/volume ratio of the reactor brought about by adding spun glass or some other high surface area material to the reactor and variations in the nature of the surface itself through the use of Teflon, syrupy phosphoric acid, or other coatings. The reduction in the rate may occasionally be referred to as negative catalysis, but it is more commonly known as *inhibition*. An additive that slows down a reaction is called an inhibitor.

In the case of chain reactions a mere trace of inhibitor can reduce reaction rates by orders of magnitude. Such inhibitors break the chain, perhaps as a result of a reaction in which a relatively nonreactive free radical is formed. Another manner in which an inhibitor may act is by combining with a catalyst and rendering it inoperative.

Other additives can promote chain reactions by acting as *initiators* of the chain. These compounds are materials that can be more readily dissociated than the primary reactants. For example, benzoyl peroxide has often been used to initiate the polymerization of olefinic monomers because of the ease with which it dissociates.

In many cases complex chain reactions are characterized by an *induction period* (i.e., a period at the beginning of the reaction during which the rate is significantly less than that which subsequently prevails). This period is often caused by the presence of small amounts of inhibitors. After the inhibitors are consumed, the reaction may then proceed at a much faster rate.

4.2.5 Branched Chain Reactions and Explosions

Chain reactions can lead to thermal explosions when the energy liberated by the reaction cannot be transferred to the surroundings at a sufficiently fast rate. An "explosion" may also occur when chain branching processes cause a rapid increase in the number of chains being propagated. This section treats the branched chain reactions that can lead to nonthermal explosions and the physical phenomena that are responsible for both branched chain and thermal explosions.

A chain branching reaction is one that leads to an increase in the number of chain carriers present in a reacting system (e.g., 4.2.9). When such reactions occur to a significant extent, the total number of radicals present in a reacting mixture can increase rapidly so that steady-state conditions are no longer maintained. The reaction proceeds at a very high velocity and, since the associated release of chemical energy is also extremely rapid, the viewer sees this phenomenon as an explosion.

The generalized mechanism by which branched chain reactions proceed provides a basis for a semiquantitative understanding of explosions resulting from chain branching.*

propagation and termination steps are written as first-order processes, but this is not essential to the argument that follows.

The rate expressions have been written in generalized fashion with the terms f_p, f_b, f_{st}, and f_{gt} containing the reaction rate constants, stoichiometric coefficients, and concentrations of the various stable species present in the reaction mixture. If one also wished to consider bimolecular radical processes, these could also be lumped into the f parameters.

Initiation:	$? \rightarrow R$	$r_i = \dfrac{d[R]}{dt}$	(4.2.73)
Propagation:			
Product formation	$R \rightarrow P + R$	$r_p = \dfrac{d[P]}{dt} = f_p[R]$	(4.2.74)
Chain branching	$R \rightarrow \alpha R$	$r_b = \dfrac{d[R]}{dt} = f_b(\alpha - 1)[R]$	(4.2.75)
Termination:			
Chain breaking at solid surface	$R \rightarrow ?$	$r_{st} = -\dfrac{d[R]}{dt} = f_{st}[R]$	(4.2.76)
Chain breaking in gas phase	$R \rightarrow ?$	$r_{gt} = -\dfrac{d[R]}{dt} = f_{gt}[R]$	(4.2.77)

where R represents the various active centers or free radical chain carriers involved in the reaction, and the detailed chemistry of the various steps is unspecified. The number of radicals formed in the chain branching step per radical that reacts is assigned the symbol α. All of the

* Adapted from Chemical Kinetics by K. J. Laidler. Copyright © 1965. Used with permission of McGraw-Hill Book Company.

In terms of the above rate expressions, the Bodenstein steady-state approximation for free radical intermediates gives:

$$r_i + f_b(\alpha - 1)[R] - f_{st}[R] - f_{gt}[R] = 0 \quad (4.2.78)$$

Note that this equation does not contain a term corresponding to the straight chain step in which the primary products are formed. This step does not have any *net* effect on the total concentration of free radicals. Thus

$$[R] = \frac{r_i}{f_{st} + f_{gt} - f_b(\alpha - 1)} \quad (4.2.79)$$

The overall rate of reaction is therefore

$$r_p = \frac{r_i f_p}{f_{st} + f_{gt} - f_b(\alpha - 1)} \quad (4.2.80)$$

Since the branching parameter α is greater than unity (usually it is 2), it is conceivable that under certain circumstances the denominator of the overall rate expression could become zero. In principle this would lead to an infinite reaction rate (i.e., an explosion). In reality it becomes very large rather than infinite, since the steady-state approximation will break down when the radical concentration becomes quite large. Nonetheless, we will consider the condition that $f_b(\alpha - 1)$ is equal to $(f_{st} + f_{gt})$ to be a valid criterion for an explosion limit.

Because there are two positive terms in the denominator of equation 4.2.85 (either of which may be associated with the dominant termination process), this equation leads to two explosion limits. At very low pressures the mean free path of the molecules in the reactor is quite long, and the radical termination processes occur primarily on the surfaces of the reaction vessel. Under these conditions gas phase collisions leading to chain breaking are relatively infrequent events, and $f_{st} \gg f_{gt}$. Steady-state reaction conditions can prevail under these conditions if $f_{st} > f_b(\alpha - 1)$.

As the pressure in the reaction vessel increases, the mean free path of the gaseous molecules will decrease and the ease with which radicals can reach the surfaces of the vessel will diminish. Surface termination processes will thus occur less frequently; f_{st} will decline and may do so to the extent that $f_{st} + f_{gt}$ becomes equal to $f_b(\alpha - 1)$. At this point an explosion will occur. This point corresponds to the first explosion limit shown in Figure 4.1. If we now jump to some higher pressure at which steady-state reaction conditions can again prevail, similar

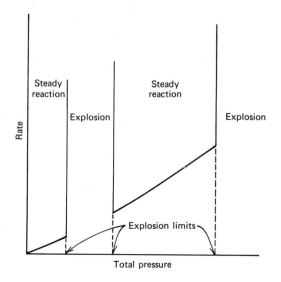

Figure 4.1

The rate of the hydrogen-oxygen reaction as a function of the total pressure. (Adapted from *Chemical Kinetics* **by K. J. Laidler. Copyright © 1965. Used with permission of McGraw-Hill Book Company.;**

semiquantitative arguments can be used to explain the phenomenon known as the second explosion limit. At these pressures the dominant processes by which chains are terminated will occur in the gas phase. The increased pressure hinders the diffusion of radicals to the vessel surfaces and provides a number density of gas phase radicals that is sufficient for radical recombination and disproportionation processes.

If the pressure in the reaction vessel is now decreased, the rate at which chains are broken will also decrease. Eventually the pressure will reach a point at which the rates of the termination processes will become equal to the rate at which the radical concentration is increasing because of the chain branching process; that is

$$[f_{gt} + f_{st}][R] = f_b(\alpha - 1)[R] \quad (4.2.81)$$

At this point an explosion will occur corresponding to the second explosion limit in Figure 4.1.

There is a third explosion limit indicated in Figure 4.1 at still higher pressures. This limit is a thermal limit. At these pressures the reaction rate becomes so fast that conditions can no longer remain isothermal. At these pressures the energy liberated by the exothermic chain reaction cannot be transferred to the surroundings at a sufficiently fast rate, so the reaction mixture heats up. This increases the rate of the process and the rate at which energy is liberated so one has a snowballing effect until an explosion occurs.

It should be evident from this discussion that the first explosion limit will be quite sensitive to the nature of the surface of the reaction vessel and its area. If the surface is coated with a material that inhibits the surface chain termination process, the first explosion limit will be lowered. Inert foreign gases can also have the effect of lowering the first explosion limit, since they can hinder diffusion to the surface. If something like spun glass or large amounts of fine wire are inserted, one can effect an increase in the first explosion limit by changing the surface/volume ratio of the system.

4.2.6 Supplementary References

Our discussion of chain reaction processes is necessarily incomplete. For more detailed treatments consult the following references.

1. C. H. Bamford and C. F. H. Tipper (Editors), *Comprehensive Chemical Kinetics, Volume II, The Theory of Kinetics*, Elsevier, New York, 1969.
2. G. J. Minkoff and C. F. H. Tipper, *Chemistry of Combustion Reactions*, Butterworths, London, 1962.
3. N. N. Semenov, *Some Problems in Chemical Kinetics and Reactivity*, Volumes I and II, Princeton University Press, Princeton, N.J., 1958, 1959.

4.2.7 Cautionary Note on Reaction Mechanisms

It is appropriate to conclude our discussion of reaction mechanisms on a note of caution. A mechanism is merely a logical *hypothesis* as to the sequence of molecular events that occur during the course of a chemical reaction. Reaction mechanisms should *not* be regarded as experimental facts. They are plausible explanations of experimental data that are consistent with the data, but that are subject to revision as new data are obtained. Even if one proposed mechanism gives agreement with all available experimental facts, this is no sign that the mechanism is unique and that other mechanisms could not give such agreement.

4.3 MOLECULAR THEORIES OF CHEMICAL REACTION KINETICS

A "complete" theory of reaction kinetics would provide a basis for calculating the rate of an elementary reaction from a knowledge of the properties of the reacting molecules and their concentrations. In terms of the present state of our theoretical knowledge, a "complete" theory can be regarded as a goal that is far, far down the road. Existing theories are extremely unsatisfactory. Nevertheless, chemical engineers must be cognizant of the primary features of these theories.

4.3.1 Simple Collision Theory

Before a chemical reaction can occur, energy must be available and localized in such a way that it is possible to break and make certain chemical bonds in the reactant molecules. Moreover, the participants in the reaction must be in a spatial configuration such that it is possible for the atomic and electronic rearrangements to occur. The most common means by which such redistributions of energy and changes in geometric configurations can occur are molecular collision processes. In this sense all theories of reaction are collision theories. However, we will use this term in a more limited sense. We restrict its use to the theory that links chemical kinetics to the kinetic theory of gases by the use of the theoretical expression for bimolecular collision frequencies.

4.3.1.1 Rates of Bimolecular Reactions. For purposes of chemical reaction kinetics the primary result of the kinetic theory of gases is that the total number of collisions between A molecules and B molecules per unit volume per unit time is given by Z_{AB}.

$$Z_{AB} = n'_A n'_B \pi \sigma_{AB}^2 \sqrt{\frac{8k_B T}{\pi \mu_{AB}}} \qquad (4.3.1)$$

where

n'_A and n'_B are the number densities of molecules
$\quad$ A and B, respectively
T is the absolute temperature
k_B is the Boltzmann constant
σ_{AB} is the arithmetic average of the hard
$\quad$ sphere diameters of molecules A
$\quad$ and B

$$\sigma_{AB} = \frac{\sigma_A + \sigma_B}{2} \qquad (4.3.2)$$

and μ_{AB} is the reduced mass of the system expressed in terms of the molecular masses m_A and m_B as

$$\mu_{AB} = \frac{m_A m_B}{m_A + m_B} \qquad (4.3.3)$$

If A and B are identical, then $m_A = m_B = m$, $\mu = m/2$, $\sigma_A = \sigma_B = \sigma_{AB} = \sigma$, and $n'_A = n'_B = n'$. To avoid double counting of collisions, an additional factor of $1/2$ must also be introduced into equation 4.3.1 to obtain an expression for the number of collisions between identical molecules per unit volume per unit time.

$$Z_{AA} = (n')^2 \frac{\pi}{2} \sigma^2 \sqrt{\frac{8k_B T}{\pi(m/2)}}$$

$$= 2(n')^2 \sigma^2 \sqrt{\frac{\pi k_B T}{m}} \qquad (4.3.4)$$

For a typical gas at standard conditions, Z_{AA} and Z_{AB} are both of the order of 10^{28} collisions per cubic centimeter per second.

Derivations of the above expressions are contained in most physical chemistry texts and other books dealing with kinetic theory (35–39). Different authors may write equations 4.3.1 and 4.3.4 in slightly different forms because of different assumptions involved in their derivations and the degree of mathematical rigor that they choose to use. They may precede these expressions by some numerical factor that will not differ appreciably from unity. Such differences are insignificant for purposes of chemical kinetics. From a practical viewpoint the uncertainties involved in predicting reaction rates do not justify using anything more complex than the simple hard sphere model of the kinetic theory.

Bimolecular processes are the primary vehicle by which chemical change occurs. The frequency with which these encounters occur is given by equations 4.3.1 and 4.3.4. However, only an extremely small number of the collisions actually lead to reaction; for predictive purposes one needs to know what fraction of the collisions are effective in that they lead to reaction.

Since chemical reactions involve the making and breaking of chemical bonds with their associated energy effects and geometric requirements, it is not unreasonable to assume that these factors play an important role in determining the probability that a bimolecular collision will lead to chemical reaction. In addition to these factors there are restrictions on bimolecular combination or association reactions and quantum mechanical requirements that can influence this probability.

For complex organic molecules, geometric considerations alone lead one to the conclusion that only a small fraction of bimolecular collisions can lead to reaction. One can represent the fraction of the collisions that have the proper geometric orientation for reaction by a *steric factor* (P_s). Except for the very simplest reactions, this factor will be considerably less than unity. On the basis of simple collision theory, it is not possible to make numerical estimates of P_s, although it may occasionally be possible to make use of one's experience with similar reactions to determine whether P_s for a given reaction will be large or small. This failure to be able to predict values of P_s is one of the major weak points of the collision theory. It results from the inability of the theory to consider the individual geometric shapes of various reactant molecules. Since P_s must be determined empirically, it may be considerably in error because it will contain all of the errors associated with the assumption of hard sphere molecules as well as any quantum mechanical restrictions on the reaction in question.

Of all the factors that determine the effectiveness of a collision, the energy requirement is by far the most important. Reaction cannot occur unless sufficient energy is provided and localized so that the appropriate bonds can be broken and new bonds formed. It is reasonable to suppose that there will be a threshold energy requirement below which these processes cannot occur. It is usually assumed that a collision is effective for reaction purposes only if the relative kinetic energy along the line of collision centers is greater than or equal to this threshold value. Kinetic energy associated with motion perpendicular to this line corresponds to a sideswipe that leads to changes in rotational levels but cannot be expected to be effective in promoting reaction.

The principles of kinetic theory may be used to arrive at an expression for the number of collisions whose relative kinetic energy along the line of centers is greater than ε_c. The result is the following expression for the number of such collisions per unit volume per unit time.

$$Z'_{AB} = n'_A n'_B \pi \sigma^2_{AB} \left(\frac{8 k_B T}{\pi \mu_{AB}} \right)^{1/2} e^{-\varepsilon_c/k_B T} \quad (4.3.5)$$

If this equation is now written in terms of energy per mole instead of per molecule, it becomes

$$Z'_{AB} = n'_A n'_B \pi \sigma^2_{AB} \left(\frac{8 k_B T}{\pi \mu_{AB}} \right)^{1/2} e^{-E_c/RT} \quad (4.3.6)$$

an expression that becomes identical with equation 4.3.1 if we set E_c equal to zero. Equation 4.3.5

provides the surprisingly simple result that the fraction of the collisions that will involve at least an energy ε_c directed along the line of centers is given by $e^{-\varepsilon_c/k_B T}$ or $e^{-E_c/RT}$. If the steric factor defined previously is introduced, the rate of reaction between unlike molecules becomes

$$r_{AB} = P_s n'_A n'_B \pi \sigma^2_{AB} \left(\frac{8 k_B T}{\pi \mu_{AB}} \right)^{1/2} e^{-E_c/RT} = k n'_A n'_B$$

$$(4.3.7)$$

Comparison of this equation with the Arrhenius form of the reaction rate constant reveals a slight difference in the temperature dependence of the rate constant, and this fact must be explained if one is to have faith in the consistency of the collision theory. Taking the derivative of the natural logarithm of the rate constant in equation 4.3.7 with respect to temperature, one finds that

$$\frac{d \ln k}{dT} = \frac{1}{2T} + \frac{E_c}{RT^2} \quad (4.3.8)$$

while the Arrhenius form is given by equation 3.3.54.

$$\frac{d \ln k}{dT} = \frac{E_A}{RT^2} \quad (4.3.9)$$

These last two equations can be reconciled only if the experimental activation energy is temperature dependent in such a way as to accommodate the variation in collision frequency with temperature; that is,

$$E_A = E_c + \frac{RT}{2} \quad (4.3.10)$$

Usually the last term on the right side of equation 4.3.10 is only a very small fraction of the total. For example, if the observed activation energy (E_A) is 10 kcal at 500°K, the $RT/2$ term is only 0.5 kcal or 5%. Such discrepancies are usually within the scatter of the data.

Illustration 4.4 indicates how collision frequency calculations are used to obtain a reaction rate expression.

ILLUSTRATION 4.4 COMPARISON OF EMPIRICAL RATE EXPRESSION WITH COLLISION FREQUENCY EXPRESSION

At 700 °K the rate expression for the decomposition of HI,

$$2HI \rightarrow H_2 + I_2$$

is given by (40)

$$r = 1.16 \times 10^{-3}[HI]^2 \text{ kmoles/m}^3\text{·sec}$$

when [HI] is expressed in kilomoles per cubic meter. Compare this reaction rate expression with that predicted by the analog of equation 4.3.7, which corresponds to collisions of the type A- 4.

Data

$$M_{HI} = 127.9 \qquad \sigma_{HI} = 2.0 \times 10^{-10} \text{ m}$$

The measured activation energy is equal to 186.1 kJ/mole.

Solution

Using equation 4.3.4, the $A-A$ analog of equation 4.3.7 is

$$r_{AA} = P_s 2(n'_A)^2 \sigma_A^2 \sqrt{\frac{\pi k_B T}{m_A}} e^{-E_c/RT}$$

or

$$r_{AA} = P_s 2(n'_A)^2 \sigma_A^2 \sqrt{\frac{\pi R T}{M_A}} e^{-E_c/RT}$$

From equation 4.3.10, using $R = 8.31$ J/mole·°K,

$$E_c = 186\,100 - 8.31(700)/2 = 183\,200 \text{ J}$$

In SI units $M_A = 0.1279$ kg/mole. If n'_A is measured in molecules per cubic meter and r_{AA} in molecules per cubic meter per second, the last equation becomes

To convert from number densities of molecules to reactant concentrations in kilomoles per cubic meter,

$$C_A = \frac{n'_A}{1000 N_0}$$

where N_0 is Avogadro's number.

A similar conversion is necessary to measure r_{AA} in kilomoles per cubic meter per second. In these units

$$r_{AA} = 6.35 \times 10^{-31}(6.023 \times 10^{23}) \times 10^3 P_s C_A^2$$
$$= 3.83 \times 10^{-4} P_s C_A^2$$

If the steric factor is comparable to unity the calculated rate is within an order of magnitude of the experimental value of $1.16 \times 10^{-3}[HI]^2$. Authors of other textbooks have reported even better agreement between experimental values of this reaction rate expression and those calculated from collision theory. Changes in the values used for the activation energy of the reaction and the molecular diameter are often sufficient to bring the calculated values into much closer agreement with the experimental values. Since measurements of different properties lead to significant differences in calculated values of hard sphere molecular diameters, these quantities are not accurately known. Moreover, uncertainties of several percent in reaction activation energies are not at all unusual.

Consideration of a variety of other systems leads to the conclusion that very rarely does the collision theory predict rate constants that will be comparable in magnitude to experimental values. Although it is not adequate for predictions of reaction rate constants, it nonetheless provides a convenient physical picture of the reaction act and a useful interpretation of the concept of activation energy. The major short-

$$r_{AA} = P_s 2(n'_A)^2 (2.0 \times 10^{-10})^2 \sqrt{\frac{\pi(8.31)700}{0.1279}} e^{-183200/8.31(700)}$$

$$= 6.35 \times 10^{-31} P_s (n'_A)^2$$

comings of the theory lie in its failure to relate the steric factor and the activation energy to molecular parameters from which *a priori* predictions can be made.

4.3.1.2 Termolecular Reactions. If one attempts to extend the collision theory from the treatment of bimolecular gas phase reactions to termolecular processes, the problem of how to define a termolecular collision immediately arises. If such a collision is defined as the simultaneous contact of the spherical surfaces of all three molecules, one must recognize that two hard spheres will be in contact for only a very short time and that the probability that a third molecule would strike the other two during this period is vanishingly small.

In order to have a finite probability that termolecular collisions can occur, we must relax our definition of a collision. We will assume that the approach of rigid spheres to within a distance ℓ of one another constitutes a termolecular collision that can lead to reaction if appropriate energy and geometry requirements are met. This approach is often attributed to Tolman (41). The number of ternary collisions per unit volume per unit time between molecules A, B, and C such that A and C are both within a distance ℓ of B is given by Z_{ABC}.

One may estimate the relative frequency of bimolecular and termolecular collisions using equations 4.3.1 and 4.3.11.

$$\frac{Z_{ABC}}{Z_{AB}} = 4\pi\sigma_{BC}^2 \ell \sqrt{\mu_{AB}} \left(\frac{1}{\sqrt{\mu_{AB}}} + \frac{1}{\sqrt{\mu_{BC}}} \right) n_C' \tag{4.3.13}$$

If A, B, and C have similar molecular weights, σ_{BC} is of the order of 2 Å (2×10^{-10} m), and the gas is at atmospheric pressure so that $n_C' = 3 \times 10^{25}$ molecules/m^3, then

$$\frac{Z_{ABC}}{Z_{AB}} = \mathcal{O}[4(\pi)(2 \times 10^{-10})^2(1 \times 10^{-10})(3 \times 10^{25})$$

$$= \mathcal{O}[10^{-3}] \tag{4.3.14}$$

so that in a gas at standard conditions there are approximately 1000 bimolecular collisions for every termolecular collision.

When the fact that ternary collisions are relatively rare occurrences is combined with the fact that there will probably be severe geometric restrictions on such reactions, one concludes that these reactions must have relatively low activation energies or else their reaction rates would be vanishingly small. This expectation is confirmed by experimental data on such reactions.

Termolecular reactions are quite rare and the best-known examples are the recombination re-

---

$$Z_{ABC} = 8\sqrt{2}\pi^{3/2}\sigma_{AB}^2\sigma_{BC}^2 \ell \sqrt{k_B T} \left(\frac{1}{\sqrt{\mu_{AB}}} + \frac{1}{\sqrt{\mu_{BC}}} \right) n_A' n_B' n_C' \tag{4.3.11}$$

---

where the symbols correspond to those used previously and ℓ is a somewhat indefinite parameter that should have a value typical of the length of a chemical bond (i.e., $\ell \cong 1$ Å). If one assumes that the same considerations that govern the efficiency of bimolecular collisions are relevant to termolecular processes, then the reaction rate of these processes is given by

$$r = Z_{ABC} P_s e^{-E/RT} \tag{4.3.12}$$

actions of small atoms and radicals (e.g., H + H + M → H$_2$ + M).

4.3.1.3 Unimolecular Reactions. Experimental evidence for unimolecular processes posed a dilemma for early kineticists. They observed that several species that appeared to be stable at low temperatures underwent first-order decomposition or isomerization reactions at higher temperatures. For reaction to take place, molecules have to possess sufficient energy to permit the necessary rearrangement of chemical bonds. If this energy is not supplied in the form of electromagnetic radiation from external sources, the source of this energy can only be molecular collisions. It was difficult for early workers to reconcile the fact that collisions are most frequently bimolecular processes with the fact that the reactions of interest were first-order. In 1922, however, Lindemann (42) resolved this dilemma by showing how collisional processes can give rise to first-order kinetics under certain circumstances. He proposed a mechanism that was supported by several experiments suggested by his hypothesis. His mechanism is the basis for all modern theories of unimolecular reactions, although several important modifications and additions to the theory have been made since 1922.

According to Lindemann's theory, at any given time a small fraction of the reactant molecules possess enough energy distributed among their various degrees of freedom so that it is possible for them to be converted directly to product molecules without receiving additional energy from any other molecules. This process does, however, require localization of sufficient energy in the appropriate vibrational degree of freedom for reaction to occur. This fraction is referred to as "activated" or "energized" molecules. Lindemann assumed that decomposition of the activated molecules is not instantaneous at the moment the energizing collision occurs, but that a certain time is required for the energy to redistribute itself among the different vibrational modes of motion. When the energy in a certain vibrational mode of motion exceeds the bond strength, the bond breaks and reaction occurs.

If the stoichiometric equation for a unimolecular reaction is $A \rightarrow B + C$, and if the energized molecules are denoted by A^*, the Lindemann mechanism consists of the following sequence of events.

I. Activation by collision:

$$A + A \overset{k_1}{\rightarrow} A^* + A \qquad (4.3.15)$$

II. Deactivation by collision:

$$A^* + A \overset{k_2}{\rightarrow} A + A \qquad (4.3.16)$$

This process is the reverse of the previous one. It is expected to occur at the first collision of A^* after it has been formed. The rate constant k_2 will be much greater than k_1, since it is not restricted by the large energy requirement associated with the activation process.

III. Unimolecular decomposition:

$$A^* \overset{k_3}{\rightarrow} B + C \qquad (4.3.17)$$

Since collisional processes occur so rapidly, the concentration of the A^* molecules builds up to its steady-state value in a small fraction of a second and the steady-state approximation for A^* is appropriate for use.

$$\frac{d[A^*]}{dt} = k_1[A]^2 - k_2[A][A^*] - k_3[A^*] \approx 0 \qquad (4.3.18)$$

Thus

$$A^* = \frac{k_1[A]^2}{k_2[A] + k_3} \qquad (4.3.19)$$

The rate of the overall reaction is identical with the rate of formation of B.

$$\frac{d[B]}{dt} = k_3[A^*] = \frac{k_1 k_3 [A]^2}{k_2[A] + k_3} \qquad (4.3.20)$$

This equation shows that the reaction rate is neither first-order nor second-order with respect to species A. However, there are two limiting cases. At high pressures where [A] is large, the bimolecular deactivation process is much more rapid than the unimolecular decomposition (i.e., $k_2[A][A^*] \gg k_3[A^*]$). Under these conditions the second term in the denominator of equation 4.3.20 may be neglected to yield a first-order rate expression.

$$\frac{d[B]}{dt} = \frac{k_1 k_3}{k_2} [A] \qquad (4.3.21)$$

However, as the pressure is decreased, one eventually reaches a point where the rate of the decomposition reaction becomes much larger than the collisional deactivation process, so that $k_3 \gg k_2[A]$. In this situation the overall rate expression becomes second-order in A.

$$\frac{d[B]}{dt} = k_1 [A]^2 \qquad (4.3.22)$$

The lifetime of the energized molecule relative to the time between collisions determines the reaction order. If the lifetime is short compared to the time between collisions, virtually all energized molecules will react before they can be deactivated by subsequent collisions. In this case each energizing collision will lead to reaction, and the rate limiting step in the overall reaction is the rate of activation by collision, a second-order process. On the other hand, if the lifetime of the energized molecule is long compared to the time between collisions, practically all of the energized molecules become deactivated by subsequent collisions without reacting. A quasi-equilibrium situation will exist between the concentrations of energized and "normal" molecules. The concentration of the former will be proportional to that of the latter. Since the reaction rate is proportional to the concentration of energized molecules, it will also be proportional to the concentration of normal molecules (i.e., first-order).

As the system pressure is decreased at constant temperature, the time between collisions will increase, thereby providing greater opportunity for unimolecular decomposition to occur. Consequently, one expects the reaction rate expression to shift from first-order to second-order at low pressures. Experimental observations of this transition and other evidence support Lindemann's theory. It provides a satisfactory qualitative interpretation of unimolecular reactions, but it is not completely satisfactory from a quantitative viewpoint, and certain important modifications have been made in the years since Lindemann set forth this mechanism. The interested reader may wish to consult textbooks that emphasize gas phase reaction kinetics (43–50).

One aspect of extensions of the theory is particularly worthy of note. In a bimolecular reaction, the act of bond breaking must occur at the instant of collision. Consequently, the distribution of energy at that instant must be exactly that required for reaction. In a unimolecular reaction, however, the entire time interval between the activating collision and the next collision is available for the energy in the activated molecule to be redistributed and cause bond breaking. This argument retains the "go-no go" concept that reaction will or will not occur, depending on whether or not a minimum energy requirement is met as a result of a collision, but it predicts higher rates than the conventional kinetic theory approach.

Hinshelwood (51) used reasoning based on statistical mechanics to show that the energy probability factor in the kinetic theory expressions $(e^{-E/RT})$ is strictly applicable only to processes for which the energy may be represented in two square terms. Each translational and rotational degree of freedom of a molecule corresponds to one squared term, and each vibrational degree of freedom corresponds to two squared terms. If one takes into account the energy that may be stored in S squared terms, the correct probability factor is

$$\int_E^\infty \frac{e^{-E/RT} E^{(S/2)-1} \, dE}{\left(\frac{S}{2} - 1\right)! (RT)^{S/2}}$$

To a very good approximation when $S/2$ is an integer and when $E \gg RT$, the integral becomes

$$\frac{e^{-E/RT}}{\left(\frac{S}{2} - 1\right)!} \left(\frac{E}{RT}\right)^{(S/2)-1}$$

For the case where $S = 2$ this expression reduces to the simple exponential form of Arrhenius. For values of S greater than 2, it yields a much larger probability of reaction than one would obtain from the normal Arrhenius form. The enhancement may be several orders of magnitude. For example, when $S = 10$ and $E/RT = 30$, the ratio of the probability factor predicted by Hinshelwood's approach to that predicted by the conventional Arrhenius method is $(30)^4/4! = 3.375 \times 10^4$. The drawback of the approach is that one cannot accurately predict S a priori. When one obtains an apparent steric factor in excess of unity, this approach can often be used in interpretation of the data.

4.3.2 Transition State Theory

The transition state theory provides a useful framework for correlating kinetic data and for codifying useful generalizations about the dynamic behavior of chemical systems. This theory is also known as the "activated complex theory," the "theory of absolute reaction rates," and "Eyring's theory." This section introduces chemical engineers to the terminology, the basic aspects, and the limitations of the theory.

The transition state theory differs from collision theory in that it takes into account the internal structure of reactant molecules. It describes a reaction in terms of movements on a multidimensional surface in which the potential energy of the system is depicted as a function of the relative positions of the nuclei constituting the participants in the reaction. The progress of the reaction is characterized in terms of movement along the surface. Although one may envision an infinite number of paths linking the points corresponding to the original reactants and the final products, certain of these paths will require significantly smaller gross energy inputs to surmount the energy barrier for the reaction. The minimum energy requirement is related to the activation energy for the reaction. The atomic configuration corresponding to potential energy values near the top of this minimal barrier is referred to as the activated complex. The region in hyperspace corresponding to this configuration is referred to as the transition state. Progress along the path corresponding to the series of positions occupied by the system as it moves from reactants to the activated complex and on to the final product is measured in terms of distance along the reaction coordinate. If this quantity is used as a measure of the progress of reaction, it is possible to convert the multidimensional potential energy diagram into a two-dimensional free energy diagram, one form of which is shown in Figure 4.2. Such figures emphasize the commonly used analogy in which the reaction and its energy requirements are compared to the movement of a sphere across hilly terrain in the presence of a gravitational field.

In terms of the transition state theory the rate of reaction is controlled by the rate of passage through the region at the top of the energy barrier. This passage may correspond either to movement over the top of the barrier or, in some cases, to quantum mechanical tunneling through the barrier. Several approaches to formulating the reaction rate in terms of such movements have been described, but we will employ the one utilized by Eyring and his co-workers (52–54). The basic hypothesis on which Eyring's formulation is based is that an equilibrium exists between certain activated complex species and the reactants. This equilibrium determines the number of activated complexes corresponding to specified reactant concentrations, regardless of whether or not an overall chemical equilibrium exists between reactant and product species. This equilibrium hypothesis also permits formulation of the rate expression either in terms of the characteristic partition functions of statistical mechanics or in terms of pseudo thermodynamic parameters.

The task of expressing the energy of a system comprised of a number of nuclei and electrons as a function of their spatial configuration is

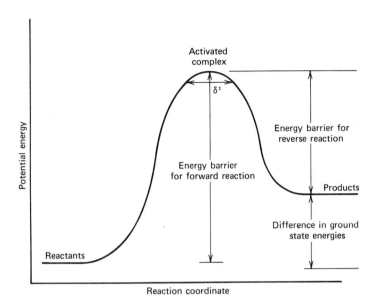

Figure 4.2
Schematic representation of reaction progress in terms of the reaction coordinate.

obviously a complex problem. In fact, it is one that is not amenable to exact solution for even the simplest possible chemical reaction. However, theoretical physicists have attacked this problem with diligence during the past several decades, and a number of useful approximate methods have been developed. In principle they permit one to generate multidimensional potential energy surfaces linking reactants and products.

Consider a reaction involving three atoms.

$$X + YZ \rightarrow XY + Z \qquad (4.3.23)$$

In general three position variables will be needed to specify the potential energy of the reaction system. These may be the X–Y, Y–Z, and X–Z internuclear distances or two internuclear distances and the included angle. Even in this relatively simple case, four dimensions would be required for generation of the potential energy surface. However, if we restrict our attention to linear configurations of these atoms, it is possible

to depict the potential energy surface in three dimensions, as shown in Figure 4.3.

Certain qualitative and semiquantitative aspects of potential energy surfaces are conveniently illustrated using this figure. The vertical elevation represents the potential energy of the system as a function of the two internuclear distances r_{xy} and r_{yz}. Such surfaces may be generated using the techniques of the theoretical physicist. They correspond to two valleys linked by a saddle-shaped pass or col. One side of each valley corresponds to a very steep hill, said terrain arising from the strong dependence of the repulsive component of interatomic forces on separation distance. The opposite sides of the valley rise less steeply and eventually lead to a plateau at large values of both internuclear separations. The valleys do not join except by way of a pass or saddle point near K and L.

If a plane is passed perpendicular to the r_{xy} axis at a sufficiently large value of this parameter, the cross section of the potential energy surface

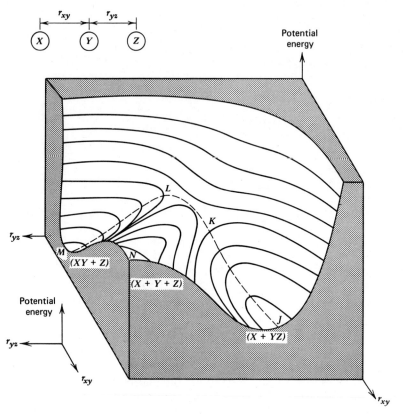

Figure 4.3
Schematic representation of the potential energy of a system comprised of three atoms in a linear configuration as a function of the internuclear separation distances. The dashed line represents the reaction $X + YZ \rightarrow XY + Z$.

one obtains is essentially the potential energy curve governing the formation of the YZ molecule whose constituent atoms exert both attractive and repulsive forces. At smaller values of r_{xy}, atom X interacts with atoms Y and Z to perturb the shape of the potential function so that the depth of the potential well is less. Point J is located on the floor of one of the valleys and has a large value of r_{xy} and a very small value for r_{yz}. It corresponds to an X atom far removed from a YZ molecule.

Similarly, it can be shown that a cross section perpendicular to the r_{yz} axis at large r_{yz} gives

the potential energy curve characteristic of the formation of XY from its constituent atoms. Point M on the figure corresponds to a Z atom and an XY molecule. At point N both r_{xy} and r_{yz} are large, so atoms X, Y, and Z are well separated and no molecules are present. At points K and L both internuclear distances are small, as in a collision between an X atom and a YZ molecule (point K) or between a Z atom and an XY molecule (point L). The region in the neighborhood of these points is referred to as the *transition state*, and the configuration assumed by the atoms in this state is referred to as the *activated complex.* We shall denote this complex by the symbol $XYZ^{\ddagger}$.

The reaction $X + YZ \rightarrow XY + Z$ will correspond to motion from point J in one valley to point M in the second valley. For this reaction to take place with a minimum amount of energy the system will travel along the floor of the first valley, over the col, and down into the second valley. This path is indicated by a dashed line. Energy considerations dictate that the majority of the reaction systems will follow this path. The elevation of the saddle point above the floor of the first valley is thus related to the activation energy for the reaction.

Another common representation of the progress of the system as it moves from reactants to products is that shown in Figure 4.2. Here the potential energy of the system is plotted versus distance along the dashed line shown in Figure 4.3. This variable is referred to as the *reaction coordinate.* In both figures the energy differences between the transition state and the reactant and product states are related to the activation energies for the forward and reverse reactions, respectively. Energy differences between the reactant and product states are related to the normal thermodynamic energy functions characteristic of the reaction. For exothermic reactions the potential energy of the products lies below that of the reactants, and for endothermic reactions the relative elevations are reversed. Thus, Figure 4.2 indicates an endothermic reaction.

It is important to recognize that the mechanism indicated by the figures is quite different from one corresponding to an initial complete dissociation of the YZ molecule followed by combination of the X and Y atoms. This latter mechanism corresponds to movement from one valley at J up to the plateau at N and back down to the second valley at M. The activation energy for the latter mechanism would be equal to the energy required to dissociate the YZ molecule. By moving along the valley as in the mechanism indicated by the dashed line in the

figure, the system can obviously achieve reaction with significantly lower energy requirements. In physical terms the process of forming the $X–Y$ bond continuously contributes to the energy requirement for breaking the original $Y–Z$ bond.

For reactions involving more than three atoms the number of dimensions required to depict the potential energy surface exceeds human capacity for visualizing the surface. Thus it may be more convenient to consider such reactions as taking place between various moieties that play the same role as the atoms X, Y and Z in the discussion above.

The semiempirical nature of the methods used to construct multidimensional potential energy surfaces makes the quantitative validity of the results questionable. Hence the present state of the theoretical calculation of activation energies is unsatisfactory.

The development of reaction potential energy surfaces was an essential precursor to the formulation of the transition state theory. The first calculation of reaction rates in terms of a specific potential energy surface is attributed to Pelzer and Wigner (55), who studied the reaction between atomic and molecular hydrogen. Eyring and his co-workers (52–54) later developed and extended the basic concepts, leading to codification of the theory and thereby forging the close link between his name and the theory. At about the same time Evans and Polanyi (56–58) presented a somewhat similar formulation of the problem. These and other analyses of the problem lead to very similar conclusions. The chemical engineer should be familiar with these conclusions, even if he or she does not choose to become acquainted with the analytical details.

In Eyring's formulation of the problem he assumes that an "equilibrium" exists between the activated complex species and the reactant molecules. This "equilibrium" is said to exist at all times, regardless of whether or not a true chemical equilibrium has been established between the reactants and products. Although the basic assumption is not unreasonable, neither does it lend itself to simple verification. It does, however, lead to results that are consistent with those obtained by procedures that do not require one to assume equilibrium between reactants and the activated complex.

The reaction rate is then taken to be the product of the frequency at which activated complexes cross the energy barrier and their concentration *at the top of the barrier*. If $(C^{\ddagger})'$ represents the concentration *lying within a region of length $\delta^{\ddagger}$ at the top of the barrier* (see Figure 4.2) and if $\bar{v}^{\ddagger}$ is the mean speed at which molecules move from left to right across the barrier, the rate is given by

$$\bar{r} = \frac{(C^{\ddagger})'\bar{v}^{\ddagger}}{\delta^{\ddagger}} \qquad (4.3.24)$$

These activated complexes differ from ordinary molecules in that in addition to the three normal translational degrees of freedom, they have a fourth degree of translational freedom corresponding to movement along the reaction coordinate. This degree of freedom replaces one vibrational degree of freedom that would otherwise be observed.

The motion of activated complexes within the transition state may be analyzed in terms of classical or quantum mechanics. In terms of classical physics, motion along the reaction coordinate may be analyzed in terms of a one-dimensional velocity distribution function. In terms of quantum mechanics, motion along the reaction coordinate within the limits of the transition state corresponds to the traditional quantum mechanical problem involving a particle in a box.

Such analyses lead to the conclusions that

$$\bar{v}^{\ddagger} = \left[\frac{k_B T}{2\pi m^{\ddagger}}\right]^{1/2} \qquad (4.3.25)$$

and

$$(C^{\ddagger})' = (XYZ^{\ddagger})\frac{\delta^{\ddagger}}{h}(2\pi m^{\ddagger}k_B T)^{1/2} \qquad (4.3.26)$$

where $m^{\ddagger}$ is the effective mass of the activated

complex, h is Planck's constant, and $(XYZ^{\ddagger})$ is the total concentration of activated complexes. Combination of equations 4.3.24 to 4.3.26 gives

$$\bar{r} = \frac{k_B T}{h}(XYZ^{\ddagger}) \qquad (4.3.27)$$

Thus the effective frequency with which activated complexes are transformed into reaction products is $k_B T/h$. At a temperature of 300 °K, this group has a value of 6×10^{12} sec^{-1}, which is comparable in magnitude to normal molecular vibration frequencies.

If one describes the equilibrium between reactants and the activated complex by

$$K^{\ddagger} = \frac{(XYZ^{\ddagger})}{(X)(YZ)} \qquad (4.3.28)$$

then

$$\bar{r} = \frac{k_B T}{h}K^{\ddagger}(X)(YZ) \qquad (4.3.29)$$

which is a second-order form.

The problem of predicting a rate constant thus reduces to one of evaluating $K^{\ddagger}$. There are two basic approaches that are used in attacking this problem: one is based on statistical mechanics and the other on thermodynamics. From statistical mechanics it is known that for a reaction of the type $X + YZ \rightleftharpoons XYZ^{\ddagger}$ the equilibrium constant is given by

$$K^{\ddagger} = \frac{Q_{XYZ}\ e^{-E_0/RT}}{Q_X Q_{YZ}} \qquad (4.3.30)$$

where the Q's are molecular partition functions for the various species and E_0 is the energy increase accompanying the reaction at absolute zero when 1 mole of activated complex is formed. Combination of equations 4.3.29 and 4.3.30 indicates that the second-order rate constant is given by

$$k = \frac{k_B T}{h}K^{\ddagger} = \frac{k_B T}{h}\frac{Q_{XYZ^{\ddagger}}e^{-E_0/RT}}{Q_X Q_{YZ}} \qquad (4.3.31)$$

This relation indicates that the rate constant can be determined from a knowledge of the partition functions of the activated complex and the reactant species. For stable molecules or atoms the partition functions can be calculated from experimental data that do not require kinetic measurements. However, they do require that molecular constants such as the vibrational frequencies and moments of inertia be evaluated from spectroscopic data. Evaluation of the partition function for the activated complex ($Q_{XYZ\ddagger}$) presents a more difficult problem, since the required moments of inertia and vibrational frequencies cannot be determined experimentally. In principle theoretical calculations would permit one to determine moments of inertia from the various internuclear distances and vibrational frequencies from the curvature of the potential energy surface in directions normal to the reaction coordinate. In practice one would seldom (if ever) have available a sufficiently accurate potential energy surface for the reaction whose rate constant is to be determined.

Moreover, $Q_{XYZ\ddagger}$ is a partition function from which the degree of freedom corresponding to motion along the reaction coordinate has been removed. The contribution of this motion has already been taken into account in the analysis leading to equation 4.3.27. Readers who wish to acquaint themselves with the manner in which partition functions are calculated are encouraged to consult standard texts in physical chemistry and statistical mechanics.

Although equation 4.3.29 refers to a second-order reaction between an atom X and a molecule YZ, the theory is readily generalized to other reaction stoichiometries. An expression characterizing the equilibrium between reactants and the activated complex is used to eliminate the latter from equation 4.3.27, and the desired result is obtained.

Instead of formulating the reaction rate expression in terms of molecular partition functions, it is often convenient to employ an approach utilizing pseudo thermodynamic functions. From equation 4.3.29, the second-order rate constant is given by

$$k = \frac{k_B T}{h} K^{\ddagger} \qquad (4.3.32)$$

An equilibrium constant is simply related to a standard Gibbs free energy change, as indicated by equation 2.4.7.

$$\Delta G^{\ddagger} = -RT \ln K^{\ddagger} \qquad (4.3.33)$$

Hence, the reaction rate constant can be written as

$$k = \left(\frac{k_B T}{h} \right) e^{-\Delta G^{\ddagger}/RT} \qquad (4.3.34)$$

Since the factor $(k_B T/h)$ is independent of the nature of the reaction, this approach to the transition state theory argues that the free energy of activation ($\Delta G^{\ddagger}$) determines the reaction rate at a given temperature.

The free energy of activation can also be expressed in terms of an entropy and an enthalpy of activation in conventional thermodynamic fashion.

$$\Delta G^{\ddagger} = \Delta H^{\ddagger} - T \Delta S^{\ddagger} \qquad (4.3.35)$$

These quantities provide another manner of expressing the rate constant in thermodynamic terms.

$$k = \frac{k_B T}{h} e^{\Delta S^{\ddagger}/R} e^{-\Delta H^{\ddagger}/RT} \qquad (4.3.36)$$

If k is expressed in liters per mole per second, the standard state for the free energy and entropy of activation is 1 mole/liter. If the units of k are cubic centimeters per molecule per second, the corresponding standard state concentration is 1 molecule/cm^3. The magnitudes of $\Delta G^{\ddagger}$ and $\Delta S^{\ddagger}$ reflect changes in the standard state, so it is not useful to say that a particular reaction is characterized by specific numerical values of these parameters unless the standard states associated with them are clearly identified. These standard states are automatically determined by the units chosen to describe the reactant concentrations in the phenomenological rate expressions.

In terms of the collision theory a bimolecular reaction rate is written as

$$r = P_s Z_{AB} e^{E/RT} \qquad (4.3.37)$$

where P_s is the steric factor, Z_{AB} is the collision frequency, and E is the activation energy of the reaction. In terms of the thermodynamic formulation of the transition state theory a bimolecular reaction rate is given by

$$r = \frac{k_B T}{h} e^{\Delta S^{\ddagger}/R} e^{-\Delta H^{\ddagger}/RT}(A)(B) \quad (4.3.38)$$

Comparison of these expressions indicates that the activation entropy is related to the steric factor for the reaction. One may interpret the steric factor in terms of the degree of order of molecular configurations required to bring about the reaction, and this viewpoint is generally regarded as more satisfactory from an intellectual viewpoint than is that which regards P_s as an *a posteriori* correction factor necessary to obtain agreement between theory and experiment.

Experimental values of $\Delta S^{\ddagger}$ are readily calculated from measured values of reaction rate constants and activation energies. These experimental values provide significant insight into the nature of the transition state and the structure of the activated complex. Loosely bound complexes have higher entropies than tightly bound ones. A positive entropy of activation implies that the entropy of the complex is greater than that of the reactants. More often there is a decrease in entropy associated with formation of the activated complex. For bimolecular reactions the complex is formed by association of two molecules, and there is a loss of three translational and at least one rotational degree of freedom. Hence $\Delta S^{\ddagger}$ is usually negative and, in some cases, is not too different from the overall entropy change conventionally associated with the reaction. In these cases for reactions of the type $A + B \rightarrow AB$, it indicates that the activated complex closely resembles the product molecule in its structure. For a long time such reactions were regarded as abnormal because they had unusually low steric factors. However, the transition state theory clearly indicates that these low steric factors are merely a result of the large degree of order (large entropy decrease) necessary to obtain the proper molecular configuration for complex formation.

The thermodynamic formulation of the transition state theory is useful in considerations of reactions in solution when one is examining a particular class of reactions and wants to extrapolate kinetic data obtained for one reactant system to a second system in which the same function groups are thought to participate (see Section 7.4). For further discussion of the predictive applications of this approach and its limitations, consult the books by Benson (59) and Laidler (60). Laidler's kinetics text (61) and the classic by Glasstone, Laidler, and Eyring (54) contain additional useful background material.

Although the collision and transition state theories represent two important methods of attacking the theoretical calculation of reaction rates, they are not the only approaches available. Alternative methods include theories based on nonequilibrium statistical mechanics, stochastic theories, and Monte Carlo simulations of chemical dynamics. Consult the texts by Johnson (62), Laidler (60), and Benson (59) and the review by Wayne (63) for a further introduction to the theoretical aspects of reaction kinetics.

The various theories can provide useful insight into the way in which reactions occur, but we must again emphasize that they must be regarded as inadequate substitutes for experimental rate measurements. Experimental work to determine an accurate reaction rate expression is an essential prerequisite to the reactor design process.

LITERATURE CITATIONS

1. Wayne, R. P., "The Detection and Estimation of Intermediates" in *Comprehensive Chemical Kinetics, Volume I, The Practice of Kinetics*, edited by C. H. Bamford and C. F. H. Tipper, Elsevier, New York, 1969.

2. Gardiner, W. C., Jr., *Rates and Mechanisms of Chemical Reactions*, W. A. Benjamin, New York, 1969.

3. Daniels, F., and Johnston, E. H., *J. Am. Chem. Soc., 43* (53), 1921.

4. Ogg, R. A., *J. Chem. Phys.*, *15* (337), 1947; *18* (572), 1950.

5. Edwards, J. O., Greene, E. F., and Ross, J., *J. Chem. Education*, *45* (381), 1968. Adapted with permission.

6. Liebhafsky, H. A., and Mohammed, A., *J. Am. Chem. Soc.*, *55* (3977), 1933.

7. King, E. L., in *Catalysis*, edited by P. H. Emmett, Volume 2, Part 2, pp. 337–456, Reinhold, New York, 1955.

8. Gould, E. S., *Mechanism and Structure in Organic Chemistry*, p. 136, Holt, Rinehart and Winston, New York, 1959.

9. Melville, H. W., and Gowenlock, B. G., *Experimental Methods in Gas Reactions*, Macmillan, London, 1963.

10. Anderson, R. B., *Experimental Methods in Catalytic Research*, Academic Press, New York, 1970.

11. Friess, S. L., Lewis, E. S., and Weissberger, A., Editors, *Investigation of Rates and Mechanisms of Reactions*, Volume VIII, Parts I and II of *Technique of Organic Chemistry*, Second Edition, Interscience, New York, 1961, 1963.

12. Lewis, E. S. (Editor), *Investigation of Rates and Mechanisms of Reactions*, Third Edition, Volume VI, Part I of *Techniques of Chemistry*, Wiley, New York, 1974.

13. Hammes, G. G. (Editor), *Investigation of Rates and Mechanisms of Reactions*, Third Edition, Volume VI, Part II of *Techniques of Chemistry*, Wiley, New York, 1974.

14. Morris, J. C., *J. Am. Chem. Soc.*, *63* (2535), 1941; *66* (584), 1944.

15. Zemany, P. D., and Burton, M., *J. Phys. and Colloid Chem.*, *55* (949), 1951.

16. Wall, L. A., and Moore, W. J., *J. Phys. and Colloid Chem.*, *55* (965), 1951; *J. Am. Chem. Soc.*, *73* (2840), 1951.

17. Tolman, R. C., *Phys. Rev.*, *23* (699), 1924; *The Principles of Statistical Mechanics*, p. 163, Clarendon Press, Oxford, 1938.

18. Denbigh, K. G., *The Thermodynamics of the Steady State*, pp. 31–38, Methuen and Co., London, 1958. Adapted with permission.

19. Hirschfelder, J., *J. Chem. Phys.*, *9* (645), 1941.

20. Semenov, N. N., *Some Problems in Chemical Kinetics and Reactivity*, Volume I, pp. 29 and 64, translated by M. Boudart, Princeton University Press, Princeton, N.J., 1958.

21. Benson, S. W., *Thermochemical Kinetics, Methods for the Estimation of Thermochemical Data and Rate Parameters*, Second Edition, Wiley, New York, 1976.

22. Semenov, N. N., op. cit., Chapter I.

23. Laidler, K. J., *Chemical Kinetics*, Chapter 3, McGraw-Hill, New York, 1965.

24. Bodenstein, M., and Lind, S. C., *Z. Physik. Chem.*, *57* (168), 1907.

25. Christiansen, J. A., *Kgl. Danske Videnskab. Selskab., Mat.-Fys. Medd.*, *1* (14), 1919.

26. Herzfeld, K. F., *Ann. Physik*, *59* (635), 1919.

27. Polanyi, M., *Z. Elektrochem.*, *26* (50), 1920.

28. Jost, W., and Jung, G., *Z. Physik. Chem.*, *B3* (83), 1929.

29. Rice, F. O., and Herzfeld, K. F., *J. Am. Chem. Soc.*, *56* (284), 1934.

30. Goldfinger, P., Letort, M., and Niclause, M., *Contribution à l'étude de la structure moléculaire*, Victor Henri Commemorative Volume, p. 283, Desoer, Liege, 1948.

31. Laidler, op. cit., Chapters 7 and 8.

32. Frost, A. A., and Pearson, R. G., *Kinetics and Mechanism*, Second Edition, Chapter 10, Wiley, New York, 1961.

33. Benson, op. cit., Chapter XIII.

34. Dainton, F. S., *Chain Reactions*, Methuen and Co. London, 1956.

35. Kennard, E. H., *Kinetic Theory of Gases*, McGraw-Hill, New York, 1938.

36. Present, R. D., *Kinetic Theory of Gases*, McGraw-Hill, New York, 1958.

37. Guggenheim, E. A., *Elements of the Kinetic Theory of Gases*, Pergamon Press, New York, 1960.

38. Golden, S., *Elements of the Kinetic Theory of Gases*, Addison Wesley, Reading, Mass., 1964.

39. Kauzmann, W., *Kinetic Theory of Gases*, W. A. Benjamin, New York, 1966.

40. Moelwyn-Hughes, E. A., *Physical Chemistry*, p. 1109, Pergamon Press, New York, 1957.

41. Tolman, R. C., cited in Frost and Pearson, op. cit., pp. 68–69.

42. Lindemann, F. A., *Trans. Faraday Soc.*, *17* (598), 1922.

43. Benson, op. cit., Chapter 11.

44. Laidler, op. cit., pp. 143–175.

45. Wayne, op. cit., pp. 256–289.

46. Bunker, D. L., *Theory of Elementary Gas Reaction Rates*, pp. 48–74, Pergamon Press, Oxford, 1960.

47. Johnston, H. S., *Gas Phase Reaction Rate Theory*, pp. 263–297, Ronald Press, New York, 1966.

48. Kondratiev, V. N., *Chemical Kinetics of Gas Reactions*, pp. 283–321, Pergamon Press, Oxford, 1964.

49. Gardiner, op. cit., pp. 111–135.

50. Slater, N. B., *Theory of Unimolecular Reactions*, Cornell University Press, Ithaca, N. Y., 1959.

51. Hinshelwood, C. N., *The Kinetics of Chemical Change*, p. 81, Clarendon Press, Oxford, 1940.

52. Eyring, H., *J. Chem. Phys.*, *3* (107), 1935.

53. Wynne-Jones, W. F. K., and Eyring, H., *J. Chem. Phys.*, *3* (492), 1935.

54. Glasstone, S., Laidler, K. J., and Eyring, H., *The Theory of Rate Processes*, McGraw-Hill, New York, 1941.

55. Pelzer, H., and Wigner, E., *Z. Physik. Chem.*, *B15* (445), 1932.

56. Evans, M. G., and Polanyi, M., *Trans. Faraday Soc.*, *31* (875), 1935.

57. Evans, and Polanyi, op. cit., *33* (448), 1937.

58. Polanyi, M., *J. Chem. Soc.*, *1937*, 629.

59. Benson, S. W., *Thermochemical Kinetics*, Wiley, New York, 1968.

60. Laidler, K. J., *Theories of Chemical Reaction Rates*, McGraw-Hill, New York, 1969.

61. Laidler, op. cit., Chapter 3, 1965.

62. Johnston, H. S., *Gas Phase Reaction Rate Theory*, Ronald Press, New York, 1966.

63. Wayne, R. P., in *Comprehensive Chemical Kinetics*, *Volume 2*, *The Theory of Kinetics*, edited by C. H. Bamford and C. F. H. Tipper, Elsevier, Amsterdam, 1969.

PROBLEMS

1. Houser and Lee [*J. Phys. Chem.*, 71 (3422), 1967] have studied the pyrolysis of ethyl nitrate using a stirred flow reactor. They have proposed the following mechanism for the reaction.

Initiation:

$$C_2H_5ONO_2 \xrightarrow{k_1} C_2H_5O\cdot + NO_2$$

Propagation:

$$C_2H_5O\cdot \xrightarrow{k_2} CH_3\cdot + CH_2O$$

$$CH_3\cdot + C_2H_5ONO_2 \xrightarrow{k_3} CH_3NO_2 + C_2H_5O\cdot$$

Termination:

$$2C_2H_5O\cdot \xrightarrow{k_4} CH_3CHO + C_2H_5OH$$

(a) What rate expression is consistent with this mechanism?

(b) Since these investigators carried out their study in a stirred flow reactor, they were able to measure the reaction rate directly as a function of ethyl nitrate concentration. They have reported the following data.

Rate $(moles/m^3 \cdot ksec)$	Ethyl nitrate concentration $(moles/m^3)$
13.4	0.0975
12.2	0.0759
12.1	0.0713
23.0	0.2714
20.9	0.2346

Is the proposed mechanism consistent with this data? What are the order of the reaction and the rate constant that they observed experimentally?

2. Under appropriate conditions the decomposition of ethyl bromide may be explained by the following mechanism.

Initiation:

$$C_2H_5Br \xrightarrow{k_1} C_2H_5\cdot + Br\cdot \qquad E_{a,1}$$

Propagation:

$$Br\cdot + C_2H_5Br \xrightarrow{k_2} HBr + C_2H_4Br\cdot \qquad E_{a,2}$$

$$C_2H_4Br\cdot \xrightarrow{k_3} C_2H_4 + Br\cdot \qquad E_{a,3}$$

Termination:

$$Br\cdot + C_2H_4Br\cdot \xrightarrow{k_4} C_2H_4Br_2 \qquad E_{a,4}$$

(a) Derive an expression for the rate of disappearance of ethyl bromide.
(b) What is the apparent order of the reaction?
(c) What is the apparent activation energy for the reaction in terms of the activation energies of the individual steps?

Note that the amount of ethyl bromide that disappears by the initiation reaction is small compared to that which disappears via reaction 2.

3. Khan and Martell [*J. Am. Chem. Soc.*, 91 (4668), 17, 1969] have reported the results of a kinetic study of the uranyl ion catalyzed oxidation of ascorbic acid. The stoichiometric equation for this reaction may be represented as

$$H_2A + O_2 \xrightarrow{UO_2^{++}} A + H_2O_2$$

The rate expression for this reaction is of the form.

$$-\frac{d(H_2A)}{dt} = k[H_2A][H^+][O_2][UO_2^{++}]$$

Is the following mechanism consistent with the experimentally observed rate expression?

What is the resultant expression for the overall rate of consumption of oxygen in terms of the individual rate constants and the concentrations of stable species? Experimentally, the reaction kinetics follow the expression:

$$-\frac{d(O_2)}{dt} = k(CH_3CHO)^{3/2}(Co^{+++})^{1/2}$$

Note that the overall stoichiometry requires $[d(O_2)/dt] = [d(CH_3CHO)/dt]$. What assumptions are necessary to make the postulated mechanism agree with the observed kinetics?

$H_2A \rightleftharpoons HA^- + H^+$	fast, equilibrium constant K_1
$HA^- + UO_2^{++} \rightleftharpoons UO_2HA^+$	fast, equilibrium constant K_2
$UO_2HA^+ + H^+ \rightleftharpoons UO_2H_2A^{++}$	fast, equilibrium constant K_3
$UO_2H_2A^{++} + H^+ \rightleftharpoons UO_2H_3A^{+++}$	fast, equilibrium constant K_4
$UO_2H_3A^{+++} + O_2 \rightleftharpoons UO_2H_3A(O_2)^{+++}$	fast, equilibrium constant K_5
$UO_2H_3A(O_2)^{+++} \xrightarrow{k_6} UO_2H_3A\cdot(O_2\cdot)^{+++}$	slow, rate constant k_6
$UO_2H_3A\cdot(O_2\cdot)^{+++} \xrightarrow{k_7} UO_2^{++} + H_2O_2 + A + H^+$	fast, rate constant k_7

4. Havel [*Sb. Ved. Praci, Vysoka Chem. Technol., Pardubice* 1 (83), 1965] has suggested the following chain reaction mechanism for the Co^{+++} catalyzed oxidation of aldehyde to peracetic acid.

Initiation:

$$CH_3CHO + Co^{+++} \xrightarrow{k_1} CH_3CO\cdot + H^+ + Co^{++}$$

Propagation:

$$CH_3CO\cdot + O_2 \xrightarrow{k_2} CH_3CO_3\cdot$$

$$CH_3CO_3\cdot + CH_3CHO \xrightarrow{k_3} CH_3CO_3H + CH_3CO\cdot$$

Termination:

$$2CH_3CO_3\cdot \xrightarrow{k_4} \text{inactive products}$$

$$CH_3CO_3\cdot + CH_3CO\cdot \xrightarrow{k_5} \text{inactive products}$$

$$2CH_3CO\cdot \xrightarrow{k_6} \text{inactive products}$$

5. The following mechanism has been proposed for the oxidation of nitric oxide.

$$NO + O_2 \rightleftharpoons NO_3 \text{ equilibrium constant } K_c$$

$$NO_3 + NO \underset{k_2}{\overset{k_1}{\rightleftharpoons}} NO_3\cdot NO$$

$$NO_3 + NO \xrightarrow{k_3} 2NO_2$$

$$NO + NO_3\cdot NO \xrightarrow{k_4} 2NO_2 + NO$$

$$O_2 + NO_3\cdot NO \xrightarrow{k_5} 2NO + 2O_2$$

Derive a rate expression that is consistent with this mechanism. Treacy and Daniels [*J. Am. Chem. Soc.*, 77 (2033), 1955] have determined that the orders of the reaction with respect to oxygen and nitric oxide are one and two, respectively, at high pressures and less than one and greater than two at low pressures. Is the proposed mechanism consistent with this data?

6. It is thought that the fluorination of both *cis*- and *trans*-perfluorobutene-2 may proceed by the mechanism shown below.

Initiation:

$$C_2F_4 = C_2F_4 + F_2 \xrightarrow{k_1} C_2F_4 - \overset{F}{\overset{|}{\underset{\cdot}{C}_2F_4}} + F\cdot$$

Propagation:

$$C_2F_4 = C_2F_4 + F\cdot \xrightarrow{k_2} C_2F_4 - \overset{F}{\overset{|}{\underset{\cdot}{C}_2F_4}}$$

$$C_2F_4 - \overset{F}{\overset{|}{\underset{\cdot}{C}_2F_4}} + F_2 \xrightarrow{k_3} C_2F_5 - C_2F_5 + F\cdot$$

Termination:

$$2C_2F_4 - \overset{F}{\overset{|}{\underset{\cdot}{C}_2F_4}} \xrightarrow{k_4} \text{minor products}$$

What rate expression for the overall reaction is consistent with this mechanism?

7. Bond and Pinsky [*J. Am. Chem. Soc.*, 92 (32), 1970] have proposed the following mechanism for the decomposition of tetraborane.

$$B_4H_{10} \xrightarrow{k_1} B_3H_7 + BH_3$$

$$B_3H_7 \xrightarrow{k_2} BH_2 + B_2H_5$$

$$B_2H_5 \xrightarrow{k_3} BH_2 + BH_3$$

$$BH_2 + B_4H_{10} \xrightarrow{k_4} B_4H_9 + BH_3$$

$$B_4H_9 \xrightarrow{k_5} B_2H_5 + B_2H_4$$

$$B_2H_4 \xrightarrow{k_6} \text{solid} + H_2$$

$$2BH_2 \xrightarrow{k_7} B_2H_4$$

$$2BH_3 \xrightarrow{k_8} B_2H_6$$

(a) The stable products of the reaction are B_2H_6, H_2, and the solid. Derive the equation for the rate of disappearance of tetraborane.
(b) These investigators obtained the data below in an isothermal constant volume batch re-actor operating at 60 °C starting with an initial tetraborane pressure of 14.13 kPa.

Time, t (sec)	Tetraborane concentration (moles/m^3)
0	4.83
1800	3.68
3600	3.38
5400	2.91
7200	2.46
9000	2.18

What is the experimentally observed order of the reaction? Under what circumstances is the proposed mechanism consistent with the experimentally observed rate expression? The order is equal to $n(1/2)$ where $n = 0, 1, 2, 3$, or 4.

8. The following mechanism has been proposed for the oxidation of ammonia in the presence of ClO.

$$NH_3 + ClO \xrightarrow{k_1} NH_2 + HOCl$$

$$NH_2 + O_2 \xrightarrow{k_2} NO + H_2O$$

$$NH_2 + O_2 \xrightarrow{k_3} HNO + OH$$

$$2HNO \xrightarrow{k_4} H_2O + N_2O$$

(a) Derive an expression for the rate of formation of N_2O that contains only the concentrations of O_2, NH_3, and ClO and the reaction rate constants.
(b) What are the limiting cases of this expression if:
 (1) $k_2 \gg k_3$?
 (2) $k_3 \gg k_2$?
(c) Discuss the relative rates of formation of H_2O and N_2O in the two limiting cases mentioned in part (b). Be as quantitative as possible.

9. Two alternative mechanisms have been proposed to explain the formation of gaseous

phosgene ($COCl_2$) from carbon monoxide and chlorine and the decomposition of phosgene into these species.

Mechanism I:

$$Cl_2 \rightleftharpoons 2Cl \quad \text{equilibrium constant } K_1$$
$$Cl + CO \rightleftharpoons COCl \quad \text{equilibrium constant } K_2$$
$$COCl + Cl_2 \overset{k_3}{\rightarrow} COCl_2 + Cl$$
$$Cl + COCl_2 \overset{k_4}{\rightarrow} COCl + Cl_2$$

Mechanism II:

$$Cl_2 \rightleftharpoons 2Cl \quad \text{equilibrium constant } K_1$$
$$Cl + Cl_2 \rightleftharpoons Cl_3 \quad \text{equilibrium constant } K_3$$
$$Cl_3 + CO \overset{k_5}{\rightarrow} COCl_2 + Cl$$
$$Cl + COCl_2 \overset{k_6}{\rightarrow} Cl_3 + CO$$

Derive expressions for $d(COCl_2)/dt$ for both mechanisms. Can *simple* kinetic measurements be used to determine which mechanism is "correct"? If not, what might be done experimentally to determine which mechanism is preferred?

10. The following mechanism has been proposed for the thermal decomposition of acetone.

what change occurs in the order of the reaction? What does the overall activation energy become?

(d) The chain length is given by the ratio of the rate of chain propagation (reaction 3) to that for termination (reaction 5). What is the chain length in this case?

11. The gas phase *decomposition of ozone* is catalyzed by nitrogen pentoxide. Two mechanisms that purport to explain this process are given below. Can relatively simple kinetic measurements be used to eliminate one of the proposed mechanisms from further consideration? Explain.

Mechanism A:

$$N_2O_5 \underset{k_{-1}}{\overset{k_1}{\rightleftharpoons}} NO_2 + NO_3$$
$$NO_2 + O_3 \overset{k_2}{\rightarrow} NO_3 + O_2$$
$$NO_3 + O_3 \overset{k_3}{\rightarrow} 2O_2 + NO_2$$

Mechanism B:

$$N_2O_5 \underset{k_{-1}}{\overset{k_1}{\rightleftharpoons}} NO_2 + NO_3$$
$$NO_2 + O_3 \overset{k_2}{\rightarrow} NO_3 + O_2$$
$$NO_3 + NO_3 \overset{k_4}{\rightarrow} O_2 + 2NO_2$$

Initiation:	$CH_3COCH_3 \overset{k_1}{\rightarrow} CH_3\cdot + CH_3CO\cdot$	$E_a = 84 \text{ kcal}$
	$CH_3CO\cdot \overset{k_2}{\rightarrow} CH_3\cdot + CO$	$E_a = 10 \text{ kcal}$
Propagation:	$CH_3\cdot + CH_3COCH_3 \overset{k_3}{\rightarrow} CH_4 + CH_3COCH_2\cdot$	$E_a = 15 \text{ kcal}$
	$CH_3COCH_2\cdot \overset{k_4}{\rightarrow} CH_3\cdot + CH_2CO$	$E_a = 48 \text{ kcal}$
Termination:	$CH_3\cdot + CH_3COCH_2\cdot \overset{k_5}{\rightarrow} C_2H_5COCH_3$	$E_a = 5 \text{ kcal}$

(a) Express the overall rate in terms of the individual rate constants and concentrations of stable species.
(b) What is the overall activation energy?
(c) If reaction 1 is second-order, that is,

$$CH_3COCH_3 + CH_3COCH_3 \overset{k_1}{\rightarrow} CH_3COCH_3 + CH_3\cdot + CH_3CO\cdot$$

12. Experimental studies of the catalytic decomposition of ozone in the presence of nitrogen pentoxide follow a rate expression of the form

$$r = k(O_3)^{2/3}(N_2O_5)^{2/3}$$

A colleague has proposed the following mechanism as an explanation for these observations.

$$N_2O_5 + M \underset{k_2}{\overset{k_1}{\rightleftharpoons}} N_2O_5^* + M$$

$$N_2O_5^* \underset{k_4}{\overset{k_3}{\rightleftharpoons}} NO_2 + NO_3$$

$$NO_3 + O_3 \overset{k_5}{\rightarrow} NO_2 + 2O_2$$

$$NO_2 + O_3 \overset{k_6}{\rightarrow} O_2 + NO_3$$

where $N_2O_5^*$ is an excited nitrogen pentoxide molecule and M is any molecule. Show whether this mechanism is or is not consistent with the experimentally observed rate expression. If it is not, can you suggest an alternative mechanism including reactions 1 to 4 and 6 that will be consistent? (*Note*: It will not be necessary to introduce any chemical species in addition to those used above. Consider only bimolecular reactions.)

13. Garratt and Thompson [*J. Chem. Soc.*, *1934*, 524, 1817, 1822] have studied the photochemical and thermal decomposition of nickel tetracarbonyl. Later work by Day, Pearson and Basolo [*J. Am. Chem. Soc.*, 90 (6933), 1968] confirmed that the rate law postulated by Garratt and Thompson was obeyed for the homogeneous process. The mechanism postulated by both groups is:

$$Ni(CO)_4 \underset{k_{21}}{\overset{k_{12}}{\rightleftharpoons}} Ni(CO)_3 + CO$$

$$Ni(CO)_3 \overset{k_{23}}{\rightarrow} Ni + 3CO$$

(a) What rate law is consistent with this mechanism?
(b) Callear [*Proc. Roy. Soc. A265* (71), 1961] has also investigated the decomposition of $Ni(CO)_4$, but he used a flash photolysis

approach. He proposed the following mechanism.

$$Ni(CO)_4 \underset{k_{21}}{\overset{k_{12}}{\rightleftharpoons}} Ni(CO)_3 + CO$$

$$Ni(CO)_3 \underset{k_{32}}{\overset{k_{23}}{\rightleftharpoons}} Ni(CO)_2 + CO$$

$$Ni(CO)_2 \overset{k_{34}}{\rightarrow} Ni(s) + 2CO$$

Under what conditions are the two mechanisms equivalent?

14. Many polymerization reactions proceed by free radical mechanisms. We may characterize certain of these reactions by the following sequence of elementary reactions
(1) Dissociation of initiator (I) into two radical fragments
$$I \overset{k_D}{\rightarrow} 2R\cdot$$

(2) Addition of initial monomer (*M*) unit
$$R\cdot + M \overset{k_i}{\rightarrow} RM\cdot$$

(3) Chain propagation
$$RM\cdot + M \overset{k_p}{\rightarrow} RMM\cdot$$
$$RMM\cdot + M \overset{k_p}{\rightarrow} RMMM\cdot$$
$$\text{etc.}$$

(4) Chain termination
(a) Radical combination
$$R(M)_m\cdot + R(M)_n\cdot \overset{k_{tc}}{\rightarrow} R(M)_m(M)_nR$$

(b) Radical disproportionation
$$R(M)_m\cdot + R(M)_n\cdot \overset{k_{tD}}{\rightarrow} \text{saturated polymer} \\ + \text{unsaturated} \\ \text{polymer}$$

For these reactions k_p, k_{tc}, and k_{tD} are independent of the molecular weight of the radical. Show that the rate of polymerization in the absence of mass transfer limitations in a homogeneous system is given by

$$-\frac{d[M]}{dt} = k_p[M]\left(\frac{k_D f[I]}{k_{tc} + k_{tD}}\right)^{1/2}$$

where

$-\dfrac{d[M]}{dt}$ is the rate of polymerization of monomer

$[M]$ is the monomer concentration

$[I]$ is the initiator concentration

f is the fraction of initiator radicals reacting with monomer

Note that high molecular weights may be obtained in 1 to 10 sec. Also note that some of the initiator fragments are effectively excluded from consideration by reacting with impurities, etc. State explicitly any additional assumptions that you make. Are the following data for the polymerization of methyl methacrylate in benzene with azodiisobutyronitrile initiator consistent with the derived model?

$[M]$ (kmoles/m^3)	$[I]$ (moles/m^3)	$-\dfrac{d[M]}{dt}$ (moles/m^3 · sec)
9.04	0.235	0.193
8.63	0.206	0.170
7.19	0.255	0.165
6.13	0.228	0.129
4.96	0.313	0.122
4.75	0.192	0.0937
4.22	0.230	0.0867
4.17	0.581	0.130
3.26	0.245	0.0715
2.07	0.211	0.0415

15. The following mechanism has been proposed to explain the kinetics of the radiation curing process for elastomeric polyurethanes and polythio ethers.

Initiation:

$$RSH + h\nu \overset{k_1}{\to} RS\cdot + H\cdot$$

reaction rate $= k_1 I\,(RSH)$ where I is the intensity of the incident radiation

Chain transfer:

$$H\cdot + RSH \overset{k_2}{\to} H_2 + RS\cdot$$

Propagation:

$$RS\cdot + CH_2=CHR' \overset{k_3}{\to} RSCH_2-\dot{C}HR'$$

$$RSCH_2-\dot{C}HR' + RSH \overset{k_4}{\to} RS\cdot + RSCH_2CH_2R'$$

Termination:

$$2RS\cdot \overset{k_5}{\to} RSSR$$

(a) In terms of the usual Bodenstein steady-state approximation, derive an expression for the net rate of disappearance of thiol (RSH) in terms of the concentrations of stable species and the various rate constants. Reactions 3 and 4 occur at rates that are much greater than those for the other reactions.

(b) The following data have been reported for the Co^{60} irradiation of an equimolar solution of RSH and CH$_2$=CHR' using a constant dose rate.

Time, t (sec)	Thiol concentration (kmoles/m^3)
0	0.874
300	0.510
600	0.478
900	0.432
1280	0.382
1800	0.343
3710	0.223

Note that the overall stoichiometry of the propagation steps is:

$$RSH + CH_2=CHR' \to RSCH_2-CH_2R'$$

Is this data consistent with the mechanism proposed in part (a)?

16. From viscosity measurements, the diameter of the N$_2$O molecule has been found to be 0.33 nm. If the thermal decomposition of N$_2$O follows second-order kinetics at 930 °K and at

an initial pressure of 39.46 kPa, determine the collision frequency. If the activation energy for the process is 243 kJ, what is the apparent second-order reaction rate constant? The experimental rate constant is 0.0404 (m^3/mole-ksec). On the basis of this approach, what is the steric factor? Comment.

17. Rodebush and Klingelhoefer [*J. Am. Chem. Soc.*, 55 (130), 1933] have studied the reaction of atomic chlorine with molecular hydrogen. Chlorine atoms were formed by partial dissociation of molecular chlorine in an electrodeless discharge. A stream of this gas was then mixed with a hydrogen stream and passed through a thermostatted reaction vessel. At the far end of the vessel the reaction was effectively quenched by using a piece of silver foil to catalyze the recombination of chlorine atoms. The products of the reaction were determined by freezing out the Cl_2 and HCl in liquid air traps and titrating samples with standard thiosulfate and alkali respectively. On the basis of the data and assumptions listed below, determine:

(a) The average number of collisions that a chlorine atom undergoes with hydrogen molecules in the reaction vessel.

(b) The average number of HCl molecules formed per entering chlorine atom.
(c) The probability that a collision between a chlorine atom and a hydrogen molecule leads to reaction.

It may be assumed that each reaction of the type

$$Cl + H_2 \rightarrow HCl + H$$

is immediately followed by the much faster reaction:

$$H + Cl_2 \rightarrow HCl + Cl$$

Data

σ_{H_2} = 2.39 Å	σ_{Cl} = 2.97 Å
Hydrogen flow rate	6.3 cm^3 (STP)/min
Chlorine flow rate (as Cl_2)	9.1 cm^3 (STP)/min
Fraction Cl_2 dissociated	11%
Volume of reaction vessel	10 cm^3
Pressure in vessel	0.340 mm Hg
Temperature in vessel	0 °C
Length of run	10 min
Thiosulfate titre of products	36.5 cm^3 of 0.2N solution
Alkali titre of products	9.1 cm^3 of 0.1N solution

5 Chemical Systems Involving Multiple Reactions

5.0 INTRODUCTION

The chemical composition of many systems can be expressed in terms of a single reaction progress variable. However, a chemical engineer must often consider systems that cannot be adequately described in terms of a single extent of reaction. This chapter is concerned with the development of the mathematical relationships that govern the behavior of such systems. It treats reversible reactions, parallel reactions, and series reactions, first in terms of the mathematical relations that govern the behavior of such systems and then in terms of the techniques that may be used to relate the kinetic parameters of the system to the phenomena observed in the laboratory.

5.1 REVERSIBLE REACTIONS

Reversible reactions are those in which appreciable quantities of *all* reactant and product species coexist at equilibrium. For these reactions the rate that is observed in the laboratory is a reflection of the interaction between the rate at which reactant species are transformed into product molecules and the rate of the reverse transformation. The ultimate composition of the systems in which such reactions occur is dictated not by exhaustion of the limiting reagent, but by the constraints imposed by the thermodynamics of the reaction.

5.1.1 Mathematical Characterization of Simple Reversible Reaction Systems

The time dependence of the composition of a system in which a reversible reaction is occurring is governed by the mathematical form of the rate expressions for the forward and reverse reactions, the net rate of reaction being the difference between these two quantities.

$$r = \frac{1}{V}\frac{d\xi}{dt} = \vec{r} - \overleftarrow{r} \qquad (5.1.1)$$

In this section we discuss the mathematical forms of the integrated rate expression for a few simple combinations of the component rate expressions. The discussion is limited to reactions that occur isothermally in constant density systems, because this simplifies the mathematics and permits one to focus on the basic principles involved. We will again place a "V" to the right of certain equation numbers to emphasize that such equations are not general but are restricted to constant volume batch reactors. The use of the extent per unit volume in a constant volume system (ξ^*) will also serve to emphasize this restriction. For constant volume systems,

$$r = \frac{d\xi^*}{dt} = \vec{r} - \overleftarrow{r} \qquad (5.1.2)\text{V}$$

Several forms of the reaction rate expression will now be considered.

5.1.1.1 Opposing First-Order Reactions $A \underset{k_{-1}}{\overset{k_1}{\rightleftharpoons}} B$.

The simplest case of reversible reactions is that in which the forward and reverse reactions are both first order. This case may be represented by a rate expression of the form

$$r = k_1 C_A - k_{-1} C_B \qquad (5.1.3)$$

Equation 5.1.3 can be rewritten in terms of the extent of reaction per unit volume as:

$$\frac{d\xi^*}{dt} = k_1(C_{A0} - \xi^*) - k_{-1}(C_{B0} + \xi^*)$$

$$= k_1 C_{A0} - k_{-1} C_{B0} - \xi^*(k_1 + k_{-1})$$
$$(5.1.4)\text{V}$$

Separation of variables and integration subject

to the constraint that $\xi^* = 0$ at $t = 0$ gives

$$t = -\frac{1}{k_1 + k_{-1}} \ln\left[\frac{k_1 C_{A0} - k_{-1}C_{B0} - (k_1 + k_{-1})\xi^*}{k_1 C_{A0} - k_{-1}C_{B0}}\right] \qquad (5.1.5)V$$

or

$$\xi^* = \frac{k_1 C_{A0} - k_{-1}C_{B0}}{k_1 + k_{-1}}\left[1 - e^{-(k_1 + k_{-1})t}\right]$$
$$(5.1.6)V$$

Expressions for the time dependence of reactant and product concentrations may be obtained in the usual fashion

$$C_i = C_{i0} + \nu_i \xi^* \qquad (5.1.7)V$$

Equations 5.1.5 and 5.1.7 may be combined to give

$$-(k_1 + k_{-1})t = \ln\left[\frac{k_1 C_A - k_{-1}C_B}{k_1 C_{A0} - k_{-1}C_{B0}}\right]$$

$$= \ln\left[\frac{\dfrac{k_1}{k_{-1}}C_A - C_B}{\dfrac{k_1}{k_{-1}}C_{A0} - C_{B0}}\right]$$
$$(5.1.8)V$$

Equations 5.1.5, 5.1.6, and 5.1.8 are alternative methods of characterizing the progress of the reaction in time. However, for use in the analysis of kinetic data, they require an *a priori* knowledge of the ratio of k_1 to k_{-1}. To determine the individual rate constants, one must either carry out initial rate studies on both the forward and reverse reactions or know the equilibrium constant for the reaction. In the latter connection it is useful to indicate some alternative forms in which the integrated rate expressions may be rewritten using the equilibrium constant, the equilibrium extent of reaction, or equilibrium species concentrations.

From the requirement that the reaction rate must become zero at equilibrium, equation 5.1.4 indicates that the equilibrium extent of reaction per unit volume (ξ_e^*) is given by

$$\xi_e^* = \frac{k_1 C_{A0} - k_{-1}C_{B0}}{k_{-1} + k_1} \qquad (5.1.9)V$$

Equations 5.1.6 and 5.1.9 may be combined to give

$$\frac{\xi^*}{\xi_e^*} = 1 - e^{-(k_1 + k_{-1})t} \qquad (5.1.10)V$$

or

$$\ln(1 - \xi^*/\xi_e^*) = -(k_1 + k_{-1})t \qquad (5.1.11)V$$

This expression is useful in analyzing kinetic data using the technique developed in Section 3.3.3.2 for certain types of physical property measurements.

At equilibrium equation 5.1.3 becomes

$$r_e = k_1 C_{Ae} - k_{-1}C_{Be} = 0 \qquad (5.1.12)$$

or

$$\frac{k_1}{k_{-1}} = \frac{C_{Be}}{C_{Ae}} = K \qquad (5.1.13)$$

Subtraction of equation 5.1.12 from 5.1.3 gives

$$r = k_1(C_A - C_{Ae}) - k_{-1}(C_B - C_{Be}) \qquad (5.1.14)$$

or

$$\frac{d\xi^*}{dt} = k_1[(C_{A0} - \xi^*) - (C_{A0} - \xi_e^*)]$$
$$- k_{-1}[(C_{B0} + \xi^*) - (C_{B0} + \xi_e^*)]$$
$$(5.1.15)V$$

Simplification gives

$$\frac{d\xi^*}{dt} = (k_1 + k_{-1})(\xi_e^* - \xi^*) \qquad (5.1.16)V$$

Integration leads to 5.1.10. The form of this equation indicates that the reaction may be considered as first order in the departure from equilibrium, where the effective rate constant is the sum of the rate constants for the forward and reverse reactions.

Equations 5.1.13 and 5.1.8 may be combined to give other forms that are sometimes useful in the analysis of kinetic data

$$-(k_1 + k_{-1})t = \ln\left(\frac{KC_A - C_B}{KC_{A0} - C_{B0}}\right)$$

$$= \ln\left(\frac{\dfrac{C_{Be}C_A}{C_{Ae}} - C_B}{\dfrac{C_{Be}C_{A0}}{C_{Ae}} - C_{B0}}\right)$$

$$(5.1.17)V$$

In the analysis of kinetic data from reactions believed to be first-order in both directions, the equation that is most suitable for use depends on the pertinent equilibrium data available. Equations 5.1.17 and 5.1.11 are perhaps the most useful, but others may be more appropriate for use in some cases. The integrated forms permit one to determine the sum $(k_1 + k_{-1})$, while equilibrium data permit one to determine the equilibrium constant $K = k_1/k_{-1}$. Such information is sufficient to determine k_1 and k_{-1}.

5.1.1.2 Opposing Second Order Reactions

$$A + B \underset{k_r}{\overset{k_f}{\rightleftharpoons}} R + S.$$

From a mathematical standpoint the various second-order reversible reactions are quite similar, so we will consider only the most general case—a mixed second-order reaction in which the initial system contains both reactant and product species.

$$r = k_f C_A C_B - k_r C_R C_S \qquad (5.1.18)$$

In a constant volume system the rate may be written in terms of the extent per unit volume as

$$\frac{d\xi^*}{dt} = k_f(A_0 - \xi^*)(B_0 - \xi^*)$$

$$- k_r(R_0 + \xi^*)(S_0 + \xi^*) \quad (5.1.19)V$$

or

$$\frac{d\xi^*}{dt} = (k_f A_0 B_0 - k_r R_0 S_0)$$

$$- [k_f(A_0 + B_0) + k_r(R_0 + S_0)]\xi^*$$

$$+ (k_f - k_r)\xi^{*2} \qquad (5.1.20)V$$

where A_0, B_0, R_0, and S_0 refer to the initial concentrations of the various species. As in the previous section, one may separate variables and integrate to obtain an expression for ξ^* as a function of time. However, the resultant expression is of extremely limited utility for purposes of analyzing kinetic data.

A somewhat more useful approach is to consider the relations that should exist between the extent of reaction per unit volume at time t and the extent at equilibrium. At equilibrium equation 5.1.19 becomes

$$r_e = 0$$

$$= k_f(A_0 - \xi_e^*)(B_0 - \xi_e^*) - k_r(R_0 + \xi_e^*)(S_0 + \xi_e^*)$$

$$(5.1.21)V$$

This equation may be rewritten in terms of the equilibrium constant by recognizing that

$$K = \frac{k_f}{k_r} \qquad (5.1.22)V$$

$$0 = K(A_0 - \xi_e^*)(B_0 - \xi_e^*) - (R_0 + \xi_e^*)(S_0 + \xi_e^*)$$

$$(5.1.23)V$$

ξ_e^* is the root of the quadratic expression that corresponds to physical reality (i.e., it does not give rise to a reactant or product concentration that is negative). The equilibrium concentrations of all species may be evaluated using stoichiometric relations of the form

$$C_{ie} = C_{i0} + v_i\xi_e^* \qquad (5.1.24)V$$

At any time the concentrations of the various species are given by relations of the form

$$C_i = C_{i0} + v_i\xi^* \qquad (5.1.25)V$$

Combining equations 5.1.24 and 5.1.25 gives

$$C_i = C_{ie} + v_i(\xi^* - \xi_e^*) \quad (5.1.26)V$$

Substitution of expressions of this form into equation 5.1.18 gives

$$\frac{d\xi^*}{dt} = k_f[A_e - (\xi^* - \xi_e^*)][B_e - (\xi^* - \xi_e^*)]$$

$$- k_r[R_e + (\xi^* - \xi_e^*)][S_e + (\xi^* - \xi_e^*)]$$

$$(5.1.27)V$$

which may be rewritten as

$$\frac{d\xi^*}{dt} = k_f A_e B_e - k_r R_e S_e$$

$$- (\xi^* - \xi_e^*)[k_f(A_e + B_e) + k_r(R_e + S_e)]$$
$$+ (k_f - k_r)(\xi^* - \xi_e^*)^2 \qquad (5.1.28)V$$

The first two terms on the right side represent the net reaction rate at equilibrium, which must be zero. Hence,

$$\frac{d\xi^*}{dt} = -(\xi^* - \xi_e^*)[k_f(A_e + B_e) + k_r(R_e + S_e)]$$

$$+ (k_f - k_r)(\xi^* - \xi_e^*)^2 \qquad (5.1.29)V$$

Now

$$\frac{d\xi^*}{dt} = \frac{d}{dt}(\xi^* - \xi_e^*) \qquad (5.1.30)V$$

Hence, equation 5.1.29 becomes

The quantity $\xi^* - \xi_e^*$ is simply related to reactant concentrations by equation 5.1.26, so if the concentration of one reactant is known at various times, it is possible to evaluate the left side at these times. Alternatively, if the data take the form of physical property measurements of the type treated in section 3.3.3.2, equation 3.3.50 may be used to relate $\xi^* - \xi_e^*$ to the property values.

$$\xi^* - \xi_e^* = \xi_e^* \left(\frac{\xi^*}{\xi_e^*} - 1\right) = \xi_e^* \left(\frac{\lambda - \lambda_0}{\lambda_\infty - \lambda_0} - 1\right)$$

$$= \xi_e^* \left(\frac{\lambda - \lambda_\infty}{\lambda_\infty - \lambda_0}\right) \qquad (5.1.35)V$$

In either event one should be able to plot the left side of equation 5.1.34 versus time and obtain a straight line with a slope related to the rate constants for the forward and reverse reactions. Equation 5.1.22 gives another relation between these parameters, which permits us to determine each rate constant individually.

$$\frac{d(\xi^* - \xi_e^*)}{\{(k_f - k_r)(\xi^* - \xi_e^*) - [k_f(A_e + B_e) + k_r(R_e + S_e)]\}(\xi^* - \xi_e^*)} = dt \qquad (5.1.31)V$$

Integration gives

$$t = \frac{1}{k_f(A_e + B_e) + k_r(R_e + S_e)} \ln \left[\frac{\{(k_f - k_r)(\xi^* - \xi_e^*) - [k_f(A_e + B_e) + k_r(R_e + S_e)]\}\xi_e^*}{(\xi^* - \xi_e^*)\{(k_f - k_r)\xi_e^* + k_f(A_e + B_e) + k_r(R_e + S_e)\}}\right]$$

$$(5.1.32)V$$

If one again employs the relation between the equilibrium constant and the reaction rate constant (5.1.22), equation 5.1.32 can be written as

$$\ln \left[\frac{\{(K - 1)(\xi^* - \xi_e^*) - [K(A_e + B_e) + (R_e + S_e)]\}\xi_e^*}{(\xi^* - \xi_e^*)\{(K - 1)\xi_e^* + K(A_e + B_e) + (R_e + S_e)\}}\right] = [k_f(A_e + B_e) + k_r(R_e + S_e)]t$$

$$(5.1.33)V$$

From thermodynamic data it is possible to evaluate the equilibrium constant and all of the equilibrium concentrations in the above equation, reducing it to the following form.

$$\ln \left[\frac{M_1(\xi^* - \xi_e^*) + M_2}{(\xi^* - \xi_e^*)}\right] = [k_f(A_e + B_e) + k_r(R_e + S_e)]t \qquad (5.1.34)V$$

where M_1 and M_2 are known constants.

5.1.1.3 Second-Order Reaction Opposed by First-Order Reaction $A + B \underset{k_r}{\overset{k_f}{\rightleftharpoons}} R$.

For a rate expression of the form

$$r = k_f C_A C_B - k_r C_R \qquad (5.1.36)$$

the procedures employed in Sections 5.1.1.1 and 5.1.1.2 lead to the following relation for the time dependence of ξ^*.

$$[k_f(A_e + B_e) + k_r]t = \ln\left[\frac{\{K(\xi^* - \xi_e^*) - [K(A_e + B_e) + 1]\}\xi_e^*}{[K\xi_e^* + K(A_e + B_e) + 1](\xi^* - \xi_e^*)}\right] \qquad (5.1.37)V$$

where

$$\xi_e^* = \frac{K(A_0 + B_0) + 1 \pm \sqrt{K^2(A_0 - B_0)^2 + 2K(A_0 + B_0 + 2R_0) + 1}}{2K} \qquad (5.1.38)V$$

From thermodynamic data it is possible to evaluate all quantities pertaining to equilibrium in equation 5.1.37, reducing it to the form

$$[k_f(A_e + B_e) + k_r]t$$
$$= \ln\left[\frac{M_1(\xi^* - \xi_e^*) + M_2}{(\xi^* - \xi_e^*)}\right] \qquad (5.1.39)V$$

where M_1 and M_2 are numerical constants. This equation is of the same form as 5.1.34 and may be handled in the manner outlined previously.

5.1.2 Determination of Reaction Rate Expressions for Reversible Reactions

The problem of determining the mathematical form of the rate expression for a chemical reaction is one that involves a combination of careful experimental work and sound judgment in the analysis of the data obtained thereby. In many cases the analytical techniques discussed in Section 3.3 are directly applicable to studies of reversible reactions. In other cases only minor modifications are necessary.

5.1.2.1 General Techniques for the Interpretation of Reaction Rate Data for Reversible Reactions.

The determination of the mathematical form of a reaction rate expression is generally a two-step procedure. One first determines the dependence of the rate on the concentrations of the various reactant and product species at a fixed temperature and then evaluates the temperature dependence of the various rate constants appearing in the rate expression. For reversible reactions, at least two rate constants will be involved and the Arrhenius relation must be used to analyze data on the temperature dependence of each rate constant individually in order to determine the influence of temperature changes on the overall reaction rate.

Approaches to the determination of the concentration-dependent terms in expressions for reversible reactions are often based on a simplification of the expression to limiting cases. By starting with a mixture containing reactants alone and terminating the study while the reaction system is still very far from equilibrium, one may use an initial rate study to determine the concentration dependence of the forward reaction. In similar fashion one may start with mixtures containing only the reaction products and use the initial rates of the reverse reaction to determine the concentration dependence of this part of the rate expression. Additional simplifications in these initial rate studies may arise from the use of stoichiometric ratios of reactants and/or products. At other times the use of a vast excess of one or more of the reactants may give rise to simplifications. Approaches involving initial rates in which one focuses on one or the other of the two terms comprising the rate expression are the differential methods that are most appropriate for use with reversible reactions. When both terms are significant, one needs a knowledge of the numerical value of the thermodynamic equilibrium constant at the temperature in question in order to be able to employ the general differential approach outlined at the start of Section 3.3.1.1.

It is also possible to use integral methods to determine the concentration dependence of the reaction rate expression and the kinetic parameters involved. In using such approaches one again requires a knowledge of the equilibrium constant for use with one of the integrated forms developed in Section 5.1.1.

The generalized physical property approach discussed in Section 3.3.3.2 may be used together with one of the differential or integral methods, which are appropriate for use with reversible reactions. In this case the extent of reaction per unit volume at time t is given in terms of equation 3.3.50 as

$$\xi^* = \xi_e^* \left(\frac{\lambda - \lambda_0}{\lambda_\infty - \lambda_0} \right) \qquad (5.1.40)V$$

where λ, λ_0, and λ_∞ are the property measurements at times t, 0, and ∞, respectively.

Illustration 5.1 indicates how one may determine kinetic parameters for a reversible reaction.

ILLUSTRATION 5.1 DETERMINATION OF REACTION RATE EXPRESSION FOR THE REACTION BETWEEN SULFURIC ACID AND DIETHYL SULFATE

Hellin and Jungers (1) have reported the data below as being typical of the reaction between sulfuric acid and diethyl sulfate.

$$H_2SO_4 + (C_2H_5)_2SO_4 \rightleftharpoons 2C_2H_5SO_4H$$
$$A + B \rightleftharpoons 2R$$

Temperature: 22.9 °C
Initial H_2SO_4 concentration: 5.50 kmoles/m^3
Initial $(C_2H_5)_2SO_4$ concentration:
   5.50 kmoles/m^3

Time, t (sec)	$C_2H_5SO_4H$ concentration (kmoles/m^3)
0	0
1680	0.69
2880	1.38
4500	2.24
5760	2.75
7620	3.31
9720	3.81
10800	4.11
12720	4.45
16020	4.86
19080	5.15
22740	5.35
24600	5.42

After 11 days the $C_2H_5SO_4H$ concentration is approximately equal to 5.80 kmoles/m^3.

These individuals have suggested that the rate expression for this reaction is of the form

$$r = k_f[(C_2H_5)_2SO_4][H_2SO_4]$$
$$\quad - k_r[C_2H_5SO_4H]^2$$
$$= k_f(A)(B) - k_r(R)^2 \qquad (A)$$

Are the above data consistent with the proposed rate expression? If so, what are the values of the rate constants at 22.9 °C

Solution

The first thing to note is that stoichiometric quantities of reactants were used in this investigation. Because the reaction rate expression simplifies when stoichiometric quantities of reactants are used, the equations developed earlier in this chapter cannot be applied directly in the solution of this problem. Thus we will have to derive appropriate relations in the course of our analysis.

Each of the reactant concentrations is given by

$$A = B = A_0 - \xi^*$$

while the concentration of the product is given by

$$R = 2\xi^*$$

The long time data may be used to determine the equilibrium concentrations of all species and the equilibrium constant for the reaction.

$$R_e = 2\xi_e^* = 5.80 \text{ kmoles/m}^3$$

Hence

$$\xi_e^* = 2.90 \text{ kmoles/m}^3$$

and

$$A_e = B_e = 5.50 - 2.90 = 2.60 \text{ kmoles/m}^3$$

The equilibrium constant for this reaction is given by

$$K = \frac{k_f}{k_r} = \frac{R_e^2}{A_e B_e} = \frac{(5.80)^2}{(2.60)^2} = 4.98 \quad \text{(B)}$$

Hence

$$k_f = 4.98 k_r \quad \text{(C)}$$

Equations of the form of 5.1.26 may now be written for each species.

$$B = A = A_e - (\xi^* - \xi_e^*) \quad R = R_e + 2(\xi^* - \xi_e^*) \quad \text{(D)}$$

Substitution of these relations into equation A gives

$$\frac{d\xi^*}{dt} = k_f [A_e - (\xi^* - \xi_e^*)]^2 - k_r [R_e + 2(\xi^* - \xi_e^*)]^2$$

Expansion and the use of equation 5.1.30 gives

$$\frac{d(\xi^* - \xi_e^*)}{dt} = k_f A_e^2 - k_r R_e^2 - 2k_f A_e(\xi^* - \xi_e^*)$$
$$- 4k_r R_e(\xi^* - \xi_e^*)$$
$$+ k_f(\xi^* - \xi_e^*)^2 - 4k_r(\xi^* - \xi_e^*)^2$$

The first two terms on the right represent the net rate of reaction at equilibrium which must be zero. Hence

$$\frac{d(\xi^* - \xi_e^*)}{dt} = -(2k_f A_e + 4k_r R_e)(\xi^* - \xi_e^*)$$
$$+ (k_f - 4k_r)(\xi^* - \xi_e^*)^2$$

Separation of variables and integration gives

$$t = \frac{1}{2k_f A_e + 4k_r R_e} \left\{ \ln\left[\frac{(k_f - 4k_r)(\xi^* - \xi_e^*) - (2k_f A_e + 4k_r R_e)}{(\xi^* - \xi_e^*)}\right] \right.$$
$$\left. - \ln\left[\frac{(k_f - 4k_r)(-\xi_e^*) - (2k_f A_e + 4k_r R_e)}{-\xi_e^*}\right] \right\} \quad \text{(E)}$$

Combination of equations B and E and re-arrangement gives

$$t = \frac{1}{k_r(2KA_e + 4R_e)} \left\{ \ln\left(\frac{\xi_e^*}{\xi_e^* - \xi^*}\right) \right.$$
$$\left. + \ln\left[\frac{(K - 4)(\xi^* - \xi_e^*) - (2KA_e + 4R_e)}{(K - 4)(-\xi_e^*) - (2KA_e + 4R_e)}\right] \right\} \quad \text{(F)}$$

Now

$$2KA_e + 4R_e = 2(4.98)(2.60) + 4(5.80)$$
$$= 49.10 \text{ kmoles/m}^3$$

Substitution of numerical values in equation F gives

$$t = \frac{1}{49.10 k_r} \left\{ \ln\left(\frac{2.90}{2.90 - \xi^*}\right) \right.$$
$$\left. + \ln\left[\frac{(4.98 - 4)(\xi^* - 2.90) - 49.10}{(4.98 - 4)(-2.90) - 49.10}\right] \right\}$$

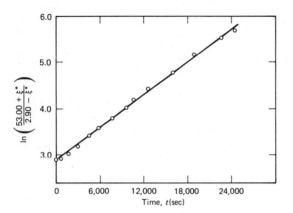

Figure 5.1

Data plot for Illustration 5.1.

or

$$49.1k_r t = \ln\left\{\frac{(2.90)}{(51.94)}\frac{(51.94 + 0.98\xi^*)}{(2.90 - \xi^*)}\right\}$$

$$= \ln\left[\frac{1}{18.28}\left(\frac{53.00 + \xi^*}{2.90 - \xi^*}\right)\right]$$

Hence a plot of $\ln[(53.00 + \xi^*)/(2.90 - \xi^*)]$ versus time should be linear if the data are to be consistent with the proposed rate expression. The experimental data are worked up in this manner and plotted in Figure 5.1.

The data appear to be consistent with the proposed rate expression. The slope of the plot is equal to $[(5.67 - 2.90)/(24000 - 0)]$, or 1.154×10^{-4} sec^{-1}. It is also equal to $49.1k_r$ kmoles/m^3. Hence

$$k_r = \frac{1.154 \times 10^{-4}}{49.1} = 2.35 \times 10^{-6} \text{ m}^3/\text{kmole·sec}$$

From equation C,

$$k_f = 4.98(2.35 \times 10^{-6})$$

$$= 1.17 \times 10^{-5} \text{ m}^3/\text{kmole·sec}$$

___

5.1.2.2 Use of Relaxation Techniques to Study Rapid Reversible Reactions. If one is interested in the kinetics of reactions that occur at very

fast rates, having half-lives of the order of a fraction of a second or less, the methods that we have previously discussed for the determination of reaction rates are no longer applicable. Instead, measurements of the response of an equilibrium system to a perturbation are used to determine its *relaxation time*. The rate at which the system approaches its new equilibrium condition is observed using special electronic techniques. From an analysis of the system behavior and the equilibrium conditions, the form of the reaction rate expression can be determined.

To illustrate how relaxation methods can be used to determine reaction rate constants, let us consider a reaction that is first-order in both the forward and reverse directions.

$$A \underset{k_{-1}}{\overset{k_1}{\rightleftharpoons}} B \qquad (5.1.41)$$

The analysis is very similar to that employed in proceeding from equation 5.1.12 to equation 5.1.16, but the physical situation is somewhat different. The reaction is first allowed to come to equilibrium with A_e and B_e representing the equilibrium concentrations of species A and B, and ξ_e^* the equilibrium extent of reaction per unit volume in a constant volume system. Under these conditions the net rate of reaction is zero.

$$r_e = \left(\frac{d\xi^*}{dt}\right)_e = 0 = k_1 A_e - k_{-1} B_e \quad (5.1.42)\text{V}$$

Now suppose that the temperature of the system is suddenly altered slightly so that it is no longer at equilibrium. The net rate of reaction is now given by

$$r = \frac{d\xi^*}{dt} = k_1 A - k_{-1} B \qquad (5.1.43)\text{V}$$

From equation 5.1.26,

$$A - A_e = -(\xi^* - \xi_e^*) \qquad (5.1.44)\text{V}$$

and

$$B - B_e = \xi^* - \xi_e^* \qquad (5.1.44a)\text{V}$$

If we denote the deviation from equilibrium conditions by

$$\Delta \xi^* = \xi^* - \xi_e^* \qquad (5.1.45)V$$

then

$$A = A_e - \Delta \xi^*, \qquad B = B_e + \Delta \xi^* \qquad (5.1.46)V$$

and

$$\frac{d\xi^*}{dt} = \frac{d \Delta \xi^*}{dt} \qquad (5.1.47)V$$

Combining equations 5.1.43, 5.1.46, and 5.1.47 gives

$$\frac{d \Delta \xi^*}{dt} = k_1(A_e - \Delta \xi^*) - k_{-1}(B_e + \Delta \xi^*)$$

$$= k_1 A_e - k_{-1} B_e - (k_1 + k_{-1}) \Delta \xi^* \qquad (5.1.48)V$$

Combination of equations 5.1.42 and 5.1.48 gives

$$\frac{d \Delta \xi^*}{dt} = -(k_1 + k_{-1}) \Delta \xi^* \qquad (5.1.49)V$$

Integration subject to the condition that $\Delta \xi^* = (\Delta \xi^*)_0$ at $t = 0$ gives

$$\ell n \left[\frac{(\Delta \xi^*)_0}{(\Delta \xi^*)} \right] = (k_1 + k_{-1})t \qquad (5.1.50)V$$

The relaxation time (t^*) for the chemical reaction is defined as the time corresponding to

$$\ell n \left[\frac{(\Delta \xi^*)_0}{(\Delta \xi^*)} \right] = 1 \qquad (5.1.51)V$$

The relaxation time is thus the time at which the distance from equilibrium has been reduced to the fraction $1/e$ of its initial value. From equation 5.1.50, it is evident that

$$t^* = \frac{1}{k_1 + k_{-1}} \qquad (5.1.52)V$$

Consequently, if one can experimentally determine the relaxation time of a system, the sum of the rate constants $(k_1 + k_{-1})$ is known. From

the equilibrium constant for the reaction one can determine the ratio of these rate constants (k_1/k_{-1}). Such information is adequate to determine the individual rate constants k_1 and k_{-1}.

For certain types of reversible reactions it is necessary to use only slight perturbations from equilibrium so that the differential equations resulting from the analysis are amenable to integration. The theoretical treatment in these cases varies slightly from that presented above and indeed varies slightly with the orders of the forward and reverse reactions. Consider the reaction represented by the following second-order mechanistic equations.

$$A + B \underset{k_{-1}}{\overset{k_1}{\rightleftharpoons}} C + D \qquad (5.1.53)$$

The reaction is assumed to occur in a constant volume system. A relaxation analysis of the type employed for the first-order reaction leads to the following analog of 5.1.49.

$$\frac{d \Delta \xi^*}{dt} = -[k_1(A_e + B_e) + k_{-1}(C_e + D_e)] \Delta \xi^*$$

$$+ (k_1 - k_{-1})(\Delta \xi^*)^2 \qquad (5.1.54)V$$

If the displacement from equilibrium is small, the term involving $(\Delta \xi^*)^2$ is small compared to the term involving $\Delta \xi^*$. This condition implies that the terms $(A_e + B_e)$ and $(C_e + D_e)$ are both very much greater in magnitude than $\Delta \xi^*$. Under these conditions equation 5.1.54 becomes

$$\frac{d \Delta \xi^*}{dt} = -[k_1(A_e + B_e) + k_{-1}(C_e + D_e)] \Delta \xi^* \qquad (5.1.55)V$$

which, after integration, becomes

$$\ell n \left[\frac{(\Delta \xi^*)_0}{(\Delta \xi^*)} \right] = [k_1(A_e + B_e) + k_{-1}(C_e + D_e)]t \qquad (5.1.56)V$$

From the basic definition of the relaxation time it is evident that

$$t^* = \frac{1}{k_1(A_e + B_e) + k_{-1}(C_e + D_e)} \qquad (5.1.57)V$$

As before, the constants k_1 and k_{-1} can be separated by making use of the fact that k_1/k_{-1} is the equilibrium constant. Alternatively, the rate constants can be separated by measuring t^* at various values of $(A_e + B_e)$ and $(C_e + D_e)$.

Chemical engineers should be aware of the existence of relaxation techniques for studies of very fast reactions. However, since relaxation time measurements call for sophisticated experimental equipment and techniques, they are seldom made outside of basic research laboratories.

5.1.3 Thermodynamic Consistency of Rate Expressions

For reversible reactions one normally assumes that the observed rate can be expressed as a difference of two terms, one pertaining to the forward reaction and the other to the reverse reaction. Thermodynamics does not require that the rate expression be restricted to two terms or that one associate individual terms with intrinsic rates for forward and reverse reactions. This section is devoted to a discussion of the limitations that thermodynamics places on reaction rate expressions. The analysis is based on the idea that at equilibrium the net rate of reaction becomes zero, a concept that dates back to the historic studies of Guldberg and Waage (2) on the law of mass action. We will consider only cases where the net rate expression consists of two terms, one for the forward direction and one for the reverse direction. Cases where the net rate expression consists of a summation of several terms are usually viewed as corresponding to reactions with two or more parallel paths linking reactants and products. One may associate a pair of terms with each parallel path and use the technique outlined below to determine the thermodynamic restrictions on the form of the concentration dependence within each pair. This type of analysis is based on the principle of detailed balancing discussed in Section 4.1.5.4.

Consider an arbitrary reaction, which we may write in the form

$$aA + bB \rightleftharpoons rR + sS \qquad (5.1.58)$$

Suppose that under conditions where the concentrations of species R and S are very small, experimental evidence indicates that the initial reaction rate is of the form

$$r_f = k[A]^\alpha[B]^\beta[R]^\rho[S]^\sigma \qquad (5.1.59)$$

where the various orders may be positive, negative, or zero. We would like to know what thermodynamics has to say about permissible forms of the concentration dependence of the reverse reaction. It is logical to expect that the form of the reverse reaction rate will also be a product of powers of concentrations, for example,

$$r_r = k'[A]^{\alpha'}[B]^{\beta'}[R]^{\rho'}[S]^{\sigma'} \qquad (5.1.60)$$

Our problem is that of determining the allowable reaction orders in the last equation.

The net rate of reaction is given by

$$r = k[A]^\alpha[B]^\beta[R]^\rho[S]^\sigma \\ - k'[A]^{\alpha'}[B]^{\beta'}[R]^{\rho'}[S]^{\sigma'} \qquad (5.1.61)$$

Hence, at equilibrium, one requires that

$$\frac{k}{k'} = [A_e]^{\alpha'-\alpha}[B_e]^{\beta'-\beta}[R_e]^{\rho'-\rho}[S_e]^{\sigma'-\sigma} \qquad (5.1.62)$$

where the subscript e's indicate any set of concentrations at which equilibrium exists. The rate constants k and k' and thus their ratio are independent of the system composition and depend only on temperature. In an ideal solution thermodynamics indicates that the equilibrium constant expressed in terms of concentrations (K_c) is also a function only of temperature. Consequently the ratio k/k' must be a function of K_c.

$$\frac{k}{k'} = f(K_c) \qquad (5.1.63)$$

Now, for equation 5.1.58, the equilibrium constant can be written as

$$K_c = \frac{[R_e]^r[S_e]^s}{[A_e]^a[B_e]^b} \qquad (5.1.64)$$

Combining equations 5.1.62 to 5.1.64 gives

$$[A_e]^{\alpha'-\alpha}[B_e]^{\beta'-\beta}[R_e]^{\rho'-\rho}[S_e]^{\sigma'-\sigma}$$
$$= f\{[A_e]^{-a}[B_e]^{-b}[R_e]^r[S_e]^s\} \quad (5.1.65)$$

This equality will be satisfied for all positive values of the concentrations of the various species if the function in question is a power function

$$[A_e]^{\alpha'-\alpha}[B_e]^{\beta'-\beta}[R_e]^{\rho'-\rho}[S_e]^{\sigma'-\sigma}$$
$$= \{[A_e]^{-a}[B_e]^{-b}[R_e]^r[S_e]^s\}^n \quad (5.1.66)$$

and if $(\alpha' - \alpha)$, $(\beta' - \beta)$, $(\rho' - \rho)$, and $(\sigma' - \sigma)$ are each the same multiple of the corresponding stoichiometric coefficient (3),

$$\frac{\alpha'-\alpha}{-a} = \frac{\beta'-\beta}{-b} = \frac{\rho'-\rho}{r} = \frac{\sigma'-\sigma}{s} = n \quad (5.1.67)$$

The exponent n may take on any *positive* value, including fractions.

In more general terms for an arbitrary reaction and rate expression the orders of the forward and reverse reactions must obey the relation

$$n = \frac{\beta_i' - \beta_i}{v_i} \qquad (5.1.68)$$

where β_i and β_i' are the orders with respect to species i for the forward reaction and reverse reaction, respectively, and v_i is the generalized stoichiometric coefficient for species i. Hence, if the concentration dependence is known for one direction, one may choose different values of n and use equation 5.1.68 to determine the orders of the opposing reaction that are consistent with this value of n. The resulting rate expressions will then be thermodynamically consistent, and the relation between the rate constants and the equilibrium constant will be given by

$$\frac{k}{k'} = K_c^n \qquad (5.1.69)$$

Obviously if one knows the complete form of the rate expression in one direction and the order of the opposing reaction with respect to one species, this information is sufficient to determine n uniquely and thus determine the complete form of the rate expression for the opposing reaction.

The following illustration indicates one application of the principle of thermodynamic consistency.

ILLUSTRATION 5.2 APPLICATION OF THE PRINCIPLE OF THERMODYNAMIC CONSISTENCY TO AN ABSORPTION PROCESS

Denbigh and Prince (4) have studied the kinetics of the absorption of gaseous NO_2 in aqueous nitric acid solutions. The following stoichiometric relation governs the process.

$$H_2O + 3NO_2(\text{or } \tfrac{3}{2} N_2O_4) \rightleftharpoons 2HNO_3 + NO$$
$$\text{(A)}$$

They postulated that the rate limiting molecular processes involved in absorption might be

$$N_2O_4 + H_2O \xrightarrow{k_1} HNO_3 + HNO_2$$

or

$$2NO_2 + H_2O \xrightarrow{k_2} HNO_3 + HNO_2$$

If both these reactions occur simultaneously in aqueous solution, it would be expected that the absorption rate could be written as

$$r_f = k_1(N_2O_4) + k_2(NO_2)^2 \qquad (B)$$

where we have incorporated the water concentration into the reaction rate constants. However, the equilibrium between NO_2 and N_2O_4 is established very rapidly so the two species concentrations are related by

$$K_1 = \frac{N_2O_4}{(NO_2)^2} \qquad (C)$$

Hence equation B may be written as

$$r_f = k_f'(N_2O_4) \qquad \text{where} \qquad k_f' = k_1 + \frac{k_2}{K_1}$$

This type of rate expression provides a good correlation of the experimental data for dilute aqueous solutions.

If it is assumed that in more concentrated solutions the rate of the forward reaction continues to follow this rate expression, what forms of the reverse rate are thermodynamically consistent in concentrated acid solution? Equilibrium is to be established with respect to equation A when written in the N_2O_4 form. It may be assumed that the dependence on NO_2 and N_2O_4 concentrations may be lumped together by equation C.

Solution

At equilibrium we require that the net rate of reaction be zero. If we postulate a net rate expression of the general power function form

$$r = k_f'[N_2O_4] - k_r'[N_2O_4]^{\alpha}[NO]^{\beta}[HNO_3]^{\gamma}$$

At equilibrium,

$$\frac{k_f'}{k_r'} = \frac{[N_2O_4]^{\alpha}[NO]^{\beta}[HNO_3]^{\gamma}}{[N_2O_4]} = K^n$$

$$= \left\{ \frac{[NO]^1[HNO_3]^2}{[N_2O_4]^{3/2}} \right\}^n$$

Thus we require that

$$\frac{\alpha - 1}{-\frac{3}{2}} = \frac{\beta}{1} = \frac{\gamma}{2} = n$$

Thermodynamically consistent forms may be obtained by choosing different values of n.

The following table indicates the reaction orders corresponding to different values of n.

n	α	β	γ
1	-1/2	1	2
2/3	0	2/3	4/3
1/2	1/4	1/2	1
0	1	0	0

On the basis of their experimental studies and a search of the literature, Denbigh and Prince indicated that the correct choice of n is $1/2$. They found that the data could be fitted by an expression of the form

$$r = k^1[N_2O_4] - C[N_2O_4]^{1/4}[NO]^{1/2}$$

where the "constant" C depends on the acid concentration. (In very concentrated solutions one must worry about the extent of dissociation of HNO_3, so some difficulties arise in the HNO_3 term.)

In addition to its constraints on the concentration dependent portions of the rate expression thermodynamics requires that the activation energies of the forward and reverse reactions be related to the enthalpy change accompanying reaction. In generalized logarithmic form equation 5.1.69 can be written as

$$\ell n\, k - \ell n\, k' = n\, \ell n\, K_c \qquad (5.1.70)$$

Differentiation with respect to reciprocal temperature and use of the Arrhenius and van't Hoff relations gives

$$E_A - E_A' = n\, \Delta H^0 \qquad (5.1.71)$$

where E_A and E_A' are the activation energies for the forward and reverse reactions, respectively and ΔH^0 is the standard enthalpy change for the reaction.

5.2 PARALLEL REACTIONS

The term *parallel reactions* describes situations in which reactants can undergo two or more reactions independently and concurrently. These reactions may be reversible or irreversible. They include cases where one or more species may react through alternative paths to give two or more different product species (simple parallel reactions),

$$A \overset{\displaystyle \nearrow C + D}{\underset{\displaystyle \searrow B}{}} \qquad (5.2.1)$$

as well as cases where one reactant may not be common to both reactions (competitive parallel reactions).

$$A \diagup \begin{matrix} \xrightarrow{+B} C \\ \xrightarrow{+B'} D \end{matrix} \qquad (5.2.2)$$

In this section we will consider the kinetic implications of both general classes of parallel reactions.

5.2.1 Mathematical Characterization of Parallel Reactions

When dealing with parallel reactions, it is necessary to describe the time-dependent behavior of the system in terms of a number of reaction progress variables equal to the number of independent reactions involved. When one tries to integrate the rate expressions to determine the time dependence of the system composition, one finds that in many cases the algebra becomes unmanageable and that closed form solutions cannot be obtained. On the other hand, it is often possible to obtain simple relations between the concentrations of the various species, thereby permitting one to determine the fractions of the original reactants transformed by each of the reactions. The technique of eliminating time as a dependent variable is an invaluable asset in the determination of these relations. It enables one to determine relative values of reaction rate constants while at the same time circumventing the necessity for obtaining a complete solution to the differential equations that describe the reaction kinetics. We will use this technique repeatedly in the sections that follow. We will start by describing the mathematical behavior of simple parallel reactions and then proceed to a discussion of competitive parallel reactions. Throughout the discussion we will restrict our analysis to constant volume systems in order to simplify the mathematics and thereby focus our attention on the fundamental principles involved.

5.2.1.1 Simple Parallel Reactions. The simplest types of parallel reactions involve the irreversible transformation of a single reactant into two or more product species through reaction paths that have the same dependence on reactant concentrations. The introduction of more than a single reactant species, of reversibility, and of parallel paths that differ in their reaction orders can complicate the analysis considerably. However, under certain conditions, it is still possible to derive useful mathematical relations to characterize the behavior of these systems. A variety of interesting cases are described in the following subsections.

5.2.1.1.1 Irreversible First-Order Parallel Reactions. Consider the irreversible decomposition of a reactant A into two sets of products by first-order reactions.

$$A \diagup \begin{matrix} \xrightarrow{k_1} R \qquad (5.2.3) \\ \xrightarrow{k_2} T \qquad (5.2.4) \end{matrix}$$

For the first reaction,

$$\frac{d\xi_1^*}{dt} = k_1(A_0 - \xi_1^* - \xi_2^*) \qquad (5.2.5)\text{V}$$

and, for the second,

$$\frac{d\xi_2^*}{dt} = k_2(A_0 - \xi_1^* - \xi_2^*) \qquad (5.2.6)\text{V}$$

where ξ_1^* and ξ_2^* are the extents per unit volume of reactions 1 and 2, respectively. These two differential equations are coupled and must be solved simultaneously. In the present case, division of the first by the second gives

$$\frac{\left(\dfrac{d\xi_1^*}{dt}\right)}{\left(\dfrac{d\xi_2^*}{dt}\right)} = \frac{d\xi_1^*}{d\xi_2^*} = \frac{k_1}{k_2} \qquad (5.2.7)\text{V}$$

Integration gives

$$\xi_1^* = \frac{k_1}{k_2} \xi_2^* \qquad (5.2.8)\text{V}$$

Combination of equations 5.2.6 and 5.2.8 gives

$$\frac{d\xi_2^*}{dt} = k_2 \left[A_0 - \left(\frac{k_1}{k_2} + 1 \right) \xi_2^* \right] \quad (5.2.9)V$$

Separation of variables and integration leads to

$$\xi_2^* = \frac{k_2}{k_2 + k_1} A_0 [1 - e^{-(k_1+k_2)t}] \quad (5.2.10)V$$

Combination of 5.2.8 and 5.2.10 gives

$$\xi_1^* = \frac{k_1 A_0}{k_2 + k_1} [1 - e^{-(k_1+k_2)t}] \quad (5.2.11)V$$

The concentrations of the various components may now be determined from basic stoichiometric principles.

$$R = R_0 + \nu_R \xi_1^*$$

$$= R_0 + \frac{k_1 A_0}{k_2 + k_1} [1 - e^{-(k_1+k_2)t}]$$

$$(5.2.12)V$$

$$T = T_0 + \nu_T \xi_2^*$$

$$= T_0 + \frac{k_2 A_0}{k_2 + k_1} [1 - e^{-(k_1+k_2)t}]$$

$$(5.2.13)V$$

$$A = A_0 + \nu_{A1}\xi_1^* + \nu_{A2}\xi_2^* = A_0[e^{-(k_1+k_2)t}]$$

$$(5.2.14)V$$

The increments in product concentrations are in constant ratio to each other, independent of time and of initial reactant concentration.

$$\frac{R - R_0}{T - T_0} = \frac{k_1}{k_2} \quad (5.2.15)V$$

This result is generally applicable to all parallel reactions in which the alternative paths have the same dependence on reactant concentrations. Also note that even in cases where the reaction is immeasurably fast, it is still possible to determine relative values of the rate constants by measuring the increments in species concentrations and using equation 5.2.15.

5.2.1.1.2 Reversible First-Order Parallel Reactions. This section extends the analysis developed in the last section to the case where the reactions are reversible. Consider the case where the forward and reverse reactions are all first-order, as indicated by the following mechanistic equations.

$$\begin{array}{c} R \\ k_1 \nearrow \quad \searrow k_{-1} \\ A \\ k_{-2} \searrow \quad \nearrow k_2 \\ S \end{array}$$

$$(5.2.16)$$

$$(5.2.17)$$

In this case the differential equation governing the first reaction may be written as

$$\frac{d\xi_1^*}{dt} = k_1(A) - k_{-1}(R)$$

$$= k_1[(A_0) - \xi_1^* - \xi_2^*] - k_{-1}[(R_0) + \xi_1^*]$$

$$(5.2.18)V$$

$$\frac{d\xi_1^*}{dt} = k_1(A_0) - k_{-1}(R_0) - (k_1 + k_{-1})\xi_1^* - k_1\xi_2^*$$

$$(5.2.19)V$$

Similarly,

$$\frac{d\xi_2^*}{dt} = k_2(A) - k_{-2}(S)$$

$$= k_2[(A_0) - \xi_1^* - \xi_2^*] - k_{-2}[(S_0) + \xi_2^*]$$

$$(5.2.20)V$$

or

$$\frac{d\xi_2^*}{dt} = k_2(A_0) - k_{-2}(S_0) - k_2\xi_1^* - (k_2 + k_{-2})\xi_2^*$$

$$(5.2.21)V$$

Elimination of time as an independent variable between equations 5.2.19 and 5.2.21 gives

$$\frac{d\xi_2^*}{d\xi_1^*} = \frac{k_2(A_0) - k_{-2}(S_0) - k_2\xi_1^* - (k_2 + k_{-2})\xi_2^*}{k_1(A_0) - k_{-1}(R_0) - (k_1 + k_{-1})\xi_1^* - k_1\xi_2^*}$$

(5.2.22)V

which is not a useful result, since this equation is not readily solvable.

In this case it is more convenient to rewrite equations 5.2.19 and 5.2.21 in terms of equilibrium concentrations and extents per unit volume. From basic stoichiometric principles

$$A = A_0 - \xi_1^* - \xi_2^*$$ (5.2.23)V

and

$$A_e = A_0 - \xi_{1e}^* - \xi_{2e}^*$$ (5.2.24)V

where the subscript e denotes an equilibrium value. These equations may be combined to give

$$A = A_e - (\xi_1^* - \xi_{1e}^*) - (\xi_2^* - \xi_{2e}^*)$$
$$= A_e - (\Delta\xi_1^*) - (\Delta\xi_2^*)$$ (5.2.25)V

where $\Delta\xi_1^*$ and $\Delta\xi_2^*$ refer to the differences between the extent per unit volume at time t and the extent at equilibrium for reactions 1 and 2, respectively.

Similarly, it may be shown that

$$R = R_e + \Delta\xi_1^* \quad \text{and} \quad S = S_e + \Delta\xi_2^*$$

(5.2.26)V

Substitution of these relations into equation 5.2.18 gives

$$\frac{d\xi_1^*}{dt} = k_1[(A_e) - \Delta\xi_1^* - \Delta\xi_2^*] - k_{-1}[(R_e) + \Delta\xi_1^*]$$

$$= k_1(A_e) - k_{-1}(R_e) - (k_1 + k_{-1})\Delta\xi_1^* - k_1\Delta\xi_2^*$$
(5.2.27)V

Now

$$\frac{d\xi^*}{dt} = \frac{d(\Delta\xi^*)}{dt}$$ (5.2.28)V

and, at equilibrium,

$$k_1 A_e = k_{-1} R_e$$ (5.2.29)

Thus

$$\frac{d(\Delta\xi_1^*)}{dt} = -(k_1 + k_{-1})\Delta\xi_1^* - k_1\Delta\xi_2^*$$

(5.2.30)V

Similarly, it can be shown that

$$\frac{d(\Delta\xi_2^*)}{dt} = -k_2\Delta\xi_1^* - (k_2 + k_{-2})\Delta\xi_2^*$$ (5.2.31)V

This pair of differential equations may be solved by differentiating the first with respect to time and eliminating $\Delta\xi_2^*$ and $d(\Delta\xi_2^*)/dt$ between the derived equation and the two original equations in order to arrive at a second-order differential equation with constant coefficients.

$$\frac{d^2\Delta\xi_1^*}{dt^2} + (k_1 + k_{-1} + k_2 + k_{-2})\frac{d(\Delta\xi_1^*)}{dt}$$

$$+ (k_1 k_{-2} + k_{-1} k_2 + k_{-1} k_{-2})\Delta\xi_1^* = 0$$

(5.2.32)V

The solution contains two exponential terms and may be written as

$$\Delta\xi_1^* = C_1 e^{m_1 t} + C_2 e^{m_2 t}$$ (5.2.33)V

where m_1 and m_2 are the roots of the quadratic equation

$$m^2 + (k_1 + k_{-1} + k_2 + k_{-2})m$$
$$+ (k_1 k_{-2} + k_{-1} k_2 + k_{-1} k_{-2}) = 0$$ (5.2.34)V

Both roots will be negative.

Now, at $t = 0$,

$$\Delta\xi_1^* = R_0 - R_e \quad \text{and} \quad \Delta\xi_2^* = S_0 - S_e$$ (5.2.35)V

These relations may be used to evaluate the constants C_1 and C_2 in equation 5.2.33. The result is

$$C_1 = \frac{1}{m_1 - m_2}[-(R_0 - R_e)(m_2 + k_1 + k_{-1}) - k_1(S_0 - S_e)] \qquad (5.2.36)\text{V}$$

and

$$C_2 = \frac{1}{m_1 - m_2}[(R_0 - R_e)(m_1 + k_1 + k_{-1}) + k_1(S_0 - S_e)] \qquad (5.2.37)\text{V}$$

Thus

$$\Delta\xi_1^* = \frac{(R_e - R_0)[(m_2 + k_1 + k_{-1})e^{m_1 t} - (m_1 + k_1 + k_{-1})e^{m_2 t}]}{m_1 - m_2}$$

$$+ \frac{(S_e - S_0)(k_1 e^{m_1 t} - k_1 e^{m_2 t})}{m_1 - m_2} \qquad (5.2.38)\text{V}$$

and

$$R = R_e + \Delta\xi_1^* = R_e + \frac{(R_e - R_0)[(m_2 + k_1 + k_{-1})e^{m_1 t} - (m_1 + k_1 + k_{-1})e^{m_2 t}]}{m_1 - m_2}$$

$$+ \frac{k_1(S_e - S_0)}{m_1 - m_2}(e^{m_1 t} - e^{m_2 t}) \qquad (5.2.39)\text{V}$$

In order to evaluate $\Delta\xi_2^*$ and hence $S(t)$ one need only substitute the relation for $\Delta\xi_1^*$ into equation 5.2.31 and solve the differential equation. The result is:

$$S = S_e + \frac{(S_e - S_0)[(m_1 + k_2 + k_{-2})e^{m_2 t} - (m_2 + k_2 + k_{-2})e^{m_1 t}]}{m_2 - m_1}$$

$$+ \frac{k_2(R_e - R_0)}{m_2 - m_1}(e^{m_2 t} - e^{m_1 t}) \qquad (5.2.40)\text{V}$$

The time dependence of the various species concentrations will depend on the relative magnitudes of the four rate constants. In some čases the curves will involve a simple exponential rise to an asymptote, as is the case for irreversible reactions. In other cases the possibility of overshoot ₍exists, as indicated in Figure 5.2. Whether or not this phenomenon will occur depends on the relative magnitudes of the rate constants and the initial conditions. However, the fact that both roots of equation 5.2.34 must be real requires that there be only one maximum in the curve for $R(t)$ or $S(t)$.

5.2.1.1.3 Higher-Order Irreversible, Simple Parallel Reactions. Many simple parallel reactions do not fit the categories discussed in the last two subsections. Of particular interest are the reactions between different chemical species to give two or more different products (e.g. the formation of otho-, meta-, and para- derivatives of an aromatic compound. This section is devoted to a discussion of the mathematical relations that govern such reactions. Consider the following two stoichiometric equations as representative of this class of reactions.

$$A + B \rightarrow R + \cdots \qquad r_1 = \frac{d\xi_1^*}{dt} = k_1(A)^{m_1}(B)^{n_1}$$

$$(5.2.41)\text{V}$$

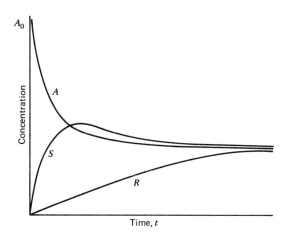

Figure 5.2
Concentration versus time curves for reversible parallel reactions indicating the possibility of a maximum in the concentration of one product species.

$$A + B \to S + \cdots \qquad r_2 = \frac{d\xi_2^*}{dt} = k_2(A)^{m_2}(B)^{n_2}$$

$$(5.2.42)V$$

The form of the solutions that one obtains for reactions of this type is dictated by the relative values of the four reaction orders. For some choices of these orders it is not possible to obtain simple closed form solutions to these rate equations unless one places additional restrictions on the initial composition of the reaction mixture. The three possible cases are discussed below.

Case I. *The Orders with Respect to Each of the Reactants Are Equal* ($m_2 = m_1$ and $n_2 = n_1$).

In this case elimination of time as an independent variable gives

$$\frac{d\xi_1^*}{d\xi_2^*} = \frac{k_1}{k_2} \qquad (5.2.43)V$$

which leads to

$$\xi_1^* = \frac{k_1}{k_2}\,\xi_2^* \qquad (5.2.44)V$$

and

$$\frac{R - R_0}{S - S_0} = \frac{k_1}{k_2} \qquad (5.2.45)V$$

The fractional yield of product R is given by

$$\frac{R - R_0}{A_0 - A} = \frac{\xi_1^*}{\xi_1^* + \xi_2^*} = \frac{k_1}{k_1 + k_2} \qquad (5.2.46)V$$

Similarly,

$$\frac{S - S_0}{A_0 - A} = \frac{k_2}{k_1 + k_2} \qquad (5.2.47)V$$

Thus the fractional yields of the different products are independent of time and of the orders of the reaction. If the increments in the products can be determined (in many cases R_0 and S_0 will be zero), then it is possible to evaluate the ratio k_1/k_2. Moreover, for this case,

$$-\frac{dA}{dt} = (k_1 + k_2)(A)^{m_1}(B)^{n_1} \qquad (5.2.48)V$$

and the methods developed in Chapter 3 may be used to evaluate m_1, n_1 and $(k_1 + k_2)$. The last result together with a knowledge of the ratio k_1/k_2 permits one to determine individual values of these parameters.

Case II. *The Orders with Respect to Each of the Constituents Are Different, but the Total Order $(m + n)$ of Each Reaction Is the Same.*

In this case elimination of time as an independent variable gives

$$\frac{d\xi_1^*}{d\xi_2^*} = \frac{k_1}{k_2}(A)^{m_1 - m_2}(B)^{n_1 - n_2} \qquad (5.2.49)V$$

or

$$\frac{d\xi_1^*}{d\xi_2^*} = \frac{k_1}{k_2}(A_0 - \xi_1^* - \xi_2^*)^{m_1 - m_2}(B_0 - \xi_1^* - \xi_2^*)^{n_1 - n_2}$$

$$(5.2.50)V$$

Only in the case where $A_0 = B_0$ is it possible to obtain a closed form solution to this differential equation. In this particular instance

$$\frac{d\xi_1^*}{d\xi_2^*} = \frac{k_1}{k_2} \qquad (5.2.51)\text{V}$$

and the equations involved in the solution become identical with those of Case I.

In the more general case, the product distribution is strongly influenced by the manner in which the reaction is carried out—in particular the manner of contacting and the type of reactor used.

Case III. *The Total Order is Not the Same for Both Reactions.*

In this case $m_1 + n_1$ differs from $m_2 + n_2$ and there are a variety of possible forms that the rate expression may take. We will consider only some of the more interesting forms. In this case elimination of time as an independent variable leads to the same general result as in the previous case (equation 5.2.50). As before, in order to obtain a closed form solution to this equation, it is convenient to restrict our consideration to a system in which $A_0 = B_0$. In this specific case equation 5.2.50 becomes

$$\frac{d\xi_1^*}{d\xi_2^*} = \frac{k_1}{k_2}(A_0 - \xi_1^* - \xi_2^*)^{m_1 + n_1 - (m_2 + n_2)} \quad (5.2.52)\text{V}$$

The relationship that must exist between ξ_1^* and ξ_2^* is most readily obtained by noting that

$$\frac{d(\xi_1^* + \xi_2^*)}{d\xi_2^*} = 1 + \frac{d\xi_1^*}{d\xi_2^*} \qquad (5.2.53)\text{V}$$

or, using equation 5.2.52,

$$\frac{d(\xi_1^* + \xi_2^*)}{d\xi_2^*} = 1 + \frac{k_1}{k_2}[A_0 - (\xi_1^* + \xi_2^*)]^{m_1 + n_1 - (m_2 + n_2)}$$

$$(5.2.54)\text{V}$$

A closed form solution to this equation exists for all values of $[(m_1 + n_1) - (m_2 + n_2)]$. However, the resultant function will depend on this difference. One case that occurs often is that in which the difference is unity.

In this case separation of variables and integration leads to

$$\xi_2^* = \frac{k_2}{k_1} \ln \left\{ \frac{1 + \dfrac{k_1}{k_2}A_0}{1 + \dfrac{k_1}{k_2}[A_0 - (\xi_1^* + \xi_2^*)]} \right\}$$

$$(5.2.55)\text{V}$$

In order to evaluate the relative values of the reaction rate constants one need only plot $(\xi_1^* + \xi_2^*)$ versus ξ_2^* and take the slope at the origin. From equation 5.2.54 this slope is equal to $1 + (k_1/k_2)A_0$ when $[m_1 + n_1 - (m_2 + n_2)] = 1$. From the slope one can determine k_1/k_2. This ratio and the relation between ξ_1^* and ξ_2^* given by equation 5.2.55 may be used with either of the original rate expressions 5.2.41 and 5.2.42 to obtain individual values of k_1 and k_2 for specified values of $m_1 + n_1$ and $m_2 + n_2$.

In this subsection we have treated a variety of higher-order simple parallel reactions. Only by the proper choice of initial conditions is it possible to obtain closed form solutions for some of the types of reaction rate expressions one is likely to encounter in engineering practice. Consequently, in efforts to determine the kinetic parameters characteristic of such systems, one should carefully choose the experimental conditions so as to ensure that potential simplifications will actually occur. These simplifications may arise from the use of stoichiometric ratios of reactants or from the degeneration of reaction orders arising from the use of a vast excess of one reactant. Such planning is particularly important in the early stages of the research when one has minimum knowledge of the system under study.

5.2.1.2 Competitive Parallel Reactions. Competitive parallel reactions are those in which two or more reactant species compete for yet another reactant; for example,

$$A \Bigg\langle \begin{matrix} \xrightarrow{+\,B} C \\ \xrightarrow{+\,B'} D \end{matrix} \qquad (5.2.56)$$

If the kinetic parameters for the upper reaction are denoted by the subscript 1 and those for the lower reaction by the subscript 2, the appropriate rate expressions for constant volume systems may be written as

$$\frac{d\xi_1^*}{dt} = k_1(A)^{m_1}(B)^{n_1}$$

$$= k_1(A_0 - \xi_1^* - \xi_2^*)^{m_1}(B_0 - \xi_1^*)^{n_1} \quad (5.2.57)V$$

and

$$\frac{d\xi_2^*}{dt} = k_2(A)^{m_2}(B')^{n_2}$$

$$= k_2(A_0 - \xi_1^* - \xi_2^*)^{m_2}(B_0' - \xi_2^*)^{n_2} \quad (5.2.58)V$$

The general relationship between ξ_1^* and ξ_2^* may be determined by eliminating time as an independent variable in the usual fashion.

$$\frac{d\xi_1^*}{d\xi_2^*} = \frac{k_1}{k_2}(A_0 - \xi_1^* - \xi_2^*)^{m_1 - m_2}\frac{(B_0 - \xi_1^*)^{n_1}}{(B_0' - \xi_2^*)^{n_2}} \quad (5.2.59)V$$

The complexity of this relation depends on the reaction orders involved. Generally, one finds that it is not easy to arrive at expressions for the time dependence of the various species concentrations. However, it is often possible to obtain relations for the relative extents of reaction that are useful for design purposes. In Chapter 9 we will see the implications of such relations in the selection of reactor type and modes of contacting.

Equation 5.2.59 simplifies greatly when $m_1 = m_2$ (i.e., both reactions are the same order with respect to A) because it is then possible to separate variables and integrate each term directly. Reactions of this type are the only ones that we will consider in more detail.

In the case where n_1 and n_2 are both unity, equation 5.2.59 becomes

$$\frac{d\xi_1^*}{d\xi_2^*} = \frac{k_1}{k_2}\frac{(B_0 - \xi_1^*)}{(B_0' - \xi_2^*)} \quad (5.2.60)V$$

Separation of variables and integration gives

$$\ln\left(\frac{B_0 - \xi_1^*}{B_0}\right) = \frac{k_1}{k_2}\ln\left(\frac{B_0' - \xi_2^*}{B_0'}\right) \quad (5.2.61)V$$

or

$$\ln\left(\frac{B}{B_0}\right) = \frac{k_1}{k_2}\ln\left(\frac{B'}{B_0'}\right) \quad (5.2.62)V$$

Thus, from simultaneous measurements of the unreacted fractions of the two competing species, one may readily determine the ratio of reaction rate constants.

In order to indicate the types of complications involved in proceeding from this point to equations that indicate the time-dependent behavior of the various species concentrations, it is instructive to consider the following rate expressions.

$$\frac{d\xi_1^*}{dt} = k_1(A)(B)$$

$$= k_1(A_0 - \xi_1^* - \xi_2^*)(B_0 - \xi_1^*) \quad (5.2.63)V$$

and

$$\frac{d\xi_2^*}{dt} = k_2(A)(B')$$

$$= k_2(A_0 - \xi_1^* - \xi_2^*)(B_0' - \xi_2^*) \quad (5.2.64)V$$

Now equation 5.2.61 can be rearranged to give the following expression.

$$\xi_2^* = B_0'\left[1 - \left(\frac{B_0 - \xi_1^*}{B_0}\right)^{k_2/k_1}\right] \quad (5.2.65)V$$

Substitution of this result in equation 5.2.63 gives

$$\frac{d\xi_1^*}{dt} = k_1\left\{A_0 - \xi_1^* - B_0'\left[1 - \left(\frac{B_0 - \xi_1^*}{B_0}\right)^{k_2/k_1}\right]\right\}(B_0 - \xi_1^*) \quad (5.2.66)V$$

In general this equation must be solved using numerical methods. Once ξ_1^* has been related to time in such fashion, equation 5.2.65 may be used to evaluate ξ_2^* as a function of time. Basic stoichiometric principles may then be used to determine the corresponding concentrations of the various product and reactant species.

When $m_1 = m_2$ and n_1 and n_2 are both two, equation 5.2.59 can be integrated to give

$$\frac{1}{B_0 - \xi_1^*} - \frac{1}{B_0} = \frac{k_1}{k_2}\left(\frac{1}{B_0' - \xi_2^*} - \frac{1}{B_0'}\right)$$

(5.2.67)V

or

$$\frac{1}{B} - \frac{1}{B_0} = \frac{k_1}{k_2}\left(\frac{1}{B'} - \frac{1}{B_0'}\right) \quad (5.2.68)\text{V}$$

When m_1 equals m_2 and n_1 and n_2 are one and two, respectively, integration of equation 5.2.59 gives

$$\ell n\left(\frac{B_0}{B}\right) = \frac{k_1}{k_2}\left(\frac{1}{B'} - \frac{1}{B_0'}\right) \quad (5.2.69)\text{V}$$

Equations 5.2.62, 5.2.68, and 5.2.69 indicate that simultaneous measurements of the concentrations of the competitive species permit one to determine the relative values of the two rate constants using plots appropriate to the rate expressions in question.

5.2.2 Techniques for the Interpretation of Kinetic Data in the Presence of Parallel Reactions

In general an analysis of a system in which noncompetitive parallel reactions are taking place is considerably more difficult than analyses of the type discussed in Chapter 3. In dealing with parallel reactions one must deal with the problems of determining reaction orders and rate constants for each of the individual reactions. The chemical engineer must be careful both in planning the experiment and in analyzing the data so as to obtain values of the kinetic constants that are sufficiently accurate for purposes of reactor design.

The first point to be established in any experimental study is that one is dealing with parallel reactions and not with reactions between the products and the original reactants or with one another. One then uses data on the product distribution to determine relative values of the rate constants, employing the relations developed in Section 5.2.1. For simple parallel reactions one then uses either the differential or integral methods developed in Section 3.3 in analysis of the data.

There are few short-cut methods for analyzing simple parallel systems. One useful technique, however, is to use stoichiometric ratios of reactants so that the ratio of the time derivatives of the extents of reaction simplifies where possible. For higher-order irreversible simple parallel reactions represented by equations 5.2.41 and 5.2.42, the degenerate form of the ratio of reaction rates becomes

$$\frac{d\xi_1^*}{d\xi_2^*} = \frac{k_1}{k_2}[A]^{m_1 + n_1 - (m_2 + n_2)} \quad (5.2.70)\text{V}$$

For these conditions the product distribution is time independent when the overall reaction orders are identical. However, when $(m_1 + n_1) > (m_2 + n_2)$, the product distribution changes as time proceeds with the ratio of the product formed by reaction 1 to that formed by reaction 2, declining as the reaction proceeds. Conversely, if $(m_2 + n_2) > (m_1 + n_1)$, this ratio will increase. Experiments of this type can provide extremely useful clues as to the form of the reaction rate expression and as to the type of experiments that should be performed next. Another technique that is often useful in studies of these systems is the use of a large excess of one reactant so as to cause a degeneration of the concentration dependent term to a simpler form.

In the case of competitive parallel reactions one has the option of studying each reaction independently by varying the composition of the initial reaction mixture. If a chemical species is absent, it obviously cannot react. What is more interesting from the viewpoint of the kineticist is the possibility of using reactions of this type to gain information about complex reaction systems. If one reaction is well characterized on the basis of previous work and if another reaction is very rapid and not amenable to investigation by conventional techniques, comparative rate studies can be very useful. In such cases it is convenient to "starve" the system with respect to the species for which other reactants compete by providing less than a stoichiometric amount of this species for any of the reactions involved. One lets the competitive parallel reactions go to completion and analyzes the resultant product mixture. Since the equations resulting from the elimination of time as an independent variable are applicable to the final product mixture, compositions of such mixtures may be used to determine relative values of the rate constants involved.

Relative values of the rate constants are useful in themselves; by measuring such values for the reactions of a series of compounds with the same reactant, one is able to determine the rank order of reactivity within the series. Such determinations are useful in the development of correlations of the effects of substituent groups on the

value that permits one to convert the relative rate constants into absolute values for each parameter.

Illustrations 5.3 and 5.4 indicate how one utilizes the concepts developed in this section in the determination of kinetic parameters for competitive parallel reactions.

ILLUSTRATION 5.3 DETERMINATION OF RELATIVE RATE CONSTANTS FOR COMPETITIVE PARALLEL SECOND-ORDER REACTIONS (FIRST-ORDER IN EACH SPECIES)

When a mixture of benzene (B) and benzyl chloride (C) are reacted with nitric acid (A) in acetic anhydride solution, the products are nitrobenzene (NB) and nitrobenzyl chloride (NC).

$$A + B \xrightarrow{k_1} NB + H_2O \qquad r_1 = k_1(A)(B) \quad (A)$$

$$A + C \xrightarrow{k_2} NC + H_2O \qquad r_2 = k_2(A)(C) \quad (B)$$

The data below are typical of this reaction at 25 °C when both B and C have initial concentrations of 1 kmole/m^3 (5). The data were recorded after very long times so all the nitric acid present initially was consumed. What is the value of k_2/k_1?

Run	Initial HNO$_3$ (kmoles/m^3)	Nitrobenzene concentration after complete reaction (kmoles/m^3)	Nitrobenzyl chloride concentration after complete reaction (kmoles/m^3)
1	0.228	0.172	0.056
2	0.315	0.235	0.080
3	0.343	0.257	0.086
4	0.411	0.307	0.104
5	0.508	0.376	0.132

rates of a given class of reactions. By measuring a series of rate constant ratios, one eventually is able to arrive at one reaction that is amenable to investigation by conventional procedures. A study of this reaction provides the key numerical

Solution

These reactions are competitive parallel reactions that are each first-order in the competitive species. Equation 5.2.61 is applicable.

$$\ln\left(\frac{B_0 - \xi_1^*}{B_0}\right) = \frac{k_1}{k_2}\ln\left(\frac{C_0 - \xi_2^*}{C_0}\right) \quad \text{(C)}$$

As originally derived, equation 5.2.61 is not restricted to use at infinite time (complete reaction) but, if one applies it at this time, he or she must be careful that species A is not in stoichiometric excess with respect to either reaction. In such a case either B or C would be completely consumed at some point, thereby invalidating the analysis.

Rearrangement of equation C gives:

$$\ln[1 - (\xi_1^*/B_0)] = \frac{k_1}{k_2}\ln[1 - (\xi_2^*/C_0)] \quad \text{(D)}$$

From the reaction stoichiometry ξ_1^* is numerically equal to the nitrobenzene concentration and ξ_2^* is equal to the nitrobenzyl chloride concentration. Since B_0 and C_0 are each unity, equation D may be written as

$$\ln(1 - NB) = \left(\frac{k_1}{k_2}\right)\ln(1 - NC)$$

We will use a numerical averaging procedure to evaluate the ratio of rate constants. Thus

Run	$1 - NB$	$1 - NC$	k_1/k_2
1	0.828	0.944	3.28
2	0.765	0.920	3.21
3	0.743	0.914	3.30
4	0.693	0.896	3.34
5	0.624	0.868	3.33
			Average 3.29 ± 0.05

ILLUSTRATION 5.4 DETERMINATION OF REACTION RATE CONSTANTS FOR COMPETITIVE DIELS-ALDER REACTIONS

Pannetier and Souchay (6) have indicated that when an equimolar gaseous mixture of butadiene and acrolein is allowed to react for 40 min at 330°C and 1 atm, 78.7% of the initial butadiene and 64.9% of the acrolein have been consumed.

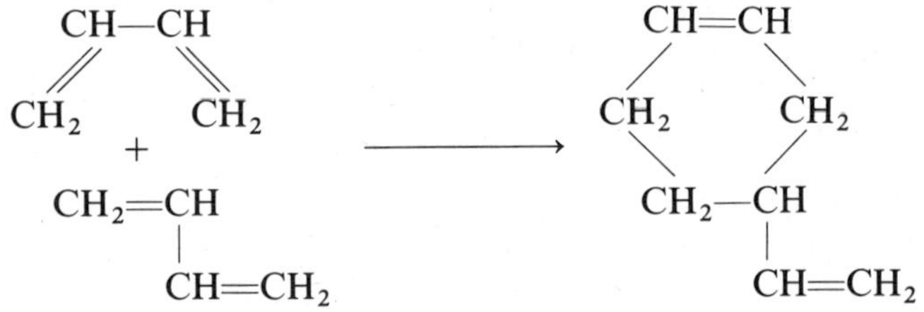

$$(2A \to C) \qquad r_1 = k_1 C_A^2 \quad \text{(A)}$$

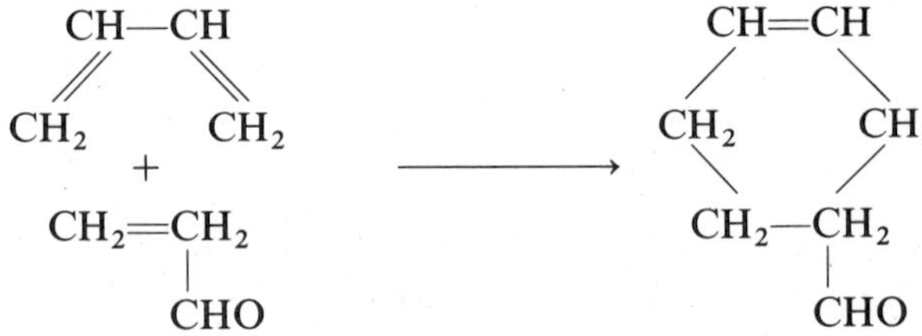

$$(A + B \to D) \qquad r_2 = k_2 C_A C_B \quad \text{(B)}$$

For the conditions cited, what are the numerical values of the two rate constants?

Solution

The form of equations A and B differs from the competitive parallel reactions considered in Section 5.2.1.2. It will be necessary to derive appropriate equations for use in the course of our analysis.

From the indicated data and basic stoichiometric principles,

$$C_A = C_{A0}(1 - 0.787) = C_{A0} - 2\xi_1^* - \xi_2^* \tag{C}$$

and

$$C_B = C_{B0}(1 - 0.649) = C_{B0} - \xi_2^* = C_{A0} - \xi_2^* \tag{D}$$

where we have noted that $C_{A0} = C_{B0}$. Hence

$$\xi_2^* = 0.649 C_{A0}$$

and

$$\xi_1^* = [(0.787 - 0.649)/2]C_{A0} = 0.069 C_{A0}.$$

In the present case elimination of time as an independent variable between equations A and B gives

$$\frac{d\xi_1^*}{d\xi_2^*} = \frac{k_1 C_A^2}{k_2 C_A C_B} = \frac{k_1 C_A}{k_2 C_B}$$

or, using equations C and D,

$$\frac{d\xi_1^*}{d\xi_2^*} = \frac{k_1}{k_2} \frac{(C_{A0} - 2\xi_1^* - \xi_2^*)}{C_{A0} - \xi_2^*}$$

$$= \frac{k_1}{k_2}\left(1 - \frac{2\xi_1^*}{C_{A0} - \xi_2^*}\right)$$

The product distribution thus changes as the reaction proceeds.

Rearrangement gives

$$\frac{d\xi_1^*}{d(C_{A0} - \xi_2^*)} - \frac{2k_1}{k_2}\left(\frac{\xi_1^*}{C_{A0} - \xi_2^*}\right) = -\frac{k_1}{k_2}$$

If we let $x = C_{A0} - \xi_2^*$ and $r = k_1/k_2$.

$$\frac{d\xi_1^*}{dx} - 2r\frac{\xi_1^*}{x} = -r$$

This first-order linear differential equation may be solved using an integrating factor approach to give

$$\xi_1^* = \frac{r}{2r - 1}\left[x - C_{A0}\left(\frac{x}{C_{A0}}\right)^{2r}\right]$$

which can be rewritten as

$$\frac{\xi_1^*}{C_{A0}} = \frac{r}{2r - 1}\left[1 - \frac{\xi_2^*}{C_{A0}} - \left(1 - \frac{\xi_2^*}{C_{A0}}\right)^{2r}\right] \tag{E}$$

This transcendental equation must be solved for r using the indicated numerical values and recognizing that $0 < r < 1$ for the product distribution cited.

$$0.069 = \frac{r}{2r - 1}[1 - 0.649 - (1 - 0.649)^{2r}]$$

The result is

$$r = \frac{k_1}{k_2} = 0.123$$

In order to evaluate each of the rate constants individually it is necessary to obtain another relation between k_1 and k_2. This will involve integration of one of the rate expressions. From equations B to D,

$$\frac{d\xi_2^*}{dt} = k_2(C_{A0} - 2\xi_1^* - \xi_2^*)(C_{A0} - \xi_2^*)$$

Rearrangement and integration gives

$$k_2 C_{A0} t = \int_0^{0.649} \frac{d\left(\frac{\xi_2^*}{C_{A0}}\right)}{\left(1 - 2\frac{\xi_1^*}{C_{A0}} - \frac{\xi_2^*}{C_{A0}}\right)\left(1 - \frac{\xi_2^*}{C_{A0}}\right)} \tag{F}$$

Equation E relates ξ_1^* and ξ_2^*, so ξ_1^* may be evaluated at various values of ξ_2^* using the value of r determined above.

$$\frac{\xi_1^*}{C_{A0}} = \frac{(0.123)}{2(0.123) - 1}\left(1 - \frac{\xi_2^*}{C_{A0}}\right)\left[1 - \left(1 - \frac{\xi_2^*}{C_{A0}}\right)^{2(0.123) - 1}\right]$$

$$= 0.163\left(1 - \frac{\xi_2^*}{C_{A0}}\right)\left[\left(1 - \frac{\xi_2^*}{C_{A0}}\right)^{-0.754} - 1\right]$$

Using the values of ξ_1^*/C_{A0} determined from this relation, it is possible to integrate equation F numerically. The result is

$$k_2 C_{A0} t = 2.37$$

The initial reactant concentration may be determined from the ideal gas law.

$$C_{A0} = \frac{y_A P}{RT} = \frac{(0.5)(1)}{(0.082)(603)}$$

$$= 1.01 \times 10^{-2} \text{ moles/liter}$$

Thus,

$$k_2 = \frac{2.37}{(1.01 \times 10^{-2})(40)}$$

$$= 5.86 \text{ liters/mole-minute}$$

Consequently,

$$k_1 = \left(\frac{k_1}{k_2}\right)k_2 = (0.123)(5.86)$$

$$= 0.72 \text{ liters/(mole-minute)}$$

These values of the reaction rate constants differ from those cited by Pannetier and Souchay (6) because these individuals erroneously treated the two reactions as if they were of the simple parallel type instead of as if there were a competition between the acrolein and butadiene molecules for other butadiene molecules.

5.3 SERIES OR CONSECUTIVE REACTIONS—IRREVERSIBLE

The term "series reactions" refers to those reactions in which one or more of the products formed initially undergoes a subsequent reaction to give yet another product. Significant amounts of both the intermediate and the final product species will be present during the normal course of the reaction. The general scheme for these reactions may be represented as

$$[A \cdots] \to [B \cdots] \to [C \cdots] \quad (5.3.1)$$

where the quantities in brackets may denote more than one molecular species. Among the many general types of reactions that fall into this category are those leading to mono-, di-, tri-, etc. substituted products, sequential partial oxidation reactions, multiple cracking reactions, and polymerization reactions.

This section discusses the kinetic implications of series reactions. We will be concerned only with those cases where the progress of the various stages of the overall transformation is not influenced by either parallel or reverse reactions. The discussion will again be limited to constant volume systems.

5.3.1 Mathematical Characterization of Series Reactions

5.3.1.1 Consecutive First-Order Reactions.

The simplest case of series reactions is that in which every reaction in the sequence obeys first-order kinetics. It may be represented in terms of the following sequence of mechanistic equations.

$$A \overset{k_1}{\to} B \overset{k_2}{\to} C \overset{k_3}{\to} D \overset{k_4}{\to} \cdots \quad (5.3.2)$$

The classic example of "reactions" of this type is a sequence of radioactive decay processes that result in nuclear transformations. The differential equations that govern kinetic systems of this type are most readily solved by working in terms of concentration derivatives. For the first reaction,

$$\frac{dA}{dt} = -k_1 A \quad (5.3.3)\text{V}$$

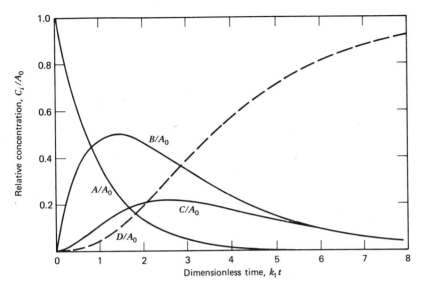

Figure 5.3
Time dependence of species concentrations for the consecutive reactions
$A \xrightarrow{k_1} B \xrightarrow{k_2} C \xrightarrow{k_3} D$ for $k_2 = 0.5\, k_1$ and $k_3 = 0.9\, k_1$.

Thus

$$A = A_0 e^{-k_1 t} \qquad (5.3.4)V$$

where A_0 is the initial concentration of reactant A. The concentration of A decreases with time in a fashion that is not influenced by subsequent reactions.

For species B,

$$\frac{dB}{dt} = k_1 A - k_2 B$$

$$= k_1 A_0 e^{-k_1 t} - k_2 B \qquad (5.3.5)V$$

This equation is linear first-order and may be solved in a variety of fashions. One may use an integrating factor approach, Laplace transforms, or rearrange the equation and obtain the sum of the homogeneous and particular solutions. The solution is

$$B = B_0 e^{-k_2 t} + \frac{k_1 A_0}{k_2 - k_1}(e^{-k_1 t} - e^{-k_2 t}) \qquad (5.3.6)V$$

The form of the equation for B is dependent on the relative values of the reaction rate constants and the initial concentrations of species A

and B. Figure 5.3 indicates the type of behavior to be expected for the case where $B_0 = 0$, while Figure 5.4 indicates two modes of behavior for $B_0 = 0.5 A_0$.

The time corresponding to a maximum concentration of species B may be found by differentiating the last equation with respect to

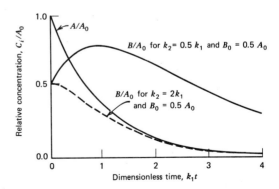

Figure 5.4
Schematic representation of the time dependence of the concentration of the first intermediate in a series of first-order reactions. Initial intermediate concentration is nonzero.

time and setting the derivative equal to zero. The result is

$$t_{max} = \frac{1}{k_2 - k_1} \ell n \left[\frac{k_2}{k_1} \left(1 + \frac{B_0}{A_0} - \frac{k_2 B_0}{k_1 A_0} \right) \right]$$

$$(5.3.7)V$$

The conditions under which the maximum can exist may be determined by examination of the initial slope. When it is positive, a maximum will occur; if it is negative, there will be no maximum. From equation 5.3.5,

$$\left(\frac{dB}{dt} \right)_0 = k_1 A_0 - k_2 B_0 \qquad (5.3.8)V$$

Hence there will be a maximum when $k_1 A_0 > k_2 B_0$. When $k_2 B_0 > k_1 A_0$, the curve for B starts out with a negative slope and decreases monotonically to zero at infinite time, as shown in Figure 5.4.

In order to arrive at an equation for the time dependence of the concentration of species C, one may proceed in a similar fashion. The result is

One may clearly extend the technique to include as many reactions as desired. The irreversibility of the reactions permits one to solve the rate expressions one at a time in recursive fashion. If the first reaction alone is other than first-order, one may still proceed to solve the system of equations in this fashion once the initial equation has been solved to determine $A(t)$. However, if any reaction other than the first is not first-order, one must generally resort to numerical methods to obtain a solution.

When one is dealing with a finite series of reactions, it is possible to use stoichiometric principles to determine the concentration of the final species. For example, if only four species (A, B, C, D) are involved in the sequence of equation 5.3.2, then

$$A_0 + B_0 + C_0 + D_0 = A + B + C + D$$

$$(5.3.10)V$$

The concentration of species D at a particular time may then be determined using this relation and the equations for species $A, B,$ and C derived above.

$$C = \begin{cases} C_0 e^{-k_3 t} \\ + B_0 \left(\dfrac{k_2 e^{-k_2 t}}{k_3 - k_2} - \dfrac{k_2 e^{-k_3 t}}{k_3 - k_2} \right) \\ + A_0 \left[\dfrac{k_1 k_2 e^{-k_1 t}}{(k_3 - k_1)(k_2 - k_1)} - \dfrac{k_1 k_2 e^{-k_2 t}}{(k_3 - k_2)(k_2 - k_1)} + \dfrac{k_1 k_2 e^{-k_3 t}}{(k_3 - k_2)(k_3 - k_1)} \right] \end{cases} \qquad (5.3.9)V$$

Obviously the curve depicting the time-dependent behavior of the concentration of species C can take on even more forms than that for species B. The shape of the curve is dependent on the initial concentrations of the various species and the three reaction rate constants. Figure 5.3 depicts the time-dependent behavior for the specific case where only species A is present initially.

5.3.1.2 Consecutive Reactions that are other than First-Order. For consecutive reactions that are not first-order, closed form analytical solutions do not generally exist. This situation is a consequence of the nonlinearity of the set of differential equations involving the time derivatives of the various species concentrations. A few two-member sequences have been analyzed. Unfortunately, the few cases that have been analyzed are seldom encountered in industrial practice. For the most part one must resort to approximate methods or to initial conditions that cause a degeneration of the reaction rate expression to a more simple form (e.g., using a stoichiometric ratio of reactants or a large excess of one species).

One general technique that is often useful in efforts to analyze the behavior of these systems is the elimination of time as an independent variable.

5.3.2 Techniques for the Interpretation of Kinetic Data in the Presence of Series Reactions

The first point that must be established in an experimental study is that one is indeed dealing with a series combination of reactions instead of with some other complex reaction scheme. One technique that is particularly useful in efforts of this type is the introduction of a species that is thought to be a stable intermediate in the reaction sequence. Subsequent changes in the dynamic behavior of the reaction system (or lack thereof) can provide useful information about the character of the reactions involved.

If one monitors the rate of disappearance of the original reactant species the general differential and integral approaches outlined in Section 3.3 may be used to determine the rate expression for the initial reaction in the sequence. Once this expression is known one of several other methods for determining either absolute or relative values of the rate constants for subsequent reactions may be used.

If the data are sufficiently accurate, one may use a general differential approach in which the expression for the net rate of formation of a stable intermediate is postulated and tested against experimental data. The difference between the rate of formation by the initial reaction for which the kinetics are known and the actual net rate of formation is tested against the proposed rate expression for the reaction responsi-

ble for the destruction of the intermediate. The process involves taking differences in the slopes of two concentration versus time curves (or a simple combination thereof). Since these differences may be very imprecise, this method is often inappropriate for use.

In some cases one is able to start with the stable intermediate and determine the rate at which it disappears using conventional methods.

For first-order and pseudo first-order reactions of the series type several methods exist for determining ratios of rate constants. We will consider a quick estimation technique and then describe a more accurate method for handling systems whose kinetics are represented by equation 5.3.2.

The first approach is based on equation 5.3.7 and makes use of the fact that the time at which the maximum concentration of the intermediate B is reached is a function only of the two rate constants and initial concentrations. For the case where no B is present initially, equations 5.3.7 and 5.3.6 can be written as

$$k_1 t_{max} = \frac{1}{\kappa - 1} \ln \kappa \qquad (5.3.11)V$$

and

$$\frac{B_{max}}{A_0} = \frac{1}{\kappa - 1} (e^{-k_1 t_{max}} - e^{-k_2 t_{max}}) \qquad (5.3.12)V$$

where

$$\kappa = \frac{k_2}{k_1} \qquad (5.3.13)$$

Combination of these relations and further algebraic manipulation gives

$$\frac{B_{max}}{A_0} = \kappa^{\kappa/(1 - \kappa)} \qquad (5.3.14)V$$

This equation indicates that the maximum becomes more pronounced as k_2/k_1 becomes smaller. If k_1 is known and the maximum concentration of intermediate is measured, k_2 may be determined from this equation. If t_{max} is known sufficiently accurately, equation 5.3.14 may be used to determine κ and 5.3.11 to determine k_1. For meaningful results this approach requires that one accurately determine the maximum concentration of the intermediate species B and the time at which this maximum is reached.

Swain (7) has discussed the general problem of determining rate constants from experimental data of this type and some of the limitations of numerical curve-fitting procedures. He suggests that a reaction progress variable for two consecutive reactions like 5.3.2 be defined as

$$\delta^* = \frac{B + 2C}{A_0} = \frac{B + 2(A_0 - A - B)}{A_0} \quad (5.3.15)\text{V}$$

so that $50\,\delta^*$ is a measure of the percent of the total irreversible reaction that has taken place (i.e., $\delta^* = 2$ corresponds to complete conversion to C). Equations 5.3.4, 5.3.6, and 5.3.13 can be

manipulated to show that

$$50\,\delta^* = 50\left[2 + \frac{(2\kappa - 1)e^{-k_1 t}}{1 - \kappa} - \frac{e^{-\kappa k_1 t}}{1 - \kappa}\right]$$

$$(5.3.16)\text{V}$$

If we let t_1 and t_2 represent the times corresponding to reaction progress variables δ_1^* and δ_2^*, respectively, the time ratio t_2/t_1 for fixed values of δ_1^* and δ_2^* will depend only on the ratio of rate constants κ. One may readily prepare a table or graph of δ^* versus $k_1 t$ for fixed κ and then cross-plot or cross-tabulate the data to obtain the relation between κ and $k_1 t$ at a fixed value of δ^*. Table 5.1 is of this type. At specified values of δ_1^* and δ_2^* one may compute the difference $\log(k_1 t)_2 - \log(k_1 t)_1$, which is identical with $\log t_2 - \log t_1$. One then enters the table using experimental values of t_2 and t_1 and reads off the value of $\kappa = k_2/k_1$. One application of this time-ratio method is given in Illustration 5.5.

Table 5.1
Series First-Order Reactions. Time-Percentage Reaction Relations for Various Relative Rate Constants

κ	$(k_1 t)_{15}$	$(k_1 t)_{35}$	$(k_1 t)_{70}$	$\log(t_{35}/t_{15})$	$\log(t_{70}/t_{15})$	$\log(t_{70}/t_{35})$
100	0.168	0.436	1.21	0.415	0.858	0.443
50	0.172	0.441	1.21	0.407	0.847	0.440
20	0.188	0.457	1.23	0.385	0.815	0.430
10	0.209	0.484	1.26	0.366	0.781	0.415
5	0.236	0.536	1.32	0.356	0.748	0.392
2	0.277	0.664	1.54	0.367	0.745	0.378
1.5	0.289	0.686	1.65	0.376	0.757	0.381
1.1	0.300	0.734	1.80	0.388	0.779	0.391
0.9	0.308	0.766	1.93	0.395	0.796	0.401
0.7	0.315	0.806	2.11	0.409	0.826	0.417
0.5	0.324	0.863	2.41	0.425	0.871	0.446
0.2	0.342	0.999	3.81	0.465	1.047	0.582
0.1	0.349	1.078	6.19	0.490	1.249	0.759
0.05	0.353	1.132	11.10	0.506	1.497	0.991
0.02	0.355	1.173	26.55	0.519	1.874	1.355
0.01	0.356	1.188	52.09	0.524	2.166	1.642

ILLUSTRATION 5.5 DETERMINATION OF RELATIVE RATE CONSTANTS USING THE TIME-RATIO METHOD

The hydrolysis of 2,7-dicyanonaphthalene has been studied by Kaufler (9).

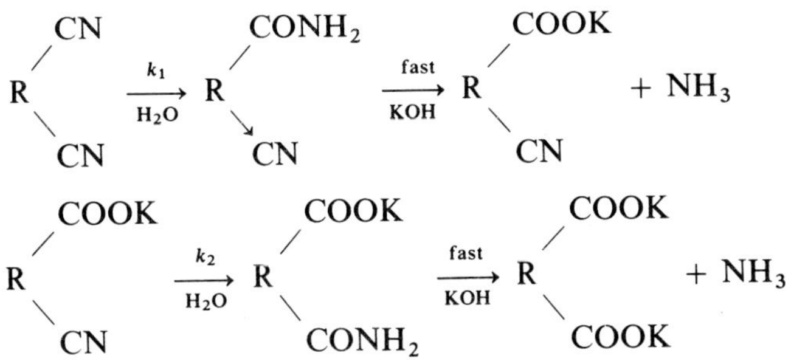

5.4 COMPLEX REACTIONS

5.4.1 General Comments

Many industrially significant reactions do not follow any of the types of rate expressions dis-

The kinetics observed are typical of two first-order reactions in series.

The following data represent the progress of the reaction as reported in Swain's (7) analysis of Kaufler's work.

Percent reaction $(50\delta^*)$	Time, t (hr)
15	0.367
35	1.067
70	4.200

Use the time ratio method to determine k_1/k_2.

Solution

From the data indicated it is possible to arrive at three time ratios and their logarithms. We will evaluate κ using each of these ratios and average the results. From the ratios and Table 5.1:

$$\log(t_{70}/t_{15}) = 1.0586 \qquad \kappa_I = 0.194$$
$$\log(t_{70}/t_{35}) = 0.5951 \qquad \kappa_{II} = 0.193$$
$$\log(t_{35}/t_{15}) = 0.4635 \qquad \kappa_{III} = 0.196$$

The average value of k_2/k_1 is thus 0.194.

cussed previously. There are a great number of possible combinations of parallel, series, and reversible reactions that we have not considered, but that may occur in nature in chemical processing operations. Benson (10) has described this situation in an apt fashion: "Kinetic systems when investigated in detail display an anarchistic tendency to become unique laws unto themselves." In this section we turn our attention to these more complex systems.

It is extremely difficult to generalize with regard to systems of complex reactions. Often it is useful to attempt to simplify the kinetics by using experimental techniques which cause a degeneration of the reaction order by using a large excess of one or more reactants or using stoichiometric ratios of reactants. In many cases, however, even these techniques will not effect a simplification in the reaction kinetics. Then one must often be content with qualitative or semi-quantitative descriptions of the system behavior.

This text has not treated all the mathematical descriptions of reacting systems that have appeared in the literature. Indeed, such coverage goes far beyond the scope and spirit of this text. For material of this type, consult the kinetics literature.

5.4.2 Competitive Consecutive Second-Order Reactions

Competitive consecutive reactions are combinations of parallel and series reactions that include processes such as multiple halogenation and nitration reactions. For example, when a nitrating mixture of HNO_3 and H_2SO_4 acts on an aromatic compound like benzene, NO_2 groups substitute for hydrogen atoms in the ring to form mono-, di-, and tri-substituted nitro compounds.

Since many reactions of this type involve a series of second-order processes, it is instructive to consider how one analyzes systems of this sort in order to determine the kinetic parameters that are necessary for reactor design purposes. We will follow a procedure described previously by Frost and Pearson (11). Consider the following mechanistic equations.

$$B + A \xrightarrow{k_1} C + Z \qquad (5.4.1)$$

$$C + A \xrightarrow{k_2} D + Z' \qquad (5.4.2)$$

These reactions may be regarded as parallel with respect to species A and series with respect to species B, C, and D. Only species A and B are present at time zero.

The equations for the time derivatives of certain species concentrations are given by

$$\frac{dA}{dt} = -k_1 BA - k_2 CA \qquad (5.4.3)V$$

$$\frac{dB}{dt} = -k_1 BA \qquad (5.4.4)V$$

$$\frac{dC}{dt} = k_1 BA - k_2 CA \qquad (5.4.5)V$$

By the use of a large excess of A one may cause these equations to degenerate to a pseudo first-order form. The analysis for these conditions is then equivalent to that presented in Section 5.3.

There are two other limiting forms of these equations that are also of interest. If $k_1 \gg k_2$, the first step is very rapid compared to the second, so that it is essentially complete before the latter starts. The reaction may then be treated as a simple irreversible second-order reaction with the second step being rate limiting. On the other hand, if $k_2 \gg k_1$, the first step controls the reaction so the kinetics observed are those for a single second-order process. However, the analysis must take into account the fact that in this case 2 moles of species A will react for each mole of B that is consumed.

For the more general case of arbitrary rate constants, the analysis is more complex. Various approximate techniques that are applicable to the analysis of reactions 5.4.1 and 5.4.2 have been described in the literature, and Frost and Pearson's text (11) treats some of these. One useful general approach to this problem is that of Frost and Schwemer (12–13). It may be regarded as an extension of the time-ratio method discussed in Section 5.3.2. The analysis is predicated on a specific choice of initial reactant concentrations. One uses equivalent amounts of reactants A and B ($A_0 = 2B_0$) instead of equimolal quantities.

A material balance on species A indicates that

$$A_0 - A = C + 2D \qquad (5.4.6)V$$

while a balance on species B gives

$$B_0 = B + C + D \qquad (5.4.7)V$$

Elimination of D between equations 5.4.7 and 5.4.6 gives

$$A_0 - A = C + 2(B_0 - B - C) \qquad (5.4.8)V$$

or

$$A_0 - 2B_0 = A - 2B - C \qquad (5.4.9)V$$

For the specified choice of initial reactant concentrations, this last equation requires that

$$C = A - 2B \qquad (5.4.10)V$$

With this substitution equation 5.4.3 becomes

$$\frac{dA}{dt} = (2k_2 - k_1)BA - k_2 A^2 \quad (5.4.11)\text{V}$$

Elimination of time as an independent variable between equations 5.4.4 and 5.4.11 gives

$$\frac{dA}{dB} = \left(1 - 2\frac{k_2}{k_1}\right) + \frac{k_2}{k_1}\left(\frac{A}{B}\right) \quad (5.4.12)\text{V}$$

This equation may be solved using the substitution

$$A = B\phi \quad (5.4.13)\text{V}$$

Hence

$$\frac{dA}{dB} = \phi + B\frac{d\phi}{dB} = \phi + \frac{d\phi}{d\ln B} \quad (5.4.14)\text{V}$$

Combination of equations 5.4.12 to 5.4.14 gives

$$\phi + \frac{d\phi}{d\ln B} = \left(1 - 2\frac{k_2}{k_1}\right) + \frac{k_2}{k_1}\phi$$

$$(5.4.15)\text{V}$$

Solution of this linear differential equation by separation of variables and subsequent mathematical manipulation leads to

$$\frac{A}{A_0} = \frac{1 - 2\kappa}{2(1 - \kappa)}\left(\frac{B}{B_0}\right) + \frac{1}{2(1 - \kappa)}\left(\frac{B}{B_0}\right)^\kappa$$

$$(5.4.16)\text{V}$$

where $\kappa = k_2/k_1$.

If this result is substituted into equation 5.4.4, one obtains

$$\frac{dB}{dt} = -k_1 B\left[\frac{(1 - 2\kappa)}{2(1 - \kappa)}\frac{B}{B_0} + \frac{1}{2(1 - \kappa)}\left(\frac{B}{B_0}\right)^\kappa\right]A_0$$

$$(5.4.17)\text{V}$$

This equation may be rewritten in terms of a dimensionless time τ^* and a dimensionless concentration for species B if we let

$$\tau^* = k_1 B_0 t \quad \text{and} \quad \beta = \frac{B}{B_0} \quad (5.4.18)$$

In this case equation 5.4.17 becomes

$$\frac{d\beta}{d\tau^*} = -\frac{B}{B_0}\left[\frac{(1 - 2\kappa)\beta}{2(1 - \kappa)} + \frac{\beta^\kappa}{2(1 - \kappa)}\right]\frac{A_0}{B_0}$$

$$(5.4.19)\text{V}$$

Recognition that $A_0/B_0 = 2$ leads to

$$\frac{d\beta}{d\tau^*} = -\left[\frac{(1 - 2\kappa)\beta^2}{1 - \kappa} + \frac{\beta^{\kappa+1}}{1 - \kappa}\right] \quad (5.4.20)\text{V}$$

Separation of variables and integration gives

$$\tau^* = -(1 - \kappa)\int_1^\beta \frac{d\beta}{(1 - 2\kappa)\beta^2 + \beta^{\kappa+1}}$$

$$= (1 - \kappa)\int_\beta^1 \frac{d\beta}{(1 - 2\kappa)\beta^2 + \beta^{\kappa+1}}$$

$$(5.4.21)\text{V}$$

In principle this integral may be evaluated in closed form for any value of κ that is a rational number (i.e., a ratio of integers). In other cases numerical integration is necessary. This integral will degenerate to the limiting forms mentioned earlier for κ values of zero and infinity. The integral also simplifies for $\kappa = 1/2$. If we let $\alpha = A/A_0$, the limiting expressions for equations 5.4.16 and 5.4.21 become:

For $\kappa = 0$ or ($k_1 \gg k_2$):

$$\tau^* = \int_\beta^1 \frac{d\beta}{\beta^2 + \beta} = \ln\left(\frac{\beta + 1}{2\beta}\right) \quad \alpha = \frac{\beta + 1}{2}$$

$$(5.4.22)\text{V}$$

For $\kappa = \frac{1}{2}$ or ($k_1 = 2k_2$):

$$\tau^* = \frac{1}{2}\int_\beta^1 \frac{d\beta}{\beta^{3/2}} = \frac{1}{\beta^{1/2}} - 1 \quad \alpha = \sqrt{\beta}$$

$$(5.4.23)\text{V}$$

For $\kappa = \infty$ or ($k_1 \ll k_2$):

$$\tau^* = \frac{1}{2}\left(\frac{1}{\beta} - 1\right) \quad \alpha = \beta \quad (5.4.24)\text{V}$$

Frost and Schwemer have developed a time-ratio technique based on equations 5.4.21 and 5.4.16 in order to facilitate the calculation of second-order rate constants for the class of reactions under consideration. Data for A/A_0 versus τ^* at various values of κ are presented in Table 5.2, and time ratios are given in Table 5.3. The latter values may be used to determine κ by using various time ratios from a single kinetic run if one recognizes that $(\tau_1^*/\tau_2^*) = (t_1/t_2)$. Once κ has been determined, Table 5.2 may be used to determine the τ^* values at a given A/A_0 and κ. Equation 5.4.18 may then be used to determine

k_1. This value and the value of κ then may be combined to evaluate k_2. The following illustration indicates the application of this technique.

ILLUSTRATION 5.6 DETERMINATION OF REACTION RATE CONSTANTS FOR COMPETITIVE CONSECUTIVE SECOND-ORDER REACTIONS

Burkus and Eckert (14) have studied the kinetics of the triethylamine catalyzed reaction of 2,6-tolylene diisocyanate with 1-butanol in toluene

Table 5.2

τ^* as a Function of κ and A/A_0 (11)

$1/\kappa$	$A/A_0 = 0.8$ 20% reaction	$A/A_0 = 0.7$ 30% reaction	$A/A_0 = 0.6$ 40% reaction	$A/A_0 = 0.5$ 50% reaction	$A/A_0 = 0.4$ 60% reaction
2.0	0.2500	0.4286	0.6667	1.000	1.500
3.0	0.2599	0.4564	0.7305	1.133	1.770
4.0	0.2656	0.4741	0.7756	1.239	2.011
5.0	0.2693	0.4865	0.8098	1.327	2.235
6.0	0.2720	0.4957	0.8368	1.404	2.449
7.0	0.2740	0.5028	0.8589	1.471	2.657
8.0	0.2755	0.5085	0.8773	1.531	2.862
9.0	0.2768	0.5131	0.8929	1.586	3.066
10.0	0.2778	0.5170	0.9064	1.637	3.270

Table 5.3

Time Ratios as a Function of κ (11) (t_{60}/t_{20} is the ratio of the time for 60% A reacting to the time for 20% reaction, or where $A/A_0 = 0.4$ and 0.8, respectively)

$1/\kappa$	t_{60}/t_{20}	t_{60}/t_{30}	t_{60}/t_{40}	t_{60}/t_{50}	t_{50}/t_{20}	t_{50}/t_{30}
2.0	6.000	3.500	2.250	1.500	4.000	2.333
3.0	6.812	3.878	2.423	1.562	4.362	2.483
4.0	7.571	4.241	2.592	1.623	4.666	2.614
5.0	8.297	4.593	2.760	1.684	4.928	2.728
6.0	9.003	4.940	2.927	1.745	5.161	2.832
7.0	9.698	5.285	3.094	1.806	5.369	2.925
8.0	10.388	5.629	3.263	1.869	5.558	3.012
9.0	11.078	5.975	3.434	1.933	5.731	3.091
10.0	11.772	6.325	3.607	1.998	5.892	3.166

solution. The reactions may be represented as

$$B + A \xrightarrow{k_1} C \qquad \qquad \text{(A)}$$

$$C + A \xrightarrow{k_2} D \qquad \qquad \text{(B)}$$

A titrimetric method was used to follow the progress of the reaction. On the basis of the data given below what are the values of the reaction rate constants, k_1 and k_2?

Data

Temperature: 39.69 °C

Initial concentration of 2,6-tolylene diisocyanate: 53.2 moles/m³

Initial concentration of 1-butanol: 106.4 moles/m³

Catalyst concentration (triethylamine): 31.3 moles/m³

Time, t (ksec)	Percent 1-butanol reacted
0.360	20
0.657	30
1.116	40
1.866	50
3.282	60

Solution

The time-ratio method of Frost and Schwemer (12–13) may be used in the solution of this problem, since *equivalent* amounts of reactants were employed ($A_0 = 2B_0$). From Table 5.3 the following values of $1/\kappa$ may be determined at the time ratios indicated.

$$t_{60}/t_{50} = \frac{3.282}{1.866} = 1.759 \qquad 1/\kappa = 6.21$$

$$t_{60}/t_{40} = \frac{3.282}{1.116} = 2.941 \qquad 1/\kappa = 6.08$$

$$t_{60}/t_{30} = \frac{3.282}{0.657} = 4.995 \qquad 1/\kappa = 6.16$$

$$t_{60}/t_{20} = \frac{3.282}{0.360} = 9.117 \qquad 1/\kappa = 6.16$$

$$t_{50}/t_{30} = \frac{1.866}{0.657} = 2.840 \qquad 1/\kappa = 6.09$$

$$t_{50}/t_{20} = \frac{1.866}{0.360} = 5.183 \qquad 1/\kappa = 6.11$$

$$\text{Average: } 1/\kappa = 6.13$$

The values of τ^* that correspond to the various conversion levels and $1/\kappa = 6.13$ may be found in Table 5.2. The rate constant k_1 may then be calculated from equation 5.4.18. Hence,

Percent conversion	τ^*	$k_1 = \tau^*/B_0 t$ $(m^3/kmole \cdot ksec)$
20	0.2722	14.21
30	0.4966	14.21
40	0.8397	14.14
50	1.413	14.23
60	2.476	14.18
	Average	14.19

Thus $k_2 = \kappa k_1 = \dfrac{14.19}{6.13} = 2.31 \text{ m}^3/\text{kmole} \cdot \text{ksec}$

LITERATURE CITATIONS

1. Hellin, M., and Jungers, J. C., *Bull. Soc. Chim.*, *1957* (386).

2. Guldberg, C. M., and Waage, P., *Études sur les affinités chimiques*, Brøgger and Christie, Christiania, Oslo, 1867.

3. Denbigh, K. G., *The Principles of Chemical Equilibrium*, Second Edition, p. 445, Cambridge University Press, 1966.

4. Denbigh, K. G., and Prince, A. J., *J. Chem. Soc.*, *1947* (790).

5. Emanuel', N. M., and Knorre, D. G., *Chemical Kinet-*

ics—*Homogeneous Reactions*, translated by R. Kondor, p. 215, Wiley, New York, 1973.

6. Pannetier, G., and Souchay, P., *Chemical Kinetics*, translated by H. D. Gesser and H. H. Emond, pp. 177–178, Elsevier, Amsterdam, 1967. Adapted with permission.

7. Swain, C. G., *J. Am. Chem. Soc.*, 66 (1696), 1944.

8. Frost, A. A., and Pearson, R. G., *Kinetics and Mechanism*, Second Edition, pp. 170–171, Wiley, New York, copyright © 1961. Used with permission.

9. Kaufler, F., *Z. physik Chem.*, 55 (502), 1906.

10. Benson, S. W., *Foundations of Chemical Kinetics*, p. 27, McGraw-Hill, New York, copyright © 1960. Used with permission.

11. Frost, A. A., and Pearson, R. G., op. cit., pp. 178–84. Used with permission.

12. Frost, A. A., and Schwemer, W. C., *J. Am. Chem. Soc.*, 74 (1268), 1952.

13. Schwemer, W. C., and Frost, A. A., *J. Am. Chem. Soc.*, 73 (4541), 1951.

14. Burkus, J., and Eckert, C. F., *J. Am. Chem. Soc.*, 80 (5948), 1958.

PROBLEMS

1. One of the classic examples of a reversible reaction that is first-order in both directions is the conversion of γ-hydroxybutyric acid into its lactone in aqueous solution.

$$CH_2OH—CH_2—CH_2—COOH \underset{k_{-1}}{\overset{k_1}{\rightleftharpoons}} H_2O + \begin{array}{c} CH_2—CH_2—CH_2—C = O \\ \overline{\qquad\qquad O \qquad\qquad} \end{array}$$

$$B \rightleftharpoons H_2O + L$$

In aqueous solution the water concentration may be considered constant, so the reverse reaction follows pseudo first-order kinetics. The data below on this reaction have been taken from Emanuel and Knorre. Use them to determine the values of both first-order rate constants.

Data

Temperature: 25 °C
Initial acid concentration: 182.3 moles/m^3

Time, t (ksec)	Lactone concentration (moles/m^3)
0	—
1.26	24.1
2.16	37.3
3.00	49.9
3.90	61.0
4.80	70.8
6.00	81.1
7.20	90.0
9.60	103.5
13.20	115.5
169.20	132.8

From *Chemical Kinetics—Homogeneous Reactions* by N. M. Emanuel' and D. G. Knorre. Copyright © 1973. Reprinted by permission of Keter Publishing House Ltd.

2. Kistiakowsky and Smith [*J. Am. Chem. Soc.*, 56 (638), 1934] have studied the kinetics of the *cis-trans* isomerization of isostilbene. The reaction is believed to be pseudo first-order in both the forward and reverse directions under the experimental conditions used by these investigators.

$$A \underset{k_2}{\overset{k_1}{\rightleftharpoons}} B$$

isostilbene   stilbene

These authors reported the following values for the equilibrium constant for this reaction.

Temperature, $T(°K)$	Equilibrium constant K
593	14.62
614	11.99

Starting with pure isostilbene, the authors reported the following data at 574 °K.

Initial pressure (mm Hg)	Time, t (sec)	Fraction reacted
11.8	1008	0.226
119	1140	0.241
113	3624	0.598
155	1800	0.360
189	1542	0.307
205	1896	0.371

Do these data fit the proposed rate expression? If so, what are the rate constants for the forward and reverse reactions?

3. The data recorded below are typical of a study of the mutarotation of π-bromonitrocamphor in chloroform solution at 14 °C. The reaction obeys the kinetics indicated by

$$A \underset{k_{-1}}{\overset{k_1}{\rightleftharpoons}} B$$

The initial concentration of B is zero. From the data below, calculate the quantity $k_1 + k_{-1}$.

Time, t (ksec)	Polarimeter reading (degrees)
0	189.0
10.8	169.0
18.0	156.0
25.2	146.0
86.4	84.5
259.2	37.3
∞	31.3

The optical rotation is an additive property of the various species present in solution.

4. Under certain conditions acetaldehyde decomposes into methane and carbon monoxide in a second-order reaction.

$$CH_3CHO \rightleftharpoons CH_4 + CO$$

For a constant volume reactor the rate of the forward reaction is given by

$$\frac{dC_{CH_3CHO}}{dt} = -k_f C^2_{CH_3CHO}$$

Give two possible rate laws for the reverse reaction.

5. In an unpublished study, Grieger and Hansel have used absorbance measurements to monitor the progress of the reaction forming an enzyme-substrate complex of imidazole (S) and metmyoglobin (E) at high pressures. Formation of the enzyme-substrate complex (ES) may be represented by the following equation

$$E + S \rightleftharpoons ES$$

where the forward reaction is first-order in each component (second-order overall) and the reverse reaction is first-order. Since the equilibrium is rapidly established, a relaxation approach was used to investigate the reaction kinetics. The data below are typical of the reaction at 103.42 MPa and 19.5 °C. What is the relaxation time characteristic of these conditions? What additional experiments must be performed in order to determine the individual reaction rate constants?

Data

Initial imidazole concentration: 70.0 moles/m³
Initial myoglobin concentration: 0.114 mole/m³
(Initial conditions refer to those used in preparing the system for study, not those at the moment the equilibrium is perturbed.)

Time, t (msec)	Absorbance (arbitrary units)
40	35.0
80	29.0
120	25.5
160	24.0
200	21.5
240	21.0
280	21.0
320	20.0
360	20.0
400	20.0

(*Note.* The absorbance could be measured only to the nearest 0.5 unit.)

6. Ziman et al. [*Kinetika i Kataliz*, 9 (117), 1968] have studied the kinetics of the catalytic oxidative dehydrogenation of various butene isomers to form 1,3-butadiene. Over a Bi-Mo catalyst the following reactions are important.

$$1\text{-butene} \underset{k_2}{\overset{k_1}{\rightleftharpoons}} trans\text{-2-butene}$$

$$1\text{-butene} \underset{k_4}{\overset{k_3}{\rightleftharpoons}} cis\text{-2-butene}$$

$$trans\text{-2-butene} \underset{k_6}{\overset{k_5}{\rightleftharpoons}} cis\text{-2-butene}$$

$$1\text{-butene} \overset{k_7}{\rightleftharpoons} 1,3\text{-butadiene} + H_2$$

$$trans\text{-2-butene} \overset{k_8}{\rightleftharpoons} 1,3\text{-butadiene} + H_2$$

$$cis\text{-2-butene} \overset{k_9}{\rightleftharpoons} 1,3\text{-butadiene} + H_2$$

At 396 °C these authors have reported the following values of the reaction rate constants in units of inverse seconds.

k_1	9.0	k_6	13.0
k_2	6.0	k_7	19.0
k_3	11.0	k_8	1.0
k_4	7.0	k_9	3.0

A value of k_5 was not reported because of the large uncertainty in this parameter. Use the principle of microscopic reversibility to determine a value of k_5 at this temperature for this catalyst.

7. The catalytic synthesis of ammonia

$$N_2 + 3H_2 \rightleftharpoons 2NH_3$$

is believed to occur by the following mechanism

$$N_2 \underset{k_{-1}}{\overset{k_1}{\rightleftharpoons}} 2N(a) \qquad (1)$$

$$H_2 \underset{k_{-2}}{\overset{k_2}{\rightleftharpoons}} 2H(a) \qquad (2)$$

$$N(a) + H(a) \underset{k_{-3}}{\overset{k_3}{\rightleftharpoons}} NH(a) \qquad (3)$$

$$NH(a) + H(a) \underset{k_{-4}}{\overset{k_4}{\rightleftharpoons}} NH_2(a) \qquad (4)$$

$$NH_2(a) + H(a) \underset{k_{-5}}{\overset{k_5}{\rightleftharpoons}} NH_3 \qquad (5)$$

The symbol (a) denotes an adsorbed species. If all steps are at equilibrium and if the second step is believed to be rate controlling, what relation must exist between the overall equilibrium constant and the observed rate constants? The rate of the forward reaction is to be taken as $k_2 C_{H_2}$ where k_2 is the rate constant observed for the forward reaction. Start by determining the appropriate form of the rate constant observed for the reverse reaction in terms of the k_i values used above.

The *allo*-ocimene may undergo subsequent reaction to form α- and β-pyronene and a dimer of *allo*-ocimene.

Fuguitt and Hawkins have reported the data below as typical of these reactions. The reported data have been modified somewhat for the purposes of this problem. By summing the amounts of dimer, α- and β-pyronene, and *allo*-ocimene present in the final mixture, one may determine the total amount of *allo*-ocimene formed in the initial reaction. These investigators have postulated that all three of the parallel reactions are first-order. Is the data consistent with this hypothesis? If so, what are the values of the three first-order rate constants?

Data

Reaction of α-Pinene at 204.5 °C

Time, t (ksec)	Percent α-pinene unreacted	Percent α and β-pyronene	Percent *allo*-ocimene	Percent dimer	Percent *dl*-limonene	Percent racemized
26.4	85.9	0.4	4.1	0.6	8.2	0.8
49.5	74.3	0.8	6.8	1.6	15.6	0.9
72.0	65.1	1.0	7.7	3.4	21.5	1.3
90.0	58.6	1.2	8.4	5.0	25.5	1.3
122.4	48.1	1.6	8.5	8.2	31.9	1.7
183.6	32.1	2.0	8.2	13.5	42.0	2.2
363.6	11.2	2.7	6.9	21.9	54.7	2.6

Reprinted with permission from *J. Am. Chem. Soc.*, **69** (319), 1947. Copyright by the American Chemical Society.

8. When optically active α-pinene is heated above 200 °C in the liquid phase, it becomes optically inactive. Initially, the reaction processes may be written as

$$d\text{-pinene} \overset{k_1}{\rightarrow} dl\text{-limonene}$$

$$d\text{-pinene} \overset{k_2}{\rightarrow} allo\text{-ocimene}$$

$$d\text{-pinene} \overset{k_3}{\rightarrow} dl \text{ pinene}$$

9. Schmid and Heinola [*J. Am. Chem. Soc.*, **90** (131), 1968] have studied the liquid phase reactions of 1-phenylpropyne (*A*) and 2,4-dinitrobenzenesulfenyl chloride (*B*) in chloroform solution at 51° C. The primary products of the reaction are: 1-phenyl-*trans*-1-chloro-2-(2, 4 dinitrophenylthio) propene (*C*) and 1-phenyl-*trans*-2-chloro-1-(2, 4-dinitrophenylthio) propene (*D*). The reaction stoichiometry may be

represented as:

$$A + B \xrightarrow[94\%]{k_1} C$$

$$A + B \xrightarrow[6\%]{k_2} D$$

Using equimolal concentrations of reactants A and B, the authors obtained the following data.

Time, t (ksec)	A (kmoles/m³)
0.0	0.181
16.2	0.141
21.6	0.131
32.4	0.119
37.8	0.111
86.4	0.0683
99.0	0.0644
108.0	0.0603

It may be assumed that the rate expressions for reactions 1 and 2 are identical except for the values of the rate constants. If the reaction order is an integer, determine the overall order of the reaction and the rate constants k_1 and k_2.

10. Regna and Caldwell [*J. Am. Chem. Soc.*, 66 (246), 1944] have studied the kinetics of the acid-catalyzed transformation of 2-ketopolyhydroxy acids into ascorbic acids and other products.

$$K \xrightarrow{k_1} A \xrightarrow{k_2} F \xrightarrow{k_3} P$$

where K is a 2-ketohydroxy acid, A is the corresponding ascorbic acid analog, F is the furfural formed by the decomposition of the ascorbic acid, and P is the polymerized furfural. At 59.9 °C, the reaction of 2-keto-$\mathscr{L}$-gluconic acid follows pseudo first-order kinetics with $k_1 = 2.53 \times 10^{-3}$ min^{-1} and $k_2 = 4.91 \times 10^{-4}$ min^{-1}. If one starts with a solution that is 0.05 molar in both 2-keto-$\mathscr{L}$-gluconic acid and its ascorbic acid derivative, what is the maximum concentration of ascorbic acid that will be obtained? How long does it take to achieve this maximum?

11. In carrying out an enzyme assay it may be convenient to introduce an auxiliary enzyme to the system to effect the removal of a product produced by the first enzymatic reaction. McClure [*Biochemistry*, 8 (2782), 1969] has described the kinetics of certain of these coupled enzyme assays. The simplest coupled enzyme assay system may be represented as

$$A \xrightarrow[\text{primary enzyme}]{k_1} B \xrightarrow[\text{auxiliary enzyme}]{k_2} C$$

If the first reaction is regarded as zero-order irreversible (i.e., the enzyme is saturated with substrate), and the second reaction is first-order in the product B, determine the time-dependent behavior of the concentration of species B if no B is present initially. How long does it take to reach 98% of the steady-state value if $k_1 = 0.833$ mole/m³-ksec and $k_2 = 0.767$ sec^{-1}? What is this steady-state value?

12. Debande and Huybrechts [*Int. J. Chem. Kinetics*, 6 (545), 1974] have studied the gas phase Diels-Alder additions of propylene (P) to cyclohexa-1, 3-diene (Chd) to give the exo- and endo-isomers of 5-methylbicyclo (2,2,2) oct-2-ene (MBO). The initial reaction rates were both found to be of the mixed second-order type with

$$r_{\text{endo}} = k_{\text{endo}} C_P C_{\text{Chd}} \quad \text{and} \quad r_{\text{exo}} = k_{\text{exo}} C_P C_{\text{Chd}}$$

and

$$\log k_{\text{endo}} = -5698/T + 5.74$$
$$\log k_{\text{exo}} = -6577/T + 6.66$$

where T is expressed in degrees Kelvin and k in cubic meters per kilomole per second.

The following biradical mechanism has been proposed for this reaction.

(endo-MBO)

(Chd) (P) (R:)

(exo-MBO)

(a) Does this mechanism give rise to a rate expression that is consistent with the initial rate data?

(b) If the reverse Diels-Alder reactions are faster than the isomerization reaction between the exo- and endo-forms, what does the rate expression reduce to?

(c) If the endo-form is the desired product, do you recommend operation at the high or low end of the temperature range 512 to 638 °K? Determine the product distribution for 30% conversion of cyclohexadiene at your specified temperature.

13. The reaction of p-chlorophenylsilane with benzyl alcohol may be characterized as follows.

$$RSiH_3 + R'OH \xrightarrow{k_1} RSiH_2OR' + H_2$$

$$RSiH_2OR' + R'OH \xrightarrow{k_2} RSiH(OR')_2 + H_2$$

where both reactions are second-order. From the data below determine the two rate constants if the reaction of 127.3 mg of p-chlorophenylsilane and 198.2 mg of benzyl alcohol in the presence of metallic copper liberates hydrogen gas as follows.

Time, t (sec)	62	116	198	396	878
$cm^3 H_2$	8.12	12.18	16.24	20.30	24.36

The initial silane concentration is equivalent to 0.5 kmole/m³.

14. Eldib and Albright [*Ind. Eng. Chem., 49* (825), 1957] have indicated that the main reactions in the catalytic hydrogenation of cottonseed oil are

If each of these reactions is regarded as irreversible pseudo first-order, derive equations for the time dependence of each species in terms of the initial concentrations and the appropriate rate constants.

For run 3, the following parameter values were estimated.

$k_1 = 0.0133 \ \text{min}^{-1}$ $k_4 = 0.0024 \ \text{min}^{-1}$

$k_2 = 0.0108 \ \text{min}^{-1}$ $k_5 = 0.0080 \ \text{min}^{-1}$

$k_3 = 0.0024 \ \text{min}^{-1}$

If one were to start with pure linoleic acid, sketch the time dependent behavior of all species in terms of normalized concentrations (C_i/C_{A0}).

6 Elements of Heterogeneous Catalysis

6.0 INTRODUCTION

Since the chemical, petrochemical, and petroleum industries rely heavily on catalytic processing operations, chemical engineers must be cognizant of the fundamental and applied aspects of catalysis. In 1964 it was estimated that 18% of the total wholesale value of manufactured goods produced in the United States directly involved catalytic technology (1). Some commercially significant catalytic processes are listed in Table 6.1. Chapter 6 discusses the chemical aspects of heterogeneous catalytic phenomena. Chapter 12 is devoted to the engineering aspects of catalytic phenomena which are involved in the design of commercial scale reactors.

When philosophical and metaphysical ideas enjoyed precedence over experimental facts as a source of scientific theory, many of the concepts that we would now interpret as catalytic in nature played a key role in scientific "theory." The "Philosopher's Stone" was a hypothetical substance that could turn base metals into gold and could promote good health and long life. The chemical action of the "Philosopher's Stone" required that a small quantity of this substance could bring about a large change, an idea inherent in the modern view of catalysis. The aura of mysticism that is associated with catalytic phenomena reaches back to the days of the alchemist. Only in recent years has catalysis been regarded more as a science than an art. Indeed, much art still remains, and our imperfect understanding of catalysts has long been reflected in the manufacturing processes employed to produce commercial catalysts.

Table 6.1
Typical Industrial Catalytic Reactions (2, 3)

Type of reaction	Reactants	Catalysts	Products
1. Ammonia synthesis	$N_2 + H_2$	Iron promoted with Al_2O_3 and K	NH_3
2. Ammonia oxidation	$NH_3 + O_2$	Pt–Rh	NO used in manufacture of HNO_3
3. Oxidation of sulfur dioxide	$SO_2 + O_2$	V_2O_5 or Pt	SO_3 used in manufacture of oleum and sulfuric acid
4. Hydrogenation of fats and edible oils	H_2 + unsaturated oil	Ni	Saturated oil
5. Cracking	Various petroleum fractions	Silica, alumina, and combinations with molecular sieves	Wide range of compounds
6. Polymerization	Ethylene	Aluminum alkyls and Ti Cl_4 MoO_3 or CrO_3 on alumina	Polyethylene
7. Dehydrogenation	Ethylbenzene	Iron oxide or chromia alumina	Styrene

In this list, the first five processes are among the most important commercial catalytic processes in terms of total processing capacity.

In 1835 Berzelius coined the term *catalysis* to describe the influence of certain substances on the nature of diverse reactions, the substances themselves apparently being unchanged by the reaction. He imbued these materials with a *catalytic force* capable of awakening the potential for chemical reaction between species that would normally be nonreactive at a given temperature. In more modern terms the following definition is appropriate.

A catalyst is a substance that affects the rate or the direction of a chemical reaction, but is not appreciably consumed in the process.

There are three important aspects of the definition. First, a catalyst may increase or decrease the reaction rate. Second, a catalyst may influence the direction or selectivity of a reaction. Third, the amount of catalyst consumed by the reaction is negligible compared to the consumption of reactants.

Catalysts that decrease reaction rates are usually referred to as *inhibitors*. They usually act by interfering with the free radical processes involved in chain reactions, and the mechanism differs from that involved in accelerating a reaction. The most familiar example of the use of inhibitors is the addition of tetraethyl lead to gasoline to improve its antiknock properties.

A catalyst *cannot* change the *ultimate equilibrium point* set by thermodynamics, but it can affect the rate at which this point is approached. However, it can facilitate approach to equilibrium with respect to a desired reaction while not influencing the rates of other less desirable reactions. In optimizing yields of desired products, chemical engineers are very concerned with the *selectivity* or *specificity* of a catalyst. For commercial applications, selectivity is often more important than activity *per se*.

The idea that a catalyst remains unaltered by the reaction it catalyzes is naive and misleading. Physical and chemical changes can and do occur either during or as a result of the catalytic process. The ratio of metal to oxygen in an oxide catalyst will frequently change with changes in temperature and in the composition of the fluids with which it is in contact. Pure metal catalysts will often change in crystal structure or surface roughness on use. Many commercial catalysts are gradually deactivated by poisoning reactions that accompany the main reaction process. The catalysts used to initiate free radical polymerization reactions end up as an integral part of the polymer formed by the reaction and are thus consumed. In neither case does a stoichiometric whole number relationship exist between the amount of the catalyst consumed and the amounts of reactants converted to products. The number of moles of reactants converted is orders of magnitude greater than the number of moles of catalyst consumed.

For reversible reactions the principle of microscopic reversibility (Section 4.1.5.4) indicates that a material that accelerates the forward reaction will also catalyze the reverse reaction. In several cases where the catalytic reaction has been studied from both sides of the equilibrium position, the observed rate expressions are consistent with this statement.

Catalytic reactions are often classified as homogeneous or heterogeneous. True *homogeneous catalysis* takes place when the catalyst and the reactants are both present in the same fluid phase. Acid and base catalyzed reactions in aqueous solution are reactions of this type. Reactions that occur between a gas and a liquid or between two liquid phases are also generally considered to fall in this category. Although the mass transfer characteristics in these cases differ from those where the reaction takes place in a single phase, the reactants are mobile in both phases. The characteristics of these reactions are much more similar to those of reactions occurring in a single fluid phase than they are to those of reactions occurring at a fluid-solid interface. The term *heterogeneous catalysis* is generally restricted to catalytic phenomena involving a solid catalyst and reactants in a gas or liquid phase. These phenomena are sometimes referred to as *contact catalysis.*

Catalysts may be employed in solid, gaseous, or liquid form. However, the overwhelming majority of industrial catalytic processes involve solid catalysts and gaseous reactants. In many cases it is possible to develop useful new catalysts on the basis of analogies drawn between the behavior of certain chemical systems in the presence of heterogeneous or homogeneous catalysts. For example, reactions that take place in the presence of strong acids in liquid solution are often catalyzed by acidic solid catalysts such as silica-alumina. In some cases inorganic salts or metal complexes that have catalytic activity in aqueous solution will exhibit activity for the same reaction when deposited on a solid support. For example, an aqueous solution of cupric chloride and palladium chloride can be used as a catalyst for oxidizing ethylene to acetaldehyde in the Wacker process (2). These same materials can be deposited on a solid support to obtain a heterogeneous catalyst for this reaction, although this is not done in the current commercial process.

6.1 ADSORPTION PHENOMENA

This section treats chemical and physical adsorption phenomena in order to provide necessary background material for the discussion of catalytic reaction mechanisms in Section 6.3.1.

Adsorption is the preferential concentration of a species at the interface between two phases. Adsorption on solid surfaces is a very complex process and one that is not well understood. The surfaces of most heterogeneous catalysts are not uniform. Variations in energy, crystal structure, and chemical composition will occur as one moves about on the catalyst surface. In spite of this it is generally possible to divide all adsorption phenomena involving solid surfaces into two main classes: physical adsorption and chemical adsorption (or chemisorption). Physical adsorption arises from *intermolecular* forces involving permanent dipole, induced dipole, and quadrupole interactions. It involves van der Waals or secondary valence forces. It is akin to condensation. Chemisorption, on the other hand, involves a chemical interaction with attendant transfer of electrons between the adsorbent and the adsorbing species (*adsorbate*). The adsorbed species are held to the surface by valence forces that are the same as those that hold atoms together in a molecule.

In practice one can usually distinguish between chemisorption and physical adsorption on the basis of experimental evidence. However, it is frequently necessary to consider several criteria in making the classification.

The most important criterion for differentiating between physical and chemical adsorption is the magnitude of the enthalpy change accompanying the adsorption process. The energy evolved when physical adsorption occurs is usually similar to the heat of liquefaction of the gas (i.e., approximately 2 to 6 kcal/mole). Like condensation, the process is exothermic. For physical adsorption the heat of adsorption often lies between the heat of vaporization and the heat of sublimation. The enthalpy change accompanying chemisorption is significantly greater than that for physical adsorption and, in some cases, may exceed 100 kcal/mole. More often it will lie between 10 and 50 kcal/mole. In practice chemisorption is regarded as an exothermic process and, in the past, the view was frequently expressed that endothermic adsorption cannot occur. However, if a molecule dissociates on chemisorption and the dissociation energy of the molecule is greater than the energy of formation of the bonds with the surface, then the process can be endothermic. Since the Gibbs free energy change accompanying any spontaneous process must be negative, the only cases in which endothermic adsorption can be observed are those for which there will be a large positive entropy change. The dissociative chemisorption of hydrogen on glass is endothermic and is an example of this relatively rare phenomenon. More often dissociative chemisorption is exothermic, like other chemisorption processes.

A second criterion that is frequently used to differentiate empirically between chemical and physical adsorption is the *rate* at which the process occurs and, in particular, the temperature dependence of the rate. For physical adsorption of a gas on a solid surface, equilibrium is usually attained very quickly and is readily reversible. The activation energy of the process is usually less than a kilocalorie. Physical adsorption usually takes place so fast that the observed rate is limited by the rate at which molecules can be transported to the surface instead of by the adsorption process itself. Chemisorption may occur at rates comparable to those of physical adsorption, or it may occur at much slower rates, depending on the temperatures involved. Most types of chemisorption are characterized by a finite activation energy and thus proceed at appreciable rates only above certain minimum temperatures. This type of phenomenon is referred to as *activated chemisorption*. However, in some systems, chemisorption occurs very rapidly even at very low temperatures and has an activation energy near zero. For example, chemisorption of hydrogen and oxygen occurs on many clean metal surfaces at liquid nitrogen temperatures. These cases are described as *nonactivated chemisorption*. It is often observed that for a given adsorbent and adsorbate, the initial chemisorption is nonactivated, while subsequent chemisorption is activated.

Another aspect of rate measurements that is useful in discriminating between the two types of adsorption involves studies of the rate of desorption. The activation energy for desorption from a physically adsorbed state is seldom more than a few kilocalories per mole, whereas that for desorption from a chemisorbed state is usually in excess of 20 kcal/mole. Consequently, the ease with which desorption occurs on warming from liquid nitrogen temperature to room temperature is often used as a quick test for determining the nature of the adsorption process.

A third empirical criterion is based on the effect of temperature on the amount adsorbed. For physical adsorption the amount of gas adsorbed always decreases monotonically as the temperature is increased. Significant amounts of physical adsorption should *not* occur at temperatures in excess of the normal boiling point at the operating pressure. Appreciable chemisorption can occur at temperatures above the boiling point and even above the critical temperature of the material. Because chemisorption can be an activated process that takes place at a slow rate, it may be difficult to determine the amount of chemisorption corresponding to true equilibrium. Moreover, the process may not be reversible. It is also possible for two or more types of chemisorption or for chemical and physical adsorption to occur simultaneously on the same surface. These facts make it difficult to generalize with regard to the effect of temperature on the amount adsorbed. Different behavior will be observed for different adsorbent-adsorbate systems.

The experimental procedure for measuring the extent of adsorption generally involves admitting a gas to a chamber that contains the sample and waiting for a fixed period of time, during which "equilibrium" is supposedly achieved. Figure 6.1 indicates one common type of behavior that can be observed if physical adsorption and one type of activated chemisorption are present. At low temperatures, physical adsorption processes dominate and the amount of chemisorption is insignificant. The amount of material physically adsorbed decreases as the temperature increases. Within the time scale of the normal adsorption experiment, the amount of material adsorbed via the chemisorption process will increase as the temperature increases in the low temperature range. Consequently its contribution to the total will become more and more significant as the temperature

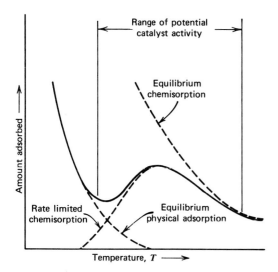

Figure 6.1
Effect of temperature on amount adsorbed for simultaneous physical adsorption and activated chemisorption. (Adapted from *Chemical Engineering Kinetics* **by J. M. Smith. Copyright © 1970. Used with permission of McGraw-Hill Book Company.)**

rises, and one eventually reaches a point at which the decrease in physical adsorption with increasing temperature is more than offset by the increase in the amount of material that can be chemisorbed in the time period involved. Consequently, the total amount adsorbed will increase with increasing temperature over a limited temperature range. At higher temperatures the rate of desorption from the chemisorbed state begins to be important, and the curve will go through a local maximum and then decrease. At the highest temperatures equilibrium can be established with respect to the chemisorption process. Note that the above discussion and the solid curve in Figure 6.1 apply to the situation where one starts at a low temperature and continuously increases the temperature. If we start at a high temperature and move in the direction of decreasing temperature, we would not retrace the solid curve but, instead, would follow the dashed line in

Figure 6.1 until rate considerations again become significant. Beyond this point the experimental curve will again lie below the true equilibrium curve.

Another characteristic that can be used to differentiate between chemical and physical adsorption is the degree of specificity in the gas-solid interaction. Physical adsorption is nonspecific. It will occur on all surfaces for a given gas, provided that the ratio of the adsorbate partial pressure to its saturation vapor pressure is sufficiently large. Since chemisorption is a chemical reaction confined to the surface of a solid and since the possibility of chemical reaction is highly specific to the nature of the species involved, chemisorption can take place only if the adsorbate is capable of forming a chemical bond with the adsorbent. Moreover, because it requires a chemical interaction, chemisorption is limited to a maximum of one layer of molecules on the surface (a monolayer). It frequently involves lower coverage. The valence forces holding the molecules on the surface fall off very rapidly with distance and become too small to form chemical bonds when the distance from the solid surface exceeds normal bond distances. Chemical and physical adsorption can occur together, but any adsorbed layers beyond the first must be physically adsorbed.

Physical adsorption is a readily reversible process, and alternate adsorption and desorption stages can be carried out repeatedly without changing the character of the surface or the adsorbate. Chemisorption may or may not be reversible. Often one species may be adsorbed and a second desorbed. Oxygen adsorbed on charcoal at room temperature is held very strongly, and high temperatures are necessary to accomplish the desorption. CO and/or CO_2 are the species that are removed from the surface. Chemical changes like these are *prima facie* evidence that chemisorption has occurred.

Although there is no single definitive test that is available to characterize the type of adsorption that takes place in an arbitrary system, the general criteria discussed above, taken collectively, provide a suitable basis for discrimination. Table 6.2 summarizes these criteria.

One of the basic tenets of virtually all mechanistic approaches to heterogeneous catalysis is that chemisorption of one or more reactant species is an essential step in the reaction. Because of the large changes in activation energies that are observed when shifting from a homogeneous reaction to one that is heterogeneously catalyzed, energetic considerations indicate that chemisorption must be involved. The forces involved in physical adsorption are so small relative to those involved in chemical bonding that it is hard to imagine that they can distort the electron clouds around a molecule to such an extent that they have an appreciable effect on its reactivity. Many heterogeneous catalytic reactions take place at temperatures above the critical temperatures of the reactants involved and, presumably, physical adsorption cannot occur under these conditions. Chemisorption has, however, been observed in many such cases.

This section has treated the qualitative features of chemisorption and physical adsorption processes. Both are important to the chemical engineer, who is interested in the design of heterogeneous catalytic reactors: chemisorption because it is an essential precursor of the reaction, and physical adsorption because it provides a means of characterizing heterogeneous catalysts in terms of specific surface areas and the distribution of pore sizes within a porous catalyst.

6.2 ADSORPTION ISOTHERMS

Plots of an amount of material adsorbed versus pressure at a fixed temperature are known as *adsorption isotherms*. They are generally classified in the five main categories described by Brunauer and his co-workers (4). In Figure 6.2 adsorbate partial pressures (P) are normalized by dividing by the saturation pressure at the temperature in question (P_0). Type I is referred to as Langmuir-type adsorption and is characterized by a monotonic approach to a limiting amount of adsorption, which presumably corresponds to formation of a monolayer. This type of behavior is that expected for chemisorption.

No other isotherms imply that one can reach a saturation limit corresponding to completion

Table 6.2
Comparison of Chemisorption and Physical Adsorption

Parameter	Chemical adsorption	Physical adsorption
Bonding forces	Primary valence forces (intramolecular forces)	Secondary valence forces (intermolecular forces)
Coverage	Monolayer	Multilayer
Adsorbent	Some solids	All solids
Adsorbate	Chemically reactive vapors	All gases below critical temperature
Reversibility	May be reversible or irreversible	Readily reversible
Rate	May be fast or slow, depending on the temperature	Rapid, may be limited by diffusion
Temperature dependence	May be complex—see Figure 6.1	Decreases with increasing temperature
Enthalpy effect	Virtually always exothermic, similar in magnitude to heats of reaction	Always exothermic, similar in magnitude to heats of condensation
Uses of adsorption studies	Determination of catalytically active surface area and elucidation of reaction kinetics	Determination of specific surface areas and pore size distributions

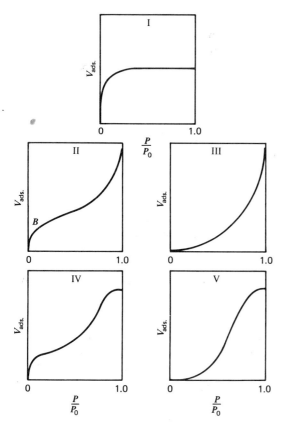

Figure 6.2
Five types of isotherms for adsorption according to Brunauer, Deming, Deming, and Teller (4).

pore condensation are often, but not always, encountered in this type of system. They arise from the effects of surface curvature on vapor pressure.

Types III and V are relatively rare. They are typical of cases where the forces giving rise to monolayer adsorption are relatively weak. Type V differs from Type III in the same manner that Type IV differs from Type II.

There are three approaches that may be used in deriving mathematical expressions for an adsorption isotherm. The first utilizes kinetic expressions for the rates of adsorption and desorption. At equilibrium these two rates must be equal. A second approach involves the use of statistical thermodynamics to obtain a pseudo equilibrium constant for the process in terms of the partition functions of vacant sites, adsorbed molecules, and gas phase molecules. A third approach using classical thermodynamics is also possible. Because it provides a useful physical picture of the molecular processes involved, we will adopt the kinetic approach in our derivations.

The Langmuir adsorption isotherm provides a simple mechanistic picture of the adsorption process and gives rise to a relatively simple mathematical expression. It can also be used to obtain a crude estimate of specific surface areas. More important, from the viewpoint of the chemical engineer, it serves as a point of departure for formulating rate expressions for heterogeneous catalytic reactions.

6.2.1 The Langmuir Adsorption Isotherm

Virtually all theoretical treatments of adsorption phenomena are based on or can be readily related to the analysis developed by Langmuir (5,6). The Langmuir isotherm corresponds to a highly idealized type of adsorption and the analysis is predicated on the following key assumptions.

1. Gas phase molecules are adsorbed at discrete points of attachment on the surface that

of a monolayer. Type II is a very common example of the behavior normally observed in physical adsorption. At values of P/P_0 approaching unity, capillary and pore condensation phenomena occur. Point B, the knee of the curve, corresponds roughly to completion of a monolayer. A statistical monolayer is built up at relatively low values of P/P_0 (0.1 to 0.3).

Type IV behavior is similar to Type II behavior except that a limited pore volume is indicated by the horizontal approach to the right-hand ordinate axis. This type of curve is relatively common for porous structures of many kinds. Hysteresis effects associated with are referred to as adsorption sites. Each site can accommodate only a single adsorbed species.

2. The energy of an adsorbed species is the same anywhere on the surface and is independent of the presence or absence of nearby adsorbed molecules. This assumption implies that the forces between adjacent adsorbed molecules are so small as to be negligible and that the probability of adsorption onto an empty site is independent of whether or not an adjacent site is occupied. This assumption usually implies that the surface is completely uniform in an energetic sense. If one prefers to use the concept of a nonuniform surface with a limited number of active centers that are the only points at which chemisorption occurs, this is permissible if it is assumed that all these active centers have the same activity for adsorption and that the rest of the surface has none.

3. The maximum amount of adsorption that is possible is that which corresponds to a monolayer.

4. Adsorption is localized and occurs by collision of gas phase molecules with vacant sites.

5. The desorption rate depends only on the amount of material on the surface.

For physical adsorption processes the third assumption is the poorest of these assumptions. For the case of chemical adsorption the worst of these assumptions is the second. There is a significant amount of experimental evidence that is contradictory to this assumption. Taylor (7) was the first to emphasize that adsorption sites may vary in energy. He noted that atoms at peaks on the surface and along crystal edges will be in high-energy states and will be the points at which adsorption first occurs. Other evidence for the lack of surface uniformity includes experimental data indicating that:

1. Adsorption of a catalyst poison to an extent that represents only a very small fraction of

a monolayer can cause an extremely disproportionate reduction in catalytic activity.

2. Heats of chemisorption usually decrease markedly as adsorption proceeds and the surface becomes covered.

In using the Langmuir adsorption isotherm as a basis for correlating heterogeneous catalytic rate data, the fact that the heat of chemisorption varies with extent of adsorption is less significant than would appear at first glance. Molecules adsorbed on high-energy sites may be bound so strongly that they are unable to serve as intermediates in the catalytic reaction. On the low-energy sites the energy change on adsorption may be insufficient to open up the catalytic path for the reaction. It is possible that only those sites of intermediate energy are effective in catalysis.

6.2.1.1 Derivation of the Langmuir Equation— Adsorption of a Single Species.

The kinetic approach to deriving a mathematical expression for the Langmuir isotherm assumes that the rate of adsorption on the surface is proportional to the product of the partial pressure of the adsorbate in the gas phase and the fraction of the surface that is bare. (Adsorption may occur only when a gas phase molecule strikes an uncovered site.) If the fraction of the surface covered by an adsorbed gas A is denoted by θ_A, the fraction that is bare will be $1 - \theta_A$ if no other species are adsorbed. If the partial pressure of A in the gas phase is P_A, the rate of adsorption is given by

$$r_{\text{adsorption}} = kP_A(1 - \theta_A) \qquad (6.2.1)$$

where k may be regarded as a "pseudo rate constant" for the adsorption process.

The rate of desorption depends only on the number of molecules that are adsorbed. Thus

$$r_{\text{desorption}} = k'\theta_A \qquad (6.2.2)$$

where k' may be regarded as a pseudo rate constant for the desorption process.

At equilibrium the rates of adsorption and desorption are equal.

$$kP_A(1 - \theta_A) = k'\theta_A \qquad (6.2.3)$$

The fraction of the sites occupied by species A is then

$$\theta_A = \frac{kP_A}{k' + kP_A} \qquad (6.2.4)$$

If one takes the ratio of the pseudo rate constant for adsorption to that for desorption as an equilibrium constant for adsorption (K), equation 6.2.4 can be written as

$$\theta_A = \frac{\dfrac{k}{k'} P_A}{1 + \dfrac{kP_A}{k'}} = \frac{KP_A}{1 + KP_A} \qquad (6.2.5)$$

The fraction of the sites that are occupied is also equal to the ratio of the volume of gas actually adsorbed to that which would be adsorbed in a monolayer (v_m).

$$\theta_A = \frac{v}{v_m} \qquad (6.2.6)$$

Both volumes are measured at standard conditions or at a fixed reference temperature and pressure.

The last two equations can be combined to give a relation between the pressure of the gas and the amount that is adsorbed.

$$v = \frac{v_m KP_A}{1 + KP_A} \qquad (6.2.7)$$

A plot of v versus P_A is of the same form as the Type I adsorption isotherm. At low values of P_A the term KP_A is small compared to unity, and the amount adsorbed will be linear in pressure. At high pressures the term KP_A is large compared to unity, and the surface coverage is nearly complete. In this case v will approach v_m.

Equation 6.2.7 may be transformed into several expressions that can be used in analyzing experimental data. However, the following form is preferred because it avoids undue emphasis on the low pressure points, which are most susceptible to error.

$$\frac{P_A}{v} = \frac{1}{v_m K} + \frac{P_A}{v_m} \qquad (6.2.8)$$

If Type I adsorption behavior is obeyed, a plot of P_A/v versus P_A should be linear with slope $1/v_m$. Once the volume corresponding to a monolayer has been determined, it can be converted to the number of molecules adsorbed by dividing by the molal volume at the reference conditions and multiplying by Avogadro's number (N_0). When this number of molecules is multiplied in turn by the area covered per adsorbed molecule (α), the total surface area of the catalyst (S) is obtained. Thus,

$$S = \left(\frac{v_m N_0}{v_{\text{reference conditions}}} \right) \alpha \qquad (6.2.9)$$

Specific surface areas are then obtained by dividing by the weight of catalyst employed in the experiments in question. It should be pointed out, however, that it is the BET adsorption isotherm that is the basis for conventional determinations of catalyst surface areas. (See Section 6.2.2.)

6.2.1.2 The Langmuir Equation for the Case Where Two or More Species May Adsorb. Adsorption isotherms for cases where more than one species may adsorb are of considerable significance when one is dealing with heterogeneous catalytic reactions. Reactants, products, and inert species may all adsorb on the catalyst surface. Consequently, it is useful to develop generalized Langmuir adsorption isotherms for multicomponent adsorption. If θ_i represents the fraction of the sites occupied by species i, the fraction of the sites that is vacant is just $1 - \sum \theta_i$ where the summation is taken over all species that can be adsorbed. The pseudo rate constants for adsorption and desorption may be expected to differ for each species, so they will be denoted by k_i and k'_i, respectively.

The rates of adsorption and desorption of each species must be equal at equilibrium. Thus:

For species A:
$$k_A P_A(1 - \theta_A - \theta_B - \theta_C \cdots) = k'_A \theta_A$$

For species B:
$$k_B P_B(1 - \theta_A - \theta_B - \theta_C \cdots) = k'_B \theta_B$$

For species C:
$$k_C P_C(1 - \theta_A - \theta_B - \theta_C \cdots) = k'_C \theta_C$$
$$\vdots$$
$$(6.2.10)$$

where P_i is the partial pressure of the ith species in the gas phase.

If these equations are solved for the fractions occupied by each species,

$$\theta_A = \frac{k_A}{k'_A} P_A(1 - \theta_A - \theta_B - \theta_C \cdots)$$
$$= K_A P_A(1 - \theta_A - \theta_B - \theta_C \cdots)$$

$$\theta_B = \frac{k_B}{k'_B} P_B(1 - \theta_A - \theta_B - \theta_C \cdots)$$
$$= K_B P_B(1 - \theta_A - \theta_B - \theta_C \cdots)$$

$$\theta_C = \frac{k_C}{k'_C} P_C(1 - \theta_A - \theta_B - \theta_C \cdots)$$
$$= K_C P_C(1 - \theta_A - \theta_B - \theta_C \cdots)$$
$$\vdots$$
$$(6.2.11)$$

where the adsorption equilibrium constants K_i have been substituted for the ratios of k_i to k'_i.

If the expressions for θ_i are added,

or
$$1 - \sum \theta_i = \frac{1}{1 + \sum K_i P_i} \qquad (6.2.15)$$

Thus
$$\theta_A = \frac{K_A P_A}{1 + \sum K_i P_i}$$
$$= \frac{K_A P_A}{1 + K_A P_A + K_B P_B + K_C P_C + \cdots} \qquad (6.2.16)$$

and
$$\theta_B = \frac{K_B P_B}{1 + K_A P_A + K_B P_B + K_C P_C + \cdots} \qquad (6.2.17)$$

Note that at low surface coverages where $(1 + K_A P_A + K_B P_B + \cdots) \approx 1$, the fraction of the sites occupied by each species will be proportional to its partial pressure.

6.2.1.3 The Langmuir Equation for the Case Where Dissociation Occurs on Adsorption.

There is evidence that certain chemical adsorption processes involve dissociation of the adsorbate to form two bonds with the adsorbent surface. On many metals hydrogen is adsorbed in atomic form.

In this case the kinetic approach to the derivation of the Langmuir equation requires that the process be regarded as a reaction between the gas molecule and *two surface sites*. Thus the

$$\theta_A + \theta_B + \theta_C + \cdots = (1 - \theta_A - \theta_B - \theta_C \cdots)(K_A P_A + K_B P_B + K_C P_C + \cdots) \qquad (6.2.12)$$

or
$$\sum \theta_i = (1 - \sum \theta_i)(\sum K_i P_i) \qquad (6.2.13)$$

Solving for $\sum \theta_i$,

$$\sum \theta_i = \frac{\sum K_i P_i}{1 + \sum K_i P_i} \qquad (6.2.14)$$

adsorption rate is written as

$$r_{\text{adsorption}} = k_A P_A(1 - \theta_A)^2 \qquad (6.2.18)$$

where we have assumed that only one species is adsorbed.

The desorption process must involve a reaction between two adsorbed atoms to regenerate the gas phase molecules. Consequently, it may be regarded as a second-order reaction between two surface species:

$$r_{\text{desorption}} = k_A' \theta_A^2 \qquad (6.2.19)$$

At equilibrium

$$k_A P_A (1 - \theta_A)^2 = k_A' \theta_A^2 \qquad (6.2.20)$$

or

$$\frac{\theta_A}{1 - \theta_A} = \sqrt{\frac{k_A P_A}{k_A'}} = \sqrt{K_A P_A} \qquad (6.2.21)$$

Solving for θ_A,

$$\theta_A = \frac{\sqrt{K_A P_A}}{1 + \sqrt{K_A P_A}} \qquad (6.2.22)$$

At low pressures where the surface is sparsely covered, the fraction of the sites occupied by fragments of species A will be proportional to the square root of the gas phase partial pressure.

If several species can adsorb but only species A dissociates, it is easily shown, using the type of analysis employed in Section 6.2.1.2, that

$$\theta_A = \frac{\sqrt{K_A P_A}}{1 + \sqrt{K_A P_A} + K_B P_B + K_C P_C \cdots} \qquad (6.2.23)$$

and that

$$\theta_B = \frac{K_B P_B}{1 + \sqrt{K_A P_A} + K_B P_B + K_C P_C \cdots} \qquad (6.2.24)$$

6.2.2 The BET Isotherm

Inasmuch as the Langmuir equation does not allow for nonuniform surfaces, interactions between neighboring adsorbed species, or multilayer adsorption, a variety of theoretical approaches that attempt to take one or more of these factors into account have been pursued by different investigators. The best-known alternative is the BET isotherm, which derives its name from the initials of the three individuals responsible for its formulation, Brunauer, Em-

mett, and Teller (8). It takes multilayer adsorption into account and is the basis of standard methods for determining specific surface areas of heterogeneous catalysts. The extended form of the BET equation (4) can be used to derive all five types of isotherms as special cases.

The BET approach is essentially an extension of the Langmuir approach. Van der Waals forces are regarded as the dominant forces, and the adsorption of all layers is regarded as physical, not chemical. One sets the rates of adsorption and desorption equal to one another, as in the Langmuir case; in addition, one requires that the rates of adsorption and desorption be identical for each and every molecular layer. That is, the rate of condensation on the bare surface is equal to the rate of evaporation of molecules in the first layer. The rate of evaporation from the second layer is equal to the rate of condensation on top of the first layer, etc. One then sums over the layers to determine the total amount of adsorbed material. The derivation also assumes that the heat of adsorption of each layer other than the first is equal to the heat of condensation of the bulk adsorbate material (i.e., van der Waals forces of the adsorbent are transmitted only to the first layer). If it is assumed that a very large or effectively infinite number of layers can be adsorbed, the following result is arrived at after a number of relatively elementary mathematical operations

$$v = \frac{v_m c P}{(P_0 - P)\left[1 + \dfrac{(c - 1)P}{P_0}\right]} \qquad (6.2.25)$$

where c is a constant exponentially related to the heats of adsorption of the first layer and the heat of liquefaction, and P_0 is the saturation pressure.

This equation can be rearranged to give a somewhat more familiar form.

$$\frac{x}{v(1 - x)} = \frac{1}{v_m c} + \left(\frac{c - 1}{v_m c}\right) x \qquad (6.2.26)$$

where x is the normalized pressure (P/P_0).

If one plots the left side of this equation versus x, a straight line should result, with

$$v_m = \frac{1}{\text{slope} + \text{intercept}} \quad (6.2.27)$$

Equation 6.2.9 may then be used in determining the specific surface area of an adsorbent. Alternative arrangements of equation 6.2.25 are possible, but they have the inherent disadvantage of placing undue emphasis on the low-pressure data points that are most susceptible to error.

If one restricts the number of layers of adsorbate that may be stacked up, as would be the case in the very narrow capillaries of a porous catalyst, the BET analysis must be modified to allow for this. If n is the number of permissible layers, it can be shown that the adsorption isotherm becomes:

$$v = \frac{v_m c x [1 - (n+1)x^n + nx^{n+1}]}{(1-x)[1 + (c-1)x - cx^{n+1}]}$$
$$(6.2.28)$$

When $n = 1$, this equation reduces to the Langmuir form. Since x is always less than unity, it does not take very large values of n to approach the limiting form indicated by equation 6.2.25. By appropriate choice of the parameters in equation 6.2.28, it is possible to generate the five types of curves shown in Figure 6.2.

Many individuals have developed more elegant theoretical treatments of adsorption processes since Brunauer, Emmett, and Teller published their classic paper. Nonetheless, the BET and Langmuir isotherms are the most significant ones for chemical engineering applications.

6.3 REACTION RATE EXPRESSIONS FOR HETEROGENEOUS CATALYTIC REACTIONS

When a heterogeneous catalytic reaction occurs, several physical and chemical processes must take place in proper sequence. Hougen and Watson (9) and others have broken down the steps that occur on a molecular scale in the following manner.

1. Mass transfer of reactants from the main body of the fluid to the gross exterior surface of the catalyst particle.
2. Molecular diffusion and/or Knudsen flow of reactants from the exterior surface of the catalyst particle into the interior pore structure.
3. Chemisorption of at least one of the reactants on the catalyst surface.
4. Reaction on the surface. (This may involve several steps.)
5. Desorption of (chemically) adsorbed species from the surface of the catalyst.
6. Transfer of products from the interior catalyst pores to the gross external surface of the catalyst by ordinary molecular diffusion and/or Knudsen diffusion.
7. Mass transfer of products from the exterior surface of the particle into the bulk of the fluid.

Several of these steps are shown in schematic fashion in Figure 6.3. Of course, if the catalyst is nonporous, steps 2 and 6 are absent. Steps 1, 2, 6, and 7 are obviously physical processes, while steps 3 to 5 are basically chemical in character. The rates of the various steps depend on a number of factors in addition to the concentration profiles of the reactant and product species.

Steps 1 and 7 are highly dependent on the fluid flow characteristics of the system. The mass velocity of the fluid stream, the particle size, and the diffusional characteristics of the various molecular species are the pertinent parameters on which the rates of these steps depend. These steps limit the observed rate only when the catalytic reaction is very rapid and the mass transfer is slow. Anything that tends to increase mass transfer coefficients will enhance the rates of these processes. Since the rates of these steps are only slightly influenced by temperature, the influence of these processes

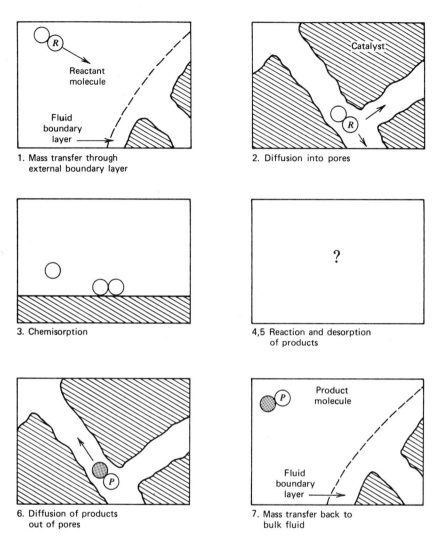

Figure 6.3
Schematic representation of heterogeneous catalytic reaction on a porous catalyst.

on the overall conversion rate will vary as the temperature changes. Their influence is often negligible at low temperatures, but may be quite significant at higher temperatures.

There are two commonly used techniques for determining the range of process variables where these steps influence conversion rates. The first approach involves carrying out a series of runs

in which one measures the conversion achieved as a function of the ratio of the weight of catalyst (W) to the molal feed flow rate (F). One then repeats the experiments at a different linear velocity through the reactor (e.g., by changing the length-to-diameter ratio of the catalyst bed in a tubular reactor). The data are then plotted as shown in Figures 6.4a and 6.4b. If the two curves do not coincide as shown in Figure 6.4b, mass transfer resistances are significant, at least in the low-velocity series. If the curves coincide as shown in Figure 6.4a, mass transfer effects probably do not influence the observed conversion rate. This diagnostic test must be applied

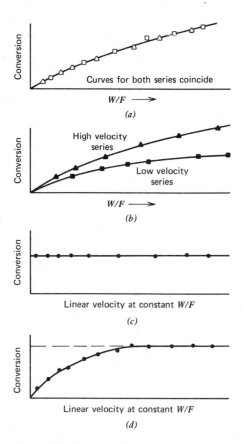

Figure 6.4
Tests for external mass transfer limitations on conversion rates. (a) External mass transfer probably does not limit conversion rate. (b) Mass transfer limitations are present. (c) External mass transfer probably does not limit conversion rate. (d) Mass transfer limitations are present at low velocities. Adapted from Chemical Engineers' Handbook, **Fourth Edition, edited by R. H. Perry, C. H. Chilton, and S. D. Kirkpatrick. Copyright © 1969. Used with permission of McGraw-Hill Book Company.)**

with caution; Chambers and Boudart (10) have shown that it often lacks sensitivity under conditions commonly employed in laboratory scale studies. Heat and mass transfer coefficients are not strongly dependent on fluid velocity for Reynolds numbers below 50. (The characteristic length dimension is taken as the pellet diameter.) Consequently, at low Reynolds numbers, the conversion will not change significantly with fluid velocity at constant W/F, even though heat or mass transfer limitations may be present. The scatter in the experimental data will be sufficient to mask these effects for many reactions. Thus, when the experimental diagnostic test indicates that mass transfer limitations are probably not present, one's conclusions should be verified by calculations of the sensitivity of the test to variations in fluid velocity. The methods developed in Sections 12.4 and 12.5 may be used for this purpose.

The second approach involves simultaneous variation of the weight of catalyst and the molal flow rate so as to maintain W/F constant. One then plots the conversion achieved versus linear velocity, as shown in Figures 6.4c and 6.4d. If the results are as indicated in Figure 6.4d, mass transfer limitations exist in the low-velocity regime. If the conversion is independent of velocity, there probably are no mass transfer limitations on the conversion rate. However, this test is also subject to the sensitivity limitations noted above.

When fluid velocities are high relative to the solid, mass transfer is rapid. However, in stagnant regions or in batch reactors where no provision is made for agitation, one may encounter cases where mass transfer limits the observed reaction rate. We should also note that in industrial practice pressure drop constraints may make it impractical to employ the exceedingly high velocities necessary to overcome the mass transfer resistance associated with highly active catalysts.

For porous catalysts the external gross surface area of a particle often represents only an insignificant fraction of the total surface area that is potentially capable of catalyzing a given reaction. For these materials, diffusional processes occurring within the pores can have marked effects on the conversion rates that can be observed in the laboratory. When working with this type of catalyst, it is necessary to recognize that steps 2 and 6 enumerated above play key roles in determining how effective the total surface area is in bringing about the desired reaction. In dealing with an active catalyst, equilibrium or complete conversion may be substantially achieved in the region near the mouth of the pore, and the observed conversion rate will not differ appreciably from that which would be observed if the catalytic surface area deep within the pores were not present. The efficiency with which the internal surface area of a porous catalyst is used is an extremely important consideration in the design of catalytic reactors. This subject is treated in detail in Section 12.3. The parameters that govern the influence of pore diffusion processes on the observed conversion rates are the degree of porosity of the catalyst, the dimensions of the pores, the degree to which the pores are interconnected, the gross dimensions of the catalyst particle itself, the rate at which the reaction occurs on the catalyst surface, and the diffusional characteristics of the reaction mixture. In order to determine if steps 2 and 6 are influencing the overall conversion rate, examine the effect of changing the gross particle dimensions, for example, by breaking up a catalyst into smaller pieces. For a given weight of catalyst the total surface area will remain essentially unchanged as a result of this process, but the effective surface area will increase significantly when pore diffusion processes have had a strong influence on the experimentally observed rate. Reducing the gross exterior dimensions of the particles decreases the average length of the pores and makes a greater fraction of the total surface area available to the reactant species. The increase in effective surface area that accompanies particle fragmentation is shown schematically in Figure 6.5.

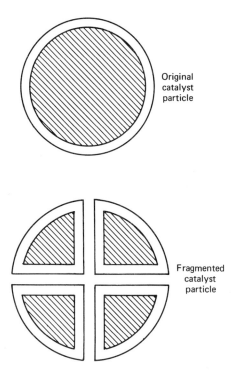

Original catalyst particle

Fragmented catalyst particle

Figure 6.5

Effect of fragmentation on catalyst utilization when intraparticle diffusion is rate controlling (shaded areas represent regions of the catalyst with insignificant concentrations of reactants).

The three remaining steps (chemisorption of reactants, reaction on the surface, and desorption of adsorbed products) are all chemical in nature. It is convenient to employ the concept of a rate limiting step in the treatment of these processes so that the reaction rate becomes equal to that of the slowest step. The other steps are presumed to be sufficiently rapid that quasi-equilibrium relations may be used. The overall rate of conversion will then be determined by the interaction of the rate of the process that is rate limiting from a chemical point of view with the rates of the physical mass transfer processes discussed above.

The remainder of this section is devoted to a discussion of the types of rate expressions that are obtained when purely chemical phenomena determine observed conversion rates.

6.3.1 Rate Expressions for Heterogeneous Catalytic Reactions Limited by the Rates of Chemical Processes

In the treatment of rate expressions for heterogeneous catalytic reactions the definition of local reaction rates in terms of interfacial areas (3.0.10) is appropriate.

$$r'' = \frac{1}{S}\frac{d\xi}{dt} \qquad (6.3.1)$$

where S is the surface area of the solid catalyst.

If an attempt is made to fit heterogeneous catalytic reaction rate data to a rate expression of the form

$$r'' = kC_A^{\beta_A}C_B^{\beta_B}C_C^{\beta_C}\cdots \qquad (6.3.2)$$

it is often found that the reaction orders are not integers, that the orders depend on temperature and perhaps on concentration, and that the observed activation energy may depend on temperature. Such expressions are of rather limited use for design purposes and can be used with confidence only under conditions that closely resemble those under which the experimental data were recorded. The basic reason this sort of approach often gives rise to such difficulties is that it assumes that the driving force for reaction is the concentrations of the species present in the fluid phase. A more logical approach is to use the surface concentrations of the adsorbed species in an expression analogous to equation 6.3.2; that is,

$$r'' = k''\theta_A^{\beta_A}\theta_B^{\beta_B}\theta_C^{\beta_C}\cdots \qquad (6.3.3)$$

where θ_i is the fraction of the surface covered by species i. If the various θ_i can be related to bulk fluid phase concentrations, one has an appropriate mathematical form to use in testing a proposed reaction rate expression. The basis for most analyses of this type is the Langmuir isotherm. When expressions for the various θ_i in terms of the appropriate partial pressures involved in the Langmuir isotherm are substituted in equation 6.3.3, a Langmuir-Hinshelwood or Hougen-Watson rate expression is

obtained. The latter two individuals were responsible for analyzing a number of possible reaction mechanisms in these terms (11, 12) and for popularizing this approach to analyzing catalytic reaction rate data. For simplicity the term "Hougen-Watson models" will be used throughout the remainder of this text.

In addition to the assumptions implicit in the use of the Langmuir isotherm the following assumption is applicable to all Hougen-Watson models: the reaction involves at least one species chemisorbed on the catalyst surface. If reaction takes place between two adsorbed species, they must be adsorbed on neighboring sites in order for reaction to occur. The probability of reaction between adsorbed A and adsorbed B is assumed to be proportional to the product of the fractions of the sites occupied by each species ($\theta_A\theta_B$). Similar considerations apply to termolecular reactions occurring on the surface.

In addition to the aforementioned assumptions, one *normally* elects one of two mutually incompatible assumptions as a basis for his or her analysis. These two limiting cases are as follows.

Category I. *Adsorption equilibrium is maintained at all times, and the overall rate of reaction is governed by the rate of chemical reaction on the surface. The expressions developed for θ_i in Section 6.2 can be used in this case.*

Category II. *The rate of chemical reaction on the surface is so rapid that adsorption equilibrium is not achieved, but a steady-state condition is reached in which the amount of adsorbed material remains constant at some value less than the equilibrium value. This value is presumed to be that corresponding to equilibrium for the surface reaction at the appropriate fractional coverages of the other species involved in the surface reaction. The rate of adsorption or desorption of one species is presumed to be much slower than that of any other species. This step is then the rate limiting step in the overall reaction.*

We will treat Category I models first.

6.3.1.1 Hougen-Watson Models for the Case of Equilibrium Adsorption.

This section treats Hougen-Watson mathematical models for cases where the rate limiting step is the chemical reaction rate on the surface. In all cases it is assumed that equilibrium is established with respect to adsorption of all species.

Case I. *Irreversible Reaction (Unimolecular)*

Consider the following mechanistic equation for a reaction occurring on a catalyst surface.

$$A \rightarrow R \qquad (6.3.4)$$

where both A and R are adsorbed in appreciable amounts. In this case the reaction rate is proportional to the fraction of the surface occupied by species A.

$$r'' = k\theta_A \qquad (6.3.5)$$

This fraction is given by equation 6.2.16, the Langmuir adsorption isotherm for the case where several species may be adsorbed. Thus

$$r'' = \frac{kK_A P_A}{1 + K_A P_A + K_R P_R} \qquad (6.3.6)$$

There are a number of limiting forms of the rate expression depending on the magnitudes of the various terms in the denominator relative to unity and to each other. Any species that is weakly adsorbed will not appear in the denominator. If species R undergoes dissociation on adsorption, the term $K_R P_R$ must be replaced by $\sqrt{K_R P_R}$ according to the discussion in Section 6.2.1.3. If an inert species I is also capable of adsorption, the term $K_I P_I$ must be added to the sum in the denominator. If the product R is strongly adsorbed, equation 6.3.6 provides a basis for explaining inhibition of the reaction by that material. For example, if the dominant term in the denominator is $K_R P_R$, the appropriate approximation for the rate expression is

$$r'' = \frac{kK_A P_A}{K_R P_R} \qquad (6.3.7)$$

The decomposition of ammonia on platinum has a rate expression of this form. The reaction is first order in ammonia and inverse first order in hydrogen.

The temperature variation of heterogeneous catalytic reaction rates can be quite complex. The apparent activation energy can in many cases be related to the activation energy of the surface reaction and the enthalpy changes accompanying the adsorption of various species. If one of the terms in the denominator of the right side of equation 6.3.6 is very much larger than any of the other terms, this is possible. If not, one cannot relate the apparent activation energy to these parameters. The dependence of the adsorption equilibrium constants on temperature is given by

$$\frac{d \ln K}{d \left(\frac{1}{T} \right)} = \frac{-\Delta H}{R} \qquad (6.3.8)$$

where ΔH is the enthalpy change accompanying chemisorption. Consequently, the apparent activation energies corresponding to some limiting forms of equation 6.3.6 are as given below.

Dominant term in denominator	Limiting form of rate expression	Apparent rate constant	Apparent activation energy
1	$r'' = kK_A P_A$	kK_A	$E + \Delta H_A$
$K_A P_A$	$r'' = k$	k	E
$K_R P_R$	$r'' = \dfrac{kK_A P_A}{K_R P_R}$	$\dfrac{kK_A}{K_R}$	$E + \Delta H_A - \Delta H_R$

where E is the activation energy of the surface reaction and ΔH_i is the enthalpy change accompanying chemisorption of species i. In the last case, the relative magnitudes of the heats of adsorption of A and R will determine whether the apparent activation energy will be greater than or less than that of the surface reaction.

Case II. *Irreversible Bimolecular Reaction Between Adsorbed Species on the Same Type of Site*

Consider the following irreversible bimolecular reaction, which takes place between two species adsorbed on a catalyst surface.

$$A + B \rightarrow R + S \qquad (6.3.9)$$

In this case the reaction rate is assumed to be proportional to the product of the fractions of the sites occupied by species A and B.

$$r'' = k\theta_A\theta_B \qquad (6.3.10)$$

In the general case where all species may be adsorbed on the same type of surface site equations 6.2.16 and 6.2.17 may be used for θ_A and θ_B. Thus

$$r'' = \frac{kK_AK_BP_AP_B}{(1 + K_AP_A + K_BP_B + K_RP_R + K_SP_S)^2} \qquad (6.3.11)$$

There are several limiting forms of this rate expression, depending on which species are strongly or weakly adsorbed.

If all partial pressures except that of one reactant, say species A, are held constant and the partial pressure of species A is varied, the rate will go through a maximum. The rate increases initially with increasing partial pressure of reactant A as the fraction of sites occupied by species A increases. However, as this fraction increases, the fraction occupied by species B declines as B molecules are displaced by A molecules. Eventually one reaches a point where the decline in the value of θ_B more than offsets the increase in θ_A and the product $\theta_A\theta_B$ goes through a maximum.

For the case of bimolecular reaction between two adsorbed A molecules, similar arguments lead to the following rate expression.

$$r'' = \frac{kK_A^2P_A^2}{(1 + K_AP_A + K_RP_R + K_SP_S)^2} \qquad (6.3.12)$$

Case III. *Irreversible Bimolecular Reactions Between Species Sorbed on Different Types of Sites*

With certain types of catalysts it is easy to postulate that more than one type of chemisorption site may exist on the solid surface. For example, in the case of metal oxide catalysts, one might speculate that certain species could chemisorb by interaction with metal atoms at the surface, while other species could interact with surface oxygen atoms. Consider the possibility that species A adsorbed on one type of site will react with species B adsorbed on a second type of site according to the following reaction.

$$A_{(\text{Type 1 site})} + B_{(\text{Type 2 site})} \rightarrow \text{Products} \qquad (6.3.13)$$

For this mechanism the rate expression is given by

$$r'' = k\theta_{A1}\theta_{B2} \qquad (6.3.14)$$

where θ_{A1} is the fraction of the sites of Type 1 occupied by species A and θ_{B2} is the fraction of the sites of Type 2 occupied by species B. If the Langmuir adsorption isotherm equation is written for each type of site, if only species A is adsorbed on sites of Type 1, and if only species B is adsorbed on sites of Type 2,

$$r'' = k\left(\frac{K_{A1}P_A}{1 + K_{A1}P_A}\right)\left(\frac{K_{B2}P_B}{1 + K_{B2}P_B}\right) \qquad (6.3.15)$$

Rate expressions of this type have been proposed to correlate the data for the reaction between hydrogen and carbon dioxide on tungsten.

Since adsorption occurs independently on each type of site and displacement of one species by another does not occur, this type of reaction will not exhibit a maximum of the type noted in

<u>Case II</u>. Instead, the rate will increase with increasing pressure of one reactant, eventually approaching an asymptotic limit corresponding to saturation of the type of site in question.

Case IV. Irreversible Bimolecular Reactions Between an Adsorbed Species and a Fluid Phase Molecule

The first three examples consider cases where only adsorbed species participate in the reaction. The possibility also exists that reaction may occur between a gaseous molecule and an adsorbed species. This mechanism is referred to as a Rideal mechanism. If species A refers to the adsorbed species and species B to a gas phase molecule, one form of this bimolecular reaction may be written as

$$A + B(g) \rightarrow R + S \qquad (6.3.16)$$

with a rate expression given by

$$r'' = k\theta_A P_B \qquad (6.3.17)$$

If we consider the general case where all species may be adsorbed, θ_A is given by equation 6.2.16, and the rate becomes

$$r'' = \frac{kK_A P_A P_B}{1 + K_A P_A + K_B P_B + K_R P_R + K_S P_S} \qquad (6.3.18)$$

If species B had been adsorbed rather than species A, the corresponding rate equation would be

$$r'' = \frac{kK_B P_A P_B}{1 + K_A P_A + K_B P_B + K_R P_R + K_S P_S} \qquad (6.3.19)$$

Both equations have the same mathematical form. In these cases there will not be a maximum in the rate as the pressure of one species varies. Instead, the rate will approach an asymptotic value at high pressures. Consequently this type of mechanism can be excluded from subsequent consideration if it can be shown that the experimentally observed rate passes through a

maximum as the concentration of a reactant is increased at constant values of other species concentrations.

If the surface is nearly covered ($\theta_A \approx 1$) the reaction will be first-order in the gas phase reactant and zero-order in the adsorbed reactant. On the other hand, if the surface is sparsely covered ($\theta_A \approx K_A P_A$) the reaction will be first-order in each species or second-order overall. Since adsorption is virtually always exothermic, the first condition will correspond to low temperature and the second condition to high temperatures. This mechanism thus offers a ready explanation of a transition from first- to second-order reaction with increasing temperature.

Case V. Reversible Reactions Between Adsorbed Species (Change in Number of Moles on Reaction)

Consider the reaction

$$A \underset{k_{-1}}{\overset{k_1}{\rightleftharpoons}} R + S \qquad (6.3.20)$$

where all species are adsorbed on surface sites. It is assumed that an adsorbed A molecule reacts with a *vacant site* to form an intermediate that then dissociates to form adsorbed R and S species. This implies that the number of sites involved in a reaction obeys a conservation principle.

The rate of this surface reaction can be written as

$$r'' = k_1 \theta_A \theta_v - k_{-1} \theta_R \theta_S \qquad (6.3.21)$$

where θ_v is the fraction of the sites that are vacant. This quantity is given by

$$\theta_v = 1 - (\theta_A + \theta_R + \theta_S) \qquad (6.3.22)$$

or

$$\theta_v = 1 - \frac{K_A P_A + K_R P_R + K_S P_S}{1 + K_A P_A + K_R P_R + K_S P_S}$$

$$= \frac{1}{1 + K_A P_A + K_R P_R + K_S P_S} \qquad (6.3.23)$$

Substitution of this result into equation 6.3.21 together with the appropriate expressions for the fractional coverages of other species gives

$$r'' = \frac{k_1 K_A P_A - k_{-1} K_R K_S P_R P_S}{(1 + K_A P_A + K_R P_R + K_S P_S)^2}$$

(6.3.24)

or

$$r'' = \frac{k_1 K_A \left(P_A - \dfrac{K_R K_S P_R P_S}{K_r K_A} \right)}{(1 + K_A P_A + K_R P_R + K_S P_S)^2}$$

(6.3.25)

where K_r is the equilibrium constant for the surface reaction.

$$K_r = \frac{k_1}{k_{-1}}$$

(6.3.26)

If we let

$$k = k_1 K_A$$

(6.3.27)

and

$$K = \frac{K_r K_A}{K_R K_S}$$

(6.3.28)

equation 6.3.25 can be rewritten as

$$r'' = \frac{k \left(P_A - \dfrac{P_R P_S}{K} \right)}{(1 + K_A P_A + K_R P_R + K_S P_S)^2}$$

(6.3.29)

The cases discussed above represent only a small fraction of the surface reaction mechanisms which might be considered. Yang and Hougen (12) have considered several additional surface reaction mechanisms and have developed tables from which rate expressions for these mechanisms may be determined. They approached this problem by writing the rate expression in the following form.

$$\text{Rate} = \frac{(\text{kinetic term}) \times (\text{driving force})}{(\text{adsorption term})^n}$$

(6.3.30)

Let us examine each of the terms in this formulation of the rate expression.

Driving Force Term: In all rate expressions the *driving force* must become zero when thermodynamic equilibrium is established. The "equilibrium" constant K appearing in equation 6.3.29 can be regarded as the appropriate ratio of partial pressures for the overall reaction.

$$K = \frac{K_r K_A}{K_R K_S} = \frac{P_{Re} P_{Se}}{P_{Ae}}$$

(6.3.31)

In the general case the value of K appearing in the driving force term is the product of the equilibrium constant for the surface reaction K_r and the product of the adsorption equilibrium constants for the reactants divided by the product of the adsorption equilibrium constants for the reaction products.

Kinetic Term: The designation *kinetic term* is something of a misnomer in that it contains both rate constants and adsorption equilibrium constants. For the cases where surface reaction controls the overall conversion rate it is the product of the surface reaction rate constant for the forward reaction and the adsorption equilibrium constants for the reactant surface species participating in the reaction. When adsorption or desorption of a reactant or product species is the rate limiting step, it will involve other factors.

Adsorption Term: When all reactants, products, and inerts are adsorbed without dissociation such that adsorption equilibrium prevails, the *adsorption term* is given by

$$(1 + K_A P_A + K_B P_B + K_R P_R + K_S P_S + K_I P_I)^n$$

where K_I and P_I are the adsorption equilibrium constant and the gas phase partial pressure of inerts, respectively. When dissociation of a species occurs, the term $K_i P_i$ must be replaced by $\sqrt{K_i P_i}$ as long as adsorption equilibrium holds for all species. When adsorption equilibrium is not maintained, other modifications in the adsorption term are necessary.

The exponent n on the adsorption term is equal to the number of surface sites participating in the reaction, whether they hold adsorbed reactants or participate as vacant sites. If more than one site is involved, we assume that the sites must be adjacent for the surface reaction to occur.

6.3.1.2 Hougen-Watson Models for Cases where Adsorption and Desorption Processes are the Rate Limiting Steps.

When surface reaction processes are very rapid, the overall conversion rate may be limited by the rate at which adsorption of reactants or desorption of products takes place. Usually only one of the many species in a reaction mixture will not be in adsorptive equilibrium. This generalization will be taken as a basis for developing the expressions for overall conversion rates that apply when adsorption or desorption processes are rate limiting. In this treatment we will assume that chemical reaction equilibrium exists between various adsorbed species *on the catalyst surface*, even though reaction equilibrium will not prevail in the fluid phase.

Consider the reversible surface reaction

$$A + B \underset{k_{-1}}{\overset{k_1}{\rightleftharpoons}} R + S \qquad (6.3.32)$$

in which we presume that species A is not in adsorptive equilibrium and that the rate limiting step in the overall conversion process is the rate of adsorption of species A. When A molecules are adsorbed, the chemical reaction equilibrium existing between surface species is disturbed, and adsorbed A molecules must then react. The reaction occurs so rapidly that adsorptive equilibrium for species A cannot be established, and θ_A is determined by surface reaction equilibrium constraints.

Since the *surface reaction* is at equilibrium,

$$k_1 \theta_A \theta_B = k_{-1} \theta_R \theta_S \qquad (6.3.33)$$

Thus

$$\theta_A = \frac{k_{-1}\theta_R\theta_S}{k_1\theta_B} = \frac{\theta_R\theta_S}{K_r\theta_B} \qquad (6.3.34)$$

where K_r is the equilibrium constant for the surface reaction.

The rates of adsorption and desorption of species B are equal. Hence,

$$k_{1B}P_B\theta_v = k_{-1B}\theta_B \qquad (6.3.35)$$

where θ_v is the fraction of the sites that are vacant, and k_{1B} and k_{-1B} are the rate constants for adsorption and desorption of species B.

Solving for θ_B,

$$\theta_B = \frac{k_{1B}P_B\theta_v}{k_{-1B}} = K_B P_B \theta_v \qquad (6.3.36)$$

A similar analysis is applicable to other species that attain adsorptive equilibrium. Thus,

$$\theta_R = K_R P_R \theta_v \quad \text{and} \quad \theta_S = K_S P_S \theta_v \qquad (6.3.37)$$

The sum of the fractions of sites occupied by the various species and the fraction vacant must be unity.

$$1 = \theta_A + \theta_B + \theta_R + \theta_S + \theta_v \qquad (6.3.38)$$

Combination of equations 6.3.34 and 6.3.36 to 6.3.38 gives

$$\theta_v = \frac{1}{1 + \dfrac{K_R K_S}{K_r K_B}\dfrac{P_R P_S}{P_B} + K_B P_B + K_R P_R + K_S P_S} \qquad (6.3.39)$$

Since the rate limiting step in the overall process is the rate of adsorption of species A, the net rate of reaction is equal to the difference between the rates of adsorption and desorption.

$$r'' = k_{1A}P_A\theta_v - k_{-1A}\theta_A \qquad (6.3.40)$$

or

$$r'' = k_{1A}\left(P_A\theta_v - \frac{\theta_A}{K_A}\right) \qquad (6.3.41)$$

Combining equations 6.3.34, 6.3.36, 6.3.37, 6.3.39, and 6.3.41 gives

$$r'' = \frac{k_{1A}\left(P_A - \dfrac{K_R K_S P_R P_S}{K_r K_A K_B P_B}\right)}{1 + \dfrac{K_R K_S}{K_r K_B}\dfrac{P_R P_S}{P_B} + K_B P_B + K_R P_R + K_S P_S} \qquad (6.3.42)$$

This Hougen-Watson model for the case where the adsorption of a single species is rate limiting is of the same mathematical form as equation 6.3.30. In this case the kinetic term is merely the rate constant for the rate controlling adsorption process. The driving force term will depend on the stoichiometry of the surface reaction, which is presumed to be at equilibrium. The adsorption term must also be modified in these cases, the change occurring in the element in the summation corresponding to the species that is not at adsorptive equilibrium. This term will also depend on the nature of the surface reaction involved.

Extension of the analysis developed above to cases where the stoichiometry of the surface reaction differs from that considered is relatively simple and straightforward. An interesting case is that where the overall conversion rate is limited by adsorption of a species that dissociates on adsorption. Consider a reaction whose stoichiometry can be represented by

$$A_2 \rightleftharpoons R + S \qquad (6.3.43)$$

where dissociation of the A_2 molecule occurs on adsorption. The mechanism of the reaction on the catalyst surface is represented by

$$2A \underset{k_{-1}}{\overset{k_1}{\rightleftharpoons}} R + S \qquad (6.3.44)$$

In this case the above procedure can be used to show that

$$r'' = \frac{k_{1A}\left(P_{A_2} - \dfrac{K_R K_S P_R P_S}{K_A K_r}\right)}{\left(1 + \sqrt{\dfrac{K_R K_S P_R P_S}{K_r}} + K_R P_R + K_S P_S\right)^2} \qquad (6.3.45)$$

Now consider the case where the controlling step is the rate of desorption of a product species R for a reversible surface reaction of the form

$$A + B \underset{k_{-1}}{\overset{k_1}{\rightleftharpoons}} R + S \qquad (6.3.46)$$

All other species are presumed to be in adsorp-

tive equilibrium; that is,

$$\theta_A = K_A P_A \theta_v \qquad \theta_B = K_B P_B \theta_v \qquad \theta_S = K_S P_S \theta_v \qquad (6.3.47)$$

The reaction *on the surface* is also assumed to be at equilibrium.

$$\theta_R = \frac{K_r \theta_A \theta_B}{\theta_S} \qquad (6.3.48)$$

The sum of the various types of fractional coverage is unity.

$$1 = \theta_A + \theta_B + \theta_R + \theta_S + \theta_v \qquad (6.3.49)$$

Combination of equations 6.3.47 to 6.3.49 and rearrangement gives

$$\theta_v = \frac{1}{1 + K_A P_A + K_B P_B + K_S P_S + \dfrac{K_r K_A K_B P_A P_B}{K_S P_S}} \qquad (6.3.50)$$

The conversion rate is equal to the net rate of desorption of species R. Hence,

$$r'' = k_{-1R}\theta_R - k_{1R}P_R\theta_v = k_{-1R}\left(\theta_R - K_R P_R \theta_v\right) \qquad (6.3.51)$$

Combination of equations 6.3.47, 6.3.48, 6.3.50, and 6.3.51 gives

$$r'' = \frac{k_{-1R}\left(\dfrac{K_r K_A K_B P_A P_B}{K_S P_S} - K_R P_R\right)}{1 + K_A P_A + K_B P_B + K_S P_S + \dfrac{K_r K_A K_B P_A P_B}{K_S P_S}} \qquad (6.3.52)$$

The analyses developed in this section are readily extended to reactions with different stoichiometries. Regardless of whether an adsorption or a desorption process is rate limiting, the resulting rate expressions may be written in the typical Hougen-Watson fashion represented by equation 6.3.30. A comprehensive summary of such relations has been developed by Yang and Hougen (12). Illustration 6.1 indicates the development of a Hougen-Watson model in an attempt to fit rate data for a catalytic reaction.

ILLUSTRATION 6.1 DEVELOPMENT OF A HOUGEN-WATSON RATE EXPRESSION FOR A HETEROGENEOUS CATALYTIC REACTION

Oldenberg and Rase (13) have studied the catalytic vapor phase hydrogenation of propionaldehyde over a commercially supported nickel catalyst. Their data indicate that the mathematical form of the reaction rate at very low conversions and 150 °C can be expressed quite well in the following manner.

$$r = \frac{kP_P}{P_{H_2}^{1/2}}$$

where the subscript P refers to propionaldehyde and the subscript H_2 to molecular hydrogen.

It has been suggested that the rate limiting step in the mechanism is the chemisorption of propionaldehyde and that the hydrogen undergoes dissociative adsorption on nickel. Determine if the rate expression predicted by a Hougen-Watson model based on these assumptions is consistent with the experimentally observed rate expression.

Solution

It is convenient to assign the following symbols before proceeding to a discussion of the mechanism proper.

P = propionaldehyde

A = propanol

σ = surface site

P_σ = adsorbed propionaldehyde—surface coverage = θ_P

H_σ = atomically adsorbed hydrogen—surface coverage = θ_H

A_σ = adsorbed alcohol—surface coverage = θ_A

The following mechanistic equations for chemisorption and reaction are appropriate.

$$P + \sigma \underset{k_{-1P}}{\overset{k_{1P}}{\rightleftharpoons}} P_\sigma \tag{A}$$

$$H_2 + 2\sigma \rightleftharpoons 2H_\sigma \tag{B}$$

K_{H_2} = adsorption equilibrium constant for hydrogen

$$P_\sigma + 2H_\sigma \rightleftharpoons A_\sigma + 2\sigma \tag{C}$$

K_{eq} = equilibrium constant for surface reaction

$$A_\sigma \rightleftharpoons A + \sigma \tag{D}$$

$1/K_A$ = reciprocal of adsorption equilibrium constant for alcohol

The rate of the overall reaction is equal to that of the rate controlling step.

$$r'' = k_{1P}P_P\theta_v - k_{-1P}\theta_P \tag{E}$$

where θ_v and θ_P are the fractions of the sites that are vacant and occupied by propionaldehyde, respectively.

From equation B,

$$K_{H_2} = \frac{\theta_H^2}{P_{H_2}\theta_v^2} \quad \text{or} \quad \theta_H = \sqrt{K_{H_2}P_{H_2}}\,\theta_v \tag{F}$$

From equation C,

$$K_{eq} = \frac{\theta_A\theta_v^2}{\theta_P\theta_H^2} \quad \text{or} \quad \theta_P = \frac{\theta_A\theta_v^2}{K_{eq}\theta_H^2} \tag{G}$$

From equation D,

$$1/K_A = \frac{P_A\theta_v}{\theta_A} \quad \text{or} \quad \theta_A = K_A P_A \theta_v \tag{H}$$

The expressions for θ_A and θ_H may now be substituted into equation G to give

$$\theta_P = \frac{K_A P_A \theta_v}{K_{eq}K_{H_2}P_{H_2}} \tag{I}$$

The sum of the fractional coverages must be unity. Hence,

$$1 = \theta_H + \theta_P + \theta_A + \theta_v \tag{J}$$

Substitution of equations F, H, and I into equation J and rearrangement gives

$$\theta_v = \cfrac{1}{1 + K_A P_A + \sqrt{K_{H_2} P_{H_2}} + \cfrac{K_A P_A}{K_{eq} K_{H_2} P_{H_2}}}$$

Substitution of this result into equation E gives

$$r'' = \cfrac{k_{1P} P_P - (k_{-1P} K_A P_A / K_{eq} K_{H_2} P_{H_2})}{1 + K_A P_A + \sqrt{K_{H_2} P_{H_2}} + \cfrac{K_A P_A}{K_{eq} K_{H_2} P_{H_2}}}$$

For an initial rate study, $P_A \approx 0$, and the last equation becomes

$$r_0'' = \frac{k_{1P} P_P}{1 + \sqrt{K_{H_2} P_{H_2}}}$$

If the second term in the denominator is large compared to unity,

$$r_0'' \approx \frac{k_{1P} P_P}{\sqrt{K_{H_2}} P_{H_2}^{1/2}}$$

Under these conditions the Hougen-Watson model is consistent with the experimentally observed rate expression.

6.3.2 Interpretation of Experimental Data

The problem of determining which, if any, of the Hougen-Watson models developed in the previous section fit experimental data for a heterogeneous catalytic reaction is generally more complex than the corresponding problem for a homogeneous phase reaction. Usually many possible controlling steps must be examined. Physical limitations on the overall reaction rate must first be eliminated by operating in a turbulent system and using sufficiently small catalyst pellets. The multitude of Hougen-Watson rate expressions for the cases where adsorption, desorption, and reaction on the surface are rate limiting is then considered. The trial-and-error procedures developed in Chapter

3 may be applied to data from catalytic reactions, (i.e. one postulates a rate expression and then uses an integral or differential method to determine if the data fits the rate expression). Because of the complexity of Hougen-Watson rate models, it is often convenient to use an initial rate approach to analyze the data. The remainder of this section is devoted to a discussion of the use of this approach to determine reaction rate expressions.

Studies of the influence of total pressure on the initial reaction rate for pure reactants present in stoichiometric proportions provide a means of discriminating between various classes of Hougen-Watson models. Isolation of a class of probable models by means of plots of initial reaction rate versus total pressure, feed composition, and temperature constitutes the first step in developing a Hougen-Watson rate model. Hougen (14) has considered the influence of total pressure for unimolecular and bimolecular surface reactions; the analysis that follows is adopted from his monograph.

In initial rate studies no products need be present in the feed, and the terms in the rate expression involving the partial pressures of these species may be omitted under appropriate experimental conditions. The use of stoichiometric ratios of reactants may also cause a simplification of the rate expression. If one considers a reversible bimolecular surface reaction between species A and B,

$$A + B \rightleftharpoons R + S \qquad (6.3.53)$$

an analysis of the type employed in Section 6.3.1.1 indicates that if the surface reaction is controlling, the conversion rate is given by:

$$r'' = \frac{k_1 \left(K_A K_B P_A P_B - \dfrac{K_R K_S P_R P_S}{K_r} \right)}{(1 + K_A P_A + K_B P_B + K_R P_R + K_S P_S)^2}$$
$$(6.3.54)$$

If an equimolar feed of species A and B is used, the dependence of the *initial* rate on total pressure (P) is given by

$$r_0'' = \frac{\frac{k_1}{4} K_A K_B P^2}{\left[1 + (K_A + K_B)\frac{P}{2}\right]^2} \quad (6.3.55)$$

In this case the initial rate increases as the square of the pressure at very low pressures and approaches a constant asymptotic limit at high pressures.

If the reaction proceeds by a mechanism requiring dissociation of one of the reactants (A_2) on adsorption, the expression for the reaction rate is given by

$$r'' = \frac{k_1 K_{A_2} K_B \left(P_{A_2} P_B - \frac{K_R K_S P_R P_S}{K_r K_{A_2} K_B}\right)}{(1 + \sqrt{K_{A_2} P_{A_2}} + K_B P_B + K_R P_R + K_S P_S)^3} \quad (6.3.56)$$

This expression assumes that the reaction on the surface is the rate limiting step with $r'' = k_1 \theta_A^2 \theta_B$.

If stoichiometric quantities of A_2 and B are used, the dependence of the initial rate on total

In the case of a Rideal-type mechanism (equation 6.3.16) a similar analysis indicates that for the case where the surface reaction is rate limiting, the initial rate starts out proportional to the square of the pressure and increases indefinitely with increasing total pressure. At high pressures it is linear in the total pressure.

In like manner one can treat cases where the rate controlling step is adsorption or desorption of some species. Hougen (14) has considered the effect of total pressure on initial reaction rates for several cases where these processes are rate limiting.

Studies of the effect of pressure on initial rates limit the possible Hougen-Watson rate expressions to certain classes. Subsequent studies using nonstoichiometric feeds and inerts and product species in the feed mixture further serve to determine the exact form of the reaction rate expression.

In determining the constants involved in a rate model it is often convenient to invert the rate expression to obtain a form that is suitable for data analysis. For example, equation 6.3.11 can be rearranged to give

$$\left(\frac{P_A P_B}{r''}\right)^{1/2} = \frac{1}{\sqrt{kK_A K_B}}(1 + K_A P_A + K_B P_B + K_R P_R + K_S P_S) \quad (6.3.58)$$

pressure is given by

$$r_0'' = \frac{k_1 K_{A_2} K_B \left(\frac{P}{2}\right)^2}{\left(1 + \sqrt{\frac{K_{A_2} P}{2}} + \frac{K_B P}{2}\right)^3} \quad (6.3.57)$$

In the low-pressure limit the rate will increase as the square of the pressure, but at very high pressures it will fall off in a manner proportional to the reciprocal of the pressure. Consequently, the initial rate increases at first, goes through a maximum, and then declines.

so a least squares analysis can be used to determine the five constants involved. Proper control of the nature of the experimental data can greatly simplify the problem of determining these parameters. However, this example illustrates one of the key difficulties involved in determining rate expressions for heterogeneous catalytic reactions, and we must temper any enthusiasm that we may have developed for Hougen-Watson rate models with a recognition of the situations one faces in the real world. The fact that various specific models can be proposed, each leading to a corresponding mathematical formulation (which is not necessarily unique), tempts one to try to correlate experimental data by each of a variety of such mathematical forms and then to "conclude" that the one which best fits the data is the "true" kinetic mechanism. Considerably more judgment than this is required and, although a given mathematical expression may provide a good fit of the data, it does not necessarily imply that the presumed mechanism is indeed the "correct" one. For rate expressions of the form of equation 6.3.58, there are five adjustable constants; when reversible reactions, the possibility of adsorption of inerts, or reactions in parallel or series are considered, still more constants are introduced. With the flexibility given by these constants it is not hard to get a good fit of data to a mathematical expression. In the words of one of my colleagues, "With six constants you can fit a charging rhinoceros."

Furthermore, one often finds that "best" fits of data may give rise to negative adsorption equilibrium constants. This result is clearly impossible on the basis of physical arguments. Nonetheless, reaction rate models of this type may be entirely suitable *for design purposes* if they are not extrapolated out of the range of the experimental data on which they are based.

We should also point out that the adsorption equilibrium constants appearing in the Hougen-Watson models cannot be determined from adsorption equilibrium constants obtained from nonreacting systems if one expects the mathematical expression to yield accurate predictions of the reaction rate. One explanation of this fact is that probably only a small fraction of the catalyst sites are effective in promoting the reaction.

6.4 PHYSICAL CHARACTERIZATION OF HETEROGENEOUS CATALYSTS

The ease with which gaseous reactant molecules achieve access to the interior surfaces of a porous substance is often of crucial significance in determining whether or not a given catalyst

formulation and preparation technique will produce a material that is suitable for industrial applications. The magnitude of the surface area alone is insufficient to predict the efficacy with which a given mass of catalyst promotes a given reaction. For fast reactions, the greater the amount of catalyst surface open to easy access by reactant molecules, the faster the rate at which reactants can be converted to products. If we are to be able to develop useful predictive models of catalyst behavior, we must know something about the interior pore structure of the catalyst and the facility with which reactant and product species can be transported within catalyst pellets.

This section discusses the techniques used to characterize the *physical properties* of solid catalysts. In industrial practice, the chemical engineer who anticipates the use of these catalysts in developing new or improved processes must effectively combine theoretical models, physical measurements, and empirical information on the behavior of catalysts manufactured in similar ways in order to be able to predict how these materials will behave. The complex models are beyond the scope of this text, but the principles involved are readily illustrated by the simplest model. This model requires the specific surface area, the void volume per gram, and the gross geometric properties of the catalyst pellet as input.

The specific surface area is usually determined by the BET technique discussed in Section 6.2.2. For the most reliable BET measurements the adsorbate gas molecules should be small, approximately spherical, inert (to avoid chemisorption), and easy to handle at the temperature in question. For economy, nitrogen is the most common choice with measurements usually made at 77 °K, the normal boiling point of liquid nitrogen. Krypton is another material that is frequently employed.

The remainder of this section is devoted to a discussion of the experimental techniques used to determine the other physical properties of catalysts that are of primary interest for reactor design purposes, the void volume and the pore size distribution.

6.4.1 Determination of Catalyst Void Volumes

The simplest method of determining the void volume or the pore volume of a given catalyst sample is to measure the increase in weight that occurs when the pores are filled with a liquid of known density. Water, carbon tetrachloride, and various hydrocarbons have been used successfully. The procedure involves boiling a known weight of dry catalyst pellets in the liquid for 20 to 30 min to displace the air in the pores. After the boiling fluid is replaced with cool liquid, the pellets are placed on an absorbent cloth and rolled to remove the excess liquid. They are then weighed. The difference between the wet and dry weights ($W_{wet} - W_{dry}$) divided by the density of the imbibed liquid (ρ_L) gives the void volume. The void volume per gram of catalyst (V_g) is obtained by

$$V_g = \frac{W_{wet} - W_{dry}}{\rho_L W_{dry}} \qquad (6.4.1)$$

This technique will determine the total volume of the pores with radii between approximately 10 and 1500 Å. It is limited in accuracy by the fact that it is difficult to dry the external surface of the particles without removing liquid from the large pores. Some liquid also tends to be retained around the points of contact between particles. These two sources of error offset one another. Any air retained within the pores after boiling will lead to erroneous results.

A much more accurate method of determining the pore volume of a catalyst sample is the helium-mercury method. One places a known weight of catalyst (W) in a chamber of known volume. After the chamber has been evacuated, a known quantity of helium is admitted. From the gas laws and measurements of the temperature and pressure, one may then proceed to determine the volume occupied by the helium (V_{He}). This volume is equal to the sum of the volume exterior to the pellets proper and the void volume within the pellets (V_{void}). The helium is then pumped out, and the chamber is filled with mercury at atmospheric pressure. Since the mercury will not penetrate the pores of most catalysts at atmospheric pressure, the mercury will occupy only the volume exterior to the pellets proper (V_{Hg}). Hence:

$$V_{void} = V_{He} - V_{Hg} \qquad (6.4.2)$$

and the void volume per gram (V_g) is

$$V_g = \frac{V_{He} - V_{Hg}}{W} \qquad (6.4.3)$$

This method determines the pore volume corresponding to pore radii below 75,000 Å. By varying the pressure on the system, it is possible to force the mercury into some of the pores and determine the void volumes corresponding to different pore radii. We will pursue this point in the next section.

The porosity of the catalyst pellet (ε_p) is defined as the void fraction.

$$\varepsilon_p = \frac{\text{void volume of catalyst particle}}{\text{total volume of catalyst particle}} \qquad (6.4.4)$$

For a particle of mass m_P,

$$\varepsilon_p = \frac{m_P V_g}{m_P V_g + \dfrac{m_P}{\rho_{skeletal}}} \qquad (6.4.5)$$

where $\rho_{skeletal}$ is the true density of the bulk solid and where the second term in the denominator is the volume occupied by the solid proper. Many commercial catalysts have porosities in the neighborhood of 0.5, indicating that the gross particle volume is about evenly split between void space and solid material. Significantly higher porosities are not encountered in commercial catalysts because of the problems of achieving sufficient mechanical strength at these porosity levels.

The total pore volume can also be determined from adsorption measurements if one knows the *volume of vapor adsorbed under saturation conditions*. For high surface area catalysts the amount of material adsorbed on particle exteriors will be negligible compared to that condensed in the pores. Hence the liquid phase volume equivalent to the amount of gas adsorbed is equal to the pore volume. The liquid density is assumed to be that corresponding to the saturation conditions in question. This technique is less accurate than that described previously.

Illustration 6.2 indicates how void volume and surface area measurements can be combined in order to evaluate the parameters involved in the simplest model of catalyst pore structure.

ILLUSTRATION 6.2 EVALUATION OF PARAMETERS IN A MATHEMATICAL MODEL OF POROUS CATALYSTS

To develop analytical models for processes employing porous catalysts it is necessary to make certain assumptions about the geometry of the catalyst pores. A variety of assumptions are possible, and Thomas and Thomas (15) have discussed some of these. The simplest model assumes that the pores are cylindrical and are not interconnected. Develop expressions for the average pore radius ($\bar{r}$), the average pore length ($\bar{L}$), and the number of pores per particle (n_p) in terms of parameters that can be measured in the laboratory [i.e., the apparent particle dimensions, the void volume per gram (V_g), and the surface area per gram (S_g)].

Solution

We will start by developing an expression for the average pore radius $\bar{r}$. If we denote the mass of an individual catalyst particle by m_P, simple geometric considerations indicate that the void volume per particle is given by

$$m_P V_g = n_p(\pi \bar{r}^2 \bar{L}) \qquad (A)$$

while the surface area per particle is given by

$$m_P S_g = n_p(2\pi \bar{r} \bar{L}) \qquad (B)$$

where we have neglected the area at the end of the cylinder, since $\bar{L} \gg \bar{r}$ for virtually all systems of interest. Division of equation A by equation B and rearrangement gives

$$\bar{r} = \frac{2V_g}{S_g} \qquad (C)$$

For a monodisperse system this result is in good agreement with the values obtained from pore size distribution measurements, but it can be significantly in error if one is dealing with a bimodal pore size distribution (see Section 6.4.2).

To evaluate the average pore length, it is necessary to recognize that the porosity ε_p will represent not only the volumetric void fraction but also at any cross section the fraction of the area occupied by the pore openings. If the average open area associated with each pore is assumed to be $\pi \bar{r}^2$, the definition of the porosity indicates that

$$\varepsilon_p = \frac{m_P V_g}{V_p} = \frac{n_p \pi \bar{r}^2}{S_x} \qquad (D)$$

where V_p and S_x are the gross geometric volume and geometric surface area, respectively.

Substitution of equation A into equation D gives

$$\frac{n_p(\pi \bar{r}^2)\bar{L}}{V_p} = \frac{n_p \pi \bar{r}^2}{S_x}$$

which can be rearranged to give

$$\bar{L} = \frac{V_p}{S_x}$$

For a spherical catalyst pellet of radius R,

$$\bar{L} = \frac{\frac{4}{3}\pi R^3}{4\pi R^2} = \frac{R}{3}$$

The number of pores per particle may be obtained by solving equation D for n_p and using equation C to eliminate $\bar{r}$

$$n_p = \frac{\varepsilon_p S_x S_g^2}{4\pi V_g^2}$$

where ε_p is determined from equation 6.4.5.

6.4.2 Determination of Pore Size Distributions

Scanning electron microscopy and other experimental methods indicate that the void spaces in a typical catalyst particle are not uniform in size, shape, or length. Moreover, they are often highly interconnected. Because of the complexities of most common pore structures, detailed mathematical descriptions of the void structure are not available. Moreover, because of other uncertainties involved in the design of catalytic reactors, the use of elaborate quantitative models of catalyst pore structures is not warranted. What is required, however, is a model that allows one to take into account the rates of diffusion of reactant and product species through the void spaces. Many of the models in common use simulate the void regions as cylindrical pores; for such models a knowledge of the distribution of pore radii and the volumes associated therewith is required.

There are two well-established experimental techniques for determining the distribution of pore radii. They are the mercury penetration technique and the desorption isotherm method.

The mercury penetration approach is based on the fact that liquid mercury has a very high surface tension and the observation that mercury does not wet most catalyst surfaces. This situation holds true for oxide catalysts and supported metal catalysts that make up by far the overwhelming majority of the porous commercial materials of interest. Since mercury does not wet such surfaces, the pressure required to force mercury into the pores will depend on the pore radius. This provides a basis for measuring pore size distributions through measurements of the volume of mercury contained within the pores as a function of applied pressure.

The smallest pores that can be observed using this approach depend on the highest pressure to which the mercury can be subjected in a particular piece of equipment. Volumes corresponding to pore radii as small as 100 to 200 Å can be measured with commercially available equipment. Beyond this point the pressures required to fill up the capillaries with smaller radii become impractical for routine use. Unfortunately, there are many catalysts of industrial significance where these very small capillaries contribute substantially to the specific surface area. Special research grade mercury porosimeters capable of measurements down to 15 Å radii have been developed but, for routine measurements, the desorption approach described below is more suitable.

The desorption isotherm approach is the second generally accepted method for determining the distribution of pore sizes. In principle either a desorption or adsorption isotherm would suffice but, in practice, the desorption isotherm is much more widely used when hysteresis effects are observed. The basis of this approach is the fact that capillary condensation occurs in narrow pores at pressures less than the saturation vapor pressure of the adsorbate. The smaller the radius of the capillary, the greater is the lowering of the vapor pressure. Hence, in very small pores, vapor will condense to liquid at pressures considerably below the normal vapor pressure. Mathematical details of the analysis have been presented by Cranston and Inkley (16) and need not concern us here.

The size of the largest pores that can be determined by this technique is limited by the rapid change in meniscus radius with pressure as the relative pressure P/P_0 approaches unity. This limit corresponds to pore radii in the neighborhood of 150 to 200 Å, corresponding to a relative pressure of 0.93 in the former case. The smallest pore radii that can be observed by this technique are those near 10 Å. Although measurements may be reported corresponding to smaller pore sizes, such results are suspect because the molecular diameter of the adsorbate molecules is comparable in magnitude to the pore radius in this regime. Moreover, the analysis assumes that the properties of the condensed phase in the capillaries are the same as those of bulk liquids, yet the concepts of surface tension and a hemispherical surface become hazy or unrealistic below radii of 10 Å, and perhaps even at larger radii. In the regions where the mercury penetration approach and the desorption method overlap, they agree reasonably well. However, in many cases of practical significance, it is necessary to employ both techniques to cover the entire range of pore sizes (10 to 10,000 Å) that may be present. This is particularly true of catalysts like pelleted alumina in which one has a bimodal pore size distribution of the type discussed below.

The catalysts employed in commercial fixed bed reactors often exhibit a bimodal pore size distribution, sometimes referred to as bidisperse or macro-micro distributions. In these systems the bulk of the catalytic reaction occurs in pores with radii below 200 Å, since this is where the bulk of the surface resides. However, the transport of reactants to these small pores occurs primarily through macropores with radii ranging from 200 to 10,000 Å. These catalyst pellets are prepared by compacting fine porous powders. The micropore region arises from the pore structure within each of the particles of the original powder, while the macropores are formed by the passageways around the compacted particles. As the pelleting pressure is increased, the macropores become successively reduced in size, but the micropores remain unaffected unless the crushing strength is exceeded. Most commercial forms of alumina catalysts exhibit this type of bimodal pore size distribution. For example, Rothfeld (17) has observed two peaks, one centered at a radius of 6250 Å and the other at 60 Å. The micropores accounted for about 99% of the surface area and 65% of the void volume. It is these pores that one

would expect to be most significant from a catalytic viewpoint. Some other industrial catalysts are deliberately fabricated with a macropore-micropore structure in order to minimize or eliminate pore diffusion limitations on the reaction rate. The macropores serve as expressways that facilitate reactant transport to the internal surface area of the catalyst.

6.5 CATALYST PREPARATION, FABRICATION, AND ACTIVATION

The task of developing a suitable catalyst for commercial applications involves many considerations, ranging from obvious factors like catalyst activity and selectivity to variables like the catalyst shape and the composition of the binder used in a pelletizing process. This section is devoted to a discussion of these considerations and of the techniques involved in manufacturing industrial catalysts.

Figure 6.6 illustrates some of the requirements to be kept in mind when developing a catalyst for industrial applications. The art of catalyst formulation involves reconciling conflicting demands that may be imposed by such requirements. The keystone of the arch is a material that has high intrinsic activity and selectivity for promoting the reaction of interest. This material may be used in bulk form or, more commonly, it will be supported on a carrier material that may or may not have catalytic activity of its own. The activity per unit volume is of practical significance because process economics are often strongly dependent on the cost of packed reactor space. Consequently, the catalyst should have a high specific surface area, and it should have a pore structure such that reactants can gain easy access to the inner surfaces of the catalyst pellet. The first of these requirements follows directly from the fact that the reaction takes place at the fluid-solid interface, while the requirement for an appropriate pore structure follows from a desire to avoid diffusional limitations on chemical conversion rates. (See Chapter 12.)

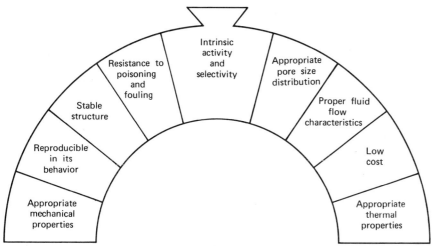

Figure 6.6
Requirements for industrial catalysts.

The lifetime of a catalyst is another factor that plays a key role in determining process economics. This lifetime is the period during which the catalyst produces the *desired product* in yields in excess of or equal to a designated value. The life of a catalyst may terminate because of unacceptable changes in activity and/or selectivity, because of physical attrition, or because of unacceptable changes in mechanical properties arising from thermal cycling or other factors in the process environment. Resistance to poisoning and fouling as well as to other catalyst deactivation processes is an attribute to be preferred in commercial catalysts.

Mechanical and geometric properties also play a strong role in determining if a given formulation will give rise to a commercially viable catalyst. Catalyst pellets must be suitably shaped so that the reactant fluid can pass through a packed bed without excessive pressure drop or maldistribution of the flow. The finished catalyst must also be mechanically robust to avoid attrition during handling and loading and to avoid crushing under sustained loads during fixed bed operation. The requirements of high strength and a stable structure depend on a firm welding together of the catalyst com-

ponents into a structure that is not greatly weakened or changed by sintering during use.

Very seldom does an industrial catalyst consist of a single chemical compound or metallic element. Most often a catalyst formulation consists of a multitude of components, each of which performs an essential task in the creation of a commercially viable catalyst. Figure 6.7,

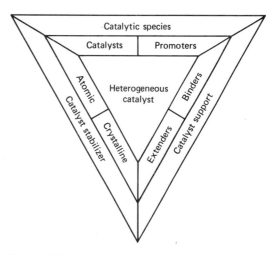

Figure 6.7
Components of industrial catalysts. (Ownership of this diagram is by ICI)

adapted from the *Catalyst Handbook* (18), indicates that the components commonly found in industrial catalysts can be classified as catalytic species, catalyst supports, and catalyst stabilizers. The support may be a ceramic matrix bound together by a hydraulic cement. The support gives body and strength to the catalyst granule and is frequently used to decrease the concentration of the usually more expensive catalyst proper. The catalytic agents consist of the catalyst proper together with any materials added to promote or modify the active surface. Stabilizers may be present in the form of gross crystals readily seen under the scanning electron microscope, or they may be present in highly dispersed form. They prevent sintering and loss of active surface area through diffusional merging of contiguous small crystals and crystallites. Catalysts in which the active constituent is a metal with a relatively low melting point normally are stabilized through the use of refractory materials such as alumina, chromia, and magnesia, which act as spacers between the readily sinterable metal crystals. Since commercial catalysts are complex multicomponent mixtures, it is appropriate to describe in broad terms some of the procedures used in the preparation of these materials.

6.5.1 Catalyst Preparation

Natural products and common industrial chemicals in massive form are seldom useful as catalysts because they have low specific surface areas, may contain various amounts of impurities that have deleterious effects on catalyst performance, do not usually have the exact chemical composition desired, or are too expensive to use in bulk form. The preparation of an industrial catalyst generally involves a series of operations designed to overcome such problems. Many catalysts can be produced by several routes. The actual choice of technique for the manufacture of a given catalyst is based on ease of preparation, homogeneity of the final catalyst, stability of the catalyst, reproducibility

of quality, and cost of manufacture. The two most commonly used techniques for catalyst preparation are the only ones we will consider.

1. Precipitation or gel formation of the active component or components through the interaction of aqueous solutions of two or more chemical compounds.
2. Impregnation of a carrier using a solution containing a compound of the desired catalyst component.

Precipitation is frequently used in the preparation of single- and multiple-component catalysts. If possible, the catalyst precursor is precipitated as a compound that will undergo some chemical change before it is converted into the substance that actually functions as the catalyst. Dehydration of a hydrated precipitate may be sufficiently drastic to accomplish this purpose. The nature of the impurities occluded or adsorbed on the precipitate must be considered in making up the starting solutions. Experience has indicated that the use of dilute solutions, nitrates or sulfates of the metal elements, and ammonia or ammonium salts as the precipitating agents generally facilitates the removal of impurities. In these cases subsequent heat treatment often suffices to drive off the impurities in volatile form. Significant variations in catalytic activity have been observed for precipitates of the same material formed under different conditions. When preparing multicomponent catalysts by coprecipitation techniques, the process variables must be manipulated to produce a uniform product. By adding a solution of two metal salts to an excess of the precipitating counter ion (carbonate, sulfide, hydroxide, etc.) simultaneous precipitation of both species is obtained. If the addition is carried out in the reverse order there is a preferential precipitation of one metal salt relative to the other because of differences in solubility products.

One of the simplest, most commonly used techniques for preparing a catalyst involves dispersing an active component (or components) on a support material. Normally one impregnates the carrier material with a solution of a soluble precursor of the catalyst and then converts this precursor to the desired product by oxidation, reduction, thermal decomposition, or some other suitable step. Where appropriate, it is preferable to use a granular support instead of a powder, since it eliminates the necessity for pelleting or extrusion in order to obtain the final product.

In many cases drying operations are critical to the production of successful commercial catalysts. Close control of the drying process is necessary to achieve the proper distribution of the catalyst precursor within the pore structure of the support. Drying also influences the physical characteristics of the finished catalyst and the ease with which subsequent pelleting or extrusion processes may be carried out.

In general all materials used to catalyze reactions of a vapor phase feed are calcined or activated at temperatures above 400 to 500 °F. This heat treatment may accomplish one or more of the following tasks.

1. Reduce the moisture content to a level consistent with the temperature at which the catalyst will be employed.
2. Decompose salts containing the catalyst precursors such as metal nitrates, formates, oxalates, or acetates. (Oxides are the usual products of the decomposition process.)
3. Form metal catalysts by reduction of metal oxides with hydrogen.
4. Increase the strength of the finished pellet or extrudate.
5. Influence the initial activity and stability of the catalyst.

The form in which the catalyst is employed is strongly dependent on the reaction involved, the scale of the process, the specific nature of the catalyst, and the type of reactor. Catalysts used for reactions carried out with a liquid feed are usually ground so as to pass through a 100 to 200-mesh screen. Catalysts employed with gaseous feeds in fixed bed reactors are most frequently used in the form of pellets, granules, or extrusions, whereas those used in fluidized beds are usually ground to pass through a 60-mesh screen. With cylindrical pellets the ratio of length to diameter is usually kept below 3:1, so that the pellets will not pack parallel to one another and thus lead to excessive pressure drop or excessive bypassing, channeling, or maldistribution of the flow. The ratio of the diameter of the reaction vessel to a characteristic pellet dimension is also significant, since it is almost impossible to achieve uniform packing if this ratio is less than 5:1.

More detailed treatments of catalyst manufacturing processes are available in the literature (19–22).

6.5.2 Catalyst Supports, Promoters, and Inhibitors

It is very seldom that a commercial catalyst consists of only a single chemical compound or element. Often the active constituent is supported on a carrier material that may or may not possess catalytic activity of its own. Enhanced catalytic activity, selectivity, or stability may also be achieved by the addition of other materials referred to as promoters or inhibitors.

Early workers viewed carriers or catalyst supports as inert substances that provided a means of spreading out an expensive material like platinum or else improved the mechanical strength of an inherently weak material. The primary factors in the early selection of catalyst supports were their physical properties and their cheapness; hence pumice, ground brick, charcoal, coke, and similar substances were used. No attention was paid to the possible influence of the support on catalyst behavior; differences in behavior were attributed to variations in the distribution of the catalyst itself.

However, it is now recognized that the catalyst carrier does more than provide a physical framework on which the catalyst is supported.

In many cases there is an interaction between the carrier and the active component of the catalyst so that the character of the active surface will change. For example, the electronic character of the supported catalyst may be influenced by the transfer of electrons across the catalyst-carrier interface. In some cases the carrier itself has a catalytic activity for the primary reaction, an intermediate reaction, or a subsequent reaction, and a "dual-function" catalyst is thereby obtained. Materials of this type are widely employed in reforming processes. There are other cases where the interaction of the catalyst and support are much more subtle and difficult to label. For example, the crystal size and structure of supported metal catalysts as well as the manner in which the metal is dispersed can be influenced by the nature of the support material.

One should distinguish between true catalyst supports and diluents. A catalyst support (or carrier) is a material on which a thin layer of catalyst is deposited; a diluent is an inert material thoroughly mixed with the catalyst to enhance the binding properties of a powdered catalyst or to assist in pelleting or extrusion fabrication procedures.

An ideal carrier is one that is:

1. Inexpensive.
2. Available in large quantities of uniform composition.
3. Sufficiently porous to permit dispersion of catalyst on its interior surfaces.
4. Free of any components that may poison the catalyst.
5. Stable under operating and regeneration conditions.
6. Strong enough to resist any thermal or mechanical shocks that it is likely to suffer as well as any disruptive action arising from fouling material that may be deposited in its pores.
7. Resistant to attrition.
8. Inert to attack from components of the feedstream and product stream and any components present in any regeneration streams to which it may be exposed.
9. Noncatalytic with regard to any undesirable side reactions.

The average pore size and the pore size distribution should be such that physical limitations are not placed on the conversion of reactants to products. The particle size of the carrier must also be suitable for the purpose intended (i.e., small for fluidized bed reactors and significantly larger for fixed bed applications).

The types of materials used commercially as catalyst supports can be categorized as having high or low specific surface areas. The low surface area (up to 1 m^2/g) materials are generally ceramic materials like silica or alumina. They are often formed by high-temperature fusion (2000 °C) in electric furnaces. The product of this operation is crushed, sieved, and fabricated into various shapes (spheres, granules, rings, cylindrical annuli, or irregular shapes). Partial sintering is then accomplished by firing at temperatures in the neighborhood of 1400 °C. The resultant products are stable mechanically and are easily impregnated with a variety of catalyst precursor solutions. The final product is quite porous with pore sizes typically in the range of 20 to 100 microns. Low surface area carriers are used where the need for a high activity per unit mass of finished catalyst is less important than other factors like catalyst cost or selectivity.

The high specific surface area supports (10 to 100 m^2/g or more) are natural or man-made materials that normally are handled as fine powders. When processed into the finished catalyst pellet, these materials often give rise to pore size distributions of the macro-micro type mentioned previously. The micropores exist within the powder itself, and the macropores are created between the fine particles when they are compressed together in a pellet press. Diatomaceous earth and pumice (or cellular lava) are naturally occurring low-cost materials that are representative of this class of catalyst support. Among the synthetic carriers that can be created by modern technology are those derived from clays, bauxite, activated carbon, and xerogels of silica gel and alumina gel.

Although a carrier is often the major constituent of a catalyst pellet, one often finds that there are materials that are added in small amounts during catalyst preparation in order to impart improved characteristics to the finished catalyst. These materials are referred to as *promoters*. They may lead to better activity, selectivity, or stability. The manner in which promoters act is not well understood, although a number of plausible explanations have been set forth. They remain one of the reasons for the "black magic" aura of catalysis.

In 1920 Pease and Taylor (23) proposed a definition of promoter action that is still appropriate today. Promotion occurs in those "cases in which a mixture of two or more substances is capable of producing a greater catalytic effect than can be accounted for on the supposition that each substance in the mixture acts independently and in proportion to the amount present." Hence "better" catalyst performance implies an improvement beyond that expected from a simple additive rule. The phenomenon is often characterized by a steep rise in activity to a sharp maximum at a low concentration of promoter followed by a decline to a lower or zero level of activity, depending on whether or not the promoter alone is capable of acting as a catalyst.

Some hypotheses that have been offered in an effort to explain promotion effects are:

1. The promoter may change the electronic structure of the solid in such a way that the activity per unit area is increased.
2. The promoter may catalyze an intermediate step in the reaction. In this sense it may act as one component of a dual function catalyst.
3. The promoter may slow down, or otherwise influence crystal formation and growth, or produce lattice defects. These effects may lead either to a higher activity per unit area or to a higher specific surface area.

Promoters may influence selectivity by poisoning undesired reactions or by increasing the rates of desired intermediate reactions so as to increase the yield of the desired product. If they act in the first sense, they are sometimes referred to as *inhibitors*. An example of this type of action involves the addition of halogen compounds to the catalyst used for oxidizing ethylene to ethylene oxide (silver supported on alumina). The halogens prevent complete oxidation of the ethylene to carbon dioxide and water, thus permitting the use of this catalyst for industrial purposes.

Another example of catalyst promotion is the use of additives to inhibit loss of active surface area during operation. Since this effect enhances catalyst stability, such additives are often called *stabilizers*. These promoters may inhibit sintering of active sites on the surface or growth of microcrystalline regions. They may form solid solutions with higher melting points than the active agents alone, thus permitting a reaction to be carried out at a higher temperature than would otherwise be possible. Alumina is used in the manufacture of iron-based ammonia synthesis catalysts for this purpose.

In commercial catalysts it is often useful to employ a multiplicity of promoters, each of which serves a particular function or acts in conjunction with other constituents to produce a more desirable catalyst.

6.6 POISONING AND DEACTIVATION OF CATALYSTS

Since the earliest days of heterogeneous catalysis, decreases in activity during use have been observed. The rates at which catalyst deactivation processes take place may be fast or slow. In some cases the decline in activity is so rapid that the catalyst ceases to function effectively after a few minutes or hours of exposure to a reactant feedstream. On the other hand, there are cases where the deactivation processes occur so slowly that the catalyst may function effectively for months or years. In the design of commercial catalytic processes, one obviously must take these factors into account so as to allow for periodic replacement or regeneration of his heterogeneous catalyst.

If the deactivation is rapid and caused by the decomposition or degradation of reactants or products on the catalyst surface, the process is termed *fouling*. In this case a deposit is formed on the surface or in the pores that physically blocks a portion of the catalyst and prevents it from catalyzing the reaction. Such fouling is particularly rapid with silica-alumina cracking catalysts and in this application, the design of a unit for catalyst regeneration is as important as the design of the reactor proper. An example of a cracking reaction leading to fouling might be $C_{10}H_{22} \rightarrow C_5H_{12} + C_4H_{10} + C$ (solid on catalyst surface). The carbon deposit may be burned off with oxygen and/or steam to regenerate the catalyst so that the active surface is no longer covered.

If the activity of the catalyst is slowly modified by chemisorption of materials that are not easily removed, the deactivation process is termed *poisoning*. It is usually caused by preferential adsorption of small quantities of impurities (poisons) present in the feedstream. Adsorption of extremely small amounts of the poison (a small fraction of a monolayer) is often sufficient to cause very large losses in catalytic activity. The bonds linking the catalyst and poison are often abnormally strong and highly specific. Consequently, the process is often irreversible. If the process is reversible, a change in the temperature or the composition of the gas to which it is exposed may be sufficient to restore catalyst activity. This process is referred to as reactivation. If the preferential adsorption of the poison cannot be readily reversed, a more severe chemical treatment or complete replacement of the catalyst may be necessary.

In addition to fouling and poisoning, there is a third catalyst deactivation process that is commonly encountered in industrial practice. This is the phenomenon referred to as *aging*. It involves a loss in specific activity due to a loss in catalyst surface area arising from crystal growth or sintering processes. This process becomes more rapid as the temperature increases. It may also be increased by the presence of certain components of the feedstream or product stream or of trace constituents of the catalyst. In some cases a flux or glaze capable of blocking catalyst pores may be produced.

Various schemes have been proposed for classifying poisons, but the one that is perhaps the most convenient for chemical engineers interested in reactor design is the classification in terms of the manner by which the poison affects chemical activity. In these terms one can distinguish between four general but not sharply differentiated classes.

1. *Intrinsic Activity Poisons.* These poisons decrease the activity of the catalyst for the primary chemical reaction by virtue of their direct electronic or chemical influence on the catalyst surface or active sites. The mechanism appears to be one that involves coverage of the active sites by poison molecules, removing the possibility that these sites can subsequently adsorb reactant species. Common examples of this type of poisoning are the actions of compounds of elements of the groups Vb and VIb (N, P, As, Sb, O, S, Se, Te) on metallic catalysts.

2. *Selectivity Poisons.* These poisons decrease the selectivity of the catalyst for the main reaction. In many cases impurities in the feedstream will adsorb on the catalyst surface and then act as catalysts for undesirable side re-

actions. The classic examples of poisons of this class are heavy metals such as Ni, Cu, V, and Fe, which are present in petroleum stocks in the form of organometallic compounds such as porphyrins. When these feedstocks are subjected to catalytic cracking, these organometallics decompose and deposit on and within the catalyst. The metals then act as dehydrogenation catalysts. The product distribution is markedly affected with lower yields of gasoline and higher yields of light gas, coke, and hydrogen. Very small poison concentrations suffice to produce large changes in selectivity. In cases where a large number of parallel and series reactions are involved, the adsorbed poison is thought to change the activation energy barrier for some of the competitive intermediate steps, resulting in increased rates of undesirable side reactions relative to that of the primary reaction.

3. *Stability Poisons.* These poisons decrease the structural stability of the catalytic agent or of the carrier by facilitating recrystallizations and other structural rearrangements. Steam acts as this type of a poison for silica-alumina gel catalysts. It acts not so much by reducing the intrinsic activity per unit surface area, but by reducing the active area, thus decreasing the activity per unit weight of catalyst. The temperature at which the reactor operates has a marked effect on stability poisoning. Sintering, localized melting, and recrystallization occur much more rapidly at high temperatures than at low temperatures.

4. *Diffusion Poisons.* This phenomenon is closely akin to catalyst fouling. Blockage of pore mouths prevents full use of the interior surface area of the pellet. Entrained dust particles or materials that can react on the catalyst to yield a solid residue give rise to this type of poisoning.

Many reviews listing specific poisons for specific catalysts are available in the literature (23–25). For a more up-to-date review of catalyst deactivation processes, consult the article by Butt (26).

LITERATURE CITATIONS

1. Haensel, V., *Industrial and Engineering Chemistry, 57* (6), p. 18, 1965.

2. Thomas, C. L., *Catalytic Processes and Proven Catalysts*, Academic Press, New York, 1970.

3. Thomson, S. J., and Webb, G., *Heterogeneous Catalysis*, Oliver and Boyd, Edinburgh, 1968.

4. Brunauer, S., Deming, L. S., Deming, W. E., and Teller, E., *J. Am. Chem. Soc., 62* (1723), 1940.

5. Langmuir, I., *J. Am. Chem. Soc., 38* (221), 1916.

6. Langmuir, I., *J. Am. Chem. Soc., 40* (1361), 1918.

7. Taylor, H. S., *Proc. Roy. Soc., A108* (105), 1925.

8. Brunauer, S., Emmett, P. H., and Teller, E., *J. Am. Chem. Soc., 60* (309), 1938.

9. Hougen, O. A., and Watson, K. M., *Chemical Process Principles, Part 3, Kinetics and Catalysis*, pp. 906–907, Wiley, New York, copyright © 1947. Used with permission.

10. Chambers, R. P., and Boudart, M., *J. Catalysis, 6* (141), 1966.

11. Hougen, O. A., and Watson, K. M., *Ind. Eng. Chem., 35* (529), 1943.

12. Yang, K., and Hougen, O. A., *Chem. Eng. Prog., 46* (3), p. 149, 1950.

13. Oldenburg, C. C., and Rase, H. F., *AIChE J., 3* (462), 1957.

14. Hougen, O. A., "Reaction Kinetics in Chemical Engineering," *Chem. Eng. Prog. Monograph Series*, No. 1, *47*, Chapter 5, copyright © 1951. Adapted with permission.

15. Thomas, J. M., and Thomas, W. J., *Introduction to the Principles of Heterogeneous Catalysis*, pp. 206–211, Academic Press, New York, 1967.

16. Cranston, R. W., and Inkley, F. A., *Adv. Catalysis, 9* (143), 1957.

17. Rothfeld, L. B., *AIChE J., 9* (19). 1963.

18. *Catalyst Handbook*, Springer-Verlag, New York, 1970.

19. Ciapetta, F. G., and Plank, C. J., "Catalyst Preparation," Chapter 7 in *Catalysis*, Volume I, edited by P. H. Emmett, Reinhold, New York, 1964.

20. Ciapetta, F. G., Helm, C. D., and Baral, L. L., "Commercial Preparation of Industrial Catalysts," Chapter 2 in *Catalysis in Practice*, edited by C. H. Collier, Reinhold, New York, 1957.

21. Innes, W. B., "Catalyst Carriers, Promoters, Accelerators, Poisons, and Inhibitors," Chapter 6 in *Catalysis*, Volume I, edited by P. H. Emmett, Reinhold, New York, 1954.

22. Anderson, J. R., *Structure of Metallic Catalysts*, Academic Press, New York, 1975.

23. Pease, R. N., and Taylor, H. S., *J. Phys. Chem.*, *24* (241), 1920.

24. Maxted, E. B., *J. Soc. Chem. Ind.*, (*Lond.*) *67* (93), 1948.

25. Maxted, E. B., *Adv. Catalysis*, *3* (129), 1951.

26. Butt, J. B., "Chemical Reaction Engineering," *Adv. Chem. Series*, *109* (259), 1972.

PROBLEMS

1. The data below were recorded at constant temperature for the adsorption of nitrogen on silica gel at $-196\ °C$. Prepare both a Langmuir and a BET plot using coordinates that would be expected to give rise to linear plots. What specific surface areas are predicted in each case? The area covered per molecule of adsorbed nitrogen is $0.162\ nm^2$. The normal boiling point of liquid nitrogen is $-196\ °C$.

Comment on your results.

Pressure (kPa)	Volume adsorbed per gram of sample (cm³ at STP)
0.8	6.1
3.3	12.7
18.7	17.0
30.7	19.7
38.0	21.5
42.7	23.0
57.3	27.7
67.3	33.5

2. Taylor and Liang [*J. Am. Chem. Soc.*, *69* (1306), 1947] have studied the extent of adsorption of hydrogen on zinc oxide at a constant gas pressure and at successively increasing temperatures. A typical curve of the amount adsorbed as a function of time is shown in Figure 6P.1. Interpret these data.

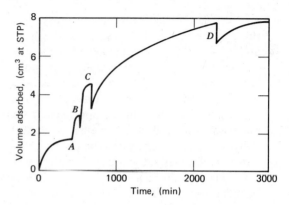

Figure 6P.1

Typical adsorption rate curve for the adsorption of hydrogen on zinc oxide at 1 atm pressure. The curve to the left of *A* applies to 0 °C; at *A* the temperature is raised to 111°, at *B* to 154°, at *C* to 184, and at *D* to 218°. [Reprinted with permission from *J. Am. Chem. Soc.*, *69* (1306), 1947. Copyright **by the American Chemical Society.]**

3. Figure 6P.2 contains the Raman spectra of pyridine adsorbed on silica gel at three temperatures as reported by Schrader and Hill [*Rev. Sci. Instrum*, *46* (1335), 1975]. Four bands are apparent at 991, 1006, 1032, and 1069 cm^{-1}. If the band intensities are proportional to surface concentrations and if each band is associated with one vibrational degree of freedom of an adsorbed species, what is your interpretation of these data?

4. Restelli and Coull [*AIChE J.*, *12* (292), 1966] have studied the transmethylation reaction of dimethylamine in a differential flow reactor using montmorillonite as a catalyst. They measured initial reaction rates under isothermal conditions for this heterogeneous catalytic process. Steady-state operating data were recorded.

The initial reaction is represented by the following stoichiometry.

$$2(CH_3)_2NH \rightleftharpoons (CH_3)_3N + CH_3NH_2$$

The feedstream consisted of essentially pure di-

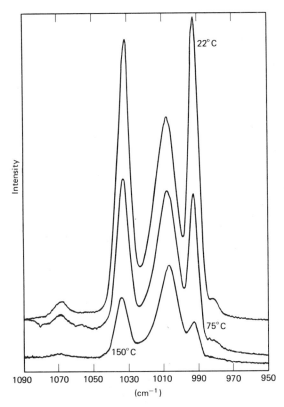

Figure 6P.2
Pyridine adsorbed on silica gel as a function of temperature (°C). $P = 1.86 \times 10^{-2}$ bar. (Reprinted with permission from *Review of Scientific Instruments*, **46 (1335), 1975. Copyright by the American Institute of Physics.)**

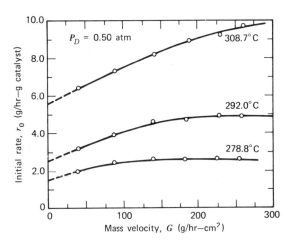

Figure 6P.3
Initial rate versus mass velocity. [Reprinted with permission from *AIChE.*, *12* **(292), 1966.]**

methylamine. What is your interpretation of the data presented below? In your discussion place particular emphasis on the various physical and chemical processes that can affect the rate of a chemical reaction.

(a) Using a catalyst size -50, $+60$ mesh with a feed pressure of 0.5 atm, the data reproduced in Figure 6P.3 were reported. G is the mass velocity under steady-state operating conditions in grams per hour per square centimeter and r_0 is the initial reaction rate in grams per hour per gram of catalyst. The

three curves correspond to three different operating temperatures.

(b) A series of experiments involved measurement of reaction rates for different catalyst sizes, maintaining all other variables constant. The measured rate was found to be constant for catalyst particle sizes from -100, $+120$ to -30, $+40$ mesh, a fourfold variation in average particle diameter. These experiments were carried out at mass velocities corresponding to the asymptotic portions of the curves in Figure 6P.3.

(c) Several additional series of experiments were also carried out at mass velocities corresponding to the asymptotic portions of the curves shown in Figure 6P.3. Initial rates were measured as a function of the dimethylamine pressure for the four temperatures indicated. The data are summarized in Figure 6P.4. These data may be fit by an expression of the form $r_0 = k_D P_D^{0.523}$, where P_D is the dimethylamine pressure.

5. Hinshelwood and Burk [*J. Chem. Soc., 127* (1105), 1925] have investigated the decomposition of ammonia on tungsten and platinum

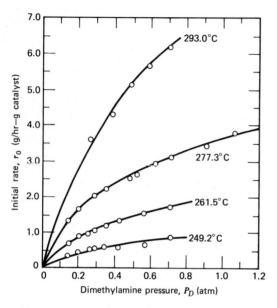

Figure 6P.4

Initial rate versus dimethylamine pressure. [Reprinted with permission from *AIChE J.*, 12 (292), 1966.]

surfaces. The data below were reported for the decomposition of ammonia on an electrically heated tungsten wire at 856 °C. What conclusions can you draw from these data relative to the rate expression and surface coverages for the decomposition of ammonia on tungsten? It is *not* necessary to determine a reaction rate constant!

6. Mardaleishvilli, Sin-Chou, and Smorodinskaya [*Kinetics and Catalysis*, 8 (664), 1967] have studied the catalytic decomposition of ammonia on quartz. The following initial rate data were obtained by these investigators at 951°C

Initial rate (arbitrary units)	Initial ammonia pressure (mm Hg)
0.0033	20
0.0085	50
0.0140	90
0.0236	140

What is the order of the reaction with respect to ammonia?

7. Hinshelwood and Burk [*J. Chem. Soc., 127* (1105), 1925] have studied the decomposition of ammonia over a heated platinum filament at 1138° C. The reaction stoichiometry is

$$2NH_3 \rightarrow N_2 + 3H_2$$

Initially, pure NH_3 was present in the reaction vessel. The data on page 207 are representative of the kinetics of the reaction. The reactor volume is a constant.

Time (sec)	Initial mixture composition		
	13.3 kPa NH_3	13.3 kPa NH_3 13.3 kPa H_2	26.6 kPa NH_3
	kPa of ammonia decomposed		
100	1.80	1.87	1.87
200	3.13	3.27	3.60
300	4.47	4.60	5.07
400	5.67	5.87	6.47
500	6.80	7.00	7.87

Time, t (sec)	Total pressure (kPa)
0	26.7
10	30.4
60	34.1
120	36.3
240	38.5
360	40.0
720	42.7

It has been postulated that the rate expression for the reaction is of the form

$$r = k \frac{P_{NH_3}}{P_{H_2}}$$

Is the experimental data consistent with this rate expression? If so, what is the value of the rate constant? What type of Hougen-Watson model gives rise to this form for the rate expression?

8. Brittan, Bliss, and Walker [*AIChE J.*, 16 (305), 1970] have studied the catalytic oxidation of carbon monoxide,

$$CO + \tfrac{1}{2}O_2 \rightleftharpoons CO_2$$

using a hopcalite catalyst (a mixture of oxides of manganese and copper and small quantities of other oxides). They used a batch circulation reactor (constant volume) and obtained the following data for one of their runs.

	System temperature 40 °C		
Time, t (ksec)	CO	Partial Pressures (kPa) O$_2$	CO$_2$
0.0	13.6	14.5	1.2
0.12	12.1	13.7	2.8
0.30	10.3	13.1	4.3
0.60	8.0	12.3	6.3
1.20	5.7	10.8	8.4
1.86	3.9	10.3	10.3
2.58	2.7	9.5	12.0
3.48	1.3	8.5	13.1
4.98	0.3	8.3	13.9

(a) In many previous studies of the catalytic oxidation reaction it has been found that the reaction is first-order with respect to carbon monoxide.

$$r = -kP_{CO}$$

Determine if the data obtained by Brittan, Bliss, and Walker are consistent with this type of rate expression.

(b) On the basis of their experimental data, Walker, Bliss, and Brittan proposed the following mechanism for this reaction.

$$CO + O_\sigma \xrightarrow{k_1} O_\sigma(CO) \qquad \text{slow}$$

$$CO + O_\sigma \xrightarrow{k_2} CO_{2\sigma} \qquad \text{fast}$$

$$O_\sigma(CO) \xrightarrow{k_3} CO_2 + \sigma \qquad \text{fast}$$

$$O_\sigma(CO) + \tfrac{1}{2}O_2 \xrightarrow{k_4} CO_2 + O_\sigma$$

$$\sigma + \tfrac{1}{2}O_2 \xrightarrow{k_5} O_\sigma$$

O_σ = atomically adsorbed oxygen

σ = vacant catalyst site

$O_\sigma(CO)$ = complex formed on surface of catalyst

Assuming that the adsorption of CO is rate controlling, derive the kinetic expression for the rates of change of the concentrations (partial pressures) of CO, O$_2$, and CO$_2$. Note that the reaction does not proceed stoichiometrically in that the hopcalite may act as a source or sink for the oxygen. Hence separate rate expressions are required for each component.

(c) Determine whether the expression obtained from this mechanism for the disappearance of carbon monoxide agrees with the experimental results.

$$k_1 + k_2 = 0.00737 \ (\text{min}^{-1})(\text{mm Hg}^{-1})$$

$$O_\sigma^0 = 11.81 \text{ equivalent mm Hg} \ (\text{initial level})$$

$$\sigma^0 = 0.95 \text{ equivalent mm Hg} \ (\text{initial level})$$

$$k_1 = 0.00732 \ (\text{min}^{-1})(\text{mm Hg})^{-1}$$
$$k_3/k_4 = 3.13 \ (\text{mm Hg})^{1/2}$$
$$k_5 = 0.0085 \ (\text{min}^{-1})(\text{mm Hg})^{-1}$$

9. Corrigan et al. [*Chem. Eng. Prog.*, *49* (603), 1953] have investigated the catalytic cracking of cumene over a silica-alumina catalyst at 950 °C.

$$\text{Cumene} \rightarrow \text{Propylene} + \text{Benzene}$$
$$\text{C} \rightarrow \text{P} + \text{B}$$

They indicated that both single- or dual-site mechanisms could be postulated as follows.

Single site	Dual site
(1) $C + \sigma \rightleftharpoons C_\sigma$	(4) $C + \sigma \rightleftharpoons C_\sigma$
(2) $C_\sigma \rightleftharpoons B_\sigma + P$	(5) $C_\sigma + \sigma \rightleftharpoons B_\sigma + P_\sigma$
(3) $B_\sigma \rightleftharpoons B + \sigma$	(6) $B_\sigma \rightleftharpoons B + \sigma$
	(7) $P_\sigma \rightleftharpoons P + \sigma$.

(a) Assuming that the reaction is reversible and that steps 2 and 5 are rate limiting for the single- and dual-site mechanisms, respectively, derive Hougen-Watson rate expressions for these mechanisms.
(b) These investigators reported the *initial* rate data shown in Figure 6P.5. Is either of the mechanisms considered in part (a) consistent with these data?

10. The water gas reaction takes place on a platinum catalyst in a batch reactor at 1000 °C. The reaction is

$$CO_2 + H_2 \rightarrow H_2O + CO$$

No products other than H_2O or CO are formed. In a series of experiments in which the initial CO_2 pressure was held constant (and in substantial excess over hydrogen), the rate of formation of CO was found to be directly proportional to the pressure of H_2. In another series of experiments the initial partial pressure of H_2 was held constant at 100 mm Hg, and the following partial pressures of CO were found after a reaction time of 120 sec.

Initial pressure CO_2 (mm Hg)	CO pressure after 120 sec (mm Hg)
25	3.8
50	7.0
75	10.0
100	12.4
125	14.4
150	16.2
175	17.5
200	18.0
225	17.8
250	17.3
300	15.4

The CO was found to have a negligible retarding influence on the rate of reaction. In all runs the water formed was removed immediately by condensation in a dry ice trap.
(a) Derive the general Hougen-Watson rate expression that applies to this kind of reaction.
(b) What is the simplest form to which the general equation reduces that still adequately fits both series of experiments?
(c) Why does the rate go through a maximum with an increase in the initial pressure of

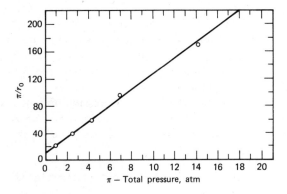

Figure 6P.5

Data plot for problem 9. π/r_0 **versus pressure at 950 °F. [Used with permission. From Corrigan et al.,** *Chem. Eng. Prog.*, *49* **(603), 1953.]**

CO_2? Explain qualitatively, in terms of the mathematical model, and in mechanistic terms.

11. An addition reaction of the type

$$A + B \rightarrow C$$

is being carried out over a relatively inactive metal catalyst at 400 °K. At this temperature the rate appears to be first-order in the gas phase concentrations of both A and B.

$$r = kP_A P_B$$

In studies of the temperature dependence of the reaction, the rate goes through a maximum (at fixed P_A and P_B) at a temperature near 400 °K. How do you interpret these facts in terms of the Hougen-Watson framework?

nickel catalyst. Instead of using a more complex Hougen-Watson model for the kinetic equation, they used an equation of the following form to correlate the data.

$$r_0 = kP_a^m P_h^n$$

where

r_0 = initial reaction rate, g-moles propionaldehyde/ksec-g unreduced catalyst

P_a, P_h = partial pressures of propionaldehyde, hydrogen, kPa.

m, n = exponents in rate equation

For their runs at approximately 150 °C, the unreduced catalyst weight was 0.7961 g and the following data were obtained.

Partial pressures (kPa)		Space velocity g-moles propionaldehyde/ ksec-g unreduced catalyst	Percent conversion
propionaldehyde	Hydrogen		
68.9	68.9	0.697	2.00
103.4	103.4	0.697	2.63
137.9	137.9	0.697	3.04
155.1	155.1	0.697	3.20
172.4	172.4	0.697	3.37
189.6	189.6	0.697	3.48
206.8	206.8	0.697	3.60
112.4	94.5	0.819	2.47
99.3	107.6	0.819	2.23
88.9	117.9	0.819	1.85
76.5	130.3	0.819	1.57
67.6	139.3	0.819	1.33
112.4	94.5	0.819	2.47
104.8	102.0	0.358	5.45
106.2	100.7	1.106	1.77
109.6	97.2	1.958	0.92

12. Oldenburg and Rase [*AIChE J.*, 3 (462), 1957] studied the catalytic vapor phase hydrogenation of propionaldehyde by making low conversion runs on a commercial supported

Determine the values of k, m, and n in the simplified rate equation. Assume that the catalyst activity change is negligible over the course of the experiments.

Hint 1. The reactor can be treated as a differential reactor, since the conversions are so low.

Hint 2. There is a way of using all of the data to obtain these three constants by taking each of the two sets of data and plotting the appropriate straight line for each. Round off your values of m and n to the nearest half integer.

13.* Yang and Hougen [*Chem. Eng. Progr., 46* (146), 1950] have suggested that initial rate data can be analyzed in terms of total pressure to facilitate the acceptance or rejection of a mechanism. For their analyses, experimental conditions are set such that the two reactants are present in constant composition (often equimolar amounts).

Mathur and Thodos [*Chem. Eng. Sci., 21* (1191), 1966] used the initial rate approach to analyze the kinetics of the catalytic oxidation of sulfur dioxide. They summarized the most plausible rate controlling steps for the reaction as:

1. For sulfur dioxide chemisorbed, oxygen dissociated and chemisorbed, sulfur trioxide chemisorbed:
   (a) Adsorption of oxygen controlling.
   (b) Adsorption of sulfur dioxide controlling.
   (c) Desorption of sulfur trioxide controlling.
   (d) Surface reaction between sulfur dioxide and atomic oxygen controlling.
2. For sulfur dioxide chemisorbed, oxygen in gas phase and not adsorbed, sulfur trioxide chemisorbed.

   For a Single-Site Mechanism
   (e) Adsorption of sulfur dioxide controlling.
   (f) Desorption of sulfur trioxide controlling.
   (g) Surface reaction between chemisorbed sulfur dioxide and gaseous oxygen controlling.

For a Dual-Site Mechanism
(h) Adsorption of sulfur dioxide controlling.
(i) Desorption of sulfur trioxide controlling.
(j) Surface reaction between chemisorbed sulfur dioxide and gaseous oxygen controlling.

I. Derive equations relating the *initial* reaction rate (r_0) to the total pressure (π) for each of the above cases when the sulfur dioxide and oxygen are initially present in equimolar amounts. Do this using the Hougen-Watson mechanistic models. Show your derivations.

Hint. For the single-site mechanism assume (as Mathur and Thodos did) that the surface reaction step can be written as:

$$SO_2 \cdot \sigma + (\tfrac{1}{2}O_2)_{gas} \rightarrow SO_3 \cdot \sigma$$

For the dual-site mechanism assume (as Mathur and Thodos did) that the reaction step can be written as:

$$SO_2 \cdot \sigma + \sigma + (\tfrac{1}{2}O_2)_{gas} \rightarrow SO_3 \cdot \sigma + \sigma$$

II. Using the data on page 211, prepare plots of r_0 versus π for each temperature.

Reject any mechanisms that are grossly inadequate by *qualitative* reasoning. Do not attempt to do a complicated quantitative or statistical analysis.

Hint 1. Although sulfur dioxide and oxygen do not appear to be present in equimolar amounts for several of the data points, you may assume, as Mathur and Thodos did, that they are close enough for the purposes of this *qualitative* analysis.

Hint 2. Look for the presence or absence of a maximum. Assume that the data have been taken over a wide enough range for a maximum to show up if it were going to.

* Adapted from G. P. Mathur and G. Thodos, *Chem. Eng. Sci., 21* (1191) copyright © 1966. Used with permission of Pergamon Press, Ltd.

Total pressure atm	Superficial mass velocity (lb-moles/hr-ft^2)		Initial rate lb-moles SO$_2$ hr-lb catalyst	Temperature, T (°F)	Grams catalyst
	SO$_2$	O$_2$			
1.54	1.14	1.33	0.0322	649	3.145
2.36	1.84	2.04	0.0489	649	3.145
3.72	2.61	2.92	0.0541	649	3.145
7.12	4.93	4.99	0.0757	649	3.145
9.84	4.29	6.01	0.0585	649	3.145
1.68	1.29	1.31	0.0320	649	2.372
2.36	1.83	1.90	0.0532	649	2.372
3.72	2.59	2.90	0.0642	649	2.372
7.12	4.87	4.94	0.0642	649	2.372
9.71	4.21	5.95	0.0514	649	2.372
1.54	1.13	1.32	0.0202	701	3.145
2.36	1.84	2.04	0.0904	701	3.145
7.12	4.94	4.98	0.0496	701	3.145
1.54	1.23	1.32	0.0289	701	2.372
2.36	1.83	1.90	0.0688	701	2.372
3.04	2.10	2.48	0.0845	701	2.372
3.72	2.59	2.90	0.0894	701	2.372
7.12	4.85	4.94	0.0520	701	2.372
1.54	1.13	1.33	0.0578	752	3.145
2.36	1.85	2.04	0.0620	752	3.145
3.72	2.62	2.93	0.0659	752	3.145
5.08	3.05	2.88	0.0817	752	3.145
7.12	4.93	4.99	0.0947	752	3.145
1.68	1.29	1.31	0.0459	752	2.372
2.36	1.83	1.90	0.0592	752	2.372
3.72	2.59	2.90	0.0679	752	2.372
5.08	3.04	3.88	0.0852	752	2.372
7.12	4.89	4.93	0.1031	752	2.372

14. The kinetics of the irreversible reaction $A \rightarrow B + C$ has been studied in a laboratory scale recycle reactor. A nonporous solid catalyst is being used to obtain an appreciable reaction rate. The data reported by your technician are shown in Figure 6P.6. Each of the sets of data was recorded holding the partial pressures of the other species constant. The homogeneous gas phase reaction may be neglected. The reaction rate is expressed in terms of the number of gram moles of A converted per minute per gram of catalyst.

(a) What is your interpretation of these data in terms of the Hougen-Watson models?
(b) How would you replot these data in order to determine the rate constants and adsorption equilibrium constants involved in the model you propose in part (a)? These plots are to be straight lines.

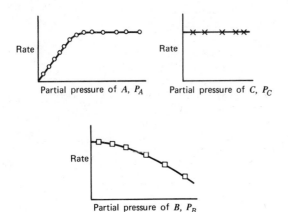

Figure 6P.6
Rate versus partial pressure plots.

15.* In a discussion of catalyst testing procedures, Dowden and Bridger [*Adv. Catalysis, 9*
(669), 1957] have reported the effect of particle
size and mass velocity on the rate of oxidation
of SO_2 to SO_3. They studied this reaction at
400 and at 470 °C using commercial catalyst
pellets (5.88 mm diameter) and two sizes of
crushed pellets (2.36 and 1.14 mm diameter).
In all runs the feedstream composition was kept
constant.

* This problem has been contributed by Professor C. N.
Satterfield of MIT.

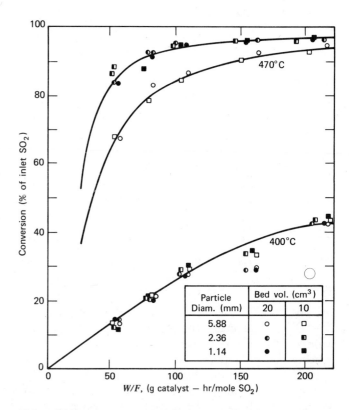

Figure 6P.7
**Effect of particle size and linear velocity on conversion.
[Reprinted with permission from the contribution of D. A.
Dowden and G. W. Bridger to** *Advances in Catalysis, 9* **(669).
Copyright © 1957, Academic Press.]**

The effect of mass velocity on the conversion rate was studied by using a tube of fixed diameter that was filled with a sample of a given catalyst diameter to give beds with volumes of either 10 or 20 cm³. At a constant ratio of catalyst weight to reactant feed, this method of varying the bed volume has the effect of varying the mass velocity through the bed.

The data in Figure 6P.7 indicate the conversion achieved for different operating conditions. What is your interpretation of these data?

16. Carlton and Oxley [*A.I.Ch.E. J.*, *13* (86), 1967] have studied the heterogeneous catalytic decomposition of nickel tetracarbonyl over the temperature range from 100 to 225° C.

$$Ni(CO)_4 \rightarrow Ni + 4 CO$$

The equilibrium conversion for this reaction is shown as a function of temperature and pressure in Figure 6P.8. The conversion is expressed in terms of the percent nickel tetracarbonyl decomposed at equilibrium.

A schematic diagram of the system used in this kinetic study is shown in Figure 6P.9.

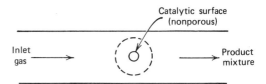

Figure 6P.9
Schematic diagram of reactor.

The pressure range investigated was 20 to 200 mm Hg. In some of the 50 mm total pressure runs Ar, He, and CO were added to the feedstream. In none of the studies did the conversion of the inlet $Ni(CO)_4$ exceed 8%.

The effect of temperature on the decomposition rate is shown in Figure 6P.10. In Figure 6P.11 the effect of various additives to the inlet stream is shown. In all cases shown in this figure, a constant mass flow rate of

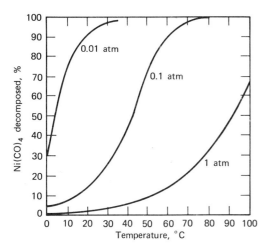

Figure 6P.8
Equilibrium decomposition of nickel tetracarbonyl. [Reprinted with permission from *AIChE J.*, *13* (86), 1967.]

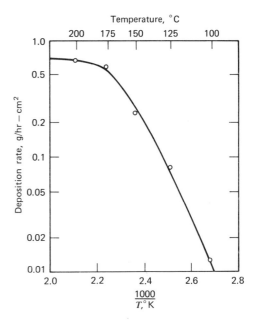

Figure 6P.10
Effect of temperature on decomposition rate at a Reynolds number of about 15 and a pressure of 20 torr. [Reprinted with permission from *AIChE J.*, *13* (86), 1967.]

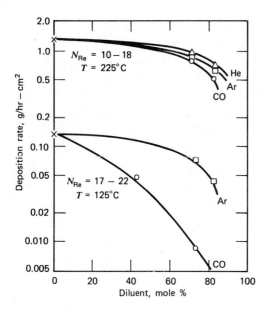

Figure 6P.11
Effect of diluent on deposition rate of nickel at 50 torr total pressure. [Reprinted with permission from *AIChE J.*, *13* (86), 1967.]

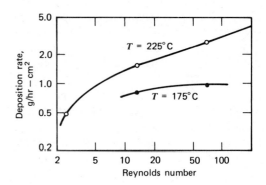

Figure 6P.12
Effect of flow rate on nickel deposition at 50 torr total pressure. [Reprinted with permission from *AIChE J.*, *13* (86), 1967.]

$Ni(CO)_4$ was used. In yet another series of runs the effect of the gas Reynolds number on the decomposition rate was studied. These data are shown in Figure 6P.12.

Interpret the data shown in Figure 6P.8 and Figures 6P.10 to 6P.12.

7 Liquid Phase Reactions

7.0 INTRODUCTION

The dynamic behavior of reactions in liquids may differ appreciably from that of gas phase reactions in several important respects. The short-range nature of intermolecular forces leads to several major differences in the macroscopic properties of the system, often with concomitant effects on the dynamics of chemical reactions occurring in the liquid phase.

The various effects are often classified as physical or chemical, depending on the role played by solvent molecules in the course of the elementary reaction acts. The effects associated with the presence of solvent molecules are *chemical* in nature when the solvent molecules themselves (or fragments thereof) participate in the microscopic reaction acts that comprise the reaction mechanism. In some cases the solvent molecules are regenerated by the sequence of reactions and, like catalysts, do not enter into the overall stoichiometry of the reaction. In other cases there may be a net increase or decrease in the amount of solvent present. However, in the majority of the situations likely to be encountered in industrial practice, the changes in solvent concentration arising from its participation in the reaction stoichiometry will not be appreciable. When solvent molecules play a chemical role, this effect is superimposed on a number of *physical* effects arising from the interplay of intermolecular forces in liquid solution. The most significant of the physical effects is ionization, which affords the possibility of alternative reaction mechanisms to those normally occurring in vapor phase reactions, with concomitant changes in the energy requirements for the molecular processes constituting the reaction mechanism. The *net* energy requirements for the conversion of reactants to products, however, remain unchanged.

This chapter discusses the aspects of the kinetic behavior of reactions in liquid solutions that are most germane to the education of a chemical engineer. Particular emphasis is placed on catalysis by acids, bases, and enzymes and a useful technique for correlating kinetic data.

Solvent molecules may play a variety of roles in liquid phase reactions. In some cases they merely provide a physical environment in which encounters between reactant molecules take place much as they do in gas phase reactions. Thus they may act merely as space fillers and have negligible influence on the observed reaction rate. At the other extreme, the solvent molecules may act as reactants in the sequence of elementary reactions constituting the mechanism. Although a thorough discussion of these effects would be beyond the scope of this textbook, the paragraphs that follow indicate some important aspects with which the budding kineticist should be familiar.

The most important physical effects associated with the presence of the solvent are electrostatic in nature. The production of ions from neutral species in the gas phase involves a large energy requirement (a few hundred kilocalories/gram mole). Consequently, ionic species are seldom involved in gas phase reaction mechanisms. In solution the interaction energy between the solvent molecules and the ions may often be of the same order of magnitude as the energy required for formation of ions from their neutral precursors. The interactions between reactant and solvent molecules thus may be strong enough to stabilize a charge on a fragment of a reactant molecule. Ionic species are then more likely to be present in much higher concentrations in the liquid phase than in the vapor phase and are much more likely to be involved as reaction intermediates in liquid phase reactions.

Electrostatic effects other than ionization are also important. Interactions between reacting ions depend on the local electrical environment of the ions and thus reflect the influence of the dielectric constant of the solvent and the presence of other ions and various solutes that may be present. In dilute solutions the influence of ionic strength on reaction rates is felt in the primary and secondary salt effects (see below).

Another important physical effect arising from the presence of solvent molecules is the efficiency of energy transfer. In condensed phases the rapid energy transfer in the abundant collisions between reactant and solvent molecules maintains vibrational thermal equilibrium at all times. Other solvent effects on reaction rates and mechanisms include effects arising from the acidity of the medium; effects of selective solvation by one component in a mixed solvent or of different degrees of solvation by different solvents of the reactants, reaction intermediates (including activated complexes), and reaction products; effects arising from hydrogen bonding; effects of changing from a protic solvent (one that may be regarded as a hydrogen bond donor, such as H_2O, NH_3, and alcohols) to a dipolar aprotic solvent (not a hydrogen bond donor, such as acetone, SO_2, or nitrobenzene); and cage effects. The various solvent effects may, of course, be superimposed or overlap one another such that they are inextricably linked. The magnitude of these solvent effects may range from insignificance to several powers of ten. For example, at 25 °C the reaction

$$CH_3I + Cl^- \rightarrow CH_3Cl + I^-$$

occurs more than 10^6 times as fast in dimethylacetamide as in acetone (1). On the other hand, the decomposition of nitrogen pentoxide in several different solvents is characterized by rate constants that lie within a factor of two of the gas phase rate constant (2).

Since the forces giving rise to the formation of chemical bonds are very short-range forces, reactions in liquid solutions will require some sort of encounter or "collision" between reactant molecules. These encounters will differ appreciably from gas phase collisions in that they will occur in close proximity to solvent molecules. Indeed, in liquids any individual molecule will always be interacting with several surrounding molecules at the same time, and the notion of a bimolecular collision becomes rather arbitrary. Nontheless, a number of approaches to formulating expressions for collision frequencies in the liquid phase have appeared in the chemical literature through the years. The simplest of these approaches presumes that the gas phase collision frequency expression is directly applicable to the calculation of liquid phase collision frequencies. The rationale for this approach is that for several second-order gas phase reactions that are also second-order in various solvents, the rate constants and preexponential factors are pretty much the same in the gas phase and in various solvents. For further discussion of the collision theory approach to reactions in liquids, consult the monograph by North (3).

There is another aspect of collisions in liquid solution that is of particular interest with regard to chemical reactions. Collisions in solution are often repeated, so that multiple collisions of the same two molecules occur. Consider the molecules labeled A and B in Figure 7.1. Each molecule is surrounded by several neighboring molecules. In view of the short-range order typical of liquids, the neighboring molecules will all be located at approximately the same distance from the molecule in question. This distance will be somewhat larger than a typical hard sphere molecular diameter, but considerably less than twice this diameter. Consequently, geometric and molecular force considerations indicate that the passage of any specified reference molecule between any two neighbors will be restricted by the repulsive portion of the intramolecular potential. To escape from the "cage" formed by the surrounding molecules, the reference molecule must surmount the energy barrier presented by the repulsive forces.

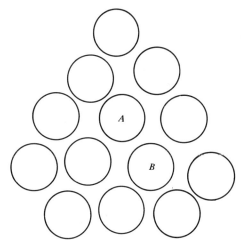

Figure 7.1
Schematic representation of the cage effect.

This molecule will undergo many collisions with its nearest molecules before it escapes from the cage. In the case of two solute molecules hemmed in by solvent molecules, multiple collisions will occur before one or both of the solute molecules can diffuse out of the cage. In liquid solution then, the total number of collisions is comparable in magnitude to the number of gas phase collisions, but repeated collisions are favored over fresh collisions.

It has been estimated (4) that in most common solvents at room temperature two reactant molecules within a cage of solvent molecules will collide from 10 to a 1000 times before they separate. The number of collisions per encounter will reflect variations in solvent viscosity, molecular separation distances, and the strength of the pertinent intermolecular forces. High viscosities, high liquid densities, and low temperatures favor many collisions per encounter.

The tendency for liquid phase collisions to occur in groups or sets does not have a very large effect for ordinary reactions that involve a significant activation energy, since no individual collision within the set is more likely to lead to reaction than any other. On the other hand, some reactions have zero or minimal

energy requirements (e.g., free radical recombinations). Such reactions will occur at virtually every collision. For these reactions the rate is limited by diffusional processes within the liquid phase in that virtually every collision between reactants leads to reaction. Since these reactions have negligible activation energies, their reaction rate constants are expected to be inversely proportional to the time elapsing between sets or groups of collisions.

The cage effect described above is also referred to as the Franck-Rabinowitch effect (5). It has one other major influence on reaction rates that is particularly noteworthy. In many photochemical reactions there is often an initiation step in which the absorption of a photon leads to homolytic cleavage of a reactant molecule with concomitant production of two free radicals. In gas phase systems these radicals are readily able to diffuse away from one another. In liquid solutions, however, the pair of radicals formed initially are caged in by surrounding solvent molecules and often will recombine before they can diffuse away from one another. This phenomenon is referred to as primary recombination, as opposed to secondary recombination, which occurs when free radicals combine after having previously been separated from one another. The net effect of primary recombination processes is to reduce the photochemical yield of radicals formed in the initiation step for the reaction.

7.1 ELECTROSTATIC EFFECTS IN LIQUID SOLUTION

In dilute solutions it is possible to relate the activity coefficients of ionic species to the composition of the solution, its dielectric properties, the temperature, and certain fundamental constants. Theoretical approaches to the development of such relations trace their origins to the classic papers by Debye and Hückel (6–8). For detailed treatments of this subject, refer to standard physical chemistry texts or to treatises on electrolyte solutions [e.g., that by Harned and Owen (9)]. The Debye-Hückel theory is useless for quantitative calculations in most of the reaction systems encountered in industrial practice because such systems normally employ concentrated solutions. However, it may be used together with the transition state theory to predict the qualitative influence of ionic strength on reaction rate constants.

For a bimolecular reaction between species A and B, the analysis gives an equation of the form

$$k = k_0 \frac{\gamma_A \gamma_B}{\gamma_{AB}^{\ddagger}} \qquad (7.1.1)$$

where k_0 is the rate constant in infinitely dilute solution and γ_A, γ_B, and $\gamma_{AB}^{\ddagger}$ are the activity coefficients of species A, B, and the activated complex, respectively.

If limiting forms of the Debye-Hückel expression for activity coefficients are used, this equation becomes

$$\ln k = \ln k_0 + 2z_A z_B \sqrt{\mu} \left[\frac{q^3 \sqrt{\dfrac{8\pi N_0}{1000}}}{2(k_B T D)^{3/2}} \right]$$

$$(7.1.2)$$

where
z_A and z_B represent the number (and sign) of charges on the ions A and B
q is the charge on an electron
N_0 is Avogadro's number
k_B is the Boltzmann constant
T is the absolute temperature
D is the dielectric constant of the medium
μ is the ionic strength of the solution

$$\mu = \tfrac{1}{2}(\textstyle\sum C_i z_i^2) \qquad (7.1.3)$$

where C_i is the molar concentration of species i. In dilute aqueous solution at $25\,°C$

$$\log k = \log k_0 + 2z_A z_B \sqrt{\mu}(0.509) \quad (7.1.4)$$

This equation is known as the Brønsted-Bjerrum equation. It may be derived in several ways. It predicts that a plot of log k versus the square root of the ionic strength should be linear over the range of ionic strengths where the Debye-Hückel limiting law is applicable. The slope of such lines is nearly equal to the product of the ionic charges of the reactants $z_A z_B$.

There are three general classes of ionic reactions to which these equations may be applied.

1. Reactions between ions of the same sign ($z_A z_B$ is positive) for which the rate constant increases with increasing ionic strength.
2. Reactions between ions of opposite sign ($z_A z_B$ is negative) for which the rate constant decreases with increasing ionic strength.
3. Reactions between an uncharged species and an ion that this equation predicts to be independent of ionic strength ($z_A z_B = 0$)

Several tests of the validity of the last equation have been made through the years. Figure 7.2, taken from Laidler (10), indicates the variation of $\log(k/k_0)$ with ionic strength for a variety of reactions. All factors considered, the agreement of theory and experiment are quite remarkable for a great many reactions. Note that the bulk of the data are reported in a range of ionic strengths that would normally be expected to be outside the range of validity of the Debye-Hückel limiting law (below $\sqrt{\mu} = 0.1$ for 1:1 electrolytes and below $\sqrt{\mu} = 0.03$ for higher valence ions). Equation 7.1.2 appears to be capable of extrapolation to much higher ionic strengths for the reactions indicated. Also note that the magnitude of the effect can be quite significant, particularly for reactions that involve ions with multiple charges. Even in the case of reactions where $z_A z_B = \pm 1$, the use of "inert" salts to increase the ionic strength can cause the rate to increase or decrease by as much as 50%.

Equation 7.1.2 characterizes what is known as the primary salt effect (i.e. the influence of ionic strength on the reaction rate through the activity coefficients of the reactants and the activated complex). Much early work on ionic reactions is relatively useless because this effect was not understood. Now it is common practice in studies of ionic reactions to add a considerable

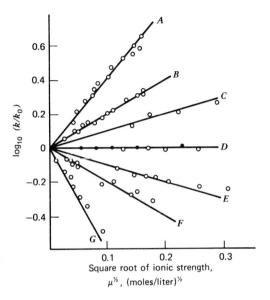

Figure 7.2
Plots of $\log_{10} (k/k_0)$ against the square root of the ionic strength, for ionic reactions of various types. The lines are drawn with slopes equal to $z_A z_B$. (10) The molecular species involved in the rate controlling steps are as follows.

A	$Co(NH_3)_5Br^{2+} + Hg^{2+}$	$(z_A z_B = 4)$
B	$S_2O_8^{2-} + I^-$	$(z_A z_B = 2)$
C	$CO(OC_2H_5)N:NO_2^- + OH^-$	$(z_A z_B = 1)$
D	$(Cr(urea)_6)^{3+} + H_2O$	$(z_A z_B = 0)$
	(open circles)	
	$CH_3COOC_2H_5 + OH^-$	$(z_A z_B = 0)$
	(closed circles)	
E	$H^+ + Br^- + H_2O_2$	$(z_A z_B = -1)$
F	$Co(NH_3)_5Br^{2+} + OH^-$	$(z_A z_B = -2)$
G	$Fe^{2+} + Co(C_2O_4)_3^{3-}$	$(z_A z_B = -6)$

(From *Chemical Kinetics* by K. J. Laidler. Copyright © 1965. Used with permission of McGraw-Hill Book Company.)

excess of inert salt (e.g., NaCl) to the reaction mixture. This practice ensures that the ionic strength does not vary substantially during the course of the reaction as it might if pure water were used. To the uninitiated observer, the primary salt effect might appear to be a form of catalysis, since the rate is affected by the

addition of substances that do not appear in the stoichiometric equation for the reaction. However, it is preferable not to regard it as a form of catalysis, since added salts affect the rate by changing the environment in which the reaction occurs through their influence on electrostatic forces instead of by opening up alternative reaction paths.

For more detailed treatments of ion-molecule reactions in liquid solution, consult the books by Amis (11), Amis and Hinton (12), and the review article by Clark and Wayne (13).

7.2 PRESSURE EFFECTS ON REACTIONS IN LIQUID SOLUTION

The effect of external pressure on the rates of liquid phase reactions is normally quite small and, unless one goes to pressures of several hundred atmospheres, the effect is difficult to observe. In terms of the transition state approach to reactions in solution, the equilibrium existing between reactants and activated complexes may be analyzed in terms of Le Chatelier's principle or other theorems of moderation. The concentration of activated complex species (and hence the reaction rate) will be increased by an increase in hydrostatic pressure if the volume of the activated complex is less than the sum of the volumes of the reactant molecules. The rate of reaction will be decreased by an increase in external pressure if the volume of the activated complex molecules is greater than the sum of the volumes of the reactant molecules. For a decrease in external pressure, the opposite would be true. In most cases the rates of liquid phase reactions are enhanced by increased pressure, but there are also many cases where the converse situation prevails.

In order to properly account for the effect of pressure on liquid phase reaction rates one should eliminate the pressure dependence of the concentration terms by expressing the latter in terms of mole ratios. It is then customary to express the general dependence of the rate constant on pressure as

$$\left(\frac{\partial \ln k}{\partial P}\right)_T = -\frac{\Delta V^{\ddagger}}{RT} \qquad (7.2.1)$$

where $\Delta V^{\ddagger}$ is the volume of activation for the reaction. If pressure-dependent concentration units (e.g., moles per liter) are employed to determine the rate constant, this equation must be corrected by a term accounting for the compressibility of the solution. In many cases the correction term is negligible, but this situation does not always prevail. Nonetheless, much of the high-pressure rate constant data available in the literature do not take this factor into account, and a good design engineer must properly evaluate the validity of the reported values before using them in calculations. Since high pressures usually enhance reaction rates, activation volumes are negative for most reactions.

Equation 7.2.1 implies that the rate constant for a reaction increases with increasing pressure if $\Delta V^{\ddagger}$ is negative, which is the most common situation. In this case the transition state has a smaller volume than the initial state. On the other hand, pressure increases bring about a decrease in the reaction rate if the formation of the activated complex requires a volume increase.

Substitution of typical numerical values of $\Delta V^{\ddagger}$ into equation 7.2.1 indicates that at room temperature $\Delta V^{\ddagger}/RT \simeq 10^{-3}$ atm^{-1}, so that one needs to go to pressures of several hundred or several thousand atmospheres to observe significant effects or to obtain accurate values of $\Delta V^{\ddagger}$ from plots of $\ln k$ versus pressure.

Activation volumes may be used to elucidate the mechanisms of classes of reactions involving the same functional groups, and changes in activation volumes can be used to characterize the point at which a change in reaction mechanism takes place in a series of homologous reactions.

For a detailed treatment of the kinetics of reactions at high pressures, consult the review article by Eckert (14).

7.3 HOMOGENEOUS CATALYSIS IN LIQUID SOLUTION

A catalyst has been defined previously as a substance that influences the *rate* or the *direction* of a chemical reaction *without being appreciably consumed*. Another definition of a catalyst that is particularly appropriate for reactions in liquid solution is the following: "A substance is said to be a catalyst for a reaction in a homogeneous system when its concentration occurs in the velocity expression to a higher power than it does in the stoichiometric equation" (15). In the overwhelming majority of cases, catalysts influence reaction rates by opening up alternative sequences of molecular reactions linking the reactant and product states. Catalyst species participate in elementary reaction steps, forming reaction intermediates that, in turn, react to yield eventually the reaction products and regenerate the original catalyst species.

Homogeneous catalytic processes are those in which the catalyst is dissolved in a liquid reaction medium. There are a variety of chemical species that may act as homogeneous catalysts (e.g., anions, cations, neutral species, association complexes, and enzymes). All such reactions appear to involve a *chemical interaction* between the catalyst and the *substrate* (the substance undergoing reaction). The bulk of the material in this section will focus on acid-base and enzyme catalysis. Students interested in learning more about these subjects and other aspects of homogeneous catalysis should consult appropriate texts (11–12, 16–29) or the original literature.

7.3.1 Acid-Base Catalysis

In acid-base catalysis there is at least one step in the reaction mechanism that consists of a generalized acid-base reaction (a proton transfer between the catalyst and the substrate). The protonated or deprotonated reactant species or intermediate then reacts further, either with another species in the solution or by a decomposition process.

The relative ease with which proton transfer is accomplished is responsible for the importance of the generalized acid-base concept in solution chemistry. The Brønsted concept of acidity is most useful in this respect. Brønsted defined an acid as a species that tends to give up a proton and a base as a species that tends to accept a proton. In this sense any proton transfer process having the general form

$$HX + Y \rightarrow HY + X \qquad (7.3.1)$$

may be regarded as an acid-base reaction between a Brønsted acid HX and a Brønsted base Y to form a conjugate base X and a conjugate acid HY. The various species involved in these reactions may be ions or neutral molecules.

The unique properties of the proton have been attributed by some authors to the fact that it has no electronic or geometric structure. The absence of any electron shell implies that it will have a radius that is about 10^5 times smaller than any other cation and that there will be no repulsive interactions between electron clouds as a proton approaches another reactant species. The lack of any geometric or electronic structure also implies that there will not be any steric limitations with regard to orientation of the proton. However, it still must attack the other reactant molecule at the appropriate site.

The small size of the proton relative to its charge makes the proton very effective in polarizing the molecules in its immediate vicinity and consequently leads to a very high degree of solvation in a polar solvent. In aqueous solutions, the primary solvation process involves the formation of a covalent bond with the oxygen atom of a water molecule to form a hydronium ion H_3O^+. Secondary solvation of this species then occurs by additional water molecules. Whenever we use the term hydrogen ion in the future, we are referring to the H_3O^+ species.

Early studies of the catalytic hydrolysis of esters indicated that in many cases, for strong acids, the observed rate constants were independent of the anion, and it became generally accepted that the active catalyst was a hydrogen ion. In other reactions, it became necessary to consider the effects of the hydroxide ion concentration and also the rate of the uncatalyzed reaction. The result was a three-term expression for the apparent rate constant:

$$k_{apparent} = k_0 + k_{H^+}[H^+] + k_{OH^-}[OH^-] \qquad (7.3.2)$$

where the first term corresponds to the uncatalyzed reaction, the second to catalysis by hydrogen ions, and the third to catalysis by hydroxide ions. The three rate constants appearing in this expression vary with temperature and with the nature of the reaction involved. Depending on the reaction conditions and the reaction involved, one or two of these terms may be negligible compared to the other(s). When only the hydronium ion is effective in catalyzing the reaction, the process is referred to as *specific acid catalysis*. When only hydroxide ions are effective, the process is classified as *specific base catalysis*. When both terms are significant, the catalysis is characterized as specific acid and specific base catalysis.

Although the concepts of specific acid and specific base catalysis were useful in the analysis of some early kinetic data, it soon became apparent that any species that could effect a proton transfer with the substrate could exert a catalytic influence on the reaction rate. Consequently, it became desirable to employ the more general Brønsted-Lowry definition of acids and bases and to write the reaction rate constant as

$$k = k_0 + k_{H^+}(H_3O^+) + k_{OH^-}(OH^-) \\ + \sum_i k_{HX_i}(HX_i) + \sum_j k_{X_j}(X_j) \qquad (7.3.3)$$

where HX_i and X_j represent all other acid and base species present in the solution apart from

H_3O^+ and OH^-. Reactions that are dependent on the concentrations of HX_i and X_j are categorized as involving *general acid* and *general base* catalysis. Table 7.1, adapted from Ashmore (30), indicates a number of catalytic reactions of the specific and general acid-base types in order to provide some orientation as to the types of reactions in the various categories. A thorough discussion of these reactions is obviously

beyond the scope of this text. Consult the books by Bell (22, 23), Frost and Pearson (20), Laidler (16), Bender (17), and Hine (31) for a further introduction to this topic.

Since the rates of acid and base catalyzed reactions are sensitive to variations in the solution pH, it is instructive to consider the types of behavior that can be observed in aqueous solution in the laboratory. The disso-

Table 7.1

Examples of Acid and Base Catalysis in Aqueous Solution (30)

Type of catalysis	Brief title of reaction	Equation of reaction
Specific acid	Inversion of cane sugar	$C_{12}H_{22}O_{11} + H_2O = C_6H_{12}O_6 + C_6H_{12}O_6$
	Hydrolysis of acetals	$R_1CH(OR_2)_2 + H_2O = R_1\cdot CHO + 2R_2OH$
	Hydration of unsaturated aldehydes	$CH_2{:}CH\cdot CHO + H_2O = CH_2OH\cdot CH_2\cdot CHO$
Specific base	Cleavage diacetone-alcohol	$CH_3CO\cdot CH_2\cdot C(OH)(CH_3)_2 = 2(CH_3)_2\cdot CO$
	Claisen condensation	$C_6H_5\cdot CHO + CH_3\cdot CHO = C_6H_5\cdot CH{:}CHO + H_2O$
	Aldol condensation	$2R\cdot CH_2\cdot CHO = R\cdot CH_2\cdot CH(OH)\cdot CHR\cdot CHO$
Specific acid and base	Hydrolysis of γ-lactones	$CH_2\cdot CH_2\cdot CH_2\cdot C{:}O + H_2O = CH_2OH\cdot CH_2\cdot CH_2\cdot COOH$ (cyclic O)
	Hydrolysis of amides	$R\cdot CO\cdot NH_2 + H_2O = R\cdot COONH_4$
	Hydrolysis of esters	$R_1\cdot COOR_2 + H_2O = R_1\cdot COOH + R_2OH$
General acid	Decomposition of acetaldehyde hydrate	$CH_3\cdot CH(OH)_2 = CH_3\cdot CHO + H_2O$
	Hydrolysis of o-esters	$HC\cdot(OC_2H_5)_3 + H_2O = H\cdot COOC_2H_5 + 2C_2H_5OH$
	Formation of nitro-compound	$CH_2{:}NO_2^- + acid = CH_3\cdot NO_2 + base^-$
General base	Decomposition of nitramide	$NH_2NO_2 = N_2O + H_2O$
	Bromination of nitromethane	$CH_3NO_2 + Br_2 = CH_2BrNO_2 + HBr$
	Aldol with acetaldehyde	$2CH_3\cdot CHO = CH_3\cdot CH(OH)\cdot CH_2\cdot CHO$
General acid and base	Halogenation, exchange, racemization of ketones	$R\cdot CO\cdot CH_3 + X_2 = R\cdot CO\cdot CH_2X + XH$
	Addition to carbonyl	$R_1\cdot CO\cdot R_2 + NH_2OH = R_1R_2\cdot COH\cdot NHOH$

ciation constant for water may be written as

$$K_w = [H^+][OH^-] \qquad (7.3.4)$$

Combining equations 7.3.4 and 7.3.2 indicates that in aqueous media for specific acid-base catalysis,

$$k_{apparent} = k_0 + k_{H^+}[H^+] + \frac{k_{OH^-}K_w}{[H^+]} \qquad (7.3.5)$$

Alternatively,

$$k_{apparent} = k_0 + \frac{k_{H^+}K_w}{[OH^-]} + k_{OH^-}[OH^-] \qquad (7.3.6)$$

In many cases one or two of the three terms on the right side of these equations is negligible compared to the others. In a 0.1N strong acid, for example, the second term is $k_{H^+} \times 10^{-1}$, while the last term is $k_{OH^-} \times 10^{-13}$, since $K_w \simeq 10^{-14}$. Consequently, unless the ratio (k_{OH^-}/k_{H^+}) is 10^9 or more, the third term will be negligible compared to the second. This large a ratio is not often encountered in practice. By the same sort of argument, it can be shown that in a 0.1N strong base, catalysis by hydrogen ions will usually be unimportant relative to catalysis by hydroxide ions. Normally one finds that there will be a lower limit on hydrogen ion concentration, above which catalysis by hydroxide ions may be regarded as insignificant. Similarly, there is usually a range of pH values in the high pH region where catalysis by hydrogen ions will be negligible and the hydroxide ions will be responsible for the catalytic effect. Within the pH range where the hydrogen ions are solely responsible for the catalytic effect, the apparent rate constant will be linear in the hydrogen ion concentration. Consequently, it is a straightforward task to determine k_{H^+} from a plot of the apparent rate constant versus hydrogen ion concentration. Within the pH range where hydroxide ions are solely responsible for the catalysis, one may use an analogous procedure to determine k_{OH^-}.

There are several types of pH-dependent kinetic behavior that can be interpreted in

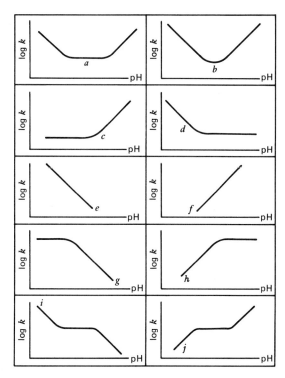

Figure 7.3
Schematic representation of log k-pH profiles for various types of acid-base catalysis.

terms of one or more of the various forms of the specific acid-base catalysis equation (equation 7.3.2). Skrabal (32) has classified the various possibilities that may arise in reactions of this type, and Figure 7.3 is based on this classification. The various forms of the plots of log k versus pH reflect the relative importance of each of the various terms in equation 7.3.2 as the pH shifts. Curve a represents the most general type of behavior. It consists of a region where acid catalysis is superimposed on the noncatalytic reaction, a region where neither acid nor base catalysis is significant, and a region where base catalysis is superimposed on the spontaneous reaction. At low pH, the bulk of the reaction may be attributed to the acid catalyzed reaction and

$$k_{apparent} \simeq k_{H^+}[H_3O^+] \qquad (7.3.7)$$

Thus

$$\begin{aligned} \log k_{\text{apparent}} &= \log k_{\text{H}^+} + \log[\text{H}_3\text{O}^+] \\ &= \log k_{\text{H}^+} - \text{pH} \end{aligned} \qquad (7.3.8)$$

The slope of the left arm of curve a is thus -1. At intermediate pH values, the rate is independent of pH, so k_0 may be determined directly from kinetic measurements in this pH regime. At high pH the only significant contribution to the rate constant is that arising from specific base catalysis. In terms of equation 7.3.5,

$$k_{\text{apparent}} \simeq \frac{k_{\text{OH}^-} K_w}{[\text{H}^+]} \qquad (7.3.9)$$

or

$$\begin{aligned} \log k_{\text{apparent}} &= \log(k_{\text{OH}^-} K_w) - \log \text{H}_3\text{O}^+ \\ &= \log(k_{\text{OH}^-} K_w) + \text{pH} \end{aligned}$$
$$(7.3.10)$$

Hence, at high pH, the slope of the $\log k_{\text{apparent}}$ versus pH plot is $+1$.

The various other forms of the log rate constant versus pH plots may be analyzed in similar terms. In curve b the horizontal portion of the curve is missing, indicating that the spontaneous reaction at no time contributes significantly to the observed reaction rate. If either k_{H^+} or k_{OH^-} is sufficiently small, the contribution to k_{apparent} from the catalytic effect in question will be negligible, and the corresponding arm of the curve will be missing. Hence, curve c reflects a combination of specific base catalysis and the intrinsic reaction, while curve d represents a combination of specific acid catalysis and the intrinsic reaction. If the horizontal region and one arm of the curve are missing, one has the degenerate forms of the rate constant expression indicated by equations 7.3.7 and 7.3.9. Curve e corresponds to specific acid catalysis and curve f to specific base catalysis. More complex dependencies of $\log k$ on pH have also been observed in the laboratory and, in these more complicated systems, considerable ambiguity may arise in the interpretation of the kinetic data.

We have indicated how to determine the various kinetic constants appearing in the expression for specific acid and base catalysis. Let us now consider how to evaluate the various contributions to the rate constant in the case of general acid-base catalysis. For reactions of this type in a solution of a weak acid or base and its corresponding salt, the possible catalysts indicated by equation 7.3.3 are the hydronium ion, the hydroxide ion, the undissociated weak acid (or base), and the conjugate base (or acid). In the case of acetic acid the general acid would be the neutral CH_3COOH species and the conjugate base would be the acetate ion $(\text{CH}_3\text{COO}^-)$. In this case the apparent rate constant can be written as

$$\begin{aligned} k_{\text{apparent}} = k_0 &+ k_{\text{H}^+}(\text{H}_3\text{O}^+) + k_{\text{OH}^-}(\text{OH}^-) \\ &+ k_{\text{HA}}(\text{HA}) + k_{\text{A}^-}(\text{A}^-) \end{aligned} \qquad (7.3.11)$$

The five constants appearing in this expression may be determined by a systematic variation of the experimental conditions so as to make one or more of the contributing terms negligible within a given set of experiments. By carrying out experiments in the absence of acetate ions and acetic acid, it is possible to determine the constants k_0, k_{H^+}, and k_{OH^-} using the procedure outlined previously for specific acid and specific base catalysis. Strong acids and strong bases whose corresponding anions and cations are known to exhibit no catalytic activity are used as the source of hydronium and hydroxide ions. One may then proceed to determine the constants characteristic of generalized acid and base catalysis by using a buffer solution of the weak acid and its conjugate base. In a series of buffer solution experiments the absolute concentrations of undissociated acetic acid species and acetate ions may be varied while maintaining a constant ratio (q) of these concentrations. In these experiments the ionic strength should also be maintained constant in order to ensure that the activity coefficients of the various species do not change. From a consideration of the dissociation equilibrium, it is evident that

$$(H^+) = K_a \frac{(HA)}{(A^-)} = K_a q \qquad (7.3.12)$$

where K_a is the dissociation constant for the acid. The hydrogen ion concentration will not vary if the ratio $(HA)/(A^-)$ is maintained constant. Thus the first three terms on the right side of equation 7.3.11 may be represented by a constant k_1 under these conditions. The equation may then be rewritten as

$$k_{apparent} = k_1 + (HA)(k_{HA} + k_{A^-}/q) \qquad (7.3.13)$$

By plotting the measured rate constant versus the undissociated acid concentration, one obtains for this type of catalysis a straight line with intercept k_1 and slope $\alpha_1 = (k_{HA} + k_{A^-}/q)$. If the procedure is repeated for other ratios, enough information is obtained to permit evaluation of k_{HA} and k_{A^-}. The hydrogen and hydroxide ion concentrations corresponding to a given ratio q may be determined from equation 7.3.12 and the dissociation constant for water.

It should be evident that the calculations and the necessary experimental program are simplified if one or more of the contributions is negligible under appropriate conditions. In a relatively simple system in which either k_{HA} or k_{A^-} is zero, the existence of general acid or general base catalysis may be deduced from the pH dependence of the reaction.

A reaction catalyzed by undissociated acid will have the dependence of log k on pH shown in Figure 7.3g. Specific acid and specific base catalysis are presumed to be absent. If specific and general acid and base catalysis are both operative, one is able to obtain a variety of interesting log k versus pH curves, depending on the relative contributions of the different terms in various pH ranges. Curves i and j of Figure 7.3 are simple examples of these types.

Since specific acid and specific base catalysis and generalized acid and generalized base catalysis by cations and anions all involve ionic species, these processes are influenced by the ionic strength of the solutions in which they take place. There are two types of salt effects that are significant in acid-base catalysis. The first of these is the primary salt effect, discussed in Section 7.1. This effect is significant for reactions in solution involving ionic species. It operates by influencing the activity coefficients of the reactants and the activated complex. The second effect is referred to as the secondary salt effect. It operates by changing the actual concentrations of the catalytically active ions. It may cause either an increase or a decrease in the rate constant, and it may be either larger or smaller than a primary salt effect. In some instances both effects occur simultaneously, and a cancellation of effects can make the reaction rate appear to be independent of ionic strength.

The secondary salt effect is important when the catalytically active ions are produced by the dissociation of a weak electrolyte. In solutions of weak acids and weak bases, added salts, even if they do not exert a common ion effect, can influence hydrogen and hydroxide ion concentrations through their influence on activity coefficients.

If one considers a reaction catalyzed by hydrogen ions formed by the dissociation of a weak acid HA, the hydrogen ion concentration is governed by the following relation:

$$K_a = \frac{\gamma_{H^+}\gamma_{A^-}}{\gamma_{HA}} \frac{(H^+)(A^-)}{(HA)} \qquad (7.3.14)$$

where K_a is the equilibrium constant for the reaction and the γ_i's are the activity coefficients of the various species. Taking logarithms and rearranging gives

$$\log(H^+) = \log K_a + \log\left(\frac{HA}{A^-}\right) + \log\left(\frac{\gamma_{HA}}{\gamma_{H^+}\gamma_{A^-}}\right) \qquad (7.3.15)$$

The concentration of hydrogen ions depends on the ionic strength through the ratio of activity coefficients appearing in the dissociation equilibrium expression. Any change in ionic strength affects the γ terms and thus the hydrogen ion concentration. Consequently, if a reaction is catalyzed by hydrogen (or hydroxide ions), the rate becomes dependent on ionic strength through this secondary salt effect.

For the range of concentrations where the Debye-Hückel theory is applicable it is possible to place the theory of the secondary salt effect on a more quantitative basis by using Debye-Hückel relations for the various activity coefficients appearing in equation 7.3.15 or its more general analog. For an acid of charge z and its conjugate base of charge $(z - 1)$, the analogous equation is

$$\log(H^+) = \log K_a + \log\left[\frac{(HA^z)}{(A^{z-1})}\right]$$
$$+ \log\left(\frac{\gamma_{HA^z}}{\gamma_{H^+}\gamma_{A^{z-1}}}\right) \qquad (7.3.16)$$

If the Debye-Hückel limiting law is used to evaluate the various activity coefficients in aqueous solution at 25 °C, the last equation becomes

$$\log(H^+) = \log K_a + \log\left[\frac{(HA^z)}{(A^{z-1})}\right]$$
$$- 0.509\sqrt{\mu}(2)(z - 1) \qquad (7.3.17)$$

In the most common case $z = 0$, and the secondary salt effect implies that the hydrogen ion concentration will increase with increasing ionic strength. However, the direction of the effect is determined by the sign of the quantity $(z - 1)$.

The existence of the primary and secondary salt effects indicates the importance of maintaining control over ionic strength in kinetics studies. One may choose to keep the ionic strength low so as to minimize its effects, or one may make a series of measurements at various ionic strengths in order to permit extrapolation to the limit of infinitely dilute solution. Another useful alternative is to maintain the ionic strength constant at a value that is suffi-

ciently large that any variations caused by the progress of the reaction will be negligible. This approach is similar to the method of excess concentrations discussed in Chapter 3. It is particularly useful in attempts to determine the rate expression for reactions that would involve significant changes in ionic strength if carried out in the absence of extraneous ionic species. Unfortunately, the rate constants determined in this fashion may be quite different from those in highly dilute solutions. Nonetheless it is good laboratory practice to add small quantities of electrolytes to those reaction systems believed to involve ionic species in order to determine the possible presence of ionic strength effects. Some judgment must be used in the selection of the added ions in order to choose species that are noncatalytic in themselves and thus influence reaction rates only through the charges resulting from electrolytic dissociation.

7.3.2 Catalysis by Enzymes

Enzymes are protein molecules that possess exceptional catalytic properties. They are essential to plant and animal life processes. Enzymes are remarkable catalysts in at least three respects: activity, specificity, and versatility.

The high activity of enzymes becomes apparent when the rates of enzyme catalyzed reactions are compared to those of the corresponding nonenzymatic reaction or to the same reaction catalyzed by an inorganic species. Rate enhancements on the order of 10^8 to 10^{11} are not unusual in the presence of enzymes. Enzyme efficiencies are often measured in terms of turnover numbers. This number is defined as the number of molecules that are caused to react in 1 min by one molecule of catalyst. For many common reactions the turnover number is in excess of 10^3 and, in some cases, it may exceed 10^6. High enzyme turnover numbers are largely the result of greatly reduced activation energies for the enzymatic reaction relative to other modes of effecting the reactions in question.

Enzymes are often considered to function by general acid-base catalysis or by covalent catalysis, but these considerations alone cannot account for the high efficiency of enzymes. Proximity and orientation effects may be partially responsible for the discrepancy, but even the inclusion of these effects does not resolve the disparity between observed and theoretically predicted rates. These and other aspects of the theories of enzyme catalysis are treated in the monographs by Jencks (33) and Bender (34).

Enzyme specificities are categorized in terms of the manner in which enzymes interact with various substrates. Some enzymes will cause only a single substrate to react. This type of specificity is known as *absolute specificity*. An example is urease, which catalyzes only the hydrolysis of urea. Other enzymes will react only with substrates having certain functional groups in certain positions relative to the bond to be attacked. This situation is called *group specificity*. An example of enzymes of this type is pepsin, which requires an aromatic group to be present in a certain position relative to a peptide linkage in order to effect its hydrolysis. *Reaction specificity* is the least specific type of enzyme catalysis. It requires only that a certain type of bond be present in the substrate. Enzymes such as the lipases will catalyze the hydrolysis of any organic ester. Many enzymes exhibit *stereochemical specificity* in that they catalyze the reactions of one stereochemical form but not the other. Proteolytic enzymes, for example, catalyze only the hydrolysis of peptides composed of amino acids in the *L* configuration.

Enzyme specificity is often explained in terms of the geometric configuration of the active site of the enzyme. The active site includes the side chains and peptide bonds that either come into direct contact with the substrate or perform some direct function during catalysis. Each site is polyfunctional in that certain parts of it may hold the substrate in a position where the other parts cause changes in the chemical bonding of the substrate, thereby producing a reaction intermediate in the sequence of steps leading to product formation. The detailed configuration of the enzyme molecule—including the conformation of the protein in folds or coils as well as the chemical structure near the active site—is quite important, and it is said that the geometric configuration is such that only molecules with certain structural properties can fit. This is the famous "lock and key" hypothesis for enzyme activity that dates back to the work of Fischer in 1894. More sophisticated models that purport to explain enzyme specificity have been proposed through the years, but the basic concept that specificity results from steric or geometric considerations remains unchanged.

The third remarkable aspect of enzyme catalysis is the versatility of these species. They catalyze an extremely wide variety of reactions—oxidation, reduction, polymerization, dehydration, dehydrogenation, etc. Their versatility is a reflection of the range and complexity of the chemical reactions necessary to sustain life in plants and animals.

7.3.2.1 Rate Expressions for Enzyme Catalyzed Single-Substrate Reactions.

The vast majority of the reactions catalyzed by enzymes are believed to involve a series of bimolecular or unimolecular steps. The simplest type of enzymatic reaction involves only a single reactant or substrate. The substrate forms an unstable complex with the enzyme, which subsequently undergoes decomposition to release the product species or to regenerate the substrate.

Reaction rate expressions for enzymatic reactions are usually derived by making the Bodenstein steady-state approximation for the intermediate enzyme-substrate complexes. This is an appropriate assumption when the substrate concentration greatly exceeds that of the enzyme (the usual laboratory situation) or when there is both a continuous supply of reactant and a continuous removal of products (the usual cellular situation).

The "classic" mechanism of enzymatic catalysis can be written as

$$E + S \underset{k_2}{\overset{k_1}{\rightleftharpoons}} ES \qquad (7.3.18)$$

$$ES \overset{k_3}{\rightarrow} E + P \qquad (7.3.19)$$

where

E represents the enzyme

S represents the substrate

ES represents the enzyme-substrate complex

P represents the product of the reaction

The stoichiometry of the reaction may be written as

$$S \rightarrow P \qquad (7.3.20)$$

In a sense this mechanism is akin to Linde-

Combining equations 7.3.21 and 7.3.23 gives

$$V = \frac{k_1 k_3 (E)(S)}{k_2 + k_3} \qquad (7.3.24)$$

Since most kinetic studies of this type involve initial rate experiments, it is usually necessary to rederive this expression in terms of the initial concentrations and the initial rate. From material balance considerations

$$(E_0) = (E) + (ES) \qquad (S_0) = (S) + (ES) \qquad (7.3.25)$$

For the conditions commonly encountered in the laboratory, $S_0 \gg E_0$. Since ES cannot exceed E_0, this implies that $S_0 \approx S$.

Solving equation 7.3.25 for E and substituting this result into equation 7.3.22 gives

$$\frac{d(ES)}{dt} = k_1[(E_0) - (ES)](S_0) - k_2(ES) - k_3(ES) \cong 0 \quad (7.3.26)$$

mann's picture of unimolecular decomposition reactions (see Section 4.3.1.3). An initial reaction produces a reactive intermediate that subsequently decomposes irreversibly to yield products or is reversibly decomposed into enzyme and substrate.

The net rate of an enzymatic reaction is usually referred to as its velocity and is assigned the symbol V. In this case

$$V = \frac{d(P)}{dt} = k_3(ES) \qquad (7.3.21)$$

The concentration of the complex can be obtained by making the usual steady-state approximation

$$\frac{d(ES)}{dt} = k_1(E)(S) - k_2(ES) - k_3(ES) \approx 0$$

$$(7.3.22)$$

or

$$(ES) = \frac{k_1(E)(S)}{k_2 + k_3} \qquad (7.3.23)$$

or

$$(ES) = \frac{k_1(E_0)(S_0)}{k_2 + k_3 + k_1(S_0)} \qquad (7.3.27)$$

Note the similarity of this expression to that for θ_A, derived by the Langmuir adsorption isotherm. $(ES)/(E_0)$ plays a role analogous to θ_A, while S_0 plays a role akin to the gas pressure. Although the expression is formally similar, we do not mean to imply that the two types of catalytic reactions proceed by similar molecular steps.

This last result may be substituted into equation 7.3.21 to give

$$V_0 = \frac{k_1 k_3 (E_0)(S_0)}{k_2 + k_3 + k_1(S_0)} \qquad (7.3.28)$$

This equation predicts that the initial rate will be proportional to the initial enzyme concentration if the initial substrate concentration is held constant. If the initial enzyme concentration is held constant, the initial rate will be proportional to the substrate concentration at low substrate concentrations, and substantially independent of substrate concentration at high substrate levels. The maximum reaction rate is equal to $k_3 E_0$, and this product is often assigned the symbol V_{max}. The group $(k_2 + k_3)/k_1$ is often assigned the symbol K and is known as the Michaelis constant. Equation 7.3.28 can be written in terms of these parameters as

$$V_0 = \frac{V_{max}(S_0)}{K + (S_0)} \qquad (7.3.29)$$

The Michaelis constant is numerically equal to the value of the initial substrate concentration that gives an initial velocity that is half that of the maximum.

Although equation 7.3.28 and, in particular, equation 7.3.29 are known as Michaelis-Menten rate expressions, these individuals used a somewhat different approach to arrive at this mathematical form for an enzymatic rate expression (35).

In attempting to determine if a given set of experimental data is of the same mathematical form as equation 7.3.29, there are three routes that permit the graphical determination of the parameters V_{max} and K. The most frequently used plot is known as a *Lineweaver-Burk* or *reciprocal plot*. It is based on rearrangement of equation 7.3.29 into the following form.

$$\frac{1}{V_0} = \frac{1}{V_{max}} + \frac{K}{V_{max}(S_0)} \qquad (7.3.30)$$

If the data fit this model, a plot of $1/V_0$ versus $1/(S_0)$ should be linear with a slope K/V_{max} and intercept $1/V_{max}$. It is analogous to that used in determining the constants in the Langmuir equation for adsorption on solid surfaces. Other forms that may be used to prepare linear plots are

$$V_0 = V_{max} - K \frac{V_0}{(S_0)} \qquad (7.3.31)$$

and

$$\frac{(S_0)}{V_0} = \frac{K}{V_{max}} + \frac{(S_0)}{V_{max}} \qquad (7.3.32)$$

Equation 7.3.31 gives rise to what is known as an Eadie or Hofstee plot, while equation 7.3.32 gives rise to a Hanes plot. The Eadie plot has the advantage of spreading the points out more evenly and of determining K and V_{max} separately. The three types of plots are shown schematically in Figure 7.4. The Lineweaver-Burk and Eadie plots are the ones used most frequently in data analysis.

Although the Michaelis-Menten equation is applicable to a wide variety of enzyme catalyzed reactions, it is not appropriate for reversible reactions and multiple-substrate reactions. However, the generalized steady-state analysis remains applicable. Consider the case of reversible decomposition of the enzyme-substrate complex into a product molecule and enzyme with mechanistic equations.

$$E + S \underset{k_2}{\overset{k_1}{\rightleftharpoons}} ES \underset{k_4}{\overset{k_3}{\rightleftharpoons}} E + P \qquad (7.3.33)$$

In this case the net rate of reaction is given by

$$V = \frac{d(P)}{dt} = k_3(ES) - k_4(E)(P) \qquad (7.3.34)$$

The steady-state approximation for the intermediate complex is

$$0 \approx \frac{d(ES)}{dt} = k_1(E)(S) - k_2(ES)$$
$$- k_3(ES) + k_4(E)(P) \qquad (7.3.35)$$

If the conservation equation for total enzyme concentration (7.3.25) is employed, the last equation becomes

$$k_1[(E_0) - ES](S) - k_2(ES) - k_3(ES)$$
$$+ k_4[(E_0) - (ES)](P) \cong 0 \qquad (7.3.36)$$

or

$$(ES) = \frac{k_1(E_0)(S) + k_4(E_0)(P)}{k_2 + k_3 + k_1(S) + k_4(P)} \qquad (7.3.37)$$

Combining equations 7.3.25, 7.3.34, and 7.3.37 gives

$$V = \frac{k_1 k_3(E_0)(S) - k_2 k_4(E_0)(P)}{k_2 + k_3 + k_1(S) + k_4(P)} \qquad (7.3.38)$$

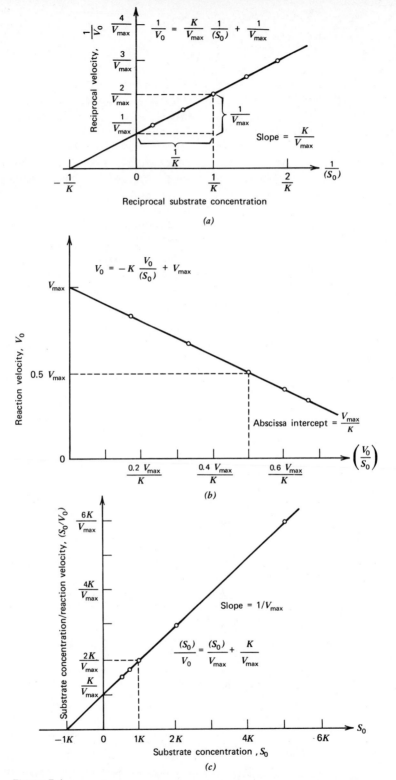

Figure 7.4
Methods of plotting data obtained from enzyme catalyzed reactions.
(a) Lineweaver-Burk plot. (b) Eadie or Hofstee plot. (c) Hanes plot.
(Adapted from *Enzyme Kinetics* by K. Plowman. Copyright © 1972.
Used with permission of McGraw-Hill Book Company.)

If one defines the following enzyme reaction parameters,

$$V_{1\,max} = k_3(E_0) \qquad V_{4\,max} = k_2(E_0)$$

$$K_s = \frac{k_2 + k_3}{k_1} \qquad K_p = \frac{k_2 + k_3}{k_4} \qquad (7.3.39)$$

the four kinetic constants ($V_{1\,max}$, $V_{4\,max}$, K_s, and K_p) can be determined from initial rate studies of the forward and reverse reactions. Equation 7.3.38 can be rephrased in terms of these parameters as

$$V = \frac{K_p V_{1\,max}(S) - K_s V_{4\,max}(P)}{K_s K_p + K_p(S) + K_s(P)} \qquad (7.3.40)$$

At equilibrium the net reaction velocity must be zero. In terms of the enzymatic kinetic constants, equation 7.3.40 then indicates that

$$\frac{(P)}{(S)} = K_{eq} = \frac{K_p V_{1\,max}}{K_s V_{4\,max}} \qquad (7.3.41)$$

where K_{eq} is the thermodynamic equilibrium constant for the overall reaction. Equation 7.3.41 is known as the Haldane relation. It indicates that the enzymatic kinetic parameters are not all independent, but are constrained by the thermodynamics of the overall reaction.

In the above example the enzyme combined with either the substrate or the product to form the same complex. This assumption is not realistic, but it still leads to the correct form of the rate expression.

The following mechanism for a reaction of identical stoichiometry introduces a second complex into the sequence of elementary reactions.

$$E + S \underset{k_2}{\overset{k_1}{\rightleftharpoons}} ES \underset{k_4}{\overset{k_3}{\rightleftharpoons}} EP \underset{k_6}{\overset{k_5}{\rightleftharpoons}} E + P \qquad (7.3.42)$$

where ES and EP are enzyme-substrate and enzyme-product complexes, respectively.

For this mechanism, the steady-state rate equation is:

A comparison of equations 7.3.43 and 7.3.38 shows that they are of the same mathematical form. Both can be written in terms of four measurable kinetic constants in the manner of equation 7.3.40. Only the relationship between the kinetic constants and the individual rate constants differs. Thus, no distinction can be made between the two mechanisms using steady-state rate studies. In general, the introduction of unimolecular steps involving only isomerization between unstable intermediate complexes does not change the form of the rate expression.

7.3.2.2 Inhibition Effects in Enzyme Catalyzed Reactions. Enzyme catalyzed reactions are often retarded or inhibited by the presence of species that do not participate in the reaction in question as well as by the products of the reaction. In some cases the reactants themselves can act as inhibitors. Inhibition usually results from the formation of various enzyme-inhibitor complexes, a situation that decreases the amount of enzyme available for the normal reaction sequence. The study of inhibition is important in the investigation of enzyme action. By determining what compounds behave as inhibitors and what type of kinetic patterns are followed, it may be possible to draw important conclusions about the mechanism of an enzyme's action or the nature of its active site.

Inhibitors may act reversibly or irreversibly, but this classification is not particularly useful. It may even be misleading, because it suggests that reversible and irreversible inhibitors act in different ways when, in fact, both act by combining with the enzyme to give inactive complexes, but with quite different "dissociation constants." The irreversible inhibitors give complexes that have very small dissociation constants; the reversible inhibitors have significantly higher dissociation constants.

$$V = \frac{(E_0)k_1 k_3 k_5(S) - (E_0)k_2 k_4 k_6(P)}{(k_2 k_5 + k_2 k_4 + k_3 k_5) + k_1(k_3 + k_4 + k_5)(S) + k_6(k_2 + k_3 + k_4)(P)} \qquad (7.3.43)$$

A much more useful classification of inhibitors can be made on the basis of the mechanisms by which they act. *Competitive inhibitors* combine with the enzyme at the same site as the substrate does, thus blocking the first step in the sequence. *Noncompetitive inhibitors* combine with the enzyme at some other site to give a complex that can still combine with the substrate, but the resultant ternary complex is unreactive. *Uncompetitive inhibition* results when the inhibitor and substrate combine with enzyme forms as in the following mechanism.

$$E + A \underset{k_2}{\overset{k_1}{\rightleftharpoons}} EA \overset{k_3}{\rightarrow} E + P \qquad (7.3.44)$$

$$EA + I \underset{k_5}{\overset{k_4}{\rightleftharpoons}} EAI \rightarrow \text{no reaction} \quad (7.3.45)$$

These three classes of inhibition can be distinguished by virtue of the effect of variations in inhibitor concentration on the slopes and intercepts of reciprocal plots. For competitive inhibition only the slope varies. For uncompetitive inhibition only the intercept varies, while for noncompetitive inhibition both the slope and the intercept vary.

If more than one substrate participates in an enzymatic reaction, the kinetic effects of an inhibitor can be quite complex. In this case, rules formulated by Cleland (36) are useful in gaining a qualitative picture of the inhibition patterns to be expected of a given mechanism.

1. If the inhibitor combines with an enzyme form different from one with which the variable substrate combines, the vertical intercept of the corresponding reciprocal plot will be affected.
2. If the inhibitor combines with an enzyme form that is the same as or is connected by a series of reversible steps to the same form with which the variable substrate combines, the slope of the corresponding reciprocal plot is affected.
3. These effects can occur separately, in which case the inhibition is either competitive or uncompetitive; or they can occur jointly, resulting in noncompetitive inhibition.

These rules can be used in conjunction with experimental inhibition studies to assess the plausibility of possible enzymatic mechanisms.

7.3.2.3 The Influence of Environmental Factors on Enzyme Kinetics.
Because enzymes are proteins, they are unusually sensitive to changes in their environment. This is true not only with regard to variations in inhibitor concentrations, but also with respect to variations in pH and temperature. Most enzymes are efficient catalysts only within relatively narrow ranges of pH and temperature.

When the rate of an enzyme catalyzed reaction is studied as a function of temperature, it is found that the rate passes through a maximum. The existence of an optimum temperature can be explained by considering the effect of temperature on the catalytic reaction itself and on the enzyme denaturation reaction. In the low temperature range (around room temperature) there is little denaturation, and increasing the temperature increases the rate of the catalytic reaction in the usual manner. As the temperature rises, deactivation arising from protein denaturation becomes more and more important, so the observed overall rate eventually will begin to fall off. At temperatures in excess of 50 to 60 °C, most enzymes are completely denatured, and the observed rates are essentially zero.

Enzyme activity generally passes through a maximum as the pH of the system in question is varied. However, the optimum pH varies with substrate concentration and temperature. Provided that the pH is not changed too far from the optimum value corresponding to the maximum rate, the changes of rate with pH are reversible and reproducible. However, if the solutions are made too acid or too alkaline, the activity of the enzyme may be irreversibly destroyed. Irreversible deactivation is usually attributed to denaturation of the proteinaceous enzyme. The range of pH in which reversible behavior is observed is generally small and this behavior is almost certainly due to changes in the amounts and activities of the various ionic forms of the enzyme, the substrate, and the enzyme complex. The maximum in the activity of the enzyme is a reflection of the maximum in the concentration of the catalytically active species.

7.4 CORRELATION METHODS FOR KINETIC DATA—LINEAR FREE ENERGY RELATIONS

A primary objective of the practicing kineticist is to be able to relate the rate of a reaction to the structure of the reactants. Although the collision and transition state theories provide useful frameworks for the discussion of the microscopic events involved in chemical re-actions, they do not in any meaningful way permit one to predict reaction rates on an a priori basis for even relatively simple reactions. However, chemical engineers should be aware that methods exist for correlating empirical rate data for homologous reactions. These correla-tions attempt to describe quantitatively the influence of variations in chemical structure on rate constants for a series of reactions involving the same functional groups. This section briefly describes the most useful correlation method in order to indicate to beginning kineticists an empirical approach which has given good results in the past. The useful correlation methods are based on what are referred to as linear free energy relations. These relations presume that when a selected molecule undergoes reactions with two different homologous compounds, the activation energy changes associated with the rate processes will be influenced in a similar fashion by the changes in structure. In essence one treats the molecule as if its structure can be arbitrarily broken up into a reaction center X and a nonreacting residue with structural ele-ments that can influence the rate of reaction at X. Similar treatments are used in the correlation of free energy changes associated with reaction equilibria and hence of the equilibrium constant

itself. The development and improvement in these quantitative correlations for reaction rate and equilibrium constant data has been one of the most striking developments in the evolution of physical organic chemistry. In the area of kinetics these relationships constitute a suitable framework for the extrapolation and interpola-tion of rate data, and they can also provide a useful insight into the events that take place on reaction at the molecular level.

It is instructive to consider the rationale underlying the various linear free energy corre-lations and to indicate in qualitative fashion how substituents may influence reaction rates. The relation between an equilibrium constant and the standard free energy change accom-panying a reaction is given by

$$\Delta G^0 = -RT \ln K_a \qquad (7.4.1)$$

Since the transition state formulation of a reaction rate expression treats the activated complex as being in equilibrium with the reactants, the resultant expression for the reac-tion rate constant depends similarly on the free energy difference between reactants and the activated complex. In this case equation 4.3.34 can be rewritten as

$$\Delta G^{\ddagger} = -RT \left[\ln k - \ln \left(\frac{k_B T}{h} \right) \right] \qquad (7.4.2)$$

where the second logarithmic term is constant at a fixed temperature. Thus, estimates of reaction rate and equilibrium constants may be regarded as equivalent to estimates of free energy differences between different species. If one examines a series of reactions, differences in $\ln k$ or $\ln K$ are then simply related to differences in the associated free energy changes.

Consider two molecular processes involving the reaction of structurally similar reactants with a common reagent; for example,

$$Y_1 N X_1 + A \xrightarrow{k_1} Y_1 N X_2 + B \qquad (7.4.3)$$

$$Y_2 N X_1 + A \xrightarrow{k_2} Y_2 N X_2 + B \qquad (7.4.4)$$

where Y_1 and Y_2 are monovalent substituent atoms or groups; X_1 and X_2 are monovalent atoms or groups that may be regarded as the reactant and product group, respectively; N is the core of the molecule that links the substituent groups to the reactant and product groups; and A and B are other reactant and product molecules. A specific example of the above pair of generalized reactions is the following.

$$H - \langle \bigcirc \rangle - CH_2Cl + KI \rightleftharpoons H - \langle \bigcirc \rangle - CH_2I + KCl \tag{7.4.5}$$

$$CH_3 - \langle \bigcirc \rangle - CH_2Cl + KI \rightleftharpoons CH_3 - \langle \bigcirc \rangle - CH_2I + KCl \tag{7.4.6}$$

where

$$
\begin{aligned}
X_1 &= Cl & X_2 &= I \\
Y_1 &= H & Y_2 &= CH_3 \\
A &= KI & B &= KCl
\end{aligned}
$$

When one looks at the general reaction systems from the standpoint of the kineticist, species $YNX_1A^{\ddagger}$ would represent the transition state configuration. In this case equation 7.4.2 indicates that

$$RT \ln \left(\frac{k_1}{k_2} \right) = \Delta G_2^{\ddagger} - \Delta G_1^{\ddagger} \tag{7.4.7}$$

Now the standard free energy content of a molecule has often been expressed as the sum of a number of contributions from the constituent parts of a molecule plus various contributions arising from the interactions of the parts with each other and with surrounding molecules (e.g., those of the solvent). Inasmuch as each chemical species represents a structurally unique combination of its constituent groups, it is possible to correlate standard free energies and many other thermodynamic properties for all known compounds by using a sufficiently large number of contributing parameters. Under a given set of standard conditions the free

energy of our representative Y_1NX_1 molecule will include terms due to the individual Y_1, N, and X_1 groups and terms arising from the interactions of these groups. For a representative molecule the standard free energy under these conditions may be written as

$$G_{Y_1NX_1}^0 = G_{Y_1} + G_N + G_{X_1} + G_{Y_1N} \\ + G_{X_1N} + G_{Y_1X_1N} \tag{7.4.8}$$

where the singly subscripted variables refer to individual group contributions, the G_{Y_1N} and G_{X_1N} terms to interactions between the substituent group and core group and between the reactive group and the core group, respectively, and $G_{Y_1X_1N}$ represents the interactions between the substituent and reactive groups through the molecular core.

The standard free energy change for reaction 7.4.3 is given by

$$\Delta G_1^0 = G_{Y_1NX_2}^0 + G_B^0 - G_A^0 - G_{Y_1NX_1}^0 \tag{7.4.9}$$

Substitution of equation 7.4.8 and analogous equations for other compounds into equation 7.4.9 and simplification gives

$$\Delta G_1^0 = G_{X_2N} + G_{Y_1X_2N} + G_{X_2} + G_B^0 - G_A^0 \\ - (G_{X_1N} + G_{Y_1X_1N} + G_{X_1}) \tag{7.4.10}$$

If we derive an analogous equation for reaction 7.4.4, we find that

$$\Delta G_2^0 = G_{X_2N} + G_{Y_2X_2N} + G_{X_2} + G_B^0 - G_A^0 \\ - (G_{X_1N} + G_{Y_2X_1N} + G_{X_1}) \tag{7.4.11}$$

Two equations of the form of equation 7.4.1 can be used with equations 7.4.10 and

7.4.11 to show that

$$RT \ln \left(\frac{K_2}{K_1}\right) = G_{Y_1 X_2 N} - G_{Y_2 X_2 N}$$

$$- (G_{Y_1 X_1 N} - G_{Y_2 X_1 N})$$

$$(7.4.12)$$

which implies that the ratio of the equilibrium constants for the two different reactions involving the same functional group but different substituents depends only on the terms for the free energy of interaction between the substituent and reactant groups. In similar fashion one may write for the ratio of reaction rate constants

$$\ln \left(\frac{k_2}{k_1}\right) = \frac{1}{RT} \left[G_{Y_1 X_2 N} - G_{Y_2 X_2 N} - (G_{Y_1 X_1 N} - G_{Y_2 X_1 N}) \right] \qquad (7.4.13)$$

It is generally thought that the interaction energies of groups that are not directly bonded to one another result from the following effects.

1. *Inductive or polar effects.* These effects involve electron displacements that are transmitted along a chain of atoms without any reorganization of the formal chemical bonds in the molecule. For example, the introduction of a methyl group in a pyridine ring involves a displacement of electrons to the nitrogen atom from the methyl group. This effect falls off rapidly with separation distance.

2. *Resonance or electromeric effects.* Certain molecular structures are characterized by the possibility of having two or more compatible electronic structures and the molecules exist in a resonance state intermediate between the several extremes. These effects are particularly characteristic of aromatic structures and other molecules containing conjugate double bonds.

3. *Steric effects.* These effects arise as a consequence of the molecular geometry of the species involved in the reaction. They include

interference with internal rotations, steric compressions or strains, etc.

These various effects combine to result in different rate and equilibrium constants for homologous reactions. The substituents influence these parameters in part by displacements of electron density of the first two types and in part by geometric effects of the last type. When some of the possible complications are considered, it is not surprising that there is no generally useful method for correlating the G_{YXN} terms with satisfactory accuracy. However, there are a number of special cases in which useful correlations can be developed. This is particularly true for those cases where N is a relatively rigid group, such as an aromatic ring. In this case the various X and Y groups will be the same distance apart in the species involved in reactions 7.4.3 and 7.4.4. Furthermore, if the X and Y groups are sufficiently far apart that there are no direct steric interactions and no direct resonance interactions between the X and Y groups, only the polar interactions contribute significantly to the G_{YXN} terms. This situation occurs in many *meta* and *para* substituted benzene derivatives. In this case G_{YXN} can be assumed to be proportional to the product of parameters for the substituent group Y and the reactant group X. These parameters are referred to as *polar substituent constants* σ_{YN} and σ_{XN}. The proportionality constant depends on the solvent, the temperature, and the nature of the core group N linking the substituents. It provides a measure of how effectively the influence of one group is transmitted to the other. This proportionality constant may be written as $(-\tau_N RT \ln 10)$, so that the interaction term can be written as

$$G_{YXN} = -(\tau_N RT \ln 10)\sigma_{YN}\sigma_{XN} \quad (7.4.14)$$

Combining equations 7.4.13 and 7.4.14 then gives

$$\ell n \left(\frac{k_1}{k_2} \right) = (\tau_N \, \ell n \, 10)[\sigma_{Y_1 N}\sigma_{X_2 N} - \sigma_{Y_2 N}\sigma_{X_2 N} - (\sigma_{Y_1 N}\sigma_{X_1 N} - \sigma_{Y_2 N}\sigma_{X_1 N})] \qquad (7.4.15)$$

or, on simplification,

$$\log \left(\frac{k_1}{k_2} \right) = \tau_N[(\sigma_{Y_1 N} - \sigma_{Y_2 N})(\sigma_{X_2 N} - \sigma_{X_1 N})] \qquad (7.4.16)$$

Within a specific sequence of reactions involving the same groups X_1 and X_2 and for uniform conditions of solvent, temperature, etc., the value of $\tau_N(\sigma_{X_2 N} - \sigma_{X_1 N})$ will be a constant $(\rho_{X_1 X_2 N})$, which characterizes the functional group interactions involved. Hence

$$\rho_{X_1 X_2 N} \equiv \tau_N(\sigma_{X_2 N} - \sigma_{X_1 N}) \qquad (7.4.17)$$

and equation 7.4.16 becomes

$$\log \left(\frac{k_1}{k_2} \right) = \rho_{X_1 X_2 N}(\sigma_{Y_1 N} - \sigma_{Y_2 N}) \qquad (7.4.18)$$

Since only *differences* between σ's for the substituents are involved, one may arbitrarily set the absolute value of one sigma value without loss of generality. Normally one chooses a hydrogen atom as a reference substituent for which σ is defined as zero. All other rate constants may then be compared to the one characterizing the reaction of the reference substance (k_0). In these terms equation 7.4.18 becomes

$$\log \left(\frac{k_1}{k_0} \right) = \rho_{X_1 X_2 N}\sigma_{Y_1 N} \qquad (7.4.19)$$

A corresponding equation exists for the ratio of equilibrium constants. The utility of equations of this form lies in the fact that the ρ value is characteristic of a particular reaction of functional groups, while the σ value is characteristic of the nonreactive functional groups. Once σ values have been determined for one class of reactions, they may be used for another class of reactions. Hence a knowledge of k_0 and ρ for

the second class permits one to obtain estimates of the rate constants for reactions involving the various substituted compounds. As we will see in the next subsection, if equations like equation 7.4.19 are to be appropriate for use, they require that a certain free energy difference or free energy contribution be a linear function of some property of a substituent group. Consequently, these equations are referred to as linear free energy relations.

7.4.1 The Hammett Equation

The Hammett equation is the best-known and most widely studied of the various linear free energy relations for correlating reaction rate and equilibrium constant data. It was first proposed to correlate the rate constants and equilibrium constants for the side chain reactions of para and meta substituted benzene derivatives. Hammett (37–39) noted that for a large number of reactions of these compounds plots of $\log k$ (or $\log K$) for one reaction versus $\log k$ (or $\log K$) for a second reaction of the corresponding member of a series of such derivatives was reasonably linear. Figure 7.5 is a plot of this type involving the ionization constants for phenylacetic acid derivatives and for benzoic acid derivatives. The point labeled p-Cl has for its ordinate $\log K_a$ for p-chlorophenylacetic acid and for its abscissa $\log K_a$ for p-chlorobenzoic acid. The points approximate a straight line, which can be expressed as

$$\log K_A = \rho \log K'_A + C \qquad (7.4.20)$$

where K_A and K'_A are the ionization constants for substituted phenylacetic and benzoic acids with a given substituent, ρ is the slope of the line, and C is the intercept. This relation may be used for any substituent including the reference substituent, normally taken as a hydrogen

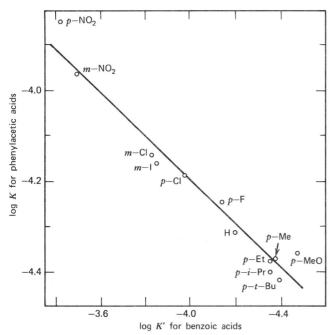

Figure 7.5
**Log-log plot of ionization constants of benzoic and phenyl-
acetic acids in water at 25°. (From** *Physical Organic Chemistry*
**by J. S. Hine. Copyright © 1962. Used with permission of
McGraw-Hill Book Company.)**

atom. If the reference substance is denoted by the subscript zero, then

$$\log K_0 = \rho \log K'_0 + C \quad (7.4.21)$$

where K_0 and K'_0 are the ionization constants of phenylacetic and benzoic acids, respectively. Elimination of the intercept C gives

$$\log \frac{K_A}{K_0} = \rho \log \frac{K'_A}{K'_0} \quad (7.4.22)$$

Equations of this type can be written for any pair of the many reactions for which linear log-log plots could be made. Consequently, it is convenient to choose a reference reaction to which others can be compared. The large amount of accurate data on the ionization of benzoic acid derivatives at 25 °C made this reaction an appropriate choice, and a new con-

stant σ was then defined as

$$\sigma_A = \log \left(\frac{K'_A}{K'_0} \right) \quad (7.4.23)$$

This constant characterizes the ionization of a particular substituted benzoic acid in water at 25 °C relative to that of benzoic acid itself. This definition reduces equation 7.4.22 to

$$\log \frac{K_A}{K_0} = \rho \sigma_A \quad (7.4.24)$$

which is of the same form as the general linear free energy relation introduced earlier in equation 7.4.19, although it predates the latter by many years. From this equation and from the definition of σ, it is evident that ρ is taken as unity for the standard reaction (i.e., the ionization of substituted benzoic acids in water at

25 °C). The value of σ may be determined from its definition in terms of the ionization constant if the appropriate benzoic acid derivative has been measured. Such σ values may then be used to determine ρ values for other reactions, and these ρ values in turn lead to the possibility of determining new σ values. Equation 7.4.24 implies that a plot of $\log K_A$ or $\log k_A$ versus σ_A should be linear for a given series of reactions involving the same reactive groups. Extensive experimental evidence attests to this relation. Table 7.2 contains some values that have been reported in the literature. Extensive tabulations

of σ and ρ values are available (37, 41-45). Note that one requires different σ values for meta and para substituents, but that only one ρ value is required. Typical σ values range from about -1 to $+2$, while ρ values range from -6 to $+4$. Hence the Hammett equation may be used to correlate data covering several orders of magnitude in the rate and equilibrium constant values. Illustration 7.1 involves the use of Table 7.2 to estimate a reaction rate constant.

The substituent constants have also been associated with the ability of the substituent group to alter the charge density at the reaction site.

Table 7.2
Hammett Substituent Constants (46)

Substituent	σ Meta	σ Para	Substituent	σ Meta	σ Para
CH_3	-0.069	-0.170	O^-	-0.708	-1.00
CH_2CH_3	-0.07	-0.151	OH	$+0.121$	-0.37
$CH(CH_3)_2$	-0.068	-0.151	OCH_3	$+0.115$	-0.268
$C(CH_3)_3$	-0.10	-0.197	OC_2H_5	$+0.1$	-0.24
C_6H_5	$+0.06$	-0.01	OC_6H_5	$+0.252$	-0.320
$C_6H_4NO_2\text{-}p$	......	$+0.26$	$OCOCH_3$	$+0.39$	$+0.31$
$C_6H_4OCH_3\text{-}p$	......	-0.10	F	$+0.337$	$+0.062$
$CH_2Si(CH_3)_3$	-0.16	-0.21	$Si(CH_3)_3$	-0.04	-0.07
$COCH_3$	$+0.376$	$+0.502$	PO_3H^-	$+0.2$	$+0.26$
COC_6H_5	......	$+0.459$	SH	$+0.25$	$+0.15$
CN	$+0.56$	$+0.660$	SCH_3	$+0.15$	0.00
CO_2^-	-0.1	0.0	$SCOCH_3$	$+0.39$	$+0.44$
CO_2H	$+0.35$	$+0.406$	$SOCH_3$	$+0.52$	$+0.49$
CO_2CH_3	$+0.321$	$+0.385$	SO_2CH_3	$+0.60$	$+0.72$
$CO_2C_2H_5$	$+0.37$	$+0.45$	SO_2NH_2	$+0.46$	$+0.57$
CF_3	$+0.43$	$+0.54$	SO_3^-	$+0.05$	$+0.09$
NH_2	-0.16	-0.66	$S(CH_3)_2^+$	$+1.00$	$+0.90$
$N(CH_3)_2$	-0.211	-0.83	Cl	$+0.373$	$+0.227$
$NHCOCH_3$	$+0.21$	0.00	Br	$+0.391$	$+0.232$
$N(CH_3)_3^+$	$+0.88$	$+0.82$	I	$+0.352$	$+0.276$
N_2^+	$+1.76$	$+1.91$	IO_2	$+0.70$	$+0.76$
NO_2	$+0.710$	$+0.778$			

(Note that σ values for charged substituents may be particularly solvent dependent.)

From *Physical Organic Chemistry* by J. S. Hine. Copyright © 1962. Used with permission of McGraw-Hill Book Company.

The groups with positive σ values are regarded as electron withdrawers, while those with negative σ's refer to electron donor substituents. Reactions involving a transition state with a highly electron deficient center would be expected to be sensitive to the stabilizing effect of substituents able to donate charge to the center. The opposite situation exists for reactions involving a reaction center with an excess electron density. In this case the stabilizing effect occurs in the presence of substituents that act to withdraw charge from the center. Reactions with positive ρ values are accelerated by electron withdrawal from the ring (positive σ), while those with negative ρ values are retarded by electron withdrawal.

In its original form the Hammett equation was appropriate for use with *para* and *meta* substituted compounds where the reaction site is separated from the aromatic group by a nonconjugating side chain. Although there have been several extensions and modifications that permit the use of the Hammett equation beyond these limitations, it is not appropriate for use with *ortho* substituted compounds, since steric effects are likely to be significant with such species. The results obtained using free radical reactions are often poor, and the correlation is more appropriate for use with ionic reactions. For a detailed discussion of the Hammett equation and its extensions, consult the texts by Hammett (37), Amis and Hinton (12), and Johnson (47).

Before terminating our discussion of the Hammett equation, we should note that the existence of linear correlations of the type indicated by equation 7.4.20 implies a linear free energy relationship. The rate or equilibrium constants can be eliminated from this equation using equation 7.4.1; that is,

$$\frac{\Delta G_A^0}{2.303RT} = \rho \frac{(\Delta G_A^0)'}{2.303RT} - C \quad (7.4.25)$$

Thus a linear relationship between the free energies for one homologous series of reactions

and those for another must exist if the Hammett equation is obeyed.

ILLUSTRATION 7.1 USE OF THE HAMMETT EQUATION FOR THE DETERMINATION OF A REACTION RATE CONSTANT

Kindler [*Ann.*, *450*(1), 1926] has studied the alkaline hydrolysis of the ethyl esters of a number of substituted benzoic acids. The m-nitro compound was found to have a rate constant 63.5 times as fast as the unsubstituted compound. What relative rate constant is predicted for the reaction of p-methoxybenzoate by the Hammett equation? The value based on experimental results is 0.214.

Solution

From Table 7.2 the σ value for the m-nitro group is 0.710. Substitution of this value and the ratio of reaction rate constants into equation 7.4.19 gives

$$\log 63.5 = \rho(0.710)$$

or

$$\rho = 2.54$$

for the alkaline hydrolysis of ethyl benzoates. For p-methoxy substitution, Table 7.2 indicates that $\sigma = -0.268$. In this case equation 7.4.19 becomes

$$\log\left(\frac{k}{k_0}\right) = 2.54(-0.268) = -0.681$$

or

$$\left(\frac{k}{k_0}\right) = 0.209$$

This value compares quite favorably with the experimental results.

7.4.2 Other Correlations

In order to correlate rate constant data for aliphatic and ortho substituted aromatic compounds, one must allow not only for the polar effects correlated by the Hammett equation but also for resonance and steric effects. Taft (48-49) has shown that it is possible to extend the range of linear free energy relations significantly by assuming that the polar, steric, and resonance effects may be treated independently. Other useful extensions of the Hammett equation include those of Swain and Scott (50-51), Edwards (52-53), Grunwald and Winstein (54), and Hansson (55). In order to obtain substituent values for the parameters appearing in the various equations, consult the text by Wells (41), the original references, or texts in physical organic chemistry. The practicing design engineer should be aware of these and other correlations in order to minimize the experimental work necessary to generate required kinetic parameters.

LITERATURE CITATIONS

1. Parker, A. J., *Adv. Phys. Org. Chem.*, 5 (192), 1967.

2. Laidler, K. J., *Reaction Kinetics, Volume 2, Reactions in Solution*, p. 2, Pergamon Press, Oxford, 1963.

3. North, A. M., *The Collision Theory of Chemical Reactions in Liquids*, Methuen, London, 1964.

4. Rabinowitch, E., and Wood, W. C., *Trans. Faraday Soc.*, 32 (1381), 1936.

5. Franck, J., and Rabinowitch, E., *Trans. Faraday Soc.*, 30 (120), 1934.

6. Debye, P., and Hückel, E., *Phys. Z.*, 24 (305), 1923.

7. Debye, P., and Hückel, E., ibid., 25 (145), 1924.

8. Debye, P., and Hückel, E., *Trans. Faraday Soc.*, 23 (334), 1927.

9. Harned, H. S., and Owen, B. B., *The Physical Chemistry of Electrolytic Solutions*, Third Edition, Reinhold Publishing, New York, 1958.

10. Laidler, K. J., *Chemical Kinetics*, p. 221, McGraw-Hill, New York, 1965.

11. Amis, E. S., *Solvent Effects on Reaction Rates and Mechanisms*, Academic Press, New York, 1966.

12. Amis, E. S., and Hinton, J. F., *Solvent Effects on Chemical Phenomena*, Academic Press, New York, 1973.

13. Clark, D. and Wayne, R. P., in *Comprehensive Chemical Kinetics, Volume 2, The Theory of Kinetics*, pp. 302–376, Elsevier, Amsterdam, 1969.

14. Eckert, C. A., *Ann. Rev. Physical Chem.*, 23 (239), 1972.

15. Bell, R. P., "Acid-Base Catalysis," p. 3, Oxford University Press, Oxford, copyright © 1941. Reprinted with permission.

16. Laidler, K. J., op. cit., 1965.

17. Bender, M. L., *Mechanisms of Homogeneous Catalysis from Protons to Proteins*, Wiley Interscience, New York, 1971.

18. Jencks, W. P., *Catalysis in Chemistry and Enzymology*, McGraw-Hill, New York, 1969.

19. Ashmore, P. G., *Catalysis and Inhibition of Chemical Reactions*, Butterworths, London, 1963.

20. Frost, A. A., and Pearson, R. G., *Kinetics and Mechanism*, Second Edition, Wiley, New York, 1961.

21. Moelwyn-Hughes, E. A., *The Chemical Statics and Kinetics of Solutions*, Academic Press, New York, 1971.

22. Bell, R. P., "Acid-Base Catalysis," Oxford University Press, Oxford, 1941.

23. Bell, R. P., *The Proton in Chemistry*, Methuen, London, 1959.

24. Laidler, K. J., *The Chemical Kinetics of Enzyme Action*, Clarendon Press, Oxford, 1958.

25. Plowman, K. M., *Enzyme Kinetics*, McGraw-Hill, New York, 1972.

26. Gould, R. F., Editor, "Homogeneous Catalysis: Industrial Applications and Implications," *ACS Adv. in Chem.*, 70, Washington, 1968.

27. Schrauzer, G. N., *Transition Metals in Homogeneous Catalysis*, Marcel Dekker, New York, 1971.

28. Jones, M. M., *Ligand Reactivity and Catalysis*, Academic Press, New York, 1968.

29. Basolo, F., and Pearson, R. G., "Mechanisms of Inorganic Reactions," Wiley, New York, 1960.

30. Ashmore, P. G., op. cit., pp. 30–31.

31. Hine, J. S., *Physical Organic Chemistry*, Second Edition, McGraw-Hill, 1962.

32. Skrabal, A., *Z. Elektrochem.*, 33 (322), 1927.

33. Jencks, W. P., *Catalysis in Chemistry and Enzymology*, McGraw-Hill, New York, 1969.

34. Bender, M. L., *Mechanisms of Homogeneous Catalysis from Protons to Proteins*, Wiley Interscience, New York, 1971.

35. Michaelis, L., and Menten, M. L., *Biochem. Z.*, 49 (333), 1913.

36. Cleland, W. W., *Biochim. Biophys. Acta*, 67 (188), 1963. Adapted with permission.

37. Hammett, L. P., *Physical Organic Chemistry*, Second Edition, McGraw-Hill, New York, 1970.

38. Hammett, L. P., *Chem. Rev.*, *17* (125), 1935.

39. Hammett, L. P., *Trans. Faraday Soc.*, *34* (156), 1938.

40. Hine, J. S., op. cit., p. 85.

41. Wells, P. R., *Linear Free Energy Relationships*, Academic Press, New York, 1968.

42. Gordon, A. J., and Ford, R. A., *The Chemist's Companion*, pp. 145–149, Wiley, New York, 1972.

43. Jaffé, H. H., *Chem. Rev.*, *53* (191), 1953.

44. McDaniel, D. H., and Brown, H. C., *J. Org. Chem.*, *23* (420), 1958.

45. Leffler, J. E., and Grunwald, E., *Rates and Equilibria of Organic Reactions*, Wiley, New York, 1963.

46. Hine, J. S., op. cit., p. 87.

47. Johnson, C. D., *The Hammett Equation*, Cambridge University Press, London, 1973.

48. Taft, R. W., Jr., *J. Am. Chem. Soc.*, *74* (2729, 3120), 1952; *75* (4231), 1953.

49. Taft, R. W., Jr., in *Steric Effects in Organic Chemistry*, M. S. Newman, Editor, Chapter 13, Wiley, New York, 1956.

50. Swain, C. G., and Scott, C. B., *J. Am. Chem. Soc.*, *75* (141), 1953.

51. Swain, C. G., Mosely, R. B., and Brown, D. E., *J. Am. Chem. Soc.*, *77* (3731), 1955.

52. Edwards, J. O., *J. Am. Chem. Soc.*, *76* (1540), 1954.

53. Edwards, J. O., *J. Am. Chem. Soc.*, *78* (1819), 1956.

54. Grunwald, E., and Winstein, S., *J. Am. Chem. Soc.*, *70* (846), 1948.

55. Hansson, J., *Svensk, Kem. Tidskr.*, *66* (351), 1954.

PROBLEMS

1. The oxidation of iodide ions by hydrogen peroxide

$$H_2O_2 + 2H^+ + 2I^- \rightarrow 2H_2O + I_2$$

has been studied by Bell et al. [*J. Phys. Chem.*, *55* (874), 1951]. At 25 °C the reaction appears to proceed by two parallel paths so that the observed rate expression is of the form

$$r = k_1(H_2O_2)(I^-) + k_2(H_2O_2)(I^-)(H^+)$$

The influence of ionic strength on the two rate constants was noted to be as follows.

μ (kmoles/m^3)	k_1 (m^3/kmoles·sec)	k_2 (m^6/kmoles^2·sec)
0.000	0.658	19.0
0.0207	0.663	15.0
0.0525	0.670	12.2
0.0925	0.679	11.3
0.1575	0.694	9.7
0.2025	0.705	9.2

Are these data consistent with the results predicted by equation 7.1.4?

2. The reaction of the nitrourethane ion with hydroxide ions can be written as

$$NO_2{=}NCOOC_2H_5^- + OH^- \rightarrow$$
$$N_2O + CO_3^= + C_2H_5OH$$

Near room temperature the reaction is essentially irreversible and second-order. In the limit of zero ionic strength at 293 °K, $k = 2.12 \, \text{m}^3/\text{kmole} \cdot \text{ksec}$. Determine the initial reaction rate in a solution that is 0.05 kmole/m^3 each in potassium nitrourethanate, NH$_4$OH, and KCl. The ionization constant for NH$_4$OH at 293 °K is 1.7×10^{-5} kmoles/m^3. The effect of the slight ionization of NH$_4$OH on ionic strength and the variation of activity coefficients with temperature between 293 and 298 °K may be neglected.

3. The following reaction takes place in aqueous solution.

$$[Co(NH_3)_5Br]^{++} + OH^- \rightarrow$$
$$[Co(NH_3)_5OH]^{++} + Br^-$$

It may be regarded as bimolecular and irreversible. Determine the ratio of the reaction rate in a system initially containing 10 moles/m^3 [Co(NH$_3$)$_5$Br](NO$_3$)$_2$, 100 moles/m^3 NH$_4$OH, and 50 moles/m^3 KNO$_3$ to the rate in a system that initially contains 100 moles/m^3 NH$_4$OH and 100 moles/m^3 [Co(NH$_3$)$_5$Br](NO$_3$)$_2$.

Take into account both primary and secondary salt effects, but neglect the contributions to the ionic strength of species resulting from the dissociation of NH$_4$OH.

4. Chen and Laidler [*Canadian J. Chem.*, 37 (599), 1959] have studied the reaction of the quinoid form of bromphenol blue with hydroxide ions to give the carbinol form of the dye. The following values of the second-order rate constant at 25 °C were reported.

Pressure (psia)	$k \times 10^4$ (liters/mole-sec)
14.7	9.298
4000	11.13
8000	13.05
12000	15.28
16000	17.94

What is the activation volume for this reaction?

5. Brönsted and Guggenheim [*J. Am. Chem. Soc.*, 49 (2554), 1927] have studied the mutarotation of glucose as catalyzed by acids and bases. The reaction takes place slowly in pure water, is weakly catalyzed by hydrogen ions, and is strongly catalyzed by hydroxide ions. When strong acids and bases are employed as catalysts, the apparent first-order rate constants can be written as

$$k = k_0 + k_{H^+}(H^+) + k_{OH^-}(OH^-)$$

where k_{OH^-}/k_{H^+} is of the order of several thousands. In the pH range 4 to 6, the contribution of the spontaneous reaction dominates, since both catalytic terms are negligible in this region. In dilute solutions the primary salt effect is not applicable to solutions of strong acids and bases.

On the basis of the data below determine the parameters k_0 and k_{H^+} at 18 °C:

Solution	$k \times 10^3$ (min^{-1})
1.0 × 10^{-4}N HClO$_4$	5.20
1.0 × 10^{-4}N HCl	5.31
1.0 × 10^{-5}N HCl	5.23
Distilled water (contaminated by CO$_2$ from the air)	5.42
1.0 × 10^{-4}N HClO$_4$ + 0.1N NaCl	5.25
1.0 × 10^{-4}N HClO$_4$ + 0.2N NaCl	5.24
1.0 × 10^{-4}N HClO$_4$ + 0.05M Ba(NO$_3$)$_2$	5.43
1.0 × 10^{-3}N HClO$_4$	5.42
2.0 × 10^{-2}N HClO$_4$	8.00
4.0 × 10^{-2}N HClO$_4$	11.26
3.85 × 10^{-2}N HClO$_4$	10.80
2.50 × 10^{-2}N HClO$_4$ + 0.2N NaCl	8.89

Since the spontaneous reaction term dominates in the pH range 4 to 6, studies of the reaction in this range are particularly suitable for measuring the small catalytic effects of weak acids and weak bases. If one employs the more general expression for k involving contributions from the undissociated acid and the anion resulting from dissociation, determine the coefficients of these terms from the data below.

Sodium propionate normality	Propionic acid normality	Hydrogen Ion concentration (kmoles/m^3)	$k \times 10^3$ (min^{-1})
0.010	0.010		5.65
0.040	0.020		6.53
0.050	0.050		6.81
0.075	0.075		7.57
0.100	0.100		8.21
0.125	0.125		9.15
0.150	0.150		9.85
0.040	0.020	0.00001	6.53
0.040	0.060	0.00002	6.60
0.040	0.110	0.00006	6.76
0.040	0.160	1.00010	6.77

6. Bell and Baughan [*J Chem. Soc. 1937*, 1947] have investigated the generalized acid-base

catalysis of the depolymerization of dimeric dihydroxyacetone. In terms of the general formulation of the first-order rate constant

$$k = k_0 + k_{H^+}(H^+) + k_{OH^-}(OH^-)$$
$$+ k_A(A) + k_B(B)$$

where k_0 is the rate constant for the water catalyzed reaction and k_x is the catalytic constant for species x. This equation can also be written as

$$k = k'' + \alpha(B)$$

where

$$k'' = k_0 + k_{H^+}(H^+) + k_{OH^-}(OH^-)$$

and

$$\alpha = k_B + [k_A(A)/(B)]$$

For the reactions in question no term may be neglected and it was necessary to carefully plan the experimental program to facilitate evaluation of all five kinetic parameters. On the basis of the data below determine these parameters when the weak acid employed is acetic acid.

$\dfrac{CH_3COOH}{CH_3COO^-}$	H_3O^+ (moles/m^3)	CH_3COO^- (moles/m^3)	k (ksec^{-1})
0.980	2.69×10^{-2}	100.4	0.707
		50.5	0.583
		31.6	0.530
		12.9	0.473
1.304	3.56×10^{-2}	104.5	0.633
		78.5	0.558
		53.3	0.490
		20.4	0.423
3.38	9.27×10^{-2}	102.7	0.433
		75.0	0.363
		40.9	0.278
		12.7	0.207
4.83	0.132	102.6	0.390
		87.1	0.363
		68.3	0.297
		51.2	0.262
		30.3	0.208
		10.3	0.154

7. The kinetic data below were reported for an enzyme catalyzed reaction of the type $E + S \rightleftharpoons ES \rightleftharpoons E + P$. Since the data pertain to initial reaction rates, the reverse reaction may be neglected. Use a graphical method to determine the Michaelis constant and V_{max} for this system at the enzyme concentration employed.

Initial substrate concentration (M)	Initial rate (μmoles/liter-min)
2×10^{-3}	150
2×10^{-4}	149.8
2×10^{-5}	120
1.5×10^{-5}	112.5
1.25×10^{-6}	30.0

What would the initial rate be at a substrate level of 2.0×10^{-5} M if the enzyme concentration were doubled? At a 2×10^{-3} M substrate level, how long does it take to achieve 80% conversion at the new enzyme level?

8. The turnover number of an enzyme is defined as the maximum number of moles of substrate reacted per mole of enzyme (or molecules per molecule) per minute under optimum conditions (i.e., saturating substrate concentration, optimum pH, etc). If 2 mg/cm^3 of a pure enzyme (50,000 molecular weight, Michaelis constant $K_m = 0.03$ mole/m^3) catalyzes a reaction at a rate of 2.5 μmoles/m$^3 \cdot$ksec when the substrate concentration is 5×10^{-3} moles/m^3, determine the turnover number corresponding to this definition and the actual number of moles of substrate reacting per minute per mole of enzyme.

9. In kinetic studies of enzymatic reactions, rate data are usually tested to determine if the reaction follows the Michaelis-Menten model of enzyme-substrate interaction. Weetall and Havewala [*Biotechnol. and Bioeng. Symposium* 3 (241), 1972] have studied the production of dextrose from cornstarch using conventional

glucoamylase and an immobilized version there-of. Their goal was to obtain the necessary data to be able to design a commercial facility for dextrose production. Their studies were carried out in a batch reactor at 60 °C. Compare the data below with that predicted from a Michaelis-Menten model with a rate expression of the form $r = k_3(E_0)(S)/[K_m + (S)]$

(a) Conventional enzyme ($E_G = 11,600$ units)

Time, t (sec)	Yield (mg dextrose produced/cm^3)
0	12.0
900	40.0
1800	76.5
3600	120.0
5400	151.2
7200	155.7
9000	164.9

(b) Immobilized enzyme data ($E_0 = 46,400$ units)

Time, t (sec)	Yield (mg dextrose produced/cm^3)
0	18.4
1,800	200.0
3,600	260.0
5,400	262.0
7,200	278.0
9,000	310.0
13,500	316.0
18,900	320.0
24,900	320.0

In both cases the initial substrate concentration was 0.14 kmole/m^3. In cases A and B the Michaelis constants (K_m) are reported to be 1.15×10^{-3} and 1.5×10^{-3} kmoles/m^3, respectively. The rate constant k_3 is equal to 1.25 μmoles/m^3·sec·unit in both cases.

If the data are not consistent, provide plausible explanations for the discrepancy.

10. Gould has reported the following acidity constants (in H$_2$O at 25 °C) for some substituted benzene seleninic acids ArSe O$_2$H.

Substituent	$K \times 10^5$	
None	1.6	From *Mechanism and Structure in Organic Chemistry* by E. S. Gould. Copyright © 1959. Used with permission of McGraw-Hill Book Company.
p-MeO—	0.89	
m-MeO—	2.2	
m-Cl	3.5	
m-NO$_2$	8.5	
p-Br	3.2	
p-C$_6$H$_5$O—	1.3	

(a) Show that the equilibria for the reaction series

$$ArSeO_2H + H_2O \rightleftharpoons ArSeO_2^- + H_3O^+$$

are governed by the Hammett equation and calculate the ρ value for this series. The dissociation constant for benzoic acid is 6.3×10^{-5}.

(b) Calculate the dissociation constant for p-nitrobenzeneseleninic acid and compare it with the observed value of 1.0×10^{-4}.

(c) What does the above data indicate is the σ value for the p-C$_6$H$_5$O—substituent?

8 Basic Concepts in Reactor Design and Ideal Reactor Models

8.0 INTRODUCTION

The chemical reactor must be regarded as the very heart of a chemical process. It is the piece of equipment in which conversion of feedstock to desired product takes place and is thus the single irreplaceable component of the process. Several different factors must be considered in selecting the physical configuration and mode of operation to be used to accomplish a specified task. This chapter discusses the fundamental principles that the chemical engineer engaged in the practice of reactor design utilizes in making these selections.

8.0.1 The Nature of the Reactor Design Problem

The chemical engineer is required to choose the reactor configuration and mode of operation that yields the greatest profit consistent with market constraints for the raw material and product costs, capital and operating costs, safety considerations, pollution control requirements, and esthetic constraints that may be imposed by management, society, or labor unions. Usually there are many combinations of operating conditions and reactor size and/or type that will meet the requirements imposed by nature in terms of the reaction rate expression involved and those imposed by management in terms of the required production capacity. The engineer is thus faced with the task of maintaining a careful balance between analytical reasoning expressed in quantitative terms and sound engineering judgment. In an attempt to maintain this balance, some or all of the following questions must be answered.

1. What is the composition of the feedstock and under what conditions is it available? Are any purification procedures necessary?
2. What is the scale of the process? What capacity is required?
3. Is a catalyst necessary or desirable? If a catalyst is employed, what are the ramifications with respect to product distribution, operating conditions, most desirable type of reactor, process economics, and other pertinent questions raised below?
4. What operating conditions (temperature, pressure, degree of agitation, etc.) are required for most economic operation?
5. Is it necessary or desirable to add inerts or other materials to the feedstock to enhance yields of desired products, to moderate thermal effects, or to prolong the useful life of any catalysts that may be employed?
6. Should the process be continuous or intermittent? Would batch or semibatch operation be advantageous?
7. What type of reactor best meets the process requirements? Are there advantages associated with the use of a combination of reactor types, or with multiple reactors in parallel or series?
8. What size and shape reactor(s) should be used?
9. How are the energy transfer requirements for the process best accomplished? Should one operate isothermally, adiabatically, or in accord with an alternative temperature protocol?
10. Is single pass operation best, or is recycle needed to achieve the desired degree of conversion of the raw feedstock?
11. What facilities are required for catalyst supply, activation, and regeneration?
12. What are the reactor effluent composition and conditions? Are any chemical separation steps or physical operations required in order to bring the effluent to a point where it is satisfactory for the desired end use?
13. Are there any special materials requirements imposed by the process conditions? Are the

process fluids corrosive? Are extremely high temperatures or pressures required?

The remainder of this text attempts to establish a rational framework within which many of these questions can be attacked. We will see that there is often considerable freedom of choice available in terms of the type of reactor and reaction conditions that will accomplish a given task. The development of an optimum processing scheme or even of an optimum reactor configuration and mode of operation requires a number of complex calculations that often involve iterative numerical calculations. Consequently machine computation is used extensively in industrial situations to simplify the optimization task. Nonetheless, we have deliberately chosen to present the concepts used in reactor design calculations in a framework that insofar as possible permits analytical solutions in order to divorce the basic concepts from the mass of detail associated with machine computation.

The first stage of a logical design procedure involves the determination of a reaction rate expression that is appropriate for the range of conditions to be investigated in the design analysis. One requires a knowledge of the dependence of the rate on composition, temperature, fluid velocity, the characteristic dimensions of any heterogeneous phases present, and any other process variables that may be significant. There are several potential sources of the experimental data that are essential for proper reactor design.

1. *Bench scale experiments.* The reactors used in these experiments are usually designed to operate at constant temperature, under conditions that minimize heat and mass transfer limitations on reaction rates. This facilitates an accurate evaluation of the intrinsic chemical effects.

2. *Pilot plant studies.* The reactors used in these studies are significantly larger than those in bench scale laboratory experiments. One uses essentially the reverse of the design procedures developed later in this chapter to determine the effective reaction rate from the pilot plant data. It may be difficult to separate the intrinsic chemical effects from any heat and mass transfer effects in the analysis of data of this type.

3. *Operating data from commercial scale reactors.* If one's company has access to actual operating data on another commercial installation of the same type as that contemplated, it provides the closest approximation to the conditions likely to be encountered in industrial practice. Such access may result from licensing arrangements or from previous experience within the company. Unfortunately, such data are often incomplete or inaccurate, and the problems of backing the intrinsic chemical kinetics out of the mass of data may be insurmountable. In such systems physical limitations on rates of heat and mass transfer may disguise the true kinetics to a significant degree.

In the design of an industrial scale reactor for a new process, or an old one that employs a new catalyst, it is common practice to carry out both bench and pilot plant studies before finalizing the design of the commercial scale reactor. The bench scale studies yield the best information about the intrinsic chemical kinetics and the associated rate expression. However, when taken alone, they force the chemical engineer to rely on standard empirical correlations and prediction methods in order to determine the possible influence of heat and mass transfer processes on the rates that will be observed in industrial scale equipment. The pilot scale studies can provide a test of the applicability of the correlations and an indication of potential limitations that physical processes may place on conversion rates. These pilot plant studies can provide extremely useful information on the temperature distribution in the reactor and on contacting patterns when more than a single phase reactant-catalyst system is employed.

8.0.2 Reactor Types

In terms of the physical configurations encountered, there are basically only two types of reactors: the tank and the tube.

The *ideal tank reactor* is one in which stirring is so efficient that the contents are always uniform in composition and temperature throughout. The simple tank reactor may be operated in a variety of modes: batch, semibatch, and continuous flow. These modes are illustrated schematically in Figure 8.1. In the simple *batch reactor* the fluid elements will all have the same composition, but the composition will be time dependent. The stirred tank reactor may also be operated in *semibatch* fashion. In this mode the tank is partially filled with reactant(s), and additional reactants are added progressively until the desired end composition is achieved. Alternatively, one may charge the reactants all at once and continuously remove products as they are formed. In the continuous flow mode of operation the stirred tank reactor is continuously supplied with feed; at the same time an equal volume of reactor contents is discharged in order to maintain a constant level in the tank. The composition of the effluent stream is identical with that of the fluid remaining in the tank.

The *ideal tubular reactor* is one in which elements of the homogeneous fluid reactant move through a tube as plugs moving parallel to the tube axis. This flow pattern is referred to as

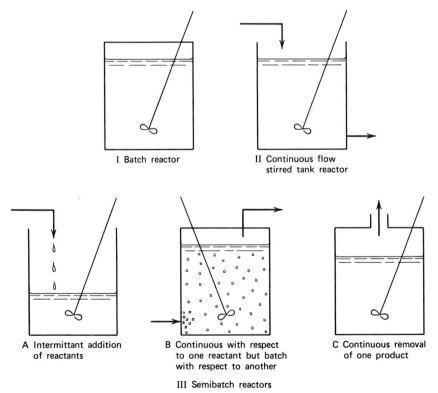

I Batch reactor

II Continuous flow
stirred tank reactor

A Intermittant addition
of reactants

B Continuous with respect
to one reactant but batch
with respect to another

C Continuous removal
of one product

III Semibatch reactors

Figure 8.1
Types of tank reactors.

plug flow or *piston flow*. The velocity profile at a given cross section is flat and it is assumed that there is no axial diffusion or back-mixing of fluid elements.

Batch reactors are often used for liquid phase reactions, particularly when the required production is small. They are seldom employed on a commercial scale for gas-phase reactions because the quantity of product that can be produced in reasonably sized reactors is small. Batch reactors are well suited for producing small quantities of material or for producing several different products from one piece of equipment. Consequently they find extensive use in the pharmaceutical and dyestuff industries and in the production of certain specialty chemicals where such flexibility is desired. When rapid fouling is encountered or contamination of fermentation cultures is to be avoided, batch operation is preferable to continuous processing because it facilitates the necessary cleaning and sanitation procedures.

When the specified production capacities are low, processes based on batch reactors will usually have lower capital investment requirements than processes calling for continuous operation, so batch reactors are often preferred for new and untried processes during the initial stages of development. As production requirements increase in response to market demands, it may become more economic to shift to continuous processing but, even in these cases, there are many industrial situations where batch operation is preferable. This is particularly true when the operating expenses associated with the reactor are a minor fraction of total product cost. At low production capacities, construction and instrumentation requirements for batch reactors are usually cheaper than for continuous process equipment. Moreover, it is generally easier to start up, shut down, and control a batch reactor than a comparable capacity continuous flow reactor.

The disadvantages associated with the use of a batch reactor include the high labor and materials handling costs involved in filling, emptying, and cleaning of these reactors. While batch reactors are being filled, emptied, or cleaned, and while the reactor contents are being heated to the reaction temperature or cooled to a point suitable for discharge, batch reactors are not producing reaction products. The sum of the nonproductive periods may often be comparable in length to the time necessary to carry out the reaction. In determining long-term production capacities for batch reactors, these dead times must be taken into account.

Continuous flow reactors are almost invariably preferred to batch reactors when the processing capacity required is large. Although the capital investment requirements will be higher, the operating costs per unit of product will be lower for continuous operation than for batch reaction. The advantages of continuous operation are that it:

1. Facilitates good quality control for the product through the provision of greater constancy in reaction conditions.
2. Facilitates automatic process control.
3. Minimizes the labor costs per unit of product.

Often the decision to select a batch or continuous processing mode involves a determination of the relative contributions of capital and operating expenses to total process costs for the proposed level of capacity. As Denbigh (1) points out, what is best for a highly industrialized country with high labor costs is not necessarily best for a lesser developed country. In many cases selectivity considerations determine the processing mode, particularly when the reaction under study is accompanied by undesirable side reactions. The yield of the desired product may differ considerably between batch and continuous operation and between the two primary types of continuous processes. When the yield is lower for a continuous process, this factor may be so important in the overall process economics as to require the use of a batch reactor.

At this point we wish to turn to a brief discussion of the types of batch and flow reactors used in industrial practice for carrying out homogeneous fluid phase reactions. Treatment of heterogeneous catalytic reactors is deferred to Chapter 12.

8.0.2.1 Batch Reactors (Stirred Tanks).

Batch reactors are usually cylindrical tanks and the orientation of such tanks is usually vertical. Cylindrical vessels are employed because they are easier to fabricate and clean than other geometries and because the construction costs for high-pressure units are considerably less than for alternative configurations. For simple stirred vertical batch reactors, the depth of liquid is usually comparable to the diameter of the reactor. For greater liquid height to diameter ratios more complex agitation equipment is necessary. Agitation can be supplied by stirrer blades of various shapes or by forced circulation with an external or built-in pump. Where more gas-liquid interfacial area is required for evaporation or gas absorption, or where it is necessary to minimize the hydrostatic head (e.g., to minimize the boiling point rise), horizontal reactors will be used. The latter orientation may also be preferable when the reactor contents are quite viscous or take the form of a slurry. Batch reactors may be fabricated from ordinary or stainless steel, but there are often advantages to using glass or polymer coatings on interior surfaces in order to minimize corrosion or sanitation problems.

Because of the large energy effects that often accompany chemical reaction, it is usually necessary to provide for heat transfer to or from the reactor contents. Heating or cooling may be accomplished using jacketed walls, internal coils, or internal tubes filled with a heat transfer fluid that is circulated through an external heat exchanger. Energy may also be supplied by electrical heating or direct firing. If the process fluid is noncorrosive and readily and safely pumped, it may be preferable to employ an external heat exchanger and circulation pump. Good temperature control can be achieved using an external reflux condenser for cases where appreciable vapor is given off. The selection of either internal or external heat transfer equipment is governed by the required area, the susceptibility of the heat transfer surface to fouling, the temperature and pressure requirements imposed by the heat transfer medium, and the potential adverse effects that might occur if the process fluid and the heat transfer medium come in direct contact through leakage.

For high-pressure operation, safety considerations are extremely important and care must be taken to ensure proper mechanical design. Closures must be designed to withstand the same maximum pressure as the rest of the autoclave. Various authors have treated the problems involved in designing medium- and high-pressure batch reactors (2-4).

8.0.2.2 Continuous Flow Reactors—Stirred Tanks.

The continuous flow stirred tank reactor is used extensively in chemical process industries. Both single tanks and batteries of tanks connected in series are used. In many respects the mechanical and heat transfer aspects of these reactors closely resemble the stirred tank batch reactors treated in the previous subsection. However, in the present case, one must also provide for continuous addition of reactants and continuous withdrawal of the product stream.

It is possible to employ either multiple individual tanks in series or units containing multiple stages within a single shell (see Figure 8.2). Multiple tanks are more expensive, but provide more flexibility in use, since they are more readily altered if process requirements change. In order to minimize pump requirements and maintenance, one often chooses to allow for gravity flow between stages. When the reactants are of limited miscibility, but differ

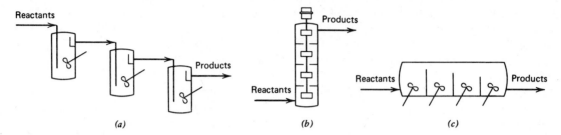

Figure 8.2
Types of staged reactors. (a) Reactor battery. (b) Vertically staged. (c) Compartmented. (Adapted from Reaction Kinetics for Chemical Engineers **by S. M. Walas. Copyright © 1959. Used with permission of McGraw-Hill Book Company.)**

in density, the vertical staged shell lends itself to countercurrent operation. This approach is useful when dealing with reversible reactions between immiscible fluids.

For purposes of calculation, each stage in a multiple-stage unit is treated as an individual reactor. The process stream flows from one reactor to the next, so that there is essentially a step change of composition between successive reactors. The step change is a direct consequence of efficient mixing. Unless the fluid phase is highly viscous, it is not difficult to approach perfect mixing in industrial scale equipment. All that is required is that the time necessary to distribute an entering element of fluid uniformly throughout the tank be very small compared to the average residence time in the tank.

Because of the dilution that results from the mixing of entering fluid elements with the reactor contents, the average reaction rate in a stirred tank reactor will usually be less than it would be in a tubular reactor of equal volume and temperature supplied with an identical feed stream. Consequently, in order to achieve the same production capacity and conversion level, a continuous flow stirred tank reactor or even a battery of several stirred tank reactors must be much larger than a tubular reactor. In many cases, however, the greater volume requirement is a relatively unimportant economic factor, particularly when one operates at ambient pres-

sure in tanks constructed of inexpensive materials such as mild steel.

In addition to lower construction costs, a continuous flow stirred tank reactor possesses other advantages relative to a tubular flow reactor, such as the facilitation of temperature control. Efficient stirring of the reactor contents insures uniform temperature and the elimination of local hot spots. The large heat capacity of the reactor contents also acts as a heat sink to moderate temperature excursions when changes occur in process conditions. The physical configuration of cylindrical tanks provides a large heat transfer area on the external surface of the tank and permits augmentation of this area through the use of submerged coils within the tank. However, the rate of heat transfer per unit volume of reaction mixture is generally lower in a conventional stirred tank reactor than in conventional tubular reactors because of the lower ratio of heat transfer surface area to volume in the tank reactor. Consequently, tubular reactors are preferred for fast reactions when energy transfer requirements are very large.

Ease of access to the interior surface of stirred tanks is an additional advantage of this type of reactor. This consideration is particularly significant in polymerization reactors, where one needs to worry about periodic cleaning of internal surfaces.

Selectivity considerations may also dictate the use of stirred tank reactors. They are preferred if undesirable side reactions predominate at high reactant concentrations, and they are also useful when one desires to "skip" certain concentration or temperature ranges where by-product formation may be excessive.

Stirred tank reactors are employed when it is necessary to handle gas bubbles, solids, or a second liquid suspended in a continuous liquid phase. One often finds that the rates of such reactions are strongly dependent on the degree of dispersion of the second phase, which in turn depends on the level of agitation.

Large stirred tank reactors are generally not suited for use at high pressures because of mechanical strength limitations. *They are used mainly for liquid phase reaction systems at low or medium pressures when appreciable residence times are required.*

8.0.2.3 *Continuous Flow Reactors—Tubular Reactors.*

The tubular reactor is so named because the physical configuration of the reactor is normally such that the reaction takes place within a tube or length of pipe. The idealized model of this type of reactor assumes that an entering fluid element moves through the reactor as a plug of material that completely fills the reactor cross section. Thus the terms piston flow or *plug flow reactor (PFR)* are often employed to describe the idealized model. The contents of a given elemental plug are presumed to be uniform in temperature and composition. This model may be used to treat both the case where the tube is packed with a solid catalyst (see Section 12.1) and the case where the fluid phase alone is present.

The majority of tubular reactors may be classified in terms of three major categories:

1. Single-jacketed tubes.
2. Shell-and-tube heat exchangers.
3. Tube furnaces, in which the tubes are exposed to thermal radiation and heat transfer from combustion gases.

The single-jacketed tube reactor is the simplest type of tubular reactor to conceptualize and to fabricate. It may be used only when the heat transfer requirements are minimal because of the low surface area to volume ratio characteristic of these reactors.

When the shell-and-tube configuration is utilized, the reaction may take place on either the tube side or the shell side. The shell-and-tube tubular reactor has a much greater area for heat transfer per unit of effective reactor volume than the single-jacketed tube. Consequently, it may be used for reactions where the energy transfer requirements are large. On occasion the reaction zone may be packed with granular solids to promote increased turbulence or better contacting of heterogeneous fluid phases or to act as a thermal sink to facilitate control of the reactor. In many cases energy economies can be achieved using countercurrent flow of a hot product stream to preheat an incoming reactant stream to the temperature where the reaction occurs at an appreciable rate. Two commercial scale processes that employ this technique are the synthesis of ammonia from its elements and the oxidation of sulfur dioxide to sulfur trioxide.

Tubular furnaces are used only when it is necessary to carry out *endothermic* reactions at fairly high temperatures on very large quantities of feedstock. Thermal reforming reactions and other reactions used to increase the yield of gasoline from petroleum-based feedstocks are commercial scale processes that employ this type of reactor. A tubular furnace is basically a combustion chamber with reactor tubes mounted on its walls and ceiling. Tube dimensions are typically 3 to 6 in. in diameter with lengths ranging from 20 to 40 ft. As many as several hundred tubes may be used, with either series or parallel connections possible depending on the required residence time.

Because there is no back-mixing of fluid elements along the direction of flow in a tubular reactor, there is a *continuous* gradient in reactant concentration in this direction. One does not encounter the step changes characteristic of multiple stirred tank reactors. Consequently, for the same feed composition and reaction temperature, the average reaction rate will generally be significantly higher in a plug flow reactor than it would be in a single stirred tank or a battery of stirred tanks with a total volume equal to that of the tubular reactor. The more efficient utilization of reactor volume is an advantage of the tubular reactor that permits one to use it in processes which demand very large capacity. Because variations in temperature and composition may occur in the axial direction in tubular reactors, these systems may be somewhat more difficult to control than continuous flow stirred tank reactors. However, the problems are usually not insurmountable, and one can normally obtain steady-state operating conditions which give rise to uniform product quality.

Other advantages of the tubular reactor relative to stirred tanks include suitability for use at higher pressures and temperatures, and the fact that severe energy transfer constraints may be readily surmounted using this configuration. The tubular reactor is usually employed for liquid phase reactions when relatively short residence times are needed to effect the desired chemical transformation. It is the reactor of choice for continuous gas phase operations.

8.0.2.4 Semibatch or Semiflow Reactors. Semibatch or semiflow operations usually take place in a single stirred tank using equipment extremely similar to that described for batch operations. Figure 8.1 indicates some of the many modes in which semibatch reactors may be operated.

One common mode of operation involves loading some of the reactants into a stirred tank as a single charge and then feeding in the remaining material gradually. This mode of operation is advantageous when large heat effects accompany the reaction. Exothermic reactions may be slowed down and temperature control maintained by regulating the rate at which one

of the reactants is fed. This point is demonstrated quite dramatically in Illustration 10.7. This mode of operation is also desirable when high reactant concentrations favor the formation of undesirable side products or when one of the reactants is a gas of limited solubility.

Another mode of semibatch operation involves the use of a purge stream to remove continuously one or more of the products of a reversible reaction. For example, water may be removed in esterification reactions by the use of a purge stream or by distillation of the reacting mixture. Continuous removal of product(s) increases the net reaction rate by slowing down the reverse reaction.

Semibatch or semiflow processes are among the most difficult to analyze from the viewpoint of reactor design because one must deal with an open system under nonsteady-state conditions. Hence the differential equations governing energy and mass conservation are more complex than they would be for the same reaction carried out batchwise or in a continuous flow reactor operating at steady state.

8.0.3 Fundamental Concepts Used in Chemical Reactor Design

The bread and butter tools of the practicing chemical engineer are the material balance and the energy balance. In many respects chemical reactor design can be regarded as a straightforward application of these fundamental principles. This section indicates in general terms how these principles are applied to the various types of idealized reactor models.

8.0.3.1 Material and Energy Balances in the Design of Industrial Reactors. The analysis of chemical reactors in terms of material and energy balances differs from the analysis of other process equipment in that one must take into account the rate at which molecular species are converted from one chemical form to another and the rate at which energy is transformed by the process. When combined with material and energy balances on the reactor, the reaction rate expression provides a means of determining the production rate and the composition of the products as functions of time. Both steady-state and time varying situations may be analyzed using the same fundamental relations. Differences in the analyses result from the retention of different terms in the basic balance equations.

A material balance on a reactant species of interest for an element of volume ΔV can be written as:

$$
\begin{pmatrix} \text{Rate of flow} \\ \text{of reactant} \\ \text{into volume} \\ \text{element} \end{pmatrix} = \begin{pmatrix} \text{Rate of flow} \\ \text{of reactant} \\ \text{out of volume} \\ \text{element} \end{pmatrix} + \begin{pmatrix} \text{Rate of disappearance} \\ \text{of reactant by chemical} \\ \text{reactions within the} \\ \text{volume element} \end{pmatrix} + \begin{pmatrix} \text{Rate of accumulation} \\ \text{of reactant within the} \\ \text{volume element} \end{pmatrix} \quad (8.0.1)
$$

or, in shorter form,

Input = output + disappearance by reaction
                    + accumulation   (8.0.2)

The flow terms represent the convective and diffusive transport of reactant into and out of the volume element. The third term is the product of the size of the volume element and the reaction rate per unit volume evaluated using the properties appropriate for this element. Note that the reaction rate per unit volume is equal to the intrinsic rate of the chemical reaction only if the volume element is uniform in temperature and concentration (i.e., there are no heat or mass transfer limitations on the rate of conversion of reactants to products). The final term represents the rate of change in inventory resulting from the effects of the other three terms.

In the analysis of batch reactors the two flow terms in equation 8.0.1 are omitted. For continuous flow reactors operating at steady state, the accumulation term is omitted. However, for the analysis of continuous flow reactors under transient conditions and for semibatch reactors it may be necessary to retain all four terms. For ideal well-stirred reactors the composition and temperature are uniform throughout the reactor and all volume elements are identical. Hence the material balance may be written over the entire reactor in these cases. For tubular flow reactors the composition is not independent of position and the balance must be written on a differential element of reactor volume and then integrated over the entire reactor using appropriate flow conditions and concentration and temperature profiles. Where nonsteady-state conditions are involved, it will be necessary to integrate over time as well as over volume in order to determine the performance characteristics of the reactor.

Since the rate of a chemical reaction is normally strongly temperature dependent, it is essential to know the temperature at each point in the reactor in order to be able to utilize equation 8.0.1 properly. When there are temperature gradients within the reactor, it is necessary to utilize an energy balance in conjunction with the material balance in order to determine the temperature and composition prevailing at each point in the reactor at a particular time.

The general energy balance for an element of volume ΔV over a time Δt can be written as:

$$
\begin{pmatrix} \text{Accumulation of} \\ \text{energy within the} \\ \text{volume element} \end{pmatrix} = \begin{pmatrix} \text{Energy transferred} \\ \text{from surroundings} \\ \text{to volume element} \\ \text{by heat and shaft} \\ \text{work interactions} \end{pmatrix} + \begin{pmatrix} \text{Energy effects} \\ \text{associated with} \\ \text{the entry of} \\ \text{matter into the} \\ \text{volume element} \end{pmatrix} - \begin{pmatrix} \text{Energy effects} \\ \text{associated with} \\ \text{the transfer of} \\ \text{matter out of the} \\ \text{volume element} \end{pmatrix} \quad (8.0.3)
$$

For completeness, the terms corresponding to the entry of material to the volume element and exit therefrom must contain, in addition to the ordinary enthalpy of the material, its kinetic and potential energy. However, for virtually all cases of interest in chemical reactor design, only the enthalpy term is significant. Since only changes in internal energy or enthalpy can be evaluated, the datum conditions for the first, third, and fourth terms must be identical in order to achieve an internally consistent equation. Although heat interactions in chemical reactors are significant, shaft work effects are usually negligible. The chemical reaction rate does not appear explicitly in equation 8.0.3, but its effects are implicit in all terms except the second. The first, third, and fourth terms reflect differences in temperature and/or in composition of the entering and leaving streams. The energy effects associated with composition changes are a direct reflection of the enthalpy change associated with the reaction.

There are a variety of limiting forms of equation 8.0.3 that are appropriate for use with different types of reactors and different modes of operation. For stirred tanks the reactor contents are uniform in temperature and composition throughout, and it is possible to write the energy balance over the entire reactor. In the case of a batch reactor, only the first two terms need be retained. For continuous flow systems operating at steady state, the accumulation term disappears. For adiabatic operation in the absence of shaft work effects the energy transfer term is omitted. For the case of semibatch operation it may be necessary to retain all four terms. For tubular flow reactors neither the composition nor the temperature need be independent of position, and the energy balance must be written on a differential element of reactor volume. The resultant differential equation must then be solved in conjunction with the differential equation describing the material balance on the differential element.

8.0.3.2 *Vocabulary of Terms Used in Reactor Design.*

There are several terms that will be used extensively throughout the remainder of this text that deserve definition or comment. The concepts involved include steady-state and transient operation, heterogeneous and homogeneous reaction systems, adiabatic and isothermal operation, mean residence time, contacting and holding time, and space time and space velocity. Each of these concepts will be discussed in turn.

Large-scale industrial reactions are almost invariably carried out on a continuous basis with reactants entering at one end of the reactor network and products leaving at the other. Usually such systems are designed for *steady-state* operation but, even during the design of such systems, adequate care must be made to provide for the *transient* condition that will invariably be incurred during start-up and shut-down periods. By the term steady-state operation we imply that conditions at any point in the reactor are time-invariant. Changes in composition occur in the spatial dimension instead of in a time dimension. It should be emphasized that operation at steady state does not imply *equilibrium*. The last term is restricted to isolated systems that undergo no net change with time. Insofar as the analysis of continuous flow reactors is concerned, the major thrust of this and succeeding chapters involves steady-state operation. The basic principles described in the previous subsection remain valid for the analysis of transient systems. However, in most transient cases it is necessary to resort to numerical solutions in order to predict the response of a continuous flow reactor network to changes in operating conditions. Batch reactors are inherently unsteady-state systems, even when the reactor contents are uniform throughout.

Both homogeneous and heterogeneous reaction systems are frequently encountered in commercial practice. The term *homogeneous* reaction system is restricted in this text to fluid systems in which the system properties *vary continuously* from point to point within the reactor. The term embraces both catalytic and noncatalytic reactions, but it requires that any catalysts be uniformly dispersed throughout the fluid phase. The term *heterogeneous* reaction system refers to systems in which there are two or more phases involved in the reaction process, either as reactants or as catalysts.

Adiabatic operation implies that there is no heat interaction between the reactor contents and their surroundings. *Isothermal* operation implies that the feed stream, the reactor contents, and the effluent stream are equal in temperature and have a uniform temperature throughout. The present chapter is devoted to the analysis of such systems. Adiabatic and other forms of nonisothermal systems are treated in Chapter 10.

The terms *holding time, contact time,* and *residence time* are often used in discussions of the performance of chemical reactors. As employed by reactor designers, these terms are essentially interchangeable. They refer to the length of time that an element of process fluid spends in the reactor in question. For a batch reactor it is usually assumed that no reaction occurs while the reaction vessel is being filled or emptied or while its contents are brought up to an ignition temperature. The holding time is thus the time necessary to "cook" the contents to the point where the desired degree of conversion is achieved. The terms "contact time" and "mean residence time" are used primarily in discussions of continuous flow processes. They represent the average length of time that it takes a fluid element to travel from the reactor inlet to the reactor outlet. For plug flow reactors, all fluid elements will have the same residence time. However, for stirred tank reactors or other reactors in which mixing effects are significant, there will be a spread of residence times for the different fluid elements. This has important implications for the conversions that will be achieved in such reactors.

Although the concept of "mean residence time" is easily visualized in terms of the average time necessary to cover the distance between reactor inlet and outlet, it is not the most fundamental characteristic time parameter for purposes of reactor design. A more useful concept is that of the reactor *space time*. For continuous flow reactors the space time (τ) is defined as the ratio of the reactor volume (V_R) to a characteristic volumetric flow rate of fluid ($\mathscr{V}$).

$$\tau = \frac{V_R}{\mathscr{V}} \qquad (8.0.4)$$

The reactor volume is taken as the volume of the reactor *physically occupied* by the reacting fluids. It does not include the volume occupied by agitation devices, heat exchange equipment, or headroom above liquids. One may arbitrarily select the temperature, pressure, and even the state of aggregation (gas or liquid) at which the volumetric flow rate to the reactor will be measured. For design calculations it is usually convenient to choose the reference conditions as those that prevail at the the inlet to the reactor. However, it is easy to convert to any other basis if the pressure-volume-temperature behavior of the system is known. Since the reference volumetric flow rate is arbitrary, care must be taken to specify precisely the reference conditions in order to allow for proper interpretation of the resultant space time. Unless an explicit statement is made to the contrary, *we will choose our reference state as that prevailing at the reactor inlet and emphasize this choice by the use of the subscript zero.* Henceforth,

$$\tau = \frac{V_R}{\mathscr{V}_0} \qquad (8.0.5)$$

where $\mathscr{V}_0$ is the volumetric flow rate at the inlet temperature and pressure and a fraction conversion of zero.

Using this convention, a space time of 30 min means that every 30 min one reactor volume of feed (measured at inlet conditions) is processed by the reactor. One reactor volume enters and one reactor volume leaves, but this statement does *not* imply that we have simply displaced the original charge from the reactor. Some or all of the original contents may leave, and some of the fresh charge may leave as well. In the latter circumstance the reactor contents become a mixture of the original contents and new material.

The space time is not necessarily equal to the average residence time of an element of fluid in the reactor. Variations in the number of moles on reaction as well as variations in temperature and pressure can cause the volumetric flow rate at arbitrary points in the reactor to differ appreciably from that corresponding to inlet conditions. Consequently, even though the reference conditions may be taken as those prevailing at the reactor inlet, the space time need not be equal to the mean residence time of the fluid. The two quantities are equal only if *all* of the following conditions are met.

1. Pressure and temperature are constant throughout the reactor.
2. The density of the reaction mixture is independent of the extent of reaction. For gas phase reactions this requirement implies that there can be no change in the number of moles on reaction. In terms of equation 3.1.45, we require that $\delta = 0$.
3. The reference volumetric flow rate is evaluated at reactor inlet conditions.

When the space time and the mean residence time differ, it is the space time that should be regarded as the independent process variable that is directly related to the constraints imposed on the system. We will see in Sections 8.2 and 8.3 that it is convenient to express the fundamental design relations for continuous flow reactors in terms of this parameter. We will also see that for these reactors the mean residence time cannot be considered as an independent variable, but that it is a parameter that can be determined only

after the nature of the changes occurring within the reactor is known.

The reciprocal of the space time is known as the *space velocity* (S).

$$S = \frac{1}{\tau} = \frac{\mathscr{V}}{V_R} \qquad (8.0.6)$$

Like the definition of the space time, the definition of the space velocity involves the volumetric flow rate of the reactant stream measured at some reference condition. A space velocity of $10 \ hr^{-1}$ implies that every hour, 10 reactor volumes of feed can be processed.

When dealing with reactions where a liquid feed must be vaporized prior to being fed to the reactor proper, one must state very clearly whether the space velocity is based on the volumetric flow rate of the feed as a liquid or as a gas. Unless an explicit statement to the contrary is made, the term space velocity in this text will refer to the ratio of the volumetric flow rate *evaluated at reactor inlet conditions* to the reactor volume.

The term *space velocity* has somewhat different connotations when dealing with heterogeneous catalytic reactors. In this case it denotes the ratio of the mass flow rate of feed to the mass of catalyst used (W).

$$WHSV = \frac{\rho \mathscr{V}}{W} \qquad (8.0.7)$$

where ρ is the mass density of the feed and WHSV is termed the *weight hourly space velocity*. Sometimes the term volumetric hourly space velocity (VHSV) is used to denote the ratio of the volumetric flow rate of a gaseous feed to the weight of the catalyst bed.

$$VHSV = \frac{\mathscr{V}}{W} \qquad (8.08)$$

On other occasions the volume of catalyst instead of the mass of catalyst may be used in the denominator. The units associated with a particular space velocity indicate the definition employed.

8.1 DESIGN ANALYSIS FOR THE BATCH REACTOR

Batch reactors are widely used in the chemical industry for producing materials that are needed in limited quantity, particularly in those cases where the processing cost represents only a small fraction of the total value of the product. Since modern industry stresses the use of continuous processes because they lend themselves most readily to mass production, chemical engineers may, in some instances, tend to overlook the economic superiority of batch operations. One should not become so fascinated with the continuous process, or the more complex and interesting design analysis associated therewith, as to lose sight of the economic penalty exacted by this degree of technical sophistication.

The starting point for the development of the basic design equation for a well-stirred batch reactor is a material balance involving one of the species participating in the chemical reaction. For convenience we will denote this species as A and we will let $(-r_A)$ represent the *rate of disappearance* of this species by reaction. For a well-stirred reactor the reaction mixture will be uniform throughout the effective reactor volume, and the material balance may thus be written over the entire contents of the reactor. For a batch reactor equation 8.0.1 becomes

limiting reagent.) Thus,

$$\text{Rate of accumulation} = \frac{dN_A}{dt} = v_A \frac{d\xi}{dt}$$

$$= -N_{A0} \frac{df_A}{dt} \quad (8.1.2)$$

where N_{A0} is number of moles of species A present when the fraction conversion is zero.

The total rate of disappearance of reactant A is given by

$$\text{Rate of disappearance} = (-r_A)V_R \quad (8.1.3)$$

We again emphasize that V_R is the volume physically occupied by the reacting fluid. Combining equations 8.1.1 to 8.1.3 gives

$$N_{A0} \frac{df_A}{dt} = (-r_A)V_R \quad (8.1.4)$$

Rearrangement and integration gives

$$t_2 - t_1 = N_{A0} \int_{f_{A1}}^{f_{A2}} \frac{df_A}{(-r_A)V_R} \quad (8.1.5)$$

where f_{A2} and f_{A1} represent the fraction conversion at times t_2 and t_1, respectively. This equation is the most general form of the basic design relationship for a batch reactor. It is valid for both isothermal and nonisothermal operation as well as for both constant volume and constant pressure operation. Both the reaction rate and the reactor volume should be retained inside the integral sign, since either or both may change as the reaction proceeds.

$$\begin{pmatrix} \text{Rate of accumulation} \\ \text{of reactant } A \text{ within} \\ \text{the reactor} \end{pmatrix} = -\begin{pmatrix} \text{Rate of disappearance of} \\ \text{reactant } A \text{ within the reactor} \\ \text{by chemical reaction} \end{pmatrix} \quad (8.1.1)$$

The accumulation term is just the time derivative of the number of moles of reactant A contained within the reactor (dN_A/dt). This term also may be written in terms of either the extent of reaction (ξ) or the fraction conversion of the limiting reagent (f_A). (A is presumed to be the

There are a number of limiting forms of equation 8.1.5 that should be mentioned briefly. If $t_1 = 0$ and $f_{A1} = 0$,

$$t = N_{A0} \int_0^{f_A} \frac{df_A}{(-r_A)V_R} \quad (8.1.6)$$

If, in addition, the reactor volume (fluid density) is constant, equation 8.1.6 becomes

$$t = C_{A0} \int_0^{f_A} \frac{df_A}{(-r_A)} = -\int_{C_{A0}}^{C_A} \frac{dC_A}{(-r_A)}$$

$$(8.1.7)$$

For reactions where the fluid volume varies linearly with the fraction conversion, as indicated by equation 3.1.40, equation 8.1.6 becomes

$$t = N_{A0} \int_0^{f_A} \frac{df_A}{(-r_A)V_{R0}(1 + \delta_A f_A)}$$

$$= C_{A0} \int_0^{f_A} \frac{df_A}{(-r_A)(1 + \delta_A f_A)} \qquad (8.1.8)$$

where V_{R0} is the volume occupied by the reacting fluid at zero fraction conversion. This equation would be appropriate for use when low-pressure gas phase reactions involving a change in the number of moles on reaction take place in a batch reactor *at constant pressure*. However, gas phase reactions are rarely carried out batchwise on a commercial scale because the quantity of product that can be produced in a reasonably sized reactor is so small. The chief use of batch reactors for gas phase reactions is to obtain the data necessary for the design of continuous flow reactors.

For simple nth-order kinetics where volumetric expansion effects may be significant, equation 8.1.8 becomes

$$t = C_{A0} \int_0^{f_A} \frac{df_A}{kC_{A0}^n \left(\dfrac{1 - f_A}{1 + \delta_A f_A}\right)^n (1 + \delta_A f_A)}$$

$$= \int_0^{f_A} \frac{(1 + \delta_A f_A)^{n-1} \, df_A}{kC_{A0}^{n-1}(1 - f_A)^n} \qquad (8.1.9)$$

Equations 8.1.4 to 8.1.8 may also be written in terms of the extent of reaction (ξ) or the extent per unit volume (ξ^*). In terms of ξ the most general design relation (equation 8.1.5) becomes

$$t_2 - t_1 = v_A \int_{\xi_1}^{\xi_2} \frac{d\xi}{r_A V_R} = \int_{\xi_1}^{\xi_2} \frac{d\xi}{r V_R} \qquad (8.1.10)$$

The use of equations like this was treated in great detail in Chapter 3. There our primary objective was the determination of the mathematical form of the reaction rate expression from data on the extent of reaction or fraction conversion versus time. At present our objective is just the reverse: to determine the time necessary to achieve a given degree of conversion using our knowledge of the mathematical form of the reaction rate expression and the reaction conditions. Although it is often more convenient to work in terms of extent of reaction when analyzing rate data, it is usually more convenient to work in terms of fraction conversion in analyzing design problems. The two concepts are simply related, and the chemical engineer should learn to work in terms of either with equal facility.

The degree of difficulty associated with evaluating the integral in any of the batch reactor design equations (equations 8.1.5 to 8.1.10) depends on the composition and temperature dependence of the reaction rate expression. For nonisothermal systems an energy balance must be employed to relate the system temperature (and through it the reaction rate) to the fraction conversion. For isothermal systems it is not necessary to utilize an energy balance to determine the holding time necessary to achieve a given fraction conversion. One merely substitutes rate constants evaluated at the temperature in question directly into the rate expression. However, an energy balance must be used to determine the heat transfer requirements necessary to maintain isothermal conditions. The reader should recognize that even if a simple closed form relation between time and fraction conversion cannot be obtained, it is still possible to evaluate the integral graphically or numerically using standard methods to assure convergence. The latter approach is invariably necessary when departures from isothermal conditions are large.

The following illustrations indicate how the basic design relations developed above are used to answer the two questions with which the reactor designer is most often faced.

1. What is the time required for converting a quantity of material to the desired level under specified reaction conditions?
2. What reactor volume is required to achieve a given production rate?

ILLUSTRATION 8.1 DETERMINATION OF HOLDING TIME REQUIREMENTS FOR THE FORMATION OF A DIELS-ALDER ADDUCT

Wassermann (6) has studied the Diels-Alder reaction of benzoquinone (B) and cyclopentadiene (C) at 25 °C.

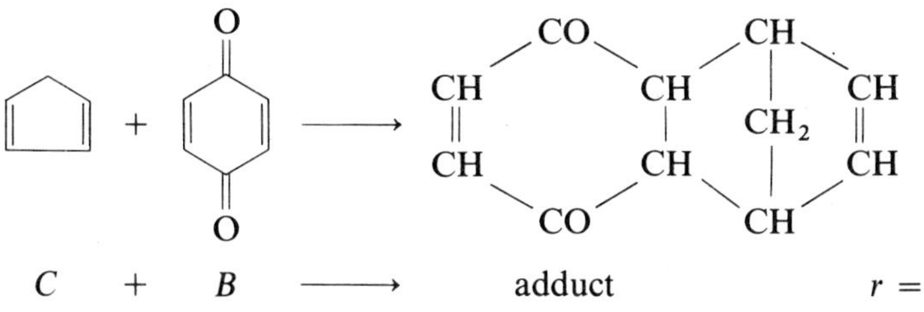

Since the reaction takes place at constant volume, the pertinent design equation is equation 8.1.7.

$$t = C_{B0} \int_0^{f_B} \frac{df_B}{k C_{B0}(1 - f_B)(C_{C0} - f_B C_{B0})}$$

Since the reaction occurs under isothermal conditions, the rate constant may be taken outside the integral sign. Integration and simplification then gives

$$t = \frac{\ln\left[\left(\dfrac{\dfrac{C_{C0}}{C_{B0}} - f_B}{1 - f_B}\right)\left(\dfrac{C_{B0}}{C_{C0}}\right)\right]}{k(C_{C0} - C_{B0})}$$

$$C \quad + \quad B \quad \longrightarrow \quad \text{adduct} \qquad\qquad r = k C_B C_C$$

Volume changes on reaction may be neglected. At 25 °C the reaction rate constant is equal to 9.92×10^{-3} m^3/kmole · sec. If one employs a well-stirred isothermal batch reactor to carry out this reaction, determine the holding time necessary to achieve 95% conversion of the limiting reagent using initial concentrations of 0.1 and 0.08 kmole/m^3 for cyclopentadiene and benzoquinone, respectively.

Solution

The limiting reagent is benzoquinone. The rate of disappearance of this species can be written in terms of the initial concentrations and the fraction conversion as

$$-r_B = k[C_{B0}(1 - f_B)](C_{C0} - f_B C_{B0})$$

Substitution of numerical values gives

$$t = \frac{\ln\left[\left(\dfrac{\dfrac{0.1}{0.08} - 0.95}{1 - 0.95}\right)\left(\dfrac{0.08}{0.10}\right)\right]}{9.92 \times 10^{-3}(0.1 - 0.08)}$$

$$= 7.91 \times 10^3 \text{ sec or } 2.20 \text{ hr}$$

ILLUSTRATION 8.2 DETERMINATION OF HOLDING TIME AND REACTOR SIZE REQUIREMENTS FOR THE PRODUCTION OF ZEOLITE A IN A BATCH REACTOR

Zeolites are hydrous aluminosilicates that are widely used as catalysts in the chemical process industry. Zeolite A is usually synthesized in the sodium form from aqueous solutions of sodium silicate and sodium aluminate. Kerr (7) and Liu (8) have studied an alternative method of synthesis from amorphous sodium aluminosilicate substrate and aqueous sodium hydroxide solution. The reaction (essentially a crystallization or recrystallization process) can be viewed as

Amorphous solid $\overset{fast}{\rightarrow}$ soluble species

Soluble species + nuclei (or zeolite crystals) $\overset{slow}{\rightarrow}$ zeolite A

Liu's description of the kinetics of the zeolite formation process can be formulated in terms of the following equation.

$$-\frac{dC_A}{dt} = \frac{k_2[\text{OH}^-]^{k_1}C_A C_Z}{k_3 C_Z + (k_3 + 1)C_A} \qquad \text{(A)}$$

where

$$[\text{OH}^-] = \text{hydroxide ion concentration} \\ (\text{kmoles/m}^3)$$

$$C_Z = \text{concentration of zeolite crystals} \\ (\text{kg/m}^3)$$

$$C_A = \text{concentration of amorphous} \\ \text{substrate (kg/m}^3)$$

and k_1, k_2, and k_3 are kinetic constants. Since the total weight of solids in a batch reactor must satisfy an overall material balance, we require that the quantity $(C_A + C_Z)$ be a constant equal to the total weight of the charge divided by the effective reactor volume. Note that the reaction is catalyzed by the presence of product in that the zeolite concentration appears in the numerator of the rate expression. Consequently, it is desirable to include some zeolite in the feed to the reactor to enhance the reaction rate. Reactions of this type are labeled as "autocatalytic." They are treated in more detail in Section 9.4.

At 100 °C the following values of the rate constants are appropriate for use.

$$k_1 = 2.36 \qquad k_2 = 0.625 \text{ ksec}^{-1} \qquad k_3 = 0.36$$

1. If one utilizes a slurry containing 1 kg/m³ of zeolite and 24 kg/m³ of amorphous substrate, determine the time necessary to achieve 98% conversion of the substrate to zeolite in a well-stirred batch reactor. The conditions to be considered are isothermal operation at 100 °C and a hydroxide concentration of 1.5 kmoles/m³.

2. Assuming the reaction conditions noted in

---

part 1, determine the reactor size and total weight of charge necessary to produce zeolite A at an average rate of 2000 kg/day. Only one reactor is to be used, and it will be necessary to shut down for 1.8 ksec between batches for removal of product, cleaning, and start-up. The zeolite to be recycled to the reactor will come from the 2000 kg produced daily.

Solution

Since the reaction takes place in slurry form, the reactor volume may be regarded as constant, and equation 8.1.7 is appropriate for use.

$$t_2 - t_1 = C_{A0} \int_0^{f_A} \frac{df_A}{(-r_A)} \qquad \text{(B)}$$

where the subscript A refers to the amorphous substrate.

The principles of stoichiometry may be used to write the reaction rate in terms of the fraction conversion. The desired conversion level is expressed in terms of the initial substrate level. Thus $C_{A0} = 24$ kg/m³. At any time the instantaneous concentration of substrate can be written as

$$C_A = C_{A0}(1 - f_A) \qquad \text{(C)}$$

while the corresponding concentration of zeolite is given by

$$C_Z = C_{Z0} + f_A C_{A0} \qquad \text{(D)}$$

The instantaneous reaction rate can be expressed in terms of the fraction conversion by combining equations A, C, and D.

$$-r_A = \frac{-dC_A}{dt} = \frac{k_2[OH^-]^{k_1}C_{A0}(1 - f_A)(C_{Z0} + f_A C_{A0})}{k_3(C_{Z0} + f_A C_{A0}) + (1 + k_3)C_{A0}(1 - f_A)}$$

Substitution of numerical values into this relation gives

$$-r_A = \frac{(0.625)(1.5)^{2.36}(24)(1 - f_A)(1 + 24f_A)}{0.36(1 + 24f_A) + (1.36)(24)(1 - f_A)}$$

$$= \frac{(39.05)(1 - f_A)(1 + 24f_A)}{33.0 - 24f_A} \qquad (E)$$

Combining equations E and B gives

$$t = 24 \int_0^{f_A} \frac{(33.0 - 24f_A)\, df_A}{39.05(1 - f_A)(1 + 24f_A)}$$

This integral may be broken up into terms that can be evaluated using standard tables.

$$t = 20.28 \int_0^{f_A} \frac{df_A}{(1 - f_A)(1 + 24f_A)}$$

$$- 14.75 \int_0^{f_A} \frac{f_A\, df_A}{(1 - f_A)(1 + 24f_A)}$$

or

$$t = \frac{20.28}{25} \ln\left(\frac{1 + 24f_A}{1 - f_A}\right)$$

$$+ \frac{14.75}{25}\left[\ln(1 - f_A) + \frac{1}{24}\ln(1 + 24f_A)\right]$$

The time required to achieve any desired degree of conversion may be calculated from this expression. For $f_A = 0.98$: $t = 3.54$ ksec $= 0.982$ hr.

For part 2, for 98% conversion, each batch will require a holding time of 3.54 ksec and a downtime of 1.80 ksec for emptying, cleaning, and start-up. The total time consumed in processing one batch is thus 5.34 ksec. In order to produce 2000 kg of zeolite A per day we require that each batch contain an amount of amorphous solids equal to

$$\frac{2000 \text{ kg/day}}{(0.98)(\text{percent conversion})} \times \frac{(5.34 \text{ ksec/batch})}{(86.4 \text{ ksec/day})} = 126 \frac{\text{kg}}{\text{batch}}$$

The weight of amorphous solids per batch must equal the product of the reactor volume and the solids concentration. Thus

$$V_R = \frac{126 \text{ kg}}{24 \text{ kg/m}^3} = 5.25 \text{ m}^3 \approx 1387 \text{ gal}$$

Note that in this case the batch reactor is operating in the production mode only two thirds of the time. This situation is not extraordinary and allowances must be made for downtime in order to meet design requirements properly.

When there are significant heating and cooling periods associated with the utilization of a batch reactor, one must take these circumstances into account in design calculations. If only one reaction is involved, the holding time computed using the methods of this section will be a conservative estimate of the length of time the reactor contents should be held at the specified temperature. However, if undesirable side reactions take place at temperatures below the operating temperature, the magnitude of the adverse effects must be considered. If they are significant it may be necessary to resort to rapid quenching techniques or to separate preheating of coreactants to the desired temperature.

8.2 DESIGN OF TUBULAR REACTORS

Tubular reactors are normally used in the chemical industry for extremely large-scale processes. When filled with solid catalyst particles, such reactors are referred to as fixed or packed bed reactors. This section treats general design relationships for tubular reactors in which isothermal homogeneous reactions take place. Nonisothermal tubular reactors are treated in Section 10.4 and packed bed reactors in Section 12.7.

8.2.1 The Plug Flow Reactor (PFR)—Basic Assumptions and Design Equations

The simplest model of the behavior of tubular reactors is the plug flow model. The essential features of this idealized model require that there be no longitudinal mixing of fluid elements as they move through the reactor and that all fluid elements take the same length of time to move from the reactor inlet to the outlet. The model may also be labeled the *slug flow* or *piston flow* model in that it may be convenient to picture the reaction as taking place within differentially thin slugs of fluid that fill the entire cross section of the tube and that are separated from one another by hypothetical pistons that prevent axial mixing. These plugs of material move as units through the reactor, and this assumption is conveniently expressed in terms of a requirement that the velocity profile be flat as one traverses the tube diameter. Each plug of fluid is assumed to be uniform in temperature, composition, and pressure, which is equivalent to assuming that radial mixing is infinitely rapid. However, there may well be variations in composition, temperature, pressure, and fluid velocity as one moves in the longitudinal direction. With respect to these variations, however, the model requires that mass transport via diffusion or turbulent mixing processes be negligible and that the plugs of material not interact with one another except for transmission of the hydrodynamic forces giving rise to the fluid motion. Some of the aforementioned requirements may be removed in more complex mathematical models of tubular reactors. In the present chapter, however, we will limit our discussion to the simplest possible model—the plug flow model. Nonetheless, when we attempt to compare the results predicted by the model with what we observe

in the real world, we should keep in mind that the model can only reflect the idealizations built into it and that deviations from ideal behavior can fall into three categories.

1. There will be velocity gradients in the radial direction so all fluid elements will not have the same residence time in the reactor. Under turbulent flow conditions in reactors with large length to diameter ratios, any disparities between observed values and model predictions arising from this factor should be small. For short reactors and/or laminar flow conditions the disparities can be appreciable. Some of the techniques used in the analysis of isothermal tubular reactors that deviate from plug flow are treated in Chapter 11.
2. There will be an interchange of material between fluid elements at different axial positions by virtue of ordinary molecular diffusion and eddy diffusion processes arising from turbulence and/or the influence of any packing in the bed. Convective mixing arising from thermal gradients in the reactor may also contribute to the exchange of matter between different fluid elements.
3. There may be radial temperature gradients in the reactor that arise from the interaction between the energy released by reaction, heat transfer through the walls of the tube, and convective transport of energy. This factor is the greatest potential source of disparities between the predictions of the model and what is observed for real systems. The deviations are most significant in nonisothermal packed bed reactors.

The tubular flow reactor is a convenient means of approaching the performance characteristics of a batch reactor on a continuous basis, since the distance-pressure-temperature history of the various plugs as they flow through the reactor corresponds to the time-pressure-temperature protocol that is used in a batch reactor. Although this analogy is often useful, it may on occasion be misleading. Batch reactors are almost invariably operated under the constraint of constant volume, whereas in a tubular flow reactor, each plug of fluid more nearly approximates constant pressure conditions. For liquid phase reactions constraints of constant volume and constant pressure may effectively be satisfied simultaneously, and the same is true of isothermal gas phase reactions that do not involve a change in the number of gas phase moles on reaction. However, when there is a change in temperature or in the number of molecules contained within the plug, the volume of the plug can change by an appreciable fraction. In order to maintain a constant mass velocity at various points along a uniform tube in cases where there is an increase in the number of moles on reaction (an increase in plug volume), an increase in the volumetric flow rate must accompany the reaction. Instead of giving rise to the pressure increase that would take place in a batch reactor, the change in the number of moles on reaction causes the fluid to accelerate. Thus the residence time of the plug will be less than that which would have been observed if the volume of the plug remained unchanged. We turn now to the problem of developing fundamental design relationships that allow for such effects.

Consider the segment of tubular reactor shown in Figure 8.3. Since the fluid composition varies with longitudinal position, we must write our material balance for a reactant species over a different element of reactor (dV_R). Moreover, since plug flow reactors are operated at steady state except during start-up and shut-down procedures, the relations of major interest are those in which the accumulation term is missing from equation 8.0.1. Thus

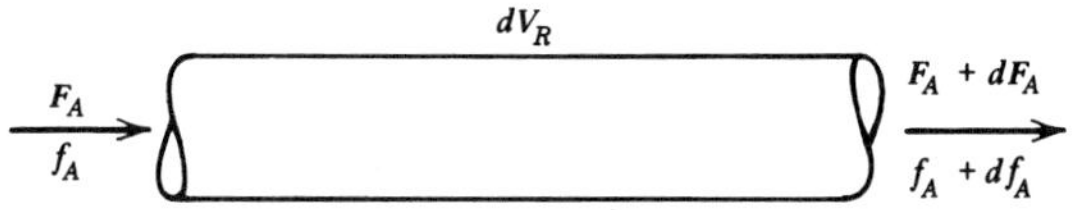

Figure 8.3
Schematic representation of differential volume element of plug flow reactor.

If F_A represents the molal flow rate of reactant A into the volume element and $F_A + dF_A$ represents the molal flow rate out of the volume element, equation 8.2.1 becomes

$$F_A = (F_A + dF_A) + (-r_A)\, dV_R \quad (8.2.2)$$

or

$$dF_A = r_A\, dV_R \quad (8.2.3)$$

At any point the molal flow rate of reactant A can be expressed in terms of the fraction conversion f_A and the molal flow rate corresponding to zero conversion F_{A0}.

$$F_A = F_{A0}(1 - f_A) \quad (8.2.4)$$

Differentiating

$$dF_A = -F_{A0}\, df_A \quad (8.2.5)$$

Combining equations 8.2.3 and 8.2.5 gives

$$\frac{dV_R}{F_{A0}} = \frac{df_A}{(-r_A)} \quad (8.2.6)$$

which may be integrated over the entire reactor volume to give

$$\frac{V_R}{F_{A0}} = \int_{f_{A\ \text{in}}}^{f_{A\ \text{out}}} \frac{df_A}{(-r_A)} \quad (8.2.7)$$

This equation is a very useful relation that indicates the reactor size necessary to accomplish a specified change in the degree of conversion for a fixed molal flow rate. It does require, however, a knowledge of the relationship between the reciprocal rate of reaction and the

$$\begin{pmatrix} \text{Rate of flow} \\ \text{of reactant into} \\ \text{volume element} \end{pmatrix} = \begin{pmatrix} \text{Rate of flow of} \\ \text{reactant out of} \\ \text{volume element} \end{pmatrix} + \begin{pmatrix} \text{Rate of disappearance} \\ \text{of reactant by chemical} \\ \text{reactions within the} \\ \text{volume element} \end{pmatrix} \quad (8.2.1)$$

fraction conversion. For nonisothermal systems this relationship can be quite complex, as we will see in Chapter 10.

It should be emphasized that for *ideal* tubular reactors, it is the *total volume* per unit of feed that determines the conversion level achieved. The ratio of the length of the tube to its diameter is irrelevant, provided that plug flow is maintained and that one uses the same flow rates and pressure-temperature profiles expressed in terms of reactor volume elements.

F_{A0} may also be written as the product of a volumetric flow rate and a reactant concentration where both are measured at some reference temperature and pressure and correspond to zero fraction conversion. Thus

$$\frac{V_R}{F_{A0}} = \frac{V_R}{C_{A0}\mathcal{V}_0} = \frac{\tau}{C_{A0}} \qquad (8.2.8)$$

where we have introduced the space time τ. Combining equations 8.2.8 and 8.2.7 gives

$$\tau = \frac{V_R}{\mathcal{V}_0} = C_{A0} \int_{f_{A\,in}}^{f_{A\,out}} \frac{df_A}{(-r_A)} \qquad (8.2.9)$$

Reactor inlet conditions are particularly useful as reference conditions for measuring the input volumetric flow rate in that they not only give physical meaning to C_{A0} and $\mathcal{V}_0$ but also usually lead to cancellation of C_{A0} with a similar term appearing in the reaction rate expression.

If the temperature is constant throughout the reactor volume, the rate constant k may be removed from the integral and an analytic solution obtained. In more general terms it will be necessary to evaluate the integrals in equations 8.2.7 and 8.2.9 using graphical or numerical methods. Figure 8.4 indicates this schematically. For nonisothermal systems one must employ an energy balance in conjunction with the basic design equation to relate the temperature (and thus the temperature dependent terms in the rate expression) to the fraction conversion. The relations are such that exact analytical solutions can rarely, if ever, be obtained. (See Chapter 10.)

For cases where $\delta = 0$, equation 8.2.9 can be written in terms of concentrations.

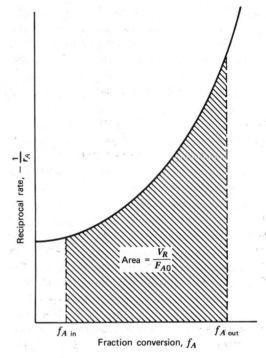

Figure 8.4
Determination of V_R/F_{A0} from plot of reciprocal rate versus fraction conversion.

$$\tau = -\int_{C_{A\,in}}^{C_{A\,out}} \frac{dC_A}{(-r_A)} \qquad (8.2.10)$$

(constant density systems only)

For variable density systems it is more convenient to work in terms of the fraction conversion (equation 8.2.9), but for constant density systems either equation 8.2.9 or equation 8.2.10 is appropriate.

For preliminary design calculations involving tubular reactors, the usual procedure is to assume plug flow with constant pressure over the length of the reactor. However, the above analysis is appropriate regardless of whether or not a pressure drop exists. The pressure enters only through its influence on the reaction rate term. For liquid phase reactions the influence of pressure variations is usually insignificant but, for gas phase reactions or when mixtures of gases and liquids are present, the pressure drop across a given segment of the reactor must be taken into account in arriving at the reactor volume required to accomplish a given task. No new principles are involved; one merely breaks the reactor up into a series of segments that can be assumed to operate at an appropriate average pressure. The pressure drop across a segment is calculated from the Bernoulli equation using appropriate empirical relations where necessary to estimate the friction factor, equivalent lengths for u-bends or other fittings, and losses associated with changes in tube diameter. The basic design equation is then applied to each segment using the average of the inlet and outlet pressures in the rate expression. One marches through the reactor assuming conversion increments that are sufficiently small that calculations allowing for the pressure drop lead to convergence of the integral. The case study in Section 13.1 indicates one possible approach to this problem.

Illustrations 8.3 and 8.4 indicate the application of the above analysis to isothermal tubular reactors with negligible pressure drop.

ILLUSTRATION 8.3 DETERMINATION OF REQUIRED PLUG FLOW REACTOR VOLUME UNDER ISOTHERMAL OPERATING CONDITIONS—CONSTANT DENSITY CASE

Inukai and Kojima (9) have studied the aluminum chloride catalyzed diene condensation of butadiene and methyl acrylate in benzene solution. The stoichiometry for this Diels-Alder reaction is

The following set of mechanistic equations is consistent with their experimental results:

$$AlCl_3 + M \rightarrow AlCl_3{\cdot}M \quad \text{(fast)} \quad \text{(A)}$$

$$B + AlCl_3{\cdot}M \xrightarrow{k_2} C{\cdot}AlCl_3 \quad \text{(slow)} \quad \text{(B)}$$

$$C{\cdot}AlCl_3 \rightarrow C + AlCl_3 \quad \text{(fast)} \quad \text{(C)}$$

where reaction B is the rate limiting step. $(AlCl_3{\cdot}M)$ and $(C{\cdot}AlCl_3)$ represent complexes formed between the dissolved $AlCl_3$ and the species in question. The concentration of the first reactive complex is essentially constant as long as enough methyl acrylate remains in solution to regenerate $(AlCl_3{\cdot}M)$ efficiently. The concentration of methyl acrylate in excess of $AlCl_3$ will not affect the rate of reaction during the early stages of reaction but, as reaction proceeds and the methyl acrylate concentration drops below the initial $AlCl_3$ concentration, the amount of complex present is limited by the amount of methyl acrylate remaining.

These investigators report that the second-order rate constant for reaction B is equal to 1.15×10^{-3} m^3/mole·ksec at 20 °C. Determine the volume of plug flow reactor that would be necessary to achieve 40% conversion of the input butadiene assuming isothermal operating conditions and a liquid feed rate of 0.500 m^3/ksec. The feed composition is as follows.

Butadiene (B)	96.5 moles/m^3
Methyl acrylate (M)	184 moles/m^3
AlCl$_3$	6.63 moles/m^3

butadiene (B)   methyl acrylate (M) ⟶ adduct (C)

Solution

The butadiene is the limiting reagent and conversions will be expressed in terms of this species. Over the composition range of interest there will always be sufficient methyl acrylate present to tie up the aluminum chloride. Consequently the concentration of the complex ($AlCl_3 \cdot M$) will remain constant throughout the length of the reactor at a value equal to the initial $AlCl_3$ concentration. For these conditions the reaction rate expression is the form

$$-r_B = kC_B C_{AlCl_3 \cdot M} = kC_{B0}(1 - f_B)C_{AlCl_3, 0} \quad (D)$$

since the volume change accompanying liquid phase reactions is negligible. Equation 8.2.9 may be used as the basic design relationship.

$$\tau = C_{B0} \int_0^{f_B} \frac{df_B}{(-r_B)}$$

$$= C_{B0} \int_0^{f_B} \frac{df_B}{kC_{B0}(1 - f_B)C_{AlCl_3, 0}} \quad (E)$$

Since the quantities $C_{AlCl_3, 0}$ and k are constant for the conditions cited, equation E may be integrated to give

$$\tau = \frac{\ln\left(\dfrac{1}{1 - f_B}\right)}{kC_{AlCl_3, 0}}$$

Substitution of numerical values gives

$$\tau = \frac{\ln\left(\dfrac{1}{1 - 0.4}\right)}{(1.15 \times 10^{-3})(6.63)} = 67.0 \text{ ksec} \approx 18.6 \text{ hr}$$

From the definition of the space time and the inlet volumetric flow rate,

$$V_R = \tau \mathcal{V}_0 = 67.0(0.500) = 33.5 \text{ m}^3$$

The space time and reactor volume required to accomplish the specified conversion in a plug flow reactor are sufficiently high that they make the use of a tubular reactor impractical for the specified operating conditions. For these conditions a cascade of stirred tank reactors would be more appropriate.

ILLUSTRATION 8.4 DETERMINATION OF REQUIRED PLUG FLOW REACTOR VOLUME UNDER ISOTHERMAL OPERATING CONDITIONS—VARIABLE DENSITY CASE

Ratchford and Fisher (10) have studied the pyrolysis of methyl acetoxypropionate at temperatures near 500 °C and a variety of pressures.

$$CH_3COOCH(CH_3)COOCH_3 \rightarrow CH_3COOH + CH_2{=}CHCOOCH_3$$

$$\text{acetic acid} \qquad\qquad \text{methyl acrylate}$$

Below 565 °C the pyrolysis reaction is essentially first order with a rate constant given by

$$k = 7.8 \times 10^9 e^{-19\,220/T} \text{ sec}^{-1}$$

where T is expressed in degrees Kelvin.

If one desires to design a pilot scale tubular reactor to operate isothermally at 500 °C, what length of 6-in. pipe will be required to convert 90% of the raw feedstock to methyl acrylate? The feedstock enters at 5 atm at a flow rate of 500 lb/hr. Ideal gas behavior may be assumed. A 6-in. pipe has an area of 0.0388 ft^2 available for flow. Pressure drop across the reactor may be neglected.

Solution

From equation 8.2.9 and the fact that the reaction is first order,

$$\tau = C_{A0} \int_0^{f_A} \frac{df_A}{kC_A} \quad (A)$$

For the case where the feed is pure methyl acetoxypropionate,

$$\delta_A = \frac{2 - 1}{1} = 1$$

Since there is a change in the number of moles of gaseous species on reaction, the concentration corresponding to a given fraction conversion is

$$C_A = C_{A0} \left(\frac{1 - f_A}{1 + \delta_A f_A} \right) = C_{A0} \frac{(1 - f_A)}{(1 + f_A)} \quad \text{(B)}$$

Combining equations A and B gives

$$\tau = \int_0^{f_A} \frac{(1 + f_A)\, df_A}{k(1 - f_A)}$$

Since the reaction takes place isothermally at 500 °C, the reaction rate constant may be moved outside the integral sign. At this temperature it is equal to 0.124 sec^{-1}. Thus

$$0.124\tau = \int_0^{f_A} \frac{df_A}{1 - f_A} + \int_0^{f_A} \frac{f_A\, df_A}{1 - f_A}$$

Evaluation of the integrals gives

$$0.124\tau = -\ell n(1 - f_A) - f_A - \ell n(1 - f_A)$$
$$= -2\, \ell n(1 - f_A) - f_A$$

For $f_A = 0.90$,

$$\tau = [-2\, \ell n(1 - 0.9) - 0.900]/0.124 = 29.9 \text{ sec}$$

If we had erred and not included the effect of volumetric expansion on reaction, we would have calculated a space time of 18.6 sec, which would have been off by 38%, and this error would propagate through the remainder of the calculations below.

If the gas behaves ideally the volumetric flow rate at the reactor inlet is given by the product of the molal flow rate [(500/146) lb moles/hr] and the molal volume at the pressure and temperature in question. The latter may be calculated by correcting the standard molal volume (359 ft^3/lb mole) for variations in temperature and pressure between the reactor inlet and standard conditions. Hence

$$\mathscr{V}_0 = \left(\frac{500}{146} \right)(359)\left(\frac{773}{273} \right)\left(\frac{1}{5} \right) \text{ ft}^3/\text{hr}$$

$$= 696 \text{ ft}^3/\text{hr} \approx 0.193 \text{ ft}^3/\text{sec}$$

From the definition of the space time,

$$V_R = \mathscr{V}_0 \tau = (0.193)(29.9) = 5.78 \text{ ft}^3$$

From geometric considerations the required length of 6-in. pipe is equal to (5.78 ft^3)/(3.88 × 10^{-2} ft^2) or 149 ft.

Illustrations 8.3 and 8.4 indicate how the plug flow design equations may be applied to homogeneous fluid phase reactions. We now wish to consider the form of the plug flow reactor design equation for heterogeneous catalytic reactions. For reactions of this type the reaction rate must be expressed per unit weight or per unit area of catalyst. Since the two are related through the specific surface area of the material, we will develop relations only in terms of the former. Consider the segment of a tubular reactor shown in Figure 8.3 in which we now presume that the differential volume element dV_R contains an amount of catalyst dW. The rate of disappearance of reactant A within the differential volume element is then equal to $(-r_{Am})\, dW$. A material balance for steady-state operating conditions based on equation 8.2.1 gives

Input   Output   Disappearance by reaction
$$F_A \quad = F_A + dF_A + (-r_{Am})\, dW$$

$$(8.2.11)$$

Introduction of the fraction conversion through equation 8.2.5 leads to the following analog of equation 8.2.7.

$$\frac{W}{F_{A0}} = \int_{f_{A\,in}}^{f_{A\,out}} \frac{df_A}{(-r_{Am})} \quad (8.2.12)$$

In this case the reaction rate will depend not only on the system temperature and pressure but also on the properties of the catalyst. It should be noted that the reaction rate term must include the effects of external and intraparticle heat and mass transfer limitations on the rate. Chapter 12 treats these subjects and indicates how equation 8.2.12 can be used in the analysis of packed bed reactors.

8.2.2 Residence Times in Plug Flow Reactors

For plug flow reactors all fluid elements take the same length of time to travel from the reactor inlet to the reactor outlet. This time is the mean residence time $\bar{t}$.

Consider the general case of a reaction accompanied by a volumetric expansion or contraction. The time necessary for a plug to travel from inlet to outlet of a tubular reactor is given by

$$\bar{t} = \int_0^{V_R} \frac{dV_R}{\mathcal{V}} \qquad (8.2.13)$$

where $\mathcal{V}$ is the volumetric flow rate. This parameter can be written in terms of the inlet reference volumetric flow rate $(\mathcal{V}_0)$, the fraction conversion, and the volumetric expansion parameter.

$$\mathcal{V} = \mathcal{V}_0(1 + \delta_A f_A) \qquad (8.2.14)$$

The increment in reactor volume can be written in terms of the fraction conversion using equation 8.2.6.

$$dV_R = \frac{F_{A0}\, df_A}{(-r_A)} \qquad (8.2.15)$$

Combining equations 8.2.13 to 8.2.15 gives

$$\bar{t} = \int_{f_A\text{ in}}^{f_A\text{ out}} \frac{F_{A0}\, df_A}{(-r_A)\,\mathcal{V}_0(1 + \delta_A f_A)} \qquad (8.2.16)$$

which can be rewritten in terms of the inlet reference concentration of the limiting reagent as

$$\bar{t} = C_{A0} \int_{f_A\text{ in}}^{f_A\text{ out}} \frac{df_A}{(-r_A)(1 + \delta_A f_A)} \qquad (8.2.17)$$

This equation is the basic relation for the mean residence time in a plug flow reactor with arbitrary reaction kinetics. Note that this expression differs from that for the space time (equation 8.2.9) by the inclusion of the term $(1 + \delta_A f_A)$ and that this term appears *inside* the integral sign. The two quantities become identical only when δ_A is zero (i.e., the fluid density is constant). The differences between the two characteristic times may be quite substantial, as we will see in Illustration 8.5. Of the two quantities, the reactor

space time is the more meaningful for reactor design purposes. A knowledge of mean residence time $\bar{t}$ does not permit us to determine the required reactor volume, but a knowledge of the space time τ readily leads to this quantity. The space time is an independent variable directly related to system parameters, whereas the mean residence time is a dependent variable found by integration of equation 8.2.17 or by tracer studies (see Chapter 11).

ILLUSTRATION 8.5 DETERMINATION OF MEAN RESIDENCE TIME IN A PLUG FLOW REACTOR UNDER ISOTHERMAL OPERATING CONDITIONS—VARIABLE DENSITY CASE

Consider the plug flow reactor used for the pyrolysis of methyl acetoxypropionate in Illustration 8.4.

$$C_6H_{10}O_4 \rightarrow C_2H_4O_2 + C_4H_6O_2$$
$$\quad A \qquad\qquad B \qquad\quad C$$

For isothermal operation at 500 °C and 5 atm, it was shown that the space time required to achieve 90% conversion was 29.9 sec. Compare this value with the mean residence time of the material in the plug flow reactor.

Solution

For first-order kinetics in a system where δ_A is nonzero, the reaction rate expression is of the form

$$-r_A = kC_{A0} \frac{(1 - f_A)}{(1 + \delta_A f_A)}$$

In the present case $\delta_A = 1$. Hence the equation for the mean residence time (equation 8.2.17) becomes

$$\bar{t} = C_{A0} \int_0^{0.9} \frac{df_A}{\left[kC_{A0} \dfrac{(1 - f_A)}{(1 + \delta_A f_A)} \right](1 + \delta_A f_A)}$$

Simplification gives

$$\bar{t} = \frac{1}{k} \int_0^{0.9} \frac{df_A}{1 - f_A} = \left. \frac{-\ln(1 - f_A)}{k} \right|_0^{0.9}$$

Substitution of numerical values gives

$$\bar{t} = \frac{1}{0.124}\left[-\ell n(1 - 0.9)\right] = 18.6 \text{ sec}$$

This value is considerably less than the reactor space time and differs from it by 38%.

8.2.3 Series-Parallel Combinations of Tubular Reactors

In order to achieve increases in production capacity or to obtain higher conversion levels, it may be necessary to provide additional reactor volume through the use of various series-parallel combinations of reactors.

Consider j plug flow reactors connected in series and let $f_1, f_2, f_3, \ldots f_i, \ldots, f_j$ represent the fraction conversion of the limiting reagent, leaving reactors $1, 2, 3, \ldots i, \ldots, j$. For each of the reactors considered above, the appropriate design equation is 8.2.7. For reactor i,

$$V_{Ri} = F_{A0} \int_{f_{i-1}}^{f_i} \frac{df_A}{(-r_A)} \qquad (8.2.18)$$

The total reactor volume is obtained by summing the individual reactor volumes.

of a reactor network must be proportioned such that equal increments in conversion occur across each leg. For this case, too, the network acts as if it were a single plug flow reactor with a volume equal to the sum of the constituent reactor volumes. Thus for any series-parallel combination of plug flow reactors, one can treat the whole system as a single plug flow reactor with a volume equal to the total volume of the individual reactors, provided that the fluid streams are distributed in a manner such that streams that combine have the same composition. Hence, for any units in parallel, the space times or V_R/F_{A0} must be identical. Other flow distributions would be less efficient.

8.3 THE CONTINUOUS FLOW STIRRED TANK REACTOR (CSTR)

Continuous flow stirred tank reactors are widely used in the chemical process industry. Although individual reactors may be used, it is usually preferable to employ a battery of such reactors connected in series. The effectiveness of such batteries depends on the number of reactors used, the sizes of the component reactors, and the efficiency of mixing within each stage.

$$V_{R,\text{ total}} = \sum_{i=1}^{j} V_{Ri} = F_{A0}\left[\int_{f_0}^{f_1} \frac{df_A}{-r_A} + \int_{f_1}^{f_2} \frac{df_A}{-r_A} + \cdots \int_{f_{j-1}}^{f_j} \frac{df_A}{(-r_A)}\right] \qquad (8.2.19)$$

From the principles of calculus, the quantity in brackets can be rewritten as a single integral. Hence

$$V_{R,\text{ total}} = F_{A0} \int_{f_0}^{f_j} \frac{df_A}{(-r_A)} \qquad (8.2.20)$$

Thus j plug flow reactors in series with a total volume $V_{R,\text{ total}}$ give the same conversion as a single reactor of volume $V_{R,\text{ total}}$.

When plug flow reactors are connected in parallel, the most efficient utilization of the total reactor volume occurs when mixing of streams of differing compositions does not occur. Consequently the feed rates to different parallel legs

Continuous flow stirred tank reactors are normally just what the name implies—tanks into which reactants flow and from which a product stream is removed on a continuous basis. CFSTR, CSTR, C-star, and back-mix reactor are only a few of the names applied to the idealized stirred tank flow reactor. We will use the letters CSTR as a shorthand notation in this textbook. The virtues of a stirred tank reactor lie in its simplicity of construction and the relative ease with which it may be controlled. These reactors are used primarily for carrying out liquid phase reactions in the organic chemicals industry, particularly for systems that are characterized by relatively slow reaction rates. If it is imperative that a gas phase reaction be carried out under efficient mixing conditions similar to those found in a stirred tank reactor, one may employ a tubular reactor containing a recycle loop. At sufficiently high recycle rates, such systems approximate CSTR behavior. This section is concerned with the development of design equations that are appropriate for use with the idealized stirred tank model.

8.3.1 Individual Stirred Tank Reactors

8.3.1.1 Basic Assumptions and Design Relationships. The most important feature of the CSTR is its mixing characteristics. The idealized model of reactor performance presumes that the reactor contents are perfectly mixed so that the system properties are uniform throughout. The effluent composition and temperature are thus identical with those of the reactor contents. This feature greatly simplifies the analysis of stirred tank reactors *vis à vis* tubular reactors for both isothermal and nonisothermal operation. It is not difficult to obtain a good approximation to CSTR behavior, provided that the fluid phase is not too viscous. The approximation to CSTR behavior is valid if the time necessary to disperse an entering element of fluid (e.g., a shot of dye or radioactive tracer) uniformly throughout the tank is very much shorter than the average residence time in the tank.

Unlike the situation in a plug flow reactor, the various fluid elements mix with one another in a CSTR. In the limit of perfect mixing, a tracer molecule that enters at the reactor inlet has equal probability of being anywhere in the vessel after an infinitesimally small time increment. Thus all fluid elements in the reactor have equal probability of leaving in the next time increment. Consequently there will be a broad distribution of residence times for various tracer molecules. The character of the distribution is discussed in Section 11.1. Because some of the

molecules have short residence times, there is a rapid response at the reactor outlet to changes in the reactor feed stream. This characteristic facilitates automatic control of the reactor.

Because the mixing process makes the properties of the entire reactor contents equal to those of the effluent stream, there will be a step change in fluid composition and temperature at the point where the feedstream enters the reactor. In the present chapter we will restrict the discussion to cases where there is no temperature change on entrance to the reactor unless otherwise noted. For these isothermal CSTR's the drop in reactant concentration as the fluid enters the reactor implies that in the vast majority of cases the volume average reaction rate will be low by comparison to that in an isothermal plug flow reactor. Consequently, when both operate at the same temperature, a CSTR will have to be significantly larger than a PFR in order to effect the same composition change in a given amount of fluid. (Exceptions occur in the case of autocatalytic reactions.) However, by using a battery of stirred tanks in series, differences in total volume requirements can be significantly reduced. Moreover, because of the simplicity of their construction, stirred tank reactors normally cost much less per unit of volume than tubular reactors. Thus, in many cases, it is more economic to employ a large stirred tank reactor or a battery of such reactors than it is to use a tubular reactor. Size comparisons of these two types can be treated quantitatively, and we will return to this subject in Sections 8.3.1.3 and 8.3.2.3. First, however, we must develop the basic design equations.

Consider the schematic representation of a continuous flow stirred tank reactor shown in Figure 8.5. The starting point for the development of the fundamental design equation is again a generalized material balance on a reactant species. For the steady-state case the accumulation term in equation 8.0.1 is zero. Furthermore, since conditions are uniform throughout the reactor volume, the material balance may be

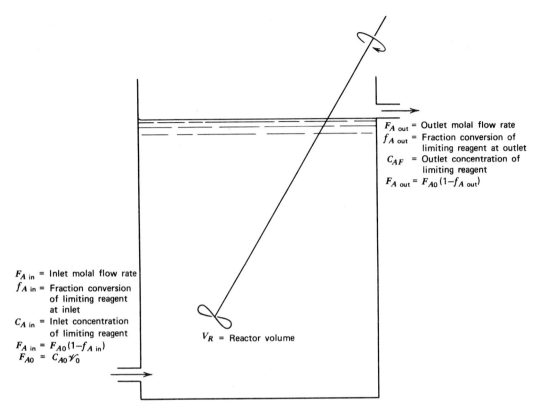

Figure 8.5
Schematic representation of CSTR indicating process variables.

written over the entire reactor. Hence,

$$\begin{pmatrix} \text{Rate of flow of} \\ \text{reactant into reactor} \end{pmatrix} = \begin{pmatrix} \text{Rate of flow of} \\ \text{reactant out of} \\ \text{reactor} \end{pmatrix} + \begin{pmatrix} \text{Rate of disappearance} \\ \text{of reactant by reaction} \\ \text{in the reactor} \end{pmatrix} \qquad (8.3.1)$$

In terms of the symbols indicated in Figure 8.5,

$$F_{A\,in} = F_{A\,out} + (-r_{AF})V_R \qquad (8.3.2)$$

where we again emphasize that the appropriate volume is that physically occupied by the reacting fluid. The quantity $(-r_{AF})$ is the rate of disappearance of reactant A evaluated at *reactor outlet conditions*.

The last equation may be rewritten in terms of the fraction conversion as

$$F_{A0}(1 - f_{A\,in}) = F_{A0}(1 - f_{A\,out}) + (-r_{AF})V_R \qquad (8.3.3)$$

where F_{A0} is again the molal flow rate corresponding to zero conversion. Rearrangement gives

$$\frac{V_R}{F_{A0}} = \frac{f_{A\,out} - f_{A\,in}}{(-r_{AF})} \qquad (8.3.4)$$

Equation 8.3.4 is an extremely useful expression relating in a simple manner the reactor volume, the molal flow rate at zero conversion, the change in fraction conversion accomplished in the reactor, and the reaction rate. A knowledge of any three of these quantities permits the fourth to be calculated directly. For reactor design purposes the two problems of primary interest can be readily solved using this equation.

1. The size reactor needed to perform a specified task under specified operating conditions may be determined.
2. For a reactor of a given size one may determine either the conversion achieved for a specified flow rate and temperature or the quantity of material that can be processed to a given degree of conversion at a specified temperature.

Equation 8.3.4 may also be used in the analysis of kinetic data taken in laboratory scale stirred tank reactors. One may directly determine the reaction rate from a knowledge of the reactor volume, flow rate through the reactor, and stream compositions. The fact that one may determine the rate directly and without integration makes stirred tank reactors particularly attractive for use in studies of reactions with complex rate expressions (e.g., enzymatic or heterogeneous catalytic reactions) or of systems in which multiple reactions take place.

Equation 8.3.4 is completely general and independent of whether the reaction occurs at constant density ($\delta_A = 0$) and of whether the feed stream and the reactor contents have identical temperatures. The effective reactor volume is independent of the particular geometry giving rise to this volume. All that is required is that the contents be well mixed.

If the molal flow rate at zero fraction conversion is written in terms of the product of a reference volumetric flow rate V_0 and a corresponding concentration (C_{A0}),

$$\frac{V_R}{V_0} = \frac{C_{A0}(f_{A\,out} - f_{A\,in})}{(-r_{AF})} \qquad (8.3.5)$$

In terms of the reactor space time

$$\tau = \frac{C_{A0}(f_{A\,out} - f_{A\,in})}{(-r_{AF})} = \frac{C_{A0}\int_{f_{A\,in}}^{f_{A\,out}} df_A}{(-r_{AF})}$$
$$(8.3.6)$$

This equation differs from that for the plug flow reactor (8.2.9) in that for a CSTR the rate is evaluated at *effluent conditions* and thus appears *outside* the integral.

It is particularly convenient to choose the reference conditions at which the volumetric flow rate is measured as the temperature and pressure prevailing at the reactor inlet, because this choice leads to a convenient physical interpretation of the parameters V_0 and C_{A0} and, in many cases, one finds that the latter quantity cancels a similar term appearing in the reaction rate expression. Unless otherwise specified, this choice of reference conditions is used throughout the remainder of this text. For constant density systems and this choice of reference conditions, the space time τ then becomes numerically equal to the average residence time of the fluid in the reactor.

Since one is almost always concerned with liquid phase reactions when dealing with stirred tank reactors, the assumption of constant fluid density is usually appropriate. In this case equation 8.3.6 can be written as

$$\tau = \frac{\int_{C_{A\,in}}^{C_{A\,out}} dC_A}{(r_{AF})} = \frac{C_{A\,in} - C_{A\,out}}{(-r_{AF})}$$

(for constant density systems only)

$$(8.3.7)$$

We now wish to consider some examples that indicate how to employ the above equations in reactor design analyses.

ILLUSTRATION 8.6 DETERMINATION OF REQUIRED CSTR VOLUME UNDER ISOTHERMAL OPERATING CONDITIONS—LIQUID PHASE REACTION

Consider the Diels-Alder reaction between 1,3-butadiene (B) and methyl acrylate (M) discussed in Illustration 8.3.

$$B + M \rightarrow C$$

If the operating conditions used in that illustration are again employed, determine the volume of a single continuous stirred tank reactor which will give 40% conversion of the butadiene when the liquid flow rate is 0.500 $m^3/ksec$.

Solution

For the conditions cited the reaction rate expression is of the form

$$-r_B = kC_{B0}(1 - f_B)C_{AlCl_3, 0} \tag{A}$$

with

$$k = 1.15 \times 10^{-3} \ m^3/mole \cdot ksec$$
$$C_{B0} = 96.5 \ moles/m^3$$
$$C_{AlCl_3, 0} = 6.63 \ moles/m^3$$

Equation 8.3.6 may be used as the basic design relationship.

$$\tau = \frac{C_{B0} \int_0^{f_B} df_B}{(-r_{BF})} = \frac{C_{B0} f_B}{(-r_{BF})} \tag{B}$$

Combining equations A and B gives

$$\tau = \frac{C_{B0} f_B}{kC_{B0}(1 - f_B)C_{AlCl_3, 0}} = \frac{f_B}{k(1 - f_B)C_{AlCl_3, 0}}$$

Substitution of numerical values gives

$$\tau = \frac{0.40}{(1.15 \times 10^{-3})(1 - 0.4)(6.63)}$$

$$= 87.4 \ ksec = 24.3 \ hr$$

From the definition of the space time and the inlet volumetric flow rate,

$$V_R = \tau \mathcal{V}_0 = (87.4)(0.500) = 43.7 \ m^3$$

This volume is appreciably larger than the volume of plug flow reactor calculated in Illustration 8.3 for the same reaction conditions and fraction conversion. However, the cost of such a reactor would be considerably less than the cost of a tubular reactor of the size determined in Illustration 8.3.

8.3.1.2 Mean Residence Time in Stirred Tank Reactors.

Stirred tank reactors differ from plug flow reactors in that not all fluid elements remain in the CSTR for the same length of time. The characteristics of the residence time distribution function are treated in Chapter 11. In this subsection we consider only the problem of determining the average residence time of a fluid element in an ideal CSTR. This problem is considerably simplified by the fact that the fluid properties are uniform throughout the reactor and equal to those prevailing at the exit. Thus the mean residence time in an ideal CSTR is given by

$$\bar{t} = \frac{V_R}{\mathcal{V}_F} \tag{8.3.8}$$

where the volumetric flow rate is evaluated at effluent conditions.

The effluent volumetric flow rate is also given by

$$\mathcal{V}_F = \mathcal{V}_0(1 + \delta_A f_{AF}) \tag{8.3.9}$$

where $\mathcal{V}_0$ is the volumetric flow rate evaluated at a composition corresponding to zero fraction conversion. Combining equations 8.3.8 and 8.3.9 gives

$$\bar{t} = \frac{V_R}{\mathcal{V}_0(1 + \delta_A f_{AF})} = \frac{\tau}{1 + \delta_A f_{AF}} \tag{8.3.10}$$

Unlike the situation in the PFR, there is always a simple relationship between the mean residence time and the reactor space time for a CSTR. Since one normally associates a liquid feed stream with these reactors, volumetric expansion effects are usually negligible ($\delta_A = 0$).

In this situation the mean residence time and the reactor space time become identical.

8.3.1.3 Relative Size Requirements for an Individual Continuous Stirred Tank Reactor and a Plug Flow Reactor.

In the development of the final reactor design for a proposed production requirement, the chemical engineer must consider a variety of reactor types and modes of operation. Several factors must be considered in the development of the final design, some of which may be only peripherally related to the kinetics of the reaction in question. Many of these factors are implicit in the questions posed in Section 8.0.1. Since a variety of operating conditions, modes of operation, and reactor types can be used to accomplish a specified task, it is not possible to generate a simple logical procedure that can be followed to arrive at a truly optimum design. A knowledge of the performance characteristics of individual ideal reactors (and combinations thereof) and sound engineering judgment based on previous design experience are both useful in arriving at a workable design. The nonanalytical reasoning that must perforce enter into design calculations is beyond the scope of this textbook. This capability comes only with the accumulation of experience and interaction with other individuals who have long been regularly engaged in the practice of reactor design. We can, however, indicate the quantitative considerations that have an important bearing on the economics of the proposed design. The choice of the reactor network employed to carry out the desired conversion will play an important role in that it will specify the size of the units needed and the distribution of products emerging from the reactor. In this section we turn our attention to the problem of determining relative size requirements for a single ideal CSTR and a PFR where both reactors operate isothermally at the same temperature. The analysis applies to systems in which only one reaction occurs to a significant extent.

For most of the commonly encountered types of reaction rate expressions the rate decreases monotonically with increasing fraction conversion. The fact that a step change in composition occurs as a fluid enters an ideal CSTR

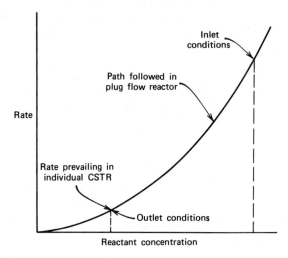

Figure 8.6a

Reaction rate versus reactant concentration plot for typical reactions—single ideal reactors.

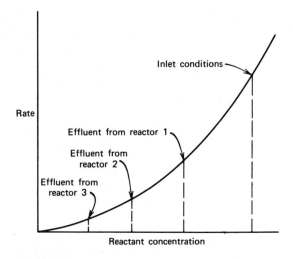

Figure 8.6b

Reaction rate versus reactant concentration plot for typical reactions—three reactors in series.

implies that for these cases the reaction rate will be much smaller in this type of reactor than it would be in a tubular reactor being used to accomplish the same composition change. Consequently, the CSTR must be significantly larger than the PFR. These considerations are evident from a brief consideration of Figure 8.6a. In a plug flow reactor one moves from right to left along the rate curve as he or she proceeds from inlet to outlet. In a CSTR, on the other hand, the reaction rate is constant throughout the reactor and equal to that prevailing at the outlet. Except in the case of autocatalytic reactions (see Section 9.4), this rate corresponds to the lowest point on the PFR rate curve and implies that a larger volume will be required to accomplish the same composition change between inlet and outlet streams.

It is possible to reduce the disparity in reactor volumes by using several tanks in series so that one obtains stepwise changes in composition as one proceeds from tank to tank. This situation is depicted in Figure 8.6b for the case of three CSTR's in series. In this case it is readily apparent that each of the CSTR's would have a larger volume than that of a PFR necessary to effect the same composition change, so that the total volume of the three CSTR's would exceed that of a single PFR used to effect the same overall change in composition. (We have shown earlier that the three PFR's in series would be equivalent to a single PFR with a volume equal to the sum of those of the individual PFR's.) As one increases the number of CSTR's in series, the disparities in total volume become less and, in the limit, as the number of CSTR's approaches infinity the battery will approach PFR behavior. For now, however, we wish to limit our considerations to a comparison of the relative sizes of a single CSTR and a PFR under isothermal operating conditions.

In order to be able to compare relative reactor sizes one needs a knowledge of the form of the reaction rate expression in either graphical or analytical terms. In Section 8.2.1 we showed that

the area under a plot of the reciprocal reaction rate $(-1/r_A)$ versus fraction conversion was equal to the ratio V_R/F_{A0} for a plug flow reactor. In the case of a CSTR equation 8.3.4 indicates that on a similar plot, the quantity V_R/F_{A0} is equal to the rectangular area shown in Figure 8.7. Thus, the ratio of the rectangular area to the area under the curve is equal to the ratio of reactor volumes if identical molal flow rates and conversion increments are employed. For any rate expression that decreases monotonically with increasing fraction conversion, the CSTR will always require a larger reactor volume than the corresponding PFR used to accomplish the same task.

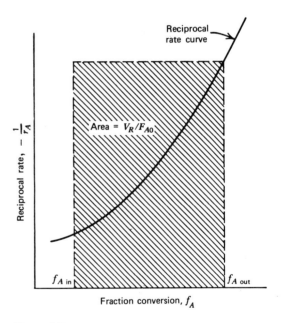

Figure 8.7

Reciprocal rate versus fraction conversion plot for determination of V_R/F_{A0} for CSTR.

If one has a knowledge of the analytical form of the reaction rate expression, equations 8.2.7 and 8.3.4 may be used to determine the relative reactor volumes required. In order to indicate the utility of these equations, let us consider the general class of reactions that follow simple nth-order kinetics with n normally lying between zero and three. Following Levenspiel's analysis (11), we will presume that the inlet fraction conversion is zero and that the volumetric expansion parameter (δ_A) has some arbitrary value. Thus

$$-r_A = \frac{kC_{A0}^n(1 - f_A)^n}{(1 + \delta_A f_A)^n} \qquad (8.3.11)$$

For these conditions the general design equation for a plug flow reactor (8.2.7) becomes

$$\left(\frac{V_R}{F_{A0}}\right)_{PFR} = \int_0^{f_A} \frac{(1 + \delta_A f_A)^n \, df_A}{kC_{A0}^n(1 - f_A)^n} \qquad (8.3.12)$$

while that for a single CSTR (equation 8.3.4) becomes

$$\left(\frac{V_R}{F_{A0}}\right)_{CSTR} = \frac{f_A}{kC_{A0}^n}\left(\frac{1 + \delta_A f_A}{1 - f_A}\right)^n \qquad (8.3.13)$$

Dividing equation 8.3.13 by equation 8.3.12 gives

$$\frac{\left(\dfrac{V_R C_{A0}^n}{F_{A0}}\right)_{CSTR}}{\left(\dfrac{V_R C_{A0}^n}{F_{A0}}\right)_{PFR}} = \frac{f_A \left(\dfrac{1 + \delta_A f_A}{1 - f_A}\right)^n}{\displaystyle\int_0^{f_A} \frac{(1 + \delta_A f_A)^n \, df_A}{(1 - f_A)^n}}$$

$$(8.3.14)$$

Levenspiel (11) has evaluated the right side of equation 8.3.14 for various values of n and δ_A. His results are presented in graphical form in Figure 8.8. For identical feed concentrations (C_{A0}) and molal flow rates (F_{A0}) the ordinate of the figure indicates the volume ratio required for a specified conversion level.

There are several useful generalizations which may be gleaned from a thorough study of Figure 8.8. They include the following.

1. If the reaction order is positive, the CSTR is larger than the PFR at all conversion levels.
2. The higher the fraction conversion involved, the greater the disparity between the sizes of the CSTR and the PFR.

3. The higher the order of the reaction, the greater the ratio of sizes at a fixed conversion level. For zero-order reactions reactor size is independent of reactor type.
4. Variations in fluid density on reaction can have significant effects on the size ratio, but the effects are secondary when compared to the variations in reaction order. For positive values of the expansion parameter δ_A, the volume ratio is increased, for negative values of δ_A, the volume ratio decreases. However, the fact that in practice CSTR's are used only for liquid phase reactions makes this point academic.

The larger volume requirement of the CSTR does not necessarily imply extra capital costs, especially for reactions that occur at ambient pressure. However, the fact that the required CSTR volume increases rapidly at high conversion levels leads to some interesting optimization problems in reactor design. The chemical engineer must find the point at which he obtains an economic trade-off between the high fraction conversion obtained in a large reactor versus the low conversions in a small reactor. In the first situation the equipment costs will be high and the product separation costs and raw material costs low; in the second case equipment costs will be low and the other costs high. The optimization problem may be further complicated by allowing the number of CSTR's employed to vary, as we will see in Section 8.3.2.3.

In Illustrations 8.3 and 8.6 we considered the reactor size requirements for the Diels-Alder reaction between 1,4-butadiene and methyl acrylate. For the conditions cited the reaction may be considered as a pseudo first-order reaction with $\delta_A = 0$. At a fraction conversion of 0.40 the required PFR volume was 33.5 m^3, while the required CSTR volume was 43.7 m^3. The ratio of these volumes is 1.30. From Figure 8.8 the ratio is seen to be identical with this value. Thus this figure or equation 8.3.14 can be used in solving a number of problems involving the

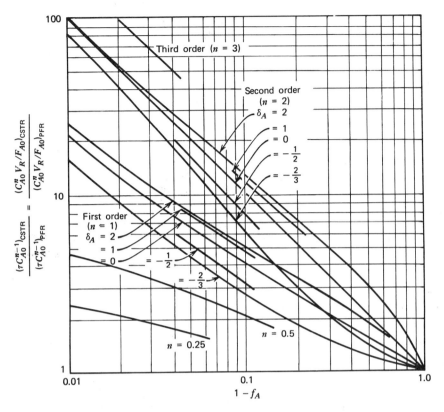

Figure 8.8
Comparison of performance of single CSTR and plug flow reactor for the nth-order reactions

$$A \rightarrow \text{products}, \quad -r_A = KC_A^n$$

The ordinate becomes the volume ratio V_{CSTR}/V_{PFR} or the space time ratio τ_{CSTR}/τ_{PFR} if the same quantities of identical feed are used. (Adapted from *Chemical Reaction Engineering*, **Second Edition, by O. Levenspiel. Copyright** © **1972. Reprinted by permission of John Wiley and Sons, Inc.)**

comparison of the performance of an individual CSTR with that of a plug flow reactor for systems that obey nth-order kinetics. Mixed second- and third-order reactions can also be handled in terms of the nth-order model if stoichiometric ratios of reactants are used. The techniques employed in solving problems using the figure will be discussed in Section 8.3.2.3 after a corresponding figure has been developed for use with multiple identical CSTR's connected in series.

8.3.1.4 CSTR Performance Under Nonsteady-State Conditions. During start-up and shutdown periods and during shifts from one steady-state operating condition to a second, the design equations developed in Section 8.3.1.1 are no longer appropriate. In these cases the starting point for the analysis must be the generalized material and energy balance equations containing accumulation terms (equations 8.0.1 and 8.0.3). In order to indicate the general approach to the nonsteady-state analysis of reactors, we wish to consider briefly the relations that govern the transient behavior of an individual CSTR operating *under isothermal conditions*. In some cases it is possible to obtain easily analytical solutions describing the approach to the steady state because of the uniform composition of the reactor contents. This situation is in distinct contrast to that prevailing in a plug flow reactor where one must invariably resort to numerical solutions of the transient material balance relations in order to describe the approach to steady-state conditions.

For nonsteady-state operating conditions the generalized material balance on reactant A is as follows.

$$\text{Input} = \text{Output} + \begin{array}{c}\text{Disappearance}\\\text{by reaction}\end{array} + \text{Accumulation}$$

$$F_{A\,\text{in}} = F_{A\,\text{out}} + (-r_{AF})V_R + \frac{dN_A}{dt} \qquad (8.3.15)$$

where the reaction rate is evaluated at the conditions prevailing at the reactor outlet and N_A is the total number of moles of species A in the reactor at time t. N_A may change as a result of changes in the volume occupied by the reacting fluid or as a result of changes in the composition of the fluid. Since

$$N_A = C_{AF}V_R \qquad (8.3.16)$$

then

$$\frac{dN_A}{dt} = V_R\frac{dC_{AF}}{dt} + C_{AF}\frac{dV_R}{dt} \qquad (8.3.17)$$

If the molal flow rates are written as the product of a concentration and a volumetric flow rate, and if equation 8.3.17 is combined with equation 8.3.15,

$$C_{A\,\text{in}}\mathcal{V}_\text{in} = C_{AF}\mathcal{V}_\text{out} + (-r_{AF})V_R$$

$$+ V_R\frac{dC_{AF}}{dt} + C_{AF}\frac{dV_R}{dt} \qquad (8.3.18)$$

For liquid phase reactions where $\delta_A = 0$, the following expression is appropriate.

$$\frac{dV_R}{dt} = \mathcal{V}_\text{in} - \mathcal{V}_\text{out} \qquad (8.3.19)$$

Combining equations 8.3.18 and 8.3.19 gives

$$\frac{dC_{AF}}{dt} - r_{AF} = (C_{A\,\text{in}} - C_{AF})\frac{\mathcal{V}_\text{in}}{V_R} \qquad (8.3.20)$$

There are several interesting forms of equation 8.3.20 that correspond to various limiting conditions. For example, if both V_R and $\mathcal{V}_\text{in}$ are time invariant, we have the situation corresponding to a shift from one steady-state operating condition to a second, and the quantity $V_R/\mathcal{V}_\text{in}$ is just the reactor space time τ. Hence, for this case,

$$\frac{dC_{AF}}{dt} - r_{AF} = \frac{C_{A\,\text{in}} - C_{AF}}{\tau} \qquad (8.3.21)$$

It is readily apparent that equation 8.3.21 reduces to the basic design equation (equation 8.3.7) when steady-state conditions prevail. Under the presumptions that $C_{A\,\text{in}}$ undergoes a step change at time zero and that the system is isothermal, equation 8.3.21 has been solved for various reaction rate expressions. In the case of first-order reactions, solutions are available for both multiple identical CSTR's in series and individual CSTR's (12). In the case of a first-order irreversible reaction in a single CSTR, equation 8.3.21 becomes

$$\frac{dC_{AF}}{dt} = \frac{C_{A\,\text{in}}}{\tau} - \frac{C_{AF}}{\tau} - kC_{AF} \qquad (8.3.22)$$

This equation is an ordinary linear differential

equation with constant coefficients if $C_{A \text{ in}}$ and the rate constant are time independent. In this case the solution may be obtained by separation of variables and integration.

$$\int_{C^*}^{C_{AF}} \frac{dC_{AF}}{\dfrac{C_{A \text{ in}}}{\tau} - \left(k + \dfrac{1}{\tau}\right)C_{AF}} = \int_0^t dt \quad (8.3.23)$$

where C^* is the reactant concentration in the tank at time zero. Integration gives

$$\frac{\ln\left[\dfrac{\dfrac{C_{A \text{ in}}}{\tau} - \left(k + \dfrac{1}{\tau}\right)C^*}{\dfrac{C_{A \text{ in}}}{\tau} - \left(k + \dfrac{1}{\tau}\right)C_{AF}}\right]}{k + \dfrac{1}{\tau}} = t \quad (8.3.24)$$

or

$$C_{AF} = \frac{C_{A \text{ in}}}{k\tau + 1} + \left(C^* - \frac{C_{A \text{ in}}}{k\tau + 1}\right) e^{-(k\tau + 1)t/\tau}$$

$$(8.3.25)$$

where the first term on the right is just the steady-state effluent concentration.

8.3.2 Stirred Tank Reactors in Series

In order to reduce the disparities in volume or space time requirements between an individual CSTR and a plug flow reactor, batteries or cascades of stirred tank reactors are employed. These reactor networks consist of a number of stirred tank reactors connected in series with the effluent from one reactor serving as the input to the next. Although the concentration is uniform within any one reactor, there is a progressive decrease in reactant concentration as one moves from the initial tank to the final tank in the cascade. In effect one has stepwise variations in composition as he moves from one CSTR to another. Figure 8.9 illustrates the stepwise variations typical of reactor cascades for different numbers of CSTR's in series. In the general nonisothermal case one will also en-

counter stepwise variations in temperature as one moves from one reactor to another in the cascade.

Each of the individual CSTR's that make up the cascade can be analyzed using the techniques and concepts developed in Section 8.3.1. The present section indicates how one may manipulate the key relations developed earlier to obtain equations that simplify the analysis of a cascade of ideal CSTR's.

We begin by indicating a few generalizations that are relevant to the treatment of batteries of stirred tank reactors. Consider the cascade of ideal CSTR's shown in Figure 8.10. For any individual reactor denoted by the subscript i the basic design equation developed earlier as equation 8.3.4 is appropriate:

$$\frac{V_{Ri}}{F_{A0}} = \frac{f_{A \text{ out}} - f_{A \text{ in}}}{-r_{AF}} = \frac{f_i - f_{i-1}}{-r_{Ai}} \quad (8.3.26)$$

where the subscripts used are as indicated in the figure and $(-r_{Ai})$ is the rate of disappearance of species A in reactor i. We wish to emphasize that F_{A0} is the molal flow rate of reactant A under conditions that would correspond to zero fraction conversion. We should also note that the conditions within any individual reactor are not influenced by what happens in reactors downstream. The conditions of the inlet stream and those prevailing within the reactor itself are the only variables that influence reactor performance under either steady-state or transient conditions.

If desired, the last equation can also be written in terms of the reactor space time for the ith reactor as

$$\tau_i = \frac{V_{Ri}}{\mathscr{V}_0} = \frac{C_{A0}(f_i - f_{i-1})}{-r_{Ai}} \quad (8.3.27)$$

where C_{A0} is the reactant concentration corresponding to zero conversion at the inlet temperature and pressure.

Equations 8.3.26 and 8.3.27 are generally applicable to all types of CSTR cascades. If one

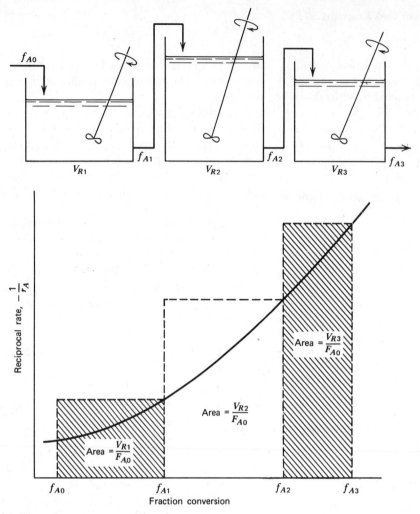

Figure 8.9
Schematic representation of reciprocal rate curve for cascade of three arbitrary size CSTR's.

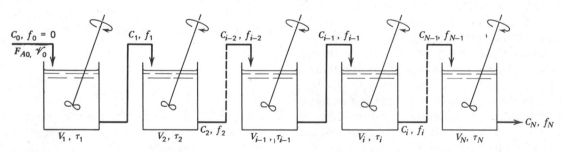

Figure 8.10
Notation for a cascade of N CSTR's in series.

recognizes that the use of such cascades is almost invariably restricted to liquid systems and that in such systems density changes caused by reaction or thermal effects are usually quite small, then additional relations or simplifications can be developed from these starting equations. In particular this situation implies that at steady state the volumetric flow rate between stages is substantially constant. It also implies that for each reactor, $\bar{\tau}_i = \tau_i$, and that the following relation between concentration and fraction conversion is appropriate:

$$C_{Ai} = C_{A0}(1 - f_i) \qquad (8.3.28)$$

where C_{Ai} is the concentration of reactant A leaving the ith reactor. We will make the assumption of constant density throughout the remainder of Section 8.3.2. We should also recall that the development of equation 8.3.26 presumed steady-state operation and thus this assumption is also made throughout Sections 8.3.2.1 to 8.3.2.3.

In view of equation 8.3.28 and these assumptions, equation 8.3.27 can be rewritten as

$$\tau_i = \frac{V_{Ri}}{\dot{V}_0} = \frac{\overset{in}{C_{A(i-1)}} - \overset{out}{C_{Ai}}}{-r_{Ai}} \qquad (8.3.29)$$

This equation is the one that is most appropriate for our use in the next two subsections, in that we will find it convenient to work in terms of reactant concentrations instead of conversion levels.

8.3.2.1 Graphical Approach to the Analysis of Batteries of Stirred Tank Reactors Operating at Steady State.
Even in reaction systems where it is not possible to determine the algebraic form of the reaction rate expression, it is often possible to obtain kinetic data that permit one to express graphically the rate as a function of the concentration of one reactant. Laboratory scale CSTR's are particularly appropriate for generating this type of kinetic data for complex reaction

systems. In this section we presume a knowledge of the reaction rate expression as a function of the concentration of reactant A, at least in graphical terms. That is, we presume

$$-r_A = g(C_A) \qquad (8.3.30)$$

where $g(C_A)$ denotes some arbitrary function. For present purposes we assume that *the CSTR's all operate at the same temperature* and that $g(C_A)$ is known at this temperature.

Equation 8.3.29 may be written in the form

$$-r_{Ai} = \frac{C_{A(i-1)} - C_{Ai}}{\tau_i} \qquad (8.3.31)$$

In graphical terms this equation indicates that a plot of $(-r_{Ai})$ versus C_{Ai} is a straight line with a slope $-1/\tau_i$ that cuts the abscissa at $C_{A(i-1)}$. An analysis of the reactor design problem involves the simultaneous solution of equation 8.3.30 and several equations of the form of equation 8.3.31 (one for each reactor). These equations are the basis for the solution of the two types of problems with which the reactor designer is most often faced in the analysis of batteries of ideal CSTR's.

1. What is the final effluent composition from a network of such reactors? (We might also require the composition of the stream leaving each reactor.)
2. What combination of ideal CSTR's is best suited to achieving a specified conversion?

Each of these problems will be considered in turn. Consider the three ideal CSTR's shown in Figure 8.11. The characteristic space times of these reactors may differ widely. Note that the direction of flow is from right to left. The first step in the analysis requires the preparation of a plot of reaction rate versus reactant concentration based on experimental data (i.e., the generation of a graphical representation of equation 8.3.30). It is presented as curve I in Figure 8.11.

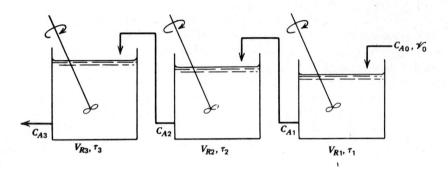

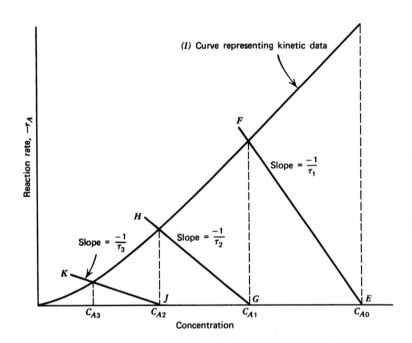

Figure 8.11
Plots used in the graphical analysis of cascades of ideal CSTR's.

Now, for the first reactor, equation 8.3.31 becomes

$$-r_1 = \frac{C_{A0} - C_{A1}}{\tau_1} \qquad (8.3.32)$$

For a specified inlet concentration this equation indicates that a plot of $-r_1$ versus C_{A1} is a straight line with slope $(-1/\tau_1)$ that passes through the point where the ordinate is zero and the abscissa is C_{A0}. However, only the point of intersection of the straight line with the curve representing equation 8.3.30 has physical meaning. It represents the conditions that must prevail in the first reactor. Hence the solution to the first part of our problem consists of drawing straight line EF through point E with slope $(-1/\tau_1)$. C_{A1} is the point of intersection of this straight line with curve I, as shown in Figure 8.11.

Now that C_{A1} is known, it is evident that a similar process can be used to find C_{A2} because, in this case, equation 8.3.31 becomes

$$-r_2 = \frac{C_{A1} - C_{A2}}{\tau_2} \qquad (8.3.33)$$

Thus we may construct straight line GH in Figure 8.11, by drawing a line of slope $(-1/\tau_2)$ through the point with an ordinate of zero and an abscissa of C_{A1}. The intersection of this line with curve I gives us the effluent concentration from the second reactor. This same procedure can be repeated for any other reactors that may be part of the cascade. The straight line JK was constructed in this fashion for the present case.

Let us turn now to the second of the problems mentioned earlier—determination of the com-

bination of CSTR's that is best suited to achieving a specified conversion level. We will begin by considering the case of two arbitrary size ideal CSTR's in series operating under isothermal conditions and then briefly treat the problem of using multiple identical CSTR's in series. Consider the two cascade configurations shown in Figure 8.12, taken from Levenspiel (13). For the first reactor, equation 8.3.26 becomes

$$\frac{V_{R1}}{F_{A0}} = \frac{f_1 - f_0}{(-r_{A1})} \qquad (8.3.34)$$

while, for the second, it becomes

$$\frac{V_{R2}}{F_{A0}} = \frac{f_2 - f_1}{(-r_{A2})} \qquad (8.3.35)$$

The relations indicated by these two equations are shown graphically in Figure 8.12 for two alternative configurations. In both cases the cascade operates between the same initial and

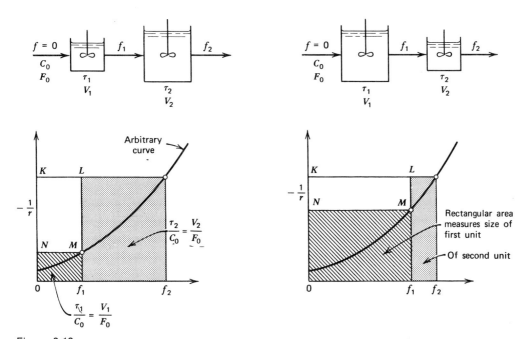

Figure 8.12

Graphical representation of variables for two CSTR's. (Adapted from *Chemical Reaction Engineering*, **Second Edition, by O. Levenspiel. Copyright © 1972. Reprinted by permission of John Wiley and Sons, Inc.)**

final conversion levels. The same arbitrary reaction rate expression is appropriate in both instances. As the composition of the effluent from the first reactor is changed, the relative size requirements for the two individual reactors also change, as does the total volume required. The size ratio is determined by the ratio of the two shaded areas and the total volume by the sum of these areas. The total reactor volume is minimized when rectangle $KLMN$ is made as large as possible so the problem of selecting the optimum sizes of the two reactor components in this sense reduces to that of selecting point f_1 so that the area of rectangle $KLMN$ is maximized. Levenspiel (13) has considered this general problem. We need consider only the results of his analysis.

For reaction rate expressions of the nth-order form it can be shown that there is always one and only one point that minimizes the total volume when $n > 0$. This situation is obtained when the intermediate fraction conversion f_1 is selected so that the slope of the curve representing the reaction rate expression at this conversion level is equal to the slope of the diagonal of rectangle $KLMN$, as shown in Figure 8.13, adopted from Levenspiel. Once this conversion level is known, equations 8.3.34 and 8.3.35 may be used to determine the required reactor sizes.

As Levenspiel points out, the optimum size ratio is generally dependent on the form of the reaction rate expression and on the conversion task specified. For first-order kinetics (either irreversible or reversible with first-order kinetics in both directions) equal-sized reactors should be used. For orders above unity the smaller reactor should precede the larger; for orders between zero and unity the larger reactor should precede the smaller. Szepe and Levenspiel (14) have presented charts showing the optimum size ratio for a cascade of two reactors as a function of the conversion level for various reaction orders. Their results indicate that the minimum in the total volume requirement is an extremely shallow one. For example, for a simple

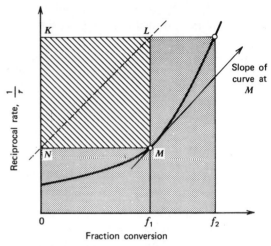

Figure 8.13

Maximization of rectangles applied to find the optimum intermediate conversion and optimum sizes of two CSTR's in series. (Adapted from *Chemical Reaction Engineering*, **Second Edition, by O. Levenspiel. Copyright © 1972. Reprinted by permission of John Wiley and Sons, Inc.)**

second-order reaction where 99% conversion is desired, the minimum total volume required is only about 3% less than if equal volume tanks had been used. For 99.9% conversion the difference is only about 4%. Generally, the savings associated with this small a reduction in total volume requirements would scarcely be adequate to cover the extra costs of engineering, installing, and maintaining two tanks of different sizes. The argument for uniformity in tank sizes becomes even stronger when cascades composed of more than two reactors are considered. Consequently, except in those rare cases where there are compelling reasons to the contrary, the reactor designer tends to employ multiple identically sized CSTR's in working up design specifications. However, it may be advantageous to run the various CSTR's at different temperatures.

For the case of multiple equal-sized reactors in series, the problem of determining the reactor sizes necessary to achieve a specified degree of conversion can be solved by a trial and error procedure. In this case the lines in figures analogous to Figure 8.11 will all be parallel to one another. Consequently, one draws a number of parallel lines equal to the number of CSTR's he or she intends to use, with the first line passing through the inlet composition. When the slope used provides the necessary match with the specified final effluent composition, this slope may be used to determine the necessary reactor volume. Illustration 8.7 indicates the use of this technique.

ILLUSTRATION 8.7 DETERMINATION OF CSTR SIZE REQUIREMENTS FOR CASCADES OF VARIOUS SIZES— GRAPHICAL SOLUTION

Consider the Diels-Alder reaction between benzoquinone (B) and cyclopentadiene (C), which was discussed in Illustration 8.1.

$$B + C \rightarrow \text{adduct}$$

If the reaction occurs in the liquid phase at 25 °C, determine the reactor volume requirements for cascades of one and three identical CSTR's. The rate at which liquid feed is supplied is 0.278 m³/ksec. Use the graphical approach outlined previously. The following constraints are applicable.

$$r = kC_B C_C \quad \text{with} \quad k = 9.92 \text{ m}^3/\text{kmole·ksec}$$

$$C_{C0} = 0.1 \text{ kmole/m}^3 \quad C_{B0} = 0.08 \text{ kmole/m}^3$$

Conversion desired—87.5%

Solution

The graphical approach requires a plot of reaction rate versus the concentration of the limiting reagent (benzoquinone). In order to prepare this plot it is necessary to relate the two reactant concentrations to one another. From the initial concentrations and the stoichiometric coefficients,

$$C_C = C_B + 0.02$$

Thus

$$r = kC_B(C_B + 0.02)$$

or, at 25 °C,

$$r = 9.92C_B^2 + 0.1984C_B \qquad (A)$$

where the rate is expressed in kilomoles per cubic meter per kilosecond when concentrations are expressed in kilomoles per cubic meter.

Equation A is presented in graphical form as curve M in Figure 8.14.

For 87.5% conversion the concentration of benzoquinone in the effluent from the last reactor in the cascade will be equal to $(1 - 0.875)(0.08)$ or 0.010 kmole/m³.

For the case where the cascade consists of only a single reactor, only a single straight line of the form of equation 8.3.31 is involved in the graphical solution. One merely links the point on curve M corresponding to the effluent concentration of benzoquinone with the point on the abscissa corresponding to the feed concentration. The slope of this line is equal to $(-1/\tau)$ or $(-\mathcal{V}_0/V_R)$. In the present instance the slope is equal to

$$(2.976 - 0) \times 10^{-3}/(0.01 - 0.08)$$

or

$$-0.0425 \text{ ksec}^{-1}$$

Thus

$$V_R = \frac{\mathcal{V}_0}{0.0425} = \frac{0.278}{0.0425} = 6.54 \text{ m}^3$$

For the case where the cascade consists of three identical reactors in series, a trial and error approach is necessary to determine the required reactor size. One starts from the inlet concentration and draws a line linking this point on the abscissa with some point J on curve M. One then draws a straight line parallel to the first, but passing through the point on the abscissa corresponding to the benzoquinone concentration at point J. This straight line intersects curve M at some point K. One then repeats the procedure by drawing yet another parallel line through the point on the abscissa corresponding to the benzoquinone concentration at K. If the intersection of this straight line with curve M occurs at a reactant concentration of 0.010

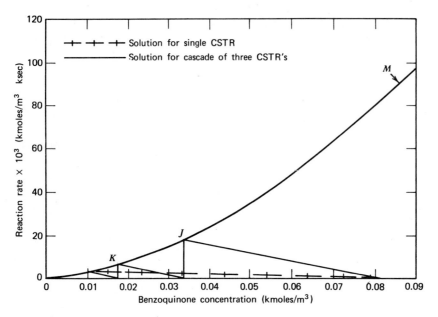

Figure 8.14
Graphical solution for cascade of CSTR's.

kmole/m^3, our initial choice of slope was correct. If not, one must choose a new point J and repeat the procedure until such agreement is obtained. Figure 8.14 indicates the construction for this case. The slopes of the straight lines in this figure are equal to $(18.2 - 0) \times 10^{-3}/(0.034 - 0.08)$ or -0.396 ksec^{-1}. Thus the volume of each individual CSTR is equal to $0.278/0.396$ or 0.70 m^3. The combined volume of all three CSTR's is thus 2.1 m^3, which is a volume reduction of more than a factor of three when compared to the single CSTR case. In Section 8.3.2.3 we will see that such large volume reductions are typical of the use of cascades of CSTR's.

8.3.2.2 *Algebraic Approach to the Analysis of Batteries of Stirred Tank Reactors Operating at Steady State.* Although the graphical approach presented in the previous subsection is quite generally applicable to the steady-state analysis of CSTR cascades, it is not highly accurate in numerical terms, particularly when the generation of the curve representing the reaction rate expression involves graphical or numerical differentiation of rate data. Algebraic methods are capable of greater accuracy than graphical ones if the functional form and constants involved in the reaction rate expression are known. In this subsection we will again consider a cascade of CSTR's where the reactor volumes are not necessarily equal to one another. We will again use the nomenclature shown in Figure 8.10 and start from equation 8.3.29, rearranged as follows.

$$C_{Ai} - r_{Ai}\tau = C_{A(i-1)} \qquad (8.3.36)$$

Starting with the first reactor and using algebraic iteration, we proceed to analyze the cascade for particular forms of the reaction rate expression.

For example, in the case of first-order kinetics

$$C_{Ai} + k_i C_{Ai} \tau_i = C_{A(i-1)} \qquad (8.3.37)$$

Thus

$$C_{Ai} = \frac{C_{A(i-1)}}{1 + k_i \tau_i} \qquad (8.3.38)$$

or

$$C_{A1} = \frac{C_{A0}}{1 + k_1 \tau_1} \qquad (8.3.39)$$

and

$$C_{A2} = \frac{C_{A1}}{1 + k_2 \tau_2} = \frac{C_{A0}}{(1 + k_1 \tau_1)(1 + k_2 \tau_2)} \qquad (8.3.40)$$

Finally,

$$C_{AN} = \frac{C_{A(N-1)}}{1 + k_N \tau_N}$$

$$= \frac{C_{A0}}{(1 + k_1 \tau_1)(1 + k_2 \tau_2) \cdots (1 + k_N \tau_N)} \qquad (8.3.41)$$

These relations are valid regardless of whether the reactors all operate at the same temperature or at different temperatures. In the case where the cascade is isothermal and all reactors have the same size,

$$C_{AN} = \frac{C_{A0}}{(1 + k\tau)^N} \qquad (8.3.42)$$

where τ is the space time for an individual CSTR.

One may use the same general approach when the reaction kinetics are other than first-order. However, except in the case of zero-order kinetics, it is not possible to obtain simple closed form expressions for C_{AN}, particularly if unequal reactor volumes are used. However, the numerical calculations for other reaction orders are not difficult to make for the relatively small number of stages likely to be encountered in industrial practice. The results for zero-order kinetics may be determined from equation

8.3.36 as

$$C_{AN} = C_{A0} - \sum_{i=1}^{N} (k_i \tau_i) \qquad (8.3.43)$$

which is again appropriate for nonisothermal, nonequal volume cases. If the cascade is isothermal and all reactor volumes are equal,

$$C_{AN} = C_{A0} - Nk\tau \qquad (8.3.44)$$

where τ is the space time or mean residence time for an individual CSTR.

Consider now the general second-order case where equation 8.3.36 becomes

$$C_{Ai} + k_i C_{Ai}^2 \tau_i = C_{A(i-1)} \qquad (8.3.45)$$

or

$$C_{Ai} = \frac{-1 + \sqrt{1 + 4 k_i \tau_i C_{A(i-1)}}}{2 k_i \tau_i} \qquad (8.3.46)$$

when we have discarded the negative root because reactant concentrations cannot be negative. Thus

$$C_1 = \frac{-1 + \sqrt{1 + 4 k_1 \tau_1 C_0}}{2 k_1 \tau_1} \qquad (8.3.47)$$

$$C_2 = \frac{-1 + \sqrt{1 + 4 k_2 \tau_2 C_1}}{2 k_2 \tau_2}$$

$$= \frac{-1 + \sqrt{1 + 4 k_2 \tau_2 \left(\dfrac{-1 + \sqrt{1 + 4 k_1 \tau_1 C_0}}{2 k_1 \tau_1} \right)}}{2 k_2 \tau_2} \qquad (8.3.48)$$

and so on. Although a general expression for C_{AN} would be rather complex algebraically, involving N nested square roots, the solution for the isothermal case is readily obtained graphically by the procedure outlined in the previous subsection. A number of other graphical solutions have been discussed in the reactor design literature by various individuals (15–18).

Illustrations 8.8 and 8.9 indicate how the techniques developed in this and the previous section may be used in the design analysis of cascades of stirred tank reactors.

ILLUSTRATION 8.8 DETERMINATION OF REACTOR SIZE REQUIREMENTS FOR A CASCADE OF CSTR's—ALGEBRAIC APPROACH

Consider the Diels—Alder reaction between benzoquinone (B) and cyclopentadiene (C), discussed in Illustrations 8.1 and 8.7.

$$B + C \rightarrow adduct$$

If one employs a feed containing equimolal concentrations of reactants, the reaction rate expression can be written as

$$r = kC_C C_B = kC_B^2$$

Determine the reactor size requirements for cascades composed of one, two, and three identical CSTR's. Use an algebraic approach and assume isothermal operation at 25 °C where the reaction rate constant is equal to 9.92 m^3/kmole·ksec. Reactant concentrations in the feed are equal to 0.08 kmole/m^3. The liquid feed rate is equal to 0.278 m^3/ksec. The desired degree of conversion is equal to 87.5%.

Solution

For the specified degree of conversion the effluent concentration of benzoquinone must be equal to $(1 - 0.875)(0.08) = 0.01$ kmole/m^3.

Case I—Single CSTR

In this case equation 8.3.45 can be solved for the reactor space time directly.

$$\tau = \frac{C_{A0} - C_{A1}}{kC_{A1}^2}$$

or

$$\frac{V_{R1}}{\dot{V}_0} = \frac{0.08 - 0.01}{9.92(0.01)^2} = 70.56 \text{ ksec}$$

Thus

$$V_{R1} = 70.56(0.278) = 19.6 \text{ m}^3$$

This value is considerably larger than that calculated in Illustration 8.7 for a nonstoichiometric feed ratio, thus indicating the potential desirability of using an excess of one reagent when it appears to be a positive power in the rate expression. In any economic analysis of a process, however, the costs of separation and recovery or disposal of the excess reagent must be taken into account.

Case II—Two Identical CSTR's in Series

In this case it will be necessary to determine the concentration in the effluent from the first reactor in order to determine the required reactor size. One way of proceeding is to write the design equation for each CSTR.

$$\frac{V_{R1}}{\dot{V}_0} = \frac{C_{B0}(f_{B1} - 0)}{kC_{B0}^2(1 - f_{B1})^2} \qquad (A)$$

$$\frac{V_{R2}}{\dot{V}_0} = \frac{C_{B0}(f_{B2} - f_{B1})}{kC_{B0}^2(1 - f_{B2})^2} \qquad (B)$$

For identical reactors equations A and B may be combined to give

$$\frac{f_{B1}}{(1 - f_{B1})^2} = \frac{f_{B2} - f_{B1}}{(1 - f_{B2})^2}$$

For $f_{B2} = 0.875$,

$$(1 - 0.875)^2(f_{B1}) = (0.875 - f_{B1})(1 - f_{B1})^2$$

This equation may be solved by trial and error or by graphical means to determine the composition of the effluent from the first reactor. Thus $f_{B1} = 0.7251$ and $C_{B1} = 0.08(1 - 0.7251) = 0.02199$ kmole/m^3. Either equation A or equation B may now be used to determine the required reactor size. Hence

$$V_{R1} = \frac{(0.278)(0.08)(0.7251)}{9.92(0.08)^2(1 - 0.7251)^2} = 3.36 \text{ m}^3$$

The total volume of the two reactors is 6.72 m^3, which is considerably less than half that required if only a single CSTR is employed.

Case III—Three Identical CSTR's in Series

In this case there are two intermediate unspecified reactant concentrations instead of just the single intermediate concentration encountered in Case II. At least one of these concentrations must be determined if one is to be able to appropriately size the reactors. In principle one may follow the procedure used in Case II where the design equations for each CSTR are written and the reactor space times then equated. This procedure gives three equations and three unknowns (V_{R1}, f_{B1}, and f_{B2}). Thus, for the first reactor,

$$\frac{V_{R1}}{\mathcal{V}_0} = \frac{C_{B0}(f_{B1} - 0)}{kC_{B0}^2(1 - f_{B1})^2} \qquad \text{(C)}$$

For the second,

$$\frac{V_{R2}}{\mathcal{V}_0} = \frac{C_{B0}(f_{B2} - f_{B1})}{kC_{B0}^2(1 - f_{B2})^2} \qquad \text{(D)}$$

and, for the third,

$$\frac{V_{R3}}{\mathcal{V}_0} = \frac{C_{B0}(0.875 - f_{B2})}{kC_{B0}^2(1 - 0.875)^2} \qquad \text{(E)}$$

Combining equations C and E gives

$$\frac{0.875 - f_{B2}}{(0.125)^2} = \frac{f_{B1}}{(1 - f_{B1})^2}$$

or

$$f_{B2} = 0.875 - \frac{(0.125)^2 f_{B1}}{(1 - f_{B1})^2} \qquad \text{(F)}$$

Combining equations C and D gives

$$\frac{f_{B1}}{(1 - f_{B1})^2} = \frac{f_{B2} - f_{B1}}{(1 - f_{B2})^2} \qquad \text{(G)}$$

Now f_{B2} may be eliminated from the last equation by using equation F.

$$\frac{f_{B1}}{(1 - f_{B1})^2} = \frac{0.875 - \dfrac{(0.125)^2 f_{B1}}{(1 - f_{B1})^2} - f_{B1}}{\left[1 - 0.875 + \dfrac{(0.125)^2 f_{B1}}{(1 - f_{B1})^2}\right]^2}$$

This equation may be solved numerically by trial and error, recognizing that f_{B1} must lie between 0 and 0.875. The appropriate value of f_{B1} is 0.6285. Equation F may now be used to determine that $f_{B2} = 0.8038$. With a knowledge of these conversions, equation C, D, or E may be used to determine the required reactor volume. Thus

$$V_{R1} = \frac{0.278(0.08)(0.6285)}{9.92(0.08)^2(1 - 0.6285)^2} = 1.60 \text{ m}^3$$

The total volume of the cascade is then $3(1.60)$ or 4.8 m³, which is again a significant reduction in the total volume requirement but not nearly as great as that brought about in going from one to two CSTR's in series.

Obviously this approach is not easily extended to cascades containing more than three reactors and, in those cases, an alternative trial and error procedure is preferable. One chooses a reactor volume and then determines the overall fraction conversion that would be obtained in a cascade of N reactors. When one's choice of individual reactor size meets the specified overall degree of conversion, the choice may be regarded as the desired solution. This latter approach is readily amenable to iterative programming techniques using a digital computer.

ILLUSTRATION 8.9 DETERMINATION OF OPTIMUM REACTOR SIZES FOR A CASCADE OF TWO CSTR's

Consider the Diels-Alder reaction between benzoquinone (B) and cyclopentadiene (C) discussed earlier in Illustrations 8.1, 8.7, and 8.8.

$$B + C \rightarrow \text{adduct}$$

We wish to determine the effect of using a cascade of two CSTR's that differ in size on the volume requirements for the reactor network. In Illustration 8.8 we saw that for reactors of equal size the total volume requirement was 6.72 m³. If the same feed composition and flow rate as in the previous illustration are employed and if the reactors are operated isothermally at 25 °C, determine the minimum total volume required and the manner in which the volume should be distributed between the two reactors. An overall conversion of 0.875 is to be achieved.

Solution

For the conditions cited the reaction rate expression is of the form

$$r = kC_B^2$$

where $k = 9.92$ m³/kmole·ksec.

From the basic design relationship for a CSTR (8.3.4),

$$\frac{V_{R1}}{F_{B0}} = \frac{f_{B1}}{kC_{B0}^2(1 - f_{B1})^2} \tag{A}$$

and

$$\frac{V_{R2}}{F_{B0}} = \frac{f_{B2} - f_{B1}}{kC_{B0}^2(1 - f_{B2})^2} = \frac{0.875 - f_{B1}}{kC_{B0}^2(1 - 0.875)^2} \tag{B}$$

Thus the total reactor volume in the cascade is given by

$$V_{R1} + V_{R2} = \frac{F_{B0}}{kC_{B0}^2}\left[\frac{f_{B1}}{(1 - f_{B1})^2} + \frac{0.875 - f_{B1}}{(0.125)^2}\right] \tag{C}$$

It is this sum that we desire to minimize. The easiest approach to finding this minimum is to plot the quantity in brackets versus f_{B1}. The minimum in this quantity then gives the minimum total volume, and the value of f_{B1} associated with the minimum may be used in equations A and B to determine the optimum distribution of the total volume between the two reactors. The minimum occurs when $f_{B1} = 0.702$.

Now

$$F_{B0} = C_{B0}\mathcal{V}_0 = (0.08)(0.278)$$
$$= 0.02224 \text{ kmole/ksec}$$

Thus,

$$V_{R1} = \frac{0.02224(0.702)}{(9.92)(0.08)^2(1 - 0.702)^2} = 2.77 \text{ m}^3$$

$$V_{R2} = \frac{0.02224(0.875 - 0.702)}{(9.92)(0.08)^2(1 - 0.875)^2} = 3.88 \text{ m}^3$$

and

$$V_{R1} + V_{R2} = 2.77 + 3.88 = 6.65 \text{ m}^3$$

This total differs from that for equal-sized reactors by only 0.07 m³ or approximately 1%. The benefits that would ensue from this small change would probably be far outweighed by the disadvantages associated with having reactors of different sizes, such as the costs of engineering design, construction, fabrication, and inventory of spare parts. In general, the economics associated with using reactors of different sizes are offset by the concomitant disadvantages. For further treatment of the problem of optimization of a two-tank CSTR cascade, consult the papers by Crooks (19) and Denbigh (20).

8.3.2.3 Size Comparisons Between Cascades of Ideal Continuous Stirred Tank Reactors and Plug Flow Reactors. In this section the size requirements for CSTR cascades containing different numbers of identical reactors are compared with that for a plug flow reactor used to effect the same change in composition.

One may define a space time for an entire cascade (τ_c) in terms of the ratio of the sum of the component reactor volumes to the inlet volumetric flow rate. Hence

$$\tau_c = \frac{\sum_{i=1}^{N} V_{Ri}}{\mathcal{V}_0} = \sum_{i=1}^{N} \tau_i \tag{8.3.49}$$

If all component reactors have the same volume

and thus the same space time (τ),

$$\tau_c = N\tau \tag{8.3.50}$$

For a first-order reaction, we showed that for a cascade composed of equal-sized reactors, equation 8.3.42 governed the effluent composition from the nth reactor.

$$\frac{C_{AN}}{C_{A0}} = \frac{1}{(1 + k\tau)^N} \tag{8.3.51}$$

where τ is the space time for an individual reactor. Combining equations 8.3.50 and 8.3.51 and rearranging gives

$$\frac{C_{A0}}{C_{AN}} = \left(1 + \frac{k\tau_c}{N}\right)^N \tag{8.3.52}$$

If the right side of this equation is expanded in a binomial series,

$$\frac{C_{A0}}{C_{AN}} = 1 + N\left(\frac{k\tau_c}{N}\right) + \frac{N(N-1)}{2!}\left(\frac{k\tau_c}{N}\right)^2$$
$$+ \frac{N(N-1)(N-2)}{3!}\left(\frac{k\tau_c}{N}\right)^3 + \cdots \tag{8.3.53}$$

and if $e^{k\tau_c}$ is expanded in series form,

$$e^{k\tau_c} = 1 + k\tau_c + \frac{(k\tau_c)^2}{2!} + \frac{(k\tau_c)^3}{3!} + \cdots \tag{8.3.54}$$

a term by term comparison in the limit as N approaches infinity indicates that

$$\frac{C_{A0}}{C_{AN}} \approx e^{k\tau_c} \quad \text{as } N \to \infty \tag{8.3.55}$$

or

$$\tau_c = \frac{1}{k}\ln\left(\frac{C_{A0}}{C_{AN}}\right) \quad \text{as } N \to \infty \tag{8.3.56}$$

This relation is identical with that which would be obtained from equation 8.2.10 for a plug flow reactor with first-order kinetics.

$$\tau_{\text{PFR}} = \frac{1}{k}\ln\left(\frac{C_{A0}}{C_{A\,\text{out}}}\right) \tag{8.3.57}$$

These relations support our earlier assertion that for the same overall conversion the total volume of a cascade of CSTR's should approach the plug flow volume as the number of reactors in the cascade is increased.

For a finite (low) number of CSTR's in series, equation 8.3.52 can be rewritten as

$$\tau_c = \frac{N}{k}\left[\left(\frac{C_{A0}}{C_{AN}}\right)^{1/N} - 1\right] \tag{8.3.58}$$

The ratio of equations 8.3.58 and 8.3.57 gives the relative *total space time* requirement for a cascade of stirred tank reactors *vis à vis* a plug flow reactor.

$$\frac{\tau_{c\text{ for }N\text{ reactors}}}{\tau_{c\text{ for }N=\infty}} = \frac{\tau_{c\text{ for }N\text{ reactors}}}{\tau_{\text{plug flow}}}$$

$$= \frac{N\left[\left(\dfrac{C_{A0}}{C_{AN}}\right)^{1/N}_{\text{cascade}} - 1\right]}{\ln\left(\dfrac{C_{A0}}{C_{A\,\text{out}}}\right)_{\text{plug flow}}} \tag{8.3.59}$$

If the effluents from the two streams are to be identical and if equimolal feed rates and compositions are employed, the ratio of space times becomes equal to the ratio of total volume requirements. Thus, for constant density systems where $C_{A\,\text{out}} = C_{AN} = C_{A0}(1 - f_A)$,

$$\frac{V_{\text{Total, cascade}}}{V_{\text{PFR}}} = \frac{N\left[\left(\dfrac{1}{1 - f_A}\right)^{1/N} - 1\right]}{-\ln(1 - f_A)} \tag{8.3.60}$$

Figure 8.15, reproduced from Levenspiel (21), is in essence a plot of this ratio versus the fraction conversion for various values of N, the number of identical CSTR's employed. The larger the value of N, the smaller the discrepancy in reactor volume requirements between the

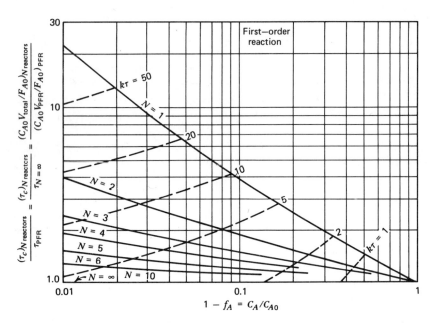

Figure 8.15

Comparison of performance of a series of *N* equal-size CSTR reactors with a plug flow reactor for the first-order reaction

$$A \rightarrow R, \qquad \delta_A = 0$$

For the same processing rate of identical feed the ordinate measures the volume ratio $V_{N\,CSTR's}/V_{PFR}$ or the space time ratio $\tau_{N\,CSTR's}/\tau_{PFR}$ directly. (Adapted from *Chemical Reaction Engineering*, Second Edition, by O. Levenspiel. Copyright © 1972. Reprinted by permission of John Wiley and Sons, Inc.)

CSTR cascade and a PFR reactor. Note how rapidly PFR behavior is approached as *N* increases. Levenspiel has also included lines of constant $k\tau$ on this figure, and these lines may be useful in solving certain types of design problems, as we will see in Illustration 8.10.

Levenspiel (22) has prepared a similar plot for second-order reactions (Figure 8.16). It is based on a generalization of equation 8.3.46 for *N* identical reactors in series and the integral form of the plug flow design equation for second-order kinetics. We again see that increasing the number of reactors in the cascade causes the total volume discrepancy between the cascade and a plug flow reactor to diminish rapidly,

with the greatest change occurring on addition of a second tank.

Although the major thrust of the material presented in this subsection has concerned the relative size requirements for CSTR cascades and plug flow reactors, the practicing chemical engineer will be more concerned with the relative economics of the two alternative reactor network configurations. In this regard it is worth repeating that the additional capital costs associated with the larger volumes of CSTR's are relatively small, particularly when the units are designed for operation at atmospheric pressure. Consequently, a plot of the total costs associated with the use of *N* reactors versus the number *N*

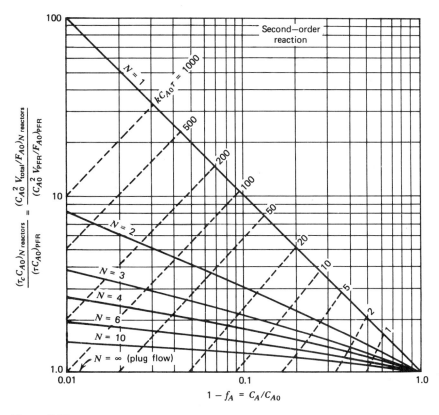

Figure 8.16

Comparison of performance of a series of N equal-size CSTR reactors with a plug flow reactor for elementary second-order reactions

$$2A \rightarrow \text{products}$$

$$A + B \rightarrow \text{products}, \qquad C_{A0} = C_{B0}$$

with negligible expansion. For the same processing rate of identical feed the ordinate gives the volume ratio $V_{N \text{ CSTR's}}/V_{\text{PFR}}$ or the space time ratio $\tau_{N \text{ CSTR's}}/\tau_{\text{PFR}}$ directly. (Adapted from _Chemical Reaction Engineering_, Second Edition, by O. Levenspiel. Copyright 1972. Reprinted by permission of John Wiley and Sons, Inc.

will usually look something like that shown in Figure 8.17. One obtains an economic trade-off between the costs associated with the high volume requirements when very few reactors are employed and the additional engineering, fabrication, installation, and maintenance costs incurred by using a larger number of reactors in

the cascade. Consequently, the reactor designer must consider cascades containing different numbers of reactors in the search for an economic optimum.

The following illustration indicates how Figures 8.15 and 8.16 are used in handling simple reactor design calculations.

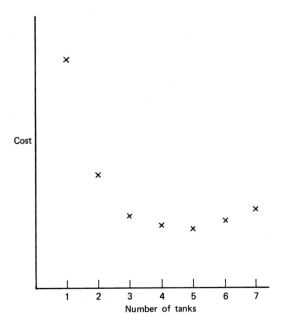

Figure 8.17
Interaction between the cost decline associated with reduced total volume and cost increases arising from design, construction, installation, and maintenance of a number of reactors.

ILLUSTRATION 8.10 USE OF THE DESIGN CHARTS FOR COMPARISON OF ALTERNATIVE REACTOR NETWORKS

Consider the Diels-Alder reaction between benzoquinone (B) and cyclopentadiene (C) discussed in the last three illustrations.

$$B + C \rightarrow \text{adduct}$$

At 25 °C the reaction is first-order in each reactant with a rate constant of 9.92 m³/kmole·ksec. A feed stream containing equimolal quantities of B and C (0.1 kmole/m³) is to be processed at a rate of 0.1111 m³/ksec. A tubular reactor (assume plug flow) with an effective volume of 2.20 m³ is to be employed in the processing operation.

Use the design charts in Figures 8.8, 8.15, and 8.16 to determine the following.

1. What degree of conversion can be obtained in the tubular reactor?
2. What reactor size would be required to achieve the same conversion if a single ideal CSTR were employed?
3. What degree of conversion would be obtained in a single CSTR equal in size to the tubular reactor?
4. If two identical ideal CSTR's in series are employed (each with a volume equal to that determined in part 2), by how large a factor can the flow rate of the feedstream be increased while maintaining the conversion level constant at the value used in parts 1 and 2?
5. If one employs these same two ideal CSTR's in series and maintains a constant feed rate, what conversion is achieved?

Solution

For an equimolal feed the reaction rate expression can be written as

$$-r_B = kC_B^2$$

A. Conversion for Plug Flow
The reactor space time is given by

$$\tau = \frac{V_R}{\mathcal{V}_0} = \frac{2.2}{0.1111} = 19.80 \text{ ksec}$$

Consequently, the characteristic dimensionless rate group for the second-order reaction and initial conditions is given by

$$kC_{B0}\tau = (9.92)(0.1)(19.80) = 19.6$$

In Figure 8.16 the line corresponding to plug flow is that where $V_{N \text{ CSTR's}}/V_{PFR} = 1$. The intersection of this line and the line for a value of $kC_{B0}\tau = 19.6$ gives the desired degree of conversion—95%.

B. Size Requirement for CSTR for Identical Processing Task
For the same feed rate and initial concentrations the ordinates of Figures 8.8 and 8.16 reduce to

the volume ratios for the two reactors. Hence, at the same f_B, we see that $V_{CSTR}/V_{PFR} = 20$. Thus $V_{CSTR} = 20(2.20) = 44 \text{ m}^3$.

C. Conversion in a CSTR of the Same Size as the Tubular Reactor

For a CSTR equal in volume to the tubular reactor, one moves along a line of constant $kC_{B0}\tau$ in Figure 8.16 in order to determine the conversions accomplished in cascades composed of different numbers of reactors but with the same overall space time. The intersection of the line $kC_{B0}\tau = 19.6$ and the curve for $N = 1$ gives $f_B = 0.80$.

D. Increase in Processing Rate Arising from the Use of a Cascade of Two CSTR's at a Specified Degree of Conversion

The values of the group $kC_{B0}\tau$ that correspond to 95% conversion and one or two CSTR's in series may be determined from Figure 8.16. They are approximately 350 and 70, respectively. Thus,

$$\frac{(kC_{B0}\tau)_{N=2}}{(kC_{B0}\tau)_{N=1}} = \frac{\tau_{N=2}}{\tau_{N=1}} = \frac{70}{350} = 0.20$$

Since

$$V_{N=2} = 2V_{N=1}$$

and

$$\frac{\tau_{N=2}}{\tau_{N=1}} = \left(\frac{V_{N=2}}{V_{N=1}}\right)\left(\frac{\mathscr{V}_{N=1}}{\mathscr{V}_{N=2}}\right) = 2\frac{\mathscr{V}_{N=1}}{\mathscr{V}_{N=2}}$$

then

$$\frac{\mathscr{V}_{N=2}}{\mathscr{V}_{N=1}} = \frac{2}{0.20} = 10$$

Consequently, the processing rate for the cascade will be an order of magnitude greater than that for a single CSTR. Note that operation of the two reactors in parallel would have merely doubled the processing capacity. Hence there is a very strong case for operating with the units in a series configuration.

E. Increase in Conversion Arising from the Use of a Cascade of Two CSTR's at a Specified Feed Rate

If the feed rate is maintained constant while the number of reactors is doubled, the overall space time for the cascade will double. In the present case $kC_{B0}\tau_{N=2} = 2(350) = 700$. From Figure 8.16, at this value of the dimensionless group and $N = 2$, it is seen that $f_B = 0.99$.

8.3.2.4 Analysis of CSTR Cascades under Nonsteady-State Conditions.

In Section 8.3.1.4 the equations relevant to the analysis of the transient behavior of an individual CSTR were developed and discussed. It is relatively simple to extend the most general of these relations to the case of multiple CSTR's in series. For example, equations 8.3.15 to 8.3.21 may all be applied to any individual reactor in the cascade of stirred tank reactors, and these relations may be used to analyze the cascade in stepwise fashion. The difference in the analysis for the cascade, however, arises from the fact that more of the terms in the basic relations are likely to be time variant when applied to reactors beyond the first. For example, even though the feed to the first reactor may be time invariant during a period of nonsteady-state behavior in the cascade, the feed to the second reactor will vary with time as the first reactor strives to reach its steady-state condition. Similar considerations apply further downstream. However, since there is no effect of variations downstream on the performance of upstream CSTR's, one may start at the reactor where the disturbance is introduced and work downstream from that point. In our generalized notation, equation 8.3.20 becomes

$$\frac{dC_{Ai}}{dt} - r_{Ai} = \frac{C_{A(i-1)} - C_{Ai}}{\tau_i} \quad (8.3.61)$$

where the reaction rate is evaluated at conditions prevailing in reactor i and where we have presumed both a constant density system and a constant reactor volume. In this case both C_{Ai} [and thus $(-r_{Ai})$] and $C_{A(i-1)}$ are generally time variant quantities.

For first-order irreversible reactions and identical space times it is possible to obtain closed form solutions to differential equations of the form of 8.3.61. In other cases it is usually necessary to solve the corresponding difference equations numerically.

If we were to extend the analysis developed in equations 8.3.22 to 8.3.25 to the case of just two CSTR's in series, the equation we would have to solve to determine the composition of the effluent from the second reactor would be the following form of equation 8.3.61.

the limit of zero recycle the system approaches plug flow behavior, and at the limit where only an infinitesimal proportion of net product stream is produced, the system will approach CSTR behavior. To develop characteristic design equations that describe the performance of recycle reactors, we will follow Levenspiel's treatment (23). The recycle ratio R is defined as

$$R = \frac{\begin{pmatrix} \text{Volume of fluid} \\ \text{returned to the} \\ \text{reactor entrance} \end{pmatrix}}{\begin{pmatrix} \text{Volume of fluid leaving} \\ \text{the system in net} \\ \text{product stream} \end{pmatrix}} \qquad (8.3.63)$$

$$\frac{dC_{A2}}{dt} - kC_{A2} = \frac{1}{\tau_2}\left[\frac{C_{A\,in}}{k\tau_1 + 1} + \left(C^* - \frac{C_{A\,in}}{k\tau_1 + 1} \right) e^{-(k\tau_1 + 1)t/\tau_1} - C_{A2} \right] \qquad (8.3.62)$$

Mathematically inclined students are encouraged to solve this equation and to move as far downstream as they desire. All students should recognize that we have seen equations of this mathematical form in Chapter 5.

8.3.3 Recycle Reactors

In cases where one desires to promote back-mixing in gaseous systems it may be desirable to use a recycle reactor (Figure 8.18). The need for back-mixing may arise from a desire to enhance reaction selectivity or to moderate thermal effects associated with the reaction.

Reaction takes place only within the plug flow element of the recycle reactor, and the gross product stream from this element is divided into two portions; one becomes the net product and the second is mixed with fresh feed. The mixture of the fresh feed and recycle stream is then fed to the plug flow element. By varying the relative quantities of the net product and recycle streams, one is able to obtain widely varying performance characteristics. At

The basic design equation for a plug flow reactor (equation 8.2.7) may be used to describe the steady-state conversion achieved in the plug flow element of the recycle reactor:

$$\frac{V_R}{F'_{A1}} = \int_{f_{A1}}^{f_{A2} = f_{A\,overall}} \frac{df_A}{(-r_A)} \qquad (8.3.64)$$

where F'_{A1} represents a hypothetical molal flow rate of A corresponding to a stream in which none of the A had reacted, and where f_{A1} is the fraction of F'_{A1} that has been converted to products prior to entering the plug flow element. It is assumed that no reaction takes place outside the plug flow element.

Material balance considerations indicate that

$$F'_{A1} = F_{A0}(1 + R) \qquad (8.3.65)$$

In other words, the molal flow rate of A corresponding to zero fraction conversion at the inlet to the PFR element is equal to the sum of the net inlet flow rate and the amount that would have entered if none of the material in the recycle stream had undergone reaction.

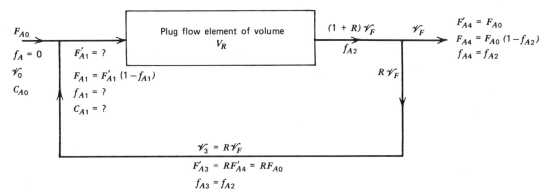

Figure 8.18
Schematic representation of recycle reactor indicating symbols used in design analysis.

The reactant concentration at the reactor inlet is given by the ratio of the molal and volumetric flow rates at the same spot. Therefore,

$$C_{A1} = \frac{F_{A1}}{\mathcal{V}_1} = \frac{F_{A0} + F_{A3}}{\mathcal{V}_0 + \mathcal{V}_3} \qquad (8.3.66)$$

where we have recognized that for a constant pressure system (no pressure drop through the reactor or recycle return line) there will be no volume change on mixing.

Now, from the definition of the recycle ratio,

$$\mathcal{V}_3 = R\mathcal{V}_4 = R\mathcal{V}_0(1 + \delta_A f_{A2}) \qquad (8.3.67)$$

and

$$F_{A3} = RF_{A4} = RF_{A0}(1 - f_{A2}) \qquad (8.3.68)$$

Combining equations 8.3.66 to 8.3.68 and 3.1.47 gives

$$C_{A1} = \frac{F_{A0}[1 + R(1 - f_{A2})]}{\mathcal{V}_0[1 + R(1 + \delta_A f_{A2})]}$$

$$= C_{A0}\left(\frac{1 - f_{A1}}{1 + \delta_A f_{A1}}\right) \qquad (8.3.69)$$

Thus

$$(1 - f_{A1})(1 + R + R\,\delta_A f_{A2})$$
$$= (1 + \delta_A f_{A1})(1 + R - f_{A2}R) \qquad (8.3.70)$$

which can be manipulated algebraically to give

the desired result

$$f_{A1} = \left(\frac{R}{R + 1}\right) f_{A2} \qquad (8.3.71)$$

Combining equations 8.3.64, 8.3.65, and 8.3.71 gives

$$\frac{V_R}{F_{A0}} = (R + 1) \int_{[Rf_{A2}/(R+1)]}^{f_{A2}} \frac{df_A}{(-r_A)} \qquad (8.3.72)$$

which may be regarded as the fundamental design relationship for a recycle reactor with a zero conversion level in the feed stream mixed with the recycle stream. It is valid for both constant and variable density systems.

Examination of the limiting forms of equation 8.3.72 for $R = 0$ and $R = \infty$ indicates that the recycle reactor can approach either plug flow or CSTR behavior. For intermediate values of the recycle ratio this equation can be integrated if the form of the reaction rate expression is known.

8.4 REACTOR NETWORKS COMPOSED OF COMBINATIONS OF IDEAL CONTINUOUS STIRRED TANK REACTORS AND PLUG FLOW REACTORS

This section indicates a few useful generalizations that are pertinent in considerations of isothermal series and parallel combinations of ideal plug flow and stirred tank reactors.

Parallel combinations are governed by the general principle enunciated in Section 8.2.3.

For most efficient utilization of the available reactor volume, all parallel streams that meet must have the same composition.

Under these circumstances each parallel leg may be considered to be operating independently insofar as the space time requirements necessary to effect a given composition change

are concerned. Total feed capacity then increases in proportion to the flow rates that can be handled by the various parallel legs. The component reactors of each of the parallel legs can then be considered as a series combination of reactors and can be optimized in terms of the general principles enunciated below.

Consider the series combination of PFR and CSTR's shown in Figure 8.19. In terms of the fundamental design equations for these idealized

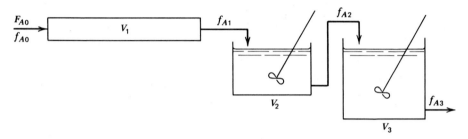

Figure 8.19a
Series combination of PFR and CSTR's.

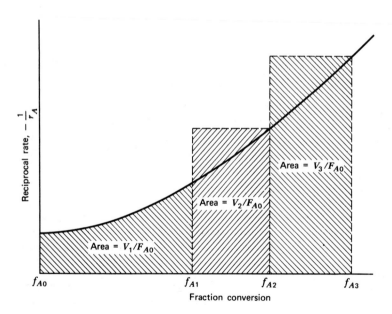

Figure 8.19b
Graphical representation of equations 8.4.1 to 8.4.3.

reactor types (8.2.7 and 8.3.4), it can be said that

$$\frac{V_1}{F_{A0}} = \int_{f_{A0}}^{f_{A1}} \frac{df_A}{(-r_A)} \qquad (8.4.1)$$

$$\frac{V_2}{F_{A0}} = \frac{f_{A2} - f_{A1}}{-r_{A2}} \qquad (8.4.2)$$

and

$$\frac{V_3}{F_{A0}} = \frac{f_{A3} - f_{A2}}{-r_{A3}} \qquad (8.4.3)$$

where the reaction rates in the last two equations are evaluated at the effluent conditions for each reactor and where F_{A0} is the molal flow rate of reactant A that would correspond to zero conversion. The above relations are depicted in graphical form in Figure 8.19b. If the mathematical form of the reaction rate expression is known, these equations can be used to specify completely the necessary design parameters.

For optimum utilization of a given set of ideal reactors operating as an isothermal reactor network, an examination of the $1/(-r_A)$ versus C_A curve is a good way to find the best arrangement of units. The following general rules have been enunciated by Levenspiel (24).

1. For a reaction whose rate-concentration curve rises monotonically (any nth-order reaction, $n > 0$) the reactors should be connected in series. They should be ordered so as to keep the concentration of reactant as high as possible if the rate-concentration curve is concave ($n > 1$), and as low as possible if the curve is convex ($n < 1$).

For all reaction orders greater than unity, the appropriate order is: plug flow, small CSTR, large CSTR. In the case of reaction orders less than unity, the reverse order should be employed. For a first-order reaction the conversion will be independent of the arrangement of the various reactors.

2. For reactions where the rate-concentration curve passes through a maximum or minimum

the arrangement of units depends on the actual shape of curve, the conversion level desired, and the units available.

(See Section 9.4 for an illustration of this type.)

8.5 SUMMARY OF FUNDAMENTAL DESIGN RELATIONS—COMPARISON OF ISOTHERMAL STIRRED TANK AND PLUG FLOW REACTORS

Table 8.1 summarizes the fundamental design relationships for the various types of ideal reactors in terms of equations for reactor space times and mean residence times. The equations are given in terms of both the general rate expression and nth-order kinetics.

If the various expressions are compared, it is evident that for constant density situations

$$t_{\text{batch, constant pressure}} = t_{\text{batch, constant volume}}$$
$$= \bar{t}_{PFR} = \tau_{PFR} \qquad (8.5.1)$$

Moreover, for negligible pressure drop through a plug flow reactor,

$$t_{\text{batch, constant pressure}} = \bar{t}_{PFR} \qquad (8.5.2)$$

As we stressed earlier, the reactor space time is the independent variable at the control of the reactor designer. This parameter is more meaningful than the mean residence time in the reactor.

If we wish to make size comparisons of batch and continuous processing equipment in terms of space time requirements, we must recognize that there are nonproductive periods associated with filling, heating, cooling, draining, cleaning, etc., and that the long-term space time requirement would be given by

$$\tau_{\text{batch}} = t + t_s \qquad (8.5.3)$$

where

t_s is the average nonproductive time per batch processed

t is given by the equations in Table 8.1

Table 8.1
Summary of Design Equations Given that $V = V_0(1 + \delta_A f_A)$

Reactor type	Measure of capacity	General design relation	Design relation for $-r_A = kC_A^n$
I. Batch			
A. Holding time			
1. Constant volume	t	$t = C_{A0} \int \dfrac{df_A}{(-r_A)}$	$t = \dfrac{1}{kC_{A0}^{n-1}} \int \dfrac{df_A}{(1 - f_A)^n}$
2. Constant pressure	t	$t = C_{A0} \int \dfrac{df_A}{(1 + \delta_A f_A)(-r_A)}$	$t = \dfrac{1}{kC_{A0}^{n-1}} \int \dfrac{(1 + \delta_A f_A)^{n-1} df_A}{(1 - f_A)^n}$
B. Equivalent space time	τ	$\tau = t + t_s$	
II. Plug flow			
A. Mean residence time	$\bar{t}^a$	$t = C_{A0} \int \dfrac{df_A}{(1 + \delta_A f_A)(-r_A)}$	$\bar{t} = \dfrac{1}{kC_{A0}^{n-1}} \int \dfrac{(1 + \delta_A f_A)^{n-1} df_A}{(1 - f_A)^n}$
B. Space time	$\tau = \dfrac{1}{S}$	$\tau = C_{A0} \int \dfrac{df_A}{(-r_A)}$	$\tau = \dfrac{1}{kC_{A0}^{n-1}} \int \dfrac{(1 + \delta_A f_A)^n df_A}{(1 - f_A)^n}$
III. CSTR			
A. Mean residence time	$\bar{t}^a$	$\bar{t} = \dfrac{C_{A0} \int df_A}{(-r_{AF})(1 + \delta_A f_{AF})}$	$\bar{t} = \dfrac{1}{kC_{A0}^{n-1}} \dfrac{(1 + \delta_A f_{AF})^{n-1}}{(1 - f_{AF})^n} \int df_A$
B. Space time	$\tau = \dfrac{1}{S}$	$\tau = \dfrac{C_{A0}}{(-r_{AF})} \int df_A$	$\tau = \dfrac{1}{kC_{A0}^{n-1}} \dfrac{(1 + \delta_A f_{AF})^n}{(1 - f_{AF})^n} \int df_A$

Adapted from *Chemical Reaction Engineering*, First Edition, by O. Levenspiel. Copyright © 1962. Reprinted by permission of John Wiley and Sons, Inc.
[a] Indicated design relations not recommended for use in reactor analysis.

8.6 SEMIBATCH OR SEMIFLOW REACTORS

For semibatch or semiflow reactors all four of the terms in the basic material and energy balance relations (equations 8.0.1 and 8.0.3) can be significant. The feed and effluent streams may enter and leave at different rates so as to cause changes in both the composition and volume of the reaction mixture through their interaction with the chemical changes brought about by the reaction. Even in the case where the reactor operates isothermally, numerical methods must often be employed to solve the differential performance equations.

It may be desirable to operate in semibatch fashion in order to enhance reaction selectivity or to control the rate of energy release by reaction through manipulation of the rate of addition of one reactant. Other situations in which semibatch operation is employed include a variety of biological fermentations where various nutrients may be added at predetermined rates to achieve optimum production capacity and cases where one reactant is a gas of limited solubility that can be fed only as fast as it will dissolve.

Semibatch operations usually employ a single well-stirred tank. In such cases it is possible to make the usual assumption that the composition and temperature of the fluid are uniform throughout the tank.

For semibatch operation, the term "fraction conversion" is somewhat ambiguous for many of the cases of interest. If reactant is present initially in the reactor and is added or removed in feed and effluent streams, the question arises as to the proper basis for the definition of f. In such cases it is best to work either in terms of the weight fraction of a particular component present in the fluid of interest or in terms of concentrations when constant density systems are under consideration. In terms of the symbols shown in Figure 8.20 the fundamental material balance relation becomes:

$$w_0 \Phi_{m0} = w_F \Phi_{mF} + (-r_{AF})V_R + \frac{d}{dt}(w_F \rho_m V_R)$$

(8.6.1)

where

    w is the weight fraction reactant

    Φ_m is the mass flow rate

    ρ_m is the mass density of the fluid in the reactor

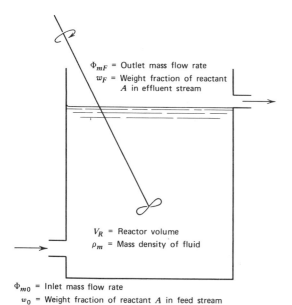

Φ_{mF} = Outlet mass flow rate
w_F = Weight fraction of reactant A in effluent stream

V_R = Reactor volume
ρ_m = Mass density of fluid

Φ_{m0} = Inlet mass flow rate
w_0 = Weight fraction of reactant A in feed stream

Figure 8.20
Schematic representation of semibatch reactor indicating process variables.

$(-r_{AF})$ is the reaction rate per unit volume expressed in terms of the mass of reactant A disappearing per unit time and evaluated at the conditions prevailing within the reactor

subscripts 0 and F indicate input and effluent conditions, respectively

If the initial condition of the reactor contents is known and if the feedstream conditions are specified, it is possible to solve equation 8.6.1 to determine the effluent composition as a function of time. The solution may require the use of material balance relations for other species or a total material balance. This is particularly true of variable volume situations where the following overall material balance equation is often useful.

$$\Phi_{m0} = \Phi_{mF} + \frac{d}{dt}(\rho_m V_R)$$ (8.6.2)

By working in terms of total mass, the reaction term disappears because the principle of conservation of mass must be satisfied.

There are a number of specific cases of equation 8.6.1 that are of potential interest for commercial applications. We wish to consider one mode of semibatch operation using Illustration 8.11 in order to indicate the general principles involved in the analysis of these systems.

Kladko (25) has presented a very interesting case study of a reactor design problem involving an exothermic isomerization reaction. Although the reaction in question was well behaved in laboratory scale apparatus, it behaved quite differently when first run on a commercial scale in a batch reactor. The system ran out of control with the temperature increasing so rapidly that the batch erupted violently through a safety valve and vented out over the building area. The fact that the strong exotherm and its concomitant effect on the reaction rate could have been predicted *a priori* on the basis of energy balance calculations indicates the necessity of considering thermal effects in reactor design calculations. These effects are the subject of Chapter 10; be particularly careful to take them into account when moving from bench or pilot scale reactor systems to commercial scale equipment. Because of the proprietary nature of the product many details of the reaction were omitted. However, by making certain assumptions or engineering estimates regarding heat capacities and molecular weights, it is possible to generate the necessary input data to permit evaluation of several alternative reactor designs. Some alternatives arc considered as illustrative examples in Chapter 10, and the next illustration indicates the type of analysis appropriate to isothermal semibatch operation.

ILLUSTRATION 8.11 ISOMERIZATION IN A SEMIBATCH REACTOR

Reagent A undergoes an essentially irreversible isomerization reaction that obeys first-order kinetics.

$$A \rightarrow B$$

Both A and B are liquids at room temperature and both have extremely high boiling points.

A 1000-gallon, glass-lined kettle is available for carrying out the reaction. The kettle may be maintained at essentially isothermal conditions by a heat transfer fluid that circulates through a jacket on its external surface. The heat transfer fluid may be cooled or heated as required by circulation through appropriate heat exchangers. Since the reaction is exothermic, Kladko and his co-workers (25) wished to consider the possibility of using cold reactant feed to provide a heat sink for some of the energy liberated by reaction. By controlling the rate of addition of feed, they could also obtain a measure of control over the rate of energy release by reaction. Hence a semibatch mode of operation appeared to be an attractive alternative. Since cold incoming reactant would crack the hot glass liner, they considered the possibility of starting with 1500 lb of product B in order to provide a thermal and

material sump. The sump not only acts as a thermal sink for the cold incoming reactant, but also dilutes it, thereby reducing the reaction rate and the rate of energy release by reaction.

If the temperature of the reactor contents is maintained constant at 163 °C, determine the total amounts of species A and B in the reactor as functions of time when it is loaded according to the schedule shown below.

Time, t (h)	Feed rate of A (lb/hr)
0–3	175
3–6	225
6–7	275
7–8	325
8–11	400
11–12	325
12–13	275
13–14	225
14–15	175
15–16	100
16–17	50
17+	0

As we will see in Illustration 10.7, this type of filling schedule is necessary to avoid dramatic exotherms that would result from sudden termination of the feed and to ensure that the heat transfer capability of the system is not exceeded.

Data and permissable assumptions

1. The reactor contents are perfectly mixed.
2. The rate expression is first-order in species A.
3. At 163 °C the reaction rate constant is 0.8 hr^{-1}.

Solution

A material balance involving the amount of species A contained within the reactor can be written as

Input = accumulation + disappearance by reaction

$$F_{A0} = \frac{dn_A}{dt} + kC_A V'_R$$

where n_A is the instantaneous number of moles of species A contained within the reactor and V_R' is the instantaneous volume occupied by the liquid solution. This equation is similar to equation 8.6.1, but it lacks the term corresponding to the effluent stream and it has been written in molal units.

Now, at any time,

$$C_A = \frac{n_A}{V_R'}$$

Thus

$$F_{A0} = \frac{dn_A}{dt} + kn_A \tag{A}$$

Equation A can be rewritten in terms of the mass of species A present in the reactor

$$\Phi_{A0} = \frac{dm_A}{dt} + km_A \tag{B}$$

where Φ_{A0} is the mass rate of flow of species A into the reactor.

Equation B may be solved in piecewise fashion to determine the mass of species A present in the reactor as a function of time. The solution can be written as

$$\int_{m_{Ai}}^{m_{A(t)}} \frac{dm_A}{\Phi_{A0} - km_A} = t - t_i$$

where one uses a constant value of Φ_{A0} appropriate to the time interval in question and m_{Ai} is the mass of A in the reactor at the start of the time interval (time t_i).

Thus

$$\frac{1}{k} \ln \left(\frac{\Phi_{A0} - km_{Ai}}{\Phi_{A0} - km_A} \right) = t - t_i$$

Table 8.I.1
Material Balance Analysis for Illustration 8.11

Time, t (hr)	m_A (lb)	m_B (lb)	Total mass (lb)	Fraction of input A that remains unconverted
0	0	1500	1500	0.000
1	120	1555	1675	0.686
2	175	1675	1850	0.500
3	199	1826	2025	0.379
4	244	2006	2250	0.325
5	265	2210	2475	0.272
6	274	2426	2700	0.228
7	312	2663	2975	0.212
8	364	2936	3300	0.202
9	439	3261	3700	0.200
10	473	3627	4100	0.182
11	488	4012	4500	0.163
12	443	4382	4825	0.133
13	388	4712	5100	0.108
14	329	4996	5325	0.086
15	268	5232	5500	0.067
16	189	5411	5600	0.046
17	119	5531	5650	0.029
18	54	5596	5650	0.013
19	24	5626	5650	0.006
20	11	5639	5650	0.003

Rearranging,

$$m_A = m_{Ai}e^{-k(t-t_i)} + \frac{\Phi_{A0}}{k}\left[1 - e^{-k(t-t_i)}\right] \quad (C)$$

Using 0.8 hr^{-1} for k and the values of Φ_{A0} given in the filling schedule, equation C can be solved in piecewise fashion. The mass of B present at time t can be found by a material balance.

$$m_B = m_{B0} + \int_0^t \Phi_{A0}\, dt - m_A$$

Values of the amounts of A and B present in the reactor at various times are given in Table 8.I.1.

LITERATURE CITATIONS

1. Denbigh, K. G., *Chemical Reactor Theory*, p. 3, Cambridge University Press, Cambridge, 1965.

2. Gooch, D. B., *Ind. Eng. Chem.*, 35 (927), 1943.

3. Clark, E. L., Golber, P. L., Whitehouse, A. M., and Storch, H. H., *Ind. Eng. Chem.*, 39 (1555), 1947.

4. Moss, F. D., *Ind. Eng. Chem.*, 45 (2133), 1953.

5. Walas, S. M., *Reaction Kinetics for Chemical Engineers*, p. 79, McGraw-Hill, New York, 1959.

6. Wassermann, A., *J. Chem. Soc.*, 1936 (1028).

7. Kerr, G. T., *J. Phys. Chem.*, 70 (1047), 1966.

8. Liu, S., *Chem. Eng. Sci.*, 24 (57), 1969.

9. Inukai, T., and Kojima, T., *J. Org. Chem.*, 32 (872), 1967.

10. Ratchford, W. P., and Fisher, C. H., *Ind. Eng. Chem.*, 37 (382), 1945.

11. Levenspiel, O., *Chemical Reaction Engineering*, pp. 125–127, Second Edition, Wiley, New York, copyright © 1972. Used with permission.

12. Mason, D. R., and Piret, E. L., *Ind. Eng. Chem.*, 43 (1210), 1951.

13. Levenspiel, O., op. cit, pp. 139–143.

14. Szepe, S., and Levenspiel, O., *Ind. Eng. Chem. Process Design Develop.*, 3 (214), 1964.

15. Eldridge, J. M., and Piret, E. L., *Chem. Eng. Prog.*, 46 (290), 1950.

16. Jones, R. W., *Chem. Eng. Prog.*, 47 (46), 1951.

17. Weber, A.P., *Chem. Eng. Prog.*, 49 (26), 1953.

18. Jenney, T. M., *Chem. Eng.*, 62 (12), p. 198, 1955.

19. Crooks, W. M., *British Chem. Eng.*, 2 (710), 1966.

20. Denbigh, K. G., *Trans. Far. Soc.*, 40 (352), 1944.

21. Levenspiel, O., op. cit., p. 136.

22. Levenspiel, O., op. cit., p. 137.

23. Levenspiel, O., op. cit., pp. 144–148.

24. Levenspiel, O., op. cit., p. 144.

25. Kladko, M., *Chem. Tech.*, 1 (141), 1971. Adapted with permission from *CHEMTECH*, the polydisciplinary magazine of the American Chemical Society.

PROBLEMS

1. Baciocchi et al. [*J. Am. Chem. Soc.*, 87 (3957), 1965] have studied the chlorination of dichlorotetramethylbenzene in acetic acid at 30° C. The reaction of interest has the following stoichiometry.

$$C_6Me_4Cl_2 + Cl_2 \rightarrow HCl + C_6Me_3(CH_2Cl)Cl_2$$

The data below are typical of those recorded in a well-stirred batch reactor. Initial concentrations were as follows.

$$Cl_2 = 19.2 \text{ moles/m}^3$$
$$C_6Me_4Cl_2 = 34.7 \text{ moles/m}^3$$

Time, t (ksec)	Fraction Cl_2 reacted
0	0
48.4	0.2133
85.1	0.3225
135.3	0.4426
171.3	0.5195
222.9	0.5955
257.4	0.6365

(a) What is the order of the reaction and the reaction rate constant?

(b) Determine the plug flow reactor volume necessary to achieve 90% conversion of the input chlorine using an input volumetric flow rate of 0.15 m^3/ksec and the same initial concentrations as used in the batch experiments.

2. The following data were recorded using a laboratory scale continuous stirred tank reactor

in which the stoichiometrically observed reaction was

$$2A \rightarrow R + S$$

Moles A fed/ksec	Moles A leaving/ksec
8.0	0.8
36	7.2
192	76.8

The effective reactor volume was 1000 cm³. The initial concentration of A was the same in all runs and equal to the solubility limit of species A in water. No R or S is present in the feed. You have been asked to scale up this reactor to produce significantly larger quantities of R. It has been suggested that you use a tubular flow reactor with an inside diameter of 2 cm and that your feed be a saturated solution of species A. No R or S is present in the feed.
(a) What is the order of the reaction?
(b) If you are to process 0.15 mole/sec of input A to 90% conversion, what length of tubular reactor do you require? The feed concentration of A will be the same as in part (a).

3. Two gaseous streams are available for use in carrying out a chemical reaction. The first contains pure A and is produced at a rate of 400 ft³/min. The second contains 50% B (remainder is an inert material) and has a flow rate of 200 ft³/min. These streams are mixed instantaneously and fed to a flow reactor. Both streams are at the same temperature (86 °C) and pressure (1 atm) and these quantities remain unchanged during the instantaneous mixing process. The gases behave ideally. A and B react to form an addition product

$$A + B \rightarrow C$$

with

$$r_C = kC_A C_B$$

The reaction is carried out isothermally in two flow reactors. Both reactors operate at a constant total pressure of 1 atm.

(a) If the reactor is a CSTR with a volume of 600 ft³ and 60% of the B is converted to C, determine
  (1) The space time for the reactor.
  (2) The average holding time in the reactor.
  (3) The outlet flow rate from the reactor (cubic feet per minute)
(b) If the reactor is a plug flow reactor with a volume of 600 ft³ and 90% of the B is converted to C, determine
  (1) The space velocity in the reactor.
  (2) The effluent volumetric flow rate.
  (3) The rate constant k.
  (4) The average holding time in the reactor.

4. A chemical reaction is being studied in a laboratory scale steady-state flow system. The reactor is a well-stirred 1000 cm³ flask containing an aqueous solution. The reactor contents (1000 cm³ of solution) are uniform throughout. The stoichiometric equation and data are given below. What is the expression for the rate of this reaction? Determine the reaction order and the activation energy.

$$\text{diacetal} + H_2O \rightarrow \text{Aldehyde} + 2 \text{ Alcohol}$$

Run	Feed Rate (cm³/sec)	Temperature, T (°C)	Concentration of Aldehyde in Effluent (kmoles/m³)
1	0.5	10	0.75
2	3.0	10	0.5
3	3.0	25	0.75

Feed concentration of diacetal is 1 kmole/m³. *Note.* Data are hypothetical. The reaction may be assumed to be zero-order in aldehyde and alcohol, and apparently zero-order in water.

5. Roper [*Chem. Eng. Sci., 2* (27), 1953] has studied the reaction of chlorine (A) with 2-ethylhexene-1 (B) in carbon tetrachloride solution. Solutions of these materials were prepared

and brought together in a mixing chamber at the inlet to a tubular flow reactor. The following data were reported at 20 °C. The initial concentrations refer to values calculated on the basis of perfect mixing.

Run	C_{A0} (moles/m^3)	C_{B0} (moles/m^3)	C_A (moles/m^3) at exit	τ (sec)
1	91	209	23	0.600
2	91	209	32	0.376
3	91	209	45	0.284
4	110	211	34	0.525
5	110	211	46	0.324
6	110	211	59	0.232

It has been suggested that the rate expression for the reaction is of the form

$$r = kC_A C_B$$

Graphically determine if this expression is consistent with the above data. If so, what is the reaction rate constant? If not, what do you recommend?

6. (a) The rate of a chemical reaction is given by

$$r = kC_A^n \qquad \text{where} \qquad r = \frac{1}{V}\frac{d\xi}{dt}$$

If 90% of the reactant A is converted to products in a reactor, and if one obtains a second reactor that is one half the size of the first, determine the increase in feed capacity that results from the following types of operation.

(1) $n = 1/2$ two plug flow reactors in series

(2) $n = 1$ two plug flow reactors in series

(3) $n = 3$ two plug flow reactors in series

(4) $n = 1/2$ two plug flow reactors in parallel

(5) $n = 1$ two plug flow reactors in parallel

(6) $n = 3$ two plug flow reactors in parallel

Assume that the exit composition from the last reactor remains unchanged in all cases and that none of the A has been converted to products prior to entering the reactor. Do not assume $\delta = 0$. Obtain a general solution. In series operation the small reactor precedes the large reactor in the sequence.

(b) Repeat part (a) for the case of continuous stirred tank reactors. Assume that there is no change in the number of moles on reaction.

7. Walter [*J. Chem. Eng. Data* 5 (468), 1960] has studied the kinetics of ethylene chloride pyrolysis over a pumice catalyst.

$$ClCH_2\text{—}CH_2Cl \rightarrow HCl + CH_2\text{=}CHCl$$

The reactor consists of a cylindrical tube 59 cm long packed with pumice stone. The catalyst charge was constant in all of the runs below and equal to W. The reactor void volume was approximately 100 cm^3. The reaction is believed to be first-order in ethylene chloride under the conditions of this study and in all cases the feedstock was pure reactant. Both the reactants and the products are gases at the conditions involved. Ideal gases may be assumed. The following data were reported at 600°C and one atmosphere.

Input flow rate (moles/ksec)	Fraction conversion
0.550	0.86
0.544	0.85
0.344	0.94

(a) What is the reaction rate constant at 600 °C in sec^{-1}?

(b) If the reaction rate constant at 500 °C is 0.141 sec^{-1}, what is the activation energy for the reaction?

8. Young and Hammett [*J. Am. Chem. Soc.*, 72 (286), 1950] have studied the alkaline bromination of acetone in a stirred flow reactor ($V_R = 118$ cm^3). The stoichiometric equation for the bromination is usually considered to be:

$$(CH_3)_2CO + 3BrO^- \rightarrow$$
$$CHBr_3 + CH_3CO_2^- + 2OH^-$$

and the reaction rate is believed to be proportional to the concentrations of acetone and hydroxyl ions and independent of the concentration of hypobromite ion. Determine if the data below are consistent with this rate expression. If they are, what is the value of the reaction rate constant?

9. The catalytic dehydrochlorination of tetrachloroethane has been studied by Shvets, Lebedev, and Aver'yanov [*Kinetics and Catalysis*, 10 (28), 1969].

$$C_2H_2Cl_4 \rightarrow C_2HCl_3 + HCl$$

The reaction is first-order with respect to tetrachloroethane with a rate constant

$$k = 10^{12} e^{-21,940/T} \text{ sec}^{-1}$$

when T is expressed in degrees Kelvin. During the reaction, small amounts of other products are produced by side reactions. The chlorine produced by the reaction can serve as a catalyst poison at concentrations above 150 ppm on a mole basis. These investigators have reported the following values for the mole ratio of chlorine to HCl.

Temperature, T (°C)	Ratio
408	1.7×10^{-4}
440	3.2×10^{-4}
455	4.0×10^{-4}

Run	Inlet hydroxide concentration (moles/m^3)	Inlet acetone concentration (moles/m^3)	Inlet hypobromite concentration (moles/m^3)	Effluent acetone concentration (moles/m^3)	Effluent hydroxide concentration (moles/m^3)	Total flow rate (cm^3/ksec)
16	1.930	1.593	6.35	0.442	4.259	37.10
17	1.930	1.593	6.35	0.447	4.249	37.17
18	1.665	1.500	6.60	0.184	4.354	19.62
19	1.665	1.500	6.60	0.176	4.370	19.60
20	1.665	1.500	6.60	0.177	4.368	19.65
21	2.410	1.048	8.40	0.206	4.119	28.27
22	2.410	1.048	8.40	0.208	4.115	28.55
23	2.124	1.599	6.40	0.604	4.129	56.42
24	2.124	1.599	6.40	0.581	4.176	55.20
25	1.980	1.492	6.94	1.028	2.911	169.05
26	1.980	1.492	6.94	1.026	2.914	169.77
27	2.666	1.649	5.92	0.909	4.151	125.83
28	2.666	1.649	5.92	0.937	4.095	126.35

It has been suggested that a pilot plant operation to determine the feasibility of developing this process be carried out in a tubular flow reactor with a volume of 0.15 m³. It is suggested that the reactor operate at 450 °C and 1 atm with a feed flow rate of 41.7 moles of pure tetrachloroethane per kilosecond. Will the catalyst be susceptible to poisoning under these operating conditions?

10. Asmus and Houser [*J. Phys. Chem.*, *73* (2555), 1969] have studied the kinetics of the pyrolysis of acetonitrile over the temperature range from 880 to 960 °C at 101 kPa using a stirred flow reactor with helium as a carrier gas. They monitored the reaction kinetics by using chromatography for quantitative analysis of the unreacted acetonitrile. The reactant mixture was sufficiently dilute that volumetric changes accompanying the reaction may be neglected. From the data below determine the apparent order of the reaction, the reaction rate constant at 880 and 940 °C, and the activation energy. Do *not* assume that the order is an integer.

Initial acetonitrile concentration (moles/m³)	Space time (sec)	Fraction acetonitrile reacted
880° C		
0.219	6.7	0.116
0.206	13.4	0.171
0.500	12.9	0.182
0.516	19.2	0.250
0.832	18.5	0.290
0.822	24.5	0.308
0.820	15.8	0.246
940° C		
0.196	6.3	0.333
0.785	5.7	0.404
0.196	12.5	0.504
0.196	18.7	0.574
0.177	2.6	0.177

11. Balasubramanian, Rihani, and Doraiswamy [*Ind. Eng. Chem. Fundamentals*, *5* (184), 1966]

have studied the reaction of ethylene and chlorine in liquid ethylene dichloride solution in a CSTR. The stoichiometry of the reaction is

$$C_2H_4 + Cl_2 \rightarrow C_2H_4Cl_2$$

Equimolar flow rates of ethylene and chlorine were used in the following experiment, which was carried out at 32 °C.

Space time (sec)	Effluent chlorine concentration (moles/cm³)
0	0.0117
300	0.0095
600	0.0082
900	0.0072
1200	0.0065
1500	0.0060
1800	0.0057

(a) Determine the overall order of the reaction and the reaction rate constant.
(b) Determine the space time necessary for 75% conversion in a CSTR.
(c) What would be the conversion in a PFR having the space time determined in part b?
In parts (b) and (c) assume that the operating temperature and the initial concentrations are the same as in part a.

12. Schultz and Linden [*Ind. Eng. Chem. Process Design and Development*, *1* (111), 1962] have studied the hydrogenolysis of low molecular weight paraffins in a tubular flow reactor. The kinetics of the propane reaction may be assumed to be first-order in propane in the regime of interest. From the data below determine the reaction rate constants at the indicated temperatures and the activation energy of the reaction.

Feed ratio $H_2/C_3H_8 = 2.0$ in all cases

Reactor pressure = 7.0 MPa in all cases

Temperature, T (°C)	Space time (sec)	Fraction propane converted
538	0	0
	42	0.018
	98	0.037
	171	0.110
593	40	0.260
	81	0.427
	147	0.635

For this problem the stoichiometry of the main reaction may be considered to be of the form

$$H_2 + C_3H_8 \rightarrow CH_4 + C_2H_6$$

13. Consider the following homogeneous gas phase reaction.

$$A + B \rightarrow C + D$$

The reaction is essentially "irreversible," and its rate in a constant volume batch reactor is given by

$$\frac{1}{V}\frac{dn_A}{dt} = -kC_A C_B$$

At the temperature of interest,

$$k = 100 \text{ m}^3/\text{kmole·sec}$$

Compounds A and B are available in the off-gas stream from an absorption column at concentrations of 20 moles/m^3 each. 14 m^3/sec of this fluid is to be processed in a long isothermal tubular reactor. If the reactor may be assumed to approximate a plug flow reactor, what volume of pipe is required to obtain 80% conversion of species A?

14. Consider the homogeneous isothermal gas phase decomposition of CH_3CHO.

$$CH_3CHO \rightarrow CH_4 + CO$$

From experiments in a constant volume batch reactor at 791 °K, it is known that the time

required for a 50% increase in total pressure is 197 sec. The initial pressure is 1 atm. The reaction is known to be second-order in acetaldehyde. You have been asked to determine the volume of a plug flow reactor necessary to process 120 liters/min of pure acetaldehyde to an 80% conversion level. The feed pressure is 1 atm. The reaction may be assumed to be essentially irreversible. The pressure drop along the length of the plug flow reactor is negligible.

15.* Ratchford and Fisher found that methylacetoxypropionate decomposes on heating to form acetic acid and methyl acrylate:

$$CH_3COOCH(CH_3)COOCH_3 \xrightarrow{k_1}$$
$$CH_3COOH + CH_2{=}CHCOOCH_3$$

The pyrolysis closely approximates a first-order irreversible reaction with a rate constant given by:

$$k_1 = 7.8 \times 10^9 \exp\left(\frac{-38,200}{RT}\right) \text{sec}^{-1}$$

where T is expressed in degrees Kelvin and R in calories per gram mole per degree Kelvin. If a plug flow reactor is used to carry out the pyrolysis, calculate the volume of pipe necessary to achieve 85% conversion of the raw material to products. The raw material enters at a temperature of 1000°R and a pressure of 5 atm with a flow rate of 1000 lb/hr.

If three perfectly mixed continuous stirred tank reactors of equal volume were used in series flow instead, what would the required volume be?

Note. In both cases ideal gases may be assumed. In both cases the reaction occurs at constant temperature.

* Adapted from *Chemical Process Principles, Part 3, Kinetics and Catalysis,* by O. A. Hougen and K. M. Watson, copyright © 1947. Reprinted by permission of John Wiley and Sons, Inc.

16. Buckles and McGrew [*J. Am. Chem. Soc.* *88* (15), 1966] have studied the dimerization of phenyl isocyanate in liquid solution in the presence of a catalyst.

$$2C_6H_5NCO \underset{k_b}{\overset{k_f}{\rightleftharpoons}} C_6H_5N \overset{\overset{O}{\underset{\parallel}{C}}}{\underset{\underset{\parallel}{C}}{}} NC_6H_5$$

phenyl
isocyanate   1,3 diphenyl 2,4-uretidinedione
(monomer)   (dimer)

The forward reaction is third-order (second-order with respect to monomer and first-order with respect to catalyst). The reverse reaction is second-order overall (first-order with respect to both catalyst concentration and dimer). The reaction is catalyzed by tributylphosphine at a concentration of 0.05 moles/liter.

The following data relative to the reaction at 25 °C are available:

$$K_e = 0.178$$
$$k_f = 1.15 \times 10^{-3} \text{ liters}^2/\text{mole}^2\text{-sec}$$
$$E_{Af} = 1.12 \text{ kcal/mole}$$
$$E_{Ab} = 11.6 \text{ kcal/mole}$$

If a monomer solution at a concentration of 1 mole/liter is fed to a CSTR at 0 °C, determine the space time necessary to achieve a conversion corresponding to 90% of the equilibrium value. If the reactor volume is 100 liters, what is the corresponding volumetric flow rate?

17. An alcohol A and an acid B are fed to a CSTR in equimolal proportions. The mechanistic and stoichiometric equation for the reaction is

$$A + B \underset{k_2}{\overset{k_1}{\rightleftharpoons}} E + H_2O$$

where E is the ester produced by the reaction. The reaction occurs at a constant temperature in an acetone solution of these species. The extent of reaction is limited by equilibrium conditions. Neglect the volume change on the reaction.

The following two data points have been reported

Space velocity	Fraction conversion of acid
33 ksec^{-1}	0.50
67.65 ksec^{-1}	0.40

What is the equilibrium degree of conversion?

18. Kem Engineer has been asked to scale up an existing process to obtain an increased production capacity for compound B. At present the process is carried out in two CSTR's in series. The reaction involved has the following stoichiometry.

$$2A \rightarrow B + C \quad \text{(liquid phase reaction)}$$

Unfortunately, the data from which the rate constants were originally determined have been lost. Kem believes that he can determine a rate constant from measurements on the plant's present system for manufacturing compound B.

Data

Volume of first CSTR	30 gal
Volume of second CSTR	40 gal
Feed to first CSTR	pure A
Fraction conversion of A in first CSTR	0.60
Overall fraction conversion of A by the two CSTR's in series	0.80
Volumetric feed rate to first CSTR	900 gal/hr

The reaction is known to be second-order in A.
(a) Is it possible for Kem to determine a rate constant for the reaction from these data? If not, what additional data does he need?

(b) The plant production capacity is to be tripled and two CSTR's in series are to be used in the new layout. The overall conversion of species A is to remain the same. It has been suggested that the 40-gal CSTR be used in the new layout as the first tank in series. What will be the size of the second CSTR?

19. A combination of two identical CSTR's in series is to be used to prepare a mixture of polysulfonated aromatic compounds. The reaction will occur isothermally in the liquid phase and may be represented as

$A + S \xrightarrow{k_1}$ ortho T $k_1 = 0.8$ liter/mole-hr

$A + S \xrightarrow{k_2}$ meta T $k_2 = 0.9$ liter/mole-hr

$A + S \xrightarrow{k_3}$ para T $k_3 = 0.3$ liter/mole-hr

Each of these reactions is first-order in A and first-order in S. The inlet concentration of A is equal to 5 moles/liter. The reactor combination is to be operated under conditions such that the fraction conversion of A based on the inlet concentration is 0.4 leaving the first reactor and 0.6 leaving the second reactor.
(a) What should the inlet concentration of S be?
(b) If the reactor volume is 100 liters, how much A can be processed per hour? Assume that A is the limiting reagent.

20. Acetaldehyde is to be decomposed in a tubular reactor operating at 520 °C and 101 k Pa. The reaction stoichiometry is

$$CH_3CHO \rightarrow CH_4 + CO$$

Under these conditions the reaction is known to be irreversible with a rate constant of 0.43 m³/kmole·sec. If 0.1 kg/sec of acetaldehyde is fed to the reactor, determine the reactor volume necessary to achieve 35% decomposition.

21. The following gas phase reaction takes place at 120 °C in a tubular reactor.

$$A + 2B \rightleftharpoons C + D$$

The initial concentrations of A and B in the feedstream are each 10 moles/m³. The remainder of the stream consists of inerts at a concentration of 30 moles/m³. The reaction is reversible and substantial amounts of all species exist at equilibrium under the pressure and temperature conditions employed. The forward reaction is first-order with respect to A and first-order with respect to B. At 120 °C the rate constant for the forward reaction is 1.4 m³/mole·ksec. The reverse reaction is first-order in C, first-order in D, and inverse first-order in B. The rate constant for the reverse reaction is 0.6 ksec^{-1}.

Determine the reactor volume necessary to convert 60% of the limiting reagent at a total input flow rate of 100 liters/hr.

22. Korbach and Stewart [*Ind. Eng. Chem. Fundamentals*, 3 (24), 1964] have studied the vapor phase hydrogenation of benzene in a batch recycle reactor.

$$3H_2(g) + C_6H_6(g) \rightleftharpoons \hexagon (g)$$

cyclohexane

This reference contains data on both the equilibrium constant for the reaction and the reaction rate.

At $\dfrac{1}{T} = 1.60 \times 10^{-3}$ $K_p = 0.0053$

$\dfrac{1}{T} = 1.80 \times 10^{-3}$ $K_p = 0.85$

where T is measured in degrees Kelvin. The rate data below were obtained using a feed ratio of 12 moles of hydrogen to 1 mole of benzene. The catalyst used was a commercial platinum-alumina material having a surface area of 500 m²/g and a bulk density of 0.79 g/cm³.

Operating pressure 7 atm

Benzene conversion	Rate at 500 °F (g moles/hr-g catalyst)	Rate at 600 °F (g moles/hr-g catalyst)
0.05	0.185	0.280
0.10	0.180	0.270
0.20	0.170	0.250
0.30	0.160	0.225
0.40	0.145	0.200
0.50	0.130	0.175
0.60	0.120	0.140
0.70	0.090	0.100
0.80	0.065	0.060
0.85	0.050	0.035
0.90	0.035	0.015
0.95	0.020	0.00
1.00	0.00	0.00

You have been asked to design a small pilot plant facility for the production of cyclohexane using some combination of existing tubular reactors. The reactor descriptions are as follows:

Reactor	Length (ft)	Inside diameter (in.)	Maximum operating temperature (°F)
A	16	1	600
B	4	2	500

If you desire to operate with the same feed ratio and total pressure as Korbach and Stewart and if you desire to produce 400 g-moles/hr of cyclohexane in reactor A, determine the fraction of the input benzene that is converted if

(a) Reactor A alone is used at 600 °F.

(b) Reactors A and B are used in series with A first. In this case what is the fraction of the benzene converted at the outlet of reactor B? Assume that each reactor operates isothermally at its maximum operating temperature.

23. An exothermic reaction with the stoichiometry $A \rightarrow 2B$ takes place in organic solution. It is to be carried out in a cascade of two CSTR's in series. In order to equalize the heat load on each of the reactors it will be necessary to operate them at different temperatures. The reaction rates in each reactor will be the same, however. In order to minimize solvent losses by evaporation it will be necessary to operate the second reactor at 120 °C where the reaction rate constant is equal to $1.5 \text{ m}^3/\text{kmole·ksec}$. If the effluent from the second reactor corresponds to 90% conversion and if the molal feed rate to the cascade is equal to 28 moles/ksec when the feed concentration is equal to 1.0 kmole/m³, how large must the reactors be? If the activation energy for the reaction is 84 kJ/mole, at what temperature should the first reactor be operated?

24. The reaction

$$A \rightarrow B + C$$

is being studied in a continuous stirred tank reactor with a volume of 0.2 m³. The reaction is first-order in A with a value of $k = 8 \text{ ksec}^{-1}$. The reactor is presently operating at steady state with a feed concentration of A equal to 1 kmole/m³ and an input volumetric flow rate of 1.6 m³/ksec. The feed concentration is suddenly changed to 2 kmoles/m³, while the contents of the reactor remain at the same temperature. The reaction takes place in the liquid phase.

(a) Calculate the effluent composition (C_A, C_B, C_C) prior to the sudden change.

(b) Calculate the new steady-state effluent composition after the change.

(c) Derive an equation for the time variation of the effluent concentration of species A.

(d) Calculate the effluent composition 60 sec after the change in the input composition.

25. The first-order isomerization of ethylbenzene to the xylenes

$$C_2H_5C_6H_5 \rightarrow \text{xylenes}$$

has been studied in a tubular flow reactor (inside diameter = 0.1 m) filled with silica-alumina catalyst by Hanson and Engel [*AIChE J.*, *13* (260), 1967]. The reaction occurred at 900 °F and 1 atm with pure ethylbenzene feed. The density of the catalyst was 2.75 g/cm^3 and the void fraction was 0.57 (volume of voids/total volume). The catalyst was found to degenerate with time, and the first-order rate constant varied accordingly. The following data were obtained.

Time, t (hr)	Reaction rate constant (liter/hr-g catalyst)
0	0.185
1	0.161
2	0.147
4	0.135
6	0.126
10	0.116
20	0.104
40	0.091
60	0.083
100	0.072

Calculate the weight of catalyst and length of reactor (in meters) required if the feed rate is 200 g-moles/hr, the conversion is 20%, and
(a) The catalyst is regenerated every 4 hr.
(b) The catalyst is regenerated every 100 hr.
Hint: The effective rate constant is

$$\frac{\int_0^{t_{regeneration}} k(t)\, dt}{\int_0^{t_{regeneration}} dt}$$

26. An autocatalytic reaction represented by the mechanistic equations

$$A + R \underset{k_{-1}}{\overset{k_1}{\rightleftharpoons}} R + R$$

is being carried out in two identical stirred tanks operating in series. The reaction is reversible and exothermic. The reaction stoichiometry is

$$A \rightarrow R$$

225 lb moles of liquid A at 70 °F is fed to the first reactor each minute. The fraction of the inlet A converted to R in the first reactor is such that the effluent from the first reactor leaves at 85 °F. The temperature of the stream leaving the second reactor is 100 °F.

The concentration of A in the entrance stream is 1.5 lb moles/ft^3. The following table provides some information about the temperature dependence of the rate constants k_1 and k_{-1}.

T(°F)	k_1 (ft^3/lb mole-min)	k_{-1} (ft^3/lb-mole-min)
70	0.55	?
85	2.90	0.078
100	?	0.625

If each reactor volume is 50 ft^3, determine the composition of the stream leaving each reactor.

27. Hydrodealkylation reactions play important roles in several commercial petrochemical processing schemes. As an example of this type of reaction, we will consider the following:

Toluene + hydrogen → benzene + methane

The rate expression for this reaction is of the following form under the conditions of interest:

$$r = kC_T C_{H_2}^{1/2}$$

where C_T and C_{H_2} are the concentration of toluene and hydrogen, respectively. From data in the literature [Shull and Hixson, *Ind. Eng. Chem., Process Design and Development*, 5(146), 1966] it is known that the reaction rate constant at 1260°F and 800 psig total pressure is 0.316 ft$^{3/2}$/lb mole$^{1/2}$-sec.

You have available in your pilot plant a small recycle reactor whose flow pattern may be represented schematically by the following diagram.

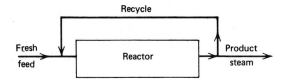

Recycle

Fresh feed → Reactor → Product steam

The volumetric flow rate of the recycle stream is many many times those of the fresh feed and product streams, and the fresh feed and recycle streams are well mixed at the juncture point. If one uses a mole ratio of 3.4 hydrogen to 1 toluene in the fresh feed stream, what fraction of the toluene is converted to benzene under the previously specified conditions? The average residence time of a fluid element is 30.1 sec. Explicitly state any assumptions that you make. In order to obtain a numerical answer, a trial and error solution will be necessary.

28. Bender and Marshall [*J. Am. Chem. Soc., 90* (201), 1968] have studied the enzymatic hydrolysis of p-nitrophenyl trimethylacetate by elastase to produce p-nitrophenol. These authors have proposed the following mechanism for this reaction.

$$E + S \xrightarrow{k_1} ES \xrightarrow{k_2} E + A$$
$$+$$
$$P$$

where

E = elastase

S = substrate = p-nitrophenyl trimethylacetate

P = product = p-nitrophenol

A = trimethylacetic acid

$k_1 = 150 \ m^3/mole \cdot ksec$

$k_2 = 2.60 \ ksec^{-1}$

(a) Derive an equation for the rate of production of species P in terms of k_1, k_2, E_0 (the total elastase concentration in its two forms, E and ES), and S the instantaneous concentration of substrate.

(b) The Badger Chemical Company is investigating the production of p-nitrophenol by this process in a continuous stirred tank

reactor with a volume of 0.3 m³. If 90% of the substrate is to be converted to the product, how large a volumetric flow rate can be processed in this reactor when the initial elastase and substrate concentrations are 10 and 100 moles/m³, respectively?

29. In recent years there has been increasing interest in the possibility of using immobilized enzymes as catalysts. The process involves attaching enzymes to solid supports and packing the supports in a tube through which liquid flows. One proposed application involves the conversion of an aqueous solution of lactose into glucose and galactose, which would permit the conversion of a waste produced in cheese manufacture to useful by-products. By averaging over the void spaces between solid particles and the particles themselves, it is possible to obtain an effective rate expression per unit volume of bed, which can be written as

$$r = \frac{kS}{S + K_M\left(1 + \dfrac{P_1}{K_1}\right)}$$

for a reaction of the type

$$S \rightarrow P_1 + P_2$$

At a given temperature the parameters k, K_M, and K_1 are constants. K_M is known as a Michaelis constant and K_1 as an inhibition constant. S and P_1 are the concentrations of reactant S and product P_1, respectively. What effective space time for a tubular reactor will be required to obtain 80% conversion of the lactose at 40 °C where $K_M = 0.0528M$, $K_1 = 0.0054M$ and $k = 5.53$ moles/(liter·min). The initial lactose concentration may be taken as 0.149M.

30.* A plant is producing nitric acid by oxidizing ammonia with air. The gases leaving the oxidation unit are cooled to condense out essentially

* Adapted with permission from C. N. Satterfield.

all of the water present. The gases leaving the cooler pass to a long pipe followed by a series of absorption towers. In these towers further oxidation of NO to NO_2 takes place. Of the NO present in the gases leaving the cooler, 90% is now being oxidized to NO_2 in the oxidizing chamber and absorption towers.

It has been suggested that the *production capacity* of the plant can be increased by introducing additional air into the gases leaving the cooler. Bucky Badger suggests that the decrease in residence time will be more than compensated for by the increased rate of reaction. He proposes that the plant throughput may be increased, keeping the conversion of the NO fixed at 90%.
(a) Is it possible to increase the production capacity as suggested?
(b) If so, what should be the number of moles of air fed per mole of gas leaving the cooler in order to maximize the production capacity? What is the percent increase in capacity under these conditions?

Data and Suggestions
The composition of the gases leaving the cooler is:

$$N_2 \quad 83.0\%$$
$$O_2 \quad 8.8\%$$
$$NO \quad 8.2\%$$

(a) Assume that this composition will be unchanged by any change in the plant production capacity.
(b) For simplification, assume the gas flow approximates plug flow at a constant temperature of 30 °C.
(c) The oxidation reaction

$$2NO + O_2 \rightarrow 2NO_2$$

is homogeneous. The reverse reaction may be neglected.
(d) Because of the large percentage of inerts in the·feed stream, the *volume change* on reaction may be neglected. This implies that our answer will only be a first approximation to the true solution.
(e) It may be assumed that absorption of the NO_2 by H_2O in the absorption towers does not affect the oxidation rate.
(f) The plant capacity is directly proportional to the quantity of NO_2 produced per unit time.
(g) The reaction rate at 30 °C is given by:

$$-\frac{dC_{NO}}{dt} = kC_{NO}^2 C_{O_2}$$

where concentrations are expressed in kilomoles per cubic meter and where

$$k = 8.0 \times 10^3 \quad m^6/kmole^2 \cdot sec$$

(h) Composition of air:

$$N_2 \quad 79\%$$
$$O_2 \quad 21\%$$

(i) Note that the addition of air increases the O_2 concentration, but also adds N_2 as an inert diluent. Thus there will be some optimum air feed rate.
(j) A useful approach is to set up an expression for the desired result in terms of the moles of air fed per 100 moles of gas leaving the cooler.

31. An addition reaction of the form $A + B \rightarrow E$ takes place in the liquid phase in a combination of two plug flow reactors in series as shown below.

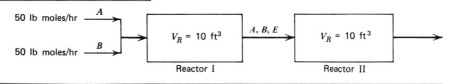

Reactor I   Reactor II

The concentrations of species A and B at the inlet to reactor I are both equal to 1.5 lb moles/ft^3. The rate expression for this reaction is given by

$$r = kC_A C_B$$

If isothermal operation at 25 °C is assumed, the value of the rate constant k is 2.0 ft^3/lb mole-hr.

(a) What fraction of the inlet A is converted to E in the first reactor?

(b) Will the fraction of the inlet A that is converted to E in the series combination of reactors be greater than, less than, or equal to twice the fraction conversion obtained in part (a)? *Why?*

(c) Would operation of the reactors in parallel give greater, lower, or the same overall conversion of A as that obtained in the present mode of operation?

Consider two conditions:

(1) Fifty percent of the A-B mixture is fed to each reactor.

(2) Sixty percent of the A-B mixture is fed to one reactor and 40% to the other.

9 Selectivity and Optimization Considerations in the Design of Isothermal Reactors

9.0 INTRODUCTION

The present chapter extends the treatment of the basic principles of reactor design to cases where multiple reactions are present. It will be shown that the choice of reactor type can have a strong influence on product distribution and thereby on the economics of the process being investigated. The material in this chapter is inextricably linked to the problem of optimization. Rigorously speaking, the choice of optimum reactor configuration should follow, not precede, investigations of the optimum operating conditions for each configuration. However, as far as selectivity considerations are concerned, it is usually possible to establish the most suitable reactor type by using relatively simple arguments based on a consideration of the various reaction rate expressions that are involved.

For reactor design purposes, the distinction between a *single* reaction and *multiple* reactions is made in terms of the number of extents of reaction necessary to describe the kinetic behavior of the system, the former requiring only one reaction progress variable. Because the presence of multiple reactions makes it impossible to characterize the product distribution in terms of a unique fraction conversion, we will find it most convenient to work in terms of species concentrations. Division of one rate expression by another will permit us to eliminate the time variable, thus obtaining expressions that are convenient for examining the effect of changes in process variables on the product distribution.

In discussions of systems in which only a single chemical reaction is involved, one may use the words yield and conversion as complementary terms. However, in dealing with multiple reactions, *conversion* refers to the proportion of a reagent that reacts, while *yield* refers to the amount of a specific product that is produced. When a number of alternative reaction paths are available to a given reactant, yield and conversion may not be simply related.

In order to avoid the possibility of obtaining yields in excess of 100% it is necessary to employ stoichiometric coefficients to normalize one's calculations properly. It is also necessary to state whether the yield is computed relative to the amount of reactant introduced into the system or relative to the amount of reactant consumed. For example, for the reaction

$$aA + bB \rightarrow rR + sS \qquad (9.0.1)$$

where we take A to be the limiting reagent, the yield of species R (Y'_R) may be defined as

$$Y'_R = \frac{a(N_R - N_{R0})}{r(N_{A0} - N_A)} \qquad (9.0.2)$$

where N_R and N_A are the moles of species R and A present after reaction and where the subscript zero refers to initial conditions. The ratio of the stoichiometric coefficients is required to give 100% yield for complete conversion to R.

The concept of yield is useful in determining the *selectivity* of a catalyst or of a given reactor and operating conditions. Different conventions have been used in assigning numerical values to selectivity, but one that is often useful is the ratio of the amount of the limiting reagent that reacts to give the desired product to the amount that reacts to give an undesirable product.

There are many industrial situations where reactor designers have opted for selective catalysts or reaction conditions even though they lead to low reactivity. Although large reactor volumes are required, the economics of these situations are more favorable than those leading to high reactivity but low selectivity. The latter situation is characterized by a smaller, cheaper reactor, but the raw material and separation costs necessary to produce a given amount of desired product are unacceptably high.

The bulk of this chapter is devoted to a discussion of optimization with regard to selectivity considerations. In the sections that follow we will take $\delta = 0$ in order to concentrate on the primary effects and to simplify the discussion. Consequently, in this chapter, the terms space time, mean residence time, and holding time may be used interchangeably.

9.1 PARALLEL REACTIONS

The possibility of a species reacting by parallel paths to yield geometric isomers or entirely different products is often responsible for low yields of a desired product. If circumstances are such that the orders of the desired and unwanted reactions are different with respect to one or more species, it is possible to promote the desired reaction by an appropriate choice of reactor type and reaction conditions.

Following the treatment of Levenspiel (1) we shall consider a set of parallel reactions in which only a single reactant species has any influence on the corresponding reaction rate expressions.

$$A \overset{k_1}{\rightarrow} V \text{ (desired or valuable product)}$$
$$(9.1.1)$$

$$A \overset{k_2}{\rightarrow} W \text{ (undesired or relatively worthless product)}$$
$$(9.1.2)$$

The corresponding rate expressions are:

$$r_V = \frac{dC_V}{dt} = k_1 C_A^{\alpha_1} \qquad (9.1.3)$$

$$r_W = \frac{dC_W}{dt} = k_2 C_A^{\alpha_2} \qquad (9.1.4)$$

Elimination of time as a variable between equations 9.1.3 and 9.1.4 gives

$$\frac{r_V}{r_W} = \frac{dC_V}{dC_W} = \frac{k_1}{k_2} C_A^{(\alpha_1 - \alpha_2)} \qquad (9.1.5)$$

It follows that one obtains maximum selectivity by choosing reaction conditions such that the ratio (r_V/r_W) always has its highest value. This ratio is often referred to as the *instantaneous selectivity*.

For a specific system at a given temperature, nature dictates the values of k_1, k_2, α_1, and α_2. The only factor that the engineer is at liberty to adjust and control is C_A. It may be maintained at a high level by using a batch or plug flow reactor, by operating at low conversions, by increasing the pressure in gas phase systems, and by avoiding the use of inert diluents in the feed. Low concentrations of a reactant are achieved using a CSTR, operating at high conversions, lowering the pressure in gaseous systems, and adding inerts to the feed stream. Note that in this case the desire for selectivity works at cross purposes to the desire for a small reactor size so that a good design with respect to one constraint may be poor with respect to the other. In this situation a detailed economic analysis is necessary to optimize the design.

Let us now consider the three possible rankings of the reaction orders in order to determine when the concentration of species A should be kept at high or low values.

Case I. $\alpha_1 > \alpha_2$
In this case the order of the desired reaction is higher than that of the unwanted reaction so the exponent on the concentration is positive. The instantaneous selectivity is promoted by employing high concentrations of reactant. Consequently, batch or plug flow reactors are most appropriate from a selectivity viewpoint.

Case II. $\alpha_2 > \alpha_1$
In this situation the order of the unwanted reaction is greater than that of the desired reaction, so the selectivity is enhanced by using low concentrations of reactant. A CSTR is appropriate from a selectivity standpoint. Unfortunately, this situation is the one in which the selectivity considerations work against the desire for a small reactor size.

Case III. $\alpha_1 = \alpha_2$

If the orders of the two parallel reactions are identical, the selectivity is a constant given by the ratio of the rate constants

$$\frac{r_V}{r_W} = \frac{k_1}{k_2} \qquad (9.1.6)$$

The relative product yields in this case are insensitive to the type of reactor used, so reactor volume considerations will govern the choice of reactor type.

In all three cases it is possible to influence the product distribution by changing process conditions that bring about changes in the ratio of the reaction rate constants. This can be accomplished by changing the operating temperature if the activation energies of the two rate constants are different. Use of a catalyst to accelerate the desired reaction and/or an inhibitor to repress the unwanted reaction are also methods by which the selectivity can be enhanced. The last approach is often the most effective means of controlling the product distribution. The use of catalysts may, however, lead to changes in the observed reaction order with concomitant implications for the type of reactor that is preferred.

In the case of other parallel reactions with different reaction rate expressions, similar analyses can be used to determine the influence of various reactant concentrations on the selectivity of a proposed process. Such analyses would lead to the following generalization, which is useful in considerations of parallel reactions where the reactant concentration level influences the product distribution.

High reactant concentrations favor the reaction of higher order, while low reactant concentrations favor the reaction of lower order.

In cases where the orders of the reactions in question are the same with respect to a particular reactant, the product distribution is independent of the concentration level of that species. It is, of course, possible to have situations in which the selectivity is enhanced by a high concentration of one reactant and by a low concentration of a second reactant. The higher the concentration dependence, the more important it is to keep a particular species at a high or a low value. Figures 9.1 and 9.2, adapted from Levenspiel (1), indicate some possible modes of contacting in batch and continuous flow reactors for some common

Desired strategy

| C_A, C_B both high | C_A, C_B both low | C_A high, C_B low |

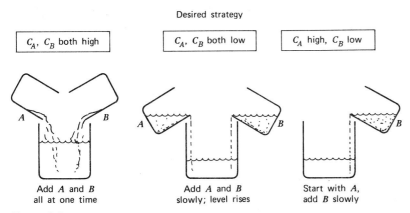

Add A and B all at one time | Add A and B slowly; level rises | Start with A, add B slowly

Figure 9.1

Contacting patterns for various combinations of high and low concentration of reactants in noncontinuous operations (batch reactors). (Adapted from *Chemical Reaction Engineering,* **Second Edition, by O. Levenspiel. Copyright © 1972. Reprinted by permission of John Wiley and Sons, Inc.)**

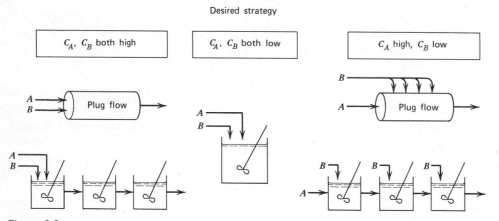

Figure 9.2

Contacting patterns for various combinations of high and low concentration of reactants in continuous flow operations. (Adapted from *Chemical Reaction Engineering*, **Second Edition, by O. Levenspiel. Copyright © 1972 Reprinted by permission of John Wiley and Sons, Inc.)**

situations. The most desirable contacting pattern and mode of operation can be determined only by considering several alternative processing modes and the possibility of recycle of reactants, either prior to or subsequent to separation of the products.

In order to determine the product distribution quantitatively, it is necessary to combine material balance and reaction rate expressions for a given reactor type and contacting pattern. On the other hand, if the reactor size is desired, alternative design equations reflecting the material balances must be employed. For these purposes it is appropriate to work in terms of the fractional yield. This is the ratio of the amount of a product formed to the amount of reactant consumed. The *instantaneous fractional yield* of a product V (denoted by the symbol y) is defined

$$y = \frac{v_A}{v_V} \frac{dC_V}{dC_A} \qquad (9.1.7)$$

while the *overall fractional yield* Y is defined as

$$Y = \frac{v_A}{v_V} \left(\frac{C_{VF} - C_{V0}}{C_{AF} - C_{A0}} \right) \qquad (9.1.8)$$

Appropriate stoichiometric coefficients are employed to ensure that y and Y lie between zero

and unity. The overall fractional yield is the average of the instantaneous fractional yield integrated over the reactor. The proper averaging technique depends on the type of reactor employed. CSTR's do not present any problems with respect to the averaging process, because the fluid composition is constant throughout the volume of the reactor. In this case the instantaneous and overall yields are identical:

$$Y_{CSTR} = y_{CSTR} = \frac{v_A r_{VF}}{v_V r_{AF}} \qquad (9.1.9)$$

where the rate expressions are evaluated at the reactor effluent composition.

For a plug flow or a batch reactor where the reactant concentration varies with position or with time, the overall yield can be determined by noting that

$$C_{VF} - C_{V0} = \int_{C_{A0}}^{C_{AF}} \frac{v_V y}{v_A} dC_A \qquad (9.1.10)$$

Substitution into the definition for the overall yield gives

$$Y = \frac{1}{C_{AF} - C_{A0}} \int_{C_{A0}}^{C_{AF}} y \, dC_A$$

$$= \frac{1}{C_{A0} - C_{AF}} \int_{C_{AF}}^{C_{A0}} y \, dC_A \qquad (9.1.11)$$

It is possible to extend this treatment to the case of multiple CSTR's operating in series by adapting the procedure outlined by Denbigh and Turner (2). Let $(\Delta C_A)_1$, $(\Delta C_A)_2$, and $(\Delta C_A)_i$, represent the changes in the concentration of species A that take place in tanks one, two, and i, respectively.

$$(\Delta C_A)_i = (C_A)_i - (C_A)_{i-1} \quad (9.1.12)$$

where $(C_A)_i$ is the concentration of reactant A in tank i. The associated changes in the concentrations of the valuable product V are given by

$$(\Delta C_V)_i = \frac{v_V}{v_A}(\Delta C_A)_i y_i \quad (9.1.13)$$

where y_i is the yield characteristic of the steady-state concentrations prevailing in the ith tank. This relation follows directly from equation 9.1.7 as applied to a finite process taking place at constant y. The overall change in the concentration of species V is obtained by summing over the pertinent number of tanks (n).

$$\Delta C_V = C_{Vn} - C_{V0} = \frac{v_V}{v_A} \sum_{i=1}^{n} [y_i(\Delta C_A)_i] \quad (9.1.14)$$

Combination of this result with the definition of the overall yield given by equation 9.1.8 gives

$$Y = \frac{\sum_{i=1}^{n} [y_i(\Delta C_A)_i]}{C_{An} - C_{A0}} \quad (9.1.15)$$

This equation is the CSTR cascade analog of equation 9.1.11 for a PFR. It indicates that the overall yield is a summation over the instantaneous yields weighted by the fraction of the concentration change that takes place in each tank.

For those cases where the rate expressions for all reactions taking place in the system under study are known, the use of the instantaneous yield in the above equations does not contribute significantly to understanding the system behavior. In such cases it is easier to determine the overall yield by substituting the appropriate ratio of reaction rate expressions for the instan-

taneous yield in equations like equation 9.1.11 and then evaluating the integral directly. Unfortunately, there are many significant industrial reactions, particularly heterogeneous catalytic reactions utilizing a complex feed stock, for which the formal rate expressions have not been determined. In these cases the concept of instantaneous yield can sometimes be quite useful. This situation occurs when the instantaneous yield depends on only a single composition variable. In such cases y can often be measured with far less effort than would be needed to determine the formal rate expressions for the various competing reactions. Plots of the instantaneous yield versus composition may be determined by carrying out a series of steady-state experiments in a CSTR. All runs are carried out at the same temperature and catalyst conditions. The shape of the curve determines which type of reactor configuration gives rise to the optimum product distribution.

Figure 9.3 contains typical instantaneous yield versus reactant concentration plots and the shaded areas indicate the composition changes of the desired product that are effected by various reactor types. From the definition of the overall yield,

$$C_{VF} - C_{V0} = -\frac{v_V}{v_A} Y(C_{A0} - C_{AF}) \quad (9.1.16)$$

Combination of this equation and equation 9.1.11 gives

$$C_{VF} - C_{V0} = -\frac{v_V}{v_A} \int_{C_{AF}}^{C_{A0}} y \, dC_A \quad (9.1.17)$$

Hence the area under the curve of y versus C_A multiplied by the ratio of stoichiometric coefficients represents the overall change in valuable product concentration between the inlet and outlet streams in a plug flow reactor or in a batch reactor. For the case of a CSTR the instantaneous yield is evaluated at the effluent composition, and the corresponding equation is

$$C_{VF} - C_{V0} = -\frac{v_V}{v_A} y_F(C_{A0} - C_{AF}) \quad (9.1.18)$$

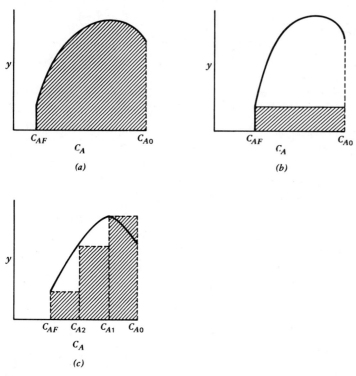

Figure 9.3
Instantaneous yield versus reactant concentration curves and their relation to overall product concentration changes. (a) plug flow or batch reactor. (b) CSTR. (c) three CSTR's in series. (Adapted from *Chemical Reaction Engineering*, **Second Edition, by O. Levenspiel. Copyright © 1972. Reprinted by permission of John Wiley and Sons, Inc.)**

so the pertinent area in this case is the rectangle shown in Figure 9.3b. For a staged cascade of stirred tank reactors a similar analysis indicates that the pertinent area is that given by the sum of the rectangles corresponding to the individual tanks.

The shape of the instantaneous yield curve determines the optimum reactor configuration and flow pattern for a particular reaction network. For cases where the instantaneous yield increases continuously with increasing reactant concentration, the optimum reactor configuration from a product selectivity viewpoint is a plug flow or batch reactor. If the instantaneous yield decreases continuously with increasing reactant concentration, the optimum product distribution is achieved using a continuous stirred tank reactor. When maxima or minima are observed in the instantaneous yield curve, the desired selectivity is enhanced by the use of a combination of plug flow and stirred tank reactors. However, considerations other than selectivity may influence the final choice of reactor type, especially when the instantaneous yield curve is relatively flat over the reactant concentration range of interest.

For cases where there are three or more reactions acting in parallel, the rate expressions may be such that the order of the reaction that leads to the desired product lies between the orders of reactions that lead to undesirable products. In this case the relative significance of the competing reactions changes as the reaction proceeds. Initially the yield is diminished because of formation of the by-product created by the highest-order reaction. This effect decreases as the reaction progresses. At high conversions, on the other hand, the yield is decreased by formation of by-products resulting from the low-order reaction. In this situation the yield is optimized by using a stirred tank reactor followed by a plug flow reactor (3). This procedure permits one to "skip" the high concentration range where one of the by-products is formed.

Illustration 9.1 indicates how the principles enunciated above may be used in optimizing the yield of a desired product when dealing with parallel reactions.

ILLUSTRATION 9.1 QUANTITATIVE TREATMENT OF IRREVERSIBLE PARALLEL REACTIONS OCCURING IN THE LIQUID PHASE

Statement

Species A is present in liquid solution at an initial concentration C_{A0}. It may undergo the reactions indicated by the following mechanistic equations:

i. $A \overset{k_1}{\to} B$ (isomerization) $r = k_1 C_A$

ii. $2A \overset{k_2}{\to}$ Products (disproportionation or dimerization) $r = k_2 C_A^2$

Neither B nor the undesirable products are present in the feedstream. Determine the maximum yields of B that can be obtained in the limit where the conversion level approaches 100% for both a plug flow reactor and a continuous flow stirred tank reactor.

Solution

Equation 9.1.11 is appropriate for determining the overall yield in a plug flow reactor.

$$Y_{PFR} = \frac{1}{C_{A0} - C_{AF}} \int_{C_{AF}}^{C_{A0}} y \, dC_A \quad (A)$$

where

$$y = \frac{\nu_A}{\nu_B} \frac{dC_B}{dC_A} = -\frac{dC_B}{dC_A} = \frac{k_1 C_A}{k_1 C_A + 2k_2 C_A^2}$$

$$= \frac{1}{1 + \dfrac{2k_2}{k_1} C_A} \quad (B)$$

Thus

$$Y_{PFR} = \left(\frac{1}{C_{A0} - C_{AF}}\right) \int_{C_{AF}}^{C_{A0}} \frac{dC_A}{1 + \dfrac{2k_2}{k_1} C_A} \quad (C)$$

or

$$Y_{PFR} = \frac{k_1}{2k_2(C_{A0} - C_{AF})} \ln \left(\frac{1 + \dfrac{2k_2}{k_1} C_{A0}}{1 + \dfrac{2k_2}{k_1} C_{AF}} \right) \quad (D)$$

In the limit, as C_{AF} approaches zero,

$$Y_{PFR} = \frac{k_1}{2k_2 C_{A0}} \ln \left(1 + \frac{2k_2}{k_1} C_{A0} \right) \quad (E)$$

Now consider the case of a stirred tank reactor. In this case the overall yield is given by equation 9.1.9.

$$Y_{CSTR} = \frac{\nu_A r_{BF}}{\nu_B r_{AF}} = \frac{k_1 C_{AF}}{k_1 C_{AF} + 2k_2 C_{AF}^2}$$

$$= \frac{1}{1 + \dfrac{2k_2 C_{AF}}{k_1}} \quad (F)$$

In the limit as C_{AF} approaches zero, the overall yield approaches unity. If the plug flow and stirred tank reactors are operated at less than complete conversion, the yields at a given conversion level can be evaluated from equations D and F if the values of the reaction rate constants and the initial concentration are known. In all cases, however, the yield from the CSTR will exceed that from the PFR.

This illustration has provided us with a concrete example that indicates in quantitative form the validity of the general rule of thumb that we have stated for analyzing parallel reactions. High concentrations favor the higher-order reaction, and low concentrations favor the lower-order reaction.

9.2 CONSECUTIVE (SERIES) REACTIONS $A \xrightarrow{k_1} B \xrightarrow{k_2} C \xrightarrow{k_3} D$

There are innumerable industrially significant reactions that involve the formation of a stable intermediate product that is capable of subsequent reaction to form yet another stable product. These include condensation polymerization reactions, partial oxidation reactions, and reactions in which it is possible to effect multiple substitutions of a particular functional group on the parent species. If an intermediate is the desired product, commercial reactors should be designed to optimize the production of this species. This section is devoted to a discussion of this and related topics for reaction systems in which the reactions may be considered as sequential or consecutive in character.

For the case where all of the series reactions obey first-order irreversible kinetics, equations 5.3.4, 5.3.6, 5.3.9, and 5.3.10 describe the variations of the species concentrations with time in an isothermal well-mixed batch reactor. For series reactions where the kinetics do not obey simple first-order or pseudo first-order kinetics, the rate expressions can seldom be solved in closed form, and it is necessary to resort to numerical methods to determine the time dependence of various species concentrations. Irrespective of the particular reaction rate expressions involved, there will be a specific time at which the concentration of a particular intermediate passes through a maximum. If interested in designing a continuous flow process for producing this species, the chemical engineer must make appropriate allowance for the flow conditions that will prevail within the reactor. That disparities in reactor configurations can bring about wide variations in desired product yields for series reactions is evident from the considerations in Illustrations 9.2 and 9.3.

ILLUSTRATION 9.2 QUANTITATIVE DEVELOPMENT OF SERIES REACTION RELATIONSHIPS FOR BATCH AND PLUG FLOW REACTORS

For the set of first-order consecutive reactions

$$A \xrightarrow{k_1} V \xrightarrow{k_2} W$$

determine the optimum holding time in a batch reactor and the optimum space time in a plug flow reactor in terms of maximizing the concentration of the intermediate V. What will the maximum concentration be in each case? It may be assumed that only species A is present initially.

Solution

For constant fluid density the design equations for plug flow and batch reactors are mathematically identical in form with the space time and the holding time playing comparable roles (see Chapter 8). Consequently it is necessary to consider only the batch reactor case. The pertinent rate equations were solved previously in Section 5.3.1.1 to give the following results.

$$\frac{C_A}{C_{A0}} = e^{-k_1 t} \tag{A}$$

$$\frac{C_V}{C_{A0}} = \frac{k_1}{k_2 - k_1} (e^{-k_1 t} - e^{-k_2 t}) \tag{B}$$

$$\frac{C_W}{C_{A0}} = 1 - \frac{C_A}{C_{A0}} - \frac{C_V}{C_{A0}}$$

The time corresponding to maximum yield of V is obtained by differentiating equation B with respect to time and setting the derivative equal to zero.

$$\frac{d(C_V/C_{A0})}{dt} = \frac{k_1}{k_2 - k_1}(-k_1 e^{-k_1 t} + k_2 e^{-k_2 t}) = 0$$

or

$$\frac{k_1}{k_2} = e^{-(k_2 - k_1)t_{\text{optimum}}}$$

Hence, for *plug flow* or a *batch reactor*,

$$t_{\text{optimum}} = \frac{\ell n(k_1/k_2)}{k_1 - k_2} = \frac{1}{k_{\text{log mean}}} \qquad \text{(C)}$$

The optimum time is also that at which the rate of formation of W is most rapid.

Equation B may be rewritten as

$$\frac{C_V}{C_{A0}} = \frac{k_1 e^{-k_1 t}}{k_2 - k_1}[1 - e^{(k_1 - k_2)t}] \qquad \text{(D)}$$

Substitution of equation C into equation D gives an expression for the maximum obtainable concentration of species V.

reactor from the standpoint of maximizing production of the intermediate. What will be the effluent concentration of V for this optimum operating condition? It may be assumed that species V and W are not present in the feedstream.

Solution

The effluent composition is readily obtained by writing a material balance on each species and solving the resultant set of equations. Hence,

Input = output + disappearance by reaction

For A:

$$C_{A0}\mathcal{V} = C_{AF}\mathcal{V} + k_1 C_{AF} V_R \qquad \text{(A)}$$

For V:

$$0 = C_{VF}\mathcal{V} + (k_2 C_{VF} - k_1 C_{AF})V_R \qquad \text{(B)}$$

For W:

$$0 = C_{WF}\mathcal{V} - k_2 C_{VF} V_R \qquad \text{(C)}$$

where the usual significance is attached to each symbol.

$$\frac{C_{V,\text{max}}}{C_{A0}} = \frac{k_1 e^{-[k_1 \ell n(k_1/k_2)]/(k_1 - k_2)}}{k_2 - k_1}[1 - e^{\ell n(k_1/k_2)}] = \frac{k_1}{k_2 - k_1}\left(\frac{k_1}{k_2}\right)^{-k_1/(k_1 - k_2)}\left(1 - \frac{k_1}{k_2}\right)$$

$$\frac{C_{V,\text{max}}}{C_{A0}} = \frac{k_1}{k_2 - k_1}\left(\frac{k_1}{k_2}\right)^{-k_1/(k_1 - k_2)}\left(\frac{k_2 - k_1}{k_2}\right) = \left(\frac{k_1}{k_2}\right)^{1 - [k_1/(k_1 - k_2)]} = \left(\frac{k_1}{k_2}\right)^{k_2/(k_2 - k_1)} \qquad \text{(E)}$$

For the conditions cited, this is the maximum possible yield of species V.

ILLUSTRATION 9.3 QUANTITATIVE DEVELOPMENT OF SERIES REACTION RELATIONSHIPS FOR A SINGLE CONTINUOUS STIRRED TANK REACTOR

For the set of first-order consecutive reactions considered in Illustration 9.2, determine the optimum space time in a single stirred tank

From equation A,

$$\frac{C_{AF}}{C_{A0}} = \frac{1}{1 + k_1 \dfrac{V_R}{\mathcal{V}}} = \frac{1}{1 + k_1 \tau} \qquad \text{(D)}$$

From equations B and D,

$$C_{VF} = \frac{k_1 C_{AF} V_R}{\mathcal{V} + k_2 V_R} = \frac{k_1 C_{AF}\tau}{1 + k_2 \tau}$$

$$= \frac{k_1 C_{A0}\tau}{(1 + k_1 \tau)(1 + k_2 \tau)} \qquad \text{(E)}$$

From equations C and E,

$$C_{WF} = k_2 C_{VF} \frac{V_R}{\mathscr{V}} = k_2 C_{VF} \tau$$

$$= \frac{k_1 k_2 C_{A0} \tau^2}{(1 + k_1 \tau)(1 + k_2 \tau)} \qquad \text{(F)}$$

The space time corresponding to a maximum concentration of species V is obtained by differentiating equation E with respect to τ and setting the derivative equal to zero.

$$\frac{dC_{VF}}{d\tau} = k_1 C_{A0} \left[\frac{(1 + k_1\tau)(1 + k_2\tau) - \tau[(1 + k_1\tau)k_2 + (1 + k_2\tau)k_1]}{(1 + k_1\tau)^2(1 + k_2\tau)^2} \right] = 0$$

Expansion of the products in the numerator leads to the result that

$$1 - k_1 k_2 \tau^2_{\text{optimum}} = 0$$

or

$$\tau_{\text{optimum}} = \frac{1}{\sqrt{k_1 k_2}} \qquad \text{(G)}$$

Substitution of this result into equation E gives an expression for the maximum possible effluent concentration of the desired species.

$$\frac{C_{VF\max}}{C_{A0}} = \frac{\left(\dfrac{k_1}{k_2}\right)^{\frac{1}{2}}}{\left[1 + \left(\dfrac{k_1}{k_2}\right)^{\frac{1}{2}}\right]\left[1 + \left(\dfrac{k_2}{k_1}\right)^{\frac{1}{2}}\right]}$$

$$= \frac{1}{\left[1 + \left(\dfrac{k_2}{k_1}\right)^{\frac{1}{2}}\right]^2} \qquad \text{(H)}$$

The ratio of the maximum yield achievable in a CSTR to that which can be obtained in a batch reactor is given by the ratio of equation H to equation E of Illustration 9.2. Hence,

$$\frac{Y_{\max,\text{CSTR}}}{Y_{\max,\text{PFR}}} = \frac{1}{\left[1 + \left(\dfrac{k_2}{k_1}\right)^{\frac{1}{2}}\right]^2 \left(\dfrac{k_1}{k_2}\right)^{k_2/(k_2 - k_1)}}$$

This relative yield is plotted in Figure 9.4 as a function of the relative rate constant k_2/k_1. The greatest disparity in the yields is achieved when the rate constants are identical. The more this ratio departs from unity, the more nearly equal the yields become. At very high and very low values of this ratio, the system behaves as if only a single reaction has any influence on the reactor design. The minimum value of the relative yield can be shown to be equal to 0.68. This is obviously a significant effect that can

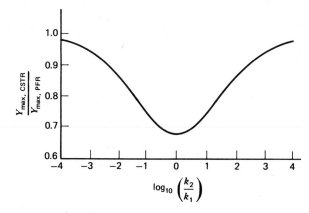

Figure 9.4
Comparison of maximum yields for series reactions in stirred tank and plug flow reactors.

strongly influence the overall process economics. Figure 9.4 indicates the range of values of k_2/k_1 for which yield considerations are significant. Successive substitution reactions usually have rate constants that lie near the minimum in the relative yield curve.

As we noted earlier, there may be heat transfer considerations or other factors that dictate the use of a CSTR even when the yield considerations are unfavorable. In such cases the yield of the desired product may be upped significantly by using a battery of CSTR's in series. If desired, one has the additional flexibility granted by using tanks with different volumes or tanks operating at different temperatures in the production line. The relative capacities of the various tanks may then be chosen to optimize product yield. This problem has been considered previously by Denbigh (4). He has shown that if both reactions are first-order and if isothermal conditions prevail, the capacities of all tanks should be the same. However, if the degradation reaction is of higher order than that producing the desired product, the capacities of the tanks should become smaller from the first tank onward. Conversely, if the degradation reaction is of lower order, then the capacities of the tanks should increase from the first tank onward.

There are a few other points worthy of note that become evident on closer inspection of the equations developed in Illustrations 9.2 and 9.3. First, except for the case where $k_2/k_1 = 1$,

the plug flow or batch reactor requires a lower space or holding time than a CSTR to achieve the maximum concentration of intermediate. The more this ratio departs from unity, the greater the difference in space times. This fact becomes evident on substitution of numerical values into equations C and G of Illustrations 9.2 and 9.3, respectively, or when plots of C_V/C_{A0} versus $k_1\tau$ are prepared for various ratios of k_2/k_1. [See, for example, Levenspiel (5).] In general, for series reactions, the maximum possible yields of intermediates are obtained when fluids of different compositions and at different stages of conversion are not allowed to mix.

Second, it is possible to plot the data in time-independent form in order to obtain curves that are useful in the determination of k_2/k_1 in kinetic studies. The experimental points are matched with one of the family of curves on the graph corresponding to the type of reactor employed in the investigation. Figure 9.5 is an

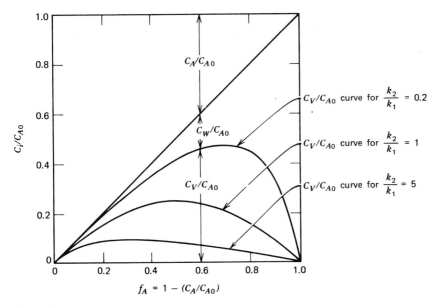

Figure 9.5
Dimensionless representation of product distribution from series reactions $A \xrightarrow{k_1} V \xrightarrow{k_2} W$ (stirred tank reactor).

example of this type of plot for a continuous stirred tank reactor.

For cases where it is possible to readily recover unused reactant A from the product mixture for recycle to the reactor inlet, the use of the definition of yield employed in Illustrations 9.2 and 9.3 is not appropriate. In this situation, a more appropriate definition is

$$Y = \frac{C_V}{C_{A0} - C_A} = \frac{C_V}{C_V + C_W} \qquad (9.2.1)$$

The equations derived earlier for the effluent concentrations in the PFR and CSTR cases may be substituted into equation 9.2.1 to obtain numerical values of the fractional yield of the intermediate V as a function of the fraction of the initial A converted. Levenspiel (6) has prepared such plots, and Figure 9.6 is reproduced from his textbook. This figure presents the fractional yield of intermediate V as a function of the ratio of rate constants (k_2/k_1) and the fraction A converted. The curves indicate that the fractional yield of the intermediate species is always higher in plug flow than when extensive back-mixing occurs, regardless of the conversion level. Moreover, the figure has important implications with regard to the conversion level of species A for which one should design. If the reaction under consideration has a value of k_2/k_1 that is greater than unity, the fractional yield of V drops very rapidly, even at low values of the conversion

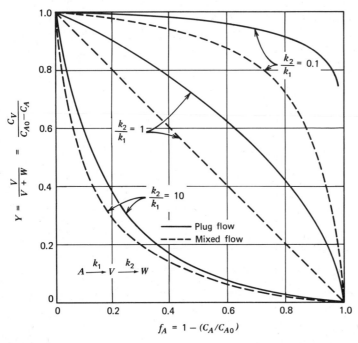

Figure 9.6

Comparison of the fractional yields of V in mixed and plug flow reactors for the consecutive first-order reactions. $A \xrightarrow{k_1} V \xrightarrow{k_2} W$. (Adapted from *Chemical Reaction Engineering,* **Second Edition, by O. Levenspiel. Copyright © 1972. Reprinted by permission of John Wiley and Sons, Inc.)**

of species A. Thus, in order to avoid excessive production of the undesirable product W, it is necessary to design for a low conversion of A per pass and recycle this species after separation from the product mixture. In such cases the load on the separation equipment will be high, and large quantities of material will have to be processed. Consequently, the costs of the separation process will strongly influence the overall process economics. As the ratio k_2/k_1 decreases below unity, the fractional yield of intermediate increases at a given conversion level. It is evident that at low values of k_2/k_1, this yield becomes relatively insensitive to fraction conversion until the conversion level begins to exceed 80 to 90%. In this situation one should design for relatively high conversion levels.

Although the bulk of the discussion in this section and the illustrative examples have been restricted to successive first-order reactions, concentration versus time curves can be developed for other cases, including those where the consecutive reactions differ in order from one another. For the batch and plug flow cases, the development requires the simultaneous solution of the pertinent differential equations. In the case of stirred tank reactors an analogous set of simultaneous algebraic equations are obtained that will generally be nonlinear. In such cases closed-form solutions are generally not available, and computers must be used to obtain numerical solutions. Fortunately, the concentration-time curves are similar in shape to those for first-order reactions, and the rules of thumb enunciated previously for that case may be regarded as general for all irreversible reactions in series. Little can be said about the product distribution curves for reactions other than first order, because they depend on the initial reactant concentration. One may conclude, as in the case of parallel reactions, that high concentrations of reactant favor the higher-order reaction and low concentrations favor the lower-order reaction. Variations in the feed concentration will shift the location of the maximum intermediate concentration, and this can be used to optimize the product distribution.

9.3 SERIES-PARALLEL REACTIONS

Reaction networks that consist of a multiplicity of reactions in which steps in series and steps in parallel are both present are often referred to as series-parallel reactions. These systems often are characterized by rate expressions that place conflicting demands on the type of contacting desired, so that it is often impossible to obtain a unique answer to a design problem. In such cases the practicing chemical engineer must exercise creativity and judgment in the choice of contacting pattern and reactor type. To illustrate the type of conflicts involved, one may consider the following combination of mechanistic equations:

$$A \rightarrow V \rightarrow W_1 \qquad (9.3.1)$$
$$A + A \rightarrow W_2 \qquad (9.3.2)$$

where

V is the desired product

W_1 and W_2 are undesirable products

To optimize the production of V, one would be persuaded by the presence of the consecutive reactions to use a plug flow or batch reactor. On the other hand, since the parallel reaction leading to W_2 is of higher order in reactant A than that leading to V, a stirred tank reactor is called for. These conflicting considerations imply that there will not be a unique solution as to the type of reactor and operating conditions to be employed. The answer will depend on the feed concentrations available and the values of the pertinent rate constants. To illustrate the general principles involved in tackling these series-parallel design problems, we will find it instructive to consider two common classes of reactions of this type: multiple substitution reactions and polymerization reactions.

9.3.1 Multiple Substitution Reactions

Multiple substitution reactions are commonly encountered in industrial practice. They may be represented in general form as:

$$A + B \xrightarrow{k_1} V \qquad (9.3.3)$$

$$V + B \xrightarrow{k_2} W \qquad (9.3.4)$$

This reaction set may be regarded as parallel reactions with respect to consumption of species B and as a series reaction with respect to species A, V, and W. Common examples include the nitration and halogenation of benzene and other organic compounds to form polysubstituted compounds. To characterize the qualitative behavior of such systems, it is useful to consider reactions 9.3.3 and 9.3.4 as mechanistic equations and to analyze the effects of different contacting patterns on the yield of species V. We shall follow the treatment of Levenspiel (7).

If one has two beakers, one containing species A, and the other containing species B, there are several ways in which their contents may be brought together. The mixing modes that represent limiting conditions are:

1. Add A slowly to B.
2. Add B slowly to A.
3. Mix A and B together rapidly.

By the use of the term slowly, we imply that the rate of mixing is slow compared to the rates of the various chemical processes involved.

In mode 1, as the first increment of A is added, it is rapidly converted to V by reaction with the B molecules. The V molecules then find themselves in the presence of excess B molecules and thus react further to yield W. The same process occurs as subsequent increments of A are supplied, the conversion rate being limited by the rate of addition of A. This mode of mixing gives rise to a situation in which one does not ever have significant amounts of V present in the vessel to which A is added. A is also absent from this vessel as long as any B remains, but it will be present after complete consumption of

B. The vessel becomes progressively richer in W until all B is consumed. Figure 9.7 is a schematic representation of the various mole numbers present in the mixing beaker.

In mode 2, a quite different situation prevails. When the first increment of B is added slowly to A, it will react to form V. The V molecules cannot react further, since all of the B molecules have been consumed. When the second increment of B is added, the V molecules formed previously will compete with unreacted A molecules for the B molecules that have been added. Since A is present in large excess during the initial stages of the addition process, it will react with most of the B to form V. The effect, however, is to increase the amount of V to the point where it competes more successfully for the B molecules. Eventually one reaches a point at which enough B will have been added so that more V will be consumed by reaction 9.3.4 than is produced by reaction 9.3.3. Beyond this point the concentration of V continues to diminish. When two moles of B have been added per mole

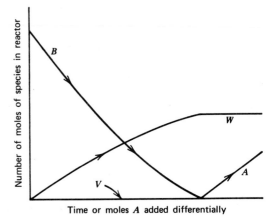

Figure 9.7

Schematic representation of product distribution versus time for parallel-series combination of reactions—Case I: series component added slowly. (Adapted from *Chemical Reaction Engineering,* **Second Edition, by O. Levenspiel. Copyright © 1972. Reprinted by permission of John Wiley and Sons, Inc.)**

of original A, the concentration of V will reach zero, and we will have a solution containing only W. Figure 9.8 is a schematic representation of the changes which occur.

In mode 3, the reaction rate is slow compared to the mixing process, and one has, in effect, a well-stirred batch reactor. Initially A and B combine to form V, which can then compete with other A molecules for the B molecules that remain. During the early stages the V molecules are at a numerical disadvantage, and more V will be produced than is consumed. As its concentration rises, V competes more successfully, eventually passing through a maximum and declining thereafter. The general behavior of this mode of mixing in terms of product distribution is quite similar to that of mode 2, which was depicted in Figure 9.8.

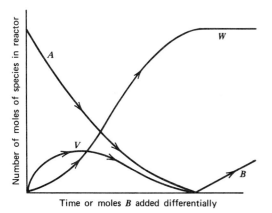

Time or moles B added differentially

Figure 9.8
Schematic representation of product distribution versus time for parallel-series combination of reactions—Case II: parallel component added slowly. (Adapted from *Chemical Reaction Engineering,* **Second Edition, by O. Levenspiel. Copyright © 1972. Reprinted by permission of John Wiley and Sons, Inc.)**

The product distributions shown in Figures 9.7 and 9.8 are obviously quite different, so we again see that the manner of contacting can have significant effects on product yields. We see that when all the A that will ever be added is present in the mixing vessel, as in modes 2 and 3, appreciable amounts of intermediate V are formed. When it is added incrementally, as in mode 1, no appreciable V is formed. Such behavior is characteristic of reactions in series, as noted in Section 9.2. As far as A, V, and W are concerned, their behavior is analogous to

$$A \overset{+B}{\to} V \overset{+B}{\to} W \qquad (9.3.5)$$

From a comparison of the qualitative results of modes 2 and 3, we see that the concentration level of B does not have a major influence on the product distribution and the reaction sequences involved. (It will, however, influence the overall conversion rate.) This behavior is characteristic of parallel reactions of the same order with respect to a particular species. From the viewpoint of species B, reactions 9.3.3 and 9.3.4 may be regarded as

$$B \overset{+A}{\underset{+V}{\diagdown}} \begin{matrix} V \\ \\ W \end{matrix} \qquad (9.3.6)$$

Levenspiel (8) has discussed these same examples and has proposed the following very useful rule.

Series-parallel reactions can be analyzed in terms of their constituent series reactions and parallel reactions in that optimum contacting for favorable product distribution is the same as for the constituent reactions.

For the reaction set 9.3.3 and 9.3.4 where V is the desired product, the rule indicates that a mixture containing A, which has reacted, should not be back-mixed with fresh A, while B may be added in any fashion which is convenient. One must apply the maxim with discretion, however, and where possible should work out the mathematics appropriate to the reaction set involved in order to obtain as much insight as possible into the factors that will influence the product distribution.

We are now prepared to develop quantitative relations for series-parallel reactions of the multiple substitution type considered above.

Case I. Quantitative Treatment for Plug Flow or Batch Reactor

The first step in the determination of the product distribution for reactions 9.3.3 and 9.3.4 is an evaluation of the instantaneous yield.

$$y = \frac{v_A}{v_V} \frac{dC_V}{dC_A}$$

$$= \frac{\dfrac{dC_V}{dt}}{\dfrac{-dC_A}{dt}} = \frac{k_1 C_A C_B - k_2 C_V C_B}{k_1 C_A C_B} \tag{9.3.7}$$

Hence, for a plug flow or batch reactor,

$$\frac{dC_V}{dC_A} = \frac{k_2 C_V}{k_1 C_A} - 1 \tag{9.3.8}$$

This differential equation can be solved using an integrating factor approach. The solution consists of two parts.

$$\frac{C_V}{C_{A0}} = \frac{1}{1 - \dfrac{k_2}{k_1}} \left[\left(\frac{C_A}{C_{A0}} \right)^{k_2/k_1} - \frac{C_A}{C_{A0}} \right] + \frac{C_{V0}}{C_{A0}} \left(\frac{C_A}{C_{A0}} \right)^{k_2/k_1} \qquad \text{for } k_2 \neq k_1 \tag{9.3.9}$$

and

$$\frac{C_V}{C_{A0}} = \frac{C_A}{C_{A0}} \left(\frac{C_{V0}}{C_{A0}} - \ell n \frac{C_A}{C_{A0}} \right) \qquad \text{for } k_2 = k_1 \tag{9.3.10}$$

One of the last two equations gives the relationship between C_V and C_A at any time in a plug flow or batch reactor. The stoichiometry of the system provides the additional information necessary to describe completely the system composition.

A material balance on species A indicates that

$$C_{A0} + C_{V0} + C_{W0} = C_A + C_V + C_W \tag{9.3.11}$$

which gives C_W. A material balance on species B gives

$$C_{B0} - C_B = (C_V - C_{V0}) + 2(C_W - C_{W0}) \tag{9.3.12}$$

and this provides a means of evaluating C_B.

Case II. Quantitative Treatment of Stirred Tank Reactor

When reactions 9.3.3 and 9.3.4 take place in a single continuous stirred tank reactor, the route to a quantitative relation describing the product distribution involves writing the design equations for species V and A.

$$\tau = \frac{C_{A0} - C_A}{-r_A} = \frac{C_{V0} - C_V}{-r_V} \tag{9.3.13}$$

or

$$\frac{C_{A0} - C_A}{k_1 C_A C_B} = \frac{C_{V0} - C_V}{-k_1 C_A C_B + k_2 C_V C_B} \tag{9.3.14}$$

Rearrangement gives

$$\frac{C_{V0} - C_V}{C_{A0} - C_A} = -1 + \frac{k_2 C_V}{k_1 C_A} \tag{9.3.15}$$

which is the difference equation analog of equation 9.3.8. Equation 9.3.15 may be solved for C_V in terms of C_A to give

$$\frac{C_V}{C_{A0}} = \frac{\dfrac{C_{V0}}{C_{A0}} + \left(1 - \dfrac{C_A}{C_{A0}} \right)}{1 + \dfrac{k_2}{k_1} \left(\dfrac{C_{A0}}{C_A} - 1 \right)} \tag{9.3.16}$$

Equations 9.3.11 and 9.3.12 are also applicable to a CSTR reactor, since they represent overall material balances. They provide the additional relations necessary to determine the complete effluent composition.

It is possible to represent the product distributions for both the batch and CSTR cases in the form of time independent plots, as shown in Figures 9.9 and 9.10. The plots are prepared using equations 9.3.9 (or 9.3.10), 9.3.11, 9.3.12, and 9.3.16. As the reaction proceeds, B is consumed, and one moves from left to right along the curve representing the appropriate value of k_2/k_1. The dashed lines of slope 2 on the figures indicate the amount of B consumed to reach a particular point on the curve being followed. This value of

consumption of B is independent of the manner in which B is added. It depends only on the total consumption. If one recognizes that the 45-degree line would represent complete conversion to the desired product, it is evident that the highest fractional yields are obtained at low fraction conversions of species A, regardless of whether k_2/k_1 is large or small. However, the larger the value of k_2/k_1, the faster the fractional yield of V drops off with increasing conversion of A. Thus, if it is possible to remove small amounts of V cheaply from large volumes of the reaction mixture, the optimum reactor configuration and mode of operation would involve the

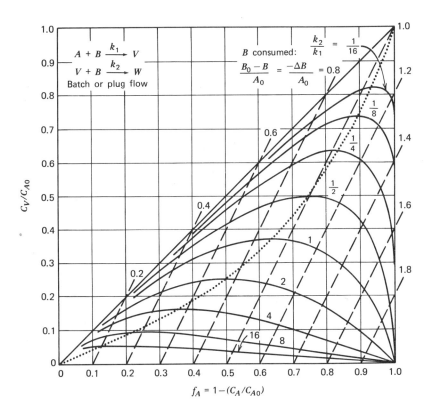

Figure 9.9
Product distribution in batch or plug flow reactors for the indicated reactions (Adapted from *Chemical Reaction Engineering*, **Second Edition, by O. Levenspiel. Copyright © 1972. Reprinted by permission of John Wiley and Sons, Inc.)**

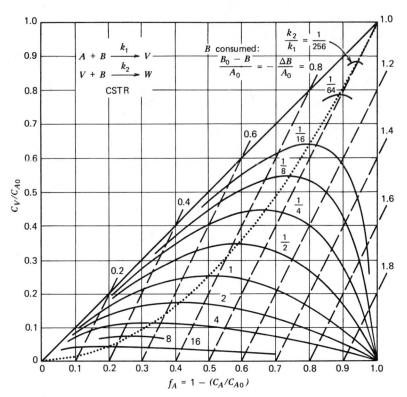

Figure 9.10

Product distribution in a continuous stirred tank reactor for the indicated reactions. (Adapted from *Chemical Reaction Engineering*, **Second Edition, by O. Levenspiel. Copyright © 1972. Reprinted by permission of John Wiley and Sons, Inc.)**

use of a plug flow reactor with low conversions of A per pass coupled with a separator to remove the product V and to recycle unconverted reactants. The exact conversion level to be employed will depend on an economic analysis of the combined reactor-separator system.

As Levenspiel points out in his discussion of this same reaction network, it is relatively easy to extend the use of figures like Figure 9.9 (PFR) to cases in which intermediates may be present in the feed to the plug flow reactor either by virtue of their presence in a recycle stream or in the raw feedstream. The progress of the reaction

will still be along the same k_2/k_1 line. However, the starting point will no longer be the point $C_V = 0$, $C_A = C_{A0}$, but the intersection of the curve corresponding to the value of k_2/k_1 characteristic of the system and a straight line passing through the point $C_V = C_A = 0$, with a slope $(-C_{V0}/C_{A0})$. The effect of the presence of intermediate V in the feed is to reduce the net fractional yield of species V.

Illustration 9.4 indicates how the concepts we have developed may be used in attempting to develop a rational reactor design for carrying out multiple substitution reactions.

ILLUSTRATION 9.4 DESIGN CONSIDERATIONS FOR A SPECIFIED PRODUCT DISTRIBUTION

Statement

For a set of parallel-series reactions represented by

$$A + B \xrightarrow{k_1} V \qquad r_1 = k_1 C_A C_B$$
$$V + B \xrightarrow{k_2} W \qquad r_2 = k_2 C_V C_B$$

it is desired to operate under conditions such that the relative yield of V is 75%, based on the amount of A that reacts. If the reaction rate constants are numerically equal, indicate briefly the constraints within which the idealized reactor types must operate. Consider conversion levels, mole ratios of the feed, and the possibility of separation and recycle of reactant or product species. It may be presumed that species A is the limiting reagent.

Solution

Figures 9.9 and 9.10 are useful for determining the conversion level at which each type of reactor should be operated. The locus of points along which the relative overall yield V is 75% is given by the straight line linking the point $f_A = 0$, $C_V/C_{A0} = 0$ to the point $f_A = 1$, $C_V/C_{A0} = 0.75$. Points above this straight line will correspond to yields in excess of the minimum specified value of 75%. The point where this straight line intersects the yield curve for the specified value of k_2/k_1 (1.0) indicates the highest conversion level at which the specified product distribution can be achieved. When such a straight line is superimposed on Figure 9.9, it is found that for a batch or plug flow reactor, the maximum possible conversion of A that is consistent with the constraint on the product distribution is $f_A = 0.42$. For a CSTR the corresponding maximum conversion level may be found using Figure 9.10. In this case $f_A = 0.25$. At the limiting conversion levels determined

above, the product mixtures will contain substantial amounts of unreacted A and B. It is thus appropriate to consider the questions of separation and recycle for each species in turn.

Reactant A

If A has significant economic value then it should be separated from the reactor effluent stream and recycled for subsequent use. Since the conversion level is higher in the plug flow reactor, the recycle rate will be much smaller and the demands on the separation equipment for reclaiming species A will also be somewhat smaller. Even when species A is of relatively little economic value, there may be circumstances when the costs associated with meeting the pollution control requirements for the process effluent will dictate separation and recycle of this reactant as the most economic alternative.

Reactant B

The product distribution is insensitive to the concentration of reactant B. If B is cheap and does not offer a potential pollution problem downstream, its concentration may be kept at any convenient level. If B is costly or must be removed for other reasons, one has the options of operating with low B concentrations at high conversions in a relatively large reactor to produce a product containing very little B, or of operating at higher B concentrations in a smaller reactor with separation and recycle of unused B. The specified product distribution requires that the mole ratio of V to W be 3:1. To produce 1 mole of W and three of V, one must consume 4 moles of A and 5 moles of B. The feed ratio employed in an actual situation may differ appreciably from 1.25 to enhance the reaction rate or to allow for discarding some A and B.

Products V and W

Separation and recycle of species V or W has no merit. Neither product enhances the reaction rate. The low conversion of A that we are forced to employ is a consequence of the fact that otherwise there would be insufficient V in the product mixture.

___

9.3.2 Polymerization Reactions

Polymerization processes represent an extremely important aspect of the chemical processing industry. Since many of the properties of polymeric materials are markedly affected by their average molecular weight and their molecular weight distribution, the design of reactors for polymerization processes offers many opportunities for the use of the principles presented earlier in this chapter.

Because tubular reactors are generally not suitable for use in effecting liquid phase or emulsion polymerizations, the principal design alternatives reduce to batch reaction or the use of single or multiple stirred tank reactors for continuous processing. In tubular reactors, the velocity profile arising from the high viscosity of the solution will imply that fluid elements near the wall have significantly longer residence times than those at the tube axis. This distribution of residence times means that the material near the wall may be polymerized to an excessive level, perhaps resulting in the precipitation of polymer and the buildup of solid material on the tube walls, eventually leading to choking or plugging of the tube.

If we represent the monomer from which a polymer is formed by M and a polymer consisting of i monomer units by P_i, a polymerization reaction can be written as

$$M + M \rightarrow P_2$$
$$P_2 + M \rightarrow P_3$$
$$P_3 + M \rightarrow P_4 \qquad (9.3.17)$$
$$\cdots \cdots \cdots \cdots$$
$$P_i + M \rightarrow P_{i+1}$$

This reaction can be regarded as a series reaction from the viewpoint of the growing polymer

molecule and as a parallel reaction from the viewpoint of the monomer molecules being consumed. Of course, if polymer molecules can react with one another,

$$P_i + P_j \rightarrow P_{i+j} \qquad (9.3.18)$$

A given sample of polymer is characterized by a distribution of molecular weights arising from the interaction of the network of reactions shown in equation 9.3.17 (and possibly equation 9.3.18), and the mixing processes taking place in the reactor in which the polymer sample was prepared.

For idealized reactor types there are two opposing factors that influence the overall molecular weight distribution.

1. *The Concentration History*. That is, in a batch reactor the monomer concentration decreases continuously whereas in a CSTR it remains constant.
2. *The Residence Time Distribution*. All fluid elements have the same residence time in a batch reactor, but there will be a wide spread in residence times in a CSTR.

In a CSTR the steady-state concentration of monomer is at a lower average value than it would be for the same feed conditions if the same reaction were carried out batchwise. In many free radical polymerization reactions, holding the monomer concentration at a constant level has the effect of reducing the variation in degree of polymerization (or molecular weight).

As the residence time of a fluid element lengthens, it is possible for the degree of polymerization of the polymer molecules contained therein to increase. Consequently, any factor that leads to a spread in residence times of individual molecules can increase the spread in the molecular weight distribution. In a CSTR some growing polymer molecules leave after a short residence time so they do not reach a very large molecular weight. Other molecules remain in the reactor for longer than the average residence time and consequently will reach a molecular weight greater than the average value.

Whether the first or the second factor dominates depends on the type of polymerization process involved. If the period during which the polymer molecule is growing is short compared to the residence time of the molecule in the reactor, the first factor dominates. This situation holds for many free radical and ionic polymerization processes where the reaction intermediates are extremely short-lived. Figure 9.11, taken from Denbigh (10), indicates the types of behavior expected for systems of this type.

For cases where the growth period is the same as the residence time in the reactor, as in polycondensation processes, the residence time distribution is the dominant factor influencing the molecular weight distribution. In this case one obtains a broader molecular weight distribution from a CSTR than from a batch reactor. Figure 9.12 [also taken from Denbigh (11)] indicates the type of behavior expected for systems of this type.

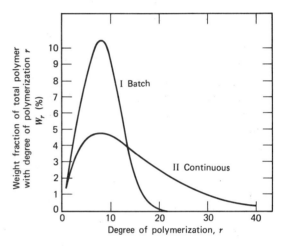

Figure 9.12

Molecular weight distribution function for the case where the length of the growth stage is long compared to the reactor residence time or for the case where there are no termination reactions. (Reprinted with permission from *Chemical Reactor Theory,* **by K. G. Denbigh and J. C. R. Turner. Copyright © 1971 by Cambridge University Press.)**

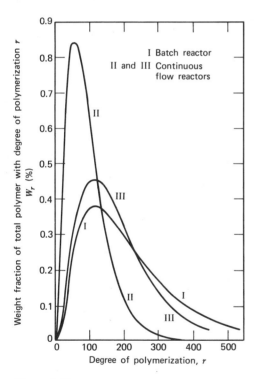

Figure 9.11

Molecular weight distribution function for the case where the length of the growth stage is short compared to the residence time in reactor. (Reprinted with permission from *Chemical Reactor Theory,* **by K. G. Denbigh and J. C. R. Turner. Copyright © 1971 by Cambridge University Press.)**

9.4 REACTOR DESIGN FOR AUTOCATALYTIC REACTIONS

9.4.1 Basic Concepts

There are many reactions in which the products formed can themselves act as catalysts for the reaction. Thus there will be a range of compositions for which the reaction rate accelerates as the reaction proceeds. This phenomenon is known as *autocatalysis*, and reactions of this type pose some interesting problems in the selection of an optimum reactor configuration.

An autocatalytic reaction is one in which the reaction rate is proportional to a product concentration raised to a positive exponent. Some of the first articles in the literature of chemical kinetics deal with reactions of this type. For example, in 1857, Baeyer (12) reported that the reaction of bromine with lactose was autocatalytic. The hydrolyses of several esters also fit into the autocatalytic category, since the acids formed by reaction give rise to hydrogen ions that serve as catalysts for subsequent reaction. Among the most significant autocatalytic reactions are the fermentation reactions that involve the action of a microorganism on an organic feedstock.

Under normal circumstances, when a material reacts, its initial rate of disappearance is high, but the rate decreases progressively as the reactant is consumed. In an autocatalytic reaction, on the other hand, the initial rate is relatively slow, because little or no product is present. The rate increases to a maximum as products are formed and then decreases to a low value as reactants are consumed or equilibrium is achieved. If there are no product species present in the initial reaction mixture, autocatalytic reactions exhibit the type of behavior shown in Figure 9.13a. If the product species that is catalytic is present in the original reaction mixture, the type of behavior that the system will exhibit is shown in Figure 9.13b.

It should be noted that we have not let the initial rate become zero in Figure 9.13 even when no catalyst is present, because the possibility

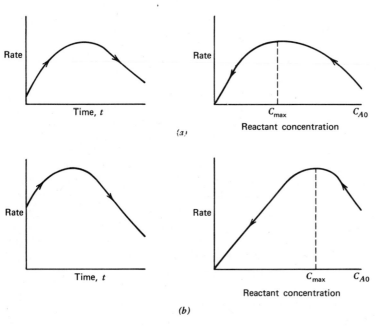

Figure 9.13

Characteristics of autocatalytic reaction. (a) No products present in initial reaction mixture. (b) Some product species present in initial mixture.

exists that there may be alternative paths from reactants to products, only one of which is autocatalytic. The observed rate will be the sum of the rates for the various paths. Reaction will proceed by the uncatalyzed path until sufficient products are produced to render conversion by this path negligible compared to conversion by the autocatalytic path. In practice one always has the possibility of spiking the original mixture with adequate catalyst but, in many cases, small amounts of product species left in the reactor from previous runs may suffice to get things going again. For hydrolysis reactions that are acid catalyzed, the hydrogen ion concentration in the feed water may be sufficient to get things started.

An experimental test for autocatalysis involves addition of the suspected autocatalytic species to the reaction mixture. If the material added is the responsible agent, one may generally expect behavior like that shown in Figure 9.14.

Illustration 9.5 indicates one type of rate expression and reaction mechanism that may be associated with an autocatalytic reaction.

ILLUSTRATION 9.5 AUTOCATALYTIC DECOMPOSITION OF 5-METHYL-2-OXAZOLIDINONE

When heated above 200 °C, pure 5-methyl-2-oxazolidinone decomposes into two products, CO_2 gas and N-(2-hydroxypropyl) imidazolidinone. The reaction stoichiometry is:

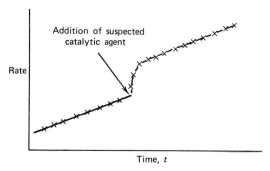

Figure 9.14

Response of autocatalytic system to addition of catalytic agent.

While we have taken some liberties with the numerical values that we will subsequently employ, the following set of mechanistic equations has been proposed in an effort to explain the observed autocatalytic kinetic behavior (13).

$$2A \xrightarrow{k_1} B + C \qquad (C)$$

$$A + B \xrightarrow{k_2} AB \qquad (D)$$

$$2AB \xrightarrow{k_3} 3B + C \qquad (E)$$

where AB is an intermediate complex. Verify that this mechanism gives rise to autocatalytic behavior.

If the following values are assumed for the reaction rate constants:

$$k_1 = 1.02 \text{ m}^3/\text{kmole·Msec}$$
$$k_2 = 150 \text{ m}^3/\text{kmole·Msec}$$
$$k_3 = 172 \text{ m}^3/\text{kmole·Msec}$$

CH₃ ... (chemical structure diagram) ... + CO₂ (A)

or

$$2A \rightarrow B + C \qquad (B)$$

where C represents CO_2.

determine the fraction conversion of A that gives rise to the maximum amount of B that can be produced in an ideal CSTR with a fixed volume

V_R. No special constraint is to be placed on the volumetric feed rate or the initial concentration of species A.

Solution

The observed rate of reaction is identical with the rate of formation of species C

$$r = k_1(A)^2 + k_3(AB)^2 \qquad (F)$$

The concentration of the reaction intermediate AB may be determined by using the customary steady-state approximation for intermediates.

$$\frac{d(AB)}{dt} = k_2(A)(B) - 2k_3(AB)^2 \approx 0 \qquad (G)$$

Hence,

$$(AB) = \sqrt{\frac{k_2(A)(B)}{2k_3}} \qquad (H)$$

Combining equations F and H gives

$$r = k_1(A)^2 + \frac{k_2}{2}(A)(B) \qquad (I)$$

Thus the mechanism gives rise to a rate expression in which species B is responsible for the autocatalytic behavior.

The design equation for a CSTR is

$$\tau = \frac{V_R}{\mathcal{V}_0} = \frac{C_{A0} \int_0^{f_A} df_A}{-r_{AF}} \qquad (J)$$

which, on rearrangement for the case where the feed contains neither B nor C, becomes

$$-r_{AF} V_R = C_{A0} \mathcal{V}_0 f_A = F_{A0} f_A \qquad (K)$$

The right side of this expression is identical with the rate of production of species B and C. Hence the maximum production rate for a fixed reactor volume occurs when the reactor contents have a composition that maximizes the specific reaction rate. Now, in terms of the fraction conversion,

$$(B) = C_{A0} f_A \qquad (L)$$

or,

$$-r_{AF} = 2\left[k_1 C_{A0}^2 (1 - f_A)^2 + \frac{k_2}{2} C_{A0}(1 - f_A)(C_{A0} f_A) \right] \qquad (M)$$

When the rate has its maximum value

$$\frac{\partial(-r_{AF})}{\partial f_A} = 2\left[-2k_1 C_{A0}^2 (1 - f_A) + \frac{k_2}{2} C_{A0}^2 (1 - 2f_A) \right] = 0 \qquad (N)$$

Thus

$$4k_1(1 - f_{A\,max}) = k_2(1 - 2f_{A\,max}) \qquad (O)$$

or

$$f_{A\,max} = \frac{k_2 - 4k_1}{2k_2 - 4k_1} \qquad (P)$$

Substituting appropriate numerical values gives

$$f_{A\,max} = \frac{150 - 4(1.02)}{2(150) - 4(1.02)} = 0.493 \qquad (Q)$$

Illustration 9.5 indicates that one may have parallel paths leading from reactants to products and that in the case of an autocatalytic reaction, one path may be preferred over a second until the product level builds up to a point where the second becomes appreciable. In this example, the magnitudes of the rate constants are such that the vast majority of the reaction occurs by the autocatalytic path. In cases such as these it is desirable to use a CSTR or recycle reactor to enhance the reaction rate by virtue of the back-mixing of product species.

An autocatalytic reaction does not necessarily imply a first-order dependence on the product species, or even an integer order with respect to this species. In the hydrolysis of an ester formed from a weak acid, an order approaching one half may be observed.

The key characteristic of all autocatalytic rate expressions is that plots of the rate versus time are of the general form shown in Figure 9.13. If converted into plots of fraction conversion versus time, these forms give rise to a characteristic S shape. These plots first rise, showing autoacceleration as the rate increases, then pass through an inflection point as the rate reaches a maximum, and finally taper off so that the fraction conversion approaches unity or its equilibrium value as the time approaches infinity.

In the more general case for liquid phase reactions or other cases where $\delta = 0$, the autocatalytic term in the reaction rate expression can be written as

$$r_{AC} = k(C_{A0} + v_A \xi^*)^{\beta_A}(C_{P0} + v_P \xi^*)^{\beta_P} \quad (9.4.1)$$

where A and P are reactant and product species, respectively. The extent of reaction per unit volume corresponding to the maximum autocatalytic rate may be determined by setting the derivative equal to zero.

$$\frac{dr_{AC}}{d\xi^*} = k[\beta_A v_A(C_{A0} + v_A \xi^*)^{\beta_A - 1}(C_{P0} + v_P \xi^*)^{\beta_P} + \beta_P v_P(C_{A0} + v_A \xi^*)^{\beta_A}(C_{P0} + v_P \xi^*)^{\beta_P - 1}] = 0$$

$$(9.4.2)$$

or

$$\beta_A v_A(C_{P0} + v_P \xi^*) + \beta_P v_P(C_{A0} + v_A \xi^*) = 0$$

$$(9.4.3)$$

Solving for the extent of reaction per unit volume that gives the maximum reaction rate

$$\xi^*_{max} = \frac{-(\beta_A v_A C_{P0} + \beta_P v_P C_{A0})}{v_A v_P(\beta_A + \beta_P)} \quad (9.4.4)$$

For the case where $v_A = -1$ and $v_P = 1$ and where no product is present initially,

$$\frac{\xi^*_{max}}{C_{A0}} = \frac{\beta_P}{\beta_P + \beta_A} \quad (9.4.5)$$

The maximum rate may be evaluated by combining equation 9.4.4 or 9.4.5 with equation 9.4.1.

9.4.2 Reactor Design Considerations

Plug flow reactors generally give greater production capacities than CSTR's of equal volume but, in the case of autocatalytic reactions, this generalization is not valid. For this class of reactions back-mixing of reacted material with fresh feed is often beneficial in optimizing the overall reactor design. This mixing can be achieved by employing a continuous stirred tank reactor or by recycling the unseparated product mixture. With the CSTR *it is possible to operate all the time at the highest reaction rate* (i.e., at the points in Figure 9.13 labeled C_{max}). One could thus require a lower volume than would be required to achieve the same conversion level in a plug flow reactor, which would operate at an average rate less than the optimum. This mode of operation would require an effluent composition corresponding to C_{max}. If lower conversions are desired, the CSTR will still require less volume than a PFR to reach this level. However, if higher conversions are desired, the optimum design solution (from the viewpoint of minimum total volume) requires the use of a CSTR to reduce the reactant concentration from its initial value to C_{max} and then a PFR to further reduce the reactant concentration to the specified level in the ultimate product.

In cases where unconverted reactants can be readily separated from the product stream it may be preferable to use only the CSTR operating at the maximum rate, regardless of the conversion level desired, because the separated reactants can be recycled. In this case one must determine the relative costs of operating in the separation and recycle mode *vis à vis* the costs of utilizing the second stage plug flow reactor and the attendant separation costs necessary to obtain the final product.

LITERATURE CITATIONS

1. Levenspiel, O., *Chemical Reaction Engineering*, Second Edition, pp. 166–167, Wiley, New York, copyright © 1972. Used with permission.

2. Denbigh, K. G., and Turner, J. C. R., *Chemical Reactor Theory*, pp. 123–124, Cambridge University Press, Cambridge, 1971.

3. Trambouze, P. J., and Piret, E. L., *AIChE J.*, 5 (384), 1959.

4. Denbigh, K. G., *Chem. Eng. Sci.*, *17* (25), 1961.

5. Levenspiel, O., op. cit., p. 180.

6. Levenspiel, O., op. cit., p. 181.

7. Levenspiel, O., op. cit., pp. 186–192.

8. Levenspiel, O., op. cit., p. 189.

9. Levenspiel, O., op. cit., pp. 191–192.

10. Denbigh, K. G., and Turner, J. C. R., op. cit., p. 115.

11. Denbigh, K. G., and Turner, J. C. R., op. cit., p. 114.

12. Baeyer, A., *Ann. Chem. Pharm.*, *103* (178), 1857.

13. Walles, W. E., and Platt, A. E., *Ind. Eng. Chem.*, *59* (6), p. 41, 1967.

PROBLEMS

1. Your plant has two liquid streams available containing solutes that are not profitably marketable at the present time. The first contains solute A, the second solute B.

Two possible reactions may occur between these species. The stoichiometric equations and rate expressions are given below.

$$A + B \rightarrow V \text{ (valuable product)}$$
$$r_V = k_1 C_A C_B C_V^{1/2} + k_2 C_A C_B$$
$$A + B \rightarrow U \text{ (undesired byproduct)}$$
$$r_U = k_3 C_A C_B$$

(a) You have two beakers containing samples of the two streams and desire to carry out a small-scale laboratory experiment in which you maximize the formation of V. In what manner would you carry out this experiment; that is, how would you mix the reactants and at what rate?

(b) If you desire to produce V in a flow reactor, what type of reactor and operating conditions do you recommend?

(c) If the activation energies for the rate constants k_1, k_2, and k_3 are 30, 25 and 25 kJ/mole, respectively, what additional statement can you make regarding recommended operating conditions?

2. Consider the following chlorination of benzene reaction system of elementary reactions.

$$Cl_2 + C_6H_6 \xrightarrow{k_1} C_6H_5Cl + HCl$$
$$Cl_2 + C_6H_5Cl \xrightarrow{k_2} C_6H_4Cl_2 + HCl$$

Assume that chlorination of dichlorobenzene does not occur to an appreciable extent. Monochlorobenzene is the desired product.

(a) Obtain an equation that relates the concentration of monochlorobenzene to the concentration of benzene, assuming that only benzene and chlorine are present initially:
  (1) For a CSTR.
  (2) For a batch reactor.

(b) Utilizing this information, for each of the above cases (1 and 2) in order to minimize the rate of production of dichlorobenzene relative to the rate of production of monochlorobenzene, would you want a high or low conversion of benzene? *Prove* this *mathematically* for each of the following cases.

$$k_1 > k_2$$
$$k_1 = k_2$$
$$k_1 < k_2$$

(c) For an industrial situation, what other consideration is particularly important in addition to minimizing the rate of production of undesired products relative to that of the desired product? Answer this question very briefly in a qualitative manner.

3. Species A and B react as follows.

$$A + 2B \xrightarrow{k_1} 2R + S \qquad (-r_A)_1 = k_1 C_A C_B^2$$

$$A + B \xrightarrow{k_2} T + U \qquad (-r_A)_2 = k_2 C_A C_B$$

An equimolar mixture of A and B is introduced into an infinitely long plug flow reactor ($C_{A0} = C_{B0} = 1.0$ kmole/m^3). The concentration of A in the effluent is 0.1 kmole/m^3. No B is present in the effluent stream. What conclusions can you draw concerning the relative magnitudes of k_1 and k_2?

4. The reactions indicated below take place in the liquid phase with the indicated rate expressions.

(i) $A \rightarrow B \quad r = k_1 C_A \quad k_1 = 20 \text{ ksec}^{-1}$

(ii) $2A \rightarrow C \quad r = k_2 C_A^2 \quad k_2 = 2 \text{ m}^3/\text{kmole·ksec}$

The feed stream consists of A dissolved in a solvent S such that the initial concentration of A is 2 kmoles/m^3. If 50% of the A fed to a flow reactor undergoes reaction, determine the effluent composition:
(a) If a CSTR is used.
(b) If a PFR is used.

5. Xylene isomerization reactions can be accomplished by contacting a hot gas stream with a solid catalyst. Under these conditions the isomerization reactions may be regarded as reversible and first-order. Unfortunately, the catalyst also catalyzes disproportionation reactions. These reactions may be regarded as essentially second-order and irreversible. If one desires to achieve an equilibrium mixture of isomers with minimal material losses due to disproportionation, what do you recommend concerning the mode in which one should operate a continuous flow reactor?

6. A process for ultra-high temperature sterilization of whey has been proposed. The purpose of the process is to destroy 99.99% of the spores in the whey. It is believed that if very high temperatures are used for short times, this result can be obtained without degrading the protein of the whey, which, besides ruining the food value, gives it a bad taste.

A PFR has been designed in which steam is injected into the whey as it enters the reactor. After it passes through the PFR, the temperature is dropped rapidly to 10 °C, where the rate of degradation is negligible. A holding time of 7 sec is needed to obtain 99.99% spore kill at a temperature of 127 °C. For the whey to meet health and flavor tests, a maximum of 10% degradation of protein is to be allowed. As a first approximation in your analysis, the reaction is to be treated as isothermal at 127 °C.

Another research group has previously conducted tests on the denaturation kinetics and has reported the data below.

Will the proposed process meet the standards set for protein degradation?

Data

Whey contains 1.5 kg/m^3 of lactalbumin and 3.0 kg/m^3 of lactoglobulin. At 85 °C the following data are available on the kinetics of protein degradation, using initial concentrations that are the same as those in whey.

Lactalbumin		Lactoglobulin	
Time, t (sec)	Percent residual protein	Time, t (sec)	Percent residual protein
0	100	0	100
120	82.5	20	83
250	72.5	35	77
360	65.0	80	67.5
725	47.5	100	58.0
850	42.0	125	54.5
960	39.0	140	52.5
1075	35.0	160	48.0
1200	30	180	44.5

Over the temperature range of interest, the activation energy for the lactalbumin degradation reaction is 25.1 kJ/mole while that for the lactoglobulin degradation reaction is 121 kJ/mole.

Note that there are problems associated with the determination of time zero for these reactions.

Assume that the orders of both reactions are integers, although not necessarily the same integer.

7. Gulyaev and Polak [*Kinetics and Catalysis,* 6 (352), 1965] have studied the kinetics of the thermal decomposition of methane with a view toward developing a method for the commercial production of acetylene in a plasma jet. The following differential equations represent the time dependence of the concentrations of the major species of interest.

$$\frac{d(CH_4)}{dt} = -k_1(CH_4)$$

$$\frac{d(C_2H_4)}{dt} = -k_3(C_2H_4) + \tfrac{1}{2}k_1(CH_4)$$

$$\frac{d(C_2H_2)}{dt} = -k_4(C_2H_2) + k_3(C_2H_4)$$

(a) Integrate these differential equations to determine the time dependent behavior of each species concentration. Assume that the only species present initially is methane and that its initial concentration is $(CH_4)_0$.

(b) (1) Determine the times corresponding to maximum concentrations of C_2H_4 and C_2H_2. It may be noted that

$$k_1 \gg k_3 \gg k_4$$

and that

$$k_1 = 4.5 \times 10^{13} e^{-45,800/T} \text{ sec}^{-1}$$
$$k_3 = 2.57 \times 10^{8} e^{-20,100/T} \text{ sec}^{-1}$$
$$k_4 = 1.7 \times 10^{6} e^{-15,100/T} \text{ sec}^{-1}$$

where T is expressed in degrees Kelvin.

(2) If the reaction at 2500 °K may be considered as occurring at constant volume, determine the times corresponding to the two maxima. Note that the time in the plasma jet will be exceedingly small.

(c) Roughly what fraction of the methane is converted to acetylene at 2500 °K if the contact time in a plasma jet is that corresponding to the maximum in acetylene concentration?

8. Consider the sequential first-order reactions $A \xrightarrow{k_1} V \xrightarrow{k_2} W$ where V is the desired product. These liquid phase reactions are to be carried out in a cascade of two equal volume CSTR's in series. If the reactors are to be sized so as to maximize the concentration of species V in the effluent from the second reactor, determine the reactor volumes necessary to process 500 gal/hr of feed containing 6 moles/gal of species A. No V or W is present in the feed. What fraction of the A ends up as V? The rate constants k_1 and k_2 are both equal to 0.5 hr^{-1}.

9. A sequential isomerization reaction is believed to be characterized by the following mechanistic equations.

$$A \xrightarrow{k_1} B \xrightarrow{k_2} C$$

These reactions are being studied in a series combination of two CSTR's. Each of these reactors has a volume of 2 liters. Liquid is being fed to the first reactor at a rate of 1 liter/min. This liquid contains only pure A and the solvent. The inlet concentration of A is 2 g moles/liter. Under the conditions of operation, $k_1 = 0.5$ min^{-1} and $k_2 = 0.25$ min^{-1}. Both reactors operate at the same temperature.

(a) What is the composition of the effluent from the first reactor? Include all species of interest.

(b) Repeat part (a) for the second reactor.

10. Kostyuk, Belen'ku, and Tezhneva [*Kinetics and Catalysis, 10 (53), 1969*] have studied the

acid catalyzed decomposition of p-diisopropyl-benzene dihydroperoxide in ethyl acetate solution. The stoichiometric equations for the two consecutive reactions that occur are:

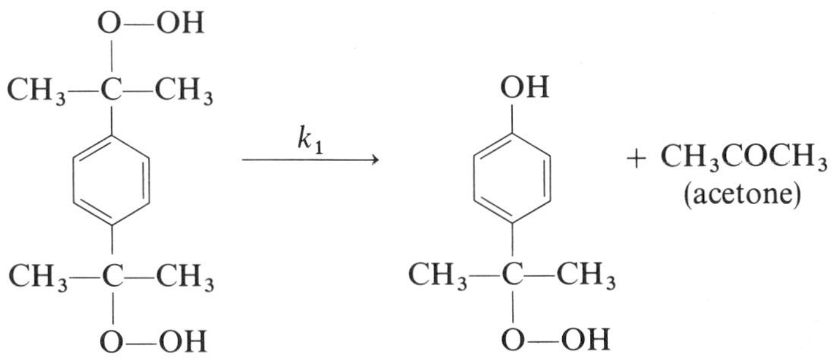

(p-hydroxycumyl hydroperoxide)

and

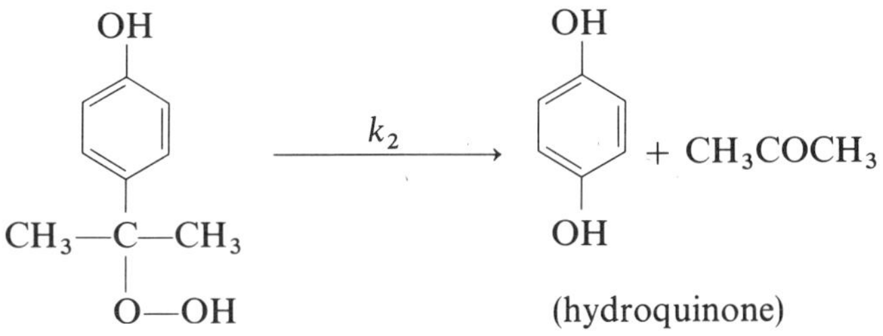

(hydroquinone)

Both reactions are first-order in catalyst concentration and in the reactant involved. If one incorporates the catalyst concentration into the rate constants k_1 and k_2, the following values are typical of the results reported by these investigators at 40 °C.

$$\frac{k_2}{k_1} = 0.95$$

$$k_1 = 3.6 \times 10^{-4} \text{ sec}^{-1}$$

Both reactions take place in liquid solution. If one desires to carry out these reactions in a continuous stirred tank reactor, determine the space time corresponding to a maximum yield of p-hydroxycumyl hydroperoxide. If the initial p-diisopropylbenzene dihydroperoxide concentration is 10 moles/m³, what are the concentrations of the various species in the effluent?

11. Vaidyanathan and Doraiswamy [*Chem. Eng. Sci.*, 23 (537), 1966] have studied the catalytic partial oxidation of benzene in a composition range where the reactions of interest all follow pseudo first-order kinetics. The pertinent stoichiometric equations are

$$C_6H_6 + \tfrac{9}{2}O_2 \rightarrow C_4H_2O_3 + 2CO_2 + 2H_2O$$
$$C_4H_2O_3 + 3O_2 \rightarrow 4CO_2 + H_2O$$
$$C_6H_6 + \tfrac{15}{2}O_2 \rightarrow 6CO_2 + 3H_2O$$

The corresponding rate expressions are

$$r_1 = k_1 P_B$$
$$r_2 = k_2 P_M$$
$$r_3 = k_3 P_B$$

where P_B refers to the partial pressure of benzene and P_M to that of maleic anhydride.

If one is interested in designing a reactor for maleic anhydride ($C_4H_2O_3$) production, determine the reactor space time for a PFR that maximizes the concentration of this species in the effluent. Start by deriving equations for P_B, P_M, and P_{CO_2} as functions of the space time. At 350 °C, the values of the rate constants are:

$$\left. \begin{array}{l} k_1 = 1.141 \times 10^{-3} \\ k_2 = 2.468 \times 10^{-3} \\ k_3 = 0.396 \times 10^{-3} \end{array} \right\} \text{g moles/(hr-g catalyst-atm)}$$

Under optimum conditions at this temperature, what fraction of the inlet benzene ends up as maleic anhydride? What is the yield of maleic anhydride based on the amount of benzene reacted? Neglect volumetric expansion.

12. The following reaction sequence takes place in the liquid phase in a continuous stirred tank reactor.

$$A \xrightarrow{k_1} B + C \xrightarrow{k_2} 2D$$

The feed concentration of A is C_{A0}, and the reactions may be assumed to be irreversible reactions proceeding by this mechanism. No B, C, or D are present in the feed. If the input volumetric flow rate is $\mathcal{V}_0$ and the reactor volume is V_R, derive equations for the effluent concentrations of B, C, and D. If species B is the desired product, determine the space time that corresponds to the maximum production of B.

13. Ziegler-Natta catalysts are used commercially for the production of stereoregular polymers, especially isotactic polypropylene and high-density linear polyethylene. The resultant polymers have number and weight average molecular weights ($\bar{M}_n$ and $\bar{M}_w$, respectively) that are defined as

$$\bar{M}_n = M \frac{\sum n N_n}{\sum N_n} \tag{I}$$

and

$$\bar{M}_w = M \frac{\sum n^2 N_n}{\sum n N_n} \tag{II}$$

where

N_n is the number of molecules containing n monomer units

M is the molecular weight of the monomer.

(The summations extend from $n = 2$ to $n = \infty$.) Keii [*Kinetics of Ziegler-Natta Polymerization,* Kodansha, Tokyo, 1972] has noted that under steady-state reaction conditions, the number of polymer molecules with degree of polymerization n desorbing per unit catalyst surface area in unit time may be written as

$$r_{\text{desorption}} = k_d N_0 \theta_n \tag{III}$$

where

k_d = desorption rate constant (assumed to be independent of n)

N_0 = number of active sites per unit surface area

θ_n = fraction of the sites covered by a polymer with degree of polymerization n

Reattachment of the n-mer does not occur.

If the reaction occurs by the so-called Rideal mechanism the net rate at which a polymer with degree of polymerization n is produced is given by

$$k_r N_0 P \theta_{n-1} - k_r N_0 P \theta_n \tag{IV}$$

where the rate constant for the reaction of gas phase monomer with adsorbed n-mer (k_r) is assumed to be independent of n and where P is the gas phase monomer pressure.

If the rate at which monomer is adsorbed to initiate polymerization is given by

$$k_a N_0 P \left(1 - \sum_{n=1}^{\infty} \theta_n \right) \tag{V}$$

and if the average rate of polymerization ($\bar{R}$) is

taken as the rate of disappearance of monomer

$$\bar{R} = \sum_{n=2}^{\infty} n \frac{dN_n}{dt}$$

show that the average rate of polymerization is given by

$$\bar{R} = \frac{N_0 k_r k_a P^2 (2k_d + k_r P)}{(k_r P + k_d)(k_a P + k_d)}$$

The breadth of the molecular weight distribution may be measured by the ratio $\bar{M}_w/\bar{M}_n$. Show that this ratio is given by

$$\frac{\bar{M}_w}{\bar{M}_n} = \frac{\dfrac{5}{\gamma} + \dfrac{2}{\gamma^2} + 4}{\left(2 + \dfrac{1}{\gamma}\right)^2}$$

where

$$\gamma \equiv \frac{k_d}{k_r P}$$

By considering the limits of $1/\gamma$ as zero and infinity corresponding to infinitely rapid desorption or slow reaction and very slow desorption or rapid reaction, respectively, show that

$$1 \le \frac{\bar{M}_w}{\bar{M}_n} \le 2$$

which is a rather narrow molecular weight distribution.

14. The following reactions take place in the liquid phase in a continuous stirred tank reactor operating at steady state.

$$A + B \rightarrow C \qquad \text{(I)}$$
$$C + A \rightarrow D \qquad \text{(II)}$$

Under the conditions of operation, reaction I is first-order in species B and is characterized by a rate constant equal to 15 ksec^{-1}. Reaction II is first-order in species C and is characterized by a rate constant equal to 3 ksec^{-1}. A stream containing 2.0 kmoles/m^3 of A and 0.5 kmoles/m^3 of B is available at a rate of 0.75 m^3/ksec.

(a) Derive equations for the effluent concentration of each species in terms of the rate constants k_I and k_{II}, the reactor volume, the initial volumetric flow rate, and the initial concentrations of species A and B.
(b) What reactor volume will give the highest concentration of species C in the effluent stream? What will this concentration be?

15. An autocatalytic reaction is to be carried out in aqueous solution in two identical continuous stirred tank reactors operating in series. The reaction stoichiometry is

$$A \rightarrow B$$

while the reaction rate expression is

$$r = k C_A C_B$$

If the system is to operate isothermally at 50 °C where the reaction rate constant is equal to 0.9 m^3/kmole·ksec, determine the reactor volume necessary to achieve an overall fraction conversion of 0.80. Species A is to be fed at a rate of 0.3 mole/sec and an initial A concentration of 2 kmoles/m^3.

16. An autocatalytic reaction represented by the mechanistic equations

$$A + R \underset{k_{-1}}{\overset{k_1}{\rightleftharpoons}} R + R$$

is being carried out in two identical stirred tanks operating in series. The reaction is reversible and exothermic. The reaction stoichiometry is

$$A \rightarrow R$$

Two hundred and twenty-five pound moles of liquid A at 70 °F are fed to the first reactor each minute. The fraction of the inlet A converted to R in the first reactor is such that the effluent from the first reactor leaves at 85 °F. The temperature of the stream leaving the second reactor is 100 °F. The mole density of A in the entrance stream is 1.5 lb moles/ft^3.

The following table provides some information about the temperature dependence of the rate constants k_1 and k_{-1} in units of cubic feet per pound mole per minute.

Temperature (°F)	k_1	k_{-1}
70	0.55	?
85	2.90	0.078
100	?	0.625

If the reactor volume is 50 ft³, determine the composition of the stream leaving each reactor.

17. Consider the following combination of reactors.

It has been suggested that the following liquid phase reaction be carried out in this reactor network.

$$A \rightarrow B + C$$

The reaction rate expression for this stoichiometric equation is

$$r = kC_A C_B$$

The feed to the CSTR contains a concentration of A of 2.0 kmoles/m³, while neither species B or C is present in this stream. Both reactors operate isothermally at the same temperature. Steady-state operation may be assumed.

The CSTR and the PFR each have a volume of 0.1 m³. The rate at which A is fed to the first reactor is 150 kmoles/ksec. Note that this is an autocatalytic reaction in which a product acts as a catalyst for subsequent reaction.

(a) If the fraction conversion leaving the CSTR is 0.25, what is the reaction rate constant?

(b) What is the effluent composition from the second reactor?

(c) It has been suggested that the rate of production of species C by this network can be increased by operating the first reactor at a fraction conversion of 0.45. To what value must the input feed rate of A be changed under these operating conditions? Will the rate of production of species C by this combination of reactors be increased under these circumstances? The composition of the feedstream is to remain unchanged.

10 Temperature and Energy Effects in Chemical Reactors

10.0 INTRODUCTION

The energy changes associated with chemical reactions play an extremely important role in the design of commercial scale reactors. Even in those cases where one desires to use a mode of operation approximating isothermal behavior, energy balance considerations are important in determining the heat transfer requirements of the process. In laboratory scale experiments it is relatively easy to maintain substantially isothermal conditions because of the large surface to volume ratio of small equipment and because the economic considerations of the heat transfer process are unimportant. In industrial scale equipment, both physical limitations on the rate of heat transfer and the economics of this process can become quite important, particularly when large enthalpy changes accompany the reaction.

Several factors govern the temperature range in which one may choose to operate a commercial reactor. The dependence of the reaction rate expression on temperature and the position of chemical equilibrium are two key factors influencing the choice of temperature level. The temperature dependence of the main reaction and of important side reactions will govern the selectivity of the conversion. Other properties of the reaction mixture may also influence the choice of operating temperature through secondary considerations. For example, the dew point of a gaseous mixture, the bubble point of a liquid mixture, the temperature at which a liquid separates into two immiscible phases, and the temperature dependence of corrosion reactions all can limit the temperature range within which one might choose to operate. Catalyst activity, selectivity, and deactivation must also be considered in selecting the operating temperature for catalytic processes.

In order to achieve adequate control during the operation of a reactor, it is necessary to maintain the temperature within at least moderate limits. Adiabatic operation is possible only when the concomitant temperatures do not rise so high that the rate becomes excessive or drop so low that the rate becomes impractically low. This mode of operation is favored when the enthalpy change accompanying the reaction is small in magnitude, when it is possible to adjust the initial temperature of the reactant mixture to a level where subsequent changes will not take the mixture out of the workable range, or when a solvent or other inert materials can be added to the reactant mixture to moderate temperature changes via sensible or latent heat effects. In some cases it is advantageous to divide the reactor into several stages, each of which operates adiabatically but between which heat exchangers are used to heat or cool the reactant mixture. In commercial scale equipment it is often impractical to try to maintain truly isothermal conditions, but one can often achieve adequate control at a satisfactory level of the reaction rate by exchanging heat with external heat reservoirs.

For nonisothermal reactors the key questions that the reactor designer must answer are: (1) How can one relate the temperature of the reacting system to the degree of conversion that has been accomplished? and (2) How does this temperature influence the subsequent performance of the system? In responding to these questions the chemical engineer must use two basic tools—the material balance and the energy balance. The bulk of this chapter deals with these topics. Some stability and selectivity considerations are also treated.

10.1 THE ENERGY BALANCE AS APPLIED TO CHEMICAL REACTORS

The principle that energy must be conserved has been expressed by innumerable textbook authors in a number of mathematical forms, depending on the particular orientation and interests of the individual concerned. One form that is especially useful for chemical engineering applications is the following.

$$\frac{dE_{sys}}{dt} = \dot{Q} - \dot{W}_s + \sum_{\substack{\text{input} \\ \text{streams}}} \left(h_{in} + \frac{v_{in}^2}{2g_c} + \frac{g}{g_c} Z_{in} \right) \dot{m}_{in} - \sum_{\substack{\text{output} \\ \text{streams}}} \left(h_{out} + \frac{v_{out}^2}{2g_c} + \frac{g}{g_c} Z_{out} \right) \dot{m}_{out} \qquad (10.1.1)$$

where

E_{sys} is the internal energy of a control volume or fixed region in space that has been chosen as the system to be investigated

t is time

$\dot{Q}$ is the *rate* at which heat is being transferred *to* the system from the surroundings across boundaries that are impermeable to the flow of matter

$\dot{W}_s$ is the rate at which *shaft work* is being done *on* the surroundings *by* the system

h is the enthalpy per unit mass

work interaction can take place only at those portions of the system boundary that are impermeable to the flow of matter. For further discussion of general forms of the energy bal-

ance equation as applied to chemical engineering systems, consult the textbooks by Reynolds (1), Smith and Van Ness (2), and Modell and Reid (3).

Equation 10.1.1 represents a very general formulation of the first law of thermodynamics, which can be readily reduced to a variety of simple forms for specific applications under either steady-state or transient operating conditions. For steady-state applications the time derivative of the system energy is zero. This condition is that of greatest interest in the design of continuous flow reactors. Thus, at steady state,

$$\dot{Q} - \dot{W}_s = \sum_{\substack{\text{outlet} \\ \text{streams}}} \left(h_{out} + \frac{v_{out}^2}{2g_c} + \frac{g}{g_c} Z_{out} \right) \dot{m}_{out} - \sum_{\substack{\text{inlet} \\ \text{streams}}} \left(h_{in} + \frac{v_{in}^2}{2g_c} + \frac{g}{g_c} Z_{in} \right) \dot{m}_{in} \qquad (10.1.2)$$

$\dfrac{v^2}{2g_c}$ is the kinetic energy per unit mass

$\dfrac{g}{g_c} Z$ is the gravitational potential energy per unit mass

$\dot{m}$ is the mass flow rate into or out of the system

subscripts in and out refer to the conditions that prevail at the various points of entry and exit, respectively.

Recall that the term *shaft work* refers to mechanical forms of energy that are interchanged between the system and the surroundings by means of a shaft that protrudes from the equipment and either rotates or reciprocates. A shaft

where

$$\sum_{\substack{\text{inlet} \\ \text{streams}}} \dot{m}_{in} = \sum_{\substack{\text{outlet} \\ \text{streams}}} \dot{m}_{out} \qquad (10.1.3)$$

In most circumstances of interest to the designer of chemical reactors, the kinetic and potential energy effects are negligible as is the shaft work term. Under these circumstances,

$$\dot{Q} = \sum_{\substack{\text{outlet} \\ \text{streams}}} h_{out}\dot{m}_{out} - \sum_{\substack{\text{inlet} \\ \text{streams}}} h_{in}\dot{m}_{in} \qquad (10.1.4)$$

For the vast majority of situations of interest in the design of industrial scale chemical reactors, equation 10.1.4 provides an adequate approximation to the true situation for continuous flow reactors operating at steady state. Since the right side of this expression represents the difference between the total enthalpy that leaves the system in time dt and the total enthalpy that enters during the same time, one often sees the statement that

$$q \approx \Delta H \qquad (10.1.5)$$

where

q is the amount of heat transferred *per unit mass* of entering material

ΔH is the difference between the enthalpies per unit mass of the exit and entrance streams

It should be emphasized that the enthalpy change includes not only sensible heat effects but also a heat-of-reaction term and in some cases pressure effects. If there are multiple inlet and outlet streams, appropriate averaging techniques must be used to employ this equation.

For batch reactors the appropriate form of the first law to use is that for closed systems.

$$\Delta E_{total} = Q - W \qquad (10.1.6)$$

where

ΔE_{total} is the change in the total energy of the system

Q is the heat transferred from the surroundings to the system

W is the work done *by the system* on the surroundings

If one again takes note of the fact that work effects and kinetic and potential energy effects are usually negligible in chemical reactors, equation 10.1.6 simplifies to

$$\Delta E = Q_V \qquad (10.1.7)$$

where

ΔE is the change in the internal energy of the system

subscript V is used to indicate that the volume is assumed to be substantially constant

The vast majority of the reactions carried out in industrial scale batch reactors involve reactants in condensed phases. Since the specific volumes of both liquids and solids are very small, the difference between internal energy and enthalpy for these materials is usually negligible. Thus one often sees the statement that for batch reactions taking place at constant volume:

$$Q \approx \Delta H \qquad (10.1.8)$$

where the Δ now represents the difference between the final and the initial values of the enthalpy of the system. For both batch and flow reactors, the enthalpy changes that occur during the course of the reaction are then important in determining the heat transfer requirements of the system under consideration. The methods used in determining the enthalpy changes that accompany chemical reactions were discussed in detail in Section 2.2.

The rate at which heat is transferred to a system can be expressed in terms of an overall heat transfer coefficient U, the area through which the heat exchange occurs and on which U is based, and the difference between the temperature of the heat source (or sink) T_m, and that of the reactor contents T.

$$\dot{Q} = UA(T_m - T) \qquad (10.1.9)$$

This expression may be combined with equations 10.1.4 and 10.1.8 in order to analyze the different situations that may arise in operating the various types of ideal reactors. These analyses are the subject of Sections 10.2 to 10.4.

10.2 THE IDEAL WELL-STIRRED BATCH REACTOR

The key assumption on which the design analysis of a batch reactor is based is that the degree of agitation is sufficient to ensure that the composition and temperature of the contents are uniform throughout the reaction vessel. Under these conditions one may write the material and energy balances on the entire contents of the reaction vessel.

If one considers a batch reactor in which the chemistry is characterized by a single extent of reaction, the material balance analysis presented in Section 8.1 indicates that the holding time necessary to change the fraction conversion from f_{A1} to f_{A2} is given by

$$t = N_{A0} \int_{f_{A1}}^{f_{A2}} \frac{df_A}{V_R(-r_A)} \qquad (10.2.1)$$

For operation in a constant pressure non-isothermal mode with a *gas phase reaction*, it is convenient to approximate the reactor volume by

$$V_R = V_{R0}(1 + \delta_A f_A) \frac{T}{T_0} \qquad (10.2.2)$$

where the temperature corresponding to a given fraction conversion (T) and that corresponding to zero conversion (T_0) are expressed in degrees absolute. The δ parameter in this equation represents the fractional volume change that would occur under isothermal operating conditions. Combining equations 10.2.1 and 10.2.2 leads to the following expression for the holding time for nonisothermal gas phase reactions carried out at constant pressure.

$$t = C_{A0} \int_{f_{A1}}^{f_{A2}} \frac{df_A}{(-r_A)(1 + \delta_A f_A)\left(\dfrac{T}{T_0}\right)} \qquad (10.2.3)$$

In order to be able to evaluate the integrals in equations 10.2.1 and 10.2.3, one must know not only the temperature dependence of the rate terms but also the relationship between the fraction conversion and the temperature of the system.

In Section 10.1 we showed that in industrial scale batch reactors

$$Q = \Delta H \qquad (10.2.4)$$

for the vast majority of the cases of interest. Strictly speaking, equation 10.2.4 applies only

to liquid phase reactions where the restrictions of constant volume and constant pressure go together and to gas phase reactions occurring at constant pressure. We should again emphasize that the enthalpy change appearing in equation 10.2.4 contains both sensible heat terms and heat of reaction terms. Consequently, one can interpret equation 10.2.4 as implying that

Heat input = change in sensible heat

+ energy consumed by reaction

$$(10.2.5)$$

or

$$Q = \int_{T_0, \xi_0}^{T_{\text{final}}, \xi_1} (m\hat{C}_p \, dT + \Delta H_R \, d\xi) \qquad (10.2.6)$$

where

$\hat{C}_p$ is an appropriate average heat capacity per unit mass

m is the total mass of the system

ΔH_R is the enthalpy change per extent of reaction ξ

Since enthalpy is a state variable, the integral on the right side of equation 10.2.6 is independent of the path of integration, and it is possible to rewrite this equation in a variety of forms that are more convenient for use in reactor design analyses. One may evaluate this integral by allowing the reaction to proceed isothermally at the initial temperature from extent ξ_0 to extent ξ_1 and then heating the final product mixture at constant pressure and composition from the initial temperature to the final temperature.

$$Q = \Delta H_{R \text{ at } T_0}(\xi_1 - \xi_0) + \int_{T_0}^{T_{\text{final}}} m\hat{C}_p \, dT \qquad (10.2.7)$$

where the appropriate average heat capacity to use is the mass average based on the composition of the *final* product mixture.

Further modification of these equations is possible by writing the last integral as

$$\int_{T_0}^{T_{\text{final}}} m\hat{C}_p \, dT = \sum \left(n_i \int_{T_0}^{T_{\text{final}}} \bar{C}_{pi} \, dT \right) \qquad (10.2.8)$$

where

the n_i are the *final mole numbers* of the various species present in the reaction mixture

the $\bar{C}_{pi}$ are the partial molal heat capacities of these species

The summation is taken over all species (including inerts) present in the system. For gaseous mixtures that follow ideal solution behavior the partial molal quantities may be replaced by the pure component values.

Since the total heat input represents the time integral of the heat transfer rate, it is evident from equations 10.1.9, 10.2.7, and 10.2.8 that

$$\int_0^t UA(T_m - T)\, dt = \Delta H_{R\,at\,T_0}(\xi_1 - \xi_0)$$
$$+ \sum \left(n_i \int_{T_0}^{T_{final}} \bar{C}_{pi}\, dT \right)$$

(10.2.9)

Since the various mole numbers can be expressed in terms of the extent of reaction, equation 10.2.9 expresses the relationship that must exist between the extent of reaction at time t and the temperature at that time. In terms of the fraction conversion where the fraction conversion at zero time is taken as zero,

$$\int_0^t UA(T_m - T)\, dt = \frac{-\Delta H_{R\,at\,T_0}}{\nu_A} n_{A0} f_A$$
$$+ \sum \left(n_i \int_{T_0}^{T_{final}} \bar{C}_{pi}\, dT \right)$$

(10.2.10)

where n_{A0} is the number of moles of the limiting reagent charged to the batch reactor.

On a differential basis,

$$UA(T_m - T) = \frac{-\Delta H_R}{\nu_A} n_{A0} \frac{df_A}{dt}$$
$$+ \left(\sum n_i \bar{C}_{pi} \right) \frac{dT}{dt} \quad (10.2.11)$$

where the summation over the mole numbers must now be evaluated at time t.

If one recognizes that the first term on the right is simply the rate at which energy is transformed by reaction ($r V_R \Delta H_R$), the last equation becomes

$$UA(T_m - T) = \Delta H_R r V_R + \left(\sum n_i \bar{C}_{pi} \right) \frac{dT}{dt}$$

(10.2.12)

For exothermic reactions ΔH_R is negative, and this term will then represent the rate of energy release by reaction.

In general, when designing a batch reactor, it will be necessary to solve simultaneously one form of the material balance equation and one form of the energy balance equation (equations 10.2.1 and 10.2.5 or equations derived therefrom). Since the reaction rate depends both on temperature and extent of reaction, closed form solutions can be obtained only when the system is isothermal. One must normally employ numerical methods of solution when dealing with nonisothermal systems.

For isothermal and adiabatic modes of operation the energy balance equations developed above will simplify so that the design calculations are not nearly as tedious as they are for the other modes of operation. In the case of adiabatic operation the heat transfer rate is zero, so equation 10.2.10 becomes

$$\Delta H_{R\,at\,T_0} \frac{n_{A0} f_A}{\nu_A} = \sum \left(n_i \int_{T_0}^{T_{final}} \bar{C}_{pi}\, dT \right) \quad (10.2.13)$$

If the partial molar heat capacities are substantially constant over the temperature range of interest, this equation may be solved to determine the relationship between the temperature and the fraction conversion.

$$T = T_0 + \frac{\Delta H_{R\,at\,T_0} n_{A0} f_A}{\nu_A \sum(n_i \bar{C}_{pi})} \quad (10.2.14)$$

where each of the n_i in the summation is evaluated at a fraction conversion equal to f_A. When this result is substituted into batch reactor design equations, one has an integral expressed in terms of a single variable, which may then be evaluated graphically or numerically.

In the case of isothermal operation the material and energy balance equations are not coupled, and design equations like 10.2.1 can be solved readily, since the reaction rate can be expressed directly as a function of the fraction conversion. For operation in this mode, an energy balance can be used to determine how the heat transfer rate should be programmed to keep the system isothermal. For this case equation 10.2.12 simplifies to the following expression for the heat transfer rate

$$\dot{Q} = UA(T_m - T) = \Delta H_R r V_R \quad (10.2.15)$$

The reaction rate can readily be determined as a function of time from the design equation, and this in turn can be used to determine how $\dot{Q}$ or T_m should be varied to obtain isothermal operating conditions.

Kladko (5) has presented a very interesting case study of the development of a reactor design for an exothermic reaction. When first carried out on a commercial scale in a batch reactor, the system exothermed rapidly, and the system ran out of control. The batch erupted violently through a safety valve and vented out over the building area. The fact that this result could have been predicted *a priori* illustrates the necessity of making the types of energy balance calculations described in this chapter when one attempts to move from bench or pilot scale reactor systems to commercial scale equipment. Because of the proprietary nature of the product, many details of the reaction were omitted but, if engineering "guesstimates" of the heat capacities and molecular weights are used, it is possible to conduct an engineering analysis of the system to complement those presented by Kladko for semibatch operation. We will use his data and our guesstimates of the system properties as the basis for several illustrations in the remainder of this chapter.

ILLUSTRATION 10.1 DETERMINATION OF REQUIRED REACTOR VOLUMES FOR ISOTHERMAL AND ADIABATIC OPERATION IN A BATCH REACTOR

Reagent A undergoes an essentially irreversible isomerization reaction that obeys first-order kinetics.

$$A \rightarrow B$$

Both A and B are liquids at room temperature and both have extremely high boiling points.

Determine the reactor volumes necessary to produce 2 million pounds of B in 7000 hr of operation:

1. If the reactor operates isothermally at 163 °C.
2. If the reactor operates adiabatically.

Use the data and assumptions listed below.

Data and Assumptions

Reaction rate expression: $r = kC_A$
Rate constant at 163 °C = 0.8 hr^{-1}
Activation energy = 28,960 cal/g mole
Heat of reaction = -83 cal/g
Molecular weight = 250

The heat capacities of species A and B may be assumed to be identical and equal to 0.5 cal/g-°C. Their densities may be assumed to be equal to 0.9 g/cm^3.

The times necessary to fill and drain the reactor may be assumed to be equal to 10 and 12 min, respectively. It may be assumed that negligible reaction occurs during the 14 min it takes to heat the feed from the temperature at which it enters the reactor to 163 °C. After 97% of the A has been isomerized, the hot product mixture is discharged to a cooling tank.

Solution

In a batch reactor maintained at constant volume the holding time is given by equation 8.1.7.

$$t = C_{A0} \int_0^{0.97} \frac{df_A}{-r_A}$$

$$= C_{A0} \int_0^{0.97} \frac{df_A}{kC_{A0}(1 - f_A)}$$

$$= \int_0^{0.97} \frac{df_A}{k(1 - f_A)} \qquad \text{(A)}$$

For isothermal operation at 163 °C the rate constant is equal to 0.8 hr^{-1}.

Thus

$$t = \frac{1}{0.8} \ell n \left(\frac{1}{1 - f_A} \right) \Big|_0^{0.97}$$

$$t = 4.38 \text{ hr}$$

The total processing time per batch is the sum of the holding time and the times necessary to fill, drain, and heat the mixture, or

$$\tau_{\text{Batch}} = 4.38 + \frac{10 + 12 + 14}{60} = 4.98 \text{ hr}$$

The total processing time per batch is thus essentially equal to 5 hr. While operating 7000 hr one will be able to process 7000/5 or 1400 batches. Each batch must contain 2,000,000/(0.97)(1400), or 1473 lb. Since the liquid has a density of 0.9 g/cm^3, the required volume will be 1473/(0.9)(62.5), or 26.2 ft^3. This volume is equivalent to 196 gal.

The maximum heat flux that will have to be maintained in order to operate isothermally will be that generated at the start of the reaction.

$$\dot{Q} = \Delta H_R r V_R = \Delta H_R k C_{A0} V_R$$

or

$$\dot{Q} = \Delta H_R k n_{A0}$$

In consistent units

$$\Delta H = -83 \frac{\text{cal}}{\text{g}} \times \frac{1 \text{ BTU}}{252 \text{ cal}} \times \frac{454 \text{ g}}{\text{lb}}$$

$$= -149.5 \frac{\text{BTU}}{\text{lb}}$$

Thus

$$\dot{Q} = (-149.5)(0.8)(1473)$$

$$= -176,000 \text{ BTU/hr}$$

This is a rather large cooling load for a system of this size. Kladko and his co-workers found it advantageous to go to a semibatch mode of operation so that incoming cold feed could be used to absorb some of the energy released by reaction and thereby drastically reduce the cooling requirements for the system. If the cold feed enters at room temperature it acts as a heat sink that is capable of absorbing nearly 80% of the energy released by reaction. This fact enables one to consider operation in the autothermal mode discussed in Section 10.5.

Table 10.I.1
Data Workup for Illustration 10.1

f_A	T (from B)	k (from C)	$k(1 - f_A)$	$\frac{1}{k(1 - f_A)}$
0.00	436	0.8	0.8	1.250
0.05	444.3	1.49	1.42	0.706
0.10	452.6	2.73	2.46	0.407
0.15	460.9	4.87	4.14	0.242
0.20	469.2	8.51	6.81	0.147
0.25	477.5	14.60	10.95	0.091
0.30	485.8	24.6	17.2	0.058
0.35	494.1	40.7	26.5	0.037
0.40	502.4	66.3	39.8	0.025
0.45	510.7	106.1	58.4	0.017
0.50	519.0	167.5	83.8	0.012
0.55	527.3	260.6	117.3	0.009
0.60	535.6	400.0	160	0.006
0.65	543.9	605.6	212	0.005
0.70	552.2	905.9	272	0.004
0.75	560.5	1339	335	0.003
0.80	568.8	1957	391	0.003
0.85	577.1	2828	424	0.002
0.90	585.4	4045	405	0.002
0.95	593.7	5729	286	0.003
0.97	597.0	6567	197	0.005

We now consider operation of the batch reactor under adiabatic conditions. We will assume that we need not worry about reaching the boiling point of the liquid and that the rate of energy release by reaction does not become sufficiently great that an explosion ensues.

For adiabatic operation,

$$T = T_0 + \frac{\Delta H_R f_A n_{A0}}{v_A \sum(n_i \bar{C}_{pi})}$$

In the present case,

$$T = 436 + \frac{(-83)f_A n_{A0}}{(-1)(0.5)n_{A0}} = 436 + 166 f_A \quad (B)$$

where T is expressed in degrees Kelvin. Note that the ultimate temperature rise will be over 160 °K.

From the Arrhenius relation,

$$\ln\left(\frac{k_2}{k_1}\right) = -\frac{E}{R}\left(\frac{1}{T_2} - \frac{1}{T_1}\right)$$

or

$$\ln\left(\frac{k}{0.8}\right) = -\frac{28,960}{1.987}\left(\frac{1}{T} - \frac{1}{436}\right)$$

Thus,

$$k = 2.61 \times 10^{14} e^{-14,570/T} \quad (C)$$

Equations B and C may now be combined with design equation A to solve for the required holding time.

The data are worked up in Table 10.I.1 in a form suitable for hand calculations, although a computerized solution would be more accurate, particularly with regard to the integration.

If we define $Z = 1/[k(1 - f_A)]$, the holding time can be represented as

$$t = \int_0^{0.97} Z \, df_A$$

and the integral may be evaluated numerically using Simpson's rule, the trapezoidal rule, or some other appropriate technique.

If one uses the trapezoidal rule for integration,

$$t = \left(\frac{Z_0}{2} + Z_1 + Z_2 + \cdots + Z_{n-1} + \frac{Z_n}{2}\right)\Delta f_A$$

or

$$t = \begin{bmatrix} \frac{1}{2}(1.250 + 0.003) + 0.706 + 0.407 + 0.242 \\ + 0.147 + 0.091 + 0.058 + 0.037 + 0.025 \\ + 0.017 + 0.012 + 0.009 + 0.006 + 0.005 \\ + 0.004 + 0.003 + 0.003 + 0.002 + 0.002 \end{bmatrix} 0.05 + \frac{1}{2}(0.003 + 0.005)0.02$$

or

$$t = 0.120 + 8 \times 10^{-5} = 0.120 \text{ hr}$$

A close inspection of the data worked up above would indicate that the increments in the fraction conversion have not been chosen small enough to cause the integral to converge. This situation results from the combination of the large temperature change that occurs and the strong dependence of the rate constant on temperature. The use of a computer would readily permit the choice of smaller increments, and a more accurate result could be thus obtained. The true answer would be 0.117 hr or 7.02 min. The labor necessary to work even a problem as simple as this one using a slide rule or a mechanical desk calculator indicates one of the reasons why the scientific basis of chemical reactor design did not develop rapidly until the advent of large-scale computers.

The time necessary to accomplish this exothermic reaction under adiabatic operating conditions is only an extremely small fraction of that necessary for isothermal operation. In fact, the times necessary to fill and drain the reactor and to heat it to a temperature where the rate becomes appreciable will be greater than that necessary to accomplish the reaction. Thus,

$$\tau_{batch,\,adiabatic} = 0.12 + \frac{10 + 12 + 14}{60} = 0.72 \text{ hr}$$

While operating 7000 hr one will be able to process 7000/0.72, or 9722 batches/year. Each batch will then contain $2 \times 10^6/0.97(9722)$, or 212 lb. The necessary reactor volume will then be 28 gal. Under these circumstances the times necessary to perform the nonreactive operations would undoubtedly be somewhat reduced. One should recognize that these steps will be the bottlenecks in this operation. The required reactor size is definitely on the small side, and it would be preferable to operate in a larger reactor and process a smaller number of batches per year with attendant reductions in labor requirements.

10.3 THE IDEAL CONTINUOUS STIRRED TANK REACTOR

The ideal continuous stirred tank reactor is the easiest type of continuous flow reactor to analyze in design calculations because the temperature and composition of the reactor contents are homogeneous throughout the reactor volume. Consequently, material and energy balances can be written over the entire reactor and the outlet composition and temperature can be taken as representative of the reactor contents. In general the temperatures of the feed and effluent streams will not be equal, and it will be necessary to use both material and energy balances and the temperature-dependent form of the reaction rate expression to determine the conditions at which the reactor operates.

The material balance on a single CSTR operating at steady state may be represented by:

$$F_{i\,in} = F_{i\,out} - r_i V_R \tag{10.3.1}$$

where the rate expression is evaluated at the effluent composition and temperature. If only one reaction is taking place one may rewrite this equation in terms of the familiar design equation

$$\tau_{CSTR} = \frac{C_{A0}(f_{A\,out} - f_{A\,in})}{-r_{AF}} \tag{10.3.2}$$

If multiple reactions are taking place and the system cannot be characterized by a single fraction conversion, an equation of the form of 10.3.1 will need to be written for each species.

The steady-state form of the energy balance for a continuous stirred tank reactor is given by equation 10.1.4.

$$\dot{Q} = \sum_{\substack{outlet \\ streams}} h_{out}\dot{m}_{out} - \sum_{\substack{inlet \\ streams}} h_{in}\dot{m}_{in}$$

$$\tag{10.3.3}$$

It should be emphasized that not all of the individual enthalpy datums may be chosen independently because of the presence of one or more reactions.

Since enthalpy changes are path-independent quantities, one is at liberty to choose a convenient path for making the calculation. If we carry out the reaction isothermally at the inlet temperature and then heat the products at constant composition to the effluent temperature, we find that

$$\dot{Q} = \Delta H_{R \text{ at } T_0} r V_R + \sum_i \left(F_{iF} \int_{T_0}^{T_{out}} \bar{C}_{pi}\, dT \right)$$

$$\tag{10.3.4}$$

where the summation involves the *outlet* molal flow rates.

From a material balance on reactant A,

$$\text{Disappearance} = \text{input} - \text{output}$$

$$-v_A r V_R = F_{A0}(1 - f_{A\,in}) - F_{A0}(1 - f_{A\,out})$$

$$= F_{A0}(f_{A\,out} - f_{A\,in}) \tag{10.3.5}$$

Thus,

$$\dot{Q} = \frac{F_{A0}(f_{A\,\text{out}} - f_{A\,\text{in}})}{-v_A} \Delta H_{R\,\text{at}\,T_0}$$
$$+ \sum \left(F_{iF} \int_{T_0}^{T_{\text{out}}} \bar{C}_{pi}\, dT \right)$$

(10.3.6)

Equation 10.3.6, the reaction rate expression, and the design equation are sufficient to determine the temperature and composition of the fluid leaving the reactor if the heat transfer characteristics of the system are known. If it is necessary to know the reactor volume needed to obtain a specified conversion at a fixed input flow rate and specified heat transfer conditions, the energy balance equation can be solved to determine the temperature of the reactor contents. When this temperature is substituted into the rate expression, one can readily solve the design equation for the reactor volume. On the other hand, if a reactor of known volume is to be used, a determination of the exit conversion and temperature will require a simultaneous trial and error solution of the energy balance, the rate expression, and the design equation.

Illustrations 10.2 and 10.3 show how the principles discussed above are applied in the design of a commercial scale reactor.

ILLUSTRATION 10.2 DETERMINATION OF HEAT TRANSFER AND VOLUME REQUIREMENTS FOR SINGLE AND MULTIPLE CONTINUOUS STIRRED TANK REACTORS

Consider the reaction system and production requirements discussed in Illustration 10.1. Consider the possibility of using one or more continuous stirred tank reactors operating in series. If each CSTR is to operate at 163 °C and if the feed stream is to consist of pure A entering at 20 °C, determine the reactor volumes and heat transfer requirements for

1. A single CSTR.
2. Three identical CSTR's in series.

Solution

The rate at which A must be processed is equal to

$$\frac{2 \times 10^6}{0.97} \text{ lb} \div 7000 \text{ hr} = 295 \frac{\text{lb}}{\text{hr}} \approx 133,700 \text{ g/hr}$$

From equation 8.3.38,

$$C_{A\,\text{out}} = \frac{C_{A\,\text{in}}}{1 + k\tau} \qquad \text{(A)}$$

For a single reactor $C_{A\,\text{out}} = 0.03 C_{A\,\text{in}}$. Thus

$$0.03 = \frac{1}{1 + 0.8\,\tau}$$

$$\tau_{1\,\text{reactor}} = \frac{1 - 0.03}{(0.03)(0.8)} = 40.4 \text{ hr}$$

The volumetric feed rate may be determined from the production requirements and the density of the material.

$$\mathscr{V}_0 = 133,700 \frac{\text{g}}{\text{hr}} \times \frac{\text{cm}^3}{0.9 \text{ g}} \times \frac{1 \text{ gal}}{3785 \text{ cm}^3} = 39.3 \frac{\text{gal}}{\text{hr}}$$

The required reactor volume may now be determined from the space time and the volumetric flow rate.

$$V_R = \mathscr{V}_0 \tau = 39.3(40.4) = 1586 \text{ gal}$$

The heat transfer in the reactor can be determined from the energy balance equation (equation 10.3.6).

$$\dot{Q} = -\frac{F_{A0}(f_{A\,\text{out}} - f_{A\,\text{in}})}{v_A} \Delta H_{R\,\text{at}\,T_0}$$
$$+ \sum \left(F_{iF} \int_{T_0}^{T_{\text{out}}} \bar{C}_{pi}\, dT \right) \qquad \text{(B)}$$

$$\dot{Q} = 133,700\,(0.97)\,(-83)$$
$$+ \int_{20}^{163} 133,700(0.5)\, dT$$

$$\dot{Q} = -10,764,000 + 9,560,000$$

$$= -1,204,000 \frac{\text{cal}}{\text{hr}} \approx -4780 \frac{\text{BTU}}{\text{hr}}$$

Thus, for a single continuous stirred tank reactor, the required reactor volume will be

1586 gal and the amount of heat that must be *removed* is equal to 4780 BTU/hr.

Now consider the case where there are three CSTR's operating in series.

If we denote the composition leaving the nth reactor by C_n, it is readily shown that

$$C_{A1} = \frac{C_{A0}}{1 + k\tau} \tag{C}$$

$$C_{A2} = \frac{C_{A1}}{1 + k\tau} = \frac{C_{A0}}{(1 + k\tau)^2} \tag{D}$$

$$C_{A3} = \frac{C_{A2}}{1 + k\tau} = \frac{C_{A0}}{(1 + k\tau)^3} \tag{E}$$

where τ is the ratio of the volume of a *single* CSTR to the input volumetric flow rate.

For 97% conversion,

$$\frac{C_{A3}}{C_{A0}} = 0.03 = \frac{1}{(1 + 0.8\tau)^3}$$

Rearrangement yields $1/(1 + 0.8\tau) = 0.3107$. Thus,

$$\tau = \frac{1 - 0.3107}{(0.8)(0.3107)} = 2.77 \text{ hr}$$

The required reactor volume per CSTR is then equal to (39.3 gal/hr)(2.77 hr) = 109 gal.

The fraction conversion in the effluent from the first reactor can be determined from equation C and the definition of the fraction conversion.

Since

$$C_{A1} = \frac{C_{A0}}{1 + k\tau} = C_{A0}(1 - f_{A1})$$

Then

$$f_{A1} = \frac{k\tau}{1 + k\tau} = \frac{0.8(2.77)}{1 + 0.8(2.77)} = 0.689$$

From equation D and the definition of fraction conversion,

$$C_{A2} = \frac{C_{A1}}{1 + k\tau} = \frac{C_{A0}(1 - f_{A1})}{1 + k\tau}$$

$$= C_{A0}(1 - f_{A2})$$

Thus

$$f_{A2} = 1 - \left(\frac{1 - f_{A1}}{1 + k\tau}\right)$$

$$f_{A2} = 1 - \frac{0.311}{1 + 0.8(2.77)}$$

$$= 1 - 0.097 = 0.903$$

The heat transfer requirements for each reactor may be determined from equations of the form of equation B.

For reactor 1,

$$Q_1 = 133{,}700(0.689)(-83)$$

$$+ \int_{20}^{163} (133{,}700)(0.5) \, dT$$

$$Q_1 = -7{,}646{,}000 + 9{,}560{,}000$$

$$Q_1 = 1{,}914{,}000 \text{ cal/hr} \approx 7595 \text{ BTU/hr}$$

For reactors 2 and 3, there will not be any sensible heat effects.

$$Q_2 = 133{,}700(0.963 - 0.689)(-83)$$

$$Q_2 = -2{,}375{,}000 \text{ cal/hr} \approx -9424 \text{ BTU/hr}$$

and

$$Q_3 = 133{,}700(0.970 - 0.903)(-83)$$

$$Q_3 = -743{,}500 \text{ cal/hr} \approx -2950 \text{ BTU/hr}$$

The equipment requirements that we have determined are well within the realm of technical feasibility and practicality. The heat transfer requirements are easily attained in equipment of this size. The fact that some of the heat transfer requirements are positive and others negative indicates that one should probably consider the possibility of at least partial heat exchange between incoming cold feed and the effluent from the second or third reactors. The heat transfer calculations show that the sensible heat necessary to raise the cold feed to a temperature where the reaction rate is appreciable represents a substantial fraction of the energy released by reaction. These calculations also indicate that it would be advisable to investigate the possibility of adiabatic operation in CSTR's operating in series.

ILLUSTRATION 10.3 ADIABATIC OPERATION OF CONTINUOUS STIRRED TANK REACTORS OPERATING IN SERIES

Consider the possibility of carrying out the reaction used as the basis for Illustrations 10.1 and 10.2 under adiabatic operating conditions. How much B will it be possible to produce from 2.1 million lb/yr of species A using a pair of 1000-gal CSTR's operating in series? Assume that you will be able to operate 7000 hr/yr. Use the data from Illustration 10.2.

The feed to the first reactor has a temperature of 20 °C.

Solution

The solution to this problem requires a trial and error iterative procedure. Since both the reactor volume and the initial volumetric flow rate are known, the space time per reactor may be calculated and we may focus our attention initially on the first reactor.

The volumetric flow rate is equal to

$$\frac{2.1 \times 10^6 \text{ lb}}{7 \times 10^3 \text{ hr}} \times 454 \frac{\text{g}}{\text{lb}} \times \frac{\text{cm}^3}{0.9 \text{ g}} \times \frac{1 \text{ gal}}{3785 \text{ cm}^3}$$

or 40.0 gal/hr.

This value corresponds to a reactor space time given by

$$\tau = \frac{V_R}{\mathcal{V}_0} = \frac{1000}{40.0} = 25 \text{ hr}$$

For a first-order reaction it was shown in Illustration 10.2 that

$$f_{A1} = \frac{k\tau}{1 + k\tau} = \frac{25k}{1 + 25k} \qquad (A)$$

where the rate constant is a temperature de-

pendent quantity given by equation C in Illustration 10.1.

$$k = 2.61 \times 10^{14} e^{-14,570/T} \qquad (B)$$

The energy balance equation for adiabatic operation becomes

$$0 = \frac{-F_{A0} f_{A1}}{v_A} \Delta H_{R \text{ at } T_0} + \sum F_i \int_{T_0}^{T_{\text{out}}} \bar{C}_{pi} \, dT$$

where the last term becomes $F_{A0} C_p (T_{\text{out}} - T_0)$.

Division by F_{A0} and rearrangement gives

$$f_{A1} = v_A \frac{C_p(T_{\text{out}} - T_0)}{\Delta H_{R \text{ at } T_0}} = \frac{0.5(T_{\text{out}} - 293)}{83} \qquad (C)$$

Equations A to C must now be solved simultaneously. Combining these equations gives

$$\frac{0.5}{83}(T - 293) = \frac{25(2.61 \times 10^{14}) e^{-14,570/T}}{1 + 25(2.6 \times 10^{14}) e^{-14,570/T}} \qquad (D)$$

This equation can be solved for T using a trial and error procedure. This gives $T = 410 \, °K$. Thus

$$f_{A1} = \frac{0.5(410 - 293)}{83} = 0.705$$

From Illustration 10.2,

$$f_{A2} = 1 - \frac{1 - f_{A1}}{1 + k\tau} = \frac{f_{A1} + k\tau}{1 + k\tau}$$

or

$$f_{A2} = \frac{0.705 + 25k}{1 + 25k} \qquad (E)$$

where k is given by equation B.

The energy balance equation for adiabatic operation becomes

$$0 = -\frac{F_{A0}(f_{A2} - f_{A1})}{v_A} \Delta H_{R \text{ at } T_0}$$

$$+ F_{A0} \int_{T_0}^{T_2} C_p \, dT - F_{A0} \int_{T_0}^{T_1} C_p \, dT$$

Division by F_{A0} and combination of the integrals gives

$$(f_{A2} - f_{A1}) \frac{\Delta H_{R \text{ at } T_0}}{v_A} = \int_{T_1}^{T_2} C_p \, dT$$

or

$$f_{A2} = f_{A1} + \frac{v_A \int_{T_1}^{T_2} C_p \, dT}{\Delta H_{R \text{ at } T_0}}$$

$$= 0.705 + \frac{0.5(T_2 - 410)}{83} \quad \text{(F)}$$

Combining equations B, E, and F gives

$$0.705 + \frac{0.5(T_2 - 410)}{83}$$

$$= \frac{0.705 + 25(2.61 \times 10^{14} e^{-14,570/T_2})}{1 + 25(2.61 \times 10^{14} e^{-14,570/T_2})} \quad \text{(G)}$$

A trial and error solution gives $T_2 = 458.5 \, °K$ (approximately 186 °C) and $f_{A2} = 0.997$.

With this arrangement one would be able to produce in excess of 2 million lb of B annually if side reactions do not occur at this higher temperature and if the vapor pressure of the solution still lies in a range that does not create problems.

10.4 TEMPERATURE AND ENERGY CONSIDERATIONS IN TUBULAR REACTORS

This section treats the material and energy balance equations for a plug flow reactor. For steady-state operation the energy balance analysis leading to equation 10.1.4 is appropriate.

$$\dot{Q} = \sum_{\substack{\text{outlet} \\ \text{streams}}} h_{\text{out}} \dot{m}_{\text{out}} - \sum_{\substack{\text{inlet} \\ \text{streams}}} h_{\text{in}} \dot{m}_{\text{in}} \quad (10.4.1)$$

while the design equation appropriate for these conditions is

$$\tau = C_{A0} \int_{f_{A \text{ in}}}^{f_{A \text{ out}}} \frac{df_A}{-r_A} \quad (10.4.2)$$

These equations must be solved simultaneously using a knowledge of the temperature dependence of reaction rate expression.

Note that in the case of the plug flow reactor, there may be a variation in the temperature of the reactor contents from point to point within the reactor. This condition is in contrast to the two ideal reactor cases discussed earlier, where the temperature of the contents of a single reactor was assumed to be constant throughout. The presence of temperature gradients in the same direction as the flow is by no means contrary to the assumption of plug flow. In this chapter we will confine our analysis to situations in which both the temperature and the composition are uniform across the cross section of the reactor. The complications arising from the existence of radial temperature gradients and variations in the composition with radial position are treated implicitly in conjunction with the discussion of the two dimensional model of fixed bed reactors in Section 12.7. For homogeneous systems the effects are usually small.

If one considers the rate at which heat is being supplied to a differential length of a tubular reactor, geometric considerations imply that

$$d\dot{Q} = U(T_m - T) \, dA = U(T_m - T) \frac{4}{D} \, dV_R$$

$$(10.4.3)$$

where D is the inside diameter of the tubular reactor.

A material balance on this differential reactor volume yields the following result.

$$F_{A0} \, df_A = -r_A \, dV_R \quad (10.4.4)$$

Combining equations 10.4.3 and 10.4.4 gives

$$d\dot{Q} = U(T_m - T) \frac{4}{D} F_{A0} \frac{df_A}{(-r_A)} \quad (10.4.5)$$

and the energy balance on this differential

volume element then becomes

$$U(T_m - T)\frac{4}{D}F_{A0}\frac{df_A}{(-r_A)} = \sum\left(F_i \int_{T_0}^{T_{\text{out of element}}} \bar{C}_{pi}\, dT\right) - \sum\left(F_i \int_{T_0}^{T_{\text{into element}}} \bar{C}_{pi}\, dT\right)$$
$$-\frac{F_{A0}\,\Delta H_{R\text{ at }T_0}}{\nu_A}\,df_A \tag{10.4.6}$$

where the first summation involves the molal flow rates of the various species leaving the volume element and the second summation involves the molal flow rates entering the volume element. The datum for the enthalpy calculations has been taken as the temperature prevailing at the inlet to the reactor. This equation may also be integrated between the reactor inlet and a point downstream where the fraction conversion is f_A to give

tion is a simple example of a Diels-Alder reaction.

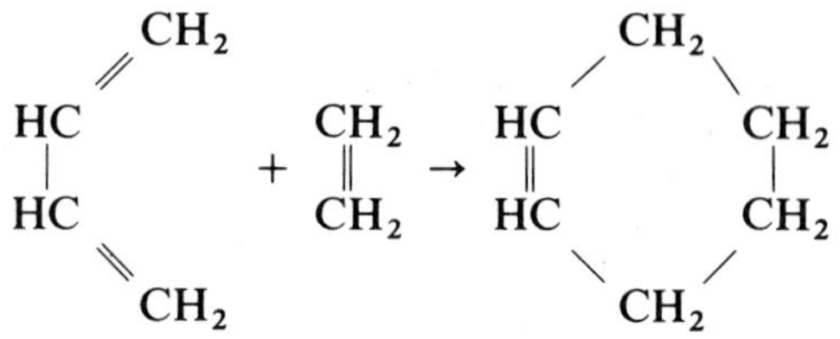

If an equimolar mixture of butadiene and

$$\int_{f_{A\text{ in}}}^{f_{A\text{ out}}} U(T_m - T)\frac{4}{D}F_{A0}\frac{df_A}{(-r_A)} = \sum\left(F_i \int_{T_0}^{T} \bar{C}_{pi}\, dT\right) - \frac{F_{A0}\,\Delta H_{R\text{ at }T_0}}{\nu_A}(f_{A\text{ out}} - f_{A\text{ in}}) \tag{10.4.7}$$

The summation involves the effluent molal flow rates. This equation and equation 10.4.2 must be solved simultaneously in order to determine the tubular reactor size and to determine the manner in which the heat transfer requirements are to be met. For either isothermal or adiabatic operation one of the three terms in equation 10.4.7 will drop out, and the analysis will be much simpler than in the general case. In the illustrations which follow two examples are treated in detail to indicate the types of situations that one may encounter in practice and to indicate in more detail the nature of the design calculations.

ILLUSTRATION 10.4 DETERMINATION OF THE VOLUME REQUIREMENTS FOR ADIABATIC OPERATION OF A TUBULAR REACTOR WITH EXOTHERMIC REACTION

Butadiene will react with ethylene in the gas phase at temperatures above 500 °C. This reac-

ethylene at 450 °C and 1 atm is fed to a reactor, determine the space times required to convert 10% of the butadiene to cyclohexene for isothermal and for adiabatic modes of operation.

Data

Wasserman (6) has reported the following data for this reaction.

$$k = 10^{7.5}e^{-27,500/RT} \text{ liters/mole-sec}$$
$$\Delta H_R = -30,000 \text{ cal/g mole}$$

The reverse reaction may be neglected.

The following values of gas phase heat capacities may be assumed to be constant over the temperature range of interest.

$$C_{p,C_4H_6} = 36.8 \text{ cal/g mole-°K}$$
$$C_{p,C_2H_4} = 20.2 \text{ cal/g mole-°K}$$
$$C_{p,C_6H_{10}} = 59.5 \text{ cal/g mole-°K}$$

Solution

From the units on the reaction rate constant, the reaction is second order. There is a volume change on reaction and $\delta = -1/2$. Thermal expansion will also occur, so equations 3.1.44 and 3.1.46 must be combined to get the reactant concentrations. Since equimolar concentrations of reactants are used, the design equation becomes

$$\tau = C_{A0} \int_0^{0.1} \frac{df_A}{kC_{A0}^2 \frac{(1-f_A)^2}{\left(1-\frac{f_A}{2}\right)^2}\left(\frac{T_0}{T}\right)^2}$$

$$= \int_0^{0.1} \left(\frac{T}{T_0}\right)^2 \frac{\left(1-\frac{f_A}{2}\right)^2 df_A}{kC_{A0}(1-f_A)^2} \quad (A)$$

The initial reactant concentration can be determined from the ideal gas law

$$C_{A0} = \frac{Y_A P_{Tot}}{RT_0} = \frac{(0.5)(1)}{(0.08206)(723)}$$

$$= 8.43 \times 10^{-3} \text{ g moles/liter}$$

For isothermal operation at 450 °C the rate constant is equal to 0.156 liters/mole-sec and

Equation 10.4.7 is appropriate for use if we set the heat transfer term equal to zero.

$$\sum\left(F_i \int_{T_0}^T \bar{C}_{pi}\, dT\right) = \frac{F_{A0}\Delta H_{R\text{ at }T_0}(f_{A\text{ out}} - f_{A\text{ in}})}{\nu_A} \quad (B)$$

From the reaction stoichiometry and the tabulated values of the heat capacities the contributions to the summation may be written as:

Butadiene

$$F_i C_{pi}(T - T_0) = F_{A0}(1 - f_A)36.8(T - T_0)$$

Ethylene

$$F_i C_{pi}(T - T_0) = F_{A0}(1 - f_A)20.2(T - T_0)$$

Cyclohexene

$$F_i C_{pi}(T - T_0) = F_{A0}f_A 59.5(T - T_0)$$

or

$$\sum F_i \int_{T_0}^T C_{pi}\, dT = (57.0F_{A0} + 2.5f_A F_{A0})(T - T_0)$$

Therefore, substitution of numerical values in equation B and rearrangement gives

$$T = 723 + \frac{30,000 f_A}{57.0 + 2.5 f_A} \quad (C)$$

$$\tau = \frac{1}{kC_{A0}} \int_0^{0.1} \frac{\left(1 - \frac{f_A}{2}\right)^2 df_A}{(1-f_A)^2}$$

$$= \frac{\left\{\dfrac{1}{1-f_A} - \left[\dfrac{1}{1-f_A} + \ell n(1-f_A)\right] - \dfrac{1}{4}\left[1 - f_A - 2\,\ell n(1-f_A) - \dfrac{1}{1-f_A}\right]\right\}_0^{0.1}}{kC_{A0}}$$

$$\tau = \frac{-\dfrac{1}{2}\ell n(1-f_A) - \dfrac{f_A(f_A - 2)}{4(1-f_A)}\Bigg|_0^{0.1}}{0.156(8.43 \times 10^{-3})} = 80.19 \text{ sec}$$

In the case of adiabatic operation the temperature must be related to the fraction conversion so that equation A can be integrated.

where T is expressed in degrees Kelvin. At a fraction conversion of 0.1 the gas temperature will have risen to 775 °K.

The design equation then becomes

$$\tau = \int_0^{0.1} \left(\frac{T}{723}\right)^2 \frac{(1 - f_A/2)^2 \, df_A}{10^{7.5} e^{-27,500/1.987T} (8.43 \times 10^{-3})(1 - f_A)^2}$$

where T is given by equation C.

Numerical evaluation of this integral gives $\tau = 47.11$ sec.

This space time is about 59% of that required for isothermal operation. This number may be somewhat low because the reaction is exothermic and the rate of the reverse reaction may be appreciable at the highest temperatures involved in our calculation.

We now wish to examine the case where we allow for heat exchange with a substantially constant temperature heat sink (e.g., an evaporating or condensing fluid or a material flowing at a velocity such that its temperature change over the reactor length is quite small compared to the driving force for heat transfer).

ILLUSTRATION 10.5 DETERMINATION OF THE VOLUME REQUIREMENTS FOR OPERATION OF A TUBULAR REACTOR UNDER NONISOTHERMAL CONDITIONS WITH HEAT EXCHANGE

Consider the reaction used as the basis for Illustrations 10.1 to 10.3. Determine the volume required to produce 2 million lb of B annually in a plug flow reactor operating under the conditions described below. The reactor is to be operated 7000 hr annually with 97% conversion of the A fed to the reactor. The feed enters at 163 °C. The internal pipe diameter is 4 in. and the piping is arranged so that the effective reactor volume can be immersed in a heat sink maintained at a constant temperature of 160 °C. The overall heat transfer coefficient based on the

inside area of the pipe may be taken as 200 kcal/(hr-m²-°K). Volumetric expansion effects are negligible.

Solution

For a plug flow reactor the appropriate design equation for the reaction at hand is:

$$\tau = C_{A0} \int_0^{0.97} \frac{df_A}{-r_{A'}} = \int_0^{0.97} \frac{df_A}{k(1 - f_A)} \quad \text{(A)}$$

From equation C of Illustration 10.1,

$$k = 2.61 \times 10^{14} e^{-14,570/T} \, \text{hr}^{-1} \quad \text{(B)}$$

The temperature may be related to the fraction conversion by means of an energy balance such as equation 10.4.7.

$$\int_0^{f_A} U(T_m - T) \frac{4}{D} F_{A0} \frac{df_A}{(-r_A)} = \sum \left(F_i \int_{T_0}^{T} C_{pi} \, dT \right) - \frac{F_{A0} \Delta H_{R \text{ at } T_0} (f_{A \text{ out}} - f_{A \text{ in}})}{\nu_A} \quad \text{(C)}$$

In the present case

$$U = 200,000 \text{ cal/m}^2 \cdot \text{hr} \cdot ^\circ\text{K}$$

$$T_m = 433 \text{ °K}$$

$$D = 4(2.54 \times 10^{-2}) = 0.1016 \text{ m}$$

$$F_{A0} = 133,700 \text{ g/hr (from Illustration 10.2)}$$

$$C_p = (0.5) \text{ cal/g} \cdot ^\circ\text{K}$$

$$T_0 = 436 \text{ °K}$$

$$\frac{\Delta H_R}{\nu_A} = 83 \text{ cal/g}$$

Division of equation C by F_{A0} and insertion of the reaction rate expression and numerical values gives

$$\int_0^{f_A} \frac{(2 \times 10^5)(433 - T)4}{(0.1016)kC_{A0}} \frac{df_A}{(1 - f_A)}$$

$$= \int_{436}^{T} (0.5) \, dT - 83 f_A \quad \text{(D)}$$

The initial reactant concentration is equal to $0.9 \ g/cm^3$ or $9 \times 10^5 \ g/m^3$ in units that are consistent with the other parameters we are using.

Substitution of this value into equation D and rearrangement gives

$$T = 436 + \frac{83f_A + \int_0^{f_A} \dfrac{(433 - T)(8 \times 10^5) \, df_A}{(0.1016)k(1 - f_A)(9 \times 10^5)}}{0.5} \tag{E}$$

Equations A, B, and E provide sufficient information to determine the necessary reactor volume. However, they do require an iterative trial and error solution.

A simple approach to solving this set of equations can be represented as follows.

1. Consider increments of reactor-volume corresponding to increments of fraction conversion $f_{A1}, f_{A2}, \ldots, f_{An}$.
2. Guess a temperature corresponding to f_{A1}.
3. Calculate values of the integrand in equation E corresponding to conversions zero and f_{A1}.
4. Use the trapezoidal rule to approximate the integral.
5. Calculate the temperature corresponding to f_{A1} and iterate until the calculated and assumed values agree.
6. Repeat the procedure for successive increments in the fraction conversion.
7. Take the functional relationship between T and f_A determined above and use it in conjunction with equation A to determine the required reactor volume.

When this procedure is followed, one finds that the space time corresponding to the operating conditions outlined above is equal to 4.642 hr. The required reactor volume is then equal to

$$V_R = \tau \mathcal{V}_0 = 4.642(39.3) = 182.4 \text{ gal}$$

or

$$V_R = 24.3 \text{ ft}^3$$

This volume corresponds to a reactor length of 69.6 ft.

The temperature of the reactor contents rises from 436 °K at the inlet to 443.20 °K at a point corresponding to $f_A = 0.22$ and then drops to a temperature of 433.19 °K at the reactor outlet.

The maximum temperature is significantly less than that which would be obtained by adiabatic operation (597 °K). This fact may be important in design considerations when side reactions have significant reaction rates at high temperatures. The required reactor volume will be greater than that for isothermal operation because, under the specified operating conditions, once a fraction conversion of 0.63 is obtained, the temperature of the reactant mixture drops below the feed temperature. In practice one might wish to dispense with all or part of the cooling capacity in the later portions of the reactor. This technique would have the effect of reducing the required reactor volume. An alternative iteration procedure for handling tubular reactor analyses involving a heat transfer term is employed in the case study in Section 13.1.

10.5 AUTOTHERMAL OPERATION OF REACTORS

In the design of processes involving exothermic reactions it is generally desirable to use the energy liberated by reaction at some other point in the process or to make it available for use elsewhere in the industrial plant. For example, it can be used to preheat the feed components, particularly when the reaction takes place at high temperatures and the feed components are supplied at much lower temperatures. The term "autothermal operation" is applied to modes of processing in which exothermic reactions are carried out such that the energy released by reaction is fed back to the incoming reactant stream. Use of this term implies that the reactor system is to a large extent self-supporting in terms of its thermal energy requirements. Consequently, it is possible to run at the desired operating temperature without using external heat sources to preheat the feed. An autothermal reaction is akin to the autocatalytic reactions discussed in Section 9.4 in the sense that a product of the reaction (in this case thermal energy) serves to enhance the reaction rate.

Since operation in an autothermal mode implies a feedback of energy to preheat the feed, provision must be made for "ignition" of the reactor in order to attain steady-state operation. The ordinary gas burner and many other rapid combustion reactions are examples of autothermal reactions in which the reactants are preheated to the reaction temperature by thermal conduction and radiation. (Back diffusion of free radicals also plays an important role in many combustion processes.)

There are several techniques by which the feedback of energy may be accomplished.

1. In a semibatch reactor, a cold feed may be heated by mixing with the reactor contents. This technique is discussed in Illustration 10.7 later in this section.
2. In reactors operating under continuous flow conditions there may be heat exchange between the effluent stream or the reactor contents and the feed stream. This exchange may occur in the reactor proper, in heat exchangers between various portions of a reactor network, or by various fluid mixing processes. In a stirred tank reactor the fresh feed is rapidly mixed with the reactor contents to promote energy transfer. Energy feedback in tubular reactors can be accomplished by recycling a fraction of the reactor effluent for mixing with the fresh feed.

In Illustration 10.2 we saw that when one uses a series of stirred tanks for carrying out an exothermic reaction under isothermal conditions there may be occasions when the heat require-

ments for the various tanks may be of opposite sign. Some tanks will require a net input of thermal energy, while others will need to be cooled. It is often useful in such situations to consider the possibility of adiabatic operation of one or more of the tanks in series, remembering the constraints that one desires to place on the temperatures of the process streams. Another means of achieving autothermal operation is to use a network consisting of a stirred tank reactor followed by a tubular reactor. This case is considered in Illustration 10.6.

ILLUSTRATION 10.6 AUTOTHERMAL OPERATION OF A REACTOR NETWORK CONSISTING OF A STIRRED TANK REACTOR FOLLOWED BY A PLUG FLOW REACTOR

Consider the reaction studied in Illustration 10.1. Autothermal operation is to be achieved using a CSTR with an effective volume of 1000 gal followed by a PFR of undetermined volume. Pure species A enters at a rate of 40.0 gal/hr and at a temperature of 20 °C. The overall fraction conversion is to be 0.97. This flow rate and conversion level will suffice to meet the annual production requirement of 2 million lb of B. Both the CSTR and the PFR are to be operated adiabatically. What PFR volume will be required, and what will be the temperature of the effluent stream?

Solution

The analysis of the performance of the CSTR is identical with that of the first CSTR considered in Illustration 10.3.

The space time for the CSTR is

$$\tau = \frac{V_R}{V_0} = \frac{1000}{40.0} = 25 \text{ hr}$$

The fraction conversion at the CSTR exit for this first-order reaction is given by

$$f_{A1} = \frac{k\tau}{1 + k\tau} = \frac{25k}{1 + 25k} \qquad \text{(A)}$$

The rate constant is given by

$$k = 2.61 \times 10^{14} e^{-14,570/T} \qquad \text{(B)}$$

The energy balance equation for adiabatic operation is

$$0 = -\frac{F_{A0}f_{A1}}{v_A} \Delta H_{R \text{ at } T_0} + F_{A0} \int_{T_0}^{T_{\text{out}}} C_p \, dT$$

or, assuming a constant heat capacity,

$$f_{A1} = \frac{v_A C_p (T_{\text{out}} - T_0)}{\Delta H_{R \text{ at } T_0}} = \frac{0.5(T_{\text{out}} - 293)}{83} \quad \text{(C)}$$

Equations A to C must now be solved simultaneously using a trial and error iterative procedure. One finds that $f_{A1} = 0.705$ and $T = 410\,°\text{K}$.

These properties are those of the stream entering the plug flow reactor. The design equation for this reactor is

$$\tau = C_{A0} \int_{0.705}^{0.97} \frac{df_A}{k C_{A0}(1 - f_A)} = \int_{0.705}^{0.97} \frac{df_A}{k(1 - f_A)} \quad \text{(D)}$$

where k is again given by equation B.

An energy balance on the PFR operating at steady state is given by equation 10.4.6. For adiabatic operation this equation becomes

$$0 = -\frac{F_{A0}(f_{A2} - f_{A1})}{v_A} \Delta H_{R \text{ at } T_0} + F_{A0} \int_{T_0}^{T_{\text{leaving PFR}}} C_p \, dT - F_{A0} \int_{T_0}^{T_{\text{entering PFR}}} C_p \, dT \quad \text{(E)}$$

Since the heat capacity of the reaction mixture is given as independent of temperature and composition, equation E simplifies to

$$0 = -\frac{(f_{A2} - f_{A1}) \Delta H_{R \text{ at } T_0}}{v_A}$$
$$+ C_p(T_{\text{leaving PFR}} - T_{\text{entering PFR}})$$

The relationship between the temperature and the fraction conversion at a particular point in the plug flow reactor can be obtained by setting $f_{A2} = f_A$ and $T_{\text{leaving PFR}}$ equal to T. Thus,

$$T = T_{\text{entering PFR}} + \frac{\Delta H_{R \text{ at } T_0}}{v_A C_p}(f_A - f_{A1})$$

or

$$T = 410 + \frac{(-83)}{(-1)(0.5)}(f_A - 0.705)$$
$$T = 410 + 166(f_A - 0.705) \quad \text{(F)}$$

At the exit from the PFR, the temperature will be 454 °K or 181 °C.

Equations B, D, and F may now be solved simultaneously to determine the required space time in the plug flow reactor.

For the present situation $\tau = 3.72$ hr so the required PFR volume is $3.72(40.0) = 148.8$ gal or 19.9 ft^3.

By operating in this mode, the necessity for using heat exchangers is eliminated or minimized. If one regards the exit temperature of 181 °C as excessive, a heat exchanger could be used between the CSTR and the PFR, or arrangements could be made for internal cooling within the PFR.

In Illustration 10.7 we will consider how to meet the vast majority of the heat transfer requirements for operation of a semibatch reactor by a semiautothermal mode of processing. By intermittent addition of a cold reactant stream to the hot contents of a well-stirred semibatch reactor, it is possible to maintain the system temperature within prescribed limits.

ILLUSTRATION 10.7   QUASIAUTOTHERMAL OPERATION OF A SEMIBATCH REACTOR USING ADDITION OF COLD FEED

At the time Kladko's article (5) was prepared, market conditions dictated that it would not be financially remunerative to develop a continuous process for producing B by the reaction discussed in Illustration 10.1 and other illustrations in this chapter. Consequently they considered the possibility of using a semibatch reactor with continuous addition of cold feed to maintain the temperature of the reactor contents within prescribed limits. The basic problem was considered earlier in Illustration 8.11. However, in addition to the material balance aspects of the design, we now wish to consider the heat transfer requirements for operation in accordance with the filling schedule outlined earlier. For the conditions indicated in Illustration 8.11 (isothermal, 163 °C), determine the direction and magnitude of the heat transfer requirements.

Solution

The material balance equations have been solved earlier with the result that

$$m_A = m_{Ai}e^{-k(t-t_i)} + \frac{\Phi_{A0}}{k}\left[1 - e^{-k(t-t_i)}\right]$$

where

m_A is the mass of A remaining in the reactor at time t

m_{Ai} is the mass of A in the reactor at time t_i

Φ_{A0} is the mass feed rate of A appropriate to the time interval between t and t_i

An energy balance on the reactor can be derived from equation 10.1.1 by omission of appropriate terms.

$$\frac{dE_{sys}}{dt} = \dot{Q} + h_{in}\dot{m}_{in}$$

Multiplying by dt and integrating between time 0 and time t gives

$$E_{sys} - E_{sys,0} = Q + h_{in}m_{in}$$

If the temperature of the inlet stream is taken as the datum temperature for enthalpy and internal energy calculations,

$$h_{in} = 0$$

and

$$E_{sys} - E_{sys,0} = Q$$

However, for condensed phases, the difference between internal energy and enthalpy is usually negligible;

$$Q \approx H_{sys} - H_{sys,0} \tag{A}$$

The enthalpy difference is given by

$$H_{sys} - H_{sys,0} = \int_{20\,°C}^{163\,°C} (m_A + m_B - m_{B0})\hat{C}\,dT$$
$$+ \Delta H_{R\text{ at }20\,°C}(m_B - m_{B,0}) \tag{B}$$

where $\hat{C}$ is the average heat capacity per unit mass and where we have assumed that the initial material sump (m_{B0}) is at 163 °C.

Equations A and B can be combined with appropriate numerical values to give

$$Q = (m_A + m_B - m_{B0})(0.5)(163 - 20)$$
$$- 83(m_B - m_{B0})$$
$$= (m_A + m_B - m_{B0})71.5 - 83(m_B - m_{B0})$$
$$= 71.5m_A - 11.5(m_B - m_{B0})$$

where

Q is expressed in calories

m is expressed in grams

If m is expressed in pounds and Q in BTU,

$$Q = 128.7m_A - 20.7(m_B - m_{B0})$$

The numerical values characterizing m_A and m_B at various times are obtained as indicated in Illustration 8.11. They may be used to determine the net heat required between time zero and time t. The pertinent data are presented in Table 10.I.2. It is evident that heat must be supplied to the reactor during the first several hours, but the reactor reaches a point beyond which cooling must be supplied.

Instantaneous heat transfer requirements may be determined by recognizing that the thermal energy input must be used either to effect a sensible heat change or the reaction. If q is the heat transfer rate,

$$q = \Phi_{A0}\int_{T_0}^{T}\hat{C}\,dT + rV_R'\,\Delta H_R$$

where V_R' is the liquid phase volume at time t.

Since $C_A = m_A/V'_R$ and the reaction obeys first-order kinetics,

$$q = \Phi_{A0} \int_{T_0}^{T} \hat{C}\, dT + km_A\, \Delta H_R$$

Substitution of numerical values gives

$$q = \Phi_{A0}(0.5)(163 - 20) + 0.8m_A(-83)$$
$$= 71.5\Phi_{A0} - 66.4m_A$$

where

$$q \text{ is expressed in calories per hour}$$
$$\Phi_{A0} \text{ is expressed in grams per hour}$$
$$m_A \text{ is expressed in grams}$$

Table 10.I.2

Material and Energy Balance Analysis for Illustration 10.7

Time, t (hr)	m_A (lb)	Total mass (lb)	$Q_{\text{total}} \times 10^{-3}$ (BTU)	$q \times 10^{-3}$ (BTU/hr)[a]
0	0	1500	0	22.52
1	120	1675	14.31	8.18
2	175	1850	18.90	1.61
3	199	2025	18.86	5.18
4	244	2250	20.93	−0.20
5	265	2475	19.41	−2.71
6	274	2700	16.10	2.65
7	312	2975	16.08	4.54
8	364	3300	17.12	7.98
9	439	3700	20.05	−0.98
10	473	4100	16.85	−5.04
11	488	4500	10.81	−16.49
12	443	4825	−2.64	−17.55
13	388	5100	−16.55	−17.41
14	329	5325	−30.02	−16.79
15	268	5500	−42.76	−19.16
16	189	5600	−56.63	−16.15
17	119	5650	−68.13	−14.22
18	54	5650	−77.84	−6.45
19	24	5650	−82.32	−2.87
20	11	5650	−84.26	−1.31

[a] Instantaneous heat transfer rates are calculated after changes are made in the mass flow rates at the times indicated.

In more conventional engineering units,

$$q = 128.7\Phi_{A0} - 119.5m_A$$

where

$$q \text{ is now expressed in BTU's per hour}$$
$$\Phi_{A0} \text{ is expressed in pounds per hour}$$
$$m_A \text{ is expressed in pounds}$$

Calculated values of this quantity are also presented in Table 10.I.2.

If one were to operate this semibatch reactor under a filling schedule, which for the first 11 hr is identical to that considered previously, and then proceed to feed A at the maximum rate of 400 lb/hr for an additional 2.875 hr, the same total amount of A would have been introduced to the reactor. However, in this case, the heat transfer requirements would change drastically. There would be a strong exotherm beginning at the moment the cold feed is stopped. The results for this case are presented in Table 10.I.3.

Table 10.I.3

Results of Analysis for Abrupt Termination of Feed at 13.875 hr

Time, t (hr)	m_A (lb)	Total mass (lb)	$Q_{\text{total}} \times 10^{-3}$ (BTU)	$q \times 10^{-3}$ (BTU/hr)[a]
11	488	4500	10.81	−6.84
12	494	4900	3.42	−7.55
13	498	5300	−4.26	−8.03
13.875	499	5650	−11.35	−59.63
14.00	452	5650	−18.38	−54.01
14.50	303	5650	−40.64	−36.21
15.00	203	5650	−55.58	−24.26
15.50	136	5650	−65.59	−16.25
16.00	91	5650	−72.31	−10.87
17	41	5650	−79.78	−4.90
18	18	5650	−83.22	−2.15
19	8	5650	−84.71	−0.96

[a] Instantaneous heat transfer rates are calculated after changes are made in the mass flow rates at the times indicated.

Consequently, if one desires to minimize the possibility that this reaction will run away as a result of exceeding the capacity of the cooling network, it would be advisable to operate in a mode such that the feed rate is gradually diminished instead of abruptly terminated at a high feed flow rate level.

10.6 STABLE OPERATING CONDITIONS IN STIRRED TANK REACTORS

In Illustrations 10.2 and 10.3 the temperatures and degrees of conversion leaving stirred tank reactors were determined by simultaneous solution of rate, material balance, and energy balance equations. In this section we will see that there may be circumstances when there will be more than one set of operating conditions that will satisfy these equations. In such cases questions arise as to the stability of the operating points corresponding to these solutions. For the elementary cases that are treated in this text, the stability can be readily determined on the basis of simple physical arguments. There are more complex cases in the general area of reactor stability, however, that require sophisticated mathematical analysis. Many of these cases have been treated in the extensive body of literature that exists on this subject. Perlmutter (7) has summarized much of this material in textbook form.

When a reactor is operating at steady state, the rate of energy release by chemical reaction must be equal to the sum of the rates of energy loss by convective flow and heat transfer to the surroundings. This statement was expressed in algebraic form in equations 10.3.4 and 10.4.6 for the CSTR and PFR, respectively. It will serve as the physical basis that we will use to examine the stability of various operating points.

Let Q_g represent the rate at which thermal energy is released by an exothermic chemical reaction in a CSTR. If Q_g is plotted versus the temperature of the reactor contents for a fixed

feed rate and feed composition, there are several curves that may result, depending on the nature of the reaction or reactions involved. Figure 10.1 shows the nature of the curve for the case of a simple first-order irreversible reaction. At low temperatures the reaction rate is negligible, so very little energy is released. At high temperatures the reaction rate constant is so large that very little unreacted material remains in the effluent stream. Consequently, still higher values of T cannot increase Q_g significantly, and the curve must approach an asymptote. At intermediate temperatures one has a competition between the increase in reaction rate arising from the effect of increased temperature and the decrease in the rate arising from depletion of the reactants. The equation of the curve is easily obtained from first principles.

$$Q_g = rV_R|\Delta H| \qquad (10.6.1)$$

where $|\Delta H|$ is the magnitude of the heat of reaction for the exothermic reaction.

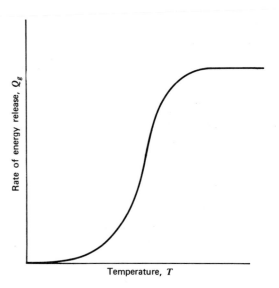

Figure 10.1

Rate of energy release by reaction versus temperature for an irreversible exothermic reaction carried out in a CSTR.

For a first-order irreversible reaction with $\nu_A = -1$

$$r = -r_A = kC_A \qquad (10.6.2)$$

If there is no volume change because of reaction, the design equation for a CSTR indicates that

$$\tau = \frac{V_R}{\mathcal{V}_0} = \frac{C_{A0} - C_{AF}}{-r_{AF}} = \frac{C_{A0} - C_{AF}}{kC_{AF}} \qquad (10.6.3)$$

or

$$C_{AF} = \frac{C_{A0}}{1 + k\tau} \qquad (10.6.4)$$

Combining equations 10.6.1, 10.6.2, and 10.6.4 gives

$$Q_g = \frac{kC_{A0}V_R|\Delta H_R|}{1 + k\tau} \qquad (10.6.5)$$

If k is now written in the Arrhenius form,

$$Q_g = \frac{Ae^{-E/RT}C_{A0}V_R|\Delta H_R|}{1 + (V_R/\mathcal{V}_0)Ae^{-(E/RT)}} \qquad (10.6.6)$$

This expression for Q_g depends on temperature in the manner shown in Figure 10.1.

Other reactions will have somewhat different forms for the curve of Q_g versus T. For example, in the case of a reversible exothermic reaction, the equilibrium yield decreases with increasing temperature. Since one cannot expect to exceed the equilibrium yield within a reactor, the fraction conversion obtained at high temperatures may be less than a subequilibrium value obtained at lower temperatures. Since the rate of energy release by reaction depends only on the fraction conversion attained and not on the position of equilibrium, the value of Q_g will thus be lower at the higher temperature than it was at a lower temperature. Figure 10.2 indicates the general shape of a Q_g versus T plot for a reversible exothermic reaction. For other reaction networks, different shaped plots of Q_g versus T will exist.

The rate at which energy is removed from the system (Q_r) is equal to the rate at which it

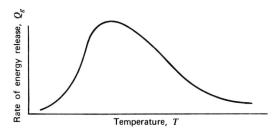

Figure 10.2
Rate of energy release by reaction versus temperature for a reversible exothermic reaction.

is being lost by heat transfer processes plus the mass flow rate times the gain in sensible heat per unit mass. Thus

$$Q_r = UA(T - T_m) + \Phi_m \hat{C}(T - T_0) \qquad (10.6.7)$$

where

T_m is the temperature of the heat transfer medium (assumed constant)

Φ_m is the mass flow rate

$\hat{C}$ is the average heat capacity per unit mass appropriate to the temperature range in question

T_0 is the feed temperature

Equation 10.6.7 can be rewritten as

$$Q_r = T(UA + \Phi_m \hat{C}) - UAT_m - \Phi_m \hat{C}T_0 \qquad (10.6.8)$$

to emphasize the fact that Q_r will be essentially linear in T if U and $\hat{C}$ are not strong functions of temperature.

At steady state the rate of transformation of energy by reaction must be equal to the rate of thermal energy loss. This implies that the intersection(s) of the curves given by equations 10.6.6 and 10.6.8 will represent the solution(s) of the combined material and energy balance equations. The positions at which the intersections occur depend on the variables appearing on the right side of equations 10.6.6 and 10.6.8. Figure 10.3 depicts some of the situations that may be encountered.

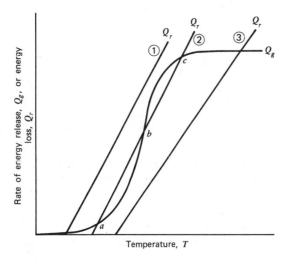

Figure 10.3
Energy release and energy loss curves for an irreversible reaction in a flow reactor.

If the rate of energy removal is represented by line 1 in Figure 10.3, the reaction mixture is cooled to such an extent that steady-state operation is possible only at a very low degree of conversion and a very low temperature. If the variables influencing the Q_g and Q_r curves are examined, one sees that this type of steady state is favored by:

1. Low magnitudes of the heat of reaction.
2. Large values of the UA product.
3. Low values of T_m or T_0.
4. Low feed rates ($\mathcal{V}_0$).
5. Long residence times ($V_R/\mathcal{V}_0$).
6. Small rate constants.

If Q_r is represented by line 3, substantially complete conversion is achieved. This type of steady state is favored by conditions that are opposite to those enumerated above.

Because of the sigmoid shape of the thermal energy generation curve, intersections at high or very low degrees of conversion are more probable than intersections at intermediate levels.

This conclusion is in agreement with observations of the performance of stirred tank reactors. Nonetheless, it is the situation where the intersection occurs at an intermediate value of the conversion (or of Q_g) that is of greatest interest from a stability analysis viewpoint.

If the Q_r curve is given by the straight line numbered 2, it is evident that there are three points of intersection with the thermal energy generation curve. This case could be achieved by reducing the cooling capacity employed in the first case, or by dropping the feed temperature and/or increasing the cooling capacity employed in case 3.

Of the three operating conditions indicated in Figure 10.3, only two (a and c) are stable; the third (b) is unstable. The stability of the points can be examined in terms of the slopes of the Q_g and Q_r curves at the point in question. When operating at point c, if a small positive temperature fluctuation occurred, one would move into a region where the rate of energy loss is greater than the rate of release of energy by chemical reaction. Consequently, the system will suffer a net loss of thermal energy and cool down until it returns to point c. On the other hand, if there were a negative temperature fluctuation, one would move to a region where more energy is transformed by reaction than is lost by heat transfer and convective flow. There would be a net gain of energy, and the system temperature would rise until point c was again reached. Because departures from this point lead to conditions tending to restore the system to the original point, c represents a *stable* operating condition. By the same reasoning it can be shown that point a and the intersections of curves 1 and 3 with the Q_g curve are stable operating points.

However, the characteristics of point b with regard to temperature fluctuations are quite different. At this point the slope of the energy release curve is greater than the slope of the energy loss curve. If a small positive temperature fluctuation were to occur, one would be in a region where thermal energy would be released faster than it could be dissipated. This would cause the system to increase further in temperature. The temperature deviation would continue to be amplified until the reactor arrived at point c, the upper stable operating point. A negative temperature fluctuation would also be amplified, the system losing energy faster than it is generated by the reaction. In this case the temperature drops until the lower stable operating point is attained. This situation corresponds to extinguishing the reaction. Because of the nature of its response to deviations in temperature, point b is referred to as an *unstable* operating point.

To operate at a point like c on curve 2 in Figure 10.3, one must provide a means of bypassing or circumventing the lower stable state a and the unstable state b. This may be accomplished by temporarily preheating the feed, operating at a reduced flow rate, or reducing the capacity of the cooling system. These procedures will permit one to move to the right of point b and "ignite" the reaction.

These stability considerations are not limited to first-order irreversible reactions. Figure 10.4 depicts the Q_g and Q_r curves for a reversible exothermic reaction. The intersections of the Q_g curve and lines 3′ and 4′ represent stable

operating conditions, as do points a and c. Point b, however, is unstable. Also note that a relatively small slope for the heat loss curve will give rise to a high reaction temperature and low yield. If the cooling capacity is increased one moves from line 4′ to line 3′ and achieves a higher degree of conversion for this equilibrium reaction. The highest yield evidently corresponds to using an energy loss curve that intersects the energy release curve near or at its peak. However, because of the shape of the curve it may be very difficult from a process control viewpoint to operate at this condition. Increases in the cooling rate can increase the slope of the energy loss curve, possibly leading to extinction of the reaction. This situation can be visualized by rotating line 2′ to the left about its intersection on the T axis.

The problem of ignition and extinction of reactions is basic to that of controlling the process. It is interesting to consider this problem in terms of the variables used in the earlier discussion of stability. When multiple steady-state solutions exist, the transitions between the various stable operating points are essentially discontinuous, and hysteresis effects can be observed in these situations.

Consider the energy release and energy loss curves shown in Figure 10.5. These curves correspond to three different operating conditions where one would be varying either the temperature of the feed stream, T_0, or the temperature of the medium to which heat is being transferred, T_m. The intercepts on the temperature axis corresponding to the three cases are T', T'' and T'''. We will assume that all other operating variables are held constant. When the T intercept is T', the reactor may operate in a stable mode at point A. However, if the cooling rate is increased (by decreasing T_m) or if the temperature of the feed stream drops, curve Q_r' will move to the left and the reaction will be extinguished. Consequently, point A represents a critical extinction point. For combustion reactions the temperature at this point may be

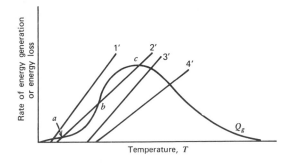

Figure 10.4
Energy release and energy loss curves for reversible exothermic reaction in a CSTR.

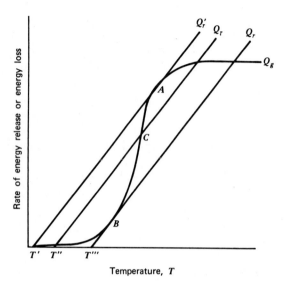

Figure 10.5
Influence of operating conditions on ignition and extinction of reactions.

referred to as a *minimum combustion temperature*. This temperature cannot be regarded as having an absolute value characteristic of the reaction. It depends on the various factors that influence the relative positions of the Q_r and Q_g curves and their point of tangency. The heat transfer coefficient, reactant feed composition, volumetric flow rate, physical properties of the reactants, and characteristic properties of the reaction all influence the absolute value of this point.

The tangent indicated at point B also represents a critical reaction condition, but of a somewhat different type. In this case the reactor temperature corresponding to point B represents the minimum temperature at which autoignition will occur. In this sense it can be regarded as a *minimum ignition temperature*. Like the critical extinction point, this temperature should not be regarded as an absolute value but as a function of various operating parameters.

If the temperature intercept of the energy loss curve lies between those corresponding to the

minimum ignition temperature and to the critical extinction point (e.g., at T''), ignition can occur only if the unstable operating point C can be bypassed.

We should note that hysteresis effects could be observed in situations like those depicted in Figure 10.5. Suppose the temperature of the feed stream is such that the temperature intercept moves from slightly below T' to slightly above T'''. Assuming that all other independent variables are held constant, the temperature of the reactor contents will rise slowly, because the conversion levels achieved will be those corresponding to the lowest stationary state condition. Once intercept T''' is exceeded, however, autoignition will occur and the temperature of the reactor contents will increase very markedly. The conversion level now will be significantly higher than it was previously. If the temperature of the feed stream is now reduced so that the temperature intercept moves to the left of T''', the temperature of the reactor contents will diminish slowly moving along the Q_g curve. The reactor will continue to operate at a high conversion level corresponding to the upper steady state. Eventually the temperature intercept will drop below T' and the reaction will be extinguished. The temperature of the reactor contents and the fraction conversion will then drop to low levels. Figure 10.6 illustrates the hysteresis phenomenon involved.

Ignition and extinction phenomena can be examined with respect to each of the various parameters of the system using the general approach outlined above. It should be noted that in these analyses and in the discussion above, one is restricted to changes that occur so slowly that the corresponding changes in reactor conditions can be regarded as a series of pseudo stationary states. It is instructive to consider the effects of opening and closing the valve on a propane torch or of varying the gas flow rate to a Bunsen burner in terms of the material we have presented in this section. The basic problem

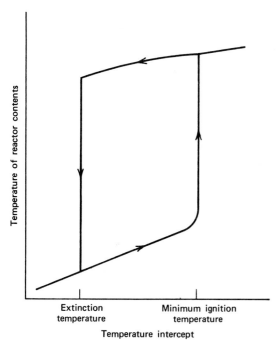

Figure 10.6
Ignition and extinction hysteresis effects in CSTR.

of obtaining a stable flame can be analyzed, in terms of the concepts developed above, even though other phenomena are undoubtedly of importance.

10.7 SELECTION OF OPTIMUM REACTOR TEMPERATURE PROFILES . . . THERMODYNAMIC AND SELECTIVITY CONSIDERATIONS

This section treats the selection of optimum temperature progressions for systems in which multiple reactions can occur. In a batch reactor this progression is the time schedule that the temperature should follow as a function of reactant conversion. In a tubular reactor it is the profile along the length of the reactor. In a series of stirred tank reactors it is the change in temperature from stage to stage. For a single reversible reaction the optimum temperature

progression is chosen as that which minimizes the required reactor volume for a given fraction conversion and feed rate. For multiple reactions optimization involves manipulation of the temperature to obtain a favorable product distribution.

10.7.1 Optimum Temperature Schedules

In order to minimize the required reactor volume for a given type of reactor and level of conversion, one must always operate with the reactor at a temperature where the rate is a maximum. For irreversible reactions the reaction rate always increases with increasing temperature, so the highest rate occurs at the highest permissible temperature. This temperature may be selected on the basis of constraints established by the materials of construction, phase changes, or side reactions that become important at high temperatures. For reversible reactions that are endothermic the same considerations apply, since both the reaction rate and the equilibrium yield increase with increasing temperature.

For reversible exothermic reactions the situation is more complicated, because kinetic and thermodynamic considerations work against one another when the temperature is raised. The rates of the forward and reverse reactions both increase with increasing temperature, but the latter increases faster than the former. Therefore, the equilibrium yield or maximum attainable conversion decreases with increasing temperature. Because of this, it is advisable to use a high temperature when the system is far from equilibrium to take advantage of the influence of temperature on the forward reaction rate. When equilibrium is approached, the temperature should be lowered in order to shift the equilibrium yield to a higher value. The optimum temperature sequence is then one that starts out high and decreases with increasing conversion.

For a single plug flow reactor optimum conditions for adiabatic operation are obtained by varying the feed temperature so that the average reaction rate has the highest possible value. For endothermic reactions this implies the maximum possible feed temperature. For exothermic reactions this means that the exit temperature should lie below that corresponding to equilibrium at the desired fraction conversion and above that corresponding to the maximum rate at this conversion. A trial and error procedure can locate an exit temperature within this range that corresponds to the minimum reactor volume. For a CSTR the reactor effluent temperature should be that which gives a maximum rate at the desired conversion level. The inlet temperature may be adjusted to give this effluent condition when the reactor is operated adiabatically.

For an exothermic reaction, adiabatic operation gives an increase in temperature with increasing conversion. However, the optimum temperature profile is one where the temperature falls with increasing conversion. Severe heat transfer requirements may be needed to make the actual temperature profile approach the desired ideal. Two ways in which this may be achieved are indicated in Figure 10.7. The case of internal heat exchange between the reaction mixture and the feed stream has been discussed by Van Heerden (8) and Kramers and Westerterp (9). The other alternative is to use staged operations with interstage cooling between adiabatic sections. This case has also been treated by Kramers and Westerterp (9). For endothermic reactions, multiple stages with reheating between stages are commonly used to keep the temperature from dropping too far.

Illustration 10.8 indicates how one determines the optimum temperature at which a single CSTR should be operated.

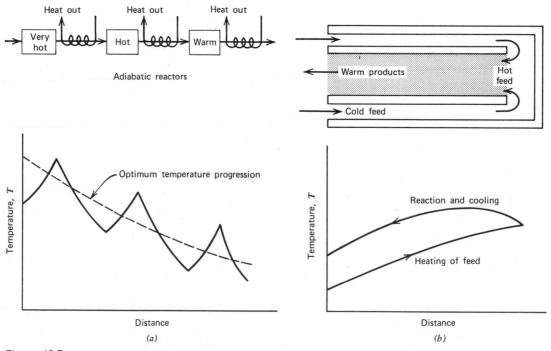

Figure 10.7

Techniques for approaching optimum temperature profiles for exothermic reaction. (a) Adiabatic operation of reactors with interstage cooling. (b) Countercurrent heat exchange. (Adapted from *Chemical Reaction Engineering,* **Second Edition, by O. Levenspiel. Copyright © 1972. Reprinted by permission of John Wiley and Sons, Inc.)**

ILLUSTRATION 10.8 DETERMINATION OF OPTIMUM TEMPERATURE FOR OPERATION OF A SINGLE CSTR IN WHICH A REVERSIBLE EXOTHERMIC REACTION IS BEING CARRIED OUT

The following reversible reaction takes place in a CSTR:

$$C \underset{k_2}{\overset{k_1}{\rightleftharpoons}} B$$

where both the forward and the reverse reaction obey first-order kinetics. The rate constants may be written in the Arrhenius form as:

$$k_1 = A_1 e^{-E_1/RT}$$
$$k_2 = A_2 e^{-E_2/RT}$$

Determine the minimum reactor volume that will be required to obtain a fraction conversion f_c if the feed is pure C and if the input volumetric flow rate is V_0. What will be the temperature of the effluent stream?

Solution

For this CSTR the appropriate design equation is

$$\tau = \frac{V_R}{V_0} = \frac{C_{C0} \int_0^{f_c} df_c}{-r_{CF}} = \frac{C_{C0} f_c}{-r_{CF}} \quad \text{(A)}$$

with

$$-r_{CF} = k_1 C_{CF} - k_2 C_{BF} \quad \text{(B)}$$

From stoichiometric considerations

$$C_{CF} = C_{C0}(1 - f_c) \quad \text{(C)}$$
$$C_{BF} = C_{C0} f_c \quad \text{(D)}$$

Combining equations A to D gives

$$\frac{V_R}{V_0} = \frac{f_c}{k_1(1 - f_c) - k_2 f_c} = \frac{f_c}{k_1 - f_c(k_1 + k_2)}$$

In order to minimize the required reactor volume one may set the temperature derivative of V_R equal to zero.

$$\left(\frac{\partial V_R}{\partial T}\right)_{f_c} = \frac{-V_0 f_c \dfrac{\partial}{\partial T}[k_1 - f_c(k_1 + k_2)]}{[k_1 - f_c(k_1 + k_2)]^2} = 0$$

or

$$\frac{\partial k_1}{\partial T} - f_c\left(\frac{\partial k_1}{\partial T} + \frac{\partial k_2}{\partial T}\right) = 0$$

However, from the Arrhenius relation,

$$\frac{d \ln k}{dT} = \frac{E_A}{RT^2} \quad \text{or} \quad \frac{dk}{dT} = \frac{kE_A}{RT^2}$$

Thus the optimum temperature will be that at which

$$\frac{k_1 E_1}{RT^2}(1 - f_c) - f_c \frac{k_2 E_2}{RT^2} = 0$$

or

$$\frac{k_1}{k_2} = \frac{E_2}{E_1}\left(\frac{f_c}{1 - f_c}\right)$$

Substituting the Arrhenius form of the reaction rate constants,

$$\frac{A_1 e^{-E_1/RT}}{A_2 e^{-E_2/RT}} = \frac{E_2}{E_1}\left(\frac{f_c}{1 - f_c}\right)$$

One can then solve for the optimum temperature:

$$\frac{1}{T}\left(-\frac{E_1}{R} + \frac{E_2}{R}\right) = \ln\left[\frac{E_2 A_2}{E_1 A_1}\left(\frac{f_c}{1 - f_c}\right)\right] \equiv \ln \beta$$

or

$$T = \frac{E_2 - E_1}{R \ln\left[\dfrac{E_2 A_2}{E_1 A_1}\left(\dfrac{f_c}{1 - f_c}\right)\right]} = \frac{E_2 - E_1}{R \ln \beta}$$

At this temperature

$$k_1 = A_1 e^{-(E_1 \ln \beta)/(E_2 - E_1)} = A_1 \beta^{-E_1/(E_2 - E_1)}$$

and

$$k_2 = A_2 \beta^{-E_2/(E_2 - E_1)}$$

Thus the minimum reactor volume is given by

$$V_R = \frac{V_0 f_c}{A_1 \beta^{-E_1/(E_2 - E_1)}(1 - f_c) - A_2 \beta^{-E_2/(E_2 - E_1)} f_c}$$

10.7.2 The Influence of Selectivity Considerations on the Choice of Reactor Operating Temperatures

When multiple reactions are possible, certain of the products have greater economic value than others, and one must select the type of reactor and the operating conditions so as to optimize the product distribution and yield. In this subsection we examine how the temperature can be manipulated with these ends in mind. In our treatment we will ignore the effect of concentration levels on the product distribution by assuming that the concentration dependence of the rate expressions for the competing reactions is the same in all cases. The concentration effects were treated in detail in Chapter 9.

Consider the following simple parallel reactions.

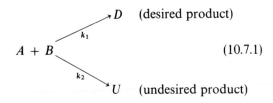

$$A + B \qquad \qquad (10.7.1)$$

It is evident that

$$\frac{dD}{dU} = \frac{k_1}{k_2} \qquad (10.7.2)$$

If reaction 1 is to be enhanced and reaction 2 depressed, the ratio of the rate constants must be made as large as possible. This ratio may be written in the Arrhenius form as

$$\frac{k_1}{k_2} = \frac{A_1 e^{-E_1/RT}}{A_2 e^{-E_2/RT}} = \frac{A_1}{A_2} e^{-(E_1-E_2)/RT}$$
$$(10.7.3)$$

It changes with temperature in a manner that depends on whether E_1 is greater than or less than E_2. When the temperature increases, the ratio increases if $E_1 > E_2$ and decreases if $E_2 > E_1$. Consequently, the reaction having the larger activation energy is the one that is most sensitive to variations in temperature. The following general rule is appropriate to competing reactions.

High temperatures favor the reaction with the higher activation energy and lower temperatures favor that with the lower activation energy.

For the parallel reactions in equation 10.7.1 one may use this general rule to select the following operating conditions as optimum from a selectivity viewpoint when the reactor operates isothermally.

If $E_1 > E_2$, use a high temperature.
If $E_2 > E_1$, use a low temperature.

On the other hand, if it is possible to use a temperature progression scheme and if one desires to obtain the maximum amount of the desired product per unit time per unit reactor volume, somewhat different considerations are applicable. If $E_1 > E_2$, one should use a high temperature throughout, but if $E_2 > E_1$, the temperature should increase with time in a batch reactor or with distance from the reactor inlet in a plug flow reactor. It is best to use a low temperature initially in order to favor conversion to the desired product. In the final stages of the reaction a higher temperature is more desirable in order to raise the reaction rate, which has fallen off because of depletion of reactants. Even though this temperature increases the production of the undesirable product, more of the desired product is formed than would otherwise be the case. Thus one obtains a maximum production capacity for the desired product.

Now consider the basic series reaction scheme

$$A \xrightarrow{k_1} D \xrightarrow{k_2} U \qquad (10.7.4)$$

where species D is the desired product. In this case the desired product distribution is favored when the ratio of the rate constants (k_1/k_2) is made very large. Therefore, for operation at a single temperature, one should use a high temperature if $E_1 > E_2$ and a low temperature if $E_2 > E_1$. On the other hand, if a temperature progression is to be employed, the relative magnitudes of the activation energies determine whether the initial temperature should be greater than or less than the final temperature. For example, if $E_2 > E_1$, the temperature should start out high to accelerate the first reaction and thus obtain a large output from a given reactor volume. However, the temperature should be progressively reduced as species D accumulates in order to take advantage of the fact that the undesirable side reaction $D \rightarrow U$ slows down faster with decreasing temperature than does the useful reaction. The problem of selecting an optimum temperature schedule has been treated by Bilous and Amundson (10).

For series-parallel combinations of reactions a number of possible situations can arise. Two general guidelines are useful to keep in mind (11). First, if one has two undesirable products being formed in parallel with a desired species or with an intermediate product that can subsequently react to form the desired species

$$A \begin{array}{c} \xrightarrow{k_1} D \\ \xrightarrow{k_2} U_1 \\ \xrightarrow{k_3} U_2 \end{array} \qquad (10.7.5)$$

there are three significant orderings of the activation energies. The type of temperature to use for each of these cases is given below:

1. $E_1 > E_2$ and $E_1 > E_3$; use high temperature.
2. $E_1 < E_2$ and $E_1 < E_3$; use low temperature.
3. $E_1 > E_2$ and $E_1 < E_3$ or $E_1 < E_2$ and $E_1 > E_3$; use intermediate temperature.

In case 3 an intermediate temperature will give the most favorable product distribution. If $E_1 > E_2$ and $E_1 < E_3$, one can show that the optimum temperature is given by

$$T = \frac{E_3 - E_2}{R \, \ell n \left\{ \left(\dfrac{E_3 - E_1}{E_1 - E_2} \right) \dfrac{A_3}{A_2} \right\}} \qquad (10.7.6)$$

A second generalization is applicable to steps occurring in series. If an early step requires a high temperature and a subsequent step is favored by a low temperature, then a temperature sequence that decreases with increasing

space time or reactor holding time should be used. Levenspiel (12) gives the following example of a situation of this type:

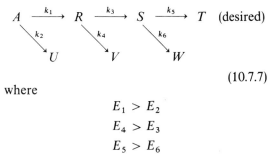

$$ (10.7.7) $$

where

$$E_1 > E_2$$
$$E_4 > E_3$$
$$E_5 > E_6$$

The optimum temperature progression to use in this sequence is a high temperature to start followed by lower temperatures where the concentration of species R is high and then increasing temperatures when the concentration of species S becomes appreciable. Levenspiel (13) has summarized the results of several analyses of the optimum temperature level and progression to be used for several general reaction schemes.

Although a determination of the operating temperature that produces the most favorable product distribution is important in working up a reactor design for a reaction scheme involving multiple reactions, other considerations may be equally important from a design viewpoint. Optimization of the reactor economics requires a favorable product distribution, high conversions, and low capital costs. If the most favorable product distribution is obtained at low temperatures, very little of any product will be formed. In this case it will be more economical to use an intermediate temperature, because the adverse effects of a change in the product distribution will be more than offset by the increase in reaction rate and conversion. This will have the effect of increasing the production capacity of a given reactor. In the case where the most favorable product distribution is favored by operation at high temperatures, the highest permissible temperature should be used, because the reaction rates will be high and the requisite reactor size small.

LITERATURE CITATIONS

1. W. C. Reynolds, *Thermodynamics*, McGraw-Hill, New York, 1965.

2. Smith, J. M., and Van Ness, H. C., *Introduction to Chemical Engineering Thermodynamics*, Third Edition, McGraw-Hill, New York, 1975.

3. Modell, M., and Reid, R. C., *Thermodynamics and its Applications*, Prentice-Hall, Englewood Cliffs, N.J., 1974.

4. Adlington, D. G., "Reactor Design," in *Chemical Engineering Practice*, Volume 8, *Chemical Kinetics*, p. 181, edited by H. W. Cremer, Butterworths, London, 1965.

5. Kladko, M., *Chemical Technology*, *1* (141), 1971.

6. Wasserman, A., *Diels-Alder Reactions*, pp. 42, 50, and 52, Elsevier, London, 1965.

7. Perlmutter, D. D., *Stability of Chemical Reactors*, Prentice Hall, Englewood Cliffs, N.J., 1971.

8. Van Heerden, C., *Ind. Eng. Chem.*, *45* (1242), 1953.

9. Kramers, H., and Westerterp, K. R., *Elements of Chemical Reactor Design and Operation*, pp. 124–128, Academic Press, New York, 1963.

10. Bilous, O., and Amundson, N. R., *Chem. Eng. Sci.*, *5* (81, 115), 1956.

11. Levenspiel, O., *Chemical Reaction Engineering*, Second Edition, p. 239, Wiley, New York, copyright © 1972. Used with permission.

12. Ibid., p. 240.

13. Ibid., p. 241.

PROBLEMS

1. Your design group has been asked to consider certain aspects of the preliminary design of a reactor for the production of vinyl chloride by reacting chlorine with ethylene.

$$C_2H_4 + Cl_2 \overset{k_1}{\to} C_2H_3Cl + HCl$$

In order to minimize the production of the undesirable addition product

$$C_2H_4 + Cl_2 \overset{k_2}{\to} C_2H_4Cl_2$$

it is suggested that the reaction be carried out in the temperature range 320 to 380 °C. The effect of the addition reaction is negligible in this range. In order to minimize the formation of multiple substitution products, an ethylene/chlorine feed ratio of 50:1 will also be used.

The following values of the pseudo first-order rate constant k_1 have been reported by Subbotin, Antonov, and Etlis [*Kinetics and Catalysis*, 7 (183), 1966].

Temperature, T (°C)	320	340	360	380
k_1 (sec^{-1})	2.23	6.73	14.3	32.6

For your preliminary considerations, assume that the following values of product and reactant heat capacities are constant over the temperature range of interest.

	C_p (cal/g mole·°K)
Ethylene	17.10
Chlorine	8.74
HCl	7.07
Vinyl chloride	10.01

The standard enthalpy change for the reaction at 25 °C may be taken as -23 kcal/g mole.

(a) Calculate the space time necessary to obtain 50% conversion of the chlorine in a PFR if the reactor is maintained isothermal at 320 °C.

(b) Calculate the space time necessary to obtain 50% conversion of the chlorine in a PFR if the feed enters at 320 °C and the reactor is operated adiabatically.

The reactor is to be designed to operate at 1 atm. Deviations from ideal gas behavior may be neglected. Note that while the volume changes associated with changes in mole numbers are negligible, you may wish to consider the effect of thermal expansion.

2. Acetic anhydride reacts with ethanol to form ethyl acetate according to the reaction

$$(CH_3CO)_2O + 2C_2H_5OH \to$$
$$2CH_3COOC_2H_5 + H_2O$$

In ethanol this reaction proceeds stepwise.

$(CH_3CO)_2O + C_2H_5OH \rightarrow$

$\qquad\qquad CH_3COOH + CH_3COOC_2H_5$

$CH_3COOH + C_2H_5OH \rightarrow$

$\qquad\qquad H_2O + CH_3COOC_2H_5$

Since the second reaction rate constant is orders of magnitude greater than the first at temperatures near room temperature, the first reaction may be regarded as the rate controlling step. Since ethanol is used as the solvent, the reaction will follow pseudo first-order kinetics. The rate of this liquid phase reaction can be expressed as

$$r = 6 \times 10^6 e^{-8857/T} C_A C_B$$

where T is temperature in degrees Kelvin, and where A and B refer to anhydride and alcohol, respectively. The concentrations are expressed in kilomoles per cubic meter, and the rate in kilomoles per cubic meter per second. On the basis of the data presented below, determine the times necessary to achieve 20% decomposition of a 1 molal anhydride solution by isothermal operation at 20 °C and by adiabatic operation starting from the same temperature.

Compound	$\Delta G^0_{f,298}$ (kJ/mole)	$\Delta H^0_{f,298}$ (kJ/mole)
Acetic anhydride (ℓ)	−509.65	−649.50
Ethanol (ℓ)	−174.81	−277.77
Ethyl acetate (ℓ)	−318.60	−463.47
Water (ℓ)	−237.30	−285.98

The density of ethanol at 20 °C is 0.789 g/cm³, and its heat capacity is 2.85 J/g·°K. These properties may be assumed to be constant over the temperature range of interest. The solution is so dilute that one may assume that its property values are essentially equal to those of pure ethanol. Batch reactor operation is assumed.

3. The decomposition of phosphine is a first-order reaction that proceeds according to the following stoichiometric equation.

$$PH_3(g) \rightarrow \tfrac{1}{4}P_4(g) + 1.5H_2(g)$$

Pure phosphine is to be admitted to a constant volume batch reactor and allowed to undergo decomposition according to the above reaction. If pure phosphine enters at 672 °C and the initial pressure is 1 atm, determine the times necessary to decompose 20% of the original phosphine for both isothermal and adiabatic operation.

The rate constant for this first-order reaction is given in the International Critical Tables as

$$\log k = -\frac{18,963}{T} + 2 \log T + 12.130 \quad (I)$$

where the temperature is expressed in degrees Kelvin and the time in seconds.

The standard enthalpy change for the gaseous reaction at 25 °C is approximately 5665 cal/g mole.

The following values may be used for the molal heat capacities (at constant pressure) of the species involved in the reaction.

$$P_4(g) \qquad C_p = 5.90 + 0.0096T$$
$$PH_3(g) \qquad C_p = 6.70 + 0.0063T$$
$$H_2(g) \qquad C_p \simeq 7.20$$

where

T is expressed in degrees Kelvin

C_p is expressed in calories per gram mole per degree Kelvin

For temperatures from 800 to 1000 °K the following values may be used to represent average C_p values.

$$P_4(g) \qquad C_p = 14.5$$
$$PH_3(g) \qquad C_p = 12.4$$

4. Consider the reaction used as the basis for Illustrations 10.1 to 10.3. Determine the volume that would be required to produce 2 million lb of B annually in a plug flow reactor operating isothermally at 163 °C. Assume that 97% of the

A fed to the reactor is to be converted to B and that the reactor can be operated for 7000 hr annually. Determine the manner in which the heat transfer requirement is distributed along the length of the reactor (i.e., what fraction of the heat evolved must be removed in the first 10% of the reactor length, the second 10%, the third 10%, etc.?).

5. A reactor designer proposes to carry out a reaction with the stoichiometry

$$A + B \rightarrow C$$

in the liquid phase in two ideal CSTR's operating in series. Since species B is very expensive, the designer has chosen the reaction conditions so that a vast excess of species A will be present and the rate expression becomes pseudo first-order in species B.

$$r = kC_B$$

The reaction is exothermic, and temperature control in both reactors is to be accomplished by heat exchange with water boiling at 1 atm ($T = 100$ °C). The contents of the first reactor will be at 106 °C; those of the second reactor will be at 117 °C. At these temperatures the values of the apparent rate constant are:

$$T = 117\,°C \qquad k = 2.79\ \text{ksec}^{-1}$$
$$T = 106\,°C \qquad k = 0.93\ \text{ksec}^{-1}$$

For the proposed design the reactor volumes are both 0.8 m^3, the input volumetric flow rate is 1.10 m^3/ksec, and the overall fraction conversion of the initial B is to be 0.80. Since the reaction is carried out in dilute liquid solution, the effective heat capacity of the liquid mixture is substantially unaffected by the reaction and may be regarded as a constant that is equal to 3.47 J/cm^3·°K. If the initial concentration of species B is 5.6 kmole/m^3 and the feed stream enters at 70 °C, determine the required heat transfer area for each reactor.

Additional data:
$\Delta H_R = -69$ kJ/mole for the reaction as written. The variation of this quantity with temperature may be neglected.
$U = 68$ J/ksec·cm^2·°K in each reactor.

6. Consider the sequential first-order reactions $A \xrightarrow{k_1} V \xrightarrow{k_2} W$ where V is the desired product. These liquid phase reactions are to be carried out in a cascade consisting of two equal-volume CSTR's in series. The reactors are operated at the same temperature and so as to maximize the concentration of species V in the effluent from the second reactor. When the feed rate is 0.5 m^3/ksec and the rate constants are each equal to 0.125 ksec^{-1}, the volume of an individual reactor that gives rise to the maximum is equal to 2 m^3. Both reactors will operate at the same temperature (93 °C).

a. If the fluid density is equal to 0.95 g/cm^3, the heat capacity of the fluid is equal to 3 J/g·°K, and the first reactor operates adiabatically, determine the temperature at which the feed must enter. The feed contains 1.5 kmoles/m^3 of reactant A.

b. At what rate must thermal energy be removed (or supplied) in order to maintain the second reactor at 93 °C?

$$\Delta H_1 = -60\ \text{kJ/mole}$$
$$\Delta H_2 = \quad 20\ \text{kJ/mole}$$

7. The hydrogenation of cottonseed oil proceeds by the following irreversible steps.

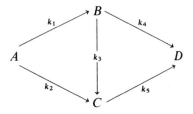

A: Linoleic acid C: Iso-oleic acid
B: *cis*-oleic acid D: Stearic acid

Relative activation energies have been estimated by comparison with similar reactions:

Reaction	E_A
1	Low
2	High
3	Low
4	High
5	High

In a batch reactor at 100 °C operated for a short time, 90 mole percent A remains unreacted, 6.2% forms B, and 3.8% forms C.

(a) It is desired to produce compound C. You are determining the optimum temperature for operating a CSTR. Assuming all reactions are first-order, explain the fact that two optimum temperatures are found.

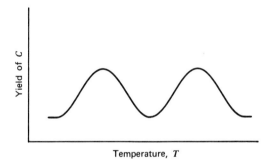

Temperature, T

(b) If B is desired, is it better to operate at low or high temperature? Why?
(c) Assume that reactions 4 and 5 have the same value of E_A. Which is larger, k_4 or k_5? Why?

8. The decomposition reaction $A \rightarrow B + C$ occurs in the liquid phase. It has been suggested that your company produce C from a stream containing equimolar concentrations of A and B by using two continuous stirred tank reactors in series. Both reactors have the same volume. The reaction is first-order with respect to A and zero-order with respect to B and C. Each reactor will operate isothermally, but each will operate at a different temperature. You have been asked to design a reactor capable of processing feed at 1.7 m³/sec.

The following additional data are available.

	A	B	C	Inerts
Specific heats (J/mole ·°K)	62.8	75.4	125.6	75.4
Species concentrations (kmole/m³)	A	B	C	Inerts
Feed	3.0	3.0	0.0	32.0
Effluent from second reactor	0.3	5.7	2.7	32.0

Feed temperature = 330 °K

$\Delta H_{\text{Reaction}} = -70{,}000$ J/mole for the reaction as written (at 330 °K)

Reactor operating temperatures: First — 330 °K, Second — 358 °K

Activation energy = 108.4 kJ/mole

At 330 °K, $k = 330$ ksec^{-1}

(a) What is the volume of each CSTR?
(b) How much thermal energy must be removed in *each* reactor if the system is to operate in the fashion described above? Be very careful to write the energy balance for the second reactor in the appropriate form.

9. The liquid phase reaction $A \rightarrow B$ is to be carried out in a plug flow tubular reactor at a constant pressure of 202.6 kPa. The feed is 600 kmoles/ksec of pure A with an inlet temperature of 200 °C. Pure A has a specific volume of 0.056 m³/kmole. The heat of reaction at 200 °C is −15 kJ/mole. The molar specific heats of A and B are both 42 J/mole·°K. The reaction rate constant in this range is given by the expression

$$k = 110 + 0.8(T - 200)$$

where

  T is measured in degrees Celsius

  k has the units ksec^{-1}

The reactor would be run adiabatically, but the maximum reaction temperature allowable is 400 °C, since above this temperature undesirable by-products are formed. Calculate the *minimum* reactor volume required to obtain 80% conversion of A. What must the heat transfer rate be in the cooling section of the reactor?

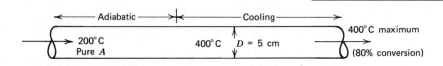

10. Consider the following reactions that take place in parallel.

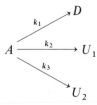

where $E_1 > E_2$ and $E_1 < E_3$. What temperature will give the maximum yield of species D?

11. An autocatalytic reaction of the type

$$A + R \underset{k_{-1}}{\overset{k_1}{\rightleftharpoons}} R + R$$

is being carried out in the combination of reactors shown below.

Pure A

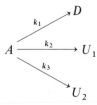

Ideal   Ideal
CSTR   plug flow reactor

The reaction is reversible and exothermic. The reaction stoichiometry is $A \rightarrow R$. One hundred pound moles of pure liquid A are fed to the

reactor each minute. The temperature of the pure A stream is 70 °F. The fraction of the A converted to R in the CSTR is 0.4, and the temperature of the effluent from the CSTR is 100 °F. The mole density of pure liquid A is 1.5 lb moles/ft^3. The following table gives the temperature dependence of the rate constants k_1 and k_{-1}.

Temperature, T (°F)	k_1 (ft^3/lb mole-min)	k_{-1} (ft^3/lb mole-min)
70	0.55	0.010
75	0.93	0.020
80	1.67	0.040
85	2.90	0.078
90	5.0	0.156
95	8.65	0.312
100	15.0	0.625

Heat capacity of pure A = 25 Btu/lb mole-°R
Heat capacity of pure R = 30 Btu/lb mole-°R
ΔH_R = −10,000 Btu/lb mole R formed for the isothermal reaction at 70 °F

(a) Determine the CSTR volume.
(b) Determine the heat transfer rate (and direction) in the first reactor.
(c) The plug flow reactor is to be operated isothermally at 100 °F. The PFR effluent conversion of the A fed to the CSTR is to be 90%. Determine the heat transfer requirements for the PFR and the necessary reactor volume.

12. A tubular reactor is being used to investigate an endothermic isomerization reaction that may be represented schematically as $A \rightarrow B$.

The following data are applicable.

$\Delta H_R = 46{,}500$ J/mole A reacted at 30 °C

$C_{p,A} = C_{p,B} = 100$ J/mole·°K

Reactor diameter = 50 cm

Pressure = 203 kPa (may be considered constant over the length of the reactor)

Feed: Pure A at a rate of 12.6 moles/sec

Initial temperature = 320 °C

If electrical heating is applied along the length of the reactor to provide energy at a constant rate of 575 J/sec·cm of reactor length and this input is evenly distributed along the length of reactor, determine the temperature profile in the reactor. Heat losses from the reactor are negligible. The experimental data on the concentration profile are as follows.

Distance from inlet (m)	Fraction A converted
0	0
0.85	0.1
1.74	0.2
3.05	0.3
4.27	0.4
6.10	0.5

13. The reaction $A + B \rightarrow C + D$ takes place in the two ideal CSTR's operating in series shown below. The reaction is first-order with respect to both A and B. Temperatures and concentrations of the species of interest are as indicated. S denotes the solvent.

The standard enthalpy change for this reaction at 200 °C is -126 kJ/mole for the reaction as written. Each CSTR has a volume of 0.2 m³. The feed rate is 66 cm³/sec. The following data on average heat capacities are available.

Species	Heat capacity (J/mole·°K)
A	54.4
B	71.1
C	96.3
D	29.2
S	41.9

At 500 °C the second-order rate constant is 1.7 m³/kmole·ksec. Determine:
(a) The composition of the effluent from the first reactor.
(b) If the heat transfer rate in the first reactor is -1.05 kJ/sec, what is the temperature of the effluent from the first reactor?
(c) What is the reaction rate constant at this temperature?

14. An irreversible liquid phase dimerization reaction of the type

$$2M \rightarrow D$$

follows second-order kinetics. You have been asked to estimate the fraction of the monomer that can be converted to dimer in an existing continuous stirred tank reactor facility designed

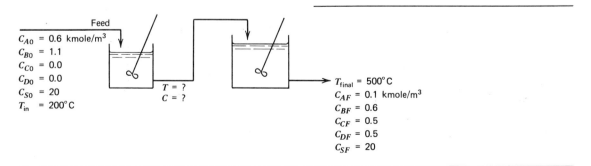

$C_{AO} = 0.6$ kmole/m³
$C_{BO} = 1.1$
$C_{CO} = 0.0$
$C_{DO} = 0.0$
$C_{SO} = 20$
$T_{in} = 200°$ C

$T = ?$
$C = ?$

$T_{final} = 500°$ C
$C_{AF} = 0.1$ kmole/m³
$C_{BF} = 0.6$
$C_{CF} = 0.5$
$C_{DF} = 0.5$
$C_{SF} = 20$

Feed

for adiabatic operation. The reactor volume is 0.4 m³ and the available input volumetric flow rate is 1.3 m³/ksec of pure monomer. Additional data are given below.

Inlet temperature = 312 °K

Standard enthalpy change on reaction at 300 °K = −42 kJ/mole D formed
Liquid heat capacity = 2.0 J/cm³·°K

Monomer feed concentration = 16 kmoles/m³

Thermal expansion effects may be neglected.

$k = 2.70 \times 10^9 e^{-12,185/T}$ m³/mole-ksec for T in degrees Kelvin

15. Consider a CSTR that is used to carry out a reversible isomerization reaction of the type

$$A \overset{k_f}{\underset{k_r}{\rightleftharpoons}} B$$

where both the forward and reverse reactions are first-order.

Data:
Feed is pure species A

$$C_{p,A} = 1255 \text{ J/mole·°K}$$
$$C_{p,B} = 1172 \text{ J/mole·°K}$$
$$k_f = 8.83 \times 10^4 e^{-6290/T} \text{ sec}^{-1}$$
$$k_r = 4.17 \times 10^{15} e^{-14947/T} \text{ sec}^{-1}$$

where T is given in degrees Kelvin

(a) Is the reaction exothermic or endothermic? What is the standard enthalpy change for the reaction?
(b) What is the equilibrium fraction conversion at 340 °K?
(c) What conversion is achieved if $\tau = 480$ sec and the reactor temperature is 340 °K?
(d) For $\tau = 480$ sec, sketch the curve of fraction conversion versus reactor temperature over the range 320 to 370 °K.
(e) Derive the equation for the curve describing the energy balance on the CSTR for adiabatic operation. Substitute variables into this expression to obtain a relation between f_A, T_{inlet}, and T_{outlet}.

(f) In order to maximize production of B when $\tau = 480$ sec, what inlet temperature should be specified?

16. A dissociation reaction of the type $A \rightarrow B + C$ is being studied in a pilot plant reactor having a volume of 0.5 m³. The reaction involves ideal gases with the following heat capacities.

$$C_{p,A} = 160 \text{ J/mole·°K}$$
$$C_{p,B} = 120 \text{ J/mole·°K}$$
$$C_{p,C} = 120 \text{ J/mole·°K}$$

Pure A is charged to the reactor at 400 kPa and 330 °K. The reaction is first-order in species A. The variation of the reaction rate constant with temperature is given below.

Temperature, T (°K)	k (ksec^{-1})
330	0.330
340	0.462
350	0.641
360	0.902
370	1.27
380	1.98

The standard heat of reaction is −11.63 kJ/mole. Determine the times necessary to achieve 90% conversion in a constant volume batch reactor under adiabatic conditions and under isothermal conditions.

17. Keairns and Manning [*AIChE J.*, *15* (660), 1969] have used the reaction between sodium thiosulfate and hydrogen peroxide in a well-stirred flow reactor to check a computer simulation of adiabatic CSTR operation. Data on their experimental conditions and the reaction parameters are listed below. The reaction may be considered second-order in sodium thiosulfate.

$$Na_2S_2O_3 + 2H_2O_2 \rightarrow \text{Products}$$
$$A + 2B \rightarrow \text{Products}$$

$\Delta H_R^0 = -548$ kJ/mole at 25 °C

$k = 6.853 \times 10^{11} e^{-E/RT} (\text{m}^3/\text{mole·ksec})$

$E = 76.480$ kJ/mole

$C_{A0} = 0.204$ kmole/m^3

$C_{B0} = 0.408$ kmole/m^3

Solution feed rate: 14.2 cm^3/sec

Inlet feed temperature: 25 °C

Reactor volume: 2790 cm^3

The heat capacity of the inlet and outlet streams may be assumed to be 4.2 J/cm^3·°K. The reaction may be treated as irreversible. If the reactor is assumed to operate adiabatically, determine the temperature of the effluent and the fraction conversion.

18. The following Diels-Alder reaction takes place in benzene solution.

Data:

Wasserman (*Diels-Alder Reactions*, pp. 42, 50, and 52, Elsevier, London, 1965) has reported the following values of the reaction rate expression and heat of reaction for reaction conditions similar to those to be used in the proposed design analysis. This expression is valid over the range from 8 to 50 °C.

$$\Delta H_R = -72.8 \text{ kJ/mole}$$
$$r = A e^{-E/RT} C_C C_B$$

where

$A = 10^{6.5} \text{m}^3/\text{kmole·sec}$

$E = 48.5$ kJ/mole

$C_C = $ cyclopentadiene concentration (kmoles/m^3)

$C_B = $ benzoquinone concentration (kmoles/m^3)

The density and heat capacity of the solution may be assumed to be constant and essentially equal to the property values for pure benzene.

cyclopentadiene   benzoquinone

5-8-endo methylene-
5-8-9-10 tetrahydro-α-
naphthaquinone

C + B $\longrightarrow$ M

Two feed streams each containing one of the reactants at a concentration of 0.2 kmoles/m^3 of benzene and each at 25 °C are to be rapidly mixed and fed to a tubular reactor.

If the flow rate of each feed stream is 0.139 m^3/ksec and if 50% conversion is to be achieved, determine the reactor volumes required for iso-thermal operation and for adiabatic operation.

$c_{\text{Benzene}} = 1.75$ J/g·°K

$\rho_{\text{Benzene}} = 880$ kg/m^3

The reverse reaction and side reactions may be neglected.

Note that on mixing the reactant concentrations will drop to half those prevailing before mixing.

11 Deviations from Ideal Flow Conditions

11.0 INTRODUCTION

The flow patterns in real reactors do not conform exactly to those postulated for ideal plug flow and continuous stirred tank reactors. Nonetheless, the conversions achieved in many real reactors so closely approximate those predicted on the basis of the idealized models that the design equations for these reactors can be used with negligible error. In other cases significant differences are noted between observed and predicted results. These differences may arise from a number of sources—from channeling of fluid as it moves through the reaction vessel, from longitudinal mixing caused by vortices and turbulence, from the presence of stagnant regions within the reactor, from bypassing or short-circuiting of portions of a packed reactor bed, from the failure of impellers or other mixing devices to provide perfect mixing, etc. In this chapter we hope to establish a rational basis for examining quantitatively and qualitatively the effect of departures from idealized flow behavior on the performance of a reactor. Read this chapter with a view toward developing a "seat of the pants" feeling for the magnitude of deviations from ideal flow conditions in various types of reactors so that you will know when these effects can be neglected and when they must be treated by the techniques developed in this chapter.

In principle, if the temperatures, velocities, flow patterns, and local rates of mixing of every element of fluid in a reactor were known, and if the differential material and energy balances could be integrated over the reactor volume, one could obtain an exact solution for the composition of the effluent stream and thus the degree of conversion that takes place in the reactor. However, most of this information is lacking for the reactors used in laboratory or commercial practice. Consequently, it has been necessary to develop approximate methods for treating nonideal flow systems in terms of data that are easily obtained experimentally. From measurements on the feed and effluent streams, one may develop parameters that can be used to characterize the magnitude of system nonidealities and to serve as input to more complex models of reactor behavior than ideal CSTR's and PFR's. We will begin by indicating how such measurements can be used to determine the residence time distribution function for a reactor, and then treat three different mathematical models that permit estimation of the conversion that will be attained. Throughout this chapter we will restrict our discussion to systems in which a single reaction takes place in a homogeneous isothermal reactor. Volume changes on reaction are also assumed to be negligible ($\delta \simeq 0$). These restrictions will permit us to focus our attention on the nonideal flow conditions.

11.1 THE RESIDENCE TIME DISTRIBUTION FUNCTION $dF(t)$

Except for the case of an ideal plug flow reactor, different fluid elements will take different lengths of time to flow through a chemical reactor. In order to be able to predict the behavior of a given piece of equipment as a chemical reactor, one must be able to determine how long different fluid elements remain in the reactor. One does this by measuring the response of the effluent stream to changes in the concentration of inert species in the feed stream—the so-called stimulus-response technique. In this section we will discuss the analytical form in which the distribution of residence times is cast, derive relationships of this type for various reactor models, and illustrate how experimental data are treated in order to determine the distribution function.

The mathematical relations expressing the different amounts of time that fluid elements spend in a given reactor may be cast in a variety of forms. [For example, see Levenspiel (1–3) and Himmelblau and Bischoff (4)]. This textbook utilizes the $F(t)$ curve, as defined by Danckwerts (5) for this purpose. For a continuous flow system $F(t)$ is the volume fraction of the fluid at the outlet that has remained in the system for a time less than t. In other words, if we were to assign "ages" to the different fluid elements leaving the system, $F(t)$ would be the volume fraction of the outlet stream having an age less than t. For constant density systems, volume fractions are identical with weight fractions, and $F(t)$ is also the weight fraction of the effluent with an age less than t. In accordance with this definition of the $F(t)$ curve, the probability that an element of volume entering the system at time $t = 0$ has left it within a period of time t is just equal to $F(t)$. The probability that it is still in the reactor and will leave at a time later than t is $[1 - F(t)]$.

It will always take some finite time for a fluid element to traverse the system, so $F(t) = 0$ at $t = 0$. Likewise, none of the material can remain in the flow reactor indefinitely, so $F(t) = 1.0$ at $t = \infty$. Figure 11.1 indicates these limiting values for an arbitrary $F(t)$ curve.

Since $F(t + dt)$ represents the volume fraction of the fluid having a residence time less than $t + dt$, and $F(t)$ represents that having a residence time less than t, the differential of $F(t)$, $dF(t)$, will be the volume fraction of the effluent stream having a residence time between t and $t + dt$. Hence $dF(t)$ is known as the residence time distribution function. From the principles of probability the average residence time ($\bar{t}$) of a fluid element is given by

$$\bar{t} = \int_{F(t)=0}^{F(t)=1} t \, dF(t) \tag{11.1.1}$$

or

$$\bar{t} = \int_{t=0}^{t=\infty} t \left(\frac{dF(t)}{dt}\right) dt \tag{11.1.2}$$

Consequently the shaded area in Figure 11.1 is equal to $\bar{t}$ and the shaded area to the right of

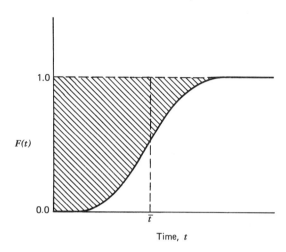

Figure 11.1
Determination of average residence time from residence time distribution function.

$t = \bar{t}$ must be equal to the unshaded area to the left of $\bar{t}$.

11.1.1 Experimental Determination of Residence Time Distribution Functions

In order to be able to make use of the residence time distribution function in the analysis of a given reactor network, one must be able to determine the function experimentally. This is done by changing some property of the fluid entering the network as a function of time and then noting the resultant response of the effluent stream. The method most commonly employed is to change the concentration of one of the nonreactive components of the feed stream. This tracer component is generally chosen on the basis of the convenience and accuracy with which it may be measured. The properties that are most often used for monitoring the concentration of these tracers are electrical conductivity, absorbance, and emission of β and γ rays. In choosing the tracer care must also be taken to ensure that portions of it do not disappear during the course of the experimental measurement (e.g., by selective adsorption on the walls of the reaction vessel or on heterogeneous catalysts present in the reactor, by settling out or being filtered out as it moves through the reactor, or by chemical reaction in the case of nonradioactive tracers).

There are three general stimulus techniques commonly used in theoretical and experimental analyses of reactor networks in order to characterize their dynamic behavior.

1. A step function change in which the input concentration is changed from one steady-state level to another.
2. A pulse input in which a relatively small amount of tracer is injected into the feed stream in the shortest possible time.
3. A sinusoidal input. The frequency of the sinusoidal variation is changed and the steady-state response of the effluent at different input frequencies is determined, thus generating a frequency-response diagram for the system.

The time variations of the effluent tracer concentration in response to step and pulse inputs and the frequency-response diagram all contain essentially the same information. In principle, any one can be mathematically transformed into the other two. However, since it is easier experimentally to effect a change in input tracer concentration that approximates a step change or an impulse function, and since the measurements associated with sinusoidal variations are much more time consuming and require special equipment, the latter are used much less often in simple reactor studies. Even in the first two cases, one can obtain good experimental results only if the average residence time in the system is relatively long.

Kramers and Alberda (6) have discussed the manner in which sinusoidal variations are analyzed, but we will discuss only the first two types of stimuli. They are sufficient for the analysis of the majority of situations that will be encountered by the chemical engineer engaged in the practice of designing chemical reactors.

The interpretation of the $F(t)$ curve as the probability that a fluid element entering the reactor at time zero has left by time t may be used to indicate how the curve may be generated from experimental data. Let us consider the case where at time zero one makes a step change in the weight fraction tracer in the feed stream from w_0^- to w_0^+. A generalized stimulus-response curve for this system is shown in Figure 11.2.

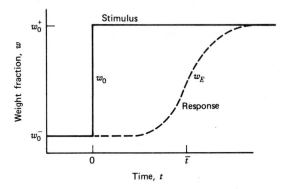

Figure 11.2

Generalized response of an arbitrary reactor to a step change in input tracer concentration.

At time t the fraction of the effluent fluid characterized by an age less than t (and thus with a composition w_0^+) is just equal to $F(t)$. At the same time, the fraction of the effluent characterized by an age greater than t (with the original inlet composition w_0^-) is equal to $1 - F(t)$. The time dependence of the weight fraction tracer in the effluent (w_E) is then given by

$$w_E(t) = w_0^+ F(t) + w_0^-[1 - F(t)] \quad (11.1.3)$$

Thus

$$F(t) = \frac{w_E(t) - w_0^-}{w_0^+ - w_0^-} \quad (11.1.4)$$

and we see that in general $F(t)$ would give the relative response of the system to a step function input.

Now consider the case where the input to the system is a pulse of tracer. The amount of tracer leaving in a time increment dt is $w_E \Phi_m \, dt$ where Φ_m is the mass flow rate. For constant density systems mass and volume fractions are identical and the basic definition of $F(t)$ indicates that

$$F(t) = \frac{\int_0^t w_E \Phi_m \, dt}{\int_0^\infty w_E \Phi_m \, dt} = \frac{\int_0^t w_E \Phi_m \, dt}{m_T} \quad (11.1.5)$$

where we have identified the term in the denominator as the total mass of tracer recovered from a pulse input. Differentiation of equation 11.1.5 gives

$$\frac{dF(t)}{dt} = \frac{w_E \Phi_m}{\int_0^\infty w_E \Phi_m \, dt} = \frac{w_E \Phi_m}{m_T} \quad (11.1.6)$$

For linear systems the relative response to a pulse input is equal to the derivative of the relative response to a step input. Illustration 11.1 indicates how the response of a reactor network to a pulse input can be used to generate an $F(t)$ curve.

ILLUSTRATION 11.1 DETERMINATION OF AN $F(t)$ CURVE FROM THE RESPONSE OF A REACTOR TO A PULSE INPUT

A slug of dye is placed in the feed stream to a stirred reaction vessel operating at steady state. The dye concentration in the effluent stream was monitored as a function of time to generate the data in the table below. Time is measured relative to that at which the dye was injected.

Time, t (sec)	Tracer concentration (g/m^3)
0	0.0
120	6.5
240	12.5
360	12.5
480	10.0
600	5.0
720	2.5
840	1.0
960	0.0
1080	0.0

Determine the average residence time of the fluid and the $F(t)$ curve for this system.

Solution

For a constant density system concentrations are directly proportional to weight fractions. Thus equation 11.1.5 becomes

$$F(t) = \frac{\int_0^t C_E \Phi_m \, dt}{\int_0^\infty C_E \Phi_m \, dt}$$

where C_E is the tracer concentration in the effluent.

One must replace the integrals by finite sums in order to be able to make use of the data given.

$$F(t) = \frac{\sum_0^t (C_E \Phi_m \, \Delta t)}{\sum_0^\infty (C_E \Phi_m \, \Delta t)}$$

The data are reported at evenly spaced time increments, and the mass flow rate is invariant for steady-state operation. Thus

$$F(t) = \frac{\sum_0^t C_E}{\sum_0^\infty C_E}$$

The average residence time is given by equation 11.1.2. Combining this equation with equations 11.1.5 and 11.1.6 gives

$$\bar{t} = \frac{\int_{t=0}^{t=\infty} t w_E \Phi_m \, dt}{\int_{t=0}^{t=\infty} w_E \Phi_m \, dt}$$

In terms of the data and the aforementioned assumptions this equation becomes

$$\bar{t} = \frac{\sum_{t=0}^{t=\infty} t C_E}{\sum_{t=0}^{t=\infty} C_E}$$

Table 11.I.1
Data Workup for Illustration 11.1

Time, t (sec)	C_E (g/m³)	$\displaystyle\sum_0^t C_E$ (g/m³)	$F(t)$	tC_E (g · sec/m³)	$t^2 C_E$ (g · sec²/m³)
0	0.0	0.0	0	0	0
120	6.5	6.5	0.13	780	93,600
240	12.5	19.0	0.38	3000	720,000
360	12.5	31.5	0.63	4500	1,620,000
480	10.0	41.5	0.83	4800	2,304,000
600	5.0	46.5	0.93	3000	1,800,000
720	2.5	49.0	0.98	1800	1,296,000
840	1.0	50.0	1.00	840	705,600
960	0.0	50.0	1.00	0	0
1080	0.0	50.0	1.00	0	0

$$\sum_0^\infty C_E = 50.0 \qquad\qquad\qquad \sum_0^\infty tC_E = 18{,}720$$

Table 11.I.1 contains a workup of the data in terms of the above analysis. In the more general case one should be sure to use appropriate averaging techniques or graphical integration to determine both $F(t)$ and $\bar{t}$. When there is an abundance of data, plot it, draw a smooth curve, and integrate graphically instead of using the strictly numerical procedure employed above.

On the basis of the tabular entries

$$\bar{t} = \frac{18{,}720}{50.0} = 374.4 \text{ sec}$$

11.1.2 $F(t)$ Curves for Ideal Flow Patterns

For a few highly idealized systems, the residence time distribution function can be determined *a priori* without the need for experimental work. These systems include our two idealized flow reactors—the plug flow reactor and the continuous stirred tank reactor—and the tubular laminar flow reactor. The $F(t)$ and response curves for each of these three types of well-characterized flow patterns will be developed in turn.

The plug flow reactor has a flat velocity profile and no longitudinal mixing. These idealizations imply that all fluid elements leaving the reactor have the same age $(\bar{t})$. The $F(t)$ function for this system must then be

$$\begin{array}{lll} F(t) = 0 & \text{for} & 0 < t < \bar{t} \\ F(t) = 1.0 & \text{for} & t > \bar{t} \end{array} \qquad (11.1.7)$$

The responses of this system to ideal step and pulse inputs are shown in Figure 11.3. Because the flow patterns in real tubular reactors will always involve some axial mixing and boundary layer flow near the walls of the vessels, they will distort the response curves for the ideal plug flow reactor. Consequently, the responses of a real tubular reactor to these inputs may look like those shown in Figure 11.3.

The next case to be considered is the ideal continuous stirred tank reactor. The key to the derivation of the $F(t)$ curve for this type of reactor is the realization that the assumption of perfect mixing implies that upon entry in the reactor an element of volume can instantaneously appear in any portion of the reactor. Therefore its past or its future history cannot be derived from its position. Furthermore, the prob-

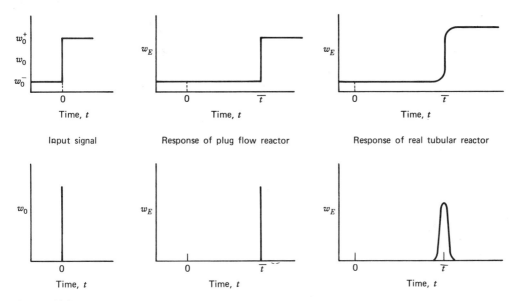

Figure 11.3
Response of ideal plug flow reactor and real tubular reactor to step and impulse inputs.

ability that it will leave the system by some future time will be independent of its past history. These statements require that the probability that a fluid element will remain in the system longer than a time $(t_1 + t_2)$ be the product of the two independent probabilities that it will remain in longer than times t_1 and t_2, respectively.

$$1 - F(t_1 + t_2) = [1 - F(t_1)][1 - F(t_2)]$$
(11.1.8)

If we now replace t_1 by t and consider the case where we let t_2 become quite small (Δt):

$$1 - F(t + \Delta t) = [1 - F(t)][1 - F(\Delta t)]$$
(11.1.9)

For a perfectly mixed reactor, all fluid elements have an equal chance of leaving the reactor, so that for a small time increment Δt, the probability that a given fluid element will leave is just the ratio of the mass of fluid leaving to the total mass contained within the reactor.

$$F(\Delta t) = \frac{\Phi_m \, \Delta t}{\rho V_R} = \frac{V_0 \, \Delta t}{V_R} = \frac{\Delta t}{\bar{t}}$$
(11.1.10)

where

ρ is the fluid density

$\bar{t}$ is the mean residence time

Combining equations 11.1.9 and 11.1.10 gives

$$F(t + \Delta t) - F(t) + F(t)\frac{\Delta t}{\bar{t}} = \frac{\Delta t}{t}$$
(11.1.11)

Dividing by Δt and taking the limit as Δt approaches zero,

$$\frac{dF(t)}{dt} + \frac{1}{\bar{t}} F(t) = \frac{1}{\bar{t}}$$
(11.1.12)

The solution to this differential equation subject to the boundary condition that $F(t) = 0$ at $t = 0$ is:

$$F(t) = 1 - e^{-t/\bar{t}}$$
(11.1.13)

This same result can also be obtained by considering the response of the effluent composition from a CSTR to a step function change in input concentration and using equation 11.1.4 to determine $F(t)$.

The relative response of a single CSTR to an ideal pulse input may be obtained by taking the time derivative of equation 11.1.13.

$$\frac{dF(t)}{dt} = \frac{e^{-t/\bar{t}}}{\bar{t}} \qquad (11.1.14)$$

From this equation it is evident that there is a wide distribution of residence times in a stirred tank reactor.

The responses of a single ideal stirred tank reactor to ideal step and pulse inputs are shown in Figure 11.4. Note that any change in the reactor inlet stream shows up immediately at the reactor outlet in these systems. This fact is used to advantage in the design of automatic control systems for stirred tank reactors.

The performance of the stirred tank flow reactors that are encountered in industrial practice may differ significantly from the ideal case discussed above. The feed stream entering the stirred region will not be instantaneously dispersed and mixed with the entire contents of the vessel. The mixing process will require a finite amount of time in order to produce a microscopically homogeneous solution, so there will be a time lag between the change in the input stream and the change in the characteristics of the effluent stream. Moreover, a fraction of the feed stream may pass through the reactor outlet without undergoing complete mixing.

These fractions may lead to an irregular response curve. This effect cannot be described quantitatively for a general system, since it will depend on the relative orientations of the inlet and outlet with respect to the impellers and any baffles present in the system. Examples of possible response curves are shown in Figure 11.5. The effluent response will also be strongly dependent on the mixing time. The approximation of an actual stirred tank reactor by an ideal CSTR improves as the ratio of the mean residence time to the mixing time increases. For most design purposes a ratio greater than 10 gives a very good approximation [7].

The final idealized flow situation that we will consider is laminar flow in a tubular reactor in the absence of either radial or longitudinal diffusion. The velocity profile in such a reactor is given by

$$u = u_0 \left[1 - \left(\frac{r}{R} \right)^2 \right] \qquad (11.1.15)$$
$$\text{for } 0 \leq r \leq R$$

where

u_0 is the centerline velocity

r is the distance from the centerline of the pipe

R is the inside radius of the pipe

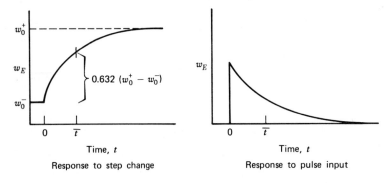

Figure 11.4
Response of ideal continuous stirred tank reactor to step and pulse Inputs.

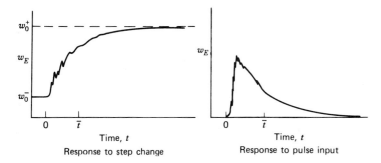

Figure 11.5
Response of real stirred tank reactor to step and pulse inputs.

The average velocity with which a fluid element moves is given by

$$\bar{u} = \frac{u_0}{2} \qquad (11.1.16)$$

Since u varies with r, the residence times of the various fluid elements will also vary with r. The time that it will take a fluid element to traverse the reactor is given by

$$t = \frac{L}{u} = \frac{L}{u_0\left[1 - \left(\dfrac{r}{R}\right)^2\right]} \qquad (11.1.17)$$

where L is the length of the reactor.

The average residence time $\bar{t}$ is

$$\bar{t} = \frac{2L}{u_0} \qquad (11.1.18)$$

Combining equations 11.1.17 and 11.1.18 gives

$$\frac{\bar{t}}{2t} = \left[1 - \left(\frac{r}{R}\right)^2\right] \qquad (11.1.19)$$

The fluid at the centerline is moving the fastest so this material will be the first to leave.

It leaves at a time t_{min} given by

$$t_{min} = \frac{L}{u_0} = \frac{\bar{t}}{2} \qquad (11.1.20)$$

Thus

$$F(t) = 0 \qquad \text{for} \qquad t < \frac{\bar{t}}{2} \qquad (11.1.21)$$

The fraction of the volumetric flow rate that takes place in the region bounded by $r = 0$ and $r = r$ will be equal to $F(t)$ where t is equal to the time necessary for a fluid element to traverse the reactor length at a given r (equation 11.1.17). Thus

$$F(t) = \frac{\text{Volumetric flow rate between } r = 0 \text{ and } r = r}{\text{Total volumetric flow rate}} \qquad (11.1.22)$$

or

$$F(t) = \frac{\int_0^r u(r)2\pi r \, dr}{\int_0^R u(r)2\pi r \, dr} \qquad (11.1.23)$$

Combining equations 11.1.15 and 11.1.23 and simplifying gives

$$F(t) = \frac{\int_0^r \left[1 - \left(\dfrac{r}{R}\right)^2\right] r \, dr}{\int_0^R \left[1 - \left(\dfrac{r}{R}\right)^2\right] r \, dr} \qquad (11.1.24)$$

Integration gives

$$F(t) = \frac{\dfrac{r^2}{2} - \dfrac{r^4}{4R^2}}{\dfrac{R^2}{2} - \dfrac{R^4}{4R^2}} = \left(\frac{r}{R}\right)^2 \left[2 - \left(\frac{r}{R}\right)^2\right]$$

(11.1.25)

From equation 11.1.19,

$$\left(\frac{r}{R}\right)^2 = 1 - \frac{\bar{t}}{2t} \qquad (11.1.26)$$

Substituting this result into equation 11.1.25 yields, for $t \geq \bar{t}/2$,

$$F(t) = \left[1 - \frac{\bar{t}}{2t}\right]\left[2 - \left(1 - \frac{\bar{t}}{2t}\right)\right]$$

$$= 1 - \left(\frac{\bar{t}}{2t}\right)^2 \qquad (11.1.27)$$

The $F(t)$ curve for a laminar flow tubular reactor with no diffusion is shown in Figure 11.6. Curves for the two other types of idealized flow patterns are shown for comparison.

11.1.3 Models for Nonideal Flow Situations

Different reactor networks can give rise to the same residence time distribution function. For example, a CSTR characterized by a space time τ_1 followed by a PFR characterized by a space time τ_2 has an $F(t)$ curve that is identical to that of these two reactors operated in the reverse order. Consequently, the $F(t)$ curve alone is not sufficient, in general, to permit one to determine the conversion in a nonideal reactor. As a result, several mathematical models of reactor performance have been developed to provide estimates of the conversion levels in nonideal reactors. These models vary in their degree of complexity and range of applicability. In this textbook we will confine the discussion to models in which a single parameter is used to characterize the nonideal flow pattern. Multiparameter models have been developed for handling more complex situations (e.g., that which prevails in a fluidized bed reactor), but these are beyond the scope of this textbook. [See Levenspiel (2) and Himmelblau and Bischoff (4).]

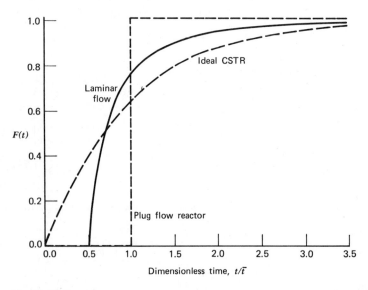

Figure 11.6
$F(t)$ curves for reactors with idealized flow patterns.

It is convenient to classify deviations from ideal flow conditions into two categories.

1. Different fluid elements may move through the reactor at different velocities. The elements remain *segregated* as they move through the reactor.
2. Fluid elements with different ages may mix on a microscopic scale. However, the *mixing* does not occur to the extent that it does in an ideal CSTR.

These two types of deviations occur simultaneously in actual reactors, but the mathematical models we will develop assume that the residence time distribution function may be attributed to one or the other of these flow situations. The first class of nonideal flow conditions leads to the segregated flow model of reactor performance. This model may be used with the residence time distribution function to predict accurately conversion levels for first-order reactions that occur isothermally. (See Section 11.2.1.) The second class may be modeled in several ways, depending on the additional assumptions one is willing to make concerning the nature of the mixing processes. Once these assumptions are made the parameters of the mathematical model can be determined from the $F(t)$ curve. The remainder of this section is

that the fluid velocity and reactant concentration are constant across the tube diameter. The magnitude of the dispersion is assumed to be independent of position within the vessel, so there will be no stagnant regions and no by-passing or short-circuiting of fluid in the model reactor. By changing the magnitude of the dispersion parameter, one may vary the performance of the reactor from that of plug flow ($\mathscr{D}_L = 0$) to that of a single continuous stirred tank reactor ($\mathscr{D}_L = \infty$).

The axial dispersion parameter $\mathscr{D}_L$ is a term that accounts for mixing both by molecular diffusion processes and by turbulent eddies and vortices. Since these two types of phenomena are to be characterized by a single parameter, and since we force the model to fit the form of Fick's law of diffusion, $\mathscr{D}_L$ should be regarded as an *effective* dispersion coefficient having the units of an ordinary molecular diffusivity (length^2/time). However, it is significantly greater in magnitude because of turbulence effects.

The response of the axial dispersion model to step or pulse tracer inputs can be determined by writing a material balance over a short tubular segment and then solving the resultant differential equations. A transient material balance on a cylindrical element of length ΔZ gives

$$\left[\left(-\mathscr{D}_L \frac{\partial C}{\partial Z} + uC\right)\pi R^2\right]_Z \Delta t = \left[\left(-\mathscr{D}_L \frac{\partial C}{\partial Z} + uC\right)\pi R^2\right]_{Z+\Delta Z} \Delta t + \pi R^2 \Delta Z \Delta C \quad (11.1.28)$$

$$\underbrace{}_{\text{Input}} \qquad\qquad \underbrace{}_{\text{Output}} \qquad\qquad \underbrace{}_{\text{Accumulation}}$$

devoted to a discussion of the interpretation of response data in terms of two of these mixing models.

11.1.3.1 The Axial Dispersion Model. The *axial dispersion* model is often used to describe the behavior of tubular reactors. This model characterizes mass transport in the axial direction in terms of an effective or apparent longitudinal diffusivity $\mathscr{D}_L$ that is superimposed on the plug flow velocity. The model also assumes

where the input and output terms allow for both dispersive and convective transport.

Dividing by the cross-sectional area, Δt, and ΔZ, and taking the limit as the last two parameters approach zero gives

$$\mathscr{D}_L \frac{\partial^2 C}{\partial Z^2} - u \frac{\partial C}{\partial Z} = \frac{\partial C}{\partial t} \quad (11.1.29)$$

The initial and boundary conditions that apply to this equation depend on whether one

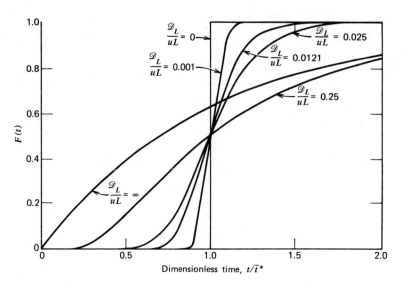

Figure 11.7
Residence time distribution curves for dispersion model.

is dealing with a pulse or step input and the characteristics of the system at the tracer injection and monitoring stations. At each of these points the tubular reactor is characterized as "closed" or "open," depending on whether or not plug flow into or out of the test section is assumed. A *closed* boundary is one at which there is plug flow outside of the test section; an *open* boundary is one at which the same dispersion parameter characterizes the flow conditions within and adjacent to the test section. There are then four different possible sets of boundary conditions on equation 11.1.29, depending on whether a completely open or completely closed vessel, a closed-open vessel, or an open-closed vessel is assumed. Different solutions will be obtained for different boundary conditions. Fortunately, for small values of the dispersion parameter, the differences between the various solutions will be small.

If we now consider a step change in tracer concentration in the feed to an "open" tube that can be regarded as extending to infinity in both directions from the injection point, the appropriate initial and boundary conditions on

equation 11.1.29 are:

$$C = \begin{cases} C_0^- & \text{for } Z > 0 \text{ at } t = 0 \\ C_0^+ & \text{for } Z < 0 \text{ at } t = 0 \end{cases} \quad (11.1.30)$$

$$C = \begin{cases} C_0^+ & \text{at } Z = -\infty \text{ for } t \geq 0 \\ C_0^- & \text{at } Z = +\infty \text{ for } t \geq 0 \end{cases} \quad (11.1.31)$$

where C_0^- and C_0^+ are the tracer concentrations before and after the step change, respectively.

In this case a closed form solution is possible with

$$\frac{C - C_0^-}{C_0^+ - C_0^-} = \frac{1}{2}\left[1 - \text{erf}\left(\frac{Z - ut}{\sqrt{4\mathscr{D}_L t}}\right)\right] \quad (11.1.32)$$

where the term in parentheses is the argument of the error function.*

The relative response at the end of a tubular reactor of length L is identical with the $F(t)$

---

* The error function is tabulated in most handbooks of mathematical tables. It is useful to note that $\text{erf}(-x) = -\text{erf}(x)$ and that

$$\text{erf } x = \frac{2}{\sqrt{\pi}}\int_0^x e^{-y^2}\, dy$$

curve at $Z = L$. If we define $\bar{t}^* = L/u$,

$$F(t) = \frac{1}{2}\left\{1 - \text{erf}\left[\frac{1}{2}\sqrt{\frac{uL}{\mathscr{D}_L}}\left(\frac{1 - t/\bar{t}^*}{\sqrt{t/\bar{t}^*}}\right)\right]\right\} \quad (11.1.33)$$

This equation is plotted in Figure 11.7 for different values of the parameter $\mathscr{D}_L/uL$. When this parameter is zero, there is no axial dispersion, and the reactor acts as a plug flow reactor.

$$C_L(t) = \frac{m_T L}{2V_R\sqrt{\pi\mathscr{D}_L t}}\,e^{-L^2\left(1 - \frac{ut}{L}\right)^2 / 4\mathscr{D}_L t}$$

$$= \frac{m_T L}{2V_R\sqrt{\pi\mathscr{D}_L t\,\dfrac{L}{u\bar{t}^*}}}\,e^{-[1 - (t/\bar{t}^*)]^2 / \left(\frac{4\mathscr{D}_L}{Lu}\right)\left(\frac{ut}{L}\right)} = \frac{m_T}{2V_R\sqrt{\pi\left(\dfrac{\mathscr{D}_L}{uL}\right)(t/\bar{t}^*)}}\,e^{-[1 - (t/\bar{t}^*)]^2 / \left(\frac{4\mathscr{D}_L}{Lu}\right)(t/\bar{t}^*)}$$

$$(11.1.37)$$

The response in this case is shown in Figure 11.7 to be the expected step-function response. At the other extreme, a value of $\mathscr{D}_L/uL$ equal to infinity corresponds to an ideal stirred tank reactor.

In the case where one injects a perfect pulse of tracer Levenspiel and Smith (8) have shown

$$\frac{C_L(t)}{\displaystyle\int_0^\infty C_L(t)\,d(t/\bar{t}^*)} = \frac{1}{2\sqrt{\pi\left(\dfrac{\mathscr{D}_L}{uL}\right)(t/\bar{t}^*)}}\,e^{-[1 - (t/\bar{t}^*)]^2 / \left(\frac{4\mathscr{D}_L}{uL}\right)\left(\frac{t}{\bar{t}^*}\right)} \quad (11.1.39)$$

that the solution to equation 11.1.29 is

$$C = \frac{m_T L}{2V_R\sqrt{\pi\mathscr{D}_L t}}\,e^{-(Z - ut)^2 / 4\mathscr{D}_L t} \quad (11.1.34)$$

where

m_T is the amount of tracer injected in consistent units

L and V_R are the length and volume of the test section, respectively

At $Z = L$,

$$C_L(t) = \frac{m_T L}{2V_R\sqrt{\pi\mathscr{D}_L t}}\,e^{-(L - ut)^2 / 4\mathscr{D}_L t} \quad (11.1.35)$$

For constant density systems the variable $\bar{t}^*$ becomes

$$\bar{t}^* = \frac{L}{u} = \frac{V_R}{V_0} \quad (11.1.36)$$

and the previous equation becomes

It should be noted that an overall material balance requires that all the tracer pass the monitoring station.

$$\int_0^\infty C_L(t)V_0\,dt = m_T \quad (11.1.38)$$

Combining equations 11.1.36 to 11.1.38 gives

If the right side of this equation is plotted versus dimensionless time for various values of the group $\mathscr{D}_L/uL$ (the reciprocal Peclet number), the types of curves shown in Figure 11.8 are obtained. The skewness of the curve increases with $\mathscr{D}_L/uL$ and, for small values of this parameter, the shape approaches that of a normal error curve. In physical terms this implies that when $\mathscr{D}_L/uL$ is small, the shape of the axial concentration profile does not change

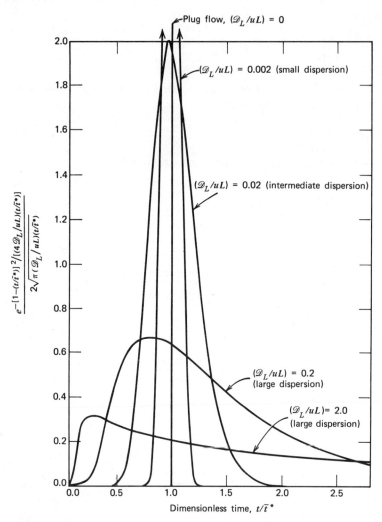

Figure 11.8

Effluent response curves for perfect impulse injection of tracer (axial dispersion model). (Adapted from *Chemical Reaction Engineering*, **Second Edition, by O. Levenspiel. Copyright © 1972. Reprinted by permission of John Wiley and Sons, Inc.)**

appreciably in the time interval required for the fluid to pass the monitoring station. However, when $\mathscr{D}_L/uL$ is of order 0.01 or greater, the shape changes significantly in this time interval.

Differentiation of the last equation with respect to $(t/\bar{t}^*)$ can be used to determine the time at which the maximum tracer concentration is observed at the monitoring station. The result is

$$\left(\frac{t}{\bar{t}^*}\right)_{max} = \sqrt{\left(\frac{\mathscr{D}_L}{uL}\right)^2 + 1} - \left(\frac{\mathscr{D}_L}{uL}\right) \quad (11.1.40)$$

or

$$\left(\frac{t}{\bar{t}^*}\right)_{max} = \frac{\mathscr{D}_L}{uL}\left[\sqrt{1 + \left(\frac{uL}{\mathscr{D}_L}\right)^2} - 1\right] \quad (11.1.41)$$

Inspection of equation 11.1.40 indicates that the time at which the maximum concentration is reached must lie between zero and $\bar{t}^*$ and that for very small values of $\mathscr{D}_L/uL$ the maximum is very close to $\bar{t}^*$.

Equations 11.1.33 and 11.1.39 provide the basis for several methods of estimating dispersion parameters. Tracer experiments are used in the absence of chemical reactions to determine the dispersion parameter $\mathscr{D}_L$; this value is then employed in a material balance for a reactive component to predict the reactor effluent composition. We will now indicate some methods that can be used to estimate the dispersion parameter from tracer measurements.

The simplest method of estimating $\mathscr{D}_L$ from experimental tracer data is based on an evaluation of the slope of the $F(t)$ curve at $t = \bar{t}^*$. If we evaluate the derivative of equation 11.1.33 at this point,

$$\left[\frac{dF(t)}{d(t/\bar{t}^*)}\right]_{(t/\bar{t}^*) = 1} = \frac{1}{2}\sqrt{\frac{uL}{\pi\mathscr{D}_L}} \quad (11.1.42)$$

This approach to determining $\mathscr{D}_L$ suffers from several disadvantages.

1. Taking the derivative of experimental data is usually a rather inaccurate procedure.
2. Only a small portion of the available data (i.e., that near $t = \bar{t}^*$) is used in this method.
3. Equation 11.1.33 is strictly applicable only to an ideal step change stimulus and the boundary conditions cited. The plant or laboratory situation may not correspond to these assumptions.

For small values of the dispersion parameter one may take advantage of the fact that equation 11.1.37 takes the shape of a normal error curve. This implies that for a step function input a plot of $(C - C_0^-)/(C_0^+ - C_0^-)$ or $F(t)$

versus time on probability paper will be linear. Since the dispersion parameter is related to the variance of the curve (see below), it is a simple matter to use this plot to determine $\mathscr{D}_L$. For a normal error curve, one standard deviation on either side of the mean comprises 68% of the total area under the curve. Thus the 16th and 84th percentile points on the F curve are two standard deviations apart, and $t_{84\%} - t_{16\%} = 2\sigma_t$. Using relations developed below for small $(\mathscr{D}_L/uL)$ (say <0.01):

$$\frac{\mathscr{D}_L}{uL} \simeq \frac{1}{2}\left[\frac{\sigma_t^2}{(\bar{t}^*)^2}\right] \quad (11.1.43)$$

Alternative methods of estimating $\mathscr{D}_L$ are based on the response of the reactor to an ideal pulse input. For example, equation 11.1.39 may be used to calculate the mean residence time and its variance. Levenspiel and Bischoff (9) indicate that for the boundary conditions cited,

$$\bar{t} = \bar{t}^*\left(1 + \frac{2\mathscr{D}_L}{uL}\right) = \frac{V_R}{\mathcal{V}_0}\left(1 + \frac{2\mathscr{D}_L}{uL}\right) \quad (11.1.44)$$

and

$$\sigma_t^2 = (\bar{t}^*)^2\left[\frac{2\mathscr{D}_L}{uL} + 8\left(\frac{\mathscr{D}_L}{uL}\right)^2\right] \quad (11.1.45)$$

For other boundary conditions or for imperfect pulse injections, modifications must be made in these expressions. For example, for a closed vessel, Levenspiel and Bischoff (9) indicate that

$$\bar{t} = \bar{t}^* \quad (11.1.46)$$

and

$$\sigma_t^2 = (\bar{t}^*)^2\left(\frac{2\mathscr{D}_L}{uL}\right)\left[1 - \frac{\mathscr{D}_L}{uL}\left(1 - e^{-\frac{uL}{\mathscr{D}_L}}\right)\right] \quad (11.1.47)$$

For the case where $(\mathscr{D}_L/uL)$ is small (say less than 0.01) the various expressions for σ_t^2 become

$$\sigma_t^2 \cong (\bar{t}^*)^2\left(\frac{2\mathscr{D}_L}{uL}\right) \quad (11.1.48)$$

This approximation is valid to within 5% at this limit. Since the axial dispersion term itself may be viewed as a perturbation or correction term for real tubular reactors, errors of this magnitude in $\mathscr{D}_L$ lead to relatively minor errors in the conversion predicted by the model.

In principle any of the equations 11.1.40, 11.1.44, 11.1.45, 11.1.47, or 11.1.48 could be used to determine the dispersion parameter. However, both equations 11.1.40 and 11.1.44 require that one accurately determine a small difference in large numbers to evaluate $\mathscr{D}_L/uL$. Hence equations 11.1.45 and 11.1.47 are preferred for evaluation of $\mathscr{D}_L/uL$ for open and closed vessels, respectively. For small $\mathscr{D}_L/uL$, equation 11.1.48 is appropriate.

The discussion thus far presumes that a perfect pulse or step input is employed when, in fact, such inputs can only be approximated. In the case of pulse inputs one is also faced with two conflicting constraints in attempting to generate a perfect pulse or delta function. Since a finite amount of tracer cannot be injected in zero time, as much material should be injected in as short a time as possible in order to approximate a delta function. However, the injection process should not disturb the system significantly. This latter requirement implies that the tracer should be injected very slowly. Fortunately, these difficulties may be circumvented by using the imperfect pulse method described by Aris (10), Bischoff (11), and Bischoff and Levenspiel (12). The technique involves monitoring the concentration at two points in the test section instead of at a single point. It does not matter where the injection point is located, provided that it is upstream of the two monitoring stations. Any type of imperfect pulse input may be employed. The variances of the concentration-time curves at the two monitoring stations are determined and their difference is taken.

$$\Delta\sigma^2 = \sigma_2^2 - \sigma_1^2 \qquad (11.1.49)$$

where the subscripts 1 and 2 refer to the upstream and downstream locations, respectively. This equation reflects the fact that variances are additive for flow through *independent* vessels or regions. This property implies that the variance of the residence time distribution can be determined for any region if the variances of the inlet and effluent streams are known. In similar fashion mean residence times are additive; thus, for the test section,

$$\bar{t}_{\text{test}} = \bar{t}_{\text{out}} - \bar{t}_{\text{in}} \qquad (11.1.50)$$

The aforementioned investigators (10–12) have derived equations relating the measured mean residence times and variances to the Peclet number or dispersion parameter for the test section. For the case where the conditions at both monitoring probes correspond to a doubly infinite pipe, it can be shown that

$$\Delta\sigma^2 = \frac{2\mathscr{D}_L}{uL} \qquad (11.1.51)$$

If the measuring points are chosen far enough away from the ends of the test vessel so that end effects are negligible, this expression may be used with confidence. Bischoff and Levenspiel (12) have presented design charts that permit one to locate monitoring stations so as to avoid end effects. For example, in a packed bed reactor with a tube to pellet diameter ratio of 15, where the packed bed is followed by an open tube, at least 8 pellet diameters are required between the measurement point and the open tube if errors below 1% are to be obtained.

In addition to the aforementioned slope and variance methods for estimating the dispersion parameter, it is possible to use transfer functions in the analysis of residence time distribution curves. This approach reduces the error in the variance approach that arises from the "tails" of the concentration versus time curves. These tails contribute significantly to the variance and can be responsible for significant errors in the determination of $\mathscr{D}_L$.

A linear system may be described by a transfer function $\hat{F}(\rho)$:

$$\hat{F}(\rho) = \frac{C_2(\rho)}{C_1(\rho)}$$

$$= \frac{\left[\int_0^\infty C_2(t)e^{-\rho t}\, dt\right] \Big/ \left[\int_0^\infty C_2(t)\, dt\right]}{\left[\int_0^\infty C_1(t)e^{-\rho t}\, dt\right] \Big/ \left[\int_0^\infty C_1(t)\, dt\right]}$$

$$(11.1.52)$$

where $C_1(t)$ and $C_2(t)$ are the tracer concentrations (as functions of time) at the upstream and downstream monitoring stations, respectively. Material balance considerations indicate that

$$\int_0^\infty C_2(t)\, dt = \int_0^\infty C_1(t)\, dt \quad (11.1.53)$$

Hence

$$\hat{F}(\rho) = \frac{\int_0^\infty C_2(t)e^{-\rho t}\, dt}{\int_0^\infty C_1(t)e^{-\rho t}\, dt} \quad (11.1.54)$$

The material balance characterizing the axial dispersion model is equation 11.1.29, which can be rewritten as

$$\left(\frac{\mathscr{D}_L}{uL}\right)\frac{\partial^2 C}{\partial(Z/L)^2} - \frac{\partial C}{\partial(Z/L)} - \left(\frac{L}{u}\right)\frac{\partial C}{\partial t} = 0 \quad (11.1.55)$$

where

L is the distance between monitoring stations

L/u is again defined as $\bar{t}^*$

For a section of a continuous system where end effects are negligible several investigators (e.g., 13–14) have shown that the transform of this equation with respect to time leads to the following expression.

$$\hat{F}(\rho) = \exp\left\{\frac{uL}{2\mathscr{D}_L}\left[1 - \left(1 + \frac{4\mathscr{D}_L}{uL}\cdot\frac{L}{u}\rho\right)^{1/2}\right]\right\}$$

$$= \exp\left\{\frac{uL}{2\mathscr{D}_L}\left[1 - \left(1 + \frac{4\mathscr{D}_L}{uL}\cdot\bar{t}^*\rho\right)^{1/2}\right]\right\}$$

$$(11.1.56)$$

Rearrangement of this equation gives

$$\frac{1}{\ln[1/\hat{F}(\rho)]} = \frac{\rho\bar{t}^*}{\{\ln[1/\hat{F}(\rho)]\}^2} - \frac{\mathscr{D}_L}{uL} \quad (11.1.57)$$

Numerical values of $\hat{F}(\rho)$ may be calculated from the experimental data for arbitrary values of ρ using equation 11.1.54. One takes these values and prepares a plot of the left side of equation 11.1.57 versus $\rho/\{\ln[1/\hat{F}(\rho)]\}^2$ for test data. This procedure should give a straight line of slope $\bar{t}^*$ and intercept $(-\mathscr{D}_L/uL)$. Hopkins et al. (15) have shown that some discretion must be used in selecting the "arbitrary" ρ values in order to minimize the error involved in determining $\mathscr{D}_L/uL$. Low values of the group $(\rho\bar{t}^*)$ lead to large errors in $\mathscr{D}_L/uL$ because the transfer function is relatively insensitive to variations in $(\mathscr{D}_L/uL)$ at low $(\rho\bar{t}^*)$. Large values of the group $(\rho\bar{t}^*)$ are also disadvantageous because of the effect of the exponential $(e^{-\rho t})$ on the individual values of the integrands in equation 11.1.54. The weighting factor $(e^{-\rho t})$ minimizes the influence of the tail, but emphasizes the values of $C(t)$ at short times when concentration fluctuations may be important. Hopkins et al. (15) recommend that values of $(\rho\bar{t}^*)$ between 2 and 5 be used when dealing with flow through packed beds.

In addition to the three methods described above, nonlinear regression methods or other transform approaches may be used to determine the dispersion parameter. For a more complete treatment of the use of transform methods, consult the articles by Hopkins et al. (15) and Ostergaard and Michelsen (14).

Illustration 11.2 indicates the use of the slope and variance methods for evaluating $\mathscr{D}_L/uL$.

ILLUSTRATION 11.2 DETERMINATION OF REACTOR DISPERSION PARAMETER FROM EXPERIMENTAL RESIDENCE TIME DATA

In Illustration 11.1 we considered the response of an arbitrary reactor to a pulse input and used this data to determine the average residence time and the $F(t)$ curve. If the pulse is assumed to be perfect, what value of $(\mathcal{D}_L/uL)$ gives a reasonable fit of the experimental data? Use the slope and variance methods to evaluate this parameter.

Solution

First consider the slope method for determining $\mathcal{D}_L/uL$. From a plot of the $F(t)$ curve the value of $dF(t)/dt$ at $t = \bar{t} = 374.4$ sec or 0.3744 ksec is equal to 2.17 ksec^{-1}. From equation 11.1.42,

$$\left[\frac{dF(t)}{d(t/\bar{t})}\right]_{t/\bar{t}=1} = \frac{1}{2}\sqrt{\frac{uL}{\pi\mathcal{D}_L}} = 0.3744 \,(2.17)$$

Thus

$$\frac{\mathcal{D}_L}{uL} = 0.121$$

The $F(t)$ curve corresponding to this value of the reactor dispersion parameter may be determined using equation 11.1.33.

Now consider the determination of this parameter using the variance of the response to a pulse input. The variance measures the spread of the distribution about the mean. For a continuous distribution it is defined as

$$\sigma^2 = \frac{\int_0^\infty (x - \mu)^2 f(x)\,dx}{\int_0^\infty f(x)\,dx}$$

where

μ is the mean

x is the property being investigated

$f(x)\,dx$ is the distribution function

In terms of the present problem for the response

to a pulse, the variance in t is given by

$$\sigma_t^2 = \frac{\int_0^\infty (t - \bar{t})^2 \dfrac{dF(t)}{dt}\,dt}{\int_0^\infty \dfrac{dF(t)}{dt}\,dt}$$

$$= \int_0^\infty t^2 \frac{dF(t)}{dt}\,dt - \bar{t}^2$$

If steady-state operation, invariant mass flow, and evenly spaced time increments are assumed, as in Illustration 11.1, this equation may be written as

$$\sigma_t^2 = \frac{\displaystyle\sum_{t=0}^{t=\infty} t^2 C_E}{\displaystyle\sum_{t=0}^{t=\infty} C_E} - \bar{t}^2$$

From the data in Table 11.I.1 it is evident that

$$\sigma_t^2 = \frac{1}{50.0}\left(\begin{array}{c} 0 + 93{,}600 + 720{,}000 + 1{,}620{,}000 + 2{,}304{,}000 \\ + 1{,}800{,}000 + 1{,}296{,}000 + 705{,}600 + 0\end{array}\right) - (374.4)^2$$

or

$$\sigma_t^2 = 30{,}609 \text{ sec}^2$$

In order to proceed from this point to a determination of the reactor dispersion parameter one must know something about the experimental setup. If we consider an open vessel, equation 11.1.45 indicates that

$$\sigma_t^2 = (\bar{t}^*)^2 \left[\frac{2\mathcal{D}_L}{uL} + 8\left(\frac{\mathcal{D}_L}{uL}\right)^2\right]$$

Since no information regarding L/u or $V_R/\mathcal{V}_0$ is provided, we assume as a first approximation that $\bar{t}^* = \bar{t} = 374.4$ sec. Thus

$$30{,}609 = (374.4)^2(2)\left[\frac{\mathcal{D}_L}{uL} + 4\left(\frac{\mathcal{D}_L}{uL}\right)^2\right]$$

or

$$4\left(\frac{\mathcal{D}_L}{uL}\right)^2 + \frac{\mathcal{D}_L}{uL} = \frac{30{,}609}{(374.4)^2(2)} = 0.1092$$

Solving for $\mathcal{D}_L/uL$ gives 0.0822.

A more refined value may be obtained using equation 11.1.44 to determine $\bar{t}^{*}$.

$$\bar{t}^{*} = \frac{\bar{t}}{1 + \dfrac{2\mathscr{D}_L}{uL}} = \frac{374.4}{1 + 2(0.0822)} = 321.5$$

As a second approximation,

$$4\left(\frac{\mathscr{D}_L}{uL}\right)^2 + \left(\frac{\mathscr{D}_L}{uL}\right) = \frac{30,609}{(321.5)^2(2)} = 0.1481$$

or

$$\frac{\mathscr{D}_L}{uL} = 0.104$$

Subsequent iterations lead to the conclusion that $\mathscr{D}_L/uL = 0.113$. This result differs from that obtained by the slope approach by 7%.

Note that if one has a value for $V_R/\mathscr{V}_0$ as in the normal situation, the iterative procedure is unnecessary. However, in this case equation 11.1.44 should not be used to determine $\mathscr{D}_L/uL$, because significant errors may be involved in accurately determining a small difference in large numbers.

11.1.3.2 The Stirred Tanks in Series Model

Another model that is frequently used to simulate the behavior of actual reactor networks is a cascade of ideal stirred tank reactors operating in series. The actual reactor is replaced by n identical stirred tank reactors whose *total* volume is the same as that of the actual reactor.

$$V_R = n V_{\text{CSTR}} \tag{11.1.58}$$

One determines the value of n that gives the best fit of the response curve of the actual reactor by the response curve of the model. Consequently, it is necessary to develop an analytical expression for the latter. For the nth stirred tank reactor in the series the time-dependent form of the material balance equation becomes (in the absence of reaction)

$$C_{n-1}\mathscr{V}_0 = C_n\mathscr{V}_0 + V_{\text{CSTR}}\frac{dC_n}{dt} \tag{11.1.59}$$

Input   Output   Accumulation

Since the *total average residence time in the actual reactor* ($\bar{t}$) is given by

$$\bar{t} = \frac{V_R}{\mathscr{V}_0} = \frac{n V_{\text{CSTR}}}{\mathscr{V}_0} \tag{11.1.60}$$

equation 11.1.59 may be rewritten as

$$\frac{dC_n}{dt} + \frac{nC_n}{\bar{t}} = \frac{nC_{n-1}}{\bar{t}} \tag{11.1.61}$$

If we are determining the response of the series of stirred tank reactors to a step change in inlet tracer concentration from 0 to C_0^+ at time zero, the initial condition for this differential equation is

$$C_n = 0 \quad \text{for} \quad t = 0 \quad \text{and} \quad n > 0$$

The solution to this equation may be obtained using the integrating factor approach to give

$$C_n = e^{-nt/\bar{t}} \int_0^t \frac{nC_{n-1}}{\bar{t}} e^{nt/\bar{t}}\, dt \tag{11.1.62}$$

The integral may be evaluated for each stage of the reactor network in turn. For the first stage, $C_{n-1} = C_0^+$, so that

$$C_1 = e^{-t/\bar{t}} \int_0^t \frac{C_0^+}{\bar{t}} e^{t/\bar{t}}\, dt \tag{11.1.63}$$

or

$$\frac{C_1}{C_0^+} = e^{-t/\bar{t}}(e^{t/\bar{t}} - 1) = 1 - e^{-t/\bar{t}} \tag{11.1.64}$$

when there is just one stage.

If there are n reactors in series, $\bar{t}_{1\text{ reactor}} = \bar{t}/n$ and

$$\frac{C_1}{C_0^+} = 1 - e^{-nt/\bar{t}} \tag{11.1.65}$$

where $\bar{t}$ is the mean residence time *for the entire network.*

Equations 11.1.62 and 11.1.64 may be combined to obtain the equation for the tracer concentration in the effluent from the second reactor when two reactors comprise the network.

$$C_2 = e^{-2t/\bar{t}} \int_0^t \frac{2C_0^+}{\bar{t}}(1 - e^{-2t/\bar{t}})e^{2t/\bar{t}}\, dt$$

$$\tag{11.1.66}$$

or

$$\frac{C_2}{C_0^+} = e^{-2t/\bar{t}} \frac{2}{\bar{t}} \int_0^t (e^{2t/\bar{t}} - 1)\, dt = e^{-2t/\bar{t}} \left\{ \left[\left(e^{2t/\bar{t}} - \frac{2t}{\bar{t}} \right) \right]_0^t \right\}$$

$$= e^{-2t/\bar{t}} \left(e^{2t/\bar{t}} - 1 - \frac{2t}{\bar{t}} \right) = 1 - e^{-2t/\bar{t}} \left(1 + 2\frac{t}{\bar{t}} \right) \qquad (11.1.67)$$

where $\bar{t}$ is the mean residence time *for the two reactor cascade.*

One may proceed stepwise in this fashion to develop a general recursion formula for the concentration leaving reactor j in an n reactor cascade.

$$\frac{C_j}{C_0^+} = 1 - e^{-nt/\bar{t}} \left[1 + \frac{nt}{\bar{t}} + \frac{1}{2!} \left(\frac{nt}{\bar{t}} \right)^2 + \frac{1}{3!} \left(\frac{nt}{\bar{t}} \right)^3 + \cdots + \frac{1}{(j-1)!} \left(\frac{nt}{\bar{t}} \right)^{j-1} \right] \qquad (11.1.68)$$

The response curve for the network as a whole is obtained by setting j equal to n in equation 11.1.68. Note that $\bar{t}$ is given by equation 11.1.60.

$$F(t) = \frac{C_n}{C_0^+} = 1 - e^{-nt/\bar{t}} \left[1 + \frac{nt}{\bar{t}} + \frac{1}{2!} \left(\frac{nt}{\bar{t}} \right)^2 + \cdots + \frac{1}{(n-1)!} \left(\frac{nt}{\bar{t}} \right)^{n-1} \right] \qquad (11.1.69)$$

Note that in this case the right side of equation 11.1.68 is zero for $t = 0$ and unity for $t = \infty$. Figure 11.9 contains several $F(t)$ curves for various values of n. As n increases, the spread in residence time decreases. In the limit, as n approaches infinity the $F(t)$ curve approaches that for an ideal plug flow reactor. If the residence time distribution function given by 11.1.69 is differentiated, one obtains an

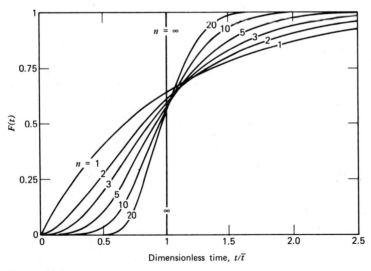

Figure 11.9
Residence time distribution curves for the n-CSTR model.

equation for the slope of the curve from which n may be determined by comparison with the experimental $F(t)$ curve.

$$\frac{dF(t)}{d(t/\bar{t})} = ne^{-nt/\bar{t}}\left[1 + \frac{nt}{\bar{t}} + \frac{1}{2!}\left[\frac{nt}{\bar{t}}\right]^2 + \cdots + \frac{1}{(n-1)!}\left[\frac{nt}{\bar{t}}\right]^{n-1}\right]$$

$$\cdot - e^{-nt/\bar{t}}\left[n + \frac{n^2 t}{\bar{t}} + \frac{n^3}{2!}\left[\frac{t}{\bar{t}}\right]^2 + \cdots + \frac{n^{n-1}}{(n-2)!}\left[\frac{t}{\bar{t}}\right]^{n-2}\right] \qquad (11.1.70)$$

or

$$\frac{dF(t)}{d(t/\bar{t})} = ne^{-nt/\bar{t}}\frac{1}{(n-1)!}\left(\frac{nt}{\bar{t}}\right)^{n-1} \qquad (11.1.71)$$

At $(t/\bar{t}) = 1$, the slope is equal to

$$\left[\frac{dF(t)}{d(t/\bar{t})}\right]_{t/\bar{t}=1} = \frac{n^n e^{-n}}{(n-1)!} = \frac{n^{n+1}e^{-n}}{n!} \qquad (11.1.72)$$

For $n > 5$ we may use Stirling's approximation and retain an accuracy of 2%.

$$n! \simeq n^n e^{-n}\sqrt{2\pi n} \qquad (11.1.73)$$

Combination of equations 11.1.72 and 11.1.73 gives

$$\left[\frac{dF(t)}{d(t/\bar{t})}\right]_{t/\bar{t}=1} \simeq \sqrt{\frac{n}{2\pi}} \qquad \text{for } n > 5 \quad (11.1.74)$$

Approximate values of the slope at $t/\bar{t} = 1$ for $n \leq 5$ are:

n	1	2	3	4	5
Slope	0.368	0.541	0.672	0.781	0.877

The model suffers from the fact that it allows only integer values of n and that it may not be possible to obtain a match of the residence time distribution function at both high and low values of $F(t)$ with the same value of n. Buffham and Gibilaro (16) have generalized the model to include noninteger values of n. The technique outlined by these individuals is particularly useful in obtaining better fits of the data for cases where n is less than 5.

An alternative to the slope approach to determining the appropriate value of n for use in model calculations is based on a determination of the variance of the response of the actual reactor to a *pulse* input. For linear systems this is equivalent to determining the variance of the derivative of the $F(t)$ curve. The response of the series of reactors to a pulse input is given by equation 11.1.71. The variance of this expression is given by its second moment.

$$\sigma^2_{(t/\bar{t})} = \int_0^\infty \left(\frac{t}{\bar{t}} - 1\right)^2 \frac{n^{n+1}}{n!} e^{-nt/\bar{t}}\left(\frac{t}{\bar{t}}\right)^{n-1} d(t/\bar{t})$$

$$(11.1.75)$$

Evaluation of this integral leads to the surprisingly simple result that

$$\sigma^2_{t/\bar{t}} = \frac{1}{n} \qquad (11.1.76)$$

Thus

$$\sigma^2_t = \frac{(\bar{t})^2}{n} \qquad (11.1.77)$$

and by determining the variance of the response of a system to a *pulse* input, one may obtain an estimate of n for use in subsequent reactor design calculations.

For cases where two monitoring stations are used, equation 11.1.49 is applicable, and

$$\frac{\Delta\sigma^2}{(\bar{t})^2} = \frac{\sigma_2^2 - \sigma_1^2}{(\bar{t})^2} = \frac{1}{j} \qquad (11.1.78)$$

where j is now the equivalent number of stirred tanks between the two stations.

Illustration 11.3 indicates how n may be determined by each of the two methods discussed above.

ILLUSTRATION 11.3 USE OF EXPERIMENTAL RESPONSE DATA TO DETERMINE THE NUMBER OF STIRRED TANK REACTORS IN SERIES

Use the data of Illustration 11.1 for the response of a reactor network to a pulse input to determine the number of identical stirred tank reactors in series that gives a reasonable fit of the experimental data. Use both the slope and variance methods described above.

Solution

From a plot of the data for Illustration 11.1,

$$\left[\frac{dF(t)}{d(t/\bar{t})} \right]_{(t/\bar{t}) = 1} = \bar{t} \left[\frac{dF(t)}{dt} \right]_{\bar{t}} = 0.3744(2.17) = 0.81$$

From the tabulated values of the slope at the point where $t/\bar{t} = 1$, it is evident that n must lie between 4 and 5.

The variance approach may also be used to determine n. From Illustration 11.2 the variance of the response data based on dimensionless time is $30609/(374.4)^2$, or 0.218. From equation 11.1.76 it is evident that n is 4.59. Thus the results of the two approaches are consistent. However, a comparison of the $F(t)$ curves for $n = 4$ and $n = 5$ with the experimental data indicates that these approaches do not provide very good representations of the data. For the reactor network in question it is difficult to model the residence time distribution function in terms of a single parameter. This is one of the potential difficulties inherent in using such simple models of reactor behavior. For more advanced methods of modeling residence time effects, consult the review article by Levenspiel and Bischoff (3) and textbooks written by these authors (2, 4).

11.2 CONVERSION LEVELS IN NONIDEAL FLOW REACTORS

The present section indicates how tracer residence time data may be used to predict the conversion levels that will be obtained in re-

actors with nonideal flow patterns. As indicated earlier, there are two types of limiting processes that can lead to a distribution of residence times within a reactor network.

1. A flow pattern in which the various fluid elements follow different paths without mutual mixing on a microscopic scale. An example of this case is laminar flow.
2. Mixing of fluid elements having different ages. Microscopic mixing produced by eddy diffusion effects is an example of this case.

Since these two types of processes have drastically different effects on the conversion levels achieved in chemical reactions, they provide the basis for the development of mathematical models that can be used to provide approximate limits within which one can expect actual isothermal reactors to perform. In the development of these models we will define a *segregated* system as one in which the first effect is *entirely responsible* for the spread in residence times. When the distribution of residence times is established by the second effect, we will refer to the system as *mixed*. In practice one encounters various combinations of these two limiting effects.

We can characterize the mixed systems most easily in terms of the longitudinal dispersion model or in terms of the cascade of stirred tank reactors model. The maximum amount of mixing occurs for the cases where $\mathscr{D}_L = \infty$ or $n = 1$. In general, for reaction orders greater than unity, these models place a lower limit on the conversion that will be obtained in an actual reactor. The applications of these models are treated in Sections 11.2.2 and 11.2.3.

In the *segregated flow* model the contents of the volume elements of the fluid do not mix with one another as they move through the reactor. Each element may be considered as a small closed system that moves through the reactor. The different systems spend varying amounts of time in the reactor, giving rise to the measured residence time distribution function. The closest approximation to this condition that one can encounter in engineering practice is a laminar flow reactor in which molecular diffusion in both longitudinal and radial directions is negligible. Another real-life situation in which a segregated flow pattern is largely responsible for the spread in residence times is one in which there is some short-circuiting or bypassing of portions of the reactor volume as the fluid moves through the reactor. The bypassed regions may be considered as "dead" spaces that contribute to the· total reactor volume but that do not contribute to the overall conversion rate in proportion to their volume. Since they are not effectively purged by the prevailing flow pattern, they often contain fluid in which the reaction has gone substantially to completion. It should be evident that this situation should be avoided in the proper design of chemical reactors. In general, for reaction orders greater than unity, the segregated flow model places an upper limit on the conversion that will be obtained in an actual reactor. For a first-order reaction occurring isothermally, the model can be used to predict accurately the conversion that will be attained in a real reactor whose $F(t)$ curve is known. The details of the analysis are discussed in Section 11.2.1.

In general, the larger the range of the distribution of residence times, the greater the discrepancy between the conversion levels predicted on the basis of the segregated flow model and those predicted by the various mixing models. For narrow distribution functions the conversions predicted by both models will be in good agreement with one another.

To illustrate the nature of the limits that the segregated flow and mixing models place on the expected conversion level, it is useful to examine what happens to two elements of fluid that have the same volume V, but that contain different reactant concentrations C_1 and C_2. We may imagine two extreme limits on the amount of mixing that may occur.

S: No mixing of the contents of the elements as they move through the reactor (complete segregation).

M: Complete mixing of the contents immediately on entrance followed by flow through the reactor in the completely mixed state.

If we assume a rate expression of the form

$$r = kC^n \tag{11.2.1}$$

the total conversion rates under the two limiting circumstances outlined above are given by

$$R_S = Vk(C_1^n + C_2^n) \tag{11.2.2}$$

and

$$R_M = 2Vk\left(\frac{C_1 + C_2}{2}\right)^n \tag{11.2.3}$$

For a differential reactor and the same residence time the ratio of these rates will be equal to the ratio of the conversion levels attained.

$$\frac{f_S}{f_M} = \frac{R_S}{R_M} = \frac{C_1^n + C_2^n}{2\left(\frac{C_1 + C_2}{2}\right)^n} \tag{11.2.4}$$

Since

$$\frac{C_1^n + C_2^n}{2} \neq \left(\frac{C_1 + C_2}{2}\right)^n \tag{11.2.5}$$

unless $n = 1$ or $n = 0$, these are the only cases for which the two extreme situations will converge to the same limit. In these cases the fraction conversion will be the same whether or not fluid elements having different reactant concentrations remain segregated or are mixed prior to or during the reaction.

For reaction orders greater than unity the segregated flow model will predict a higher conversion level than the various mixing models. For reaction orders less than unity the mixing models will predict higher conversion levels than the segregated flow model. The magnitude of the differences for representative cases are shown in Illustrations 11.5 to 11.7. The following illustration shows that for the first-order case, reactor combinations involving different mixing patterns but the same residence time distribution function will give rise to the same predicted conversion.

ILLUSTRATION 11.4 COMPARISON OF CONVERSION LEVELS ATTAINED IN TWO DIFFERENT REACTOR COMBINATIONS HAVING THE SAME RESIDENCE TIME DISTRIBUTION CURVE—FIRST-ORDER REACTION

The $F(t)$ curve for a system consisting of a plug flow reactor followed by a continuous stirred tank reactor is identical to that of a system in which the CSTR precedes the PFR. Show that the overall fraction conversions obtained in these two combinations are identical for the case of an irreversible first-order reaction. Assume isothermal operation.

Solution

Let τ_p and τ_c represent the space times of the plug flow reactor and the continuous stirred tank reactor respectively. Consider the following reactor combination

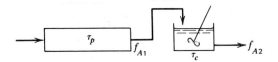

where f_{A1} and f_{A2} are the conversions at the outlet of the first and second reactors, respectively.

From the design equation for a PFR,

$$\tau_p = C_{A0} \int_0^{f_{A1}} \frac{df_A}{kC_{A0}(1 - f_A)}$$

Integration gives

$$k\tau_p = -\ln(1 - f_{A1})$$

or

$$f_{A1} = 1 - e^{-k\tau_p}$$

From the design equation for a CSTR,

$$\tau_c = \frac{C_{A0} \int_{f_{A1}}^{f_{A2}} df_A}{kC_{A0}(1 - f_{A2})} = \frac{f_{A2} - f_{A1}}{k(1 - f_{A2})}$$

Solving for f_{A2}

$$f_{A2} = \frac{f_{A1} + k\tau_c}{1 + k\tau_c}$$

or

$$f_{A2} = \frac{1 + k\tau_c - e^{-k\tau_p}}{1 + k\tau_c}$$

Consider the alternative arrangement

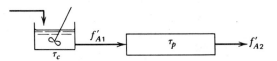

where f'_{A1} and f'_{A2} are the conversions at the outlet of the CSTR and the PFR, respectively.

From the design equation for the CSTR,

$$\tau_c = \frac{C_{A0} \int_0^{f'_{A1}} df_A}{kC_{A0}(1 - f'_{A1})} = \frac{f'_{A1}}{k(1 - f'_{A1})}$$

or

$$f'_{A1} = \frac{k\tau_c}{1 + k\tau_c} \qquad \text{(A)}$$

From the design equation for the PFR,

$$\tau_p = C_{A0} \int_{f'_{A1}}^{f'_{A2}} \frac{df_A}{kC_{A0}(1 - f_A)}$$

Integration gives

$$k\tau_p = \ln\left(\frac{1 - f'_{A1}}{1 - f'_{A2}}\right)$$

or

$$1 - f'_{A2} = (1 - f'_{A1})e^{-k\tau_p} \qquad \text{(B)}$$

Combining equations (A) and (B) and rearranging gives

$$f'_{A2} = 1 - \frac{e^{-k\tau_p}}{1 + k\tau_c} = \frac{1 + k\tau_c - e^{-k\tau_p}}{1 + k\tau_c}$$

This relation is identical to that obtained for the first arrangement. This illustration provides an example of the general principle that for irreversible first-order reactions carried out isothermally, all reactor combinations having the same residence time distribution function give rise to the same overall conversion.

You should repeat the analysis for a second-order reaction to verify that for other reaction orders the overall conversion will depend on whether the PFR precedes the CSTR or vice versa.

11.2.1 The Segregated Flow Model

The basic premise of the segregated flow model is that the various fluid elements move through the reactor at different speeds without mixing with one another. Consequently, each little fluid element will behave as if it were a batch reactor operating at constant pressure. The conversions attained within the various fluid elements will be equal to those in batch reactors with holding times equal to the residence times of the different fluid elements. The average conversion level in the effluent is then given by

$$\langle f_A \rangle = \sum \left[\binom{\text{Fraction conversion expressed as}}{\text{a function of residence time}} \times \binom{\text{Fraction of the fluid elements having}}{\text{residence times between } t \text{ and } t + dt} \right] \quad (11.2.6)$$

where the summation extends over all possible residence times. In terms of an integral

$$\langle f_A \rangle = \int_{F(t)=0}^{F(t)=1.0} f_A(t) \, dF(t) \quad (11.2.7)$$

An alternative interpretation of the segregated flow model that leads to precisely the same result involves replacing the actual reactor by a number of plug flow reactors in parallel so that the combination will give rise to the $F(t)$ curve observed experimentally. Under these conditions the degree of conversion in a fraction of the effluent stream $dF(t)$ is equal to that which would occur in an ideal PFR with a residence time t. When the different streams are recombined at the outlet, the average conversion level is again given by equations 11.2.6 and 11.2.7. In order to indicate how one makes use of these relations, consider Illustration 11.5.

ILLUSTRATION 11.5 USE OF THE SEGREGATED FLOW MODEL TO DETERMINE THE CONVERSION LEVEL OBTAINED IN A NON-IDEAL FLOW REACTOR

Use the $F(t)$ curve generated in Illustration 11.1 to determine the fraction conversion that will be achieved in the reactor if it is used to carry out a first-order reaction with a rate constant equal to $3.33 \times 10^{-3} \text{ sec}^{-1}$. Base the calculations on the segregated flow model.

Solution

Equation 11.2.7 is the key equation to the solution of this problem.

$$\langle f_A \rangle = \int_{F(t)=0}^{F(t)=1.0} f_A(t) \, dF(t) \quad (A)$$

For a first-order reaction,

$$f_A(t) = 1 - e^{-kt}$$

so that equation A becomes

$$\langle f_A \rangle = 1 - \int_{t=0}^{t=\infty} e^{-kt} \frac{dF(t)}{dt} \, dt$$

Substituting equation 11.1.6 into this relation gives

$$\langle f_A \rangle = 1 - \frac{\int_{t=0}^{t=\infty} e^{-kt} w_E \Phi_m \, dt}{\int_0^{\infty} w_E \Phi_m \, dt} \quad (B)$$

If we assume a constant density system and a constant mass flow rate and replace the integrals by finite sums, we have

$$\langle f_A \rangle = 1 - \frac{\sum_0^{\infty} (e^{-kt} w_E \, \Delta t)}{\sum_0^{\infty} w_E \, \Delta t}$$

If we now replace the mass fractions by concentrations and note that our data are recorded at evenly spaced time increments,

$$\langle f_A \rangle = 1 - \frac{\sum\limits_{0}^{\infty}(e^{-kt}C_E)}{\sum\limits_{0}^{\infty}C_E}$$

The data are worked up below.

Time, t (sec)	C_E	e^{-kt}	$C_E e^{-kt}$
0	0.0	1.00	0.0
120	6.5	0.670	4.36
240	12.5	0.449	5.61
360	12.5	0.301	3.76
480	10.0	0.202	2.02
600	5.0	0.135	0.68
720	2.5	0.091	0.23
840	1.0	0.061	0.06
960	0.0	—	0.0
1080	0.0	—	0.0
	$\sum C_E = 50.0$		$\sum C_E e^{-kt} = 16.72$

Thus

$$\langle f_A \rangle = 1 - \frac{16.72}{50.00}$$

$$= 0.666$$

As an alternative to this numerical procedure, graphical integration of the terms in equation B could be employed.

Since the reaction under consideration is a first-order reaction, this result should be in good agreement with the conversions predicted on the basis of various mixing models. That this is true can be seen from a comparison of the result obtained above with those that will be obtained in Illustrations 11.6 and 11.7.

For systems with a wide distribution of residence times, the degree of conversion can be calculated accurately only in the case of a first-order reaction. However, if it is possible to estimate in some rough fashion the extents to which mixing and segregation effects contribute to the observed residence time distribution, one can bracket the actual performance of the reactor with the solutions obtained from the completely segregated flow model and those obtained from models consisting of various combinations of ideal reactors. It is an extremely rare situation in which one knows the flow pattern in the reactor sufficiently well that an exact calculation of the conversion can be performed. Instead, one must be satisfied with bracketing the solution in terms of models based on segregated flow and various mixing effects. Two commonly used mixing models are described in the next two sections.

11.2.2 The Longitudinal Dispersion Model in the Presence of Chemical Reaction

In Section 11.1.3.1 we considered the longitudinal dispersion model for flow in tubular reactors and indicated how one may employ tracer measurements to determine the magnitude of the dispersion parameter used in the model. In this section we will consider the problem of determining the conversion that will be attained when the model reactor operates *at steady state*. We will proceed by writing a material balance on a reactant species A using a tubular reactor. The mass balance over a reactor element of length ΔZ becomes:

$$\left[\left(-\mathscr{D}_L \frac{dC}{dZ} + uC\right)\pi R^2\right]_Z = \left[\left(-\mathscr{D}_L \frac{dC}{dZ} + uC\right)\pi R^2\right]_{Z+\Delta Z} + (-r_A)\pi R^2 \,\Delta Z \quad (11.2.8)$$

$$\text{Input} \qquad\qquad\qquad \text{Output} \qquad\qquad \text{Disappearance by reaction}$$

where we have taken into account the fact that material enters and leaves the volume element by bulk flow and by longitudinal dispersion. If we divide by ΔZ and take the limit as ΔZ approaches zero, equation 11.2.8 becomes

$$\mathscr{D}_L \frac{d^2 C_A}{dZ^2} - u \frac{dC_A}{dZ} + r_A = 0 \quad (11.2.9)$$

Note the similarities and differences between this equation and the time-dependent equation used to evaluate the dispersion parameter (equation 11.1.29).

For a first-order irreversible reaction equation 11.2.9 becomes

$$\mathscr{D}_L \frac{d^2 C_A}{dZ^2} - u \frac{dC_A}{dZ} - kC_A = 0 \quad (11.2.10)$$

Wehner and Wilhelm (17) have obtained an analytical solution to this equation for the case where $\delta_A = 0$. The solution is valid both when one has plug flow and when one has dispersion in the regions adjacent to the test section.

conversions and large $\mathscr{D}_L/uL$ significantly larger reactors are required than would be predicted by a plug flow analysis. However, at large $\mathscr{D}_L/uL$, the dispersion model may well be inappropriate for use. Consequently, only the lower segment of Figure 11.10 is of general utility for design calculations.

Numerical solutions to equation 11.2.9 have been obtained for reaction orders other than unity. Figure 11.11 summarizes the results obtained by Levenspiel and Bischoff (18) for second-order kinetics. Like the chart for first-order kinetics, it is most appropriate for use when the dimensionless dispersion group is small. Fan and Bailie (19) have solved the equations for quarter-order, half-order, second-order, and third-order kinetics. Others have used perturbation methods to arrive at analogous results for the dispersion model (e.g. 20, 21).

There are several closed form approximate solutions to both the general and first-order forms of the dispersion equations (11.2.9 and 11.2.10). For example, Levenspiel and Bischoff

$$\frac{C_A}{C_{A0}} = 1 - f_A = \frac{4\beta e^{uL/2\mathscr{D}_L}}{(1 + \beta)^2 e^{\beta uL/2\mathscr{D}_L} - (1 - \beta)^2 e^{-\beta uL/2\mathscr{D}_L}} \quad (11.2.11)$$

where

$$\beta = \left[1 + 4k \left(\frac{\mathscr{D}_L}{uL} \right) \frac{L}{u} \right]^{1/2} \quad (11.2.12)$$

Levenspiel and Bischoff (18) have compared this solution with that for the plug flow case.

$$\frac{C_{A\,\text{plug}}}{C_{A0}} = 1 - f_A = e^{-k\tau} = e^{-kL/u} \quad (11.2.13)$$

They determined the ratio of the dispersion reactor volume to the plug flow reactor volume necessary to accomplish the same degree of conversion for several values of the dimensionless dispersion parameter $\mathscr{D}_L/uL$. Figure 11.10 summarizes their results. It is evident that for high

(2, 3, 4) indicate that for small dispersion numbers ($\mathscr{D}_L/uL$) equation 11.2.11 can be rewritten in the following form if one expands the exponentials and drops higher-order terms.

$$\ln \left(\frac{C_A}{C_{A0}} \right) = -k \frac{L}{u} + \left(\frac{kL}{u} \right)^2 \frac{\mathscr{D}_L}{uL}$$

$$= -k\tau + (k\tau)^2 \frac{\mathscr{D}_L}{uL} \quad (11.2.14)$$

or, using equation 11.2.13 for the same reactor space time and the same reactor volume,

$$\ln \left(\frac{C_{A\,\text{dispersion}}}{C_{A\,\text{plug}}} \right) = (k\tau)^2 \frac{\mathscr{D}_L}{uL} \quad (11.2.15)$$

This equation indicates that the conversion in

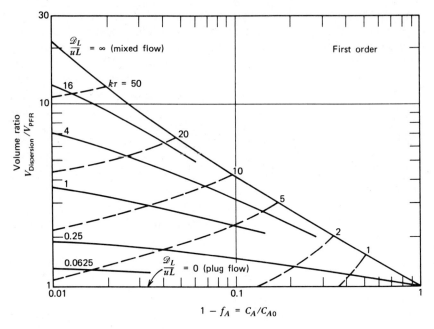

Figure 11.10

Comparison of real and plug flow reactors for the first-order reaction $A \rightarrow$ products, assuming negligible expansion ($\delta_A = 0$). (Adapted from *Chemical Reaction Engineering,* **Second Edition, by O. Levenspiel. Copyright © 1972. Reprinted by permission of John Wiley and Sons, Inc.)**

the dispersion reactor will always be less than that in the plug flow reactor ($C_{A\,\text{dispersion}} > C_{A\,\text{plug}}$). For the case where one fixes the effluent composition instead of the reactor size, equations 11.2.13 and 11.2.14 can be manipulated to show that, for small $\mathscr{D}_L/uL$,

$$\frac{L_{\text{dispersion}}}{L_{\text{plug flow}}} = \frac{V_{\text{dispersion}}}{V_{\text{plug flow}}} = 1 + (k\tau)\left(\frac{\mathscr{D}_L}{uL}\right)$$

(for same conversion)   (11.2.16)

Levenspiel and Bischoff (3) have used a derivation by Pasquon and Dente (20) to obtain an expression that gives an approximate solution to 11.2.9 for small $\mathscr{D}_L/uL$ and arbitrary kinetics.

where $r_{A\,\text{exit}}$ and r_{A0} are the rates at the exit and entrance of a plug flow reactor with the same space time as the dispersion (real) reactor.

The influence of dispersion on the yield of an intermediate produced in a series reaction has also been studied. When $\mathscr{D}_L/uL$ is less than 0.05, Tichacek's results (22) indicate that the fractional decrease in the maximum amount of intermediate formed relative to plug flow conditions is approximated by $\mathscr{D}_L/uL$ itself. Results obtained at higher dispersion numbers are given in the original article.

Douglas and Bischoff (23) have considered the influence of volumetric expansion effects on the yields obtained with dispersion.

$$C_{A\,\text{dispersion}} = C_{A\,\text{plug}} + \left(\frac{\mathscr{D}_L}{uL}\right)(r_{A\,\text{exit}}\tau)\,\ell n\left(\frac{r_{A\,\text{exit}}}{r_{A0}}\right)$$

                                                                 (11.2.17)

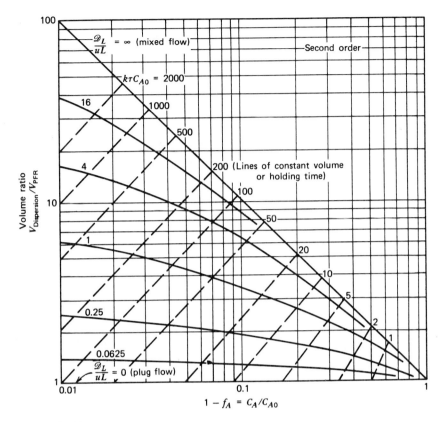

Figure 11.11
Comparison of real and plug flow reactors for the second-order reactions

$$A + B \rightarrow \text{products}, \qquad C_{A0} = C_{B0}$$

$$2A \rightarrow \text{products}$$

assuming negligible expansion ($\delta_A = 0$). (Adapted from *Chemical Reaction Engineering*, **Second Edition, by O. Levenspiel. Copyright © 1972. Reprinted by permission of John Wiley and Sons, Inc.)**

Illustration 11.6 indicates how the longitudinal dispersion model may be used to predict reactor performance.

ILLUSTRATION 11.6 USE OF THE DISPERSION MODEL TO DETERMINE THE CONVERSION LEVEL OBTAINED IN A NONIDEAL FLOW REACTOR

Use the dispersion parameter determined in Illustration 11.2 to predict the conversion that will be attained in the reactor of Illustration

11.1. Assume that the value of the first-order rate constant is 3.33×10^{-3} sec^{-1}.

Solution

Equation 11.2.11 is applicable in this case. Using the results obtained in Illustrations 11.1 and 11.2,

$$\beta = [1 + 4(3.33 \times 10^{-3})(0.113)(374.4)]^{1/2}$$
$$= 1.25$$

where we have used the value of $\mathscr{D}_L/uL$ determined by the variance approach.

From equation 11.2.11,

$$\frac{C_A}{C_{A0}} = \frac{4(1.25)e^{1/2(0.113)}}{(1 + 1.25)^2 e^{1.25/2(0.113)} - (1 - 1.25)^2 e^{-1.25/2(0.113)}} = 0.327$$

Thus

$$f_A = 1 - 0.327 = 0.673$$

This result compares to a value of 0.666 predicted on the basis of the segregated flow model. Excellent agreement should be obtained for the first-order case if the dispersion parameter gives a good fit of the experimental $F(t)$ curve. Agreement for reaction orders other than unity will not be nearly as good.

11.2.3 Determination of Conversion Levels Based on the Cascade of Stirred Tank Reactors Model

In Section 11.1.3.2 we considered a model of reactor performance in which the actual reactor is simulated by a cascade of equal-sized continuous stirred tank reactors operating in series. We indicated how the residence time distribution function can be used to determine the number of tanks that best model the tracer measurement data. Once this parameter has been determined, the techniques discussed in Section 8.3.2 can be used to determine the effluent conversion level.

If the case of an irreversible first-order reaction is considered, a material balance on the nth CSTR gives

$$C_{n-1}\mathcal{V}_{n-1} = C_n\mathcal{V}_n + kC_n\left(\frac{V_R}{n}\right) \quad (11.2.18)$$

Input   Output   Disappearance
                                  by reaction

where

V_R is the volume of the actual reactor

V_R/n is the volume of the CSTR used in the model

Equations of this form were treated in Section 8.3.2.2. For a constant density system,

$$C_n = \frac{C_{n-1}}{\left(1 + \dfrac{k\tau}{n}\right)} \quad (11.2.19)$$

where

$$\tau = \frac{V_R}{\mathcal{V}} \quad (11.2.20)$$

The final effluent concentration is related to the inlet concentration in the following manner.

$$C_n = \frac{C_0}{\left(1 + \dfrac{k\tau}{n}\right)^n} = C_0(1 - f_n) \quad (11.2.21)$$

where f_n is the fraction conversion leaving tank n.

The use of the reactor cascade model to estimate the conversion level attained in a first-order reaction is discussed in Illustration 11.7.

ILLUSTRATION 11.7 USE OF THE CASCADE OF STIRRED TANK REACTORS MODEL TO PREDICT REACTOR PERFORMANCE

Use the model based on a cascade of stirred tank reactors to predict the conversion that will be attained in the reactor of Illustration 11.1. Assume that the value of the first-order rate constant is $3.33 \times 10^{-3} \text{ sec}^{-1}$.

Solution

In Illustration 11.1 we found $\bar{t}$ or τ to be 374.4 sec, while in Illustration 11.3 we determined that the value of n must lie between 4 and 5. From equation 11.2.21, with $n = 4$,

$$f = 1 - \frac{1}{\left[1 + \dfrac{3.33 \times 10^{-3}(374.4)}{4}\right]^4}$$

$$= 1 - 0.337 = 0.662$$

For $n = 5$, $f = 0.671$, and for $n = 4.59$, $f = 0.668$.

All these values are close to those predicted by the segregated flow and dispersion models.

11.3 GENERAL COMMENTS AND RULES OF THUMB

In the previous section we indicated how various mathematical models may be used to simulate the performance of a reactor in which the flow patterns do not fit the ideal CSTR or PFR conditions. The models treated represent only a small fraction of the large number that have been proposed by various authors. However, they are among the simplest and most widely used models, and they permit one to bracket the expected performance of an isothermal reactor. However, *small variations in temperature can lead to much more significant changes in the reactor performance than do reasonably large deviations in flow patterns from idealized conditions.* Because the rate constant depends exponentially on temperature, uncertainties in this parameter can lead to design uncertainties that will make any quantitative analysis of performance in terms of the residence time distribution function little more than an academic exercise. Nonetheless, there are many situations where such analyses are useful.

Denbigh (24) has provided a set of generalizations that are useful in deciding when the conversions attained in tubular reactors will deviate significantly from those predicted on the basis of the plug flow model. For laminar flow, molecular diffusion in the longitudinal direction will not appreciably affect reactor performance if the reactor length is much greater than its diameter. Molecular diffusion in the radial direction tends to destroy any concentration gradients that have been established and thus serves to offset deviations resulting from the velocity profile. In other words, it works in favor of the idealized assumptions instead of against them. For turbulent flow, eddy diffusion is the dominant mode of dispersion, and it can significantly affect reactor performance. A crude generalization that is valid for simple reactions is that if the Reynolds number is greater than 10^4 and if the L/D ratio of the reactor is at least 50, deviations from plug flow because of longitudinal dispersion may be neglected. However, if these criteria are not met, the reactor size may have to be increased to a value appreciably greater than that required by plug flow conditions. The additional volume requirements may be determined using the models outlined previously.

Note that even though various flow models will often predict conversion levels that are within a few percent of one another, one must be extremely careful in the overall design calculations, particularly if the conversion level is high. If an ideal plug flow reactor gives a conversion of 98% and proper accounting for nonideal flow conditions by various models gives a range of conversion levels from 95 to 97%, the magnitude of subsequent separation problems will vary widely depending on the model chosen. For example, if product specifications call for a purity of 99% or better, the required separations would be quite different. In the ideal case one would have to reduce the impurity level from 2% to 1%; in the nonideal flow model it would have to be reduced from 5% to 1%. The costs of the equipment necessary to accomplish this task would vary greatly, depending on the input impurity level. Of course, it is only fair to point out that any other sources of error in the conversion estimate (e.g., temperature variations) would have the same effect.

The dispersion and stirred tank models of reactor behavior are in essence single parameter models. The literature contains an abundance of more complex multiparameter models. For an introduction to such models, consult the review article by Levenspiel and Bischoff (3) and the texts by these individuals (2, 4). The texts also contain discussions of the means by which residence time distribution curves may be used to diagnose the presence of flow maldistribution and stagnant region effects in operating equipment.

LITERATURE CITATIONS

1. Levenspiel, O., *Chemical Reaction Engineering*, First Edition, pp. 244–53, Wiley, New York, 1962.

2. Levenspiel, O., *Chemical Reaction Engineering*, Second Edition, pp. 253–314, Wiley, New York, 1972.

3. Levenspiel, O., and Bischoff, K. B., *Adv. Chem. Eng.*, *4* (95), 1963.

4. Himmelblau, D. M., and Bischoff, K. B., *Process Analysis and Simulation*, pp. 59–86, Wiley, New York, 1968.

5. Danckwerts, P. V., *Chem. Eng. Sci.*, *2* (1), 1953.

6. Kramers, H., and Alberda, G., *Chem. Eng. Sci.*, *2* (173), 1953.

7. Kramers, H., and Westerterp, K. R., *Elements of Chemical Reactor Design and Operation*, Academic Press, New York, 1963.

8. Levenspiel, O., and Smith, W. K., *Chem. Eng. Sci.*, *6* (227), 1957.

9. Levenspiel, O., and Bischoff, K. B., op. cit, p. 112.

10. Aris, R., *Chem. Eng. Sci.*, *9* (266), 1959.

11. Bischoff, K. B., *Chem. Eng. Sci.*, *12* (69), 1960.

12. Bischoff, K. B., and Levenspiel, O., *Chem. Eng. Sci.*, *17* (245), 1962.

13. Mixon, F. O., Whitaker, D. R., and Orcutt, J. C., *AIChE J.*, *13* (21), 1967.

14. Østergaard, K., and Michelsen, M. L., *Can. J. Chem. Eng.*, *47* (107), 1969.

15. Hopkins, M. J., Sheppard, A. J., and Eisenklam, P., *Chem. Eng. Sci.*, *24* (1131), 1969.

16. Buffham, B. A., and Gibilaro, L. G., *AIChE J.*, *14* (805), 1968.

17. Wehner, J. F., and Wilhelm, R. H., *Chem. Eng. Sci.*, *6* (89), 1959.

18. Levenspiel, O., and Bischoff, K. B., *Ind. Eng. Chem.*, *51* (1431), 1959.

19. Fan, L. T., and Bailie, R. C., *Chem. Eng. Sci.*, *13* (63), 1960.

20. Pasquon, I., and Dente, M., *J. Catalysis*, *1* (508), 1962.

21. Burghardt, A., and Zaleski, T., *Chem. Eng. Sci.*, *23* (575), 1968.

22. Tichacek, L. J., *AIChE J.*, *9* (394), 1963.

23. Douglas, J. M., and Bischoff, K. B., *Ind. Eng. Chem.*, *Process Design Develop.*, *3* (130), 1964.

24. Denbigh, K. G., *Chemical Reactor Theory*, pp. 50–51, Cambridge University Press, London, 1965.

PROBLEMS

1. You have been asked to carry out a residence time distribution study on a reactor network that has evolved over the years by adding whatever size and type of reactor was available at the moment. The feed stream presently contains 1 kg/m^3 of NaCl, which does not appear to affect the reactor performance or to be at all involved in the reaction under study. Your assistant recommends that you carry out a residence time distribution study by making a step change in the inlet NaCl concentration and then observing the effluent concentration of NaCl as a function of time. The data below were reported for this experiment. Use the data given to prepare an $F(t)$ curve and to determine the average residence time in the reactor network.

Time	Effluent concentration (kg/m^3)
8:00 p.m.	1.0
8:01	1.0
8:02	1.0
8:03	1.0
8:04	1.0
8:05	1.0
8:06	1.0
8:07	1.2
8:08	1.4
8:09	1.6
8:10	1.8
8.11	2.0
8:12	2.2
8:13	2.4
8:14	2.6
8:15	2.8
8:16	3.0
8:17	3.0
8:18	3.0
8:19	3.0
8:20	3.0
9:00	3.0

2. Tajbl, Simons, and Carberry [*Ind. Eng. Chem. Fundamentals*, *5* (171), 1966] have developed a stirred tank reactor for studies of catalytic reactions. Baskets containing catalyst pellets are mounted on a drive shaft that can be rotated at different speeds. The unit is designed for continuous flow operation. In order to determine if

the performance of the unit approximated that of an ideal continuous stirred tank reactor, these investigators carried out a series of tracer experiments at different agitator speeds and different volumetric flow rates. The response of a non-reacting system to a pulse input of helium injected into a steadily flowing air stream was used to characterize the reactor. The data are normalized by assuming that the concentration of the effluent at time zero is equal to the total mass of tracer injected divided by the effective volume of the reactor. On the basis of the two sets of data presented below, does the reactor appear to be a good approximation to an ideal CSTR? Be sure to use *all* the data in your analysis.

Run 1

Volumetric flow rate: 6.25 cm³/sec
Agitator speed: 1290 rpm

C_{effluent}/C_0 where C_0 is the normalizing concentration	t/τ
0.85	0.13
0.72	0.30
0.51	0.65
0.37	0.95
0.275	1.25
0.215	1.55
0.16	1.85

Run II

Volumetric flow rate: 10.6 cm³/sec
Agitator speed: 630 rpm

C_{effluent}/C_0	t/τ
0.80	0.20
0.65	0.40
0.49	0.70
0.34	1.10
0.26	1.25
0.19	1.63
0.14	2.05

3. The residence time distribution for a continuous stirred tank reactor may be represented in terms of the $F(t)$ curve as

$$F(t) = 1 - e^{-t/\tau} \quad \text{or} \quad 1 - e^{-t/\bar{t}}$$

where the mean residence time $\bar{t}$ and the space time τ are identical when there is no volume change on reaction. Show that the fraction conversion in a CSTR calculated for a first-order reaction on the basis of the design equations will be identical with that predicted by the segregated flow model and the above $F(t)$ curve. If $k = 3.0$ ksec⁻¹, the reactor volume is 1.2 m³, and the feed rate is 4.0 m³/ksec, what will this conversion be?

4. The following data have been reported as a result of an effort to determine the distribution of residence times in a packed bed reactor. Use these data to generate an $F(t)$ curve and to determine the average residence time in the reactor.

A pulse of tracer is fed to the reactor at time zero.

Time, t (sec)	Effluent tracer level (g/m³)
0	0.0
48	0.0
96	0.0
144	0.1
192	5.0
240	10.0
288	8.0
336	4.0
384	0.0
432	0.0

If one desires to utilize this reactor to carry out a first-order isomerization reaction of the type $A \rightarrow R$, and if the rate constant for the reaction is 7.5 ksec⁻¹, determine the average conversion that one expects in this reactor. Compare this value with those one would obtain in an ideal PFR and in an ideal CSTR having the same average residence time as our actual reactor.

5. The following results were reported from the operation of a flow reactor in which the reaction $A \xrightarrow{k_1} B$ was taking place. The reaction is first-order, irreversible with $k_1 = 0.0433 \text{ sec}^{-1}$. Pure A enters the reactor. The exit stream consists of 10% A and 90% B.

Time, t (sec)	$F(t)$
24	0.00
30	0.01
45	0.15
54	0.40
60	0.50
66	0.60
90	0.80
180	0.95
360	1.00

(a) Determine $\bar{t}$, the average residence time for this system.
(b) If the same conversion is to be obtained in an ideal plug flow reactor, what is the corresponding value of $\bar{t}$?
(c) What is the value of $\bar{t}$ for a CSTR for the same conversion?
(d) Comment on your results for parts (a), (b), and (c).

6. The $F(t)$ curve for a system consisting of a plug flow reactor followed by a continuous stirred tank reactor is identical to that of a system in which the CSTR precedes the PFR. Show that the overall fraction conversions obtained in these two combinations are different when the reactions are other than first-order. Derive appropriate expressions for the case of second-order irreversible reactions and indicate how the reactors should be ordered so as to maximize the conversion achieved.

7. Derive the $F(t)$ curve for a CSTR by considering its response to a step change in the input tracer concentration. Let w_0^- and w_0^+ represent the weight fraction tracer in the feed before and after the change. It may be assumed that the total mass flow rate (Φ_m) remains substantially constant.

8. Levenspiel and Smith [Chem. Eng. Sci., 6 (227), 1957] have reported the data below for a residence time experiment involving a length of 2.85 cm diameter pyrex tubing. A volume of $KMnO_4$ solution that would fill 2.54 cm of the tube was rapidly injected into a water stream with a linear velocity of 35.7 cm/sec. A photoelectric cell 2.74 m downstream from the injection point is used to monitor the local $KMnO_4$ concentration. Use slope, variance, and maximum concentration approaches to determine the dispersion parameter. What is the mean residence time of the fluid?

Time, t (sec)	$KMnO_4$ concentration (arbitrary units)
0	0
2	11
4	53
6	64
8	58
10	48
12	39
14	29
16	22
18	16
20	11
22	9
24	7
26	5
28	4
30	2
32	2
34	2
36	1
38	1
40	1
42	1

9. Bucky Badger has been investigating the residence time characteristics of a reactor that he plans to use for pilot plant work. He suggests

that the $F(t)$ curve can be approximated by the following relation.

$$F(t) = 0 \qquad\qquad 0 \leq t \leq 0.4$$
$$F(t) = 1 - e^{-1.25(t-0.4)} \qquad t \geq 0.4$$

where all times are measured in kiloseconds.

(a) Calculate the average residence time in the reactor.

(b) If Bucky uses this reactor to carry out a first-order reaction of the type $A \rightarrow P$, calculate the fraction conversion expected for isothermal operation with a reaction rate constant equal to 0.8 ksec^{-1}. Use the segregated flow model.

(c) Calculate the overall conversion expected for this reaction in a reactor network consisting of a PFR with a space time of 0.4 ksec and a CSTR with a space time of 0.8 ksec. Consider the case where the PFR is first as well as the case where it is second.

10. Use the $F(t)$ curve for two identical CSTR's in series and the segregated flow model to predict the conversion achieved for a first-order reaction with $k = 0.4$ ksec^{-1}. The space time for an individual reactor is 0.9 ksec. Check your results using an analysis for two CSTR's in series.

11. In an effort to determine the cause of low yields from a reactor network, your technicians have carried out some tracer studies in which 4.000 kg of an inert material are quickly injected at the feed port. The tracer levels leaving the reactor at various times after injection were as follows.

Time, t (sec)	Tracer concentration (kg/m^3)
12	1.960
24	1.930
120	1.642
240	1.344
600	0.736
1200	0.268
2400	0.034
3600	0.004

At any time the reactor contains 2 m^3 of fluid. The feed and effluent rates remain constant at 3.3 m^3/ksec. Does the response of the system approximate that of any simple ideal reactor? What conversion level is expected if the reaction has a first-order rate constant of 15 sec^{-1}?

12. Hopkins and co-workers [*Chem. Eng. Sci.*, *24* (1131), 1969] have indicated that the data in

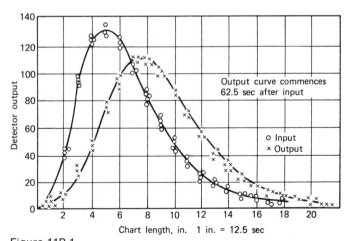

Figure 11P.1

Tracer response curve. Reprinted from M. J. Hopkins, A. J. Sheppard and P. Eisenklam, *Chem. Eng. Sci.*, *24* (1131), 1969. Used with permission of Pergamon Press, Ltd.

Figure 11P.1 can be used to determine the dimensionless dispersion parmeter ($\mathscr{D}_L/uL$) for a system of interest. Use the transfer function method to evaluate the mean residence time and ($\mathscr{D}_L/uL$) for a system subjected to the arbitrary input shown in the figure. Note that the output response has been shifted 62.5 sec to the left. Response values for the input and output streams were as follows.

Input Measurement		Output Measurement	
Time (sec)	Response (arbitrary units)	Time (sec)	Response (arbitrary units)
0	0	0	0
12.5	10.7	50	0
25.0	40.0	62.5	0
37.5	88.0	75.	4.0
50.0	126.0	87.5	10.2
61.0	133.3	100.0	22.5
75.0	126.0	112.5	50.0
85.6	104.3	123.5	75.5
100.0	80.0	137.5	99.8
111.0	63.6	148.1	110.0
125.0	46.3	162.5	110.0
134.7	36.0	171.1	103.8
150.0	25.0	187.5	85.0
160.3	18.7	197.2	70.9
175.0	13.3	212.5	58.0
186.1	10.7	222.8	45.0
200	6.7	237.5	32.5
212.5	4.7	248.6	24.0
225	2.7	262.5	16.0
236.4	0.8	275	12.0
250	0.0	287.5	8.7
		298.9	6.7
		312.5	5.2
		325.0	3.8

13. Calculate the volumes of a plug flow reactor and a laminar flow reactor required to process 0.5 m³/ksec of feed containing 1.0 kmole/m³ of species A to 95% conversion. The liquid phase reaction involved is

$$2A \xrightarrow{k_1} B + C$$

with $k_1 = 5.0$ m³/kmole·ksec.

14.(a) The following data have been reported as a result of an effort to determine the distribution of residence times in a packed bed reactor. Use these data to generate an $F(t)$ curve and to determine the average residence time in the reactor. A pulse of tracer is fed to the reactor at time zero.

Time, t (min)	Effluent tracer concentration
0	0.0
4	3.0
8	5.0
12	5.0
16	4.0
20	2.0
24	1.0
28	0.0
32	0.0

(b) If one desires to utilize this reactor to carry out a first-order isomerization reaction of the type

$$A \rightarrow R,$$

and if the rate constant for the reaction is 0.045 min⁻¹, determine the average conversion that one expects in this reactor.

(c) For the residence time distribution of part (a), what value of $\mathscr{D}_L/uL$ gives an adequate fit of the dispersion model to the data?

(d) What integer value of n gives the most accurate fit of the data in part (a) when we have n equal volume CSTR's in series? What conversion is expected on the basis of this model?

(e) It has been proposed that we model the residence time distribution by an ideal CSTR followed by a PFR. The volumetric flow rate and combined volumes of the two reactors will be the same as in part (a) What ratio of reactor volumes gives a good representation of the $F(t)$ curve? What conversion is expected on the basis of this model? Comment on the suitability of the model.

(f) Prepare a table summarizing the fraction conversions predicted by the various models.

In parts (c), (d), and (e) draw the $F(t)$ curves predicted on the basis of the various models so that they may be compared with the experimental data.

15. When catalytic reactions are carried out in a fluidized bed reactor, one may observe conversions lower than those predicted by either PFR or CSTR models. Such observations are often attributed to gas bypassing by bubbles that do not efficiently contact the catalyst. Iwasaki et al. [*Kagaku Kogaku, 4* (170), 1966] have described a method for measuring the distribution of times that reactive gases are in contact with the solid. The technique is based on pulse injections of two gases, one of which is adsorbed by the solid and the second of which is not adsorbed. Typical data are given in the table below.

Time, t (sec)	Concentration (arbitrary units)	
	Inert gas	Adsorbed gas
0.0	0.0	0.0
0.8	0.0	0.0
1.6	0.0	0.0
2.4	0.01	0.01
3.2	5.0	5.0
4.0	10.0	7.5
4.8	8.0	8.5
5.6	4.0	8.0
6.4	0.0	6.5
7.2	0.0	4.0
8.0	0.0	2.0
8.8	0.0	1.0
9.6	0.0	0.0

(a) Assuming ideal pulse injection determine
(1) The $F(t)$ curve for overall residence times for both cases.
(2) The $F(t)$ curve for the contact time distribution.
(3) The effective dispersion parameter for the inert gas case ($\mathcal{D}_L/uL$).

(b) For a first-order irreversible isomerization reaction ($A \rightarrow R$) with $k = 0.45 \text{ sec}^{-1}$, determine the conversions predicted using the following reactor models.
(1) Plug flow with $\tau = \bar{t}$ for reactive gas experiment and with $\tau =$ average contact time.
(2) CSTR with $\tau = \bar{t}$ for reactive gas experiment and with $\tau =$ average contact time.
(3) Segregated flow with reactive gas residence time curve.
(4) Segregated flow with reactive gas contact time curve.

It is assumed that the reactive gas A has adsorption characteristics similar to those of the gas used in the pulse experiment.

16. The techniques that we have developed in this chapter for the analysis of residence time distribution functions can be used in analyzing flow conditions in a stream or river where one wishes to determine the dispersion of pollutants from a given source. The following data are taken from a USGS survey of the South Platte River [R. E. Glover, "Dispersion of Dissolved or Suspended Materials in Flowing Streams," Geological Survey Professional Paper 433-B (1964)].

Average flow rate: 554 ft^3/sec
Length of reach: 19,900 ft
Naturally occurring concentration of K^+ in the stream: 8.2 mg/liter

At time zero, 1000 lb of K_2CO_3 are dumped into the upstream end of the reach. Periodically samples were taken at the downstream end of the reach and analyzed for K^+ ions with the results tabulated below:

Time, t (min)	K^+ at downstream end (g/m^3)
0	8.2
60	8.2
75	8.2
90	8.2
105	8.4
120	9.6
130	13.6
132.5	14.8
134	14.8
138	14.6
142.5	13.2
150	12.8
165	10.0
180	9.2
195	8.2
210	8.2
∞	8.2

(a) Calculate the fraction of the tracer that was recovered.
(b) Basing your calculations on the amount of tracer recovered, compute the $F(t)$ curve.
(c) Compute the average residence time $\bar{t}$ and prepare a plot of $F(t)$ versus $t/\bar{t}$.

(d) An organic species A is a potential pollutant that is present in the waste stream from a factory that empties into the upper end of the reach. The mixing cup average concentration of A at the upper end of the reach is 100 mg/liter. It is converted to a product that is harmless by a first-order reaction in solution. What fraction of the initial concentration of species A is present at the downstream end of the reach? The effective rate constant at the stream temperature is equal to 0.01 min^{-1}. Use the segregated flow model for this calculation.
(e) Based on the $F(t)$ curve, what is the apparent axial diffusion coefficient of the tracer if one assumes that the stream approximates a nonideal plug flow reactor with axial diffusion? The effective length of the reach is 19,900 ft.
(f) Based on the model of part (e), and using the axial diffusion coefficient calculated therein, will the average concentration of A present at the downstream end of the reach be greater than, less than, or equal to that predicted in part (d)?

12 Reactor Design for Heterogeneous Catalytic Reactions

12.0 INTRODUCTION

Heterogeneous catalytic reactors are the most important single class of reactors utilized by the chemical industry. Whether their importance is measured by the wholesale value of the goods produced, the processing capacity, or the overall investment in the reactors and associated peripheral equipment, there is no doubt as to the prime economic role that reactors of this type play in modern technological society. This chapter examines a number of subjects that must be considered in the design of heterogeneous catalytic reactors. Particular emphasis is placed on the concept of catalyst effectiveness factors and the implications of heat and mass transfer processes for fixed bed reactor design.

12.1 COMMERCIALLY SIGNIFICANT TYPES OF HETEROGENEOUS CATALYTIC REACTORS

The types of reactors used in industry for carrying out heterogeneous catalytic reactions may be classified in terms of a relatively small number of categories. One simple means of classification is in terms of the relative motion of the catalyst particles, or lack thereof. We consider:

1. Reactors in which the solid catalyst particles remain in a fixed position relative to one another (fixed bed, trickle bed, and moving bed reactors).
2. Reactors in which the particles are suspended in a fluid and are constantly moving about (fluidized bed and slurry reactors).

In the sections that follow, each of these reactor classes is discussed in more detail, with particular emphasis on fixed and fluidized bed reactors.

12.1.1 Heterogeneous Catalytic Reactors in which the Motion of the Catalyst Particles Relative to One Another Is Insignificant

12.1.1.1 Fixed Bed Reactors. In its most basic form, a fixed bed reactor consists of a cylindrical tube filled with catalyst pellets. Reactants flow through the catalyst bed and are converted into products. Fixed bed reactors are often referred to as packed bed reactors. They may be regarded as the workhorse of the chemical industry with respect to the number of reactors employed and the economic value of the materials produced. Ammonia synthesis, sulfuric acid production (by oxidation of SO_2 to SO_3), and nitric acid production (by ammonia oxidation) are only a few of the extremely high tonnage processes that make extensive use of various forms of packed bed reactors.

The catalyst constituting the fixed bed will generally be employed in one of the following configurations:

1. A single large bed.
2. Multiple horizontal beds supported on trays arranged in a vertical stack.
3. Multiple parallel packed tubes in a single shell.
4. Multiple beds each in their own shell.

The use of multiple catalyst sections usually arises because of the need to maintain adequate temperature control within the system. Other constraints leading to the use of multiple beds include those of pressure drop or adequate fluid distribution. In addition to the shell and tube configuration, some of the possibilities for heat transfer to or from fixed bed reactors include the use of internal heat exchangers, annular cooling spaces or cooling thimbles, and circulation of a portion of the reacting gases through an external heat exchanger.

The packing itself may consist of spherical, cylindrical, or randomly shaped pellets, wire screens or gauzes, crushed particles, or a variety of other physical configurations. The particles usually are 0.25 to 1.0 cm in diameter. The structure of the catalyst pellets is such that the internal surface area far exceeds the superficial (external) surface area, so that the contact area is, in principle, independent of pellet size. To make effective use of the internal surface area, one must use a pellet size that minimizes diffusional resistance within the catalyst pellet but that also gives rise to an appropriate pressure drop across the catalyst bed. Some considerations which are important in the handling and use of catalysts for fixed bed operation in industrial situations are discussed in the *Catalyst Handbook* (1).

The most commonly used direction of reactant flow is downward through the bed. This approach gives a stable bed that will not fluidize, dance, or lift out of the reactor. This approach minimizes catalyst attrition and potential entrainment of catalyst fines. When processing conditions are such that the reactor is subjected to wide variations in feed flow rates or when the feed is a dense fluid, it is imperative that the flow direction be downward. The attendant disadvantages of downward flow are the tendency of the bed to compress itself and the gravitation of catalyst fines (resulting from attrition) down through the bed. Both phenomena may lead to increased pressure drop and channeling or maldistribution of the flow.

Upflow has the advantage of lifting catalyst fines or fragmented particles from the bed thereby avoiding channeling and blockage of the bed. However, this mode of operation is disadvantageous because it may lead to unstable beds at high flow rates. It leads to dancing in a pulsating flow that causes catalyst abrasion and, in unusual circumstances, may lead to fluidization.

A fixed bed reactor has many unique and valuable advantages relative to other reactor types. One of its prime attributes is its simplicity, with the attendant consequences of low costs for construction, operation, and maintenance relative to moving bed or fluidized bed operation. It requires a minimum of auxiliary equipment, and is particularly appropriate for use in small commercial units when investments of large sums for control, catalyst handling, and supporting facilities would be economically prohibitive. Another major advantage of this mode of operation is implicit in the use of the term "fixed bed reactor"; i.e., there are no problems in separating the catalyst from the reactor effluent stream. (In many fluidized bed systems, catalyst recovery can be quite troublesome and require substantial equipment costs.) Another important attribute of fixed bed reactors is the wide variation in space times at which they can be operated. This flexibility is extremely important in situations where one is likely to encounter wide variations in the quantity or quality of the feedstock to be processed. For extremely high temperature or high pressure reactions employing solid catalysts, economic considerations usually dictate that the process becomes commercially viable only when a fixed bed reactor is employed.

There are some cases where the disadvantages of fixed bed operation prevent the use of such reactors. Heat transfer to or from a large fixed bed of catalyst often represents a significant problem. In some cases it is possible to circumvent this potential difficulty by appropriate variations in the physical configuration of the catalyst bed (e.g., using a shell and tube arrangement). There are also a variety of operating techniques that can be used to facilitate control over the bed temperature. These include the use of inert diluents in the feed stream to moderate the temperature changes, and the use of the "cold shot" or gas bypass technique in which a fresh cold reactant stream is mixed with a hot stream that has undergone partial reaction. Recycle of a product stream and partial temporary poisoning of the catalyst are two other expedients that can be used to help regulate the temperature in fixed bed reactors.

The problem of heat transfer is a difficult one because the rate of energy release or consumption along the length of the reactor is not uniform. The major share of the reaction takes place near the reactor inlet. In an exothermic reaction, the rate will be relatively large near the inlet because of high reactant concentrations. If the rate of energy release is greater than the rate at which heat can be transferred to a coolant fluid, the reacting fluid will continue to heat up, leading to an even faster reaction rate. This phenomenon continues as the mixture moves down the tube, until the decline in reactant concentration has a larger effect on the rate than the temperature increase. The net result is that for exothermic reactions one often observes a maximum in a plot of temperature versus reactor length.

The maintenance of uniform flow distribution in fixed bed reactors is frequently a problem. Maldistribution leads to an excessive spread in the distribution of residence times with adverse effects on the reactor performance, particularly when consecutive reactions are involved. It may aggravate problems of hot-spot formation and lead to regions of the reactor where undesired reactions predominate. Disintegration or attrition of the catalyst may lead to or may aggravate flow distribution problems.

Another disadvantage of fixed bed reactors is associated with the fact that the minimum pellet size that can be used is restricted by the permissible pressure drop through the bed. Thus if the reaction is potentially subject to diffusional limitations within the catalyst pore structure, it may not be possible to fully utilize all the catalyst area (see Section 12.3). The smaller the pellet, the more efficiently the internal area is used, but the greater the pressure drop.

One of the major disadvantages of fixed bed operation is that catalyst regeneration or replacement is relatively difficult to accomplish. If the catalyst deactivation rate is sufficiently rapid, costs associated with the catalyst regeneration or replacement step may render the entire process unattractive from a commercial standpoint. The effective catalyst life necessary to render the process economic for fixed bed operation depends on the details of the process under study, but if the lifetime is not at least several months, the costs of the shutdowns will normally be exorbitant. In situ regeneration offers a possible way around this difficulty, but if continuous operation is to be maintained, the use of in situ regeneration requires two or more reactors in parallel with concomitant higher capital costs. The most successful fixed bed operations are those where the catalyst activity does not decline markedly over long time periods. A technique that occasionally can be used to prolong the time between regenerations and shutdowns is the use of catalyst beds that are longer than are required to accomplish the desired degree of conversion. If the required bed depth is relatively short, doubling or tripling the depth of the catalyst packing will greatly prolong the time that the unit can remain on stream. When the unit is first brought on stream, the desired conversion will be accomplished in the first portion of the bed. As the catalyst activity declines, the portion of the bed in which the bulk of the reaction is accomplished will move down the bed until, finally, insufficient catalyst activity remains to accomplish the required degree of conversion. Unfortunately, this approach can be used only with certain reactions, the most common example being the ammonia synthesis reaction.

12.1.1.2 Trickle Bed Reactors (2). A trickle bed reactor utilizes a fixed bed over which liquid flows without filling the void spaces between particles. The liquid usually flows downward under the influence of gravity, while the gas flows upward or downward through the void spaces amid the catalyst pellets and the liquid holdup. Generally cocurrent downward flow of liquid and gas is preferred because it facilitates a uniform distribution of the liquid across the catalyst bed and permits the employment of higher liquid flow rates before encountering flooding constraints.

The primary uses of trickle bed reactors are for hydrodesulfurization, hydrocracking, and hydrotreating of various high-boiling petroleum fractions. The direct and capital costs are significantly less for trickle bed operation than for an equivalent hydrodesulfurization unit operating entirely in the vapor phase. The use of this reactor type makes it possible to process feedstocks with such high boiling points that straight vapor phase operation would lead to excessive undesirable side reactions.

In spite of the fact that trickle bed reactors often approach plug flow behavior, the use of a liquid phase in the feed introduces several complications in the design analysis. The nature and extent of the liquid distribution within the catalyst bed vary drastically with changes in the liquid and vapor flow rates, the properties of the reaction mixture (especially its viscosity and wetting characteristics), and the design of the reactor (especially the liquid distribution system). As the liquid distribution changes, there are concomitant changes in the contacting efficiencies between the liquid, vapor, and catalyst. Usually, gaseous reactants must first be absorbed and transported across a thin liquid film to the exterior of the catalyst pellet. They must then be subsequently transported through liquid filled pores to sites in the catalyst interior. By using a low ratio of liquid to catalyst in the reactor, one is able to minimize the extent of the homogeneous reaction. However, a balance must be struck for highly exothermic reactions, since the energy released by reaction may be sufficient to volatilize the liquid in substantial portions of the bed. In such cases, portions of the bed will not be wetted, and there will be poor contacting between liquid and catalyst.

One often finds that either external or intraparticle mass transfer effects are significant in reactors of this type. Although the treatments of these topics outlined in Sections 12.3 and 12.4 are in general applicable to trickle bed reactors, analyses specific to such reactors have been reviewed by Satterfield (2).

12.1.1.3 Moving Bed Reactors.

In moving bed reactors, a fluid phase passes upward through a packed bed of catalyst pellets. Solid is fed to the top of the bed, moves downward under the influence of gravity in a manner approximating plug flow, and is removed from the bottom. The catalyst pellets are then transferred to the top of the reactor in external equipment by pneumatic or mechanical means in continuous fashion. During the period when the catalyst pellets are not in the reactor proper, they may be regenerated or reconditioned in an auxiliary facility.

It is necessary to design special control valves to provide proper solids flow and to maintain close control over the solids level within the reactor. Moreover, care must be taken in the design of these reactors to prevent bypassing of the bed by the fluid reactant stream, and to ensure good distribution of the solids at all levels. These potential difficulties indicate why these types of reactors are less frequently used than fixed or fluidized bed reactors. When catalyst decay is slow, so that fixed bed operation is satisfactory with infrequent shutdown for catalyst regeneration, moving bed operation does not offer a viable alternative. In catalytic cracking where catalyst deactivation is so rapid that fixed beds must be regenerated after only a few minutes of time on stream, the disadvantages of these reactors are obvious. Either moving bed operation or fluidized bed operation is imperative in such cases. The primary disadvantage of moving bed operation vis à vis fluidized bed operation is in its reduced capability to facilitate heat transfer in either the reactor proper or the regeneration unit. This reduced capability makes it difficult to achieve the control of the catalyst temperature necessary to prevent high temperature deactivation processes. In instances where catalyst deactivation is not rapid but is still appreciable and can be circumvented by intermittent regeneration, moving bed operation may offer advantages over using several fixed bed reactors in parallel. The economics are dependent on the deactivation rate and other process details, and no general rule of thumb is appropriate.

The catalyst in moving bed operations usually has an important role as a heat carrying medium. The energy released by carbon burnoff in exothermic regeneration reactions can be used to supply some of the energy requirements of endothermic cracking reactions with attendant overall energy savings.

12.1.2 Heterogeneous Catalytic Reactors in which there is Significant Motion of the Catalyst Particles Relative to One Another

12.1.2.1 Fluidized Bed Reactors. A fluidized bed reactor is one in which relatively small particles of catalyst are suspended by the upward motion of the reacting fluid. In virtually all industrial applications the fluid is a gas which flows upward through the solid particles at a rate which is sufficient to lift them from a supporting grid, but which is not so large as to carry them out of the reactor or to prevent them from falling back into the fluidized phase above its free surface. The particles are in constant motion within a relatively confined region of space, and extensive mixing occurs in both the radial and longitudinal directions of the bed.

Fluidized bed reactors were first employed on a large scale for the catalytic cracking of petroleum fractions, but in recent years they have been employed for an increasingly large variety of reactions, both catalytic and non-catalytic. The catalytic reactions include the partial oxidation of naphthalene to phthalic anhydride and the formation of acrylonitrile from propylene, ammonia, and air. The noncatalytic applications include the roasting of ores and the fluorination of uranium oxide.

Typically the catalyst particles used in fluidized bed operations have dimensions in the range of 10 to 300 microns. For optimum fluidization, it is important to employ the proper particle size distribution. Beds of large uniformly sized solids often fluidize poorly with bumping, slugging, and spouting, which can cause serious structural damage in large beds. In such cases the quality of the fluidization can frequently be drastically improved by adding small amounts of fines. Fine particles with a wide size distribution will remain fluidized over a wide range of gas flow rates, permitting flexible operation with relatively large beds. Commercial scale fluidized bed reactors may be extremely large pieces of equipment. Diameters of 10 to 30 ft are not unusual for catalytic cracking units. Typically, height to diameter ratios of 2:1 or greater are employed. Large systems are required in part to accommodate heat transfer equipment, cyclone separators, and other internal equipment.

Several advantages are associated with the use of fluidized bed reactors. A remarkably uniform temperature can be maintained throughout the catalyst bed. This property is a consequence of the high degree of turbulence within the bed, the high heat capacity of the solid catalyst comprising the bed relative to the gas contained therein, and the extremely high interfacial area for heat transfer between the solid and the gas phase. These factors facilitate control over the temperature of the reactor and its contents. This in turn enhances the selectivity that can be achieved and permits very large-scale operations.

The primary advantage of fluidized bed reactors, however, is that they permit continuous, automatically controlled operations using reactant-catalyst systems that require catalyst regeneration at very frequent intervals. Fluidized bed operation permits one to easily add or remove the catalyst from the reactor or the regenerator. Regeneration can be accomplished by any convenient procedure, but the use of fluidized bed regeneration permits continuous operation and is usually most economical. Furthermore, the circulation of solids between two fluidized beds makes it possible to transport large quantities of energy between the reactor and the regenerator. This feature is particularly useful in catalytic cracking reactions where the exothermic regeneration reaction can be used thereby to supply some of the energy requirements for the endothermic cracking reaction.

Still another advantage of fluidized bed operation is that it leads to more efficient contacting of gas and solid than many competitive reactor designs. Because the catalyst particles employed in fluidized beds have very small dimensions, one is much less likely to encounter mass transfer limitations on reaction rates in these systems than in fixed bed systems.

Disadvantages are also associated with fluidized bed reactors. They cannot be used with catalyst solids that will not flow freely or that have a tendency to agglomerate. Attrition of the solids also causes some loss of material as fines, which are blown out of the reactor. Extensive solids collection systems including cyclone separators and electrostatic precipitation must often be provided to minimize catalyst losses and contamination of the environment. Another disadvantage of fluidized bed operation is that it leads to a larger pressure drop than fixed bed operation with concomitant higher operating costs. Erosion of pipes and reactor internals via abrasion by the particles can occur. In general, operating and maintenance costs will be relatively high for this mode of operation compared with similar scale operations with other reactor types. Fluidized bed operations also have the disadvantage that the fluid flow deviates markedly from plug flow, and the bypassing of solids by bubbles can lead to inefficient contacting. This problem is particularly significant when dealing with systems in which high conversions are desired. It can be circumvented to some extent by using multiple beds in series or internal staging.

In spite of the drawbacks enumerated above, fluidized bed reactors have a number of compelling advantages, as we have noted previously. By proper design it is possible to overcome their deficiencies so that their advantages predominate. This book does not discuss in detail the manner in which this problem can be solved, although the design considerations outlined in subsequent sections of this chapter are quite pertinent. For detailed treatments of fluidized bed reactor design, consult the excellent reference works by Kunii and Levenspiel (3) and by Davidson and Harrison (4).

12.1.2.2 Slurry Reactors. Slurry reactors are commonly used in situations where it is necessary to contact a liquid reactant or a solution containing the reactant with a solid catalyst. To facilitate mass transfer and effective catalyst utilization, the catalyst is usually suspended in powdered or in granular form. This type of reactor has been used where one of the reactants is normally a gas at the reaction conditions and the second reactant is a liquid, e.g., in the hydrogenation of various oils. The reactant gas is bubbled through the liquid, dissolves, and then diffuses to the catalyst surface. Obviously mass transfer limitations can be quite significant in those instances where three phases (the solid catalyst, and the liquid and gaseous reactants) are present and necessary to proceed rapidly from reactants to products.

Satterfield (5) has discussed several advantages of slurry reactors relative to other modes of operation. They include the following.

1. A well-agitated slurry may be kept at a uniform temperature throughout, eliminating "hot" spots that have adverse effects on catalyst selectivity.
2. The high heat capacity associated with the large mass of liquid facilitates control of the reactor and provides a safety factor for exothermic reactions that might lead to thermal explosions or other "runaway" events.

3. Since liquid phase heat transfer coefficients are large, heat recovery is practical with these systems.

4. The small particles used in slurry reactors may make it possible to obtain much higher rates of reaction per unit weight of catalyst than would be achieved with the larger pellets that would be required in trickle bed reactors. This situation occurs when the trickle bed pellets are characterized by low effectiveness factors (see Section 12.3).

5. Continuous regeneration of the catalyst can be obtained by continuously removing a fraction of the slurry from which the catalyst is then separated, regenerated and returned to the reactor.

6. Since fine catalyst particles are desired, the costs associated with the pelleting process are avoided, and it becomes possible to use catalysts that are difficult or impossible to pelletize.

A major deterrent to the adoption of continuous slurry reactors is the fact that published data are often inadequate for design purposes. Solubilization and mass transfer processes may influence observed conversion rates and these factors may introduce design uncertainties. One also has the problems of developing mechanical designs that will not plug up, and of selecting carrier liquids in which the reactants are soluble yet which remain stable at elevated temperatures in contact with reactants, products, and the catalyst. A further disadvantage of the slurry reactor is that the ratio of liquid to catalyst is much greater than in a trickle bed reactor. Hence, the relative rates of undesirable homogeneous liquid phase reactions will be greater in the slurry reactor, with a potential adverse effect on the process selectivity.

Slurry reactors may take on several physical forms: they may be simple stirred autoclaves; they may be simple vessels fitted with an external pump to recirculate the liquid and suspended solids through an external heat exchanger, or they may resemble a bubble-tray rectifying column with various stages placed above one another in a single shell. Since a single slurry reactor has a residence time distribution approximating a CSTR, the last mode of construction gives an easy means of obtaining stagewise behavior and more efficient utilization of the reactor volume.

For more detailed treatments of slurry reactors, see the texts of Satterfield (5) and Smith (6).

12.2 MASS TRANSPORT PROCESSES WITHIN POROUS CATALYSTS

Heterogeneous catalysis involves an interaction between a solid substrate and reactant molecules in a fluid phase. This interaction occurs at the interface between the two phases, and efficient utilization of the catalyst requires that the surface area per unit weight of catalyst be as large as possible. For commercial applications, this requirement implies that the catalyst should be porous with a large internal surface area. Typical industrial catalysts have specific surface areas ranging from 1 to 1000 m^2/g. With the exception of the use of wire gauzes for high temperature oxidation processes (e.g., NH_3 or SO_2 oxidation), nonporous catalysts have not been used extensively in commercial applications. For porous catalysts the superficial external surface of the catalyst will normally represent only an insignificant fraction of the total surface area. A spherical catalyst pellet with a diameter of 0.5 cm weighing 0.1 g would have a geometric surface area of only 7.8×10^{-5} m^2. If it had a modest specific surface area of 10 m^2/g, the internal surface area would be 1 m^2. Hence, even a thin outer shell of the catalyst, which is only 0.05 cm deep, would have a surface area associated with it that would be orders of magnitude greater than the external surface area. Consequently, to obtain high specific activities, it is necessary to use highly porous catalysts in industrial situations.

It requires only an extremely limited knowledge of fluid flow processes to recognize that the pressure drop through an ordinary catalyst bed is not sufficient to force any perceptible amount of fluid through the very small pores of catalyst particles that are required to obtain surface areas of 1 to 1000 m^2/g. If reactants are to come in contact with the interior surface of catalyst particles, they must do so by diffusion. If one has a fast reaction and relatively long catalyst pores, reactant molecules will be converted to products before they have time to diffuse very far into the pore structure. In this case, the interior of the particle will be filled primarily with product molecules, and only the outer peripheral layer of the catalyst pellet will be fully effective in promoting reaction. Section 12.3 discusses the problem of determining what fraction of the total catalyst surface area is used effectively in promoting chemical reaction and the implications of this situation for chemical reactor design. In the most general case, the problem involves a complex interaction between chemical reaction, diffusive mass transfer, and heat transfer processes.

One must understand the physical mechanisms by which mass transfer takes place in catalyst pores to comprehend the development of mathematical models that can be used in engineering design calculations to estimate what fraction of the catalyst surface is effective in promoting reaction. There are several factors that complicate efforts to analyze mass transfer within such systems. They include the facts that (1) the pore geometry is extremely complex, and not subject to realistic modeling in terms of a small number of parameters, and that (2) different molecular phenomena are responsible for the mass transfer. Consequently, it is often useful to characterize the mass transfer process in terms of an "*effective diffusivity*," i.e., a transport coefficient that pertains to a porous material in which the calculations are based on *total area* (void plus solid) *normal to the direction of transport*. For example, in a spherical catalyst pellet, the appropriate area to use in characterizing diffusion in the radial direction is $4\pi r^2$.

Unfortunately, it is not possible to obtain effective diffusivities merely by correcting bulk phase diffusivities for the reduction in cross-sectional area due to the solid phase. Two primary factors render this simple approach invalid:

1. The geometry of the pore structure makes it impossible to determine accurately the effective length of the diffusion path. Interconnections within the pore structure, the tortuous character of individual pores, and variations in cross-sectional area along the pore length all contribute to the difficulty of the task.

2. One or more of several different mechanisms may be responsible for the mass transfer process. These include ordinary bulk diffusion, Knudsen diffusion, surface diffusion, and bulk flow. For the majority of the catalysts and conditions used in industrial practice, the only significant mechanisms are bulk diffusion and Knudsen diffusion. The relative importance of these two processes depends on the relative values of the mean free path and the pore dimensions.

Ordinary or *bulk diffusion* is primarily responsible for molecular transport when the mean free path of a molecule is small compared with the diameter of the pore. At 1 atm the mean free path of typical gaseous species is of the order of 10^{-5} cm or 10^3 Å. In pores larger than 10^{-4} cm the mean free path is much smaller than the pore dimension, and collisions with other gas phase molecules will occur much more often than collisions with the pore walls. Under these circumstances the effective diffusivity will be independent of the pore diameter and, within a given catalyst pore, ordinary bulk diffusion coefficients may be used in Fick's first law to evaluate the rate of mass transfer and the concentration profile in the pore. In industrial practice there are three general classes of reaction conditions for which the bulk value of the diffusion coefficient is appropriate. For all catalysts these include liquid phase reactions and very high pressure reactions where the fluid density approaches the critical density of the material. The third class is made up of low-pressure, gas phase reactions that take place on catalysts with very large pores, say greater than 5000 Å in diameter. When bulk diffusion is operative, a number of correlations may be used to estimate the ordinary molecular diffusivity in the absence of experimental data. Reid and Sherwood (7) have summarized them in convenient form.

Knudsen diffusion will be the dominant mechanism of mass transfer whenever the mean free path between collisions is large compared with the pore diameter. This situation prevails when the gas density is low or when the pore dimensions are very small. It is not observed with liquids. The molecules hitting the walls are momentarily adsorbed and then are given off in random directions (diffusely reflected). After a collision with the pore wall, the molecule will usually fly to another spot on the wall before having a collision with a second gas phase molecule. Many collisions with the walls will take place for each collision between gas phase molecules. The molecule moves down the catalyst pore by a series of random flights, interrupted by collisions with the pore walls. The gas flux is reduced by the wall "resistance." This causes a delay because of the finite time the molecules are adsorbed and because, after a collision, the molecule is just as apt to reverse its direction as it is to continue along its original path. If the equation for the gas flow in a *straight circular* pore for the case of Knudsen flow is analyzed according to the principles of the kinetic theory of gases, it can be shown that the molar flow rate (F) can be written as

$$F = (\pi \bar{r}^2) \frac{2}{3} \bar{r} \sqrt{\frac{8RT}{\pi M}} \frac{\Delta C}{\Delta X} \qquad (12.2.1)$$

where

$\bar{r}$ is the pore radius

R is the gas constant

T is the absolute temperature

M is the molecular weight

ΔC is the concentration difference over the length of the pore

ΔX is the pore length

Since $\pi \bar{r}^2$ is the cross-sectional area for flow, and the term under the radical is the average molecular velocity ($\bar{v}$), equation 12.2.1 can be written in the form of Fick's first law as

$$\frac{F}{\pi \bar{r}^2} = D_K \frac{\Delta C}{\Delta X} \qquad (12.2.2)$$

where the Knudsen diffusivity D_K is defined as

$$D_K = \frac{2}{3} \bar{r} \sqrt{\frac{8RT}{\pi M}} = \frac{2}{3} \bar{r} \bar{v} \qquad (12.2.3)$$

The Knudsen diffusivity is thus directly proportional to the pore radius ($\bar{r}$).

Equation 12.2.3 is often written in cgs units as

$$D_K = 9.7 \times 10^3 \bar{r} \sqrt{\frac{T}{M}} \qquad (12.2.4)$$

where

D_K is expressed in square centimeters per second

$\bar{r}$ is expressed in centimeters

T is expressed in degrees Kelvin

The symbols refer to a single component. Since molecular collisions are rare events in Knudsen flow, flow and diffusion are synonymous and each component of a mixture behaves as if it alone were present. Numerical values of the Knudsen diffusivity for molecules of ordinary weight at ordinary temperatures range from 0.01 cm²/sec for pores with a radius of 10 Å up to about 10 cm²/sec for pores with 10,000 Å radii.

By comparing the relative magnitude of the mean free path (λ) and the pore diameter ($2\bar{r}$), it is possible to determine whether bulk diffusion or Knudsen diffusion may be regarded as negligible. Using the principles of the kinetic theory of gases, it can be shown that this ratio is equal to the ratio of the bulk diffusivity (D_{AB}) to the Knudsen diffusivity.

$$\frac{\lambda}{2\bar{r}} = \frac{D_{AB}}{D_K} \qquad (12.2.5)$$

Obviously, there will be a range of pressures or molecular concentrations over which the transition from ordinary molecular diffusion to Knudsen diffusion takes place. Within this region both processes contribute to the mass transport, and it is appropriate to utilize a combined diffusivity ($\mathscr{D}_c$). For species A the correct form for the combined diffusivity is the following.

$$\mathscr{D}_c = \frac{1}{1/D_K + (1 - \alpha Y_A)/D_{AB}} \qquad (12.2.6)$$

where

Y_A is the mole fraction of species A in the gas phase

D_{AB} is the bulk diffusivity in a binary gaseous mixture of A and B,

α is given by

$$\alpha = 1 + \frac{N_B}{N_A} \qquad (12.2.7)$$

where N_A and N_B are the molar fluxes of A and B relative to a fixed coordinate system.

The combined diffusivity is, of course, defined as the ratio of the molar flux to the concentration gradient, irrespective of the mechanism of transport. The above equation was derived by separate groups working independently (8–10). It is important to recognize that the molar fluxes (N_i) are defined with respect to a fixed catalyst pellet rather than to a plane of no net transport. Only when there is equimolar counterdiffusion, do the two types of flux definitions become equivalent. For a more detailed discussion of this point, the interested readers should consult Bird, Stewart, and Lightfoot (11). When there is *equimolal* counterdiffusion $N_B = -N_A$ and

equation 12.2.6 reduces to

$$\mathscr{D}_c = \frac{1}{1/D_K + 1/D_{AB}} \qquad (12.2.8)$$

Equimolar counterdiffusion takes place in catalyst pores when a reaction with a stoichiometry of the form $A \rightarrow B$ occurs under steady-state conditions.

In the case of nonequimolal counterdiffusion, equation 12.2.6 suffers from the serious disadvantage that the combined diffusivity is a function of the gas composition in the pore. This functional dependence carries over to the effective diffusivity in porous catalysts (see below), and makes it difficult to integrate the combined diffusion and transport equations. As Smith (12) points out, the variation of $\mathscr{D}_c$ with composition (Y_A) is not usually strong, and it has been an almost universal practice to use a composition independent form of $\mathscr{D}_c$ (12.2.8) in assessing the importance of intrapellet diffusion. In fact, the concept of a single effective diffusivity loses its engineering utility if the dependence on composition must be retained.

Whether Knudsen or bulk diffusion dominates the mass transport process depends on the relative magnitudes of the two terms in the denominator of equation 12.2.6. The ratio of the two diffusivity parameters is obviously important in establishing these magnitudes. In this regard, it is worth noting that D_K is proportional to the pore diameter and independent of pressure whereas D_{AB} is independent of pore size and inversely proportional to the pressure. Consequently, the higher the pressure and the larger the pore, the more likely it is that ordinary bulk diffusion will dominate.

Surface diffusion is yet another mechanism that is often invoked to explain mass transport in porous catalysts. An adsorbed species may be transported either by desorption into the gas phase or by migration to an adjacent site on the surface. It is this latter phenomenon that is referred to as surface diffusion. This phenomenon is poorly understood and the rate of mass transfer by this process cannot be predicted with any degree of accuracy. The interested student should refer to the discussion of Satterfield (13) and review articles on this subject (14–16).

Bulk or forced flow of the Hagan-Poiseuille type does not in general contribute significantly to the mass transport process in porous catalysts. For fast reactions where there is a change in the number of moles on reaction, significant pressure differentials can arise between the interior and the exterior of the catalyst pellets. This phenomenon occurs because there is insufficient driving force for effective mass transfer by forced flow. Molecular diffusion occurs much more rapidly than forced flow in most porous catalysts.

Our discussion of the various types of diffusion has presumed that they take place in a well-characterized pore structure, that is, straight cylindrical pores. However, the catalysts used in industry have extremely complex structures with interconnecting pores, tortuous pores, and wide variations in pore diameter as one moves along the length of the pore. Consequently, we need to convert the combined diffusivities that are appropriate for Knudsen and/or bulk diffusion into effective diffusivities, which are required for use in systems of unspecified pore geometry. These effective diffusivities can be calculated if we presume some sort of a realistic model for the geometry of the pore structure. The test of the model is, of course, whether or not it predicts results that are in accord with what is observed in the laboratory or the plant. The preferred method of obtaining an effective diffusivity for design calculations involves measurement of this property. It should be noted, however, that the values obtained should be interpreted with care. We briefly discuss both the calculational and the experimental routes to effective diffusivities in the paragraphs below.

If the pores of a catalyst pellet are randomly oriented, geometric considerations require that if one takes an arbitrary cross section of the porous mass, the fraction of the area occupied by the solid material will be a constant that will be identical with the volume fraction solids $[(1 - \varepsilon_p)$ where ε_p is the porosity of the pellet]. Similarly, the fraction of the cross section area through which mass transfer can take place is identical with the void fraction ε_p. Consequently, our first step in obtaining an effective diffusivity will be to multiply our combined diffusivity by ε_p, since the area available for mass transport has been reduced by this factor from the gross area of a cross section normal to the flow. The second step requires recognition that the length of the diffusion path that the molecules must traverse in real pores will be greater than that of a straight line linking the origin and termination of the diffusion path. In addition, the avenues through which diffusion proceeds are irregular in shape and vary in cross section. Since the constrictions offer resistances to diffusion that are not offset by enlargements, both of these factors cause the flux to be less than it would be in a straight cylindrical pore of the same length and mean radius. Consequently, if we introduce a length factor L' and a shape factor S' (both of which are greater than unity) to allow for these effects, the relation between the effective diffusivity and the combined diffusivity may be written as

$$\mathscr{D}_{\text{eff}} = \frac{\mathscr{D}_c \varepsilon_p}{L'S'} = \frac{\mathscr{D}_c \varepsilon_p}{\tau'} \qquad (12.2.9)$$

where the tortuosity factor τ' is used to replace the factor $L'S'$, since it is virtually impossible to separate these factors for most real porous systems. For accurate work this parameter, or $\mathscr{D}_{\text{eff}}$ itself, must be determined experimentally. There have been several attempts to relate L' and/or S' to other readily measured properties of catalyst pellets, but they are of limited applicability. The combined factor τ' has been measured for a variety of commercial catalysts. Tortuosity factors from slightly more than one to greater than ten have been reported (17). Values of 3 to 7 are easily reconciled with physical reality by multiplying together physically reasonable values of L' and S'.

In the above discussion, we have presumed that the tortuosity factor τ' is characteristic of the pore structure but not of the diffusing molecules. However, when the size of the diffusing molecule begins to approach the dimensions of the pore, one expects the solid to exert a retarding influence on the flux and this effect may also be incorporated in the tortuosity factor. This situation is likely to be significant in dealing with catalysis by zeolites (molecular sieves).

Because a priori estimates of effective diffusivities are not adequate for accurate design calculations and recourse must be made to experimental methods to evaluate these parameters, a few comments on the techniques used in these studies are in order. Since catalytic reactors are normally operated at steady state and nearly constant pressure, effective diffusivities are often measured under these constraints. The most commonly used apparatus for measuring counterdiffusion rates was first described by Wicke and Kallenbach (18). It involves sealing off the sides of a cylindrical pellet and flowing two different pure gases over the two flat faces. The rate of steady-state diffusion through the pellet is determined by analyzing each of the two pure streams for contamination by the other component. One difficulty with this approach is that since it measures diffusion through the pellet, it ignores dead-end pores, which may contribute to reaction and which are included in a measured pore-size distribution. It can be used over a wide range of pressures but cannot be readily adapted to temperatures much above ambient. The principles of kinetic theory can be used to calculate effective diffusivities for both species from measurements on only one of the two gas fluxes. In practice, however, measurements of both fluxes are desirable to facilitate the detection of leaks in the apparatus and of the presence of surface diffusion.

A second approach to measuring effective diffusivities involves filling the pore structure with one component and then measuring the rate of efflux of this component into a second component. Although somewhat simpler experimentally, this efflux method has the disadvantage that adsorption-desorption effects on the surface can conceivably complicate the results for high surface area materials. Since it is not a steady-state experiment, the equations for non-steady-state diffusion must be used in interpreting data. For gases and typical catalyst pellets, the time frame of the experiment is so short that it is difficult to obtain accurate diffusivity values. Consequently, the technique is largely restricted to liquids or to very finely porous solids, which give rise to very low effective diffusivities (e.g., zeolites).

Barrer (19) has developed another widely used nonsteady-state technique for measuring effective diffusivities in porous catalysts. In this approach, an apparatus configuration similar to the steady-state apparatus is used. One side of the pellet is first evacuated and then the increase in the downstream pressure is recorded as a function of time, the upstream pressure being held constant. The pressure drop across the pellet during the experiment is also held relatively constant. There is a time lag before a steady-state flux develops, and effective diffusion coefficients can be determined from either the transient or steady-state data. For the transient analysis, one must allow for accumulation or depletion of material by adsorption if this occurs.

The following illustration indicates how the concepts we have developed thus far in this chapter can be used in determining effective diffusivities for use in the analyses we will develop in subsequent sections.

ILLUSTRATION 12.1 ESTIMATION OF COMBINED DIFFUSIVITY FOR CUMENE IN A CRACKING CATALYST

The cumene (isopropyl benzene) cracking reaction is often used as a model reaction for determining the relative activities of cracking catalysts.

This reaction takes place on silica-alumina catalysts in the temperature range from 300 to 600 °C. It is a clean reaction with negligible production of by-products.

For the purposes of this illustrative example, we wish to calculate the combined and effective diffusivities of cumene in a mixture of benzene and cumene at 1 atm total pressure and 510 °C within the pores of a typical TCC (Thermofor Catalytic Cracking) catalyst bead. For our present purposes, the approximation to the combined diffusivity given by equation 12.2.8 will be sufficient because we will see that the Knudsen diffusion term is the dominant factor in determining the combined diffusivity.

Additional information. The following property values are associated with the TCC beads:

$$S_g = 342 \text{ m}^2/\text{g}$$
$$\varepsilon_p = 0.51$$

equivalent particle diameter = 0.43 cm

density of an individual particle $\rho_p = 1.14 \text{ g/cm}^3$

A narrow pore-size distribution and a tortuosity factor of three may be assumed. Using the methods suggested by Reid and Sherwood (7), the ordinary molecular diffusivity D_{AB} is found to be 0.150 cm²/sec.

Solution

The Knudsen diffusion coefficient may be evaluated from equation 12.2.4 if the catalyst property values are used to estimate the average pore radius. From equation C of Illustration 6.2,

$$\bar{r} = \frac{2V_g}{S_g}$$

In terms of the indicated property values,

$$V_g = \frac{\varepsilon_p}{\rho_p} = \frac{0.51}{1.14} = 0.447 \text{ cm}^3/\text{g}$$

Hence

$$\bar{r} = \frac{2(0.447)}{342 \times 10^4} = 2.61 \times 10^{-7} \text{ cm}$$
$$= 2.61 \text{ nm} = 26.1 \text{ Å}$$

and

$$D_K = 9.7 \times 10^3 (2.61 \times 10^{-7}) \sqrt{\frac{783}{120.19}}$$
$$= 6.46 \times 10^{-3} \text{ cm}^2/\text{sec}$$

Substitution of the numerical values for the bulk and Knudsen diffusion coefficients into equation 12.2.8 gives

$$\frac{1}{\mathscr{D}_c} = \frac{1}{(6.46 \times 10^{-3})} + \frac{1}{0.150} = 154.8 + 6.67$$

or

$$\mathscr{D}_c = 6.19 \times 10^{-3} \text{ cm}^2/\text{sec}$$

From the magnitudes of the diffusion coefficients, it is evident that under the conditions cited the majority of the mass transport will occur by Knudsen diffusion. Equation 12.2.9 and the tabulated values of the porosity and tortuosity may be used to determine the effective diffusivity.

$$\mathscr{D}_{\text{eff}} = \frac{\mathscr{D}_c \varepsilon_p}{\tau'} = \frac{(6.19 \times 10^{-3})(0.51)}{3}$$
$$= 1.05 \times 10^{-3} \text{ cm}^2/\text{sec}$$

12.3 DIFFUSION AND REACTION IN POROUS CATALYSTS

When diffusion and reaction occur simultaneously within a porous solid structure, concentration gradients of reactant and product species are established. If the various diffusional processes discussed in the previous section are rapid compared with the chemical reaction rate, the entire accessible internal surface of the catalyst will be effective in promoting reaction because the reactant molecules will spread essentially uniformly throughout the pore structure, before they have time to react. Here only a small concentration gradient will exist between the exterior and interior of the particle, and there will be diffusive fluxes of reactant molecules in and product molecules out that suffice to balance the reaction rate within the particle. In this case, we measure true intrinsic kinetics. On the other hand, if the catalyst is very active, many reactant molecules will have been converted to products before they have had time to diffuse very far into the pore structure. There will be steep concentration gradients of both reactant and product species near the periphery of the particle. Within the central core of the particle, the reactant concentrations will be very low and their gradients will be small. In the core the product concentrations will be high. Virtually the entire reaction will take place within a thin shell on the periphery of the particle. In this case the internal surface area is not used effectively. Still, under steady-state operating conditions, diffusion of reactants in and products out must balance the rate of reaction within the pore. Here, however, it is large peripheral concentration gradients rather than high diffusivities that are mainly responsible for

achieving the necessary mass transfer rate. Assuming that the catalyst particle is isothermal, the average reaction rate under these conditions will, in general, be less than that which would prevail if there were no mass transfer limitations. In this case, one does *not* measure true intrinsic kinetics. In this section we examine the general problem of determining what fraction of the catalyst surface area is effective when reaction and intraparticle mass transfer effects interact. Our objective is to develop expressions for the rate of reaction averaged over the entire catalyst particle, but expressed in terms of the temperature and concentrations prevailing at its exterior surface.

Quantitative analytical treatments of the effects of mass transfer and reaction within a porous structure were apparently first carried out by Thiele (20) in the United States, Damköhler (21) in Germany, and Zeldovitch (22) in Russia, all working independently and reporting their results between 1937 and 1939. Since these early publications, a number of different research groups have extended and further developed the analysis. Of particular note are the efforts of Wheeler (23–24), Weisz (25–28), Wicke (29–32), and Aris (33–36). In recent years, several individuals have also extended the treatment to include enzymes immobilized in porous media or within permselective membranes. The important consequence of these analyses is the development of a technique that can be used to analyze quantitatively the factors that determine the effectiveness with which the surface area of a porous catalyst is used. For this purpose we define an effectiveness factor η for a catalyst particle as

$$\eta = \frac{\text{actual rate for the entire catalyst particle}}{\text{rate evaluated at exterior surface conditions}} \quad (12.3.1)$$

Hence the effectiveness factor is the ratio of the actual reaction rate to that which would be observed if the total surface area throughout the catalyst interior were exposed to a fluid of the same composition and temperature as that found at the outside surface of the particle. By the proper choice of experimental conditions, the intrinsic chemical reaction rate expression can be determined. These conditions require the elimination of both external and intraparticle mass transfer resistances. Once the intrinsic chemical kinetics have been determined, the problem that we are addressing reduces to one of evaluating the effectiveness factor η. The general approach to analyzing this problem involves the development and solution of differential equations for simultaneous reaction and diffusion within a catalyst particle. In the case of nonisothermal catalyst pellets, we also require an additional differential equation for the energy transport within the particle.

12.3.1 Effectiveness Factors for Isothermal Catalyst Pellets

This section is concerned with analyses of simultaneous reaction and mass transfer within porous catalysts under isothermal conditions. Several factors that influence the final equation for the catalyst effectiveness factor are discussed in the various subsections. The factors considered include different mathematical models of the catalyst pore structure, the gross catalyst geometry (i.e., its apparent shape), and the rate expression for the surface reaction.

12.3.1.1 The Effectiveness Factor for a Straight Cylindrical Catalyst Pore: First-Order Reaction.
To illustrate the basic features of an analysis for catalyst effectiveness factors, we consider the problem of a first-order reaction taking place on the surface of a straight cylindrical catalyst pore. It is assumed that there is no change in the number of moles of gas phase species during reaction ($\delta = 0$). Figure 12.1 indicates the geometric parameters characterizing the cylindrical pore and the differential element of length over

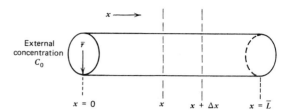

Figure 12.1
Schematic representation of straight cylindrical catalyst pore.

which we write a mass balance. If the catalyst pellet is assumed to contain n_p identical pores of this type, the analysis in Illustration 6.2 indicates that the model parameters $\bar{r}$, $\bar{L}$, and n_p (the number of pores per catalyst particle) can be related to experimentally measurable properties as follows.

$$\bar{r} = \frac{2V_g}{S_g} \qquad (12.3.2)$$

$$\bar{L} = \frac{V_p}{S_x} \qquad (12.3.3)$$

$$n_p = \frac{\varepsilon_p S_x S_g^2}{4\pi V_g^2} \qquad (12.3.4)$$

where

V_g is the void volume per gram of catalyst

S_g is the surface area per gram of catalyst

V_p is the gross volume of the catalyst particle

S_x is the gross exterior surface area of the particle (geometric area)

ε_p is the catalyst porosity

For a surface reaction that is first-order in gas phase reactant concentration, the rate *per unit surface area* may be written as

$$r = k_1'' C \qquad (12.3.5)$$

where C refers to the reactant concentration, the subscript 1 indicates that the reaction is first order, and the superscript $''$ emphasizes that the reaction rate is expressed per unit surface area.

If a material balance is written over the differential element of pore length shown in Figure 12.1, one finds that at steady state:

Input = output + disappearance by reaction

$$\pi \bar{r}^2 \left(-\mathscr{D}_c \frac{dC}{dx} \right)_x = \pi \bar{r}^2 \left(-\mathscr{D}_c \frac{dC}{dx} \right)_{x+\Delta x}$$
$$+ (2\pi \bar{r} \, \Delta x)(k_1'' C) \quad (12.3.6)$$

where we have used the symbol $\mathscr{D}_c$ to emphasize the fact that the diffusive flux may represent a combination of Knudsen and ordinary molecular diffusion and where we have assumed that the stoichiometric coefficient of the reactant for which we are writing the material balance is -1.

Dividing through by Δx, taking the limit as Δx goes to zero, and rearranging gives

$$\mathscr{D}_c \frac{d^2 C}{dx^2} = \frac{2k_1'' C}{\bar{r}} \quad (12.3.7)$$

This second-order differential equation is subject to the boundary conditions:

$$C = C_0 \quad \text{at } x = 0 \quad (12.3.8)$$

$$\frac{dC}{dx} = 0 \quad \text{at } x = \bar{L} \quad (12.3.9)$$

The second boundary condition meets the requirement that there be no flow of matter out of the closed end of the pore.

Equation 12.3.7 may be rewritten as

$$\frac{d^2 C}{d(x/\bar{L})^2} = \left(\frac{2k_1'' \bar{L}^2}{\bar{r} \mathscr{D}_c} \right) C \quad (12.3.10)$$

The term in brackets is a dimensionless group that plays a key role in determining the limitations that intraparticle diffusion places on observed reaction rates and the effectiveness with which the catalyst surface area is utilized. We define the Thiele modulus h_T as

$$h_T^2 = \frac{2k_1'' \bar{L}^2}{\bar{r} \mathscr{D}_c} \quad (12.3.11)$$

so that equation 12.3.10 may be rewritten as

$$\frac{d^2 C}{d(x/\bar{L})^2} = h_T^2 C \quad (12.3.12)$$

The solution to this equation may be written in the form

$$C = \frac{C_0 \cosh[h_T(1 - x/\bar{L})]}{\cosh h_T} \quad (12.3.13)$$

which describes the concentration profile along the length of the pore. While this equation is not our ultimate goal, a brief digression to examine a possible physical interpretation of the square of the Thiele modulus may shed some light on the discussion that follows.

Equation 12.3.11 may be rewritten as

$$h_T^2 = \frac{(2\pi \bar{r} \bar{L})k_1'' C_0}{(\pi \bar{r}^2) \left[\mathscr{D}_c \dfrac{(C_0 - 0)}{\bar{L}} \right]} \quad (12.3.14)$$

The numerator of the right side of this equation is equal to the chemical reaction rate that would prevail if there were no diffusional limitations on the reaction rate. In this situation, the reactant concentration is uniform throughout the pore and equal to its value at the pore mouth. The denominator may be regarded as the product of a hypothetical diffusive flux and a cross-sectional area for flow. The hypothetical flux corresponds to the case where there is a linear concentration gradient over the pore length equal to $C_0/\bar{L}$. The Thiele modulus is thus characteristic of the ratio of an intrinsic reaction rate in the absence of mass transfer limitations to the rate of diffusion into the pore under specified conditions.

Now at steady state, the observed rate of reaction within the pore just balances the rate of diffusion of reactant into the pore

$$r_{\text{pore}} = -\mathscr{D}_c \left(\frac{dC}{dx} \right)_{x=0} \pi \bar{r}^2 \quad (12.3.15)$$

From equation 12.3.13, the concentration gradient at the pore mouth is found to be

$$\left(\frac{dC}{dx}\right)_{x=0} = \frac{C_0}{\cosh h_T}\left\{\sinh\left[h_T\left(1 - \frac{x}{\bar{L}}\right)\right]\right\}\left(-\frac{h_T}{\bar{L}}\right)\Big|_{x=0}$$

$$= -\frac{h_T C_0}{\bar{L}}\frac{\sinh h_T}{\cosh h_T} = \frac{-h_T C_0}{\bar{L}}\tanh h_T \qquad (12.3.16)$$

According to this model, the reaction rate that would be observed within the catalyst pore is given by a combination of equations 12.3.15 and 12.3.16.

$$r_{\text{pore}} = \pi \bar{r}^2 \frac{\mathscr{D}_c h_T C_0}{\bar{L}}\tanh h_T \qquad (12.3.17)$$

If there were no limitations placed on the reaction rate by intraparticle diffusion (i.e., if the reactant concentration were C_0 throughout the pore), the reaction rate would be given by

$$r_{\text{ideal}} = 2\pi \bar{r}\bar{L}k_1''C_0 \qquad (12.3.18)$$

From the definition of the effectiveness factor and equations 12.3.17 and 12.3.18,

$$\eta = \frac{r_{\text{pore}}}{r_{\text{ideal}}} = \frac{\pi\bar{r}^2\dfrac{\mathscr{D}_c h_T C_0}{\bar{L}}\tanh h_T}{2\pi\bar{r}\bar{L}k_1''C_0}$$

$$= \left(\frac{\bar{r}\mathscr{D}_c}{2\bar{L}^2 k_1''}\right)h_T\tanh h_T \qquad (12.3.19)$$

Equation 12.3.11 indicates that the term in parentheses is just $1/h_T^2$. Hence

$$\eta = \frac{\tanh h_T}{h_T} \qquad (12.3.20)$$

Figure 12.2 is a plot of the effectiveness factor η versus the Thiele modulus h_T. For low values of h_T (slow reaction, rapid diffusion), the effectiveness factor approaches unity. For values of the Thiele modulus above 2.0, $\tanh h_T \approx 1$ and the effectiveness factor may be approximated by

$$\eta \approx \frac{1}{h_T} \qquad (12.3.21)$$

The approximation is quite good for many engineering purposes, particularly when one recognizes the disparity between the physical reality represented by a real porous catalyst and the assumptions implied by the model.

The effectiveness factor for a single pore is identical with that for the particle as a whole. Thus the reaction rate per unit mass of catalyst can be written as

$$r_{\text{mass}} = \eta k_1''C_0 S_g \qquad (12.3.22)$$

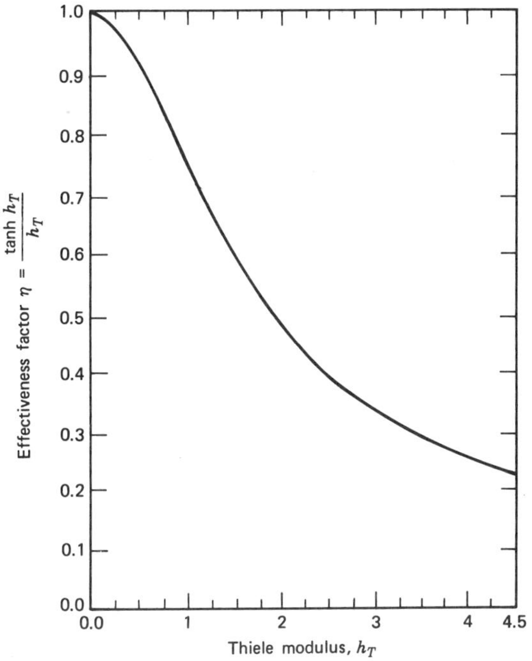

Figure 12.2
Plot of effectiveness factor versus Thiele modulus for first-order reaction.

$$\left(h_T = \bar{L}\sqrt{\frac{2k_1''}{\bar{r}\mathscr{D}_c}}\right)$$

In the laboratory, the measured rate constant for the first-order reaction would have to equal the product $\eta k_1''$ if this constant were expressed per unit area of catalyst. This relationship then gives us an alternative interpretation of the effectiveness factor in terms of our model

$$\eta = \frac{k_1'' \text{ measured}}{k_1'' \text{ intrinsic}} \qquad (12.3.23)$$

It is the ratio of the measured rate constant to the intrinsic chemical rate constant.

The Thiele modulus may be written in terms of measurable quantities using equations 12.3.2 and 12.3.3:

$$h_T = L \sqrt{\frac{2k_1''}{\bar{r}\mathscr{D}_c}}$$

$$= \frac{V_p}{S_x} \sqrt{\frac{k_1''S_g}{\mathscr{D}_c V_g}} \qquad (12.3.24)$$

For $h_T > 2$ the observed reaction rate can be approximated by combining equations 12.3.21, 12.3.22, and 12.3.24:

$$r_{\text{mass}} = \frac{S_x}{V_p} \sqrt{\frac{V_g \mathscr{D}_c}{(k_1''S_g)}} (k_1''S_g)C_0$$

$$= \frac{S_x}{V_p} \sqrt{V_g \mathscr{D}_c k_1''S_g} C_0 \qquad (12.3.25)$$

There are two points worth noting relative to equation 12.3.25. First, the ratio of the external surface area S_x to the gross geometric volume of a catalyst particle is inversely proportional to a characteristic dimension of the particle. For geometrically similar particles, this fact implies that the observed rate per unit mass of catalyst will be inversely proportional to the particle size. Second, the observed rate will be proportional to the square root of the true rate constant. Consequently, over the temperature range where intraparticle mass transfer strongly influences the reaction rate, the apparent activation energy for the reaction will be approximately one half the true activation energy for the surface reaction.

If reactor designers desire to make use of the analysis developed thus far in this subsection,

they must have a means of evaluating the effectiveness factor for the catalyst particles they plan to use in their reactor. There are two basic routes to the evaluation of η, and each is discussed in turn.

The first method of determining η is based on equation 12.3.23. It requires experimental data on the rate constants (or the reaction rate at similar external concentrations) as a function of the gross particle size of the catalyst. Experiments are carried out using smaller and smaller particles obtained by grinding or crushing the large particles that are intended for use in the commercial scale reactor. When a size regime where the rate constant is independent of particle size is reached, we are measuring the true or intrinsic rate constant, and this parameter gives us the denominator of equation 12.3.23. The numerator is, of course, the rate constant measured for the particle size one intends to use in the reactor. This method requires rate measurements on at least two different size catalysts, and preferably more.

A second approach to determining η requires rate measurements on only a single size of catalyst but it requires that we know V_p, S_x, V_g, S_g and that we know or can estimate $\mathscr{D}_c$. From a single rate experiment, we can obtain k_1'' or its equivalent $k_1''S_g$. Then there are alternative means of solving equations 12.3.11, 12.3.20, and 12.3.23 for η in terms of the input parameters. A simple trial and error technique consists of the following steps.

1. Assume a value for η.
2. Calculate h_T from equation 12.3.20 or read the value of this parameter from the curve in Figure 12.2.
3. Calculate the intrinsic rate constant from the input parameters, using equation 12.3.24.
4. Compare the measured rate constant k_1'' with the product of the assumed η and the intrinsic rate constant determined in step 3. If the two are equal, the assumption in step 1 was correct; if they are unequal, one assumes a new

value of η and iterates until consistent results are obtained.

A variation on the second approach involves a somewhat different manner of combining the equations used in the trial and error procedure. From equation 12.3.11,

$$k_1'' = \frac{\bar{r} \mathcal{D}_c}{2L^2} h_T^2 \qquad (12.3.26)$$

and, from equations 12.3.23 and 12.3.20,

$$k_1'' = \frac{k_{1\ measured}''}{\eta} = \frac{h_T k_{1\ measured}''}{\tanh h_T} \qquad (12.3.27)$$

Combining equations 12.3.26 and 12.3.27 gives

$$k_{1\ measured}'' = h_T(\tanh h_T)\left(\frac{\bar{r}\mathcal{D}_c}{2L^2}\right) \qquad (12.3.28)$$

which can be written in terms of experimentally measurable variables using equations 12.3.2 and 12.3.3:

$$k_{1\ measured}'' = h_T(\tanh h_T)\left(\frac{V_g S_x^2 \mathcal{D}_c}{S_g V_p^2}\right) \qquad (12.3.29)$$

This equation can be solved for h_T, and equation 12.3.20 may then be used to determine the effectiveness factor.

The equations for effectiveness factors that we have developed in this subsection are strictly applicable only to reactions that are first-order in the fluid phase concentration of a reactant whose stoichiometric coefficient is unity. They further require that no change in the number of moles take place on reaction and that the pellet be isothermal. The following illustration indicates how this idealized cylindrical pore model is used to obtain catalyst effectiveness factors.

ILLUSTRATION 12.2 DETERMINATION OF CATALYST EFFECTIVENESS FACTOR FOR THE CUMENE CRACKING REACTION FROM MEASUREMENT OF AN APPARENT RATE CONSTANT

The cumene cracking reaction has been studied by a number of investigators because it is a relatively clean reaction that can be used as a measure of the relative activities of different cracking catalysts. Some of the data on this reaction have been obtained using integral flow reactors. For this type of reactor, the variation of conversion with reactor space time is relatively insensitive to variations in the form of the reaction rate expression at degrees of conversion that are far from equilibrium. Consequently, data on these systems are often reported in terms of a simple first-order reaction rate constant, assuming that the volume change accompanying the reaction is negligible. An expression that is often used to determine this constant is

$$\frac{W}{F_{A0}} = \int_0^{f_A} \frac{df_A}{(-r_{Am})} = \int_0^{f_A} \frac{df_A}{kC_{A0}(1 - f_A)}$$
$$= \frac{-\ln(1 - f_A)}{kC_{A0}}$$

where W is the weight of catalyst employed and the reaction rate constant is expressed per unit weight of catalyst. (The basic equation was derived earlier as equation 8.2.12.) If the data of Corrigan and his co-workers (37) are expressed in these terms, at 510 °C the *apparent rate constant* is approximately 0.716 cm^3/sec-g catalyst) as determined from Figure 6 of the reference cited. If the catalyst employed in this study has the properties enumerated below, determine the effectiveness factor for the catalyst. The value of the combined diffusivity from Illustration 12.1 may be used in your analysis.

Data on Catalyst Properties (Silica-alumina)

Equivalent diameter	0.43 cm
Particle density	1.14 g/cm^3
Specific surface area	342 m^2/g
Porosity	0.51
Void volume per gram	0.447 cm^3/g

Solution

Since we are given the apparent rate constant rather than the true rate constant, a trial and error solution will be required. Either of the approaches described above may be used, but we will employ equation 12.3.29. Since our apparent rate constant is expressed per unit weight of catalyst rather than per unit surface area, the first step involves multiplying both sides of this equation by S_g.

$$k''_{1\ measured}S_g = h_T(\tanh h_T)\left(\frac{V_g S_x^2 \mathscr{D}_c}{V_p^2}\right)$$

where the left side is equal to 0.716 cm^3/(sec-g catalyst). The equivalent radius of the catalyst particles is 0.215 cm. Hence

$$V_p = \tfrac{4}{3}\pi R^3 = \tfrac{4}{3}\pi(0.215)^3 = 0.0416\ \text{cm}^3$$

and

$$S_x = 4\pi R^2 = 4\pi(0.215)^2 = 0.581\ \text{cm}^2$$

Thus

$$0.716 = h_T(\tanh h_T)\left[\frac{(0.447)(0.581)^2(6.19 \times 10^{-3})}{(0.0416)^2}\right]$$

or

$$h_T \tanh h_T = 1.327$$

The transcendental equation for h_T may be solved to give $h_T = 1.47$. Equation 12.3.20 then gives $\eta = (\tanh 1.47)/1.47 = 0.61$.

This value is considerably higher than the experimental value (0.17) obtained from rate measurements on different size particles, but several factors may be invoked to explain the inconsistency. There will be a distribution of both pore radii and pore lengths present in the actual catalyst rather than uniquely specified values. Alumina catalysts often have a bimodal pore-size distribution. Our estimate of an apparent first-order rate constant using the method outlined above will be somewhat in error. The catalyst surface may not be equally active throughout if selective deactivation has taken place and the peripheral region is less active than the catalyst core. Other sources of error are the

failure to allow for the change in the number of moles on reaction, and the possibility that the pellets cannot be maintained in an isothermal condition for this endothermic reaction (see Section 12.3.2). The discrepancy indicates the necessity for good experimental data as input to reactor design calculations for heterogeneous catalytic reactions. Model predictions will not reflect the real world if the assumptions on which they are based are in error. For best results, one should use experimental data taken under conditions approximating the proposed industrial application.

12.3.1.2 The Effectiveness Factor for a Straight Cylindrical Pore: Second- and Zero-Order Reactions.

This section indicates the predictions of the straight cylindrical pore model for isothermal reactions that are zero- and second-order in the gas phase concentration of reactant. Equimolal counterdiffusion is assumed ($\delta_A = 0$).

For a second-order reaction, a material balance on a differential element of pore length leads to the following differential equation

$$\mathscr{D}_c \frac{d^2C}{dx^2} = \frac{2k''_2}{\bar{r}} C^2 \qquad (12.3.30)$$

where we have used the subscript 2 to indicate that we are dealing with a second-order rate constant. This equation is subject to the boundary conditions given by equations 12.3.8 and 12.3.9.

In dimensionless form, this equation can be written as

$$\frac{d^2(C/C_0)}{d(x/\bar{L})^2} = \frac{2k''_2 C_0 \bar{L}^2}{\bar{r}\mathscr{D}_c}\left(\frac{C}{C_0}\right)^2 = h_2^2\left(\frac{C}{C_0}\right)^2 \qquad (12.3.31)$$

where the Thiele modulus for the second-order

reaction (h_2) is defined as

$$h_2 = \bar{L}\sqrt{\frac{2k_2''C_0}{\bar{r}\mathscr{D}_c}} \qquad (12.3.32)$$

This equation can be solved exactly in terms of elliptic integrals, but this solution is somewhat complex. However, by a slight modification of the boundary condition at the end of the pore, it is possible to obtain a good engineering approximation that is useful for fast reactions. In this approximation, we replace the boundary condition at the end of the pore by

$$C = C_L \quad \text{at} \quad x = \bar{L} \qquad (12.3.33)$$

To evaluate the effectiveness factor we require only the derivative of the concentration at the pore mouth. This parameter may be obtained by multiplying both sides of equation 12.3.31 by $d(C/C_0)/d(x/\bar{L})$:

$$\frac{d(C/C_0)}{d(x/\bar{L})}\frac{d^2(C/C_0)}{[d(x/\bar{L})]^2} = \frac{1}{2}\frac{d}{d(x/\bar{L})}\left[\frac{d(C/C_0)}{d(x/\bar{L})}\right]^2$$

$$= h_2^2(C/C_0)^2\frac{d(C/C_0)}{d(x/\bar{L})}$$

$$(12.3.34)$$

Integration gives

$$\frac{\left[\dfrac{d(C/C_0)}{d(x/\bar{L})}\right]^2}{2} = \frac{h_2^2(C/C_0)^3}{3} + C_1$$

$$(12.3.35)$$

where C_1 is a constant of integration.

At $x/\bar{L} = 1$ the first derivative must be zero, and C/C_0 must be C_L/C_0. Hence

$$-\frac{d(C/C_0)}{d(x/\bar{L})} = h_2\sqrt{\frac{2}{3}\left[\left(\frac{C}{C_0}\right)^3 - \left(\frac{C_L}{C_0}\right)^3\right]}$$

$$(12.3.36)$$

or

$$-\frac{dC}{dx} = \frac{C_0}{\bar{L}}h_2\sqrt{\frac{2}{3}\left[\left(\frac{C}{C_0}\right)^3 - \left(\frac{C_L}{C_0}\right)^3\right]}$$

$$(12.3.37)$$

The reaction rate within the pore is equal to the rate of diffusion into the pore as indicated by equation 12.3.15. Thus

$$r_{\text{pore}} = \pi\bar{r}^2\mathscr{D}_c\left\{h_2\frac{C_0}{\bar{L}}\sqrt{\frac{2}{3}\left[1 - \left(\frac{C_L}{C_0}\right)^3\right]}\right\}$$

$$(12.3.38)$$

where the term in braces arises from the derivative of the concentration evaluated at the pore mouth.

In the absence of diffusional limitations,

$$r_{\text{ideal}} = 2\pi\bar{r}\bar{L}k_2''C_0^2 \qquad (12.3.39)$$

The effectiveness factor is given by the ratio of equation 12.3.38 to equation 12.3.39.

$$\eta = \frac{\bar{r}\mathscr{D}_c h_2}{2\bar{L}^2 k_2'' C_0}\sqrt{\frac{2}{3}\left[1 - \left(\frac{C_L}{C_0}\right)^3\right]}$$

$$(12.3.40)$$

If we make use of equation 12.3.32, the last equation becomes

$$\eta = \frac{1}{h_2}\sqrt{\frac{2}{3}\left[1 - \left(\frac{C_L}{C_0}\right)^3\right]} \qquad (12.3.41)$$

This relation is an exact consequence of the differential equation 12.3.30, but it suffers from the disadvantage that it contains an unknown parameter, C_L, the reactant concentration at the end of the pore. However, for even moderately fast reactions we can say that C_L is significantly less than C_0 and, because of the nature of the function under the radical sign, it is a very good approximation to say that

$$\eta \cong \frac{1}{h_2}\sqrt{\frac{2}{3}} = \frac{0.8165}{h_2} \quad \text{for } h_2 > 2.5$$

$$(12.3.42)$$

Curve B of Figure 12.3 [adopted from Wheeler (38)] represents the dependence of the effectiveness factor on Thiele modulus for second-order kinetics. Values of η for first-order and zero-order kinetics in straight cylindrical pores are shown as curves A and C, respectively. Each curve is plotted versus its appropriate modulus.

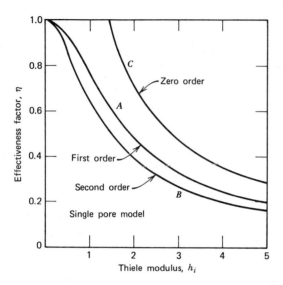

Figure 12.3

Plots of effectiveness factors versus corresponding Thiele moduli for zero-, first-, and second-order kinetics based on straight cylindrical pore model. For large h, values of η are as follows:

$$\text{Zero-order } \eta = \frac{\sqrt{2}}{h_o} \qquad h_o = \sqrt{\frac{2k_0''L^2}{\bar{r}\mathscr{D}_c C_0}}$$

$$\text{First-order } \eta = \frac{1}{h_1} \qquad h_1 = \sqrt{\frac{2k_1''L^2}{\bar{r}\mathscr{D}_c}}$$

$$\text{Second-order } \eta = \frac{\sqrt{2/3}}{h_2} \qquad h_2 = \sqrt{\frac{2k_2''C_0 L^2}{\bar{r}\mathscr{D}_c}}$$

[From A. Wheeler, *Adv. Catalysis*, *3* (249), 1951. Copyright ©1951. Used with permission of Academic Press.]

Notice that in the region of fast chemical reaction, the effectiveness factor becomes inversely proportional to the modulus h_2. Since h_2 is proportional to the square root of the external surface concentration, these two fundamental relations require that for second-order kinetics, the fraction of the catalyst surface that is effective will increase as one moves downstream in an isothermal packed bed reactor.

As Figure 12.3 indicates, it is also possible to obtain an analytical solution in terms of the straight cylindrical pore model for the case of a zero-order reaction. Here the dimensionless

form of the appropriate differential equation is

$$\frac{d^2(C/C_0)}{d(x/\bar{L})^2} = \frac{2\bar{L}^2 k_0''}{\bar{r}\mathscr{D}_c C_0} = h_0^2 \qquad (12.3.43)$$

subject to the boundary conditions of equations 12.3.8 and 12.3.9. The parameter k_0'' is the zero-order rate constant based on unit surface area of the catalyst and h_0 is the Thiele modulus for the zero-order reaction.

The solution to equation 12.3.43 can be written as

$$\frac{C}{C_0} = 1 - h_0^2 \left[\frac{x}{\bar{L}} - \frac{1}{2}\left(\frac{x}{\bar{L}}\right)^2 \right] \qquad (12.3.44)$$

The reactant concentration C will be greater than zero throughout the entire length of the pore provided that $h_0 < \sqrt{2}$. In this case the effectiveness factor will be unity because the reaction rate is independent of concentration. For values of $h_0 > \sqrt{2}$, equation 12.3.44 would call for negative values of the reactant concentration at large values of $x/\bar{L}$, a situation that is clearly impossible. Hence the boundary conditions on equation 12.3.43 must be changed so that both the reactant concentration and its gradient become zero at a point in the pore that we label with a coordinate x_c. In this situation, the concentration profile becomes

$$C = 0 \qquad \text{for } x \geq x_c \qquad (12.3.45)$$

$$\frac{C}{C_0} = 1 - 2\left[\frac{h_0 x}{\bar{L}\sqrt{2}} - \frac{1}{2}\left(\frac{h_0 x}{\bar{L}\sqrt{2}}\right)^2 \right] \qquad \text{for } x \leq x_c$$

$$(12.3.46)$$

where x_c can be determined from the point at which $dC/dx = 0$; that is,

$$x_c = \frac{\bar{L}\sqrt{2}}{h_0} \qquad \text{for } h_0 > \sqrt{2} \qquad (12.3.47)$$

Thus, for $h_0 > \sqrt{2}$, the effectiveness factor is given by

$$\eta = \frac{2\pi\bar{r}x_c k_0''}{2\pi\bar{r}\bar{L}k_0''} = \frac{x_c}{\bar{L}} = \frac{\sqrt{2}}{h_0} \qquad (12.3.48)$$

This function is plotted as curve C in Figure 12.3.

12.3.1.3 The Effectiveness Factor Analysis in Terms of Effective Diffusivities: First-Order Reactions on Spherical Pellets.

Useful expressions for catalyst effectiveness factors may also be developed in terms of the concept of effective diffusivities. This approach permits one to write an expression for the mass transfer within the pellet in terms of a form of Fick's first law based on the superficial cross-sectional area of a porous medium. We thereby circumvent the necessity of developing a detailed mathematical model of the pore geometry and size distribution. This subsection is devoted to an analysis of simultaneous mass transfer and chemical reaction in porous catalyst pellets in terms of the effective diffusivity. In order to use the analysis with confidence, the effective diffusivity should be determined experimentally, since it is difficult to obtain accurate estimates of this parameter on an a priori basis.

Consider the spherical catalyst pellet of radius R shown in Figure 12.4. The effective diffusivity approach presumes that diffusion of all types can be represented in terms of Fick's first law and an overall effective diffusion coefficient that can be taken as a constant. That is, the appropriate flux representation is

$$N_r = -\mathscr{D}_{\text{eff}}\left(\frac{dC}{dr}\right) \qquad (12.3.49)$$

where N_r is the diffusive *flux outward* in the radial direction. As before, we will assume steady-state reaction conditions, a simple irreversible first-order reaction with $\delta = 0$, and an isothermal pellet. We will focus our attention on the spherical shell of porous material contained within the region between r and $r + \Delta r$

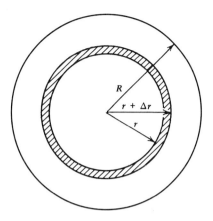

Figure 12.4
Schematic diagram of porous spherical catalyst pellet.

and write a material balance on this differential element. Reactants are transported into and out of the annular shell by diffusion and are consumed within it by reaction. For a surface reaction that is first-order in the gas phase reactant concentration, the reaction rate within the differential volume element can be written as

$$r = k_1''\rho_p S_g (4\pi r^2\, \Delta r)C \qquad (12.3.50)$$

where ρ_p is the apparent density of the catalyst particle (mass per total particle volume) and $4\pi r^2\, \Delta r$ is the differential volume element. Notice that the group $k_1''\rho_p S_g$ may be regarded as an apparent first-order rate constant per unit volume of catalyst pellet.

Assuming that the stoichiometric coefficient of the reactant is -1 and that there is no change in the number of moles on reaction, the material balance on the volume element at steady state can be written as:

(Rate of input by diffusion at $r + \Delta r$) = (rate of efflux by diffusion at r) + (rate of disappearance by chemical reaction)

$$\left[4\pi r^2\left(\mathscr{D}_{\text{eff}}\frac{dC}{dr}\right)\right]_{r+\Delta r} = \left[4\pi r^2\left(\mathscr{D}_{\text{eff}}\frac{dC}{dr}\right)\right]_r + k_1''\rho_p S_g C 4\pi r^2\, \Delta r \qquad (12.3.51)$$

The differential equation governing simultaneous chemical reaction and diffusion then becomes

$$\frac{d}{dr}\left(r^2 \mathscr{D}_{eff}\frac{dC}{dr}\right) - k_1''\rho_p S_g C r^2 = 0 \quad (12.3.52)$$

If the effective diffusivity is assumed to be a constant, equation 12.3.52 can be written as

$$\frac{d^2C}{dr^2} + \frac{2}{r}\frac{dC}{dr} - \frac{k_1''\rho_p S_g C}{\mathscr{D}_{eff}} = 0 \quad (12.3.53)$$

The boundary conditions for this differential equation require that:

1. At $r = R$ $\quad C = C_0$ $\qquad\qquad$ (12.3.54)

2. At $r = 0$ $\quad \dfrac{dC}{dr} = 0$ $\qquad\quad$ (12.3.55)

The first condition requires that the reactant concentration at the external surface of the catalyst be fixed. The second condition implies that there can be no diffusive flux through the center of the pellet, since this is a point of symmetry.

It is convenient to define a new Thiele-type modulus ϕ_s for this spherically symmetric problem as*

$$\phi_s = R\sqrt{\frac{k_1''\rho_p S_g}{\mathscr{D}_{eff}}} \quad (12.3.56)$$

Combining equations 12.3.53 and 12.3.56 gives

$$\frac{d^2C}{dr^2} + \frac{2}{r}\frac{dC}{dr} = \left(\frac{\phi_s}{R}\right)^2 C \quad (12.3.57)$$

The conventional approach to solving this linear differential equation is to introduce a

---

* We would be remiss if we did not indicate that some authors have defined yet another Thiele-type modulus (ϕ_s') for this problem as $\phi_s' = \phi_s/3$ and they have developed their analysis according to this parameter. In using plots of η versus ϕ_s, one must be careful to determine which of the two alternative definitions has been adopted for ϕ_s. Otherwise his value for η may be considerably in error. A variety of other symbols for dimensionless groups akin to the Thiele modulus have been employed by different authors.

new variable z defined such that $C = z/r$. This substitution converts equation 12.3.57 to

$$\frac{d^2z}{dr^2} = \left(\frac{\phi_s}{R}\right)^2 z \quad (12.3.58)$$

This equation is of the same form as equation 12.3.10 and the solution can be expressed in terms of exponential functions as

$$z = Cr = C_1 e^{\phi_s r/R} + C_2 e^{-\phi_s r/R} \quad (12.3.59)$$

where the constants of integration C_1 and C_2 can be evaluated from the boundary conditions. At $r = 0$, the fact that $dC/dr = 0$ implies that

$$C_1 = -C_2 \quad (12.3.60)$$

Equation 12.3.59 can thus be rewritten in terms of hyperbolic functions as

$$C = \frac{2C_1}{r}\sinh\left(\frac{\phi_s r}{R}\right) \quad (12.3.61)$$

Use of the boundary condition at the external surface of the spherical pellet ($r = R$) leads to the following expression for the concentration profile in the spherical pellet.

$$\frac{C}{C_0} = \frac{R}{r}\frac{\sinh\left(\phi_s\dfrac{r}{R}\right)}{\sinh(\phi_s)} \quad \text{(first-order reaction)}$$

$$(12.3.62)$$

Figure 12.5 contains a series of curves representing the concentration profile in the spherical pellet for different values of the Thiele modulus ϕ_s. For small values of ϕ_s, (say less than 0.5) the concentration profile is relatively flat and the reactant concentration is reasonably uniform. For large values of ϕ_s (say greater than 5), the reaction is rapid relative to diffusion and the reactant concentration at the center of the catalyst pellet is less than 7% of that at the external surface. Notice that in all cases the concentration gradient approaches zero at the center of the pellet.

We now wish to evaluate the total reaction rate for the entire catalyst pellet. This quantity may be obtained by writing an expression for

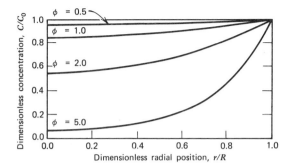

Figure 12.5
Dimensionless plot of reactant concentration versus distance from center of pellet.

the reaction rate in the spherical annulus of Figure 12.4 in terms of the local reactant concentration as given by equation 12.3.62, and then integrating this expression over the entire sphere. However, a simpler procedure involves recognizing that the overall rate of reaction within the spherical pellet must be equal to the rate of mass transfer into the pellet at steady state.

$$\text{Rate} = 4\pi R^2 \mathscr{D}_{\text{eff}} \left(\frac{dC}{dr}\right)_{r=R} \quad (12.3.63)$$

where we have eliminated the negative sign on the derivative, because we are interested in the flux *into* the pellet.

The concentration gradient at the exterior surface can be obtained by differentiating equation 12.3.62 with respect to r

$$\left(\frac{dC}{dr}\right) = C_0 \left[\phi_s \frac{\cosh\left(\phi_s \frac{r}{R}\right)}{r \quad \sinh(\phi_s)} - \frac{R}{r^2} \frac{\sinh\left(\phi_s \frac{r}{R}\right)}{\sinh(\phi_s)}\right] \quad (12.3.64)$$

and evaluating the derivative at the surface.

$$\left(\frac{dC}{dr}\right)_{r=R} = C_0 \left[\frac{\phi_s \cosh(\phi_s)}{R \sinh(\phi_s)} - \frac{1}{R}\right]$$

$$= \frac{C_0 \phi_s}{R} \left[\frac{1}{\tanh(\phi_s)} - \frac{1}{\phi_s}\right] \quad (12.3.65)$$

Combining equations 12.3.63 and 12.3.65 gives

$$\text{Rate} = 4\pi R \mathscr{D}_{\text{eff}} C_0 \phi_s \left[\frac{1}{\tanh(\phi_s)} - \frac{1}{\phi_s}\right] \quad (12.3.66)$$

If the entire active surface of the spherical pellet were exposed to reactant at a concentration C_0, the rate corresponding to this condition would be given by

Rate for $C = C_0$ throughout pellet

$$= \rho_p \frac{4}{3} \pi R^3 S_g k_1'' C_0 \quad (12.3.67)$$

The effectiveness factor is then given by the ratio of equation 12.3.66 to 12.3.67:

$$\eta = \frac{3 \mathscr{D}_{\text{eff}} \phi_s}{\rho_p R^2 k_1'' S_g} \left[\frac{1}{\tanh(\phi_s)} - \frac{1}{\phi_s}\right]$$

$$= \frac{3}{\phi_s} \left[\frac{1}{\tanh(\phi_s)} - \frac{1}{\phi_s}\right] \quad (12.3.68)$$

where we have used the defining equation for ϕ_s (12.3.56). For large values of ϕ_s, the hyperbolic tangent approaches unity and the effectiveness factor approaches $3/\phi_s$. Figure 12.6 is a plot of the function described by equation 12.3.68. It is appropriate for an isothermal first-order irreversible reaction in a sphere. At low values of ϕ_s, the effectiveness factor approaches unity.

Illustration 12.3 indicates the use of the effective diffusivity approach for estimating catalyst effectiveness factors when this parameter is determined experimentally or may be estimated.

ILLUSTRATION 12.3   DETERMINATION OF CATALYST EFFECTIVENESS FACTOR FOR THE CUMENE CRACKING REACTION USING THE EFFECTIVE DIFFUSIVITY APPROACH

Use the effective diffusivity approach to evaluate the effectiveness factor for the silica-alumina catalyst pellets considered in Illustration 12.2.

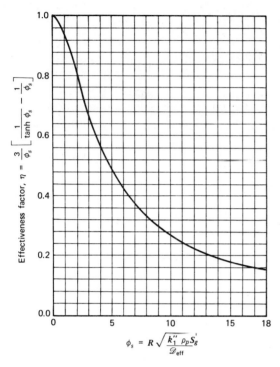

Figure 12.6
Effectiveness factor plot for spherical catalyst particles based on effective diffusivities (first-order reaction).

Assume that the effective diffusivity for cumene in these particles is equal to 1.2×10^{-3} cm^2/sec, which is a typical value for silica-alumina TCC beads (39).

Solution

The trial and error procedure outlined in Section 12.3.1.1 is quite general if appropriate modification is made for the use of ϕ_s values and effective diffusivities. Thus we will start by assuming that $\eta = 0.20$. From Figure 12.6, we find that $\phi_s = 13.4$.

Substitution of numerical values into the defining equation for ϕ_s (12.3.56) gives

$$13.4 = \left(\frac{0.43}{2}\right)\sqrt{\frac{k''_{1\,true}S_g(1.14)}{1.2 \times 10^{-3}}}$$

or

$$k''_{1\,true}S_g = 4.081$$

The measured value of $k''_1 S_g$ is 0.716 cm^3/(sec-g catalyst) and the ratio of this value to $k''_{1\,true}S_g$ should be equal to our assumed value for the effectiveness factor, if our assumption was correct. The actual ratio is 0.175, which is at variance with the assumed value. Hence we pick a new value of η and repeat the procedure until agreement is obtained. This iterative approach produces an effectiveness factor of 0.238, which corresponds to a ϕ_s value of 11.5. This result differs from the experimental value (0.17) and that calculated by the cylindrical pore model (0.61). In the above calculations, an experimental value of $\mathscr{D}_{\text{eff}}$ was not available and this circumstance is largely responsible for the discrepancy. If the combined diffusivity determined in Illustration 12.1 is converted to an effective diffusivity using equation 12.2.9, the value used above corresponds to a tortuosity factor of 2.6. If we had employed $\mathscr{D}_c$ from Illustration 12.1 and a tortuosity factor of unity to calculate $\mathscr{D}_{\text{eff}}$, we would have determined that $\eta = 0.65$, which is consistent with the value obtained from the straight cylindrical pore model in Illustration 12.2.

12.3.1.4 The Effectiveness Factor Analysis in Terms of Effective Diffusivities: Extension to Reactions Other than First-Order and Various Catalyst Geometries. The analysis developed in Section 12.3.1.3 may be extended in relatively simple straightforward fashion to other integer-order rate expressions and to other catalyst geometries such as flat plates and cylinders. Some of the key results from such extensions are treated briefly below.

For a zero-order chemical reaction, an analysis similar to that employed in Section 12.3.1.2 indicates that

$$\eta = 1 \text{ for } \phi_{so} < \sqrt{6} \qquad (12.3.69)$$

and

$$\eta = 1 - \frac{\frac{4}{3}\pi r_c^3}{\frac{4}{3}\pi R^3}$$

$$= 1 - \left(\frac{r_c}{R}\right)^3 \qquad \text{for} \qquad \phi_{s0} > \sqrt{6} \quad (12.3.70)$$

where $$\phi_{s0} = R\sqrt{\frac{k_0'' S_g \rho_p}{\mathscr{D}_{\text{eff}} C_0}} \qquad (12.3.71)$$

and where r_c is the radius at which the reactant concentration goes to zero. This radius may be determined from the constraints that both the reactant concentration and its gradient go to zero at the critical radius. r_c is a root of the equation

$$\frac{1}{2} - \frac{3}{\phi_{s0}^2} = \frac{3}{2}\left(\frac{r_c}{R}\right)^2 - \left(\frac{r_c}{R}\right)^3 \quad (12.3.72)$$

where $\phi_{s0}^2 \geq 6$ and $r_c \leq R$.

Figure 12.7 adapted from Satterfield (40) contains a plot of the effectiveness factor for a zero-order reaction versus the Thiele modulus

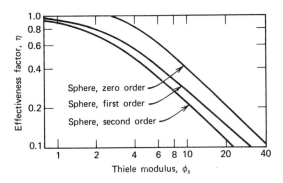

Figure 12.7
Effectiveness factor plot for *n*th-order kinetics—spherical catalyst particles

Zero-order $\phi_{s0} = R\sqrt{k_0'' S_g \rho_p / \mathscr{D}_{\text{eff}} C_0}$

First-order $\phi_{s1} = R\sqrt{k_1'' S_g \rho_p / \mathscr{D}_{\text{eff}}}$

Second-order $\phi_{s2} = R\sqrt{k_2'' S_g \rho_p C_0 / \mathscr{D}_{\text{eff}}}$

ϕ_{s0}. For purposes of comparison, plots are also presented for first- and second-order reactions in terms of the appropriate moduli.

The analysis of simultaneous diffusion and chemical reaction in porous catalysts in terms of effective diffusivities is readily extended to geometries other than a sphere. Consider a flat plate of porous catalyst in contact with a reactant on one side, but sealed with an impermeable material along the edges and on the side opposite the reactant. If we assume simple power law kinetics, a reaction in which there is no change in the number of moles on reaction, and an isothermal flat plate, a simple material balance on a differential thickness of the plate leads to the following differential equation

$$\mathscr{D}_{\text{eff}} \frac{d^2C}{dx^2} = \rho_p S_g k_n'' C^n \quad (12.3.73)$$

where n is the reaction order and x is the distance in from the open side. It is also assumed that the effective diffusivity is constant throughout the porous material. The boundary conditions for the physical situation described above require that at $x = 0$, $C = C_0$ and at $x = L$ (the plate thickness), $dC/dx = 0$.

The analyses of simultaneous reaction and mass transfer in this geometry are similar mathematically to those of the straight cylindrical pore model considered previously, because both are essentially one-dimensional models. In the general case, the Thiele modulus for semi-infinite, flat-plate problems becomes

$$\phi_{Ln} = L\sqrt{\frac{\rho_p S_g k_n'' C_0^{n-1}}{\mathscr{D}_{\text{eff}}}} \quad (12.3.74)$$

and the equations developed earlier for η in terms of the moduli h_0, h_1, and h_2 can be used to estimate η for flat-plate geometries, merely by replacing h_n by ϕ_{Ln}.

Analyses have been carried out for a variety of other catalyst geometries including among others, porous rods of infinite length (or with sealed ends), a porous rod with open ends, and the case of gases contacting the inside of a porous annulus. Refer to Satterfield (41) and Aris (35) for additional discussion along these lines and the pertinent references. However, we should point out that Luss and Amundson (42) have demonstrated that for simple first-order isothermal reactions, the spherical particle has the lowest effectiveness factor for all possible shapes having the same volume. This demonstration is largely due to the fact that the sphere has the lowest possible external surface/volume ratio. One is tempted to conclude that catalyst particles should be fabricated in shapes other than spheres to minimize intraparticle diffusional limitations on the reaction rate. However, in practice, such other considerations as the resultant pressure drop across a packed bed, the flow distribution through the bed, and the ease of manufacture weigh heavily in determining the particular geometric shape in which the catalyst will be employed.

For spherical pellets, for semi-infinite slabs, and for infinite cylinders at sufficiently large values of the Thiele modulus ϕ, the effectiveness factor becomes inversely proportional to the modulus with the constant of proportionality depending on the gross catalyst geometry and the order of the reaction. Under these conditions, the reactant concentrations rapidly approach zero as one penetrates into the catalyst, and it is only the outer periphery of the catalyst that is effective in promoting reaction. This region is very thin when ϕ is large and, in many respects, it may be regarded as a semi-infinite flat plate with a thickness V_p/S_x, irrespective of the true catalyst geometry. This limiting condition gives rise to the so-called "asymptotic forms" of the relation between the effectiveness factor and the Thiele modulus, in which the inverse proportionality noted above is observed. This approach is extremely appealing from the viewpoint of the mathematician, because it emphasizes the mathematical similarity of the differential equations and their corresponding boundary conditions when one analyzes simultaneous diffusion and chemical reaction in porous catalysts of different geometries. For discussion of the asymptotic solution approach to catalyst effectiveness factors, see the texts of Aris (36, 43) and Petersen (44).

12.3.1.5 Isothermal Effectiveness Factors—Miscellaneous Considerations.

12.3.1.5.1 Implications of Nonequimolal Counter-diffusion for Effectiveness Factors. When there is a change in the number of gas phase species on reaction, there has to be at steady state a net molar flux either into or out of a porous catalyst. When bulk diffusion is responsible for all or even a reasonably significant fraction of the total mass transfer, the pressure gradient through the pellet will usually be negligible, but the increase in the number of moles on reaction makes it more difficult for reactants to diffuse into the catalyst, thereby decreasing the effectiveness factor. A decrease in the number of moles facilitates diffusion of reactants into the pellet relative to the equimolal counterdiffusion case, thereby increasing the catalyst effectiveness factor. While we have previously neglected consideration of these phenomena, we now wish to briefly consider the influence of changes in the number of gas phase species during reaction on catalyst effectiveness factors. Treatments of this subject date back to the classic paper of Thiele (20), but the most comprehensive general treatment is that of Weekman and Goring (45), who analyzed isothermal zero-, first-, and second-order reactions in porous spherical catalysts at constant total pressure. They described the response of the effectiveness factor to changes in the number of moles on reaction in terms of a volume change modulus $\tilde{\theta}$, and the conventional Thiele modulus ϕ_s. The volume change modulus is defined as

$$\tilde{\theta} = \left(\sum \nu_i\right) Y \qquad (12.3.75)$$

where the sum of the stoichiometric coefficients is the increase in the number of moles on reaction and Y is the mole fraction reactant at the external surface of the catalyst. Their results were summarized in the form of plots of the ratio of the effectiveness factor for the volume change case (η') to that for no volume change (η) versus the volume change modulus $\tilde{\theta}$ at various values of the Thiele modulus ϕ_s. Figure 12.8 is a reproduction of their curves for first-order reactions. The more significant the departures of $\tilde{\theta}$ and ϕ_s from zero, the greater the departure of the ratio (η'/η) from unity. The departure of (η'/η) from unity also becomes more significant as the order of the reaction increases, but the effect is relatively small compared to the effects of variations in $\tilde{\theta}$ or variations in ϕ_s at low values of ϕ_s. Since the volume change modulus is directly proportional to the reactant

mole fraction at the external surface of the catalyst, one expects the influence of $\tilde{\theta}$ to be most significant near the inlet to a plug flow reactor vis à vis its effects at other points along the reactor length. Moreover, the effect should be small in cases where the reactants are highly diluted by the presence of inerts. This approach has also been extended to nonisothermal systems (46).

12.3.1.5.2 Implications of the Effectiveness Factor Concept for Kinetic Parameters Measured in the Laboratory. It is useful at this point to discuss the effects of intraparticle diffusion on the kinetic parameters that are observed experimentally. Unless we are aware that intraparticle diffusion may obscure or disguise the

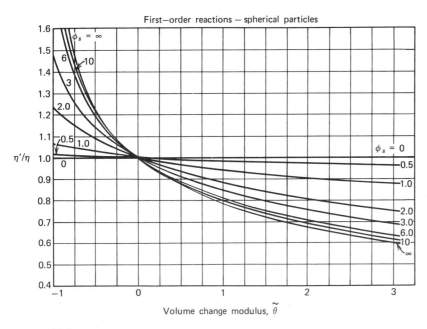

First—order reactions — spherical particles

Volume change modulus, $\tilde{\theta}$

Figure 12.8

Effectiveness factor ratios for first-order kinetics on spherical catalyst pellets.

$$\left[\left(\frac{\eta'}{\eta} = \frac{\text{factor in the presence of volume change}}{\text{factor in absence of volume change}} \right) \right]$$

From V. W. Weekman and R. L. Goring, *J. Catalysis*, 4 (260), 1965. Copyright © 1965. Used with permission of Academic Press.]

true intrinsic chemical kinetics, we may draw incorrect conclusions regarding the order and temperature dependence of the intrinsic chemical reaction. If one is interested in the fundamental nature of the chemical reaction, it is important to employ experimental conditions that eliminate the influence of both external and intraparticle mass transfer processes on the observed rate. From the viewpoint of the chemical engineer engaged in the practice of reactor design, it is even more important to recognize that in dealing with pelletized heterogeneous catalysts, the apparent kinetics may differ appreciably from the true intrinsic kinetics.

If we consider a reaction with intrinsic kinetics of simple nth order form that takes place within the pores of a catalyst pellet, the observed rate of reaction per unit mass of catalyst may be written as

$$r_{obs} = \eta k_n'' C_0^n S_g \qquad (12.3.76)$$

where C_0 is the reactant concentration in the gas phase. If we adopt the straight cylindrical pore model and write a material balance over a differential element of pore length, we find that for nth-order kinetics, the analog of equations 12.3.10, 12.3.31, and 12.3.43 becomes

$$\frac{d^2(C/C_0)}{d(x/\overline{L})^2} = \frac{2k_n''\overline{L}^2}{\overline{r}\mathscr{D}_c} \frac{C^n}{C_0}$$

$$= \left[\frac{2k_n''\overline{L}^2 C_0^{n-1}}{\overline{r}\mathscr{D}_c}\right]\left(\frac{C}{C_0}\right)^n = h^2\left(\frac{C}{C_0}\right)^n$$

$$(12.3.77)$$

where the square of the generalized Thiele modulus (the quantity in brackets) is proportional to the external concentration of reactant raised to the power $(n-1)$.

In the limit of low effectiveness factors where η becomes inversely proportional to the Thiele modulus, the apparent order of the reaction may differ from the true order. In this case, since the rate is proportional to the product of the effectiveness factor and the external concentration

raised to the nth power, it can be said that

$$r_{obs} \propto \frac{1}{h} C_0^n \propto \frac{C_0^n}{C_0^{(n-1)/2}} = C_0^{(n+1)/2} \qquad (12.3.78)$$

Thus a zero-order reaction appears to be 1/2 order and a second-order reaction appears to be 3/2 order when dealing with a fast reaction taking place in porous catalyst pellets. First-order reactions do not appear to undergo a shift in reaction order in going from high to low effectiveness factors. These statements presume that the combined diffusivity lies in the Knudsen range, so that this parameter is pressure independent.

If the dominant mode of transport within the catalyst pores is ordinary molecular diffusion, the analysis becomes somewhat more complex. The ordinary molecular diffusivity is inversely proportional to the pressure so that in this case

$$h \propto (C_0^{n-1})^{1/2} \cdot P^{1/2} \qquad (12.3.79)$$

If the reactant is by far the dominant species, equation 12.3.79 can be written as

$$h \propto (C_0^{n-1})^{1/2} (C_0 RT)^{1/2} \quad \text{or} \quad h \propto C_0^{n/2} \qquad (12.3.80)$$

and, *in the limit of low effectiveness* factors, the rate will be of the form

$$r_{obs} \propto \frac{C_0^n}{C_0^{n/2}} = C_0^{n/2} \qquad (12.3.81)$$

so that here one would again observe discrepancies between the observed and true orders of the reaction.

In a similar fashion, it is easily shown that the apparent activation energy of the reaction may differ appreciably from the intrinsic activation energy of the chemical reaction. The apparent rate constant is equal to the product of the effectiveness factor and the true rate constant and, *in the limit of low effectiveness factors*, it can be said that

$$k_{apparent} \propto \frac{k_{true}}{h} \qquad (12.3.82)$$

In terms of the generalized Thiele modulus of equation 12.3.77, the last equation becomes

$$k_{apparent} \propto \frac{k_{true}}{\sqrt{\dfrac{k_{true}}{\mathcal{D}_c}}} = \sqrt{\mathcal{D}_c k_{true}} \quad (12.3.83)$$

where we have retained only the temperature dependent terms from the Thiele modulus.

If the combined diffusivity is written as

$$\mathcal{D}_c = A e^{-E_{diffusion}/RT} \quad (12.3.84)$$

(where A is a temperature independent quantity), the Arrhenius relation can be used with equations 12.3.83 and 12.3.84 to show that

$$E_{apparent} = \frac{1}{2}(E_{diffusion} + E_{true}) \quad (12.3.85)$$

If ordinary molecular diffusion is the dominant mass transfer process, the kinetic theory of gases indicates that the diffusivity is proportional to $T^{3/2}$ and it is easily shown that

$$E_{diffusion} = \frac{3RT}{2} \quad (12.3.86)$$

If Knudsen diffusion dominates,

$$E_{diffusion} = \frac{RT}{2} \quad (12.3.87)$$

Under normal circumstances the true activation energy term in equation 12.3.85 will far exceed the diffusional activation term calculated from either equation 12.3.86 or equation 12.3.87 and, to a good approximation, it may be said that in the limit of low effectiveness factors

$$E_{apparent} \approx \frac{E_{true}}{2} \quad (12.3.88)$$

At low temperatures diffusion will be rapid compared to chemical reaction and diffusional limitations on the reaction rate will not be observed. In this temperature regime, one will observe the intrinsic activation energy of the reaction. However, since chemical reaction rates increase much more rapidly with increasing temperature than do diffusional processes, at higher temperatures one is much more likely to observe such limitations. In the regime where catalyst effectiveness factors become small one will observe an activation energy that is given by equation 12.3.85. A typical Arrhenius plot for the range of temperatures where this change takes place will resemble the one shown in Figure 12.9.

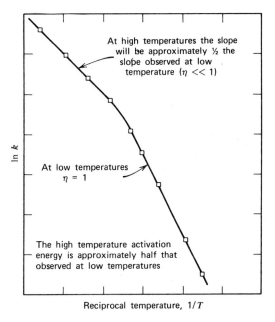

At high temperatures the slope will be approximately ½ the slope observed at low temperature ($\eta \ll 1$)

At low temperatures $\eta = 1$

The high temperature activation energy is approximately half that observed at low temperatures

$\ln k$

Reciprocal temperature, $1/T$

Figure 12.9
Schematic representation of shift in activation energy when intraparticle mass transfer effects become significant.

12.3.1.5.3 Effectiveness Factors for Hougen-Watson Rate Expressions. The discussion thus far and the vast majority of the literature dealing with effectiveness factors for porous catalysts are based on the assumption of an integer-power reaction rate expression (i.e., zero-, first-, or second-order kinetics). In Chapter 6, however, we stressed the fact that heterogeneous catalytic reactions are more often characterized by more complex rate expressions of the Hougen-Watson type. Over a narrow range of reactant concentrations, it is often possible to approximate Hougen-Watson kinetics by an integer power relation. However, if the diffusional limitations within a porous catalyst pellet are important (i.e., the effectiveness factor is low), there may be a significant gradient in reactant concentration across the pellet. The reactant partial pressure will vary from its value at the exterior surface to a value approaching zero at the center of the pellet. In this case, the range of partial pressures can be quite significant and it is not really appropriate to use the approximate integer degenerate forms of the Hougen-Watson rate expressions. Several analyses of this problem have appeared in the literature, dating back to a study by Chu and Hougen (47).

The procedure for determining catalyst effectiveness factors for generalized Hougen-Watson rate expressions is basically the same as the one for integer-power rate expressions. One sets up and solves the differential equation for simultaneous diffusion and reaction inside the catalyst pellet. However, unlike the isothermal integer rate law cases, it is not possible to obtain closed form solutions for these rate expressions, and recourse must be made to numerical solutions or various approximations that simplify the mathematical treatment. Roberts and Satterfield (48–51) have developed a generalized method for predicting catalyst effectiveness factors for reactions that are first- or second-order in the fraction of the surface covered by reactant species. Their approach allows for inhibition by both reactant and product species. They found that under certain isothermal conditions, it is possible for the effectiveness factor to exceed unity. Consult the references listed above for the generalized charts and procedural details necessary to employ their numerical solutions. Some reversible first-order reactions of the Hougen-Watson form have also been treated in the literature (52).

12.3.1.5.4 Effectiveness Factors for Reversible Reactions. The vast majority of the literature dealing with catalyst effectiveness factors pre-

sumes the reactions to be irreversible. However, in some cases it is possible to extend the analysis to certain reversible reactions. First-order reversible reactions have been treated for various catalyst geometries.

For flat-plate geometry where only one side of the plate is exposed to reactant gases, one may proceed as in previous subsections to show that for mechanistic equations of the form

$$A \underset{k''_{-1}}{\overset{k''_1}{\rightleftharpoons}} B \qquad (12.3.89)$$

the effectiveness factor is given by

$$\eta = \frac{\tanh \phi_{L,\text{rev}}}{\phi_{L,\text{rev}}} \qquad (12.3.90)$$

where

$$\phi_{L,\text{rev}} = L \sqrt{\frac{(k''_1 + k''_{-1})S_g\rho_p}{\mathscr{D}_{\text{eff},A}}} \qquad (12.3.91)$$

In terms of the equilibrium constant for the reaction, the Thiele modulus becomes

$$\phi_{L,\text{rev}} = L \sqrt{\frac{k''_1(K + 1)S_g\rho_p}{K\mathscr{D}_{\text{eff},A}}} \qquad (12.3.92)$$

Since the factor $(K + 1)/K$ is always greater than unity, $\phi_{L,\text{rev}}$ will always be greater (and η less) than the corresponding Thiele modulus for the forward reaction alone, other conditions remaining constant.

Analysis of the same reaction carried out using spherical catalyst pellets leads to similar results and conclusions, since the first-order rate constant is again replaced by the group $k''_1(K + 1)/K$.

12.3.2 The Consequences of Intraparticle Temperature Gradients For Catalyst Effectiveness Factors

When catalytic reaction rates become very large, it is possible that the energy generated (or consumed) by reaction cannot be dissipated (or supplied) at a rate that is sufficient to keep the entire catalyst pellet at the same temperature as the surrounding fluid. Temperature gradients may exist within the particle itself or within the boundary layer separating the particle from the bulk fluid. The presence of such gradients can have significant consequences for reactor design calculations. This section treats the influence of intraparticle temperature gradients in terms of their influence on catalyst effectiveness factors. This influence is most significant in cases where highly exothermic reactions are carried out on catalyst pellets that have low effective thermal conductivities. If catalyst effectiveness factors are defined in terms of the ratio of the observed rate to a rate evaluated assuming that the reactant concentration and the temperature at the pore mouths prevail throughout the pellet, it is possible to observe values of this parameter in excess of unity. When the catalyst interior is hotter than the peripheral regions, the increase in temperature as one moves inward may more than offset the decline in reactant concentrations so that one obtains faster local rates of reaction in the interior than in the peripheral regions.

12.3.2.1 Effective Thermal Conductivities of Porous Catalysts.
The effective thermal conductivity of a porous catalyst plays a key role in determining whether or not appreciable temperature gradients will exist within a given catalyst pellet. By the term "effective thermal conductivity", we imply that it is a parameter characteristic of the porous solid structure that is based on the gross geometric area of the pellet perpendicular to the direction of heat transfer. For example, if one considers the radial heat flux in a spherical pellet one can say that

$$q_r = -4\pi r^2 k_{\text{eff}} \frac{dT}{dr} \qquad (12.3.93)$$

where q_r is the rate of heat transfer in the radial direction and k_{eff} is the effective thermal conductivity.

The effective thermal conductivities of common commercial porous catalysts are quite low and fall within a surprisingly narrow range.

The heat transfer path through the solid phase offers considerable thermal resistance for many porous materials, particularly if the pellet is formed by tableting of microporous particles. Such pellets may be regarded as an assembly of particles that contact one another at only a relatively small number of points that act as regions of high thermal resistance.

An approximate relationship that is sometimes useful in predicting effective thermal conductivities is the geometric average value discussed by Woodside and Messmer (53).

$$k_{\text{eff}} = k_s^{1-\varepsilon_p} k_f^{\varepsilon_p} = k_s \left(\frac{k_f}{k_s}\right)^{\varepsilon_p} \qquad (12.3.94)$$

where k_f and k_s are the thermal conductivities of the bulk fluid and solid, respectively, and ε_p is the porosity of the pellet.

In spite of the difficulties of predicting k_{eff} on an a priori basis, it is still possible to choose a value that is reasonably correct because the possible range of values is roughly only 0.1 to 0.4 BTU/(hr-ft-°F) or 1.6 to 6.4×10^{-3} J/(sec-cm-°C). Typically, the thermal conductivity of reactant gases at room temperature will range from 8 to 24×10^{-5} J/(sec-cm-°C), and these values are an order of magnitude less than those of most porous catalysts under vacuum. Consequently, in this case the bulk of the energy transport will occur through the solid phase. On the other hand, for liquids, the thermal conductivity will be an order of magnitude greater and, when the catalyst pores are filled with liquid, both phases will make significant contributions to the heat transfer process. Butt (54) has developed a useful model for the thermal conductivity of porous catalysts that has shown good agreement with experimental data.

12.3.2.2 Effectiveness Factors for Nonisothermal Catalyst Pellets.
Here we indicate how previous effectiveness factor analyses may be extended to situations where the pellet is not isothermal. Consider the case of a spherical

pellet within which a catalytic reaction is taking place. If we examine an infinitesimally thin spherical shell with internal radius r similar to that shown in Figure 12.4 and write a steady-state energy balance over the interior core of the pellet, it is obvious that the heat flow outward by conduction across the sphere of radius r must be equal to the energy released by reaction within the central core. The latter quantity is just the product of the reaction rate and (minus) the enthalpy change accompanying the reaction (ΔH). Hence

$$-4\pi r^2 k_{\text{eff}} \left.\frac{dT}{dr}\right|_r = -\Delta H \mathscr{R} \quad (12.3.95)$$

where $\mathscr{R}$ is the reaction rate in the interior core. Since the reaction rate within the core must also be equal to the rate of transfer of reactant *into* the core by diffusion, the previous equation may also be written as

$$-4\pi r^2 k_{\text{eff}} \left.\frac{dT}{dr}\right|_r = -\Delta H \left(4\pi r^2 \mathscr{D}_{\text{eff}} \left.\frac{dC}{dr}\right|_r\right)$$

$$(12.3.96)$$

Negative signs are required on the right sides of the last two equations so that for exothermic reactions, the temperature will be hotter in the core than at the periphery. Integration of the last equation between radius r and the gross pellet radius R gives

$$k_{\text{eff}}(T - T_0) = \Delta H \mathscr{D}_{\text{eff}}(C - C_0) \quad (12.3.97)$$

where T_0 and C_0 are the temperature and reactant concentration at the external surface of the pellet. Rearrangement gives

$$T - T_0 = \frac{\Delta H \mathscr{D}_{\text{eff}}}{k_{\text{eff}}}(C - C_0) \quad (12.3.98)$$

This equation is a general result first derived by Damköhler (55). It is applicable for any form of the reaction rate expression since this quantity was eliminated through the require-

ment that the rate of mass transport equal the reaction rate.

The maximum temperature difference between the center of the pellet and the external surface $(T_c - T_0)$ occurs when the reactant concentration vanishes at $r = 0$.

$$(T_c - T_0)_{\text{max}} = -\frac{\Delta H \mathscr{D}_{\text{eff}} C_0}{k_{\text{eff}}} \quad (12.3.99)$$

Evaluation of the maximum temperature difference provides a useful criterion for determining if departures from isothermal behavior are significant. Substitution of the following property values into the above relation leads to a temperature difference of 200 °C between the center of the pellet and the exterior surface.

$$\Delta H = -80,000 \text{ J/mole}$$
$$\mathscr{D}_{\text{eff}} = 10^{-1} \text{ cm}^2/\text{sec}$$
$$C_0 = 4 \times 10^{-5} \text{ moles/cm}^3 \text{ (1 atm)}$$
$$k_{\text{eff}} = 16 \times 10^{-4} \text{ J/(cm-sec-°C) (alumina)}$$

Obviously, if the reactant concentration does not go to zero at the center, the temperature difference will be less but, in many cases, it will still be quite large. Large values of this temperature difference can lead to effectiveness factors for exothermic reactions that are considerably in excess of unity.

In Section 12.3.1.3 it was shown that for a first-order reaction of the type $A \rightarrow B$, a material balance on the infinitesimally thin spherical shell of Figure 12.4 leads to the following differential equation.

$$\frac{d^2C}{dr^2} + \frac{2}{r}\frac{dC}{dr} = \frac{k_1'' \rho_p S_g C}{\mathscr{D}_{\text{eff}}} \quad (12.3.100)$$

with boundary conditions

$$C = C_0 \quad \text{at } r = R$$

and

$$\frac{dC}{dr} = 0 \quad \text{at } r = 0 \quad (12.3.101)$$

At steady state, an energy balance on this same volume element gives

Output − input = energy released by reaction (by conduction)

$$\left(-4\pi r^2 k_{\text{eff}} \frac{dT}{dr}\right)_{r+\Delta r} - \left(-4\pi r^2 k_{\text{eff}} \frac{dT}{dr}\right)_r = -4\pi r^2 \, \Delta r \, (k_1'' CS_g \rho_p) \, \Delta H \qquad (12.3.102)$$

Rearranging and taking the limit as $\Delta r \to 0$ gives

$$\frac{d}{dr}\left(r^2 \frac{dT}{dr}\right) = r^2 \frac{k_1'' CS_g \rho_p \, \Delta H}{k_{\text{eff}}} \qquad (12.3.103)$$

or

$$\frac{d^2 T}{dr^2} + \frac{2}{r}\frac{dT}{dr} = \frac{k_1'' CS_g \rho_p \, \Delta H}{k_{\text{eff}}} \qquad (12.3.104)$$

The boundary conditions on the last equation require that

$$T = T_0 \quad \text{at} \quad r = R$$

and

$$\frac{dT}{dr} = 0 \quad \text{at} \quad r = 0 \qquad (12.3.105)$$

Since the reaction rate constant appearing in equations 12.3.100 and 12.3.104 depends exponentially on temperature, these equations are coupled in a nonlinear fashion and cannot be considered independently.

Exact analytical solutions of the coupled equations for simultaneous mass transfer, heat transfer, and chemical reaction cannot be obtained. However, various authors have employed linear approximations (56–57), perturbation techniques (58), or asymptotic approaches (59) to obtain approximate analytical solutions to these equations. Numerical solutions have also been obtained (60–61). Once the solution for the concentration profile has been determined, equation 12.3.98 may be used to determine the temperature profile. The effectiveness factor may also be determined from the concentration profile, using the approach we have used throughout this chapter. It is important to remember that here the effectiveness factor is defined as the ratio of the actual rate to the one that would occur if all portions of the interior of the pellet were exposed to reactant at the same concentration and the same temperature as exist at the outside surface of the pellet.

Regardless of the approach used to generate values of the effectiveness factor, this parameter can be expressed in terms of three dimensionless groups.

I A Thiele type modulus.

$$\phi_s = R \sqrt{\frac{k_1'' S_g \rho_p}{\mathscr{D}_{\text{eff}}}} \qquad (12.3.106)$$

where the reaction rate constant is evaluated at the external surface temperature.

II An Arrhenius number.

$$\gamma = \frac{E}{R_g T_0} \qquad (12.3.107)$$

where R_g is the gas constant and E the intrinsic activation energy of the reaction.

III An energy generation function

$$\beta = \frac{(-\Delta H)\mathscr{D}_{\text{eff}} C_0}{k_{\text{eff}} T_0} \qquad (12.3.108)$$

where β is positive for an exothermic reaction.

The parameter γ reflects the sensitivity of the chemical reaction rate to temperature variations. The parameter β represents the ratio of the maximum temperature difference that can exist within the particle (equation 12.3.99) to the external surface temperature. For isothermal pellets, β may be regarded as zero ($k_{\text{eff}} = \infty$). Weisz and Hicks (61) have summarized their numerical solutions for first-order irreversible reactions on spherical catalyst pellets in terms of four figures in which η is plotted versus ϕ_s. Each figure represents a fixed Arrhenius number and contains a series of curves for values of the heat generation parameter ranging from -0.8 to $+0.8$. Figure 12.10 is adapted from their work and represents an Arrhenius number that corresponds to a set of variables that may typically be encountered in industrial practice (i.e., $E = 28,000$ cal/g mole; $T = 700\,°K$). Their original article also contains curves for Arrhenius numbers of 10, 30, and 40.

Inspection of Figure 12.10 indicates that for exothermic reactions ($\beta > 0$), the effectiveness factor can exceed unity by a considerable amount. This situation occurs when circumstances are such that the increase in rate caused by the increase in temperature as one moves toward the center of the pellet more than offsets the drop in reactant concentration accompanying the movement. The overall rate of reaction is therefore greater than it would be if the same temperature and reactant concentration prevailed throughout the catalyst pellet. While values of η that exceed unity imply efficient catalyst utilization, there may be concomitant disadvantages. The large temperature increase at the center of the pellet may hasten catalyst deactivation processes in the interior core, or it may lead to adverse effects on catalyst selectivity if the activation energies of reactions leading to undesired products are greater than that of the reaction of interest.

The numerical solutions indicate that under conditions where $\eta \gg 1$, there is a large temperature gradient at the periphery of the pellet, but the gradient flattens out considerably as one approaches the center of the pellet. For fast exothermic reactions (large ϕ_s, $\beta > 0$), the interior of the pellet will be at a relatively uniform high temperature, but there will be a fairly sharp decrease in temperature as one nears the exterior surface. In this regime η becomes inversely proportional to ϕ_s, as in the isothermal case, and the bulk of the reaction takes place in the nonisothermal shell of catalyst at its periphery.

The shapes of the curves in Figure 12.10 that correspond to highly exothermic reactions are such that a single value of ϕ_s may give rise to as many as three values of η at fixed values of γ and β. These three values correspond to three different sets of circumstances in which the rate of energy release by reaction equals the rate of heat removal. In many respects the situation is similar to the problem of stability for a CSTR treated in Section 10.6. The intermediate value of η for given values of ϕ_s, γ and β corresponds to unstable conditions. The high and low values of η represent stable conditions and both can be realized in practice. Which condition will prevail in the real world depends on the direction from which the steady-state condition is approached. The highest value of η corresponds to a steep temperature profile in the pellet, and physical transfer processes limit the overall reaction rate. The lowest value of η will be close to unity and, in this case, the temperature gradients within the pellet will be quite small. In the last instance the overall rate is controlled by the rate of the chemical reaction on the catalyst surface.

We would be remiss in our obligations if we did not point out that the regions of multiple solutions are seldom encountered in industrial practice, because of the large values of β and γ required to enter this regime. The conditions under which a unique steady state will occur have been described in a number of publications, and the interested student should consult the literature for additional details. It should also be stressed that it is possible to obtain effectiveness factors greatly exceeding unity at relatively low values of the Thiele modulus. An analysis that presumed isothermal operation would indicate that the effectiveness factor would be close to unity at the low moduli involved. Consequently, failure to allow for temperature gradients within the catalyst pellet could lead to major errors.

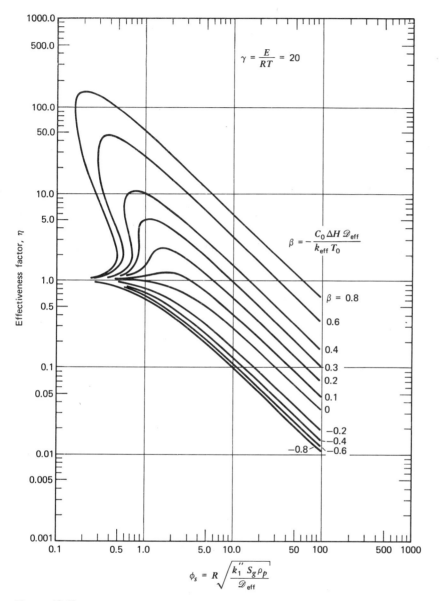

$$\gamma = \frac{E}{RT} = 20$$

$$\beta = -\frac{C_0 \, \Delta H \, \mathscr{D}_{\text{eff}}}{k_{\text{eff}} \, T_0}$$

$\beta = 0.8$

0.6

0.4
0.3
0.2
0.1
0

−0.2
−0.4
−0.8 −0.6

$$\phi_s = R \sqrt{\frac{k_1'' \, S_g \rho_p}{\mathscr{D}_{\text{eff}}}}$$

Figure 12.10

Effectiveness factor chart for first-order reaction in spherical pellets for $\gamma = 20$. [From P. B. Weisz and J. S. Hicks, *Chemical Engineering Science*, *17* (265), 1962. Copyright ⓒ 1962. Reprinted with permission of Pergamon Press.]

Under isothermal conditions, we have seen that the apparent activation energy of the reaction is approximately one half the intrinsic value when η is sufficiently low. When η exceeds unity, an opposite effect occurs (i.e., the apparent activation energy will exceed the true activation energy).

Consideration of Figure 12.10 indicates that for endothermic reactions ($\beta < 0$), nonisothermal conditions do not have as significant an effect on η as they do for exothermic cases. For endothermic reactions, η is always less than unity and the temperature within the catalyst interior is less than that at the exterior surface. Both concentration and temperature decrease as one moves radially inward. A good approximation for the effectiveness factor can be obtained by considering only the temperature gradient through the pellet and neglecting concentration gradients, even when they are appreciable. This approach has been described by Maymo, Cunningham, and Smith (62).

The following illustration indicates how experimental and calculated values of catalyst effectiveness factors may be determined for a specific case.

ILLUSTRATION 12.4 EFFECTIVENESS FACTOR DETERMINATION FOR A NONISOTHERMAL CATALYST PELLET— EXOTHERMIC REACTION

Cunningham et al (63) have studied the rate of hydrogenation of ethylene at 1 atm on a copper-magnesium oxide catalyst. They used flow reactors to study the reaction kinetics over both finely divided catalyst particles and spherical pellets, made by compressing these particles in a steel mold. They also measured the temperature difference between the center of the pellet and the external surface.

The reactor feed mixture was prepared so as to contain less than 17% ethylene (remainder hydrogen) so that the change in total moles within the catalyst pore structure would be small. This reduced the variation in total pressure and its effect on the reaction rate, so as to permit comparison of experiment results with theoretical predictions [e.g., those of Weisz and Hicks (61)]. Since the numerical solutions to the nonisothermal catalyst problem also presumed first-order kinetics, they determined the Thiele modulus by forcing the observed rate to fit this form even though they recognized that a Hougen-Watson type rate expression would have been more appropriate. Hence their Thiele modulus was defined as

$$\phi = R \sqrt{\frac{\rho_p \mathscr{R}_m}{C_0 \mathscr{D}_{eff}}} \qquad (A)$$

where $\mathscr{R}_m$ was the reaction rate per unit weight of catalyst for very small catalyst particles, and C_0 was the reactant concentration at the external surface of the pellet.

Using this definition of the Thiele modulus, the reaction rate measurements for finely divided catalyst particles noted below, and the additional property values cited below, determine the effectiveness factor for 0.5 in. spherical catalyst pellets fabricated from these particles. Comment on the reasons for the discrepancy between the calculated value of η and the ratio of the observed rate for 0.5 in. pellets to that for fine particles.

Data for Illustration 12.4

Particle size	$-100 + 150$ Tyler mesh
Pellet size	0.5 in. diameter ≈ 1.27 cm
Effective diffusivity for ethylene in pellet (estimated)	3.0×10^{-2} cm^2/sec
Effective thermal conductivity of pellets (measured)	3.5×10^{-4} cal/(cm-sec-°C)

Pellet density	1.16 g/cm^3
Void volume gram	0.236 cm^3/g
Specific surface area	90 m^2/g
Enthalpy change for reaction	$-32{,}700$ cal/mole
Temperature	80 °C
Total pressure	1 atm
Activation energy for reaction based on small particle rate measurements	17,800 cal/mole
Measured rate for small particles (extrapolated)	8×10^{-7} moles/(sec-g catalyst)
Measured rate for 0.5 in. pellets	1.8×10^{-6} moles/(sec-g catalyst)

Solution

The Arrhenius number is given by equation 12.3.107.

$$\gamma = \frac{E}{R_g T_0} = \frac{17{,}800}{(1.987)(353)} = 25.4$$

The heat generation function can be determined from equation 12.3.108.

$$\beta = \frac{(-\Delta H)\mathscr{D}_{\text{eff}}C_0}{k_{\text{eff}} T_0}$$

The external ethylene surface concentration at 80 °C, 1 atm and an ethylene mole fraction of 0.17 may be estimated using the ideal gas law as 5.87×10^{-6} moles/cm^3. Hence

$$\beta = \frac{(32{,}700)(3.0 \times 10^{-2})}{(3.5 \times 10^{-4})(353)}(5.87 \times 10^{-6}) = 0.047$$

This value of β corresponds to a maximum temperature difference between the center and the surface of the pellet of 16.2 °C. Measured values of this difference for this catalyst were 14 to 15 °C, as reported by the authors.

The pseudo Thiele modulus may be calculated from equation A.

$$\phi = \frac{1.27}{2}\sqrt{\frac{(1.16)(8 \times 10^{-7})}{(5.87 \times 10^{-6})(3.0 \times 10^{-2})}} = 1.46$$

From Figure 12.10 it is estimated that the catalyst effectiveness factor will be approximately 1.0. This figure corresponds to $\gamma = 20$, but consideration of the curve for $\gamma = 30$ in the original article leads to the same conclusion.

On the basis of the experimental measurements,

$$\eta = \frac{1.8 \times 10^{-6}}{8 \times 10^{-7}} = 2.25$$

The value calculated above is at variance with the value listed in Table 3 of reference 63, because the individuals cited used $\beta = 0.27$ based on a calculation in which the total gas phase concentration was used. Consequently, they predicted an effectiveness factor of 20.

There are several factors that may be invoked to explain the discrepancy between predicted and measured results, but the discrepancy highlights the necessity for good pilot plant scale data to properly design these types of reactors. Obviously, the reaction does not involve simple first-order kinetics or equimolal counterdiffusion. The fact that the catalyst activity varies significantly with time on-stream and some carbon deposition is observed indicates that perhaps the coke residues within the catalyst may have effects like those to be discussed in Section 12.3.3. Consult the original article for further discussion of the nonisothermal catalyst pellet problem.

12.3.3 The Influence of Catalyst Poisoning Processes on Catalyst Effectiveness Factors

In the design of commercial scale heterogeneous catalytic reactors, the activity of the catalyst will almost invariably change with time. We now wish to focus our attention on the implications of poisoning reactions for efficient use of catalyst surface areas. Since reactant molecules must interact with unpoisoned catalyst sites before reaction can occur, the poisoning process may have two effects on the reaction rate one observes.

1. It always decreases the total number of catalytic sites or the fraction of the total surface area that has the capability of promoting reaction.
2. It may increase the average distance a reactant molecule must diffuse through the pore structure before undergoing reaction.

The poisoning reaction must be viewed like any other chemical interaction between a gas phase reactant and the solid surface. The manner in which the poisoned sites are distributed throughout the pore structure is a direct consequence of the relative values of the rate at which the poison diffuses into the pore and the rate at which the poisoning reaction takes place. The problem of determining this distribution is the same as the more general problem we have been analyzing in the last several subsections: "What do the concentration profiles of reactant and product species look like for a given catalyst particle?" If the poisoning reaction is slow relative to diffusion of the poison, the poisonous reactants will be distributed uniformly throughout the porous catalyst and we will have *homogeneous* poisoning of the catalyst. This situation obviously corresponds to that in which the effectiveness factor for the poisoning reaction is essentially unity. On the other hand if the poisoning reaction is rapid relative to diffusion of the poison, the other periphery of the catalyst particle will be completely poisoned before the interior core suffers significant activity losses.

This situation is termed "*pore-mouth*" poisoning. As poisoning proceeds the inactive shell thickens and, under extreme conditions, the rate of the catalytic reaction may become limited by the rate of diffusion past the poisoned pore mouths. The apparent activation energy of the reaction under these extreme conditions will be typical of the temperature dependence of diffusion coefficients. If the catalyst and reaction conditions in question are characterized by a low effectiveness factor, one may find that poisoning only a small fraction of the surface gives rise to a disproportionate drop in activity. In a sense one observes a form of *selective poisoning*.

The two limiting cases for the distribution of deactivated catalyst sites are representative of some of the situations that can be encountered in industrial practice. The formation of coke deposits on some relatively inactive cracking catalysts would be expected to occur uniformly throughout the catalyst pore structure. In other situations the coke may deposit as a peripheral shell that thickens with time on-stream. Poisoning by trace constituents of the feed stream often falls in the pore-mouth category.

By analyzing both of the limiting situations described above, Wheeler (64–65) was able to show that the interaction of the poisoning process with the influence of intraparticle diffusion on the rates of the primary and poisoning reactions can lead to a variety of interesting relations between observed catalytic activity and the fraction of surface poisoned. Each of the limiting cases is analyzed below in terms of the mathematical model set forth by Wheeler. In both cases the reaction is presumed to proceed isothermally in a manner that is first-order in the gas phase concentration of reactant. For convenience, the straight cylindrical pore model of the catalyst is used, but the qualitative results may be regarded as essentially independent of the model employed.

12.3.3.1 Uniform Distribution of Activity Loss.
If the species giving rise to the poisoning reaction must make several collisions with the catalyst

surface before adsorption can occur, then these molecules will have a chance to diffuse deep into the catalyst pore structure before chemisorbing on the surface. Here deactivation of the catalyst will occur uniformly throughout the pore structure (homogeneous poisoning). If we denote the fraction of the surface that is poisoned by α, the fraction that is intrinsically capable of promoting reaction is equal to $(1 - \alpha)$. The reaction rate per unit surface area will then be $k_1''(1 - \alpha)C$.

A material balance over a differential element of pore length leads to the following analog of equation 12.3.10.

$$\frac{d^2C}{d(x/\bar{L})^2} = \left[\frac{2k_1''(1 - \alpha)\bar{L}^2}{\bar{r}\mathscr{D}_c}\right]C \quad (12.3.109)$$

The appropriate Thiele modulus for use with this differential equation is

$$h_p^2 = \frac{2k_1''\bar{L}^2(1 - \alpha)}{\bar{r}\mathscr{D}_c} = h_T^2(1 - \alpha) \quad (12.3.110)$$

where

h_T is the Thiele modulus for the unpoisoned case

h_p is the Thiele modulus for the poisoned case

Following the procedure used in Section 12.3.1.1 for the first-order case, it is readily shown that the effectiveness factor for the poisoned surface is given by

$$\eta_{\text{poisoned}} = \frac{\tanh h_p}{h_p} \quad (12.3.111)$$

The ratio ($\mathscr{F}$) of the reaction rate for the poisoned pore to that for the unpoisoned pore is given by

$$\mathscr{F} = \frac{\eta_{\text{poisoned}}k_1''(1 - \alpha)C_0}{\eta_{\text{unpoisoned}}k_1''C_0} \quad (12.3.112)$$

or, using equations 12.3.111 and 12.3.20,

$$\mathscr{F} = \left(\frac{\tanh h_p}{\tanh h_T}\right)\frac{h_T(1 - \alpha)}{h_p} \quad (12.3.113)$$

Using equation 12.3.110, the last equation

becomes

$$\mathscr{F} = \left[\frac{\tanh(h_T\sqrt{1 - \alpha})}{\tanh h_T}\right]\frac{h_T(1 - \alpha)}{h_T\sqrt{1 - \alpha}}$$
$$= \left[\frac{\tanh(h_T\sqrt{(1 - \alpha)})}{\tanh h_T}\right]\sqrt{1 - \alpha}$$

$$(12.3.114)$$

When the Thiele modulus for the unpoisoned pore is small (i.e., the surface is completely available), the hyperbolic tangent terms become equal to their arguments and $\mathscr{F}$ is given by

$$\mathscr{F} = 1 - \alpha \quad \text{for small } h_T \quad (12.3.115)$$

This relation is plotted as curve A in Figure 12.11 and represents the "classical case" of nonselective poisoning in which the apparent fraction of the activity remaining is equal to the fraction of the surface remaining unpoisoned. This same result is evident from equation 12.3.112 by recognizing that both effectiveness factors are unity for this situation.

On the other hand, when the Thiele modulus for the unpoisoned reaction is very large, both hyperbolic tangents in equation 12.3.114 become equal to unity and $\mathscr{F}$ is given by

$$\mathscr{F} = \sqrt{1 - \alpha} \quad \text{for very large } h_T \quad (12.3.116)$$

This relation is plotted as curve B in Figure 12.11. Smith (66) has shown that the same limiting forms for $\mathscr{F}$ are observed using the concept of effective diffusivities and spherical catalyst pellets. Curve B indicates that, for fast reactions on catalyst surfaces where the poisoned sites are uniformly distributed over the pore surface, the apparent activity of the catalyst declines much less rapidly than for the case where catalyst effectiveness factors approach unity. Under these circumstances, the catalyst effectiveness factors are considerably less than unity, and the effects of the portion of the poison adsorbed near the closed end of the pore are not as apparent as in the earlier case for small h_T. With poisoning, the Thiele modulus h_p decreases, and the reaction merely penetrates deeper into the pore.

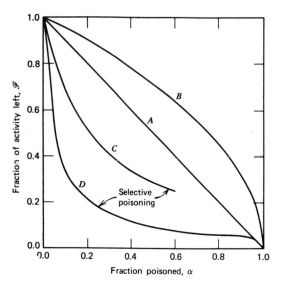

Figure 12.11
Poisoning curves for porous catalysts. Curve A is for a porous catalyst with h_r very small and poison distributed homogeneously. Curve B is for large h_r with the poison distributed homogeneously. Curves C and D correspond to preferential adsorption of poison near the pore mouths. For curve C, $h_r = 5$, and for curve D, $h_r = 20$.

12.3.3.2 Pore-Mouth or "Selective" Poisoning.
Since many metallic catalysts have high adsorption affinities, we often find that certain poison molecules are adsorbed in an immobile form after only a very few collisions with the catalyst surface. In this situation, the outer periphery of the catalyst particle will be completely poisoned while the inner shell will be completely free of poison. The thickness of the poisoned shell grows with prolonged exposure to poison molecules until the pellet is completely deactivated. During the poisoning process, the boundary between active and deactivated regions is relatively sharp.

If a fraction α of the total catalyst surface has been deactivated by poison, the pore-mouth poisoning model assumes that a cylindrical region of length $(\alpha \bar{L})$ nearest the pore mouth will have

no catalytic activity, while a region of length $[(1 - \alpha)\bar{L}]$ will be catalytically active. Figure 12.12 illustrates this situation. The reactant concentration at the pore mouth is C_0, and that at $x = \alpha \bar{L}$ is C_c (an unknown).

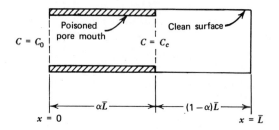

Figure 12.12
Schematic representation of the preferential adsorption of poison near the mouth of a pore. [From A. Wheeler, _Adv. Catalysis_, 3 (249), 1951. Used with permission of Academic Press.]

Under steady-state conditions, the reaction rate is equal to the rate of diffusion of reactant through the poisoned region. The latter may be written as

$$\text{Rate of diffusion} = -\pi \bar{r}^2 \mathcal{D}_c \frac{dC}{dx}$$

$$= \pi \bar{r}^2 \mathcal{D}_c \left(\frac{C_0 - C_c}{\alpha \bar{L}} \right) \tag{12.3.117}$$

where the concentration derivative may be replaced by a linear gradient in the case of a straight cylindrical pore. The rate of reaction in the unpoisoned segment of the pore can be written as the product of an effectiveness factor for a pore of length $[(1 - \alpha)\bar{L}]$ exposed to a reactant concentration C_c at its mouth, and the rate of reaction within such a pore, assuming that the reactant concentration is C_c throughout the pore length. Thus, for a first-order reaction,

Rate of reaction within the pore

$$= \eta 2\pi \bar{r}(1 - \alpha)\bar{L} k_1'' C_c \tag{12.3.118}$$

where η can be determined from equations 12.3.11 and 12.3.20 to be

$$\eta = \frac{\tanh\left[(1-\alpha)\bar{L}\sqrt{\dfrac{2k_1''}{\bar{r}\mathcal{D}_c}}\right]}{(1-\alpha)\bar{L}\sqrt{\dfrac{2k_1''}{\bar{r}\mathcal{D}_c}}} \qquad (12.3.119)$$

Thus the rate of reaction within the poisoned pore can be written as

$$\frac{\tanh\left[(1-\alpha)\bar{L}\sqrt{\dfrac{2k_1''}{\bar{r}\mathcal{D}_c}}\right]2\pi\bar{r}(1-\alpha)\bar{L}k_1''C_c}{(1-\alpha)\bar{L}\sqrt{\dfrac{2k_1''}{\bar{r}\mathcal{D}_c}}}$$

or

$$r_{poisoned} = \tanh[(1-\alpha)h_T]\sqrt{2k_1''\bar{r}\mathcal{D}_c}\,\pi\bar{r}C_c$$
$$(12.3.120)$$

$$r_{poisoned} = \frac{\tanh\left[(1-\alpha)h_T\right]\sqrt{2k_1''\bar{r}\mathcal{D}_c}\,\pi\bar{r}C_0}{1+\alpha h_T\tanh\left[(1-\alpha)h_T\right]}$$
$$(12.3.122)$$

For an unpoisoned pore, $\alpha = 0$ and the rate is given by

$$r_{unpoisoned} = (\tanh h_T)\sqrt{2k_1''\bar{r}\mathcal{D}_c}\,\pi\bar{r}C_0 \quad (12.3.123)$$

The fraction of the surface that is then apparently available for reaction is given by the ratio of the poisoned rate to the unpoisoned rate.

$$\mathscr{F} = \frac{r_{poisoned}}{r_{unpoisoned}} \qquad (12.3.123a)$$

Combination of equations 12.3.122, 12.3.123 and 12.3.123a leads to the desired result.

$$\mathscr{F} = \left\{\frac{\tanh\left[(1-\alpha)h_T\right]}{\tanh(h_T)}\right\}\left\{\frac{1}{1+\alpha h_T\tanh\left[(1-\alpha)h_T\right]}\right\} \quad (12.3.124)$$

where h_T is the Thiele modulus for the unpoisoned reaction.

The unknown concentration C_c can be eliminated between equations 12.3.117 and 12.3.120 to give

$$C_c = \frac{C_0}{1+\tanh\left[(1-\alpha)h_T\right]\alpha L\sqrt{\dfrac{2k_1''}{\bar{r}\mathcal{D}_c}}}$$

$$= \frac{C_0}{1+\alpha h_T\tanh\left[(1-\alpha)h_T\right]}$$
$$(12.3.121)$$

The reaction rate within the pore may now be obtained by combining equations 12.3.120 and 12.3.121.

In order to demonstrate the selective effect of pore-mouth poisoning, it is instructive to consider the two limiting cases of reaction conditions corresponding to large and small values of the Thiele modulus for the poisoned reaction. For the case of active catalysts with small pores, the arguments of all the hyperbolic tangent terms in equation 12.3.124 will become unity and

$$\mathscr{F} \approx \frac{1}{1+\alpha h_T}\quad[\text{for } h_T(1-\alpha) > 2] \quad (12.3.125)$$

This equation indicates that a small amount of poisoned surface can lead to a sharp decline in apparent activity. For example, if only 10% of the catalyst surface has been deactivated in the case where the Thiele modulus for the unpoisoned reaction is 40, $\mathscr{F} = 0.200$ so that the apparent activity has dropped by some 80%. In this situation, both the primary reaction and the poisoning reaction try to take place on the same surface near the pore mouth. As this region is poisoned, the reactants are forced to diffuse further into the pore before encountering a catalytically active surface. This has a concomitant strong adverse effect on the observed reaction rate. Curve D in Figure 12.11 is representative of this type of behavior.

Now consider the other extreme condition where diffusion is rapid relative to chemical reaction [i.e., $h_T(1 - \alpha)$ is small]. In this situation the effectiveness factor will approach unity for both the poisoned and unpoisoned reactions, and we must retain the hyperbolic tangent terms in equation 12.3.124 to properly evaluate $\mathscr{F}$. Curve C in Figure 12.11 is calculated for a value of $h_T = 5$. It is apparent that in this instance the activity decline is not nearly as sharp at low values of α as it was at the other extreme, but it is obviously more than a linear effect. The reason for this result is that the regions of the catalyst pore exposed to the highest reactant concentrations do not contribute proportionately to the overall reaction rate because they have suffered a disproportionate loss of activity when pore-mouth poisoning takes place.

For situations where the reaction is very slow relative to diffusion, the effectiveness factor for the poisoned catalyst will be unity, and the apparent activation energy of the reaction will be the true activation energy for the intrinsic chemical reaction. As the temperature increases, however, the reaction rate increases much faster than the diffusion rate and one may enter a regime where $h_T(1 - \alpha)$ is larger than 2, so the apparent activation energy will drop to that given by equation 12.3.85 (approximately half the value for the intrinsic reaction). As the temperature increases further, the Thiele modulus $[h_T(1 - \alpha)]$ continues to increase with a concomitant decrease in the effectiveness with which the catalyst surface area is used and in the depth to which the reactants are capable of

penetrating. In instances where $h_T(1 - \alpha)$ is large (say greater than 6) and α is appreciable (say between 0.2 and 0.8), we find that the deactivated portion of the pore represents by far the major portion of the depth to which reactants are capable of penetrating in appreciable concentrations. In this case the overall reaction rate becomes limited by the rate of diffusion through the poisoned region, and the activation energy will reflect the temperature dependence of the combined diffusivity, i.e., it will be a kilocalorie or so. Thus as the temperature increases, there is a transition from the activation energy of the intrinsic chemical reaction to an activation energy approaching zero.

12.3.4 The Influence of Intraparticle Mass Transfer Limitations on Catalyst Selectivity

The present section deals with the influence of intraparticle mass transfer processes on catalyst selectivity following the basic pattern established by Wheeler (64, 65) in his classic papers.

12.3.4.1 Independent Parallel Reactions of Different Species on the Same Catalyst. One often requires a catalyst that promotes the reactions of one component of a feedstock but does not promote the reactions of other constituents of the mixture. For example, one might desire to dehydrogenate six-membered rings, but not five-membered rings. This type of selectivity behavior may be represented by mechanistic equations of the form

$$A \overset{k_1}{\to} U + V \qquad (12.3.126)$$

$$X \overset{k_2}{\to} W + Z \qquad (12.3.127)$$

The corresponding rate equations for constant volume systems can be written as

$$\frac{d(A)}{dt} = -k_1 \eta_1 (A) \qquad (12.3.128)$$

$$\frac{d(X)}{dt} = -k_2 \eta_2 (X) \qquad (12.3.129)$$

where the effective rate constants are $k_1\eta_1$ and $k_2\eta_2$, respectively.

The selectivity of the catalyst performance can be described by the ratio of equations 12.3.128 and 12.3.129.

$$\frac{d(A)}{d(X)} = \frac{k_1\eta_1(A)}{k_2\eta_2(X)} \quad (12.3.130)$$

For an isothermal system separation of variables and integration gives

$$\ln\frac{(A)}{(A_0)} = \frac{k_1\eta_1}{k_2\eta_2}\ln\frac{(X)}{(X_0)} \quad (12.3.131)$$

or

$$\ln(1 - f_A) = \frac{k_1\eta_1}{k_2\eta_2}\ln(1 - f_X) \quad (12.3.132)$$

where f_A and f_X are the fraction conversions of A and X, respectively. The relative conversions are dependent only on the ratio of the apparent rate constants, and this group may be defined as the selectivity S:

$$S \equiv \frac{(X)}{(A)}\frac{d(A)}{d(X)} = \frac{(X)}{(A)}\frac{d(V)}{d(W)} = \frac{k_1\eta_1}{k_2\eta_2} \quad (12.3.133)$$

When the effectiveness factors for both reactions approach unity, the selectivity for two independent simultaneous reactions is the ratio of the two intrinsic reaction-rate constants. However, at low values of both effectiveness factors, the selectivity of a porous catalyst may be greater than or less than that for a plane-catalyst surface. For a porous spherical catalyst at large values of the Thiele modulus ϕ_s, the effectiveness factor becomes inversely proportional to ϕ_s, as indicated by equation 12.3.68. In this situation, equation 12.3.133 becomes

$$S = \frac{k_1}{k_2}\frac{\phi_{s,2}}{\phi_{s,1}} = \sqrt{\frac{k_1\mathscr{D}_{A,\text{eff}}}{k_2\mathscr{D}_{X,\text{eff}}}} \quad (12.3.134)$$

where we have used the definition of ϕ_s given by equation 12.3.56. If the ratio of the effective diffusivities is less than the ratio of the intrinsic

rate constants (e.g., when the larger molecule is the more reactive), then the selectivity ratio will decline as the effectiveness factors decrease. However, if the more reactive species also has a higher effective diffusivity, it is possible to obtain an *enhanced* selectivity for the porous catalyst. This situation may readily occur in dealing with molecular sieve catalysts in which certain reactants can be essentially excluded from the catalytically active regions by virtue of their molecular size or shape.

12.3.4.2 Independent Parallel Reactions of the Same Species.

Independent parallel reactions of the same species may be represented in *stoichiometric* form as

$$A \begin{array}{c} \nearrow V \\ \searrow W \end{array} \quad (12.3.135)$$

These are the types of reactions discussed in Section 9.1, and that discussion is relevant to our present considerations. A selective catalyst will be one that will promote one of these reactions relative to the other.

If the two competing reactions have the same concentration dependence, then the catalyst pore structure does not influence the selectivity because at each point within the pore structure the two reactions will proceed at the same *relative* rate, independent of the reactant concentration. However, if the two competing reactions differ in the concentration dependence of their rate expressions, the pore structure may have a significant effect on the product distribution. For example, if V is formed by a first-order reaction and W by a second-order reaction, the observed yield of V will increase as the catalyst effectiveness factor decreases. At low effectiveness factors there will be a significant gradient in the reactant concentration as one moves radially inward. The lower reactant concentration within the pore structure would then favor the lower-order reaction. For the case cited above, the yield of V relative to W will increase. When the desired reaction is of lower order than the undesired reaction, catalyst properties that lead to low effectiveness factors give a higher selectivity than those that lead to an effectiveness factor of unity. If the desired product is formed by the higher-order reaction, we should minimize the concentration gradient in the pellet and thus should use catalysts with effectiveness factors approaching unity.

12.3.4.3 Consecutive Reactions Where an Intermediate Is the Desired Product.

Consecutive reactions in which an intermediate species (V) is the desired product are often represented as a series of pseudo first-order reactions

$$A \overset{k_1}{\rightarrow} V \overset{k_2}{\rightarrow} W \qquad (12.3.136)$$

Good yields of the desired product can be obtained only when $k_1 > k_2$. Among the industrially significant reactions of this type are those involving partial oxidation and multiple substitution. Reactions represented by equation 12.3.136 were discussed in detail in Section 9.2, and that material is also pertinent here.

For a catalyst-reactant system in which the effectiveness factors for the first and second reactions both approach unity, the analysis presented in Sections 9.2 and 9.3 is appropriate. If yields are based on the initial concentration of reactant A and if no V or W is present in the original feed, equations 9.3.9 and 9.3.10 can be rearranged to express the yield of V as a function of the fraction A reacted:

Figure 12.13 contains a plot of the yield of the intermediate V as a function of the fraction A reacted for a value of k_1/k_2 equal to 5. In this case, we see that the maximum yield of V based on the initial concentration of A is equal to 66.9%.

When consecutive reactions take place within a porous catalyst, the concentrations of A and V within the pellet will be significantly different from those prevailing at the external surface. The intermediate V molecules formed within the pore structure have a high probability of reacting further before they can diffuse out of the pore.

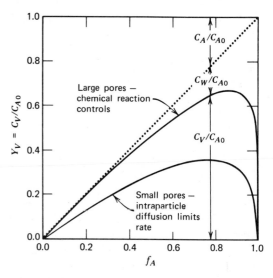

Figure 12.13
Yield of intermediate versus fraction conversion for $k_1''/k_2'' = 5$. Upper curve: $\eta = 1$; lower curve: $\eta < 0.3$.

$$Y_V = \left(\frac{1}{1 - \dfrac{k_2}{k_1}} \right) (1 - f_A) \left[(1 - f_A)^{(k_2/k_1)-1} - 1 \right] \qquad \text{for} \qquad k_2 \neq k_1 \qquad (12.3.137)$$

and

$$Y_V = (1 - f_A) \ln \left(\frac{1}{1 - f_A} \right) \qquad \text{for} \qquad k_2 = k_1 \qquad (12.3.138)$$

Consequently, catalysts with narrow pores should give lower yields of V than those with large pores.

The concentration profile of reactant A within the cylindrical pore is the same as that which prevails in the absence of the second reaction and is given by equation 12.3.13. The fact that V is a potentially reactive intermediate in no way influences the rate of reaction of A. The concentration of V within the pore under steady-state operating conditions is governed by an equation obtained in a manner similar to that used to arrive at equation 12.3.7, but in this case, the right side must allow for both the production

12.3.139 gives

$$\mathscr{D}_V \frac{d^2 C_V}{dx^2} = \frac{2}{\bar{r}}\left[k_2'' C_V - \frac{k_1'' C_{Ag}\cosh\{h_T(1 - x/\bar{L})\}}{\cosh h_T} \right]$$

(12.3.140)

where C_{Ag} is the gas phase concentration of reactant A at the pore mouth, and

$$h_T = \sqrt{\frac{2k_1''}{\bar{r}\mathscr{D}_A}}\, \bar{L}$$

(12.3.141)

($\mathscr{D}_A$ is the *combined diffusivity* of species A.)

Equation 12.3.140 can be rewritten in dimensionless form as

$$\frac{d^2(C_V/C_{Ag})}{d(x/\bar{L})^2} = \frac{2k_2''\bar{L}^2}{\bar{r}\mathscr{D}_V} \frac{C_V}{C_{Ag}} - \frac{2k_1''\bar{L}^2}{\bar{r}\mathscr{D}_V} \frac{\cosh\left[h_T(1 - x/\bar{L})\right]}{\cosh h_T}$$

(12.3.142)

or

$$\frac{d^2(C_V/C_{Ag})}{d(x/\bar{L})^2} = \left\{ h_T^2 \left(\frac{k_2''}{k_1''}\right)\left(\frac{C_V}{C_{Ag}}\right) - h_T^2 \frac{\cosh\left[h_T(1 - x/\bar{L})\right]}{\cosh h_T} \right\} \frac{\mathscr{D}_A}{\mathscr{D}_V}$$

(12.3.143)

and disappearance of V by surface reactions. Hence

$$\mathscr{D}_V \frac{d^2 C_V}{dx^2} = \frac{2}{\bar{r}}(k_2'' C_V - k_1'' C_A) \quad (12.3.139)$$

where C_V and C_A are the concentrations of V and A respectively, at a distance x from the pore mouth, and $\mathscr{D}_V$ represents the *combined diffusivity* for species V. This equation indicates that at steady state the rate of diffusive flow into the pore must equal the net rate of reaction within the pore. Combination of equations 12.3.13 and

The boundary conditions on this second-order differential equation are:

$$\text{At } x = 0 \qquad C_V = C_{Vg} \quad (12.3.144)$$

$$\text{At } x = \bar{L} \qquad \frac{dC_V}{dx} = 0 \quad (12.3.145)$$

The solution to equation 12.3.143 may be obtained by taking the sum of the homogeneous and particular solutions and evaluating the constants of integration. It may be written in the following form.

$$\frac{C_V}{C_{Vg}} = \left[1 + \frac{C_{Ag}}{C_{Vg}} \frac{\dfrac{k_1''}{k_2''}}{\left(\dfrac{k_1''\mathscr{D}_V}{k_2''\mathscr{D}_A} - 1\right)} \right] \frac{\cosh\left[h_T \sqrt{\dfrac{k_2''\mathscr{D}_A}{k_1''\mathscr{D}_V}}(1 - x/\bar{L}) \right]}{\cosh\left(h_T \sqrt{\dfrac{k_2''\mathscr{D}_A}{k_1''\mathscr{D}_V}} \right)} - \frac{C_{Ag}}{C_{Vg}} \frac{\left(\dfrac{k_1''}{k_2''}\right)\cosh\left[h_T(1 - x/\bar{L})\right]}{\left(\dfrac{k_1''\mathscr{D}_V}{k_2''\mathscr{D}_A} - 1\right)\cosh h_T}$$

(12.3.146)

The rate of production of V within a single pore must be equal to the net rate of diffusion *out* of the pore.

$$r_{V,\text{pore}} = \pi \bar{r}^2 \mathscr{D}_V \left(\frac{dC_V}{dx} \right)_{x=0} \tag{12.3.147}$$

Evaluation of the derivative and substitution of this result into equation 12.3.147 gives

$$r_{V,\text{pore}} = \frac{\pi \bar{r}^2 \mathscr{D}_V C_{Ag}}{L} \left\{ -\left[\frac{C_{Vg}}{C_{Ag}} + \frac{\dfrac{k_1''}{k_2''}}{\left(\dfrac{k_1'' \mathscr{D}_V}{k_2'' \mathscr{D}_A} - 1 \right)} \right] h_T \sqrt{\frac{k_2'' \mathscr{D}_A}{k_1'' \mathscr{D}_V}} \tanh\left(h_T \sqrt{\frac{k_2'' \mathscr{D}_A}{k_1'' \mathscr{D}_V}} \right) \right.$$
$$\left. + \left[\frac{\dfrac{k_1''}{k_2''}}{\dfrac{k_1'' \mathscr{D}_V}{k_2'' \mathscr{D}_A} - 1} \right] h_T \tanh h_T \right\} \tag{12.3.148}$$

The relative rate at which the external gas phase concentrations change is given by the ratio of the rate at which V is produced within a pore (12.3.148) to the rate at which A is consumed within a pore (12.3.17):

$$-\frac{dC_{Vg}}{dC_{Ag}} = \left[\frac{(k_1''/k_2'')}{\left(\dfrac{k_1'' \mathscr{D}_V}{k_2'' \mathscr{D}_A} - 1 \right)} - \frac{\left(\dfrac{C_{Vg}}{C_{Ag}} + \dfrac{\dfrac{k_1''}{k_2''}}{\dfrac{k_1'' \mathscr{D}_V}{k_2'' \mathscr{D}_A} - 1} \right) \sqrt{\dfrac{k_2'' \mathscr{D}_A}{k_1'' \mathscr{D}_V}} \tanh\left(h_T \sqrt{\dfrac{k_2'' \mathscr{D}_A}{k_1'' \mathscr{D}_V}} \right)}{\tanh h_T} \right] \frac{\mathscr{D}_V}{\mathscr{D}_A} \tag{12.3.149}$$

This equation gives the differential yield of V for a porous catalyst at a point in a reactor. For equal combined diffusivities and the case where h_T approaches zero (no diffusional limitations on the reaction rate), this equation reduces to equation 9.3.8, since the ratio of the hyperbolic tangent terms becomes $\sqrt{k_2'' \mathscr{D}_A / k_1'' \mathscr{D}_V}$. As h_T increases from about 0.3 to about 2.0, the selectivity of the catalyst falls off continuously. The selectivity remains essentially constant when both hyperbolic tangent terms approach unity. This situation corresponds to low effectiveness factors and, in this case, equation 12.3.149 becomes

$$-\frac{dC_{Vg}}{dC_{Ag}} = \frac{\dfrac{k_1'' \mathscr{D}_V}{k_2'' \mathscr{D}_A}}{\dfrac{k_1'' \mathscr{D}_V}{k_2'' \mathscr{D}_A} - 1} - \frac{C_{Vg}}{C_{Ag}} \sqrt{\frac{k_2'' \mathscr{D}_V}{k_1'' \mathscr{D}_A}} - \frac{\sqrt{\dfrac{k_1'' \mathscr{D}_V}{k_2'' \mathscr{D}_A}}}{\dfrac{k_1'' \mathscr{D}_V}{k_2'' \mathscr{D}_A} - 1} \tag{12.3.150}$$

or

$$-\frac{dC_{Vg}}{dC_{Ag}} = \frac{(k_1''\mathcal{D}_V/k_2''\mathcal{D}_A) - \sqrt{k_1''\mathcal{D}_V/k_2''\mathcal{D}_A}}{(\sqrt{k_1''\mathcal{D}_V/k_2''\mathcal{D}_A} - 1)(\sqrt{k_1''\mathcal{D}_V/k_2''\mathcal{D}_A} + 1)} - \frac{C_{Vg}}{C_{Ag}}\sqrt{\frac{k_2''\mathcal{D}_V}{k_1''\mathcal{D}_A}} \qquad (12.3.151)$$

Simplification gives

$$-\frac{dC_{Vg}}{dC_{Ag}} = \frac{\sqrt{k_1''\mathcal{D}_V/k_2''\mathcal{D}_A}}{\sqrt{k_1''\mathcal{D}_V/k_2''\mathcal{D}_A} + 1} - \frac{C_{Vg}}{C_{Ag}}\sqrt{\frac{k_2''\mathcal{D}_V}{k_1''\mathcal{D}_A}} \qquad (12.3.152)$$

This equation may be solved using an integrating factor approach to determine the absolute yield of V as a function of the fraction A reacted in an isothermal reactor. For the case where no V or W are present in the original gas stream, integration gives:

$$\frac{C_{Vg}}{C_{A0}} = \frac{\dfrac{k_1''}{k_2''}\left(\dfrac{\mathcal{D}_V}{\mathcal{D}_A} - \sqrt{\dfrac{k_2''\mathcal{D}_V}{k_1''\mathcal{D}_A}}\right)\left[\left(\dfrac{C_{Ag}}{C_{A0}}\right)^{\sqrt{k_2''\mathcal{D}_V/k_1''\mathcal{D}_A}-1} - 1\right]\dfrac{C_{Ag}}{C_{A0}}}{\left(\dfrac{k_1''\mathcal{D}_V}{k_2''\mathcal{D}_A} - 1\right)\left(1 - \sqrt{\dfrac{k_2''\mathcal{D}_V}{k_1''\mathcal{D}_A}}\right)} \qquad (12.3.153)$$

where C_{A0} is the concentration of reactant at the entrance to the catalytic reactor and C_{Ag} is the local gas phase concentration. This equation gives the yield of V based on the initial reactant concentration. As a function of the fraction conversion of A

$$Y_V = \frac{\dfrac{k_1''}{k_2''}\left(\dfrac{\mathcal{D}_V}{\mathcal{D}_A} - \sqrt{\dfrac{k_2''\mathcal{D}_V}{k_1''\mathcal{D}_A}}\right)[(1 - f_A)^{\sqrt{k_2''\mathcal{D}_V/k_1''\mathcal{D}_A}-1} - 1](1 - f_A)}{\left(\dfrac{k_1''\mathcal{D}_V}{k_2''\mathcal{D}_A} - 1\right)\left(1 - \sqrt{\dfrac{k_2''\mathcal{D}_V}{k_1''\mathcal{D}_A}}\right)} \qquad (12.3.154)$$

The lower curve in Figure 12.13 is a plot of the yield of V versus the fraction conversion of species A for $\mathcal{D}_V = \mathcal{D}_A$, $k_1''/k_2'' = 5$, and low effectiveness factors for the first reaction. In general, for the low effectiveness factor regime and comparable combined diffusivities, the maximum yield of an intermediate in a series of consecutive reactions will only be about half that achieved when the effectiveness factor approaches unity (64). At effectiveness factors below 0.3, the observed selectivity is nearly independent of the Thiele modulus and the effectiveness factor. The major decline in selectivity takes place as the effectiveness factor drops from unity to 0.3. If one is interested in the intermediate produced by series reactions, and the reactions in question are being carried out on catalyst pellets characterized by an effectiveness factor for the initial reaction that lies between 0.3 and 1, it is possible to increase the yield of this intermediate by going to smaller catalyst pellets or by altering the pore structure so as to increase the effective diffusivity. However, if the effectiveness factor lies well below 0.3, it is necessary to go to very large reductions in pellet size or a much more open pore structure to bring about a significant improvement in selectivity.

Wheeler's classic analysis (64, 65) has also been extended to nonisothermal situations (67–70).

Generally, under either isothermal or nonisothermal conditions, intraparticle diffusional limitations are undesirable because they reduce the selectivity below that which can be achieved in their absence. The exception to this generalization is a set of endothermic reactions that take place in nonisothermal pellets where the second reaction has an activation energy that is greater than that of the first.

The problem treated in Section 13.2 indicates' how the concepts discussed above can be applied to a practical problem.

12.4 MASS TRANSFER BETWEEN THE BULK FLUID AND EXTERNAL SURFACES OF SOLID CATALYSTS

When a solid acts as a catalyst for a reaction, reactant molecules are converted into product molecules at the fluid-solid interface. To use the catalyst efficiently, we must ensure that fresh reactant molecules are supplied and product molecules removed continuously. Otherwise, chemical equilibrium would be established in the fluid adjacent to the surface, and the desired reaction would proceed no further. Ordinarily, supply and removal of the species in question depend on two physical rate processes in series. These processes involve mass transfer between the bulk fluid and the external surface of the catalyst and transport from the external surface to the internal surfaces of the solid. The concept of effectiveness factors developed in Section 12.3 permits one to average the reaction rate over the pore structure to obtain an expression for the rate in terms of the reactant concentrations and temperatures prevailing at the exterior surface of the catalyst. In some instances, the external surface concentrations do not differ appreciably from those prevailing in the bulk fluid. In other cases, a significant concentration difference arises as a consequence of physical limitations on the rate at which reactant molecules can be transported from the bulk fluid to the exterior surface of the catalyst particle. Here, we discuss

this transport process and its implications for chemical reactor design.

In trying to analyze the flow fields within a heterogeneous catalytic reactor, we encounter an extremely complex problem, regardless of whether we consider a fixed bed, a trickle bed, a fluidized bed, or any other commonly used industrial configuration. We can, in principle, set up differential equations for the conservation of mass, momentum, and energy that take into account the chemical reaction at the fluid-solid interface. Appropriate equations are needed both for processes taking place external to the catalyst particle and for those occurring within the particle. The sets of equations are coupled through the conditions at the exterior surface of the particle, and here one must match the local fluxes of energy and chemical species as well as the local temperatures and compositions. Although these equations are readily written, no completely general method of solution exists. Sometimes it is possible to obtain analytical solutions by making appropriate assumptions so as to simplify the problem. While the assumptions are in many cases inconsistent with physical reality as expressed by the situations normally encountered in industrial practice, the predictions of conversion may well be within the accuracy of the kinetic information. Some of these analytical approaches are described in the text by Petersen (71) and interested students should consult this source. There are several intrinsic problems and/or limitations with such analyses, particularly when one attempts to employ conventional film theory. We shall find it to be more convenient to utilize average mass transfer coefficients in our analyses and to make use of the semiempirical correlations of these parameters that have been developed over the years.

12.4.1 External Mass Transfer in Packed Bed Reactors

The velocity patterns within a fixed bed reactor reflect the interactions between fluid elements flowing over different particles, variations in the available cross section for flow, the intrinsic physical properties of the fluid, and the average rate at which the fluid is supplied. The problems of analyzing the various physical transport processes in a fixed bed reactor are thus extremely complex. The conventional engineering approach to analyzing such systems involves the definition of *average* heat and mass transfer coefficients for the bed. Such coefficients are presumed to apply to the entire external surface of a given catalyst particle, even though experimental studies have shown that this situation does not correspond to physical reality. This assumption is sometimes stated by saying that the surface is *uniformly accessible*. It implies that the equations describing the transport processes can be treated as unidimensional.

The errors that result from the use of average transport coefficients are not particularly serious. The correlations that are normally employed to predict these parameters are themselves determined from experimental data on packed beds. Therefore, the applications of the correlations and the data on which they are based correspond to similar physical configurations.

Depending on the driving force we choose to employ in our analysis, there are several definitions of mass transfer coefficients that may be considered appropriate for use. If we consider an arbitrary interface between a fluid and the external surface of a catalyst particle, we might choose to define a mass transfer coefficient based on a concentration driving force (k_c) as

$$k_c = \frac{J_i}{C_{i,B} - C_{i,ES}} \qquad (12.4.1)$$

where

J_i is the molar flux of species i toward the surface relative to the molar average velocity

$C_{i,B}$ and $C_{i,ES}$ are the concentrations of species i in the bulk and at the surface, respectively.

In gas phase systems, it is convenient to define a mass transfer coefficient based on a partial pressure driving force (k_G) as

$$k_G = \frac{J_i}{P_{i,B} - P_{i,ES}} \qquad (12.4.2)$$

where $P_{i,B}$ and $P_{i,ES}$ are the partial pressures of species i in the bulk and at the surface, respectively. If the gas behaves ideally, then

$$k_c = k_G R_g T \qquad (12.4.3)$$

where R_g is the gas constant.

To explore further the usages of these and other mass transfer coefficients, consult standard references on mass transfer (72, 73) or transport phenomena (74).

Dimensional analysis of the variables characteristic of mass transfer under flow conditions suggests that the following dimensionless groups are appropriate for correlating mass transfer data.

$$\text{Reynolds number} = N_{Re} = \frac{D_p G}{\mu} \quad (12.4.4)$$

$$\text{Schmidt number} = N_{Sc} = \frac{\mu}{\rho \mathcal{D}} \quad (12.4.5)$$

$$\text{Sherwood number} = N_{Sh} = \frac{k_c D_p}{\mathcal{D}} \quad (12.4.6)$$

where

D_p is the equivalent diameter of the catalyst particle

G is the mass velocity based on the total (superficial) cross-sectional area of the reactor

μ is the fluid viscosity

ρ is the fluid density

$\mathcal{D}$ is the molecular diffusivity of the species being transferred in the system of interest

The Sherwood number is also known as the Nusselt number for mass transfer. Notice that the diameter of the catalyst pellet is used in the Reynolds and Sherwood numbers as the characteristic length dimension of the system. For flow through packed beds, the transition between laminar and turbulent flow occurs at a Reynolds number of approximately 40.

The equivalent particle diameter appearing in these dimensionless groups is the diameter of a sphere having the same external surface area as the particle in question. Thus for a cylinder of length L_c and radius r_c, the equivalent particle diameter is given by

$$4\pi \left(\frac{D_p}{2}\right)^2 = 2\pi r_c L_c + 2\pi r_c^2 \quad (12.4.7)$$

or

$$D_p = \sqrt{2 r_c L_c + 2 r_c^2} \quad (12.4.8)$$

The most convenient mathematical form for correlating mass transfer data is in terms of the well-known Chilton-Colburn (75, 76) j_D factor.

$$j_D = \frac{k_c \rho}{G} N_{Sc}^{2/3} \quad (12.4.9)$$

The functional dependence of j_D on Reynolds number has been the subject of study by many investigators [e.g., Thodos and his co-workers (77, 78), and Wilson and Geankoplis (79)]. A variety of equations have been proposed as convenient representations of the experimental data. Many of these correlations also employ the bed porosity (ε_B) as an additional correlating parameter. This porosity is the ratio of the void volume *between* pellets to the total bed volume.

Some of the more useful relations are summarized in Table 12.1.

At high Reynolds numbers the correlations for liquids and gases are quite similar, but axial mixing becomes increasingly significant in gases at low Reynolds numbers. The values of j_D predicted by equations 12.4.10 and 12.4.11 differ at most by 15% for $55 < Re < 1500$, but for low Reynolds numbers j_D for liquids will be less than j_D for gases. These equations apply to packed beds in which a single fluid fills the voids between particles. The problem of predicting appropriate mass transfer coefficients in trickle bed reactors is much more complex, and it is beyond the scope of this text. For an introduction to the problems involved, consult Satterfield's monograph (80). For other catalyst geometries (e.g., woven screens) or flow conditions (e.g., pulsatile flow), j_D factors are available and may be obtained from the current literature or standard handbooks. In addition, it is possible to estimate mass transfer coefficients using data on heat transfer coefficients, since the j_D factor is equal to an analogous factor (j_H) defined for the correlation of heat transfer data (see Section 12.5).

12.4.2 External Mass Transfer in Fluidized Bed Reactors

Since the catalysts employed in fluidized bed reactors have characteristic dimensions in the 10 to 300 micron range, the external surface

Table 12.1
Correlations for Mass Transfer Factors (j_D) in Packed Beds

Fluid	Reference	Range of N_{Re} and/or ε_B where applicable	Correlation	Equation number
Gas	Petrovic and Thodos (78)	$3 < N_{Re} < 2000$	$\varepsilon_B j_D = \dfrac{0.357}{N_{Re}^{0.359}}$	12.4.10
Liquids	Wilson and Geankoplis (79)	$55 < N_{Re} < 1500$ $0.35 < \varepsilon_B < 0.75$	$\varepsilon_B j_D = \dfrac{0.250}{N_{Re}^{0.31}}$	12.4.11
		$0.0016 < N_{Re} < 55$	$\varepsilon_B j_D = \dfrac{1.09}{N_{Re}^{2/3}}$	12.4.12

Table 12.2

Correlations for Mass and Heat Transfer Factors in Fluidized Beds

Reference	Range of variables where applicable	Correlation	Equation number
Chu et al. (81)	$1 < N'_{Re} < 30$	$j_D = 5.7(N'_{Re})^{-0.78}$	12.4.13
	$30 < N'_{Re} < 10^4$	$j_D = 1.77(N'_{Re})^{-0.44}$	12.4.14
Riccette and Thodos (82)	$100 < N'_{Re} < 7000$	$j_D = \dfrac{1}{(N'_{Re})^{0.40} - 1.5}$	12.4.15
Sen Gupta and Thodos (83)	$\dfrac{\sqrt{A_p}\,G}{\mu} > 50$	$\dfrac{\varepsilon_B j_D}{\psi} = \dfrac{0.300}{\left(\dfrac{\sqrt{A_p}\,G}{\mu}\right)^{0.35} - 1.90}$	12.4.16

where

$$N'_{Re} = \frac{D_p G}{\mu(1 - \varepsilon_B)} \qquad (12.4.17)$$

G = superficial mass velocity based on total cross-sectional area

ε_B = void fraction of bed arising from spaces between particles

A_p = surface area of a single particle

ψ = area availability factor (1.0 for spheres and 1.16 for cylinders in fluidized beds)

area per unit weight of catalyst is significantly greater than that employed in fixed bed reactors. This fact ensures that even in relatively low density fluidized beds, overall mass transfer rates will be high compared to those in fixed beds. The large extent of turbulent mixing within such systems also serves to enhance the mass transfer coefficient, but it is the high external surface to volume ratio that is most significant in ensuring rapid mass transfer.

Table 12.2 summarizes the more useful correlations for j_D and j_H that have been proposed by various investigators. The indicated correlations are quite consistent with one another in the regions where they overlap. Although the mass transfer coefficients determined from these correlations are not particularly large, they lead to high mass transfer rates when they are multiplied by the external surface area of the bed. For more detailed treatments of mass transfer

in fluidized beds, see the text of Kunii and Levenspiel (84) and the review by Beek (85).

12.4.3 Implications of External Mass Transfer Processes for Reactor Design Calculations

The only instances in which external mass transfer processes can influence observed conversion rates are those in which the intrinsic rate of the chemical reaction is so rapid that an appreciable concentration gradient is established between the external surface of the catalyst and the bulk fluid. The rate at which mass transfer to the external catalyst surface takes place is greater than the rate of molecular diffusion for a given concentration or partial pressure driving force, since turbulent mixing or eddy diffusion processes will supplement ordinary molecular diffusion. Consequently, for porous catalysts one does not encounter external mass transfer limitations, except in those circumstances where intraparticle diffusional limitations are also present.

In the presence of intraparticle mass transfer limitations, the rate per particle is expressed in terms of the species concentrations prevailing at the exterior of the catalyst. However, when external mass transfer limitations are also present, these concentrations will differ from those prevailing in the bulk. Since bulk concentrations are what one measures in the laboratory, exterior surface concentrations must be eliminated to express the observed conversion rate in terms of measurable concentrations. In the paragraphs that follow, the manner in which one eliminates surface concentrations is indicated in some detail for a specific case.

Consider an irreversible reaction of the form

$$A \rightarrow B \qquad (12.4.18)$$

that takes place on a solid catalyst surface. For the moment, we shall presume that the intrinsic chemical reaction rate follows an arbitrary kinetic expression with a functional form

$$r_m = \Phi(\text{species concentrations, temperature}) \qquad (12.4.19)$$

where the rate is expressed per unit mass of catalyst. If we take intraparticle mass transfer limitations into account, the previous relation may be written as

$$r_m = \eta\, \Phi'(\text{species concentrations, temperature}) \qquad (12.4.20)$$

where the new function Φ' must be expressed in terms of the temperature and species concentrations prevailing at the exterior surface of the catalyst pellet. For reactions that obey rate expressions that differ from simple first-order kinetics, η itself depends on the species concentrations, and the apparent order of the reaction observed in the laboratory may differ from the intrinsic order, as we have noted previously in Section 12.3.1.5.2.

Under steady-state operating conditions, the observed reaction rate must be exactly counterbalanced by the rate at which reactants are supplied to the exterior surface of the particle. On a unit mass basis, the latter rate can be written as

$$r_m = k_G(P_{A,B} - P_{A,ES})a_m \qquad (12.4.21)$$

where

$P_{A,B}$ and $P_{A,ES}$ are the reactant concentrations in the bulk fluid and at the external surface of the catalyst, respectively,

k_G is the mass transfer coefficient in appropriate units

a_m is the *external* surface area per unit mass of catalyst.

In this equation the entire exterior surface of the catalyst is assumed to be uniformly accessible. Because equimolar counterdiffusion takes place for stoichiometry of the form of equation 12.4.18, there is no net molar transport normal to the surface. Hence there is no convective transport contribution to equation 12.4.21. Let us now consider two limiting conditions for steady-state operation. First, suppose that the intrinsic reaction as modified by intraparticle diffusion effects is extremely rapid. In this case $P_{A,ES}$ will approach zero, and equation 12.4.21 indicates that the observed rate per unit mass of catalyst becomes

$$r_m = k_G a_m P_{A,B} \qquad (12.4.22)$$

The reaction will then appear to follow first-order kinetics, *regardless of the functional form of the intrinsic rate expression and of the effectiveness factor*. This first-order dependence is characteristic of reactions that are mass transfer limited. The term "diffusion controlled" is often applied to reactions that occur under these conditions.

Now consider the other extreme where the external mass transfer rate is sufficiently rapid to ensure that $P_{A,ES}$ approaches $P_{A,B}$. In this case, the concentration dependence of the reaction is given by equation 12.4.20. The apparent orders of the reaction will be given by the functional form of $\eta\Phi'$. One may or may not observe first-order kinetics. The apparent order will depend on the interaction of the intrinsic surface reaction and intraparticle diffusion effects.

To illustrate the masking effects that arise from intraparticle and external mass transfer effects, consider a surface reaction whose intrinsic kinetics are second-order in species A. For this rate expression, equation 12.4.20 can be written as

$$r_m = \eta k_{\text{intrinsic}} P_A^2 \qquad (12.4.23)$$

In the limit of low effectiveness factors where η is proportional to $(1/h)$, equation 12.3.78 indicates that the observed rate becomes

$$r_m = k_{\text{observed}} P_A^{3/2} \qquad (12.4.24)$$

when the dominant mode of transport is Knudsen diffusion.

Equations 12.4.22 and 12.4.24 indicate that the observed reaction order will differ from the intrinsic reaction order in the presence of intraparticle and/or external mass transfer limitations. To avoid drawing erroneous conclusions about intrinsic reaction kinetics, we must be careful to either eliminate these limitations by proper choice of experimental conditions or to properly take them into account in our data analysis.

Let us now consider how the external surface concentrations can be eliminated when our reaction follows simple irreversible first-order kinetics. In this instance equation 12.4.20 becomes

$$r_m = \eta k_{\text{true}}'' S_g P_{A,ES} \qquad (12.4.25)$$

At steady state, this rate is balanced by the mass transfer rate (equation 12.4.21).

$$r_m = \eta k_{\text{true}}'' S_g P_{A,ES}$$
$$- k_G(P_{A,B} - P_{A,ES})a_m \qquad (12.4.26)$$

The exterior surface concentration is then given by

$$P_{A,ES} = \frac{k_G a_m P_{A,B}}{k_G a_m + \eta k_{\text{true}}'' S_g}$$
$$= \frac{P_{A,B}}{1 + \dfrac{\eta k_{\text{true}}'' S_g}{k_G a_m}} \qquad (12.4.27)$$

With respect to the bulk concentration, the observed rate per unit weight of catalyst then becomes

$$r_m = \frac{\eta k_{\text{true}}'' S_g P_{A,B}}{1 + \dfrac{\eta k_{\text{true}}'' S_g}{k_G a_m}}$$
$$= \frac{P_{A,B}}{\dfrac{1}{\eta k_{\text{true}}'' S_g} + \dfrac{1}{k_G a_m}} \qquad (12.4.28)$$

The apparent rate constant per unit mass that one would observe in the laboratory would therefore obey the following relation.

$$\frac{1}{k_{\text{apparent}}} = \frac{1}{\eta k_{\text{true}}'' S_g} + \frac{1}{k_G a_m} \qquad (12.4.29)$$

When $k_G a_m$ is very large compared to $\eta k_{\text{true}}'' S_g$, the exterior surface concentration is not appreciably different from the bulk fluid concentration, and the apparent rate constant is equal to $\eta k_{\text{true}}'' S_g$, which reflects the interaction of the intrinsic chemical reaction and intraparticle diffusion processes. When $k_G a_m$ is very small in comparison with $\eta k_{\text{true}}'' S_g$, the exterior surface concentration will be significantly less than the bulk fluid value, and the apparent rate constant becomes equal to $k_G a_m$.

The above discussion indicates an approach that may be used in deriving an expression for the reaction rate in terms of the physical and chemical parameters of the system. However, for most practical catalyst systems, it will not be possible to arrive at closed-form expressions for the reaction rate per unit mass of catalyst. Consequently, this approach is of extremely limited utility for reactor design purposes. The most common approach to the analysis of external mass transfer limitations in heterogeneous catalytic reactors is usually couched in terms of calculations of the difference in reactant concentrations between the bulk fluid and the exterior surface. The following illustration indicates the manner in which we calculate such differences.

ILLUSTRATION 12.5 MASS TRANSFER IN A FIXED BED REACTOR—CONCENTRATION GRADIENTS BETWEEN THE BULK FLUID AND THE EXTERNAL CATALYST SURFACE

Olson, Schuler, and Smith (86) have studied the catalytic oxidation of sulfur dioxide in a differential fixed bed reactor.

$$SO_2 + \tfrac{1}{2}O_2 \rightarrow SO_3$$

The catalyst employed consisted of platinum deposited on the external surface of $1/8 \times 1/8$ in. cylindrical alumina pellets. The data below were obtained in run C-127S. Determine the composition at the exterior surface of the catalyst.

Data

Temperature = 458 °C
Pressure = 790 mm Hg
Superficial mass velocity = 245 lb/(hr-ft^2)
SO$_2$ feed rate = 0.0475 g mole/min
Air feed rate = 0.681 g mole/min
Reaction rate = 0.0940 g mole SO$_3$ produced/ (hr-g catalyst)
External surface area/mass catalyst = 5.12 ft^2/lb
Viscosity of the reaction mixture = 0.032 cp
Pellet density = 112.8 lb/ft^3 = 1.81 g/cm^3
Bed porosity (estimated) = 0.40

The methods outlined in Reid and Sherwood (87) have been used to estimate the diffusivities characteristic of binary mixtures of the various components of the fluid in the reactor.

$$\mathscr{D}_{N_2-O_2} = 0.900 \text{ cm}^2/\text{sec}$$
$$\mathscr{D}_{N_2-SO_2} = 0.602 \text{ cm}^2/\text{sec}$$
$$\mathscr{D}_{N_2-SO_3} = 0.515 \text{ cm}^2/\text{sec}$$
$$\mathscr{D}_{O_2-SO_2} = 0.598 \text{ cm}^2/\text{sec}$$
$$\mathscr{D}_{O_2-SO_3} = 0.471 \text{ cm}^2/\text{sec}$$
$$\mathscr{D}_{SO_2-SO_3} = 0.303 \text{ cm}^2/\text{sec}$$

Solution

For a differential reactor, the change in composition across the reactor will be very small, and the bulk fluid composition may be estimated from the inlet molal flow rates. Assuming that the inlet air is 79% nitrogen and 21% oxygen, the calculations below indicate the bulk fluid mole fractions and partial pressures of the various components of the reaction mixture.

$$y_{SO_2} = \frac{0.0475}{0.0475 + 0.681} = 0.0652$$

$$P_{SO_2} = y_{SO_2} P_{Total} = (0.0652)\frac{790}{760}$$
$$= 0.0678 \text{ atm}$$

$$y_{O_2} = \frac{0.21(0.681)}{0.0475 + 0.681} = 0.1963$$

$$P_{O_2} = (0.1963)\left(\frac{790}{760}\right) = 0.2040 \text{ atm}$$

$$y_{N_2} = \frac{(0.79)(0.681)}{(0.0475 + 0.681)} = 0.7385$$

$$P_{N_2} = (0.7385)\left(\frac{790}{760}\right) = 0.7677 \text{ atm}$$

The average molecular weight of the mixture is then

$$\bar{M} = 0.0652(64) + (0.1963)(32) + 0.7385(28)$$
$$= 31.1 \text{ g/mole}$$

The diffusivity of species A in a gas mixture is

given by

$$\mathscr{D}_{Am} = \frac{1 - y_A \left(\dfrac{\sum\limits_{i=1}^{n} N_i}{N_A} \right)}{\sum\limits_{j=1}^{n} \left[\dfrac{1}{\mathscr{D}_{Aj}} \left(y_j - y_A \dfrac{N_j}{N_A} \right) \right]} \quad (A)$$

where the N_i are the molar fluxes, ratios of which can be determined from the reaction stoichiometry. $\mathscr{D}_{Am}$ is the pseudo binary diffusivity of component A in the mixture, and $\mathscr{D}_{Aj}$ is the diffusivity characteristic of a binary mixture of species A and j.

Equation A may now be used to determine the diffusivity of sulfur dioxide in the gas mixture. The flux ratios may be determined from the reaction stoichiometry.

$$N_{O_2} = \tfrac{1}{2} N_{SO_2} \qquad N_{SO_2} = N_{SO_2}$$
$$N_{N_2} = 0 \qquad N_{SO_3} = -N_{SO_2}$$

Thus

$$\sum_{i=1}^{n} N_i = \frac{1}{2} N_{SO_2}$$

and

Similarly, we find the following values for the diffusivities of the other species in the reaction mixture.

$$\mathscr{D}_{O_2-m} = 0.502 \ \text{cm}^2/\text{sec}$$
$$\mathscr{D}_{N_2-m} = 1.818 \ \text{cm}^2/\text{sec}$$
$$\mathscr{D}_{SO_3-m} = 0.484 \ \text{cm}^2/\text{sec}$$

The ideal gas law may be used to determine the molar density of the reaction mixture.

$$\rho = \frac{P}{RT} = \frac{\left(\dfrac{790}{760} \right)}{82.06(731)}$$

$$= 1.73 \times 10^{-5} \ \text{g moles/cm}^3$$

Since the molecular weight of the mixture is 31.1, the corresponding mass density is $(1.73 \times 10^{-5}) \times (31.1)$ or $5.39 \times 10^{-4} \ \text{g/cm}^3$.

Schmidt numbers may now be calculated for each component.

$$(N_{Sc})_i = \frac{\mu}{\rho \mathscr{D}_{im}} = \frac{3.2 \times 10^{-4}}{5.39 \times 10^{-4} \mathscr{D}_{im}}$$

$$= \frac{0.594}{\mathscr{D}_{im}}$$

$$\mathscr{D}_{SO_2-m} = \frac{1 - y_{SO_2} \left(\dfrac{\tfrac{1}{2} N_{SO_2}}{N_{SO_2}} \right)}{\left[\dfrac{y_{SO_2} - (y_{SO_2})1}{\mathscr{D}_{SO_2-SO_2}} + \dfrac{y_{O_2} - (y_{SO_2})\tfrac{1}{2}}{\mathscr{D}_{SO_2-O_2}} + \dfrac{y_{N_2} - y_{SO_2}(0)}{\mathscr{D}_{N_2-SO_2}} + \dfrac{y_{SO_3} - y_{SO_2}(-1)}{\mathscr{D}_{SO_2-SO_3}} \right]}$$

or

$$\mathscr{D}_{SO_2-m} = \frac{1 - \tfrac{1}{2} y_{SO_2}}{\left(\dfrac{y_{O_2} - \tfrac{1}{2} y_{SO_2}}{\mathscr{D}_{SO_2-O_2}} + \dfrac{y_{N_2}}{\mathscr{D}_{N_2-SO_2}} + \dfrac{y_{SO_3} + y_{SO_2}}{\mathscr{D}_{SO_2-SO_3}} \right)}$$

Substitution of numerical values $(y_{SO_3} \simeq 0)$ gives

$$\mathscr{D}_{SO_2-m} = \frac{1 - \dfrac{0.0652}{2}}{\left(\dfrac{0.1963 - \dfrac{0.0652}{2}}{0.598} + \dfrac{0.7385}{0.602} + \dfrac{0.0652}{0.303} \right)} = 0.564 \ \text{cm}^2/\text{sec}$$

Thus

$$(N_{Sc})_{O_2} = \frac{0.594}{0.502} = 1.18$$

$$(N_{Sc})_{SO_2} = \frac{0.594}{0.564} = 1.05$$

$$(N_{Sc})_{N_2} = \frac{0.594}{1.818} = 0.327$$

$$(N_{Sc})_{SO_3} = \frac{0.594}{0.484} = 1.23$$

The particle diameter to be employed in the mass transfer calculations is that of a sphere with the same area as that of the pellets. From equation 12.4.8 with $r_c = 1/16$ and $L_c = 1/8$,

$$D_p = \sqrt{2(\tfrac{1}{16})(\tfrac{1}{8}) + 2(\tfrac{1}{16})^2} = 0.153 \text{ in.} \approx 0.389 \text{ cm}$$

The superficial mass velocity can be expressed in cgs units as

$$G = \frac{245 \text{ lb}}{\text{hr ft}^2} \times \frac{1 \text{ hr}}{3600 \text{ sec}} \times \frac{454 \text{ g}}{\text{lb}} \times \frac{\text{ft}^2}{144 \text{ in.}^2} \left(\frac{\text{in.}}{2.54 \text{ cm}}\right)^2$$

$$= 3.33 \times 10^{-2} \text{ g/cm}^2 \cdot \text{sec}$$

The Reynolds number in consistent units is then

$$N_{Re} = \frac{D_p G}{\mu} = \frac{(0.389)(3.33 \times 10^{-2})}{3.2 \times 10^{-4}} = 41$$

The j factor correlations in Table 12.1 may now be used to determine j_D. At the Reynolds number in question, equation 12.4.10 is appropriate.

$$j_D = \frac{0.357}{\varepsilon_B N_{Re}^{0.359}} = \frac{0.357}{(0.40)(41)^{0.359}} = 0.235$$

From the definition of the mass transfer j-factor,

$$k_{c,i} = \frac{G}{\rho}(N_{Sc})_i^{-2/3} j_D \tag{B}$$

At steady state, the rate of mass transfer must equal the reaction rate. Hence,

$$N_i a_m = \text{reaction rate/mass catalyst}$$

$$= -v_i r_m \tag{C}$$

where a_m is the external area per unit mass of catalyst, and N_i the molar flux of species i to the external surface of the catalyst. Since there is a change in the number of moles on reaction, we do not have equimolal counterdiffusion, and the flux of species i relative to a fixed coordinate system becomes

$$N_i = y_i \left(\sum_{j=1}^{c} N_j\right) + k_{c,i}(C_{i,B} - C_{i,ES})$$

where the subscripts B and ES refer to bulk and external surface values, respectively, and where the summation involves all c components of the mixture. The various fluxes are related by the reaction stoichiometry, and it becomes convenient to write the last equation as

$$N_i = \frac{k_{c,i}(C_{i,B} - C_{i,ES})}{1 - y_i\left(\dfrac{\sum_{j=1}^{c} N_j}{N_i}\right)}$$

where the term in the denominator represents the drift factor commonly encountered in mass transfer calculations. Now

$$N_j = \frac{v_j}{v_i} N_i$$

Hence

$$N_i = \frac{k_{c,i}(C_{i,B} - C_{i,ES})}{1 - \dfrac{y_i}{v_i}\left(\sum_{j=1}^{c} v_j\right)} \tag{D}$$

Combining equations C and D gives

$$(C_{i,B} - C_{i,ES}) = -\frac{v_i r_m}{a_m k_{c,i}}\left(1 - \frac{y_i}{v_i}\sum_{j=1}^{c} v_j\right) \quad (E)$$

which, with equation B, gives

$$C_{i,B} - C_{i,ES} = -\frac{v_i r_m \rho N_{Sc}^{2/3}}{a_m j_D G}\left(1 - \frac{y_i}{v_i}\sum_{j=1}^{c} v_j\right)$$

In consistent units

$$r_m = \frac{0.0940 \text{ g mole } SO_3 \text{ produced}}{\text{g catalyst} - \text{hr}} \times \frac{\text{hr}}{3600 \text{ sec}}$$

$$= 2.61 \times 10^{-5} \frac{\text{g moles}}{\text{sec} - \text{g catalyst}}$$

$$a_m = 5.12 \frac{\text{ft}^2}{\text{lb}} \times \frac{1 \text{ lb}}{454 \text{ g}} \times 9.29 \times 10^2 \frac{\text{cm}^2}{\text{ft}^2}$$

$$= 10.48 \text{ cm}^2/\text{g catalyst}$$

Hence

without significant error. Hence

	$y_{i,B} - y_{i,ES}$	$y_{i,B}$	$y_{i,ES}$
SO_2	0.0099	0.0652	0.0553
N_2	−0.0017	0.7385	0.7402
O_2	0.0045	0.1963	0.1918
SO_3	−0.0114	≃0.00	0.0114

The difference in mole fractions is most significant in the case of SO_2 where this difference is 15% of the bulk phase level. This result indicates that external mass transfer limitations are indeed significant, and that this difference should be taken into account in the analysis of kinetic data from this system. Note that there is a difference in nitrogen concentration between the bulk fluid and the external surface because there is a change in the number of moles on reaction, and there is a net molar flux toward

$$C_{i,B} - C_{i,ES} = -v_i \frac{2.61 \times 10^{-5}(5.39 \times 10^{-4})(N_{Sc})_i^{2/3}}{(10.48)(0.235)(3.33 \times 10^{-2})}\left[1 - \frac{y_i}{v_i}\left(-\frac{1}{2}\right)\right]$$

$$= -1.72 \times 10^{-7}(N_{Sc})_i^{2/3}v_i[1 + (y_i/2v_i)] \text{ g moles/cm}^3$$

From the ideal gas law,

$$C_i = \frac{y_i P_{\text{total}}}{RT}$$

Thus

the catalyst surface. Hence one must establish a concentration difference for nitrogen, so that the backward diffusion of nitrogen exactly counterbalances the convective transport to the catalyst surface. We should also indicate that

$$y_{i,B} - y_{i,ES} = \frac{-(82.06)(731)}{790/760}(1.72 \times 10^{-7})(N_{Sc})_i^{2/3}\left(v_i + \frac{y_i}{2}\right)$$

$$= -9.93 \times 10^{-3}(N_{Sc})_i^{2/3}\left(v_i + \frac{y_i}{2}\right)$$

Substitution of the previously calculated Schmidt numbers and composition values, and an appropriate average value for y_i permits one to determine the difference in mole fractions. In the present instance, we may use $y_{i,B}$ values

the mole fractions at the external surface sum to only 0.9987. In large measure this discrepancy arises because we have used feed composition values in our calculations rather than local average bulk fluid compositions.

We would be remiss if we did not indicate that a significant temperature difference also exists between the bulk fluid and the external surface. This ΔT has a far greater effect on the observed rate than does the SO_2 concentration difference. Illustration 12.6 indicates how the temperature difference may be calculated.

In view of the fact that our results are reasonably sensitive to the estimate of the bed porosity used in the analysis, these results are not bad. If one had employed a value of 0.3 or 0.5 rather than 0.4 for ε_B, j_D would change significantly and this would have a major influence on the calculated concentration (or mole fraction) differences. Unfortunately, bed porosity data were not noted in the article cited. In an experimental program being conducted as an aspect of a reactor design, this parameter could easily be determined.

Yoshida, Ramaswami, and Hougen (88) have developed a nomograph that in essence is based on the procedure employed in the last example. Although it employs a j_D correlation that predates those given in Table 12.1 and thus is less accurate, it eliminates many of the detailed calculations involved in determining when significant external concentration differences exist. For all practical purposes, the results obtained by using the nomograph are indistinguishable from those obtained using the procedure employed in Illustration 12.5.

Before terminating the discussion of external mass transfer limitations on catalytic reaction rates, we should note that in the regime where external mass transfer processes limit the reaction rate, the apparent activation energy of the reaction will be quite different from the intrinsic activation energy of the catalytic reaction. In the limit of complete external mass transfer control, the apparent activation energy of the reaction becomes equal to that of the mass transfer coefficient, typically a kilocalorie or so per gram mole. This decrease in activation energy is obviously much greater than the decrease encountered when intraparticle diffusion processes limit the catalytic reaction rate.

12.5 HEAT TRANSFER BETWEEN THE BULK FLUID AND EXTERNAL SURFACES OF SOLID CATALYSTS

The energy effects that accompany chemical reactions can lead to significant temperature differences between a bulk fluid and the external surface of a catalyst on which reaction is taking place. At steady state, the rate of energy release (or consumption) by reaction must equal the rate at which heat is transferred to the fluid. The temperature gradient necessary to sustain the heat transfer may be appreciable, even in situations where concentration differences between the bulk fluid and the external catalyst surface are negligibly small. Consequently, these temperature differences can act to obscure the reaction kinetics in many more situations than can the mass transfer effects previously discussed. In laboratory studies aimed at a determination of the rate expression for the intrinsic chemical reaction, external gradients in temperature and concentration can be made negligible by operating under the following constraints.

1. Using a reactant stream that is diluted with inerts to reduce the reaction rate so that the energy evolved per unit volume is greatly reduced below that encountered in the absence of inerts.
2. Employing high mass velocities to minimize resistances to heat and mass transfer.

In such studies one may also eliminate intraparticle gradients of temperature and composition by using very fine catalyst particles or by confining the catalytic species to the exterior surface of a nonporous or impervious pellet. Unfortunately, the conditions that are optimum for the elucidation of the intrinsic chemical kinetics are often inappropriate for use in commercial scale reactors, and the design engineer must be able to take into account both exterior and intraparticle gradients in concentration and temperature. We discuss here the methods used to evaluate exterior temperature gradients in both fixed and fluidized bed reactors.

Heat transfer between a fluid and a catalyst particle occurs primarily through the same combination of molecular and convective processes that are responsible for mass transfer in such systems. At sufficiently high temperatures, one must also allow for radiation contributions to the energy transfer processes within the reactor. Radiation effects are not considered in the correlations discussed below. They are not significant at temperatures below 400 °C for packed bed reactors comprised of pellets with characteristic dimensions of less than 1/4 in. (89). At higher temperatures, they may be included in an energy balance on a segment of the reactor if appropriate view factors and other geometric considerations are specified. For present purposes, however, their inclusion would greatly complicate the analysis that follows, without shedding additional light on the fundamental principles involved.

The approach we use in analyzing the heat exchange between a solid catalyst particle and the surrounding fluid to determine if a significant temperature difference exists is in many respects quite similar to that employed in the analysis of mass transfer in Section 12.4. Both analyses are based largely on the Chilton-Colburn analogies for the corresponding transport processes. However, our earlier comment that external mass transfer limitations are less significant than intraparticle limitations cannot be generalized to the heat transfer process. At high Reynolds numbers, the heat transfer coefficients, like the mass transfer coefficients, will be large and the corresponding temperature and concentration differences between the bulk fluid and the exterior surface of the catalyst will be small. However, at low Reynolds num-

bers where both transfer coefficients are small, it is quite possible for substantial temperature differences to exist between the bulk fluid and the external surface, even when the corresponding concentration differences are small and when intraparticle concentration and temperature gradients are negligible ($\eta \simeq 1$). The major resistance to heat transfer is the laminar film adjacent to the solid surface rather than the pellet itself.

Experimental data on heat transfer in fixed and fluidized bed reactors are correlated in terms of a j factor for heat transfer.

$$j_H \equiv \left(\frac{h}{C_p G}\right)\left(\frac{C_p \mu}{\kappa}\right)^{2/3} = \frac{h}{C_p G} N_{Pr}^{2/3} \quad (12.5.1)$$

where

μ is the fluid viscosity

C_p is the constant pressure heat capacity *per unit mass of fluid*

κ is the thermal conductivity of the fluid

h is the heat transfer coefficient between the catalyst particle and the bulk fluid

N_{Pr} is the Prandtl number

The functional dependence of j_H on Reynolds number has been the subject of study by many investigators, and a variety of equations have been proposed for correlation of the available data for fixed bed (77, 88) and fluidized bed reactors (81–83). Boundary layer theory indicates that the Chilton-Colburn analogy, $j_H = j_D$, represents an asymptotic solution for forced convection in three-dimensional flows in any geometry, provided that the Peclet number is large (90). [The Peclet number for mass transfer is the product of the Reynolds and Schmidt numbers $(D_p G/\mathscr{D}\rho)$, while the Peclet number for heat transfer is the product of the Reynolds and Prandtl numbers $(D_p G C_p/\kappa)$.] The relation $j_D = j_H$ agrees well with experimental data for many flow geometries. Although the literature correlations frequently indicate that the ratio j_H/j_D is slightly greater than unity for flow through packed beds (e.g., 79), the deviations are usually small. They may be explained by the fact that measured rates of vaporization or sublimation cancel out of the expression for the ratio j_H/j_D when one attempts to calculate these parameters from conventional experimental data. The deviation from unity then merely reflects an average of values determined from a large number of readings with a particular humidity chart (91).

For most purposes, the correlations for j_D presented in Tables 12.1 and 12.2 also suffice for estimating j_H. There is, however, one additional correlation for fluidized beds that is worth noting. On the basis of data for the fluidization of 20 to 40 mesh silica and alumina gel particles in air at Reynolds number values $(D_p G/\mu)$ ranging from 9 to 55, Kettenring et al. (92) suggest that

$$\frac{hD_p}{\kappa} = 0.0135(N_{\text{Re}})^{1.3} \qquad (12.5.2)$$

rate at which thermal energy is exchanged between the fluid and the solid catalyst.

$$r_m(-\Delta H) = ha_m(T_{ES} - T_B) \qquad (12.5.4)$$

Noting that $r_{Am} = v_A r_m$, equations 12.5.3 and 12.5.4 may be combined to give

$$\frac{k_{c,A}(C_{A,B} - C_{A,ES})a_m(-\Delta H)}{-v_A\left(1 - \dfrac{y_A}{v_A}\sum\limits_{j=1}^{c} v_j\right)} = ha_m(T_{ES} - T_B)$$

$$(12.5.5)$$

or

$$T_{ES} - T_B = \frac{k_{c,A}(\Delta H)(C_{A,B} - C_{A,ES})}{h\left(v_A - y_A\sum\limits_{j=1}^{c} v_j\right)}$$

$$(12.5.6)$$

If the expressions for h and $k_{c,A}$, derived from equations 12.5.1 and 12.4.9, are substituted into the last equation, we find that

$$T_{ES} - T_B = \left(\frac{j_D}{j_H}\right)\left(\frac{N_{\text{Pr}}}{N_{\text{Sc}}}\right)^{2/3}\left(\frac{\Delta H/v_A}{\rho C_p}\right)\frac{(C_{A,B} - C_{A,ES})}{\left[1 - y_A \sum\limits_{j=1}^{c}(v_j/v_A)\right]} \qquad (12.5.7)$$

Notice, however, that none of these correlations are appropriate for use when radiation makes a significant contribution to the heat transfer process.

At steady state, the rate at which reactants are supplied to the external surface of the catalyst by mass transfer must be equal to the rate at which they are consumed by the catalytic reaction. Per unit mass of catalyst, the rate of disappearance of species A is then given by

$$-r_{A,m} = \frac{k_{c,A}(C_{A,B} - C_{A,ES})a_m}{\left[1 - \dfrac{y_A}{v_A}\left(\sum\limits_{j=1}^{c} v_j\right)\right]} \qquad (12.5.3)$$

where the term in brackets is the drift factor. The rate of reaction multiplied by the energy release per unit extent of reaction must also equal the

The temperature difference is thus directly proportional to the heat of reaction per mole of diffusing species and to the difference in concentration between the bulk fluid and the exterior surface of the solid. If we recognize that $j_D \simeq j_H$ from the Chilton-Colburn relation, and that the ratio of Prandtl and Schmidt numbers is close to unity for many simple gas mixtures, the previous relation may be approximated as

$$T_{ES} - T_B \simeq \left(\frac{\Delta H}{v_A \rho C_p}\right)\frac{(C_{A,B} - C_{A,ES})}{\left[1 - y_A \sum\limits_{j=1}^{c}(v_j/v_A)\right]}$$

$$\text{(for gases)} \qquad (12.5.8)$$

Thus if the method developed in Section 12.4 is used to evaluate the concentration difference, this equation may be used to determine an approximate value for the temperature difference. (In many cases the drift factor term in brackets is essentially unity.)

In the limit where the external surface concentration becomes very small compared to the bulk fluid concentration, we obtain the maximum temperature difference.

$$\Delta T_{max} \simeq \frac{\left(\dfrac{\Delta H C_{A,B}}{v_A \rho C_p}\right)}{\left[1 - y_A \sum\limits_{j=1}^{c} (v_j/v_A)\right]} \quad \text{(for gases)} \tag{12.5.9}$$

In many cases of interest, the maximum temperature difference will be over 100 °C, and values of this magnitude have been observed experimentally for several catalytic reactions.

Combination of equations 12.5.8 and 12.5.9 gives

$$T_{ES} - T_B = \left(1 - \frac{C_{A,ES}}{C_{A,B}}\right)(\Delta T)_{max} \tag{12.5.10}$$

Because of the large magnitude of ΔT_{max}, it is quite possible that significant temperature differences will exist, even when the ratio $C_{A,ES}/C_{A,B}$ is close to unity. For example, if the exterior surface concentration is 95% of the bulk fluid level and the adiabatic temperature rise for the

for heat transfer coefficients. Hence

$$T_{ES} - T_B = \frac{r_m(-\Delta H)}{h a_m} \tag{12.5.11}$$

which may also be written in terms of the j_H factor as

$$T_{ES} - T_B = \frac{r_m(-\Delta H)}{C_p G j_H N_{Pr}^{-2/3} a_m} \tag{12.5.12}$$

Yoshida, Ramaswami, and Hougen (88) have also developed nomographs for the solution of equation 12.5.12. Illustration 12.6 indicates an analytical approach by which the temperature difference can be evaluated using the correlations given in Tables 12.1 and 12.2.

ILLUSTRATION 12.6 HEAT TRANSFER IN A FIXED BED REACTOR—TEMPERATURE GRADIENTS BETWEEN THE BULK FLUID AND THE EXTERNAL CATALYST SURFACE

In Illustration 12.5, we considered the problem of estimating the concentration differences that exist between the bulk fluid and a catalyst used for the oxidation of sulfur dioxide. If the reported temperature is that of the bulk fluid, determine the external surface temperature corresponding to the conditions cited. Additional useful data are:

Heat of reaction evaluated for the temperature range of interest $= -96.95$ kJ/mole

Heat capacity (C_p) of the reaction mixture $= 1.07$ J/g-°K

Thermal conductivity of the reaction mixture $= 3.70 \times 10^{-4}$ J/(cm-sec-°K)

---

reaction is 300 °C, equation 12.5.10 indicates that the catalyst surface is 15 °C hotter than the bulk fluid. This difference would be sufficient to lead to a significant enhancement in the reaction rate relative to bulk fluid conditions.

Another approach to estimating the temperature difference involves the solution of equation 12.5.4 for the temperature difference using measured reaction rates and empirical correlations

Solution
The data on the thermal conductivity and the heat capacity permit us to calculate the Prandtl number for the conditions of interest.

$$N_{Pr} = \frac{C_p \mu}{\kappa}$$

$$= \frac{(1.07 \text{ J/g-°K})(3.2 \times 10^{-4} \text{ g/cm-sec})}{3.70 \times 10^{-4} \text{ J/cm-sec-°K}}$$

or

$$N_{Pr} = 0.93$$

For the conditions cited, it was shown that j_D was 0.235 and, by the Chilton-Colburn analogy, $j_H = 0.235$. From the definition of the heat transfer factor j_H, it follows that

$$h = j_H C_p G(N_{Pr})^{-2/3}$$

Hence

$$h = (0.235)(1.07)(3.33 \times 10^{-2})(0.93)^{-2/3}$$
$$= 8.79 \times 10^{-3} \text{ J/cm}^2\text{-sec-}°\text{K}$$

At steady state, equation 12.5.11 must be satisfied. Thus

$$T_{ES} - T_B = \frac{(2.61 \times 10^{-5})(96,950)}{(8.79 \times 10^{-3})(10.48)} = 27.5 \,°\text{C}$$

or

$$T_{ES} = 27.5\,°\text{C} + 458 = 486\,°\text{C}$$

Thus, for this reaction, a substantial temperature difference exists between the solid surface and the bulk fluid. It has a far greater influence on the observed rate than the corresponding concentration difference. Differences of this magnitude can clearly lead to complications in the analysis of data obtained in laboratory scale reactors. If such differences will exist at the operating conditions to be employed in a commercial scale reactor, the design engineer must be sure to take them into account in his analysis.

12.6 "GLOBAL" REACTION RATES

The rates at which chemical transformations take place are in some circumstances strongly influenced by mass and heat transfer processes (see Sections 12.3 to 12.5). In the design of heterogeneous catalytic reactors, it is essential to utilize a rate expression that takes into account the influence of physical transport processes on the rate at which reactants are converted to products. Smith (93) has popularized the use of the term *global reaction rate* to characterize the overall rate of transformation of reactants to

products in the presence of heat and mass transfer limitations. We shall find this term convenient for use throughout the remainder of this chapter. Global rate expressions then include both external heat and mass transfer effects on the reaction rate, and the efficiency with which the internal surface area of a porous catalyst is used. Here we indicate how to employ the concepts developed in Sections 12.3 to 12.5 to arrive at global rate expressions.

In Section 6.3 we indicated that a catalytic reaction taking place on the surface of a porous solid involves the following sequence of physical and chemical processes.

1. Mass transfer of reactants from the bulk fluid to the gross exterior surface of the catalyst particle.
2. Transport of reactants by diffusion from the exterior surface into the pores of the catalyst.
3. Chemisorption of one or more reactant species.
4. Reaction on the surface.
5. Desorption of products.
6. Transport of products by diffusion from the interior pores to the gross external surface of the catalyst.
7. Mass transfer of products from the external surface to the bulk fluid.

At steady state, the rates of each of the individual steps will be the same, and this equality is used to develop an expression for the global reaction rate in terms of bulk-fluid properties. Actually, we have already employed a relation of this sort in the development of equation 12.4.28 where we examined the influence of external mass transfer limitations on observed reaction rates. Generally, we must worry not only about concentration differences between the bulk fluid and the external surface of the catalyst, but also about temperature differences between these points and intraparticle gradients in temperature and composition.

In laboratory experiments carried out to obtain rate data for subsequent design calculations, we normally try to operate under conditions that minimize the effects of heat and mass transfer on conversion rates so as to facilitate the extraction of the rate expression for the intrinsic chemical reaction from the observed reaction rate (the global rate). We use very fine particles and high fluid velocities to minimize these limitations so that differences in composition and temperature between the bulk fluid and the fluid within the pore structure of the catalyst are insignificant. In the absence of such differences, the global reaction rate corresponds to the intrinsic reaction rate, evaluated at the bulk fluid composition and temperature. When such differences are significant, the global rate may be appreciably higher or appreciably lower than the intrinsic rate corresponding to bulk fluid properties.

Laboratory reactors and industrial scale equipment are seldom operated under similar flow and heat transfer conditions. To obtain a global rate that is useful for design purposes, one must combine the intrinsic chemical reaction rate expression with expressions for heat and mass transfer rates corresponding to industrial operating conditions. As a general rule, the global rate reduces to the intrinsic expression evaluated at bulk fluid properties when one of the chemical steps at the catalyst surface is slow. However, if economic considerations require the use of fluid velocities or catalyst pellet sizes that differ from those employed in the laboratory, the global rate may be quite different from the intrinsic rate at bulk fluid conditions. In such cases the reaction may be mass or heat transfer limited, or controlled by pore diffusion processes. Figure 12.14 indicates some of the reactant concentration profiles that may result from control by various steps in the sequence enumerated earlier. In commercial scale fixed bed reactors, external resistances to mass transfer are usually small compared to the intraparticle

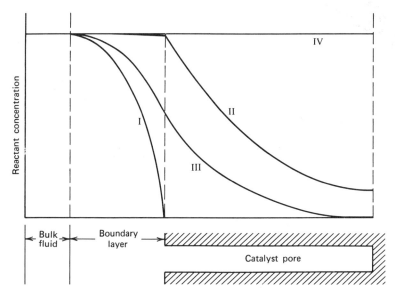

Figure 12.14
Schematic representation of reactant concentration profiles in various global rate regimes. I: External mass transfer limits rate. II: Pore diffusion limits rate. III: Both mass transfer effects are present. IV: Mass transfer has no influence on rate.

resistance for normal operating conditions. However, we have seen earlier that significant external temperature differences can exist even when the external concentration difference is small. In fact, for gaseous reactants the external temperature difference will probably be much greater than any intraparticle differences.

Let us now turn our attention to the problem of determining the global reaction rate at some arbitrary point in a heterogeneous catalytic reactor from a knowledge of the following parameters.

1. The intrinsic rate expression.
2. The properties of the bulk fluid for the region in question (temperature and composition).
3. The flow conditions (mass velocity, reactor diameter, etc.).
4. The heat of reaction at the bulk fluid temperature.
5. The physical properties of the fluid (Schmidt and Prandtl numbers, heat capacity, etc.).
6. The shape and size of the catalyst.
7. The physical properties of the catalyst (specific surface area, porosity, effective thermal conductivity, effective diffusivity, pellet density, etc.).
8. The porosity of the bed (ε_B).

Values of all of these parameters must be available or estimated if we are to determine the global reaction rate. Some of these quantities can be evaluated from standard handbooks of physical property data, or generalized correlations such as those compiled by Reid and Sherwood (87). Others can be determined only by experimental measurements on the specific reactant/catalyst system under consideration.

Determination of the global reaction rate that applies to a given segment of reactor volume involves the use of equations developed earlier.

The rate of mass transfer to the external surface of the catalyst is given by equation E of Illustration 12.5.

$$r_m = \frac{k_c a_m (C_{A,B} - C_{A,ES})}{-\left[v_A - y_A \sum_{j=1}^{c} (v_j)\right]} \quad (12.6.1)$$

where the mass transfer coefficient (k_c) is that for species A.

The reaction rate averaged over the catalyst pellet can be expressed in terms of the effectiveness factor η as

$$r_m = \eta \Psi_m(C_{ES}, T_{ES}) \quad (12.6.2)$$

where the functional form of Ψ_m is presumed to be known. The subscript m is employed to emphasize the fact that $\Psi_m(C_{ES}, T_{ES})$ is the reaction rate per unit mass of catalyst in the absence of intraparticle diffusional limitations. To determine a global rate expression from equations 12.6.1 and 12.6.2 we also need to know the dependence of η on the Thiele modulus ϕ_s, the definition of the Thiele modulus for the intrinsic reaction kinetics involved, the Arrhenius number γ (12.3.107) and the energy generation function β (12.3.108). In order to relate bulk fluid and external surface temperatures, equations 12.5.4 and 12.5.5 are employed

$$r_m(-\Delta H) = h a_m(T_{ES} - T_B) \quad (12.6.3)$$

or, using equation 12.6.1,

$$\frac{k_c a_m (C_{A,B} - C_{A,ES})(-\Delta H)}{-\left(v_A - y_A \sum_{j=1}^{c} v_j\right)} = h a_m(T_{ES} - T_B)$$

$$(12.6.4)$$

Equations 12.6.2 to 12.6.4 and the relation between η, ϕ_s, γ, and β are sufficient to calculate the global rate at specified values of T_B and C_B. Unfortunately, information on the last relation is rather limited. The curves presented in Figure 12.10 and reference 61 give the desired relation for first-order kinetics, but numerical solutions for other reaction orders are not available to this extent; we will presume that numerical solutions may be generated if needed for design purposes.

For isothermal systems, it is occasionally possible to eliminate the external surface concentrations between equations 12.6.1 and 12.6.2 and arrive at a global rate expression involving only bulk fluid compositions (e.g., equation 12.4.28 was derived in this manner). In general, however, closed form solutions cannot be achieved and an iterative trial and error procedure must be employed to determine the global rate. One possible approach is summarized below.

1. Use the j factor correlations of Section 12.4 to determine appropriate heat and mass transfer coefficients.
2. Assume a value for T_{ES}.
3. Determine C_{ES} from 12.6.4.
4. From the effectiveness factor relation, determine $\eta\,(T_{ES}, C_{ES})$.
5. Calculate the reaction rate per unit mass using equation 12.6.2. (This implies a knowledge of the intrinsic rate expression.)
6. Determine T_{ES} from equation 12.6.3 using the reaction rate determined in step 5.
7. Compare the initial guess of T_{ES} with the calculated value and iterate until agreement is reached.
8. Substitution of the final values of T_{ES} and/or C_{ES} into any of the various equations for r_m gives the global rate.

This procedure obviously requires machine computation capability if it is to employed in reactor design calculations. Fortunately, there are many reactions for which the global rate reduces to the intrinsic rate, which avoids the necessity for calculations of this type. On the other hand, several high tonnage processes (e.g., SO_2 oxidation) are influenced by heat and mass transfer effects and one must be fully cognizant of their implications for design purposes.

12.7 DESIGN OF FIXED BED REACTORS

Fixed or packed bed reactors have many advantages relative to other types of heterogeneous catalytic reactors and, consequently, are employed much more widely in the chemical industry than any other basic reactor type. This section indicates in some detail the procedures involved in the design of a tubular fixed bed reactor. This task involves the use of global rate information obtained through the techniques developed earlier to predict the composition of the effluent from a reactor for a given set of design parameters.

The design problem can be approached at various levels of sophistication using different mathematical models of the packed bed. In cases of industrial interest, it is not possible to obtain closed form analytical solutions for any but the simplest of models under isothermal operating conditions. However, numerical procedures can be employed to predict effluent compositions on the basis of the various models. In the subsections that follow, we shall consider first the fundamental equations that must be obeyed by all packed bed reactors under various energy transfer constraints, and then discuss some of the simplest models of reactor behavior. These discussions are limited to pseudo steady-state operating conditions (i.e., the catalyst activity is presumed to be essentially constant for times that are long compared to the fluid residence time in the reactor).

Models of fixed bed reactors can be categorized in a couple of ways. One basis lies in the number of spatial coordinates employed in the equations used to describe the model. One-dimensional models take into account variations in composition and temperature along the length of the reactor, while two-dimensional models also allow for variations in these properties in the radial direction. A second basis for categorizing reactor models lies in the manner in which one envisions the reaction as being distributed throughout the catalyst bed. In this sense, the models are viewed as either pseudo homogeneous or heterogeneous. For the pseudo homogeneous models, one envisions the reactions as taking place throughout the reactor volume, and not as localized at the catalyst surface. The rate expressions for use with these models are obtained by taking the product of the global reaction rate per unit mass of catalyst and the *bulk density* (ρ_B) of the catalyst (i.e., the total mass of catalyst divided by the *total volume* of the reactor).

$$r_v = r_m \rho_B \qquad (12.7.1)$$

The units of r_v are moles converted/(volume-time), and r_v is identical with the rates employed in homogeneous reactor design. Consequently, the design equations developed earlier for homogeneous reactors can be employed in these terms to obtain estimates of fixed bed reactor performance. Two-dimensional, pseudo homogeneous models can also be developed to allow for radial dispersion of mass and energy.

Heterogeneous models of fixed bed reactors explicitly account for the presence of the solid catalyst by writing material and energy balance equations for both the solid phase and the fluid phase. Several types of heterogeneous models can be employed, depending on the complications one is willing to introduce. The basic heterogeneous model considers only transport by plug flow, but differentiates between bulk fluid properties and those prevailing at the external surface of the catalyst pellet. Models may also be developed that allow for intraparticle gradients and for radial variations in system properties. The necessity for using a heterogeneous model can be circumvented through the use of *global rate* expressions in terms of r_v in the corresponding pseudo homogeneous models described in Section 12.7.2. I have chosen to utilize this approach for the level of this text. [See Froment (94, 95) for critical reviews of several heterogeneous models.]

12.7.1 General Considerations

Prediction of the performance of a fixed bed reactor under specified operating conditions is a problem whose solution is readily expressed in words. Unfortunately, this solution is not so easily expressed in the quantitative terms the engineer needs to prepare equipment specifications and cost estimates. In principle, the solution involves the use of a heterogeneous reaction rate expression in a set of material balance equations for various chemical species and a corresponding conservation equation for energy. The resultant differential equations and the attendant boundary conditions are then evaluated over the catalyst bed to determine the conditions at the reactor outlet. Before discussing these equations in detail, however, we consider briefly a few subjects that are relevant to the general problem of fixed bed reactor design. These include pressure drop, dispersion, and heat transfer in fixed beds.

12.7.1.1 Pressure Drop in Fixed Bed Reactors. In order to simplify a reactor design analysis, we often assume that the pressure drop across a reactor is negligible and has little influence on reactor size requirements. However, there are some situations in reactor design where it is important to consider the implications of pressure drop, particularly when dealing with the flow of compressible fluids through a fixed or fluidized bed reactor. Because of its significance in a wide variety of chemical engineering operations, flow through packed beds has been studied by many investigators through the years. The most useful approaches to predicting pressure drop through packed beds are those based on empirical correlations for the friction factor (f_M), defined in the following manner.

$$\frac{\mathscr{P}_0 - \mathscr{P}_L}{L} = \frac{4f_M}{D_p^*}\left(\tfrac{1}{2}\rho v_0^2\right) \qquad (12.7.2)$$

where

L is the length of the packed column

ρ is the fluid density

v_0 is the superficial velocity (the average linear velocity the fluid would have in the absence of packing)

D_p^* is the equivalent particle diameter defined as

$$D_p^* = \frac{6}{a_v}$$

where

a_v is the area per unit volume of an individual particle

$\mathscr{P}$ represents the combined effects of static pressure (P) and gravitational force and is defined as

$$\mathscr{P} = P + \rho g z \qquad (12.7.3)$$

where z is the distance upward (as opposed to gravity) from a chosen reference plane.

Subscripts 0 and L refer to inlet and outlet $\mathscr{P}$ values, respectively.

Several correlating equations for the friction factor have been proposed for both the laminar and turbulent flow regimes, and plots of f_M (or functions thereof) versus Reynolds number are frequently presented in standard fluid flow or chemical engineering handbooks (e.g., 96, 97). Perhaps the most useful of the correlations is that represented by the Ergun equation (98)

outlet densities. This approximation is useful in computer-aided design problems.

12.7.1.2 Dispersion in Packed Bed Reactors.

In Chapter 11, we indicated that deviations from plug flow behavior could be quantified in terms of a dispersion parameter that lumped together the effects of molecular diffusion and eddy diffusivity. A similar dispersion parameter is used to characterize transport in the radial direction, and these two parameters can be used to describe radial and axial transport of matter in packed bed reactors. In packed beds, the dispersion results not only from ordinary molecular diffusion and the turbulence that exists in the absence of packing, but also from lateral deflections and mixing arising from the presence of the catalyst pellets. These effects are the dominant contributors to radial transport at the Reynolds numbers normally employed in commercial reactors.

If there is no correlation between the directions in which a given fluid molecule is deflected at successive layers of pellets, the *radial distance* through which the molecule is transported as it

$$\left(\frac{[\mathscr{P}_0 - \mathscr{P}_L]\rho}{G^2}\right)\left(\frac{D_p^*}{L}\right)\left(\frac{\varepsilon_B^3}{1 - \varepsilon_B}\right) = \frac{150(1 - \varepsilon_B)}{(D_p^* G/\mu)} + 1.75 \qquad (12.7.4)$$

where each term in parentheses is a dimensionless group. At high Reynolds numbers, the first term on the right drops out, and we obtain the Burke-Plummer equation for the turbulent flow regime. Similarly, at low Reynolds numbers, the second term drops out, and the Ergun equation reduces to the Blake-Kozeny relation for the laminar flow regime.

Notice that the superficial mass velocity G is constant throughout the bed, but that ρ will vary for compressible fluids. When the pressure drop is small compared with the absolute pressure, equation 12.7.4 may be used for gases by employing the arithmetic average of the inlet and

moves longitudinally through the bed can be described using Einstein's random walk approach. Each deflection will be comparable in magnitude to the characteristic dimensions of the pellet. If a large number of molecules are considered, the effect is that of radial diffusion characterized by an apparent diffusion coefficient in the radial direction that is given by (99):

$$\mathscr{D}_R = \frac{u_z D_p}{11.2} \qquad (12.7.5)$$

where u_z is the superficial velocity in the axial direction. For gases, the magnitude of $\mathscr{D}_R$ is usually much greater than the molecular diffusion coefficient. This disparity is even greater for liquids. The dimensionless group $u_z D_p / \mathscr{D}_R$ is known as the Peclet number for radial transport.

$$N_{Pe,r} \equiv \frac{u_z D_p}{\mathscr{D}_R} \qquad (12.7.6)$$

At high Reynolds numbers where molecular diffusion effects are negligible, experimental evidence confirms the general validity of equation 12.7.5. Figure 12.15 indicates how the Peclet number for radial mixing varies with the fluid Reynolds number. Above a Reynolds number of 40, the radial Peclet number is approximately 10.

The Peclet number for axial dispersion is defined in a manner similar to the radial parameter

$$N_{Pe,z} \equiv \frac{u_z D_p}{\mathscr{D}_L} \qquad (12.7.7)$$

where $\mathscr{D}_L$ is the longitudinal dispersion parameter. Using a mathematical model that regards the interstices in the packing as a series of mixing chambers, Aris and Amundson (101) have esti-

mated that the longitudinal Peclet number should have a value of approximately 2. Experimental results indicate that for gases $N_{Pe,z}$ is equal to two for modified Reynolds numbers $(D_p G / \mu)$ above 10. For liquids, the longitudinal Peclet number is less, particularly at low Reynolds numbers.

It should be emphasized that u_z is the apparent superficial velocity in the longitudinal direction (i.e., it is based on total area, voids plus solid). Similarly, in packed beds both $\mathscr{D}_L$ and $\mathscr{D}_R$ are normally based on total cross section.

The extent to which longitudinal dispersion must be taken into account depends on the ratio of the reactor length to the size of the particles. If the ratio is 100 or more, as is the usual case in commercial reactors, the effect of longitudinal dispersion is negligible in comparison with mass transport due to bulk flow. Only in very short reactors at high conversions and low velocities is the effect significant. In such cases, however, dispersion models of the reactor would not be realistic, and channeling and nonuniform flow distribution effects would be significant.

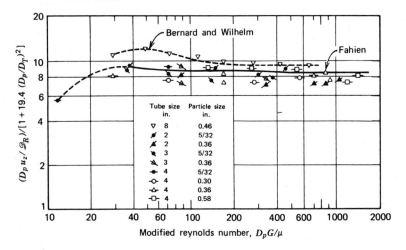

Figure 12.15

Correlation of average Peclet number with Reynolds number and the ratio of particle to tube diameter (D_p/D_T). [From Chemical Engineering Kinetics **by J. M. Smith. Copyright © 1970. Used with permission of McGraw-Hill Book Company.]**

12.7.1.3 Heat Transfer in Fixed Bed Reactors.

In reactor design, it is important to be able to describe in quantitative terms the heat transfer processes taking place within the reactor. In the design of homogeneous reactors, one normally assumes that radial mixing is sufficiently rapid that all the resistance to energy transfer is concentrated at the wall, and that the heat transfer can be described by a single heat transfer coefficient. However, in fixed bed reactors, the catalyst pellets inhibit radial mixing of the fluid, and significant gradients in temperature can exist in both the radial and axial directions. Here, we treat heat transfer between the containing wall of the reactor' and its contents and heat transfer within the packed bed. The latter process is described in terms of an effective thermal conductivity that lumps together contributions from a number of heat transfer mechanisms. Both this parameter and the wall heat transfer coefficient can be estimated on the basis of correlations that summarize extensive experimentation in this area.

12.7.1.3.1 Heat Transfer to the Containing Wall.

Heat transfer between the container wall and the reactor contents enters into the design analysis as a boundary condition on the differential or difference equation describing energy conservation. If the heat flux through the reactor wall is designated as $\dot{q}_w$, the heat transfer coefficient at the wall is defined as

$$h_w \equiv \frac{\dot{q}_w}{T_W - T_B} \qquad (12.7.8)$$

where

T_W is the wall surface temperature

T_B is the mean temperature of the bulk fluid in the volume element at the axial position in question

Normally, such heat transfer coefficients for packed beds are significantly greater than those for empty tubes at the same gas flow rate. Early reports of such data were usually reported as ratios of the coefficient in the packed bed to that in the empty tube. Typical ratios range from 5 to 7.8, depending on the ratio of the pellet diameter to the tube diameter. Jakob (102) has proposed the following correlation for wall heat transfer coefficients as a generalization of an earlier correlation by Colburn (103):

$$\frac{h_w D_T}{\kappa_g} = f^* D_T^{0.17} \left(\frac{D_p G}{\mu}\right)^{0.83} N_{Pr} \qquad (12.7.9)$$

where

D_T is the tube diameter in feet

D_p is the particle diameter in feet

f^* is a coefficient given by Figure 12.16 with D_p in inches and D_T in feet

κ_g is the thermal conductivity of the fluid in BTU's per hour-foot-degree Fahrenheit

h_w is the wall heat transfer coefficient in BTU's per hour-square foot-degree Fahrenheit

μ is the fluid viscosity in pounds mass per hour per foot

G is the superficial mass velocity in pounds mass per hour per square foot

This correlation indicates that the ratio of the fixed bed coefficient to the empty tube coefficient goes through a maximum, as D_p/D_T is varied.

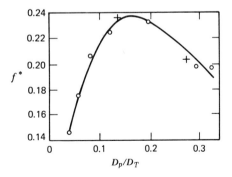

Figure 12.16

Coefficient f^* for use in equation 12.7.9. (From *Heat Transfer* by M. Jakob. Copyright © 1957. Reprinted by permission of John Wiley and Sons, Inc.)

Other useful correlations of heat transfer coefficients in packed beds have been proposed. Among these is a simple relation of Calderbank and Pogorski (104)

$$\frac{h_w D_p}{\kappa_g} = 3.6 \left(\frac{D_p G}{\mu \varepsilon_B}\right)^{0.365} \quad (12.7.10)$$

where the symbols have their usual significance. Leva (105) has also proposed two correlations for heating:

$$h_w = 0.813 \frac{\kappa_g}{D_T} e^{-6 D_p / D_T} \left(\frac{D_p G}{\mu}\right)^{0.90}$$

$$\text{for } \frac{D_p}{D_T} < 0.35 \quad (12.7.11)$$

$$h_w = 0.125 \frac{\kappa_g}{D_T} \left(\frac{D_p G}{\mu}\right)^{0.75}$$

$$\text{for } 0.35 < \frac{D_p}{D_T} < 0.6 \quad (12.7.12)$$

and one for cooling.

$$\frac{h_w D_T}{\kappa_g} = 3.50 \left(\frac{D_p G}{\mu}\right)^{0.70} e^{-4.6 D_p / D_T} \quad (12.7.13)$$

12.7.1.3.2 Energy Transfer Within a Packed Bed. Energy transfer within packed beds is an extremely complex process because it involves multiple mechanisms within a complex geometric structure. It may involve conduction through the pellets, conduction between pellets at points in contact, conduction through the fluid, radiation, turbulent transport by eddy diffusion processes, and conventional convective transport. These mechanisms combine and interact in a complex fashion to produce the observed energy transfer phenomena. Consequently, reactor designers find it convenient to use effective thermal conductivities (κ^*), which encompass all contributions to the transport of thermal energy except convection by plug flow. These transport coefficients are employed in pseudo homogeneous models of packed bed reactors that picture the reactor contents as a

homogeneous body through which heat is transferred by conduction. Consider the volume element shown in Figure 12.17. It consists of a cylindrical shell of radius R, thickness ΔR, and length Δz, which is concentric with the tube axis. If an energy balance is written for the volume element, the following input terms are necessary:

Input of energy by convective transport in the axial direction

$$2\pi R (\Delta R) \rho u_z \hat{C}_p (T - T_0)\big|_z \Delta t$$

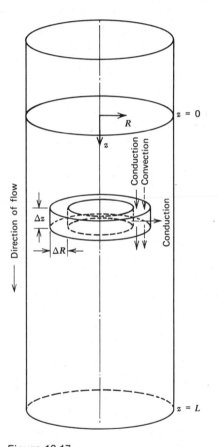

Figure 12.17

Annular ring over which energy balance is made. (From *Transport Phenomena* **by R. B. Bird, W. E. Stewart, and E. N. Lightfoot. Copyright © 1960. Reprinted with permission of John Wiley and Sons, Inc.)**

Input of thermal energy by effective radial conduction

$$-\kappa_R^* 2\pi R(\Delta z) \left(\frac{\partial T}{\partial R}\right)\bigg|_R \Delta t$$

Input of thermal energy by effective longitudinal conduction

$$-\kappa_z^* 2\pi R(\Delta R) \left(\frac{\partial T}{\partial z}\right)\bigg|_z \Delta t$$

where T_0 is the datum temperature for enthalpy

and let C_p refer to the heat capacity of the gas at temperature T.

Similar terms exist for output of thermal energy by these mechanisms. In addition to the input and output terms, we also need terms for the generation of thermal energy by chemical reaction

$$r_v(-\Delta H)2\pi R(\Delta R)(\Delta z)\,\Delta t$$

and the accumulation of energy within the volume element

$$[(\varepsilon_{\text{Total}})(\rho C_p)_{\text{gas}} + (1 - \varepsilon_{\text{Total}})(\rho C_p)_{\text{solid}}](\Delta T)2\pi R(\Delta R)(\Delta z)$$

calculations, $\hat{C}_p$ is an appropriate average heat capacity between the datum temperature and the temperature in question for the gaseous mixture, u_z is the superficial velocity in the z-direction corresponding to the point (R, z), ρ is the fluid density, κ_R^* and κ_z^* are the effective thermal conductivities in the radial and axial directions respectively, and Δt is a short time increment. For convenience, we shall choose T_0 to be close to T so that we may employ a constant pressure heat capacity value corresponding to temperature T. Thus we may delete the average symbol

where r_v is the pseudo homogeneous reaction rate per unit volume (solid plus fluid), ΔH is the enthalpy change for reaction at the indicated conditions, and the term in brackets is an effective mean heat capacity per unit volume. The porosity in the last term represents the ratio of the sum of external and intraparticle void volumes to the total volume. The quantity ΔT represents the temperature change occurring within the volume element in time Δt.

A typical balance equation can then be written as:

Input $\quad 2\pi R(\Delta R)\rho u_z C_p(T - T_0)\big|_z \Delta t - \kappa_R^* 2\pi R(\Delta z)\left(\dfrac{\partial T}{\partial R}\right)\bigg|_R \Delta t - \kappa_z^* 2\pi R(\Delta R)\left(\dfrac{\partial T}{\partial z}\right)\bigg|_z \Delta t$

$+$ generation $\quad +r_v(-\Delta H)2\pi R(\Delta R)(\Delta z)(\Delta t)$

$=$ output $\quad = 2\pi R(\Delta R)\rho u_z C_p(T - T_0)\big|_{z+\Delta z} \Delta t - \kappa_R^* 2\pi R(\Delta z)\left(\dfrac{\partial T}{\partial R}\right)\bigg|_{R+\Delta R} \Delta t$

$\qquad\qquad\qquad - \kappa_z^* 2\pi R\,\Delta R\left(\dfrac{\partial T}{\partial z}\right)\bigg|_{z+\Delta z} \Delta t$

$+$ accumulation $\quad +[\varepsilon_{\text{Total}}(\rho C_p)_{\text{gas}} + (1 - \varepsilon_{\text{Total}})(\rho C_p)_{\text{solid}}](\Delta T)2\pi R(\Delta R)(\Delta z)$

$$(12.7.14)$$

·Division by $2\pi(\Delta R)(\Delta z)(\Delta t)$ and rearrangement gives:

$$[\varepsilon_{\text{Total}}(\rho C_p)_{\text{gas}} + (1 - \varepsilon_{\text{Total}})(\rho C_p)_{\text{solid}}]R\frac{\Delta T}{\Delta t} = Rr_v(-\Delta H) + \frac{\rho u_z C_p R(T - T_0)|_z - \rho u_z C_p R(T - T_0)|_{z+\Delta z}}{\Delta z}$$

$$+ \frac{\kappa_R^* R\left(\dfrac{\partial T}{\partial R}\right)\bigg|_{R+\Delta R} - \kappa_R^* R\left(\dfrac{\partial T}{\partial R}\right)\bigg|_R}{\Delta R}$$

$$+ \frac{\kappa_z^* R\left(\dfrac{\partial T}{\partial z}\right)_{z+\Delta z} - \kappa_z^* R\left(\dfrac{\partial T}{\partial z}\right)_z}{\Delta z} \qquad (12.7.15)$$

and in the limit, as the various deltas approach zero,

$$R[\varepsilon_{\text{Total}}(\rho C_p)_{\text{gas}} + (1 - \varepsilon_{\text{Total}})(\rho C_p)_{\text{solid}}]\frac{\partial T}{\partial t} = Rr_v(-\Delta H) - R\frac{\partial}{\partial z}(\rho u_z C_p T)$$

$$+ \frac{\partial}{\partial R}\left[\kappa_R^* R\left(\frac{\partial T}{\partial R}\right)\right] + R\frac{\partial}{\partial z}\left[\kappa_z^*\left(\frac{\partial T}{\partial z}\right)\right] \qquad (12.7.16)$$

The effective thermal conductivities in the axial and radial directions depend on such variables as temperature, gas flow rate, the thermal conductivities of the gas and solid phases, particle diameter and porosity, packing geometry, and the emissivity of the solid. In general, κ_R^* should depend on radial position, but this effect is almost always neglected. Moreover, transport of energy in the direction of flow by convective flow is normally so much greater than the effective thermal conduction that the last term in (12.7.16) is virtually always neglected. It is significant only in very short beds at low velocities. These assumptions lead to the following energy balance equation for pseudo homogeneous models of fixed bed reactors.

$$\frac{\partial}{\partial z}(\rho u_z C_p T) = r_v(-\Delta H) + \kappa_R^*\left(\frac{\partial^2 T}{\partial R^2} + \frac{1}{R}\frac{\partial T}{\partial R}\right) \qquad (12.7.18)$$

This equation may be used as an appropriate form of the law of energy conservation in various pseudo homogeneous models of fixed bed reactors. Radial transport by effective thermal conduction is an essential element of two-dimensional reactor models but, for one-dimensional models, the last term must be replaced by one involving heat losses to the walls.

The last equation has been used by numerous investigators to evaluate effective thermal conductivities from experimental data. Figure 12.18 reproduced from Froment (94) indicates the

$$[\varepsilon_{\text{Total}}(\rho C_p)_{\text{gas}} + (1 - \varepsilon_{\text{Total}})(\rho C_p)_{\text{solid}}]\frac{\partial T}{\partial t} = r_v(-\Delta H) - \frac{\partial}{\partial z}(\rho u_z C_p T) + \kappa_R^*\left(\frac{\partial^2 T}{\partial R^2} + \frac{1}{R}\frac{\partial T}{\partial R}\right) \qquad (12.7.17)$$

For steady-state operation, the time derivative vanishes and we have the form that is of greatest interest for purposes of this text.

range of effective thermal conductivities typical of various Reynolds numbers. The data presented are most appropriate for use at tempera-

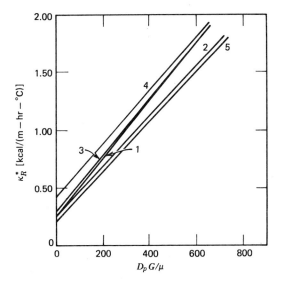

Figure 12.18
Heat transfer in packed beds. Effective thermal conductivity as a function of Reynolds number. Curve 1: Coberly and Marshall. Curve 2: Campbell and Huntington. Curve 3: Calderbank and Pogorski. Curve 4: Kwong and Smith. Curve 5: Kunii and Smith. (From G. F. Froment, "Chemical Reaction Engineering," *Adv. Chem. Ser., 109,* **1970.)**

tures below 300 °C where radiation heat transfer is negligible.

Argo and Smith (106, 107) have presented a detailed discussion of heat transfer in packed beds and have proposed the following relation for the effective thermal conductivity in packed beds:

$$\kappa^* = \varepsilon_B \left[\kappa_g + \frac{D_p C_p G}{N_{Pe,r}\varepsilon_B} + \frac{4\sigma}{2-\sigma} D_p(0.173) \frac{T^3}{100^4} \right]$$

$$+ (1 - \varepsilon_B) \frac{h'\kappa_s D_p}{2\kappa_s + h'D_p} \qquad (12.7.19)$$

where σ is the emissivity of the solid, κ_g and κ_s are the thermal conductivities of fluid and pellet, respectively, and h' is a heat transfer coefficient that characterizes the thermal interaction between adjacent particles. The terms on the right of equation 12.7.19 represent, in order, the effects of molecular conduction through the

fluid, the effects of transport by turbulent motion of fluid eddies, the effects of radiation transport through the void spaces, and the effects of conduction through the solid. The last term lumps together a number of effects associated with energy transfer from one pellet to adjacent pellets by radiation, point contacts, and convective transport to and from the fluid. The numerical constant in the radiation term implies that British Engineering Units (with time in hours) must be used for the various parameters and, in particular, the bed temperature T must be expressed in degrees Rankine.

Most of the available data have been recorded under conditions such that only the terms for eddy transport and conduction through the solid are significant. Equation 12.7.19 requires that κ^* increase with particle diameter, mass velocity, and the conductivity of the solid. It is consistent with data for low conductivity solids, but some discrepancies arise for very high conductivity solids (108). At Reynolds numbers greater than 40, the contribution of the molecular conduction term is negligible.

The parameter h', which combines several effects, must be at least as large as the convective heat transfer coefficient (h_c), calculated from the j_H factor correlations discussed in Section 12.5. It is given by an expression of the following form

$$h' = h_c + h_R + h_p \qquad (12.7.20)$$

where h_R and h_p represent the contributions of radiation between particles and conduction for pellets in contact. These quantities can be estimated as:

$$h_R = \left(\frac{2\kappa_s + h'D_p}{D_p \kappa_s} \right) \left(\frac{4\sigma}{2-\sigma} \right) D_p(0.173) \frac{T^3}{100^4} \qquad (12.7.21)$$

and

$$h_p = \left(\frac{2\kappa_s + h'D_p}{D_p \kappa_s} \right) \kappa_p \qquad (12.7.22)$$

where

$$\log \kappa_p = -1.76 + 0.0129 \frac{\kappa_s}{\varepsilon_B} \qquad (12.7.23)$$

in which κ_s is the thermal conductivity of the bulk solid. Equations 12.7.20 to 12.7.23 constitute a transcendental relation for h' that can be solved by trial and error procedures. Again because of the numerical constant in the radiation equation, we must employ British Engineering Units (with time in hours) in these relations.

Illustration 12.7 indicates how to estimate an effective thermal conductivity for use with two-dimensional, pseudo homogeneous packed bed models.

ILLUSTRATION 12.7 DETERMINATION OF THE EFFECTIVE THERMAL CONDUCTIVITY OF A PACKED BED OF CATALYST PELLETS

In Illustrations 12.5 and 12.6, some data on the catalytic oxidation of SO_2 were used to determine composition and temperature differences between the bulk fluid and the fluid at the pellet-gas interface. Use the data and results of these illustrations and the new data given below to predict the effective thermal conductivity of the bed.

Data and Results from Illustrations 12.5 and 12.6

$T = 458\ ^\circ C = 731\ ^\circ K = 1316\ ^\circ R$

$P = 790$ mm Hg

$G = 245$ lb/hr-ft^2 = 3.33×10^{-2} g/sec-cm^2

$\kappa_g = 3.70 \times 10^{-4}$ J/(cm-sec-$^\circ$K)

$\quad = 0.0214$ BTU/(hr-ft$^\circ$F)

$\mu = 0.032$ cp

$\rho_p = 1.81$ g/cm^3

$\varepsilon_B = 0.40$

$D_p = 0.389$ cm $= 1.276 \times 10^{-2}$ ft

$N_{Re} = 41$

$j_D = 0.235$

$C_p = 1.07$ J/g-$^\circ$K $= 0.255$ BTU/(lb$^\circ$R)

$N_{Pr} = 0.92$

$h_c = 8.79 \times 10^{-3}$ J/(cm^2-sec-$^\circ$K)

Additional Data

$D_T = 1.770$ in. $= 0.1475$ ft

$\sigma = 0.65$

$\kappa_s = 1.8$ BTU/(hr-ft-$^\circ$F)

Solution

Equations 12.7.20 to 12.7.23 may be used to evaluate the parameter h'. We begin by converting the convective heat transfer coefficient h_c, determined in Illustration 12.6, to British Engineering Units.

$$h_c = 8.79 \times 10^{-3}\ \text{J/cm}^2\text{-sec-}^\circ\text{K}$$
$$\approx 15.56\ \text{BTU/ft}^2\text{-hr-}^\circ\text{R}$$

From equation 12.7.20 it is evident that h' must be at least as large as h_c, and this fact provides a useful starting point for solving these transcendental equations. We begin by assuming that $h' = 35$ BTU/(hr-ft^2-$^\circ$R). From equation 12.7.23,

$$\log \kappa_p = -1.76 + 0.0129 \frac{(1.8)}{(0.40)}$$

or

$$\kappa_p = 0.0199$$

From equation 12.7.22,

$$h_p = \frac{2(1.8) + 35(0.01276)}{(0.01276)(1.8)}(0.0199)$$

$$= 3.51\ \text{BTU/(hr-ft}^2\text{-}^\circ\text{R)}$$

From equation 12.7.21,

$$h_R = \frac{2(1.8) + 35(0.01276)}{(0.01276)(1.8)}\left(\frac{4(0.65)}{2 - 0.65}\right)(0.01276)(0.173)\frac{(1316)^3}{(100)^4}$$

$$= 17.07\ \text{BTU/hr-ft}^2\text{-}^\circ\text{R}$$

From equation 12.7.20,

$$h' = 15.56 + 3.51 + 17.07$$
$$= 36.14 \text{ BTU/(hr-ft}^2\text{-°F)}$$

Since the relations for h_R and h_p are not very sensitive to the value of h', we need not iterate again and we can use the above value for subsequent purposes.

The radial Peclet number may now be estimated using Figure 12.15 at $N_{Re} = 41$,

$$\frac{D_p u_z}{\mathscr{D}_R} = 9\left[1 + 19.4\left(\frac{D_p}{D_T}\right)^2\right]$$

$$= 9\left[1 + 19.4\left(\frac{0.01276}{0.1475}\right)^2\right] = 10.3$$

We now have numerical values for all of the quantities appearing in equation 12.7.19. However, consistent units must be employed and, because of the 0.173 factor appearing in the radiation term, British Engineering Units are appropriate. The various terms in this equation may be evaluated as follows:

$$\kappa_g = 0.0214 \text{ BTU/(hr-ft-°F)}$$

$$\frac{D_p C_p G}{N_{Pe,r}\varepsilon_B} = \frac{(1.276 \times 10^{-2})(0.255)(245)}{(10.3)(0.40)} = 0.1935 \text{ BTU/(hr-ft-°F)}$$

$$\frac{4\sigma}{2 - \sigma} D_p(0.173)\frac{T^3}{100^4} = \frac{4(0.65)}{2 - 0.65}(1.276 \times 10^{-2})(0.173)\frac{(1316)^3}{(100)^4}$$

$$= 0.0969 \text{ BTU/(hr-ft-°F)}$$

$$\frac{(1 - \varepsilon_B)h'\kappa_s D_p}{2\kappa_s + h'D_p} = \frac{(1 - 0.40)(36.1)(1.8)(1.276 \times 10^{-2})}{2(1.8) + (36.1)(1.276 \times 10^{-2})} = 0.1225 \text{ BTU/(hr-ft-°F)}$$

Thus

$$\kappa^* = 0.40[0.0214 + 0.1935 + 0.0969] + 0.1225$$
$$= 0.247 \text{ BTU/(hr-ft-°F)}$$

The result is the average effective thermal conductivity of the bed. The reader should be aware, however, that the correlation employed is probably only good to 10 to 20% at best. In using such values in design calculations, one should examine the sensitivity of the results to variations in parameters that are characterized by such large uncertainties.

12.7.2 Pseudo Homogeneous* Models of Packed Bed Reactors

Pseudo homogeneous models of fixed bed reactors are widely employed in reactor design calculations. Such models assume that the fluid within the volume element associated with a single catalyst pellet or group of pellets can be characterized by a given bulk temperature, pressure, and composition and that these quantities vary continuously with position in the reactor. In most industrial scale equipment, the reactor volume is so large compared to the volume of an individual pellet and the fraction of the void volume associated therewith that the assumption of continuity is reasonable.

Pseudo homogeneous models require global rate expressions of the type discussed in Section 12.6, and the analysis that follows presumes that such expressions are available. To solve the differential equations representing material and energy balances, we use appropriate global rate expressions and boundary conditions to generate relations between the effluent conditions

---

* The term pseudo homogeneous model has been popularized by Froment (94, 95).

(composition and temperature) and the reactor volume. Usually, it is necessary to transform the coupled differential equations into difference equations and to use numerical methods to solve the transformed equations. The conventional approach to this problem involves the use of a stepwise procedure in which one starts at the reactor inlet and marches longitudinally through the reactor in appropriate volume increments. One-dimensional or two-dimensional models may be employed depending on the degree of sophistication desired, the cost constraints on computer aided design, and the level of approximation required. Either model can be handled by modern computing machines, but the one-dimensional model is used most often in preliminary design calculations because it provides a good approximation to the desired result with limited consumption of computer time, and it can be used to determine the effect of changes in design parameters and operating conditions on effluent conditions (see Section 12.7.2.1). The two-dimensional model is more complex, but provides essential information about radial temperature profiles within the bed. Such information is particularly useful in evaluating the potential for catalyst deactivation or in dealing with systems in which selectivity considerations are significant (see Section 12.7.2.2).

The computational effort required to carry out the design analysis is determined mainly by the magnitude and spatial distribution of the temperature variations that are taken into account. The maximum temperature difference between the inlet and outlet of the reactor occurs when the reactor operates adiabatically. In this case, heat transfer to the reactor wall is neglected so there is no temperature variation in the radial direction. However, the temperature does vary in the axial direction, so the material and energy balance equations are coupled through the dependence of the reaction rate on temperature. If the reactor is well insulated, and/or of large

diameter, the adiabatic model may provide a good representation of actual operating experience.

However, the most complex analysis is that in which heat transfer through the reactor walls is taken into account. This type of operation must be employed when it is necessary to supply or remove energy from the system so as to moderate the temperature excursions that would otherwise follow. It is frequently employed in industrial reactors and, to model such systems, one must often resort to two-dimensional models of the reactor that allow the concentration and temperature to vary in both the radial and axial directions. In the analysis of such systems, we make incremental calculations across the diameter of a given longitudinal segment of the packed bed reactor, and then proceed to repeat the process for successive longitudinal increments.

The various energy transfer constraints enter into the analysis primarily as boundary conditions on the difference equations, and we now turn to the generation of the differential equations on which the difference equations are based. Since the equations for the one-dimensional model are readily obtained by omitting or modifying terms in the expressions for the two-dimensional model, we begin by deriving the material balance equations for the latter. For purposes of simplification, it is assumed that only one independent reaction occurs within the system of interest. In cases where multiple reactions are present, one merely adds an appropriate term for each additional independent reaction.

Consider the element of reactor volume shown in Figure 12.17. The cylindrical shell is concentric with the cylinder axis, about which there is general symmetry. A material balance on the volume element requires consideration of several terms. The amount of a particular species A that enters the volume element during a time increment Δt consists of material entering by longitudinal bulk flow,

$$[C_A u_z 2\pi R(\Delta R)]_z (\Delta t)$$

material entering as a consequence of axial dispersion,

$$\left[-\mathscr{D}_L 2\pi R(\Delta R) \left(\frac{\partial C_A}{\partial z} \right) \right]_z (\Delta t)$$

and material entering by radial dispersion,

$$\left[-\mathscr{D}_R 2\pi R(\Delta z) \left(\frac{\partial C_A}{\partial R} \right) \right]_R \Delta t$$

where $\mathscr{D}_L$ and $\mathscr{D}_R$ are effective dispersion parameters that lump together molecular diffusion and eddy diffusion arising from turbulence and packing effects. As used in the pseudo homogeneous models, both diffusivities are based on

the total area (void plus solid) normal to the direction of transport. These parameters are those involved in the definition and discussion of the Peclet numbers presented in Section 12.7.1.2.

The amount of A that leaves the volume element during the time increment Δt consists of three similar terms. The amount of A that accumulates within the volume element during time Δt is $(\Delta C_A)2\pi R(\Delta R)(\Delta z)$, and the amount of A that is generated by chemical reaction within the volume element is $v_A r_v 2\pi R(\Delta R)(\Delta z)(\Delta t)$, where r_v is the reaction rate expressed in pseudo homogeneous form [i.e., the number of moles transformed per unit time per unit of total reactor volume (voids plus solid)].

The various terms may be combined in a generalized material balance to give:

Input	$[C_A u_z 2\pi R(\Delta R)]_z \, \Delta t - \left[\mathscr{D}_L 2\pi R(\Delta R) \left(\dfrac{\partial C_A}{\partial z} \right) \right]_z \Delta t - \left[\mathscr{D}_R 2\pi R(\Delta z) \left(\dfrac{\partial C_A}{\partial R} \right) \right]_R \Delta t$
+generation	$+ v_A r_v 2\pi R(\Delta R)(\Delta z)(\Delta t)$
=output	$[C_A u_z 2\pi R(\Delta R)]_{z+\Delta z} \, \Delta t - \left[\mathscr{D}_L 2\pi R(\Delta R) \left(\dfrac{\partial C_A}{\partial z} \right) \right]_{z+\Delta z} \Delta t$
	$- \left[\mathscr{D}_R 2\pi R(\Delta z) \left(\dfrac{\partial C_A}{\partial R} \right) \right]_{R+\Delta R} \Delta t$
+accumulation	$+(\Delta C_A)2\pi R(\Delta R)(\Delta z)$

$$(12.7.24)$$

Division by $2\pi(\Delta R)(\Delta z)(\Delta t)$ and rearrangement gives:

$$\frac{\left[\mathscr{D}_L R \left(\dfrac{\partial C_A}{\partial z} \right) \right]_{z+\Delta z} - \left[\mathscr{D}_L R \left(\dfrac{\partial C_A}{\partial z} \right) \right]_z}{\Delta z}$$

$$+ \frac{\left[\mathscr{D}_R R \left(\dfrac{\partial C_A}{\partial R} \right) \right]_{R+\Delta R} - \left[\mathscr{D}_R R \left(\dfrac{\partial C_A}{\partial R} \right) \right]_R}{\Delta R} - \frac{(C_A u_z R)_{z+\Delta z} - (C_A u_z R)_z}{\Delta z} + v_A r_v R = R \frac{\Delta C_A}{\Delta t}$$

$$(12.7.25)$$

and, in the limit, as Δz, ΔR, and Δt, all approach zero

$$\mathscr{D}_L R \frac{\partial^2 C_A}{\partial z^2} + \mathscr{D}_R \frac{\partial}{\partial R}\left(R \frac{\partial C_A}{\partial R}\right) - R \frac{\partial(C_A u_z)}{\partial z} + v_A R r_v = R\left(\frac{\partial C_A}{\partial t}\right) \tag{12.7.26}$$

where we have presumed that $\mathscr{D}_L$ and the tube diameter are constant over the length of the reactor, and that $\mathscr{D}_R$ is constant over the reactor diameter. It is necessary to retain u_z within the partial derivative to cover those cases where there is a change in fluid density on reaction. When there is an increase or decrease in the number of moles on reaction, or a change in fluid temperature or pressure, the velocity component in the longitudinal direction will vary along the length of the reactor. For gas phase reactions, the effect can be quite significant.

The last equation can be rewritten as

$$\mathscr{D}_L \frac{\partial^2 C_A}{\partial z^2} + \mathscr{D}_R \left(\frac{\partial^2 C_A}{\partial R^2} + \frac{1}{R}\frac{\partial C_A}{\partial R}\right) - \frac{\partial(C_A u_z)}{\partial z} + v_A r_v = \frac{\partial C_A}{\partial t} \tag{12.7.27}$$

At steady state, the right side vanishes, and one obtains an equation that describes in pseudo homogeneous fashion the concentration profile in the reactor.

$$\mathscr{D}_L \frac{\partial^2 C_A}{\partial z^2} + \mathscr{D}_R \left(\frac{\partial^2 C_A}{\partial R^2} + \frac{1}{R}\frac{\partial C_A}{\partial R}\right) - \frac{\partial(C_A u_z)}{\partial z} + v_A r_v = 0 \tag{12.7.28}$$

For isothermal systems this equation, together with an appropriate expression for r_v, is sufficient to predict the concentration profiles through the reactor. For nonisothermal systems, this equation is coupled to an energy balance equation (e.g., the steady-state form of equation 12.7.16) by the dependence of the reaction rate on temperature.

$$\kappa_L^* \frac{\partial^2 T}{\partial z^2} + \kappa_R^* \left[\left(\frac{\partial^2 T}{\partial R^2}\right) + \frac{1}{R}\left(\frac{\partial T}{\partial R}\right)\right] - \frac{\partial}{\partial z}(\rho u_z C_p T) - (\Delta H) r_v = 0 \tag{12.7.29}$$

Equations 12.7.28 and 12.7.29 provide a two-dimensional pseudo homogeneous model of a fixed bed reactor. The one-dimensional model is obtained by omitting the radial dispersion terms in the mass balance equation and replacing the radial heat transfer term by one that accounts for thermal losses through the tube wall. Thus the material balance becomes

$$\mathscr{D}_L \frac{\partial^2 C_A}{\partial z^2} - \frac{\partial(C_A u_z)}{\partial z} + v_A r_v = 0 \tag{12.7.30}$$

The corresponding energy conservation equation is

$$\kappa_L^* \frac{\partial^2 T}{\partial z^2} - G \frac{\partial}{\partial z}(C_p T) - \frac{4}{D_T} h_w(T - T_w) - r_v(\Delta H) = 0 \tag{12.7.31}$$

where T_w is the tube wall temperature.

The boundary conditions that govern both the one- and two-dimensional models are usually stated in the following manner.

$$\left.\begin{array}{r} C_A = C_{A0} \\ T = T_0 \end{array}\right\} \text{ at } z = 0 \text{ for all } R \qquad \begin{array}{l} (12.7.32) \\ (12.7.33) \end{array}$$

$$\left.\begin{array}{r} \left(\dfrac{\partial C_A}{\partial R}\right) = 0 \\[2mm] \left(\dfrac{\partial T}{\partial R}\right) = 0 \end{array}\right\} \text{ at } R = 0 \text{ for all } z \qquad \begin{array}{l} (12.7.34) \\[4mm] (12.7.35) \end{array}$$

$$\left(\frac{\partial C_A}{\partial R}\right) = 0 \qquad \text{at } R = R_0 \text{ for all } z \qquad (12.7.36)$$

$$(u_z C_{A0})|_{z=0^-} = -\left(\mathscr{D}_L \frac{\partial C_A}{\partial z}\right)_{z=0^+} + (u_z C_A)|_{z=0^+} \qquad \text{at } z = 0 \text{ for all } R \qquad (12.7.37)$$

In the formulation of the boundary conditions, it is presumed that there is no dispersion in the feed line and that the entering fluid is uniform in temperature and composition. In addition to the above boundary conditions, it is also necessary to formulate appropriate equations to express the energy transfer constraints imposed on the system (e.g., adiabatic, isothermal, or nonisothermal-nonadiabatic operation). For the one-dimensional models, boundary conditions 12.7.34 and 12.7.35 hold for all R, and not just at $R = 0$.

When the velocity u_z varies with *radial* position, equation 12.7.28 must be solved by a stepwise numerical procedure. Experimental evidence indicates that the axial velocity does indeed vary with radial position in fixed bed reactors. The velocity profile is relatively flat in the center of the tube. As one moves radially outward, the velocity increases gradually until a maximum is reached at a point about one pellet diameter from the tube wall. It then falls rapidly, until it reaches zero at the wall. If the ratio of the tube diameter to the pellet diameter

exceeds 30, a plug flow model of the velocity profile provides a good representation of the actual profile across the central core of the tubular reactor. Departures from the average velocity are significant only in the thin annular region near the tube walls. In single tube commercial-scale reactors this criterion is usually met [e.g., if the pellets have a characteristic dimension of 0.6 cm ($\simeq 1/4$ in.), the required tube diameter is only 18 cm ($\simeq 7$ in.)]. However, in the multitube reactors required to meet the large heat transfer requirements of very rapid highly exothermic reactions, the assumption of a relatively flat velocity profile is not a good one. Such systems are difficult to model in mathematical terms. In order to generate the necessary data for the design of such reactors, the practicing engineer frequently has to resort to pilot scale reactors that consist of one or more tubes of the same diameter and length as those to be employed in the commercial-scale multitube reactor.

We turn now from a discussion of general principles to specific models of reactor behavior.

12.7.2.1 The One-Dimensional Pseudo Homogeneous Model of Fixed Bed Reactors. The design of tubular fixed bed catalytic reactors has generally been based on a one-dimensional model that assumes that species concentrations and fluid temperature vary *only* in the axial direction. Heat transfer between the reacting fluid and the reactor walls is considered by presuming that all of the resistance is contained within a very thin boundary layer next to the wall and by using a heat transfer coefficient based on the temperature difference between the fluid and the wall. Per unit area of the tube wall, the heat flow rate from the reactor contents to the wall is then $h_w(T - T_w)$. The correlations for h_w presented in Section 12.7.1.3 may be used to evaluate this parameter.

The one-dimensional model is advantageous because it provides a rapid means of (1) obtaining an estimate of the reactor size necessary to achieve a given conversion, and (2) examining the influence of several design variables on the behavior of the reactor. Unfortunately, the model provides no information concerning the possibility of achieving an excessive temperature at the center of the tube that may be markedly different from the mean temperature at the same longitudinal position. Such temperatures may be unacceptable for reasons of reactor stability, process selectivity, or catalyst deactivation. To ensure that such excessive temperatures are not achieved along the reactor axis, one should determine if they are present using a two-dimensional model to analyze those sets of operating conditions that the one-dimensional model indicates are promising. The two-dimensional model is discussed in Section 12.7.2.2.

The equation describing the steady-state material balance for tubular packed bed reactors can be obtained from the more general relation (12.7.28) by omitting the terms corresponding to radial transport of matter. Hence the material balance relation becomes

$$\mathscr{D}_L \frac{\partial^2 C_A}{\partial z^2} - \frac{\partial(C_A u_z)}{\partial z} + v_A r_v = 0 \quad (12.7.38)$$

Further simplification of the material balance equation occurs if the axial dispersion term is neglected. In this case,

$$\frac{\partial(C_A u_z)}{\partial z} - v_A r_v = 0 \quad (12.7.39)$$

If one recognizes that

$$F_A = C_A u_z \left(\frac{\pi D_T^2}{4} \right) \quad (12.7.40)$$

and that

$$V_R = \frac{\pi D_T^2}{4} z \quad (12.7.41)$$

equation 12.7.39 can be written as

$$\frac{dF_A}{dV_R} - v_A r_v = \frac{dF_A}{dV_R} - r_{Av} = 0 \quad (12.7.42)$$

Now $dF_A = -F_{A0}\, df_A$, so the previous equation can be written as

$$\frac{V_R}{F_{A0}} = \int \frac{df_A}{-r_{Av}} \quad (12.7.43)$$

which is identical with the equation for an ideal plug-flow reactor, derived as equation 8.2.7. If both sides of this relation are multiplied by the bulk density of the catalyst, we again generate equation 8.2.12.

$$\frac{\rho_B V_R}{F_{A0}} = \frac{W}{F_{A0}} = \int \frac{\rho_B\, df_A}{-r_{Av}} = \int \frac{df_A}{-r_{Am}} \quad (12.7.44)$$

The *general* energy balance for the one-dimensional, packed bed reactor *cannot* be obtained by omitting the radial derivatives of the temperature in equation 12.7.29, because these terms ultimately lead to heat transfer through the walls of the reactor. They can be omitted *only for adiabatic operation*. In the general case of a one-dimensional model, the rate of heat loss through the wall must be considered. For a differential length of reactor (Δz), this loss is given by

$$h_w \pi D_T (T - T_w)(\Delta z)$$

where

D_T is the tube diameter

T_w is the local wall temperature

h_w is a wall heat transfer coefficient which can be evaluated from the empirical correlations presented in Section 12.7.1.3.1.

An energy balance over the differential length of reactor for steady-state operating conditions is then given by:

Input by convection   Transformation by reaction   Input by conduction

$$\rho u_z \left(\frac{\pi D_T^2}{4}\right) C_p (T - T_0)_z + r_v(-\Delta H)\left(\frac{\pi D_T^2}{4}\right)\Delta z \qquad + \qquad \left[-\kappa_L^* \left(\frac{\pi D_T^2}{4}\right)\frac{\partial T}{\partial z}\right]_z$$

Output by convection   Losses through wall   Output by conduction

$$= \rho u_z \left(\frac{\pi D_T^2}{4}\right) C_p (T - T_0)\bigg|_{z+\Delta z} + h_w \pi D_T (T - T_w)\,\Delta z + \left[-\kappa_L^* \left(\frac{\pi D_T^2}{4}\right)\frac{\partial T}{\partial z}\right]_{z+\Delta z} \qquad (12.7.45)$$

which leads to the following differential equation.

$$-\frac{\partial}{\partial z}\left[\kappa_L^* \left(\frac{\partial T}{\partial z}\right)\right] + \frac{\partial}{\partial z}(\rho u_z C_p T) + \frac{4}{D_T} h_w (T - T_w) = r_v(-\Delta H) \qquad (12.7.46)$$

This equation can also be rewritten in terms of the superficial mass velocity G, which does not vary along the length of the reactor.

$$-\frac{\partial}{\partial z}\left[\kappa_L^* \left(\frac{\partial T}{\partial z}\right)\right] + G\frac{\partial}{\partial z}(C_p T) + \frac{4}{D_T} h_w (T - T_w) = r_v(-\Delta H) \qquad (12.7.47)$$

As indicated earlier, the axial conduction term is almost always negligible compared to the convective énthalpy transport term. Therefore, equation 12.7.47 is usually simplified to give

$$G\frac{\partial}{\partial z}(C_p T) + \frac{4}{D_T} h_w (T - T_w) = r_v(-\Delta H) \qquad (12.7.48)$$

Equations 12.7.48 and 12.7.39 provide the simplest one-dimensional mathematical model of tubular fixed bed reactor behavior. They neglect longitudinal dispersion of both matter and energy and, in essence, are completely equivalent to the plug flow model for homogeneous reactors that was examined in some detail in Chapters 8 to 10. Various simplifications in these equations will occur for different constraints on the energy transfer to or from the reactor. Normally, equations 12.7.48 and 12.7.39

are coupled through the dependence of the reaction rate on temperature. However, for isothermal operation, the equations are not coupled, and $\partial T/\partial z$ becomes zero. For adiabatic operation, the heat losses through the wall are negligible, and the energy balance becomes

$$G\frac{\partial}{\partial z}(C_p T) = r_v(-\Delta H) \qquad \text{(adiabatic case)} \qquad (12.7.49)$$

Depending on the operational constraints, one of the two equations 12.7.48 or 12.7.49 or the choice of isothermal behavior must be used, together with the general material balance relation (equation 12.7.39), to determine the composition and temperature profiles along the length of the reactor.

In many instances, the pressure drop through the reactor will be relatively small so that we may employ a mean value for the total pressure in our calculations. However, the Ergun equation presented in Section 12.7.1.1 may be used, if necessary, to evaluate the pressure at various points in the bed. Solutions of the appropriate material and energy balance equations for this simplified one-dimensional model can be obtained by straightforward numerical procedures employing digital computation to solve the corresponding difference equations. While the correspondence of these solutions to the behavior of actual reactors must be viewed as only an approximation, this model provides a means of simulating the steady-state response of the system under investigation to changes in various process parameters at minimum cost in computer time and programming effort. The simulations enable us to answer questions such as:

1. What tube length will be required to achieve a given conversion?
2. What will be the corresponding effluent temperature?
3. What does the longitudinal temperature profile look like for given inlet temperatures and/or wall temperature profiles? Are the hot spots excessive for reasons of selectivity, catalyst deactivation, etc?
4. How do changes in the tube diameter influence the design calculations?
5. How do changes in catalyst pellet size affect system behavior?
6. How can one avoid excessive sensitivity of the performance to changes in process parameters?

The last question pertains to the problem of parametric sensitivity. There is extensive literature dealing with this topic, but it is beyond the scope of this book.

A more general one-dimensional model of tubular, packed bed reactors is contained within equations 12.7.38 and 12.7.47. These equations include all of the elements of the simple model discussed above and, in addition, account for the longitudinal dispersion of both thermal energy and matter. The dispersion that represents the combined effects of molecular diffusion, normal fluid turbulence, and the influence of the packing is accounted for by superimposing "effective" heat and mass transfer terms on the plug-flow equations. Because of the assumptions involved in the definitions of the effective thermal conductivity and dispersion terms, they implicitly contain the effect of radial velocity profiles within the tubes. Normally, one uses values averaged over the tube diameter, as expressed by Peclet numbers for heat and mass transfer. However, the velocity profile is not considered explicitly. For design purposes, we take the longitudinal Peclet number based on a pellet diameter, to be approximately 2. Effective thermal conductivities may be estimated using the method described in Section 12.7.1.3.2.

In many respects, the solutions to equations 12.7.38 and 12.7.47 do not provide sufficient additional information to warrant their use in design calculations. It has been clearly demonstrated that for the fluid velocities used in industrial practice, the influence of axial dispersion of both heat and mass on the conversion achieved is negligible provided that the packing depth is in excess of 100 pellet diameters (109). Such shallow beds are only employed as the first stage of multibed adiabatic reactors. There is some question as to whether or not such short beds can be adequately described by an effective transport model. Thus for most preliminary design calculations, the simplified one-dimensional model discussed earlier is preferred. The discrepancies between model simulations and actual reactor behavior are not resolved by the inclusion of longitudinal dispersion terms. Their effects are small compared to the influence of radial gradients in temperature and composition. Consequently, for more accurate simulations, we employ a two-dimensional model (Section 12.7.2.2).

Nevertheless, one feature of the one-dimensional model containing dispersion terms is of considerable interest. These terms increase the order of the partial differential equations and, under certain conditions, lead to nonuniqueness of the steady-state profile through the reactor (110). For certain ranges of operating conditions and parameter values, three or more steady-state profiles can be obtained for the same feed conditions. The two outlying steady-state profiles will be stable (at least to small perturbations), while the intermediate one will be unstable. The profile generated as a solution to equations 12.7.38 and 12.7.47 will depend on the initial guesses of T and C involved in the trial and error solution.

In physical terms, this assessment means that the steady state that would be achieved in a real reactor will depend on the initial profile in the reactor. For all situations where the initial values differ from the feed conditions, we must solve the time dependent differential equations describing the reactor to determine which profile will prevail. In order to determine if such computations are necessary, one must know whether or not multiple steady states are possible. Several criteria that permit us to make this decision on an a priori basis have been discussed by various authors (111–113). For details of these analyses, consult the indicated references. For present purposes we are interested only in the general conclusions derived from such studies. The range within which such multiple steady states are possible is very narrow, and such conditions are rarely encountered in either laboratory or industrial scale reactors. They are most likely to be realized for highly exothermic reactions carried out in extremely short beds under adiabatic operating conditions. The length of virtually all industrial scale reactors precludes the necessity of including longitudinal dispersion terms in a one-dimensional model of reactor behavior and of worrying about their implications for multiple steady states. The effects of radial gradients are much more likely to be significant.

The following illustrations indicate the manner in which the one-dimensional model is employed in reactor design analyses.

ILLUSTRATION 12.8 PRODUCTION OF SULFUR TRIOXIDE IN AN ADIABATIC FIXED BED REACTOR

Industrial scale production of sulfuric acid is dependent on the oxidation of sulfur dioxide to sulfur trioxide in fixed bed catalytic reactors.

$$SO_2 + \tfrac{1}{2}O_2 \rightarrow SO_3$$

Through the years, several catalyst formulations have been employed, but one of the traditional catalytic agents has been vanadium pentoxide. Calderbank (114) has indicated that for a catalyst consisting of V_2O_5 supported on silica gel, the kinetic data are represented by a rate expression of the form

$$r_m = \frac{k_1 P_{SO_2} P_{O_2} - k_2 P_{SO_3} P_{O_2}^{1/2}}{P_{SO_2}^{1/2}} \tag{A}$$

that may be regarded as a degenerate form of typical Hougen-Watson kinetics. The rate constants are given by

$$\ln k_1 = 12.07 - \frac{31,000}{RT} \tag{B}$$

and

$$\ln k_2 = 22.75 - \frac{53,600}{RT} \tag{C}$$

where

T is expressed in degrees Kelvin

R is expressed in calories per mole-degree Kelvin

k_1 is expressed in moles per second-gram of catalyst-atmosphere$^{3/2}$

k_2 is expressed in moles per second-gram of catalyst-atmosphere

For our present purposes, the global rate expression may be presumed to be identical with that of equation A.

The reaction is highly exothermic and must be regarded as reversible. Consequently, although high temperatures enhance the initial rate, they limit the conversion that can be achieved. This limitation can be circumvented by cooling a hot effluent to a temperature where the equilibrium is more favorable, and then contacting this stream with additional catalyst. Determine the catalyst requirements for a two-stage adiabatic fixed bed reactor with interstage cooling. Specified production requirements are 50 tons of H_2SO_4/day.

Additional Data and Specifications

Feed composition (mole percent)

SO_2	8.0
O_2	13.0
N_2	79.0

Total pressure	1 atm
First-stage inlet temperature	370 °C
First-stage effluent temperature	560 °C
Second-stage inlet temperature	370 °C
Overall conversion of SO_2	99%

Heat capacities (calories per gram mole-degree Kelvin)

N_2	$6.42 + 1.34 \times 10^{-3}T$
O_2	$6.74 + 1.64 \times 10^{-3}T$
SO_2	$9.52 + 3.64 \times 10^{-3}T$
SO_3	$12.13 + 8.12 \times 10^{-3}T$

where

   T is expressed in degrees Kelvin

Heat of reaction at temperature T (kilocalories per gram mole) $= -24.60 + 1.99 \times 10^{-3} \ T$ for T in degrees Kelvin.

Bulk density of catalyst $= 0.6$ g/cm^3

Reactor diameter $= 6$ ft

Notes

The first-stage effluent temperature has been limited to 560 °C in order to prevent excessive catalyst activity losses. The heat of reaction data is slightly inconsistent with the reported activation energies, but use of this expression demonstrates the ease with which temperature dependent properties may be incorporated in the one-dimensional model.

Solution

Since the reaction rate per unit mass of catalyst and the reaction rate per unit volume of bed are simply related by the bulk density of the catalyst, the procedures employed in Illustration 10.4 can easily be adopted in solving this problem. However, we wish to indicate a slightly different approach that is appropriate for use in hand calculations. By virtue of the arguments presented earlier, longitudinal dispersion of energy and matter are presumed to be negligible compared to convective transport. The appropriate equations for the one-dimensional, plug flow model, assuming adiabatic operation, are then equation 12.7.49 for the energy balance

$$G \frac{\partial}{\partial z}(C_p T) = r_v(-\Delta H) = r_m \rho_B(-\Delta H) \quad \text{(D)}$$

and equation 12.7.39 for the material balance.

$$\frac{\partial}{\partial z}(C_A u_z) = \nu_A r_v = \nu_A r_m \rho_B \quad \text{(E)}$$

Since there is a change in the number of moles on reaction, the volumetric expansion parameter δ will be nonzero. Consequently,

$$u_z = u_0(1 + \delta f)\frac{T}{T_0}$$

$$C_A = C_{A0}\frac{(1 - f)}{(1 + \delta f)}\frac{T_0}{T}$$

and

$$\frac{\partial}{\partial z}(C_A u_z) = \frac{\partial}{\partial z}[C_{A0} u_0 (1 - f)] = -C_{A0} u_0 \left(\frac{\partial f}{\partial z}\right) \tag{F}$$

Combining equations E and F and recognizing that $v_A = -1$ gives

$$\frac{C_{A0} u_0}{\rho_B} \left(\frac{\partial f}{\partial z}\right) = r_m \tag{G}$$

The partial pressures of the various species are numerically equal to their mole fractions since the total pressure is one atmosphere. These mole fractions can be expressed in terms of a single reaction progress variable–the degree of conversion–as indicated in the following mole table.

Initial moles		Moles at a fraction conversion f	Mole fraction at fraction conversion f
SO_2	8.0	$8.0(1 - f)$	$8.0(1 - f)/(100 - 4.0f)$
N_2	79.0	79.0	$(79.0)/(100 - 4.0f)$
O_2	13.0	$13.0 - \dfrac{8.0f}{2}$	$(13.0 - 4.0f)/(100 - 4.0f)$
SO_3	0.0	$8.0f$	$8.0f/(100 - 4.0f)$
Total	100.0	$100 - 4.0f$	1.000

In terms of the fraction conversion, the catalytic reaction rate expression then becomes

$$r_m = \frac{k_1 \left[\dfrac{(8.0)(1 - f)}{100 - 4.0f}\right]\left(\dfrac{13.0 - 4.0f}{100 - 4.0f}\right) - k_2 \left(\dfrac{8.0f}{100 - 4.0f}\right)\left(\dfrac{13.0 - 4.0f}{100 - 4.0f}\right)^{1/2}}{\left[\dfrac{8.0(1 - f)}{100 - 4.0f}\right]^{1/2}} \tag{H}$$

The mass velocity G will be constant over the reactor length, and this quantity may be determined from the specified production rate and the reactor dimensions. To produce 50 tons/day of 100% H_2SO_4, the number of pound moles of SO_2 that must be oxidized per second is:

$$50\,\frac{\text{tons}}{\text{day}} \times \frac{\text{day}}{24\ \text{hr}} \times \frac{\text{hr}}{3600\ \text{sec}} \times \frac{2000\ \text{lb}}{\text{ton}} \times \frac{\text{lb moles } H_2SO_4}{98\ \text{lb } H_2SO_4} = 1.18 \times 10^{-2}\,\frac{\text{lb moles}}{\text{sec}}$$

For 99% conversion of the SO_2, the inlet molal flow rate must be

$$\frac{1.18 \times 10^{-2}}{(0.99)(0.08)} = 0.149 \text{ lb moles/sec}$$

The average molecular weight of the inlet gas is $(0.08)(64) + 0.13(32) + 0.79(28) = 31.4$. The mass velocity G is then $(0.149)(31.4)/[\pi(6)^2/4] = 0.165$ lb/(ft²-sec).

We now wish to examine the heat capacity per unit mass to determine if it varies significantly with conversion. At the inlet conditions, the molal heat capacity of the gaseous feed will be equal to $\sum(y_i C_{pi})$. Hence at $f = 0$, with C_p in units of calories per gram degree Kelvin and T in degrees Kelvin.

$$C_p = \frac{0.79(6.42 + 1.34 \times 10^{-3}T) + 0.08(9.52 + 3.64 \times 10^{-3}T) + 0.13(6.74 + 1.64 \times 10^{-3}T)}{31.4}$$

$$= 0.214 + 4.98 \times 10^{-5} \, T \tag{I}$$

If complete conversion were to take place ($f = 1$), the heat capacity of the effluent mixture could be evaluated with the aid of the mole table on page 511.

$$C_{p(f=1)} = \frac{\frac{79}{96}(6.42 + 1.34 \times 10^{-3}T) + \frac{9}{96}(6.74 + 1.64 \times 10^{-3}T) + \frac{8}{96}(12.13 + 8.12 \times 10^{-3}T)}{\frac{79}{96}(28) + \frac{9}{96}(32) + \frac{8}{96}(80)}$$

$$= 0.212 + 5.91 \times 10^{-5}T \tag{J}$$

Equations I and J indicate that for the temperature range of interest (640 to 830 °K), the heat capacity per unit mass is substantially independent of the conversion level. Furthermore, the temperature dependent contribution to the heat capacity will not vary much over the temperature range involved. Hence without introducing errors comparable to those inherent in the use of a one-dimensional model, we may take the heat capacity as constant at 0.250 cal/g-°K or 0.250 BTU/(lb-°F).

Combining equations G and D indicates that

$$GC_p \frac{\partial T}{\partial z} = (-\Delta H)C_{A0}u_0 \frac{\partial f}{\partial z}$$

where we have used our assumption that C_p is a constant. Integration of this equation between the reactor inlet and location z gives

$$GC_p(T - T_0) = (-\Delta H)C_{A0}u_0(f - f_0) \tag{K}$$

The product $C_{A0}u_0$ is just the molal mass velocity of reactant A, which is equal to

$$\frac{0.08(0.149)}{\pi(6)^2/4} = 4.22 \times 10^{-4} \text{ lb moles/ft}^2\text{-sec}$$

Substitution of numerical values in equation K gives

$$0.165(0.250)1.8(T - T_0) = 10^3(24.60 - 1.99 \times 10^{-3}T)\left(\frac{454}{252}\right)(4.22 \times 10^{-4})(f - f_0) \tag{L}$$

where the factor 1.8 is necessary to convert T in degrees Kelvin to consistent units.

Hence the relation between the temperature and the fraction conversion at any point in the adiabatic reactor is given by

$$T - T_0 = (251.9 - 0.0204T)(f - f_0)$$

or

$$T = \frac{T_0 + 251.9(f - f_0)}{1 + 0.0204(f - f_0)} \tag{M}$$

Equations G and H may be combined to give

$$\frac{C_{A0}u_0}{\rho_B} \frac{\partial f}{\partial z} = \frac{k_1[(8.0)(1 - f)]^{1/2}(13.0 - 4.0f)}{(100 - 4.0f)^{3/2}} - \frac{k_2(8.0f)(13.0 - 4.0f)^{1/2}}{[8.0(1 - f)]^{1/2}(100 - 4.0f)} \tag{N}$$

Appropriate numerical values for use in this equation are

$$C_{A0}u_0 = 4.22 \times 10^{-4} \text{ lb moles}/(\text{ft}^2\text{-sec})$$
$$\rho_B = 0.60 \text{ g/cm}^3 \approx 37.4 \text{ lb/ft}^3$$

$$k_1 = e^{12.07}e^{-31,000/RT} \frac{\text{g moles}}{\text{g catalyst atm}^{3/2}\text{sec}} \approx 1.746 \times 10^5 e^{-31,000/RT} \frac{\text{lb moles}}{\text{lb catalyst atm}^{3/2}\text{sec}}$$

$$k_2 = e^{22.75}e^{-53,600/RT} \frac{\text{g moles}}{\text{g catalyst atm-sec}} \approx 7.589 \times 10^9 e^{-53,600/RT} \frac{\text{lb moles}}{\text{lb catalyst atm-sec}}$$

Substituting numerical values and writing equation N in terms of finite increments gives

$$\frac{\Delta f}{\Delta z} = \frac{4.38 \times 10^{10}(1 - f)^{1/2}(13 - 4f)}{e^{31,000/(1.987T)}(100 - 4f)^{3/2}} - \frac{1.9 \times 10^{15}f(13 - 4f)^{1/2}}{e^{53,600/(1.987T)}(1 - f)^{1/2}(100 - 4f)} \tag{O}$$

Equations M and O may now be solved numerically to determine the bed depth corresponding to a given conversion.

If we rewrite equation O as

$$\Delta z = \bar{\beta}(\Delta f) \tag{P}$$

where

$$\beta = \frac{(100 - 4f)}{\dfrac{4.38 \times 10^{10}(1 - f)^{1/2}(13 - 4f)e^{-31,000/1.987T}}{(100 - 4f)^{1/2}} - \dfrac{1.90 \times 10^{15}f(13 - 4f)^{1/2}}{(1 - f)^{1/2}}e^{-53,600/1.987T}}$$

the coefficient β may be evaluated at the beginning and the end of the conversion increment, and an average value employed to determine Δz.

At the inlet to the first reactor $f = 0$ and $T_0 = 370\,°C = 643.16\,°K$. Hence

$$\beta_0 = \frac{(100)^{3/2}}{(4.38 \times 10^{10})(13)e^{-31,000/(1.987)(643.16)}} = 60.2$$

If we chose our first conversion increment as $\Delta f = f_1 - f_0 = 0.05$, equation M indicates that

$$T_1 = \frac{643.16 + 251.9(0.05)}{1 + (0.0204)(0.05)} = 655.09\,°K$$

Thus

$$\beta_1 = \frac{[100 - 4(0.05)]}{\dfrac{4.38 \times 10^{10}(1 - 0.05)^{1/2}[13 - 4(0.05)]}{[100 - 4(0.05)]^{1/2}\, e^{31,000/(1.987)(655.09)}} - \dfrac{1.90 \times 10^{15}(0.05)[13 - 4(0.05)]^{1/2}}{(1 - 0.05)^{1/2}\, e^{53,600/(1.987)(655.09)}}} = 40.2$$

and

$$\Delta z = z_1 - 0 = \left(\frac{60.2 + 40.2}{2}\right)(0.05)$$

or

$$z_1 = 2.51 \text{ ft}$$

Proceeding to the next increment for Δf again equal to 0.05, we find $f_2 = 0.10$ and

$$T_2 = \frac{643.16 + 251.9(0.10)}{1 + 0.0204(0.10)} = 666.99\ ^\circ\text{K}$$

Thus

$$\beta_2 = \frac{[100 - 4(0.1)]}{\dfrac{4.38 \times 10^{10}(1 - 0.1)^{1/2}[13 - 4(0.1)]}{[100 - 4(0.1)]^{1/2}\, e^{31,000/(1.987)(666.99)}} - \dfrac{1.90 \times 10^{15}(0.1)[13 - 4(0.1)]^{1/2}}{(1 - 0.1)^{1/2}\, e^{53,600/(1.987)(666.99)}}} = 27.4$$

Hence

$$\Delta z = z_2 - z_1 = \left(\frac{40.2 + 27.4}{2}\right)0.05 = 1.69 \text{ ft.}$$

or

$$z_2 = 2.51 + 1.69 = 4.20 \text{ ft.}$$

The above procedure may be repeated to determine the depth of catalyst necessary to achieve a given conversion level. Table 12.I.1 summarizes the results of such calculations. Notice that if the first stage effluent is to be kept below 560 °C, equation M indicates that the conversion leaving this stage will be 0.81.

A similar procedure can be used to analyze the second stage of the reactor network. In this case, equation M becomes

$$T = \frac{643.16 + 251.9(f - 0.81)}{1 + 0.0204(f - 0.81)}$$

This equation and equation O suffice to determine the temperature and conversion profiles in the second stage of the reactor. The results of these calculations are also summarized in Table 12.I.1. From the table, it is evident that the bulk of the catalyst (77%) must be present in the second stage. Furthermore, a substantial fraction (27%) of the total is required merely to go from 95 to 99% conversion. Notice that the catalyst requirements for the second stage cannot be significantly reduced by changing the second stage inlet temperature. The effluent conditions are approaching the equilibrium conditions for 99% conversion ($T = 691\ ^\circ\text{K}$), and this restraint is largely responsible for the larger catalyst depth requirement in the second stage. Thermodynamic equilibrium does not limit the conversion in the first reactor, but catalyst deactivation processes do. We could, of course, consider the possibility of using a third stage to reduce catalyst requirements, but we would have

Table 12.I.1
Temperature and Conversion Profiles for Packed Bed Reactor Network

	Stage 1			Stage 2	
f_A	Temperature, $T(^\circ K)$	Catalyst depth (ft)	f_A	Temperature, $T(^\circ K)$	Catalyst depth (ft)
0	643	0	0.81	643	0
0.05	655	2.51	0.83	648	3.36
0.10	667	4.20	0.85	653	6.37
0.15	679	5.36	0.87	657	9.12
0.20	691	6.17	0.89	662	11.65
0.25	703	6.74	0.91	667	14.05
0.30	714	7.15	0.93	672	16.39
0.35	726	7.45	0.95	676	18.81
0.40	738	7.68	0.97	681	21.68
0.45	750	7.85	0.98	684	23.65
0.50	761	7.98	0.985	685	25.11
0.55	773	8.09	0.9875	685	26.21
0.60	785	8.18	0.99	686	28.76
0.65	796	8.25			
0.70	808	8.32			
0.75	820	8.38			
0.80	831	8.46			
0.81	'833	8.48			

to be willing to incur the associated capital charges.

The total weight of catalyst required is given by the product of the bulk density and the total reactor volume

$$\text{Weight of catalyst} = 37.4 \left[\frac{\pi(6)^2}{4} (8.48 + 28.76) \right]$$

$$= 39{,}400 \text{ lb or } 19.7 \text{ tons}$$

In practice, one could oversize both stages and operate with reduced inlet temperatures that could subsequently be raised as catalyst deactivation proceeds.

Before proceeding to the next illustration, it is instructive to see if our assumption of constant total pressure was indeed appropriate. If one assumes 1/4 in. spherical pellets, a nominal gas viscosity of 0.09 lb/hr-ft and a bed porosity of 0.4, the Reynolds number of the gas is given by

$$N'_{Re} = \frac{D_p G}{\mu} = \frac{\left(\frac{1}{4}\right)\left(\frac{1}{12}\right)(0.165)(3600)}{(0.09)}$$

$$= 137.5$$

From equation 12.7.4,

$$\frac{(\mathscr{P}_0 - \mathscr{P}_L)\rho}{G^2} \left(\frac{D_p}{L}\right)\left(\frac{\varepsilon_B^3}{1 - \varepsilon_B}\right) = \frac{150(1 - 0.4)}{137.5} + 1.75 = 2.40$$

where

$$\rho \cong \frac{PM}{RT} = \frac{(1)(31.4)}{0.73(670)(1.8)} = 3.6 \times 10^- \text{ lb/ft}^3$$

Hence

$$\mathscr{P}_0 - \mathscr{P}_L = 2.40 \left(\frac{1 - 0.4}{0.4^3}\right) \left[\frac{37.24}{\frac{1}{4}\left(\frac{1}{12}\right)}\right] \frac{(0.165)^2}{3.6 \times 10^{-2}} \frac{\text{lb}_m}{\text{ft-sec}^2}$$

$$= 3.05 \times 10^4 \frac{\text{lb}_m}{\text{ft-sec}^2} \times \left(\frac{\text{lb}_f \text{ sec}^2}{32.2 \text{ lb}_m\text{-ft}}\right) \times \left(\frac{\text{ft}^2}{144 \text{ in.}^2}\right)$$

$$= 6.57 \text{ psi}$$

This pressure drop is a significant fraction of the total pressure. Consequently, the analysis should be repeated, breaking up the reactor into segments that could be treated as having a constant average pressure. As an alternative, we could go to larger diameter, shorter beds of catalyst to reduce the pressure drop, while maintaining the same conversion according to the one-dimensional homogeneous model. For example, by going to a nine foot diameter bed, the total bed length could be reduced from 37.24 to 16.6 ft and the pressure drop to 0.7 psi. Total catalyst requirements would be unchanged. Such trade-offs indicate why large diameter beds with small length-to-diameter ratios are often employed in catalytic reactors. Section 13.1 indicates how to treat the pressure variation along the length of a tubular reactor.

ILLUSTRATION 12.9 PRODUCTION OF SULFUR TRIOXIDE IN A FIXED BED REACTOR WITH THERMAL LOSSES

While adiabatic operation may be approached by the use of efficient insulation techniques, heat losses from insulated reactors can be appreciable.

Indeed, it may be prohibitively expensive to accomplish large reductions in thermal energy losses beyond a certain point, because of the material and labor costs involved.

Repeat the analysis of Illustration 12.8, assuming that the heat transfer to the surroundings can be characterized by an overall heat transfer coefficient based on the temperature difference between the reactor contents and ambient conditions (70 °F). When based on the inside area of the reactor tube, the heat transfer coefficient has a numerical value of 1.2 BTU/(hr-ft^2-°F) or 6.0×10^{-4} (BTU/sec-ft^2-°K). The second-stage inlet temperature may be taken as 410 °C instead of 370 °C, as in Illustration 12.8.

Solution

The finite difference form of the material balance equation developed in Illustration 12.8 (equation P) is again applicable:

$$\Delta z = \bar{\beta} \Delta f \tag{A}$$

where

$$\frac{1}{\beta} = \frac{4.38 \times 10^{10}(1 - f)^{1/2}(13 - 4f)e^{-31,000/1.987T}}{(100 - 4f)^{3/2}} - \frac{1.90 \times 10^{15}f(13 - 4f)^{1/2}e^{-53,600/1.987T}}{(1 - f)^{1/2}(100 - 4f)}$$

However, the energy balance equation appropriate for use in this illustration differs from that employed in the previous case because thermal losses through the reactor walls must be accounted for. It will be of the same general form as equation 12.7.48, but with the wall heat transfer coefficient replaced by an overall heat transfer coefficient ($\mathscr{U}$) and with a corresponding change made in the temperature driving force.

$$G \frac{\partial}{\partial z}(C_p T) + \frac{4}{D_T} \mathscr{U}(T - T_{ambient}) = (-\Delta H)r_v$$

The right side of this equation can be rewritten in terms of the fraction conversion, using equations E and G of Illustration 12.8.

However, solution of the difference equations requires a knowledge of the temperature at the end of the conversion increment. Consequently, a trial and error procedure is indicated. One assumes a value for the temperature at the end of the increment, computes Δz from equation A, and checks the temperature assumption using equation B. Under normal circumstances, the

$$G \frac{\partial}{\partial z}(C_p T) + \frac{4}{D_T} \mathscr{U}(T - T_{ambient}) = r_m \rho_B(-\Delta H) = (-\Delta H)C_{A0} u_0 \left(\frac{\partial f}{\partial z}\right)$$

Substitution of appropriate numerical values gives

$$(0.165)(0.250)(1.8)\frac{\partial T}{\partial z} + \frac{4}{6}(6.0 \times 10^{-4})(T - 294.27)$$

$$= 10^3(24.60 - 1.99 \times 10^{-3}T)\left(\frac{454}{252}\right)(4.22 \times 10^{-4})\left(\frac{\partial f}{\partial z}\right)$$

where

T is measured in degrees Kelvin

z is measured in feet

Evaluation of the numerical constants gives

$$dT = (251.9 - 0.0204T)\,df$$
$$- 5.387 \times 10^{-3}(T - 294.27)\,dz$$

or, in finite difference form,

$$\Delta T = (251.9 - 0.0204T)\,\Delta f$$
$$- 5.387 \times 10^{-3}(T - 294.27)\,\Delta z \quad \text{(B)}$$

Equations A and B may now be solved to deter-

thermal loss term will be small compared to the heat of reaction term and we may obtain a reasonable first estimate of ΔT by neglecting the thermal loss term. From the previous illustration, we shall use $T_1 = 655.09\ °K$ as our first estimate of the temperature at the end of the first conversion increment ($\Delta f = 0.05$). We will indicate the calculations for the first two conversion increments in some detail. From Illustration 12.8, $\beta_0 = 60.2$, $\beta_1 = 40.2$, and $\Delta z = 2.51$ ft, for an assumed temperature of 655.09 °K ($\Delta T = 11.93\ °K$). Substitution of these values in equation B gives

$$\Delta T = \left[251.9 - 0.0204\left(\frac{655.09 + 643.16}{2}\right)\right](0.05) - 5.387 \times 10^{-3}\left(\frac{655.09 + 643.16}{2} - 294.27\right)(2.51)$$

$$= 7.13\ °K$$

mine the temperature and conversion profiles for the reactor. As before, one may begin by choosing a small increment in the conversion.

Thus

$$T_1 = 643.16 + 7.13 = 650.29\ °K$$

Since the preliminary estimate of ΔT is somewhat larger than the value calculated from equation B, we should iterate using the last value of T_1 as our initial estimate. For a temperature of 650.29,

$$\beta_1 = 47.9$$

Thus from equation A,

$$\Delta z = \left(\frac{60.2 + 47.9}{2}\right)(0.05) = 2.70 \text{ ft}$$

and, from equation B,

$$\Delta T = \left[251.9 - 0.0204\left(\frac{650.29 + 643.16}{2}\right)\right](0.05)$$

$$- 5.387 \times 10^{-3}\left[\left(\frac{650.29 + 643.16}{2}\right) - 294.27\right](2.70)$$

$$= 6.81 \text{ °K}$$

or $T_1 = 649.97$ °K, which is in better agreement with the initial estimate of the temperature. To achieve the desired degree of agreement, one may iterate as often as necessary.

For the second conversion increment ($\Delta f = 0.05$, $f = 0.10$), we will assume $T = 657$ °K ($\Delta T = 7.03$ °K). Here,

$$\beta_2 = 39.0$$

and, from equation A,

$$\Delta z = \left(\frac{48.5 + 39.0}{2}\right)(0.05) = 2.19 \text{ ft}$$

where a revised value of β_1 corresponding to 649.97 °K has been employed.

From equation B,

$$\Delta T_2 = \left[251.9 - 0.0204\left(\frac{657 + 649.97}{2}\right)\right](0.05)$$

$$- 5.387 \times 10^{-3}\left[\left(\frac{657 + 649.97}{2}\right) - 294.27\right](2.19)$$

$$= 7.69 \text{ °K}$$

Thus

$$T_2 = 649.97 + 7.69 = 657.66 \text{ °K}$$

A second iteration using this temperature as an initial guess gives

$$T_2 = 657.69 \text{ °K and } \Delta z = 2.17 \text{ ft.}$$

In the above manner, we may proceed downstream in the reactor until we either reach the desired conversion level, run into thermodynamic limitations on the reaction rate, or exceed the effluent temperature constraint (see Table

12.I.2). In the present case, thermodynamic constraints on the rate indicate that the first stage effluent should correspond to a conversion in the vicinity of 0.87 and an effluent temperature near 824.4 °K. Beyond this point, the system is so close to thermodynamic equilibrium that substantial increments in the reactor length do not produce noticeable increments in the conversion.

Here, one may employ a higher temperature for the second-stage feed and still achieve the desired conversion level. The conversion and temperature profiles as calculated from equations A and B are given in Table 12.I.2.

The total weight of catalyst required is equal to

$$37.4\left[\frac{\pi}{4}(6)^2(12.14 + 13.45)\right] = 27,060 \text{ lb}$$

$$\approx 13.53 \text{ tons}$$

Table 12.I.2
Conversion and Temperature Profiles for Packed Bed Reactor of Illustration 12.9

	First stage			Second stage	
f	Temperature, $T(^\circ K)$	z (ft)	f	Temperature, $T(^\circ K)$	z (ft)
0.00	643.16	0	0.87	683.16	0
0.05	650.0	2.70	0.89	685.6	1.12
0.10	657.7	4.87	0.91	687.9	2.29
0.15	666.3	6.55	0.93	689.9	3.56
0.20	675.6	7.84	0.95	691.5	5.04
0.25	685.5	8.81	0.97	691.8	7.12
0.30	695.8	9.54	0.98	690.6	8.81
0.35	706.5	10.09	0.985	688.8	10.22
0.40	717.4	10.50	0.9875	686.9	11.36
0.45	728.6	10.81	0.99	683.1	13.45
0.50	739.8	11.05			
0.55	751.2	11.24			
0.60	762.7	11.39			
0.65	774.2	11.51			
0.70	785.7	11.62			
0.75	797.2	11.72			
0.80	808.7	11.82			
0.85	820.0	11.97			
0.86	822.3	12.02			
0.87	824.4	12.14			

This value is considerably less than that obtained for pure adiabatic operation (19.7 tons). The heat losses tend to partially remove thermodynamic constraints on the reaction rate and permit a closer approach to the optimum temperature profile corresponding to minimum catalyst requirements.

A few words concerning the results of our analyses in Illustrations 12.8 and 12.9 are in order. Obviously, better estimates of the catalyst requirements could be obtained by using smaller conversion increments. We have not attempted to fully optimize the reactor stages in terms of catalyst minimization. Furthermore, we have again neglected pressure drop in each stage. Further calculations would remedy each of the aforementioned shortcomings of the analysis. They are readily accomplished with the aid of machine computation.

12.7.2.2 The Two-Dimensional, Pseudo Homogeneous Model of Fixed Bed Reactors.

Two-dimensional models permit more realistic simulation of fixed bed reactor behavior than the one-dimensional models discussed previously. Experimental measurements indicate that the fluid temperature and composition are not uniform across a section of the tube normal to the flow. The one-dimensional models discussed earlier neglect the radial resistance to heat and mass transfer and thus predict uniform temperature and composition for each longitudinal position. This assumption is obviously a vast oversimplification when reactions with large heat effects are considered. Whenever there is extensive heat exchange between the packed bed reactor and its surroundings, one requires at least a two-dimensional model to simulate the reactor performance. In such cases, the design engineer requires a model that predicts the detailed temperature and composition patterns in the reactor to be able to avoid hot spots along the reactor axis when they would be detrimental for reasons of selectivity, catalyst deactivation, etc.

A complete two-dimensional model would account for the radial distribution of velocity, for the radial concentration and temperature gradients in terms of Peclet numbers that themselves varied with radial location, and for axial dispersion of heat and mass. With machine computation, such considerations may be handled if the cost in computer time and programming effort can be withstood. However, the most significant aspects of two-dimensional models can be demonstrated by considering simple two-dimensional models. Such models assume that the mass velocity G and the radial Peclet numbers for heat and mass transfer are constant across the tube diameter and that the effective thermal conductivity and effective dispersion contributions to longitudinal transport of energy and mass are insignificant (compared to the convective transport terms). The appropriate material and energy balance equations that provide a two-dimensional description of the reactor performance at steady state have been derived earlier in discussing general aspects of heat transfer and of mass conservation in packed beds. They are equation 12.7.18

$$G \frac{\partial(C_p T)}{\partial z} - \kappa_R^* \left[\frac{\partial^2 T}{\partial R^2} + \frac{1}{R}\left(\frac{\partial T}{\partial R}\right) \right] + r_v(\Delta H) = 0$$

$$(12.7.18)$$

for energy conservation and a simplified form of equation 12.7.28

$$\mathscr{D}_R \left[\frac{\partial^2 C_A}{\partial R^2} + \frac{1}{R}\frac{\partial C_A}{\partial R} \right] - \frac{\partial(C_A u_z)}{\partial z} + v_A r_v = 0$$

$$(12.7.50)$$

for mass conservation.

These coupled partial differential equations can be solved for the temperature and composition at any point in the catalyst bed by using numerical procedures to solve the correspon-

ding difference equations. As boundary conditions, one needs to know the temperature and composition profile across the tube diameter at the reactor inlet. In addition, the solution must satisfy the requirements that

$$\frac{\partial C_A}{\partial R} = 0 \quad \text{at} \quad R = 0 \text{ for all } z \quad (12.7.51)$$

$$\frac{\partial T}{\partial R} = 0 \quad \text{at} \quad R = 0 \text{ for all } z \quad (12.7.52)$$

(for reasons of symmetry) and that

$$\frac{\partial C_A}{\partial R} = 0 \quad \text{at} \quad R = \frac{D_T}{2} = R_0 \text{ (tube wall)}$$

$$(12.7.53)$$

since the tube is not permeable to the fluid. The boundary condition at the wall on the energy conservation equation can be framed as a requirement that the heat flux at the wall be of the following form:

$$\dot{q} = -\kappa_R^* \left(\frac{\partial T}{\partial R}\right)_{R_0} = h_w(T_{R_0} - T_w)$$

$$(12.7.54)$$

where

T_w is the inside wall surface temperature

T_{R_0} is the temperature of the surrounding fluid

h_w is a heat transfer coefficient

This approach in essence assumes a temperature discontinuity at the wall. Alternative methods of formulating this constraint have also been proposed (94, 95, 115), together with empirical methods of evaluating the heat transfer coefficients introduced by each method.

Various numerical procedures may be employed to solve the difference equations corresponding to equations 12.7.18 and 12.7.50 Many sophisticated numerical procedures may be employed, but they are more properly treated in textbooks dealing with numerical methods or more advanced texts in chemical reactor design.

For a general introduction to some of the techniques employed to solve difference equations of the types encountered in chemical engineering, consult the text by Lapidus (116). Detailed numerical examples of one method of numerical solution to the two-dimensional reactor problem are contained in the texts of Smith (117) and Jensen and Jeffreys (118).

12.8 DESIGN OF FLUIDIZED BED CATALYTIC REACTORS

The design of a commercial scale fluidized bed reactor for a new process is in many respects one of the most challenging technical assignments that a practicing chemical engineer can be assigned. Even a superficial study of an operating unit reveals its intrinsic complexity. It is obvious that such systems pose tremendous analytical problems for someone trying to develop realistic mathematical models of the interactions between the chemical reaction phenomena and the various physical transport processes that occur in these reactors. Not only do we have to worry about the possibilities of intraparticle and external mass and heat transfer limitations of the types discussed previously, but we must also be concerned with the complexity of the fluid-solid contacting process and the manner in which the gas and solid catalyst are distributed throughout the reactor.

Within a fluidized bed, low density regions or "gas bubbles" are formed at apparently unpredictable rates. These bubbles may subsequently coalesce, split, or grow, perhaps even to the extent that their size approaches the physical dimensions of the reactor. Indeed, in some cases it is necessary to provide for reactor internals to limit bubble size and the problems of slugging, poor contacting, and mechanical vibration, which become serious when the bubbles become too large. Experimental studies clearly demonstrate that there is significant movement of gas from the high density regions through the bubbles and back to the high density regions. The

extent of this interchange between the gas bubbles and the remainder of the bed has obvious implications with respect to the degree of conversion that can be expected in a fluidized bed reactor. The bubbles are essentially empty, although from time to time a quantity of particles may rain through the bubble from the roof. In some cases, this movement may cause the bubble to split. The lowest third of a sphere enclosing the bubble contains a wake of particles that travel upward with the bubble. Solid particles also move in streamlines around each rising bubble, so that a spout of material is also drawn up behind the bubble. The solids moving upward in the wake behind the bubble and in the spout below provide rapid mixing of solids from bottom to top.

These factors combine to make it extremely difficult not only to develop predictive analytical models of the performance of such systems, but also to scale up experimental data from laboratory operations, particularly when the commercial scale equipment may contain baffles, heat exchangers and other internal fixtures within the bed. Some useful models have been developed for fluidized beds, but the design of industrial scale reactors is generally dominated by empirical correlations and component designs that have proven useful in past generations of equipment. This feature is particularly true of the equipment used for distribution of the incoming gas. This component has a marked effect on the solids recirculation pattern and on potential channeling through the bed.

The velocity at which gas flows through the dense phase corresponds approximately to the velocity that produces incipient fluidization. The bubbles rise, however, at a rate that is nearly an order of magnitude greater than the minimum fluidization velocity. In effect, then, as a consequence of the movement of solids within the bed and the interchange of fluid between the bubbles and the dense regions of the bed, there are wide disparities in the residence times of various fluid elements within the reactor and in the times that the fluid elements are effectively in contact with the catalyst particles.

From the standpoint of attempting to develop mathematical models for the simulation of fluidized bed reactors, one must determine if the phenomena mentioned above and other aspects of the behavior of fluidized beds can be described in terms of a relatively small number of parameters. In particular, those aspects of the behavior of fluidized beds that have a significant influence on the conversion achieved must be adequately characterized by the model if it is to be useful for purposes of reactor design. Many models, especially if they are sufficiently complex and contain many "constants," can describe a given set of data equally well. However, if the models are to be useful for design purposes, it should be possible to determine the parameters appearing therein from simple laboratory experiments or from correlations of experimental data. Indeed, the utility of the model is judged by the success obtained when extrapolating data from one set of experimental reaction conditions (usually bench or pilot scale experiments) to predict the behavior of a large-scale unit.

The physical situation in a fluidized bed reactor is obviously too complicated to be modeled by an ideal plug flow reactor or an ideal stirred tank reactor although, under certain conditions, either of these ideal models may provide a fair representation of the behavior of a fluidized bed reactor. In other cases, the behavior of the system can be characterized as plug flow modified by longitudinal dispersion, and the unidimensional pseudo homogeneous model (Section 12.7.2.1) can be employed to describe the fluidized bed reactor. As an alternative, a cascade of CSTR's (Section 11.1.3.2) may be used to model the fluidized bed reactor. Unfortunately, none of these models provides an adequate representation of reaction behavior in fluidized beds, particularly when there is appreciable bubble formation within the bed. This situation arises mainly because a knowledge of the residence time distribution of the gas in the bed is insuf-

ficient to permit a prediction of degrees of conversion, since there is not an equal distribution of catalyst between the bubble and dense phases. It is the presence of bubbles and their effect on gas/solid contacting and mixing that lies at the root of the problem and that causes the discrepancy between the residence time distribution (easily measured) and the contact time distribution (not easily measured).

Because of the inadequacies of the aforementioned models, a number of papers in the 1950s and 1960s developed alternative mathematical descriptions of fluidized beds that explicitly divided the reactor contents into two phases, a bubble phase and an emulsion or dense phase. The bubble or lean phase is presumed to be essentially free of solids so that little, if any, reaction occurs in this portion of the bed. Reaction takes place within the dense phase, where virtually all of the solid catalyst particles are found. This phase may also be referred to as a particulate phase, an interstitial phase, or an emulsion phase by various authors. Figure 12.19 is a schematic representation of two phase models of fluidized beds. Some models also define a cloud phase as the region of space surrounding the bubble that acts as a source and a sink for gas exchange with the bubble.

On the basis of different assumptions about the nature of the fluid and solid flow within each phase and between phases as well as about the extent of mixing within each phase, it is possible to develop many different mathematical models of the two phase type. Pyle (119), Rowe (120), and Grace (121) have critically reviewed models of these types. Treatment of these models is clearly beyond the scope of this text. In many cases insufficient data exist to provide critical tests of model validity. This situation is especially true of large scale reactors that are the systems of greatest interest from industry's point of view. The student should understand, however, that there is an ongoing effort to develop mathematical models of fluidized bed reactors that will be useful for design purposes. Our current

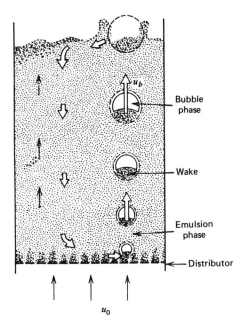

Figure 12.19
Basic two-phase model of fluidized bed. Open arrows indicate movement of solids. (Adapted from *Fluidization Engineering* by D. Kunii and O. Levenspiel. Copyright © 1969. Reprinted by permission of John Wiley and Sons, Inc.)

capabilities in this area have not been developed to the extent that they have in fixed bed reactor design, but fluidized beds are inherently much more complex systems. Indeed, the design of fluidized bed reactors is largely dominated by empirical procedures, particularly with regard to the design and operation of the gas distributor and the construction of reactor internals. Treatments of mathematical models of fluidized bed reactor performance are appropriate subject material for advanced texts in reactor design [see Kunii and Levenspiel (122) and Davidson and Harrison (123)].

LITERATURE CITATIONS

1. Goodman, D. R., and Oxon, B. A., "Handling and Using Catalysts on the Plant," in the *Catalyst Handbook*, Springer-Verlag, New York, 1970.

2. Satterfield, C. N., *AIChE J.*, *21* (209), 1975.

3. Kunii, D., and Levenspiel, O., *Fluidization Engineering*, John Wiley and Sons, New York, 1969.

4. Davidson, J. F., and Harrison, D., *Fluidized Particles*, Cambridge University Press, New York, 1963.

5. Satterfield, C. N., *Mass Transfer in Heterogeneous Catalysis*, pp. 107–128, MIT Press, Cambridge, Mass., 1970. Copyright © 1970 by The Massachusetts Institute of Technology. Adapted by permission of The MIT Press.

6. Smith, J. M., *Chemical Engineering Kinetics*, Second Edition, pp. 383–395, McGraw-Hill, New York, 1970.

7. Reid, R. C., and Sherwood, T. K., *The Properties of Gases and Liquids*, Second Edition, McGraw-Hill, New York, 1966.

8. Evans, R. B., III, Watson, G. M., and Mason, E. A., *J. Chem. Phys.*, *35* (2076), 1961.

9. Scott, D. S., and Dullien, F. A. L., *AIChE J.*, *8* (113), 1962.

10. Rothfeld, L. B., *AIChE J.*, *9* (19), 1963.

11. Bird, R. B., Stewart, W. E., and Lightfoot, E. N., *Transport Phenomena*, pp. 496–504, Wiley, New York, 1960.

12. Smith, J. M., op. cit., p. 403.

13. Satterfield, C. N., op. cit., pp. 47–54, 1970.

14. Barrer, R. M., *Appl. Mater. Res.*, *2*(3), p. 129, 1963.

15. Field, G. J., Watts, H., and Weller, K. R., *Rev. Pure Appl. Chem.*, *13* (2), 1963.

16. Dacey, J. R., *Ind. Eng. Chem.*, *57* (6), p. 27, 1965.

17. Satterfield, C. N., op. cit., pp. 37–40, 1970.

18. Wicke, E., and Kallenbach, R., *Kolloid Z.*, *97* (135), 1941.

19. Barrer, R. M., *J. Phys. Chem.*, *57* (35), 1953.

20. Thiele, E. W., *Ind. Eng. Chem.*, *31* (916), 1939.

21. Damköhler, G., *Der Chemie-Ingenieur*, *3* (430), 1937.

22. Zeldovitch, Ya. B., *Acta Physicochim. U.R.S.S.*, *10* (583), 1939, in English. Also *Z. Fiz, Chim*, *13* (163), 1939.

23. Wheeler, A., *Adv. Catalysis*, *3* (249), 1951.

24. Wheeler, A., *Catalysis*, *2* (104), edited by P. H. Emmett, Reinhold, New York, 1955.

25. Weisz, P. B., *Adv. Catalysis*, *13* (137), 1962.

26. Weisz, P. B., and Hicks, J. S., *Chem. Eng. Sci.*, *17* (265), 1962.

27. Weisz, P. B., *Chem. Eng. Prog. Symp. Ser.*, *55* (25), p. 29, 1959.

28. Weisz, P. B., and Prater, C. D., *Adv. Catalysis*, *6* (143), 1954.

29. Wicke, E., and Kallenbach, R., *Kolloid Z.*, *97* (135), 1941.

30. Wicke, E., and Brotz, W., *Chem.-Ing.-Tech.*, *21* (219), 1949.

31. Wicke, E., and Hedden, K., *Z. Elektrochem.*, *57* (636), 1953.

32. Wicke, E., *Chem.-Ing.-Tech.*, *29* (305), 1957.

33. Aris, R., *Chem. Eng. Sci.*, *6* (262), 1957.

34. Aris, R., *Ind. Eng. Chem., Fundamentals*, *4* (227), 1965.

35. Aris, R., *Ind. Eng. Chem., Fundamentals*, *4* (487), 1965.

36. Aris, R., *The Mathematical Theory of Diffusion and Reaction in Permeable Catalysts*, Volumes I and II, Clarendon Press, Oxford, 1975.

37. Corrigan, T. E., Garver, J. C., Rase, H. F., and Kirk, R. S., *Chem. Eng. Prog.*, *49* (603), 1953.

38. Wheeler, A., op. cit., p. 136, 1955.

39. Satterfield, C. N., op. cit., p. 152, 1970.

40. Satterfield, C. N., op. cit., p. 134, 1970.

41. Satterfield, C. N., op. cit., pp. 135–138, 1970.

42. Luss, D., and Amundson, N. R., *AIChE J.*, *13* (759), 1967.

43. Aris, R., *Introduction to the Analysis of Chemical Reactors*, Prentice-Hall, Englewood Cliffs, N.J., 1965.

44. Petersen, E. E., *Chemical Reaction Analysis*, p. 64, Prentice-Hall, Englewood Cliffs, N.J., 1965.

45. Weekman, V. W., Jr., and Goring, R. L., *J. Catalysis*, *4* (260), 1965.

46. Weekman, V. W., Jr., *J. Catalysis*, *5* (44), 1966.

47. Chu, C., and Hougen, O. A., *Chem. Eng. Sci.*, *17* (167), 1962.

48. Roberts, G. W., and Satterfield, C. N., *Ind. Eng. Chem., Fundamentals*, *4* (288), 1965.

49. Roberts, G. W., and Satterfield, C. N., *Ind. Eng. Chem., Fundamentals*, *5* (317), 1966.

50. Knudsen, C. W., Roberts, G. W., and Satterfield, C. N., *Ind. Eng. Chem., Fundamentals*, *5* (325), 1966.

51. Satterfield, C. N., op. cit., pp. 176–200, 1970.

52. Kao, H. S-P., and Satterfield, C. N., *Ind. Eng. Chem., Fundamentals*, *7* (664), 1968.

53. Woodside, W., and Messmer, J. H., *J. Appl. Phys.*, *32* (1688), 1961.

54. Butt, J. B., *AIChE J.*, *11* (106), 1965.

55. Damköhler, G., *Z. Phys. Chem.*, *A193* (16), 1943.

56. Schilson, R. E., and Amundson, N. R., *Chem. Eng. Sci.*, *13*, pp. 226, 237, 1961.

57. Beek, J., *AIChE J.*, *7* (337), 1961.

58. Tinkler, J. D., and Pigford, R. L., *Chem. Eng. Sci.*, *15* (326), 1961.

59. Petersen, E. E., *Chem. Eng. Sci.*, *17* (987), 1962.

60. Tinkler, J. D., and Metzner, A. B., *Ind. Eng. Chem.*, *53* (663), 1961.

61. Weisz, P. B., and Hicks, J. S., *Chem. Eng. Sci.*, *17* (265), 1962.

62. Maymo, J. A., Cunningham, R. E., and Smith, J. M., *Ind. Eng. Chem., Fundamentals*, *5* (280), 1966.

63. Cunningham, R. A., Carberry, J. J., and Smith, J. M., *AIChE J.*, *11* (636), 1965.

64. Wheeler, A., *Adv. Catalysis*, *3* (249), copyright © 1951. Used with permission of A. Wheeler and Academic Press.

65. Wheeler, A., *Catalysis*, *2* (105), edited by P. H. Emmett, Reinhold, New York, 1955. Adapted from the contribution of A. Wheeler to *Catalysis*, *2*, edited by P. H. Emmett. Copyright © 1955 by Litton Educational Publishing, Inc. Adaptation with permission of Van Nostrand Reinhold Company.

66. Smith, J. M., op. cit., pp. 457–462.

67. Carberry, J. J., *Chem. Eng. Sci.*, *17* (675), 1962.

68. Ostergaard, K., *Acta, Chem. Scand.*, *15* (2037), 1961.

69. Butt, J. B., *Chem. Eng. Sci.*, *21* (275), 1966.

70. Hutchings, J., and Carberry, J. J., *AIChE J.*, *12* (20), 1966.

71. Petersen, E. E., *Chemical Reaction Analysis*, pp. 129–164, Prentice-Hall, Englewood Cliffs, N.J., 1965.

72. Treybal, R. E., *Mass Transfer Operations*, Second Edition, McGraw-Hill, New York, 1968.

73. McCabe, W. L., and Smith, J. C., *Unit Operations of Chemical Engineering*, McGraw-Hill, New York, 1967.

74. Bird, R. B., Stewart, W. E., and Lightfoot, E. N., *Transport Phenomena*, Wiley, New York, 1960.

75. Chilton, T. H., and Colburn, A. P., *Ind. Eng. Chem.*, *26* (1183), 1934.

76. Colburn, A. P., *Trans. AIChE*, *29* (174), 1933.

77. Acetis, J. and Thodos, G., *Ind. Eng. Chem.*, *52* (1003), 1960.

78. Petrovic, L. J., and Thodos, G., *Ind. Eng. Chem., Fundamentals*, *7* (274), 1968.

79. Wilson, E. J., and Geankoplis, C. J., *Ind. Eng. Chem., Fundamentals*, *5* (9), 1966.

80. Satterfield, C. N., *Mass Transfer in Heterogeneous Catalysis*, pp. 79–97, MIT Press, Cambridge, 1970.

81. Chu, J. C., Kalil, J., and Wetteroth, W.A., *Chem. Eng. Prog.*, *49* (141), 1953.

82. Riccetti, R. E., and Thodos, G., *AIChE J.*, *7* (442), 1961.

83. Sen Gupta, A., and Thodos, G., *Chem. Eng. Prog.*, *58* (7), p. 58, 1962.

84. Kunii, D., and Levenspiel, O., loc. cit., Chapter 7.

85. Beek, W. J., "Mass Transfer in Fluidized Beds," in *Fluidization* by J. F. Davidson and D. Harrison, Academic Press, New York, 1971.

86. Olson, R. W., Schuler, R. W., and Smith, J. M., *Chem. Eng. Prog.*, *46* (614), 1950.

87. Reid, R. C., and Sherwood, T. K., *The Properties of Gases and Liquids*, Second Edition, McGraw-Hill, New York, 1966.

88. Yoshida, F., Ramaswami, D., and Hougen, O. A., *AIChE J.*, *8* (5), 1962.

89. Argo, W. B., and Smith, J. M., *Chem. Eng. Prog.*, *49* (1443), 1953.

90. Stewart, W. E., *AIChE J.*, *9* (528), 1963.

91. Sherwood, T. K., *Trans. AIChE*, *39* (583), 1943.

92. Kettenring, K. N., Manderfield, E. L., and Smith, J. M., *Chem. Eng. Prog.*, *46* (139), 1950.

93. Smith, J. M., loc. cit., Chapters 10–12.

94. Froment, G. F., "Analysis and Design of Fixed Bed Catalytic Reactors," in *Chemical Reaction Engineering*, *Adv. Chem. Series*, *109* (1), American Chemical Society, Washington, D.C., 1972.

95. Froment, G. F., Fifth European Symposium on Chemical Reaction Engineering, Elsevier, Amsterdam, 1972.

96. Bird, R. B., Stewart, W. E., and Lightfoot, E. N., *Transport Phenomena*, Wiley, New York, 1960.

97. Perry, R. H., and Chilton, C. H., *Chemical Engineer's Handbook*, Fifth Edition, McGraw-Hill, New York, 1973.

98. Ergun, S., *Chem. Eng. Prog.*, *48* (93), 1952.

99. Wehner, J. F., and Wilhelm, R. H., *Chem. Eng. Sci.*, *6* (89), 1956.

100. Smith, J. M., loc. cit., p. 504.

101. Aris, R., and Amundson, N. R., *AIChE J.*, *3* (280), 1957.

102. Jakob, M., *Heat Transfer*, p. 553, Wiley, New York, 1957.

103. Colburn, A. P., *Ind. Eng. Chem.*, *23* (910), 1931.

104. Calderbank, P. H., and Pogorski, L. A., *Trans. Inst. Chem. Engrs.* (London) *35* (195), 1957.

105. Leva, M., *Ind. Eng. Chem.*, *42* (2498), 1950.

106. Argo, W. B., and Smith, J. M., *Chem. Eng. Prog.*, *49* (443), 1953.

107. Smith, J. M., loc. cit., pp. 513–522.

108. Beek, J., *Adv. Chem. Eng.*, *3*, pp. 229–230, 1962.

109. Carberry, J. J., Wendel, M., *AIChE J.*, *9* (132), 1963.

110. Raymond, L. R., Amundson, N. R., *Can. J. Chem. Eng.*, *42* (173), 1964.

111. Luss, D., Amundson, N. R., *Chem. Eng. Sci.*, *22* (253), 1967.

112. Luss, D., *Chem. Eng. Sci.*, *23* (1249), 1968.

113. Hlavacek, V., Hofmann, H., *Chem. Eng. Sci.*, *25* pp. 173, 187, 1970.

114. Calderbank, P. H., *Chem. Eng. Prog.*, *49* (585), 1953.

115. Froment, G. F., *Ind. Eng. Chem.*, *59* (18), 1967.

116. Lapidus, L., *Digital Computation for Chemical Engineers*, McGraw-Hill, New York, 1960.

117. Smith, J. M., op. cit., pp. 536–547.

118. Jenson, V. G., and Jeffreys, G. V., *Mathematical Methods in Chemical Engineering*, pp. 412–428, Academic Press, New York, 1963.

119. Pyle, D. L., in "Chemical Reaction Engineering," *ACS Adv. Chem. Series*, *109*, p. 106, 1972.

120. Rowe, P. N., Fifth European Symposium on Chemical Reaction Engineering, Elsevier, Amsterdam, 1972.

121. Grace, J. R., AIChE Symposium Series, *67* (116), p. 159, 1971.

122. Kunii, D., and Levenspiel, O., *Fluidization Engineering*, Wiley, New York, 1969.

123. Davidson, J. F., and Harrison, D., *Fluidized Particles*, Cambridge University Press, London, 1963.

PROBLEMS

1. Thiophene (C_4H_4S) is representative of the organic sulfur compounds that are hydrogenated in the commercial hydrodesulfurization of petroleum naphtha. Estimate both the combined and effective diffusivities for thiophene in hydrogen at 660 °K and 3.04 MPa in a catalyst with a BET surface area of 168 m^2/g, a porosity of 0.40, and an apparent pellet density of 1.40 g/cm^3. A narrow pore sized distribution

and a tortuosity factor of 2.5 may be assumed. The following Lennard-Jones parameters for thiophene and hydrogen may be useful in your estimate of the ordinary molecular diffusivity.

	σ(nm)	ε/k_B (°K)
Thiophene	0.562	454
Hydrogen	0.2827	59.7

Use your best judgment in estimating the binary molecular diffusivity. However, it may be assumed that we are still in the region where the binary molecular diffusivity is inversely proportional to the pressure.

2. Villet and Wilhelm [*Ind. Eng. Chem.*, *53*(837), 1961] have studied the Knudsen diffusion of hydrogen in porous silica-alumina cracking catalyst pellets. They used apparatus of the type depicted in Figure 12P.1.

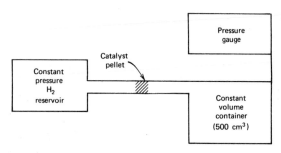

Figure 12P.1
Schematic diagram of diffusivity measurement apparatus.

The entire apparatus was immersed in a constant temperature bath at 25 °C. The upstream hydrogen pressure was maintained constant at 77.00 cm Hg. The time variation of the pressure in the constant volume container is given in the next column.

Time, t (min)	Pressure (cm Hg)
0	61.00
20	61.10
40	61.21
60	61.32
80	61.43
100	61.55

For our purposes, the pellet may be considered as cylindrical with a diameter of 1/8 in. and a length of 0.12 cm. Its porosity is 0.464 and the specific surface area is 243 m²/g.

What is the effective diffusivity of the hydrogen under these conditions?

It may be assumed that the accumulation of hydrogen within the pellet is negligible and that it may be treated as being in a quasi-steady-state condition. The finite difference form of Fick's first law may be used to determine the flow rate of hydrogen through the pellet. The diffusion constant appearing in this equation may be considered as an effective Knudsen diffusion coefficient.

3. The ortho-para conversion of molecular hydrogen is catalyzed by NiO. A supported catalyst is available with a specific surface area of 305 m²/g and a void volume of 0.484 cm³/g. A spherical catalyst pellet has an apparent density of 1.33 g/cm³ and a diameter of 0.5 cm. If the system is not far from equilibrium, an apparent first-order rate constant (k_r) can be defined in the following manner.

Rate of approach to equilibrium $= k_r(C - C_{eq})$

When the hydrogen pressure is 1 atm, and the temperature is 77 °K, the experimentally observed (apparent) rate constant is 0.159 cm³/sec-g catalyst. Determine the mean pore radius, the effective diffusivity of hydrogen, and the catalyst effectiveness factor.

4. A catalyst for cracking cumene is available commercially in the form of 0.35 cm diameter pellets. These pellets have a specific surface area of 420 m^3/g and a void volume of 0.42 cm^3/g. If the apparent first-order rate constant for this reaction is 1.49 cm^3/sec-g catalyst at 412 °C, determine the effectiveness factor of the catalyst.

5. Barnett et al. [*AIChE J.*, 7 (211), 1961] have studied the catalytic dehydrogenation of cyclohexane to benzene over a platinum-on-alumina catalyst. A 4 to 1 mole ratio of hydrogen to cyclohexane was used to minimize carbon formation on the catalyst. Studies were made in an isothermal, continuous flow reactor. The results of one run on 0.32 cm diameter catalyst pellets are given below.

> Temperature, 705 °K
> Pressure, 1.480 MPa
> H$_2$ feed rate, 8 moles/ksec
> Cyclohexane feed rate, 2 moles/ksec
> Conversion of cyclohexane, 15.5%
> Quantity of catalyst, 10.4 g

Catalyst Properties:

> Pore volume, 0.48 cm^3/g
> Surface area, 240 m^2/g
> Pellet density, 1.332 g/cm^3
> Pellet porosity, 0.59 cm^3 voids/cm^3

If the effectiveness factor of the catalyst is known to be 0.42, estimate the tortuosity factor of the catalyst assuming that the reaction obeys first-order kinetics and that Knudsen diffusion is the dominant mode of molecular transport.

6. Gupta and Douglas [*AIChE J.*, 13 (883), 1967] have studied the catalytic hydration of isobutylene to *t*-butanol, using a cation exchange resin catalyst in a stirred tank reactor.

$$(CH_3)_2C{=}CH_2 + H_2O \rightleftharpoons (CH_3)_3COH$$

The water is present in such large excess relative to the isobutylene and *t*-butanol con-

centrations that the reaction may be regarded as pseudo first-order in both directions.

Determine the effectiveness factor for the ion exchange resin at 85 °C, assuming that the reaction is reversible even though the authors presumed the reaction to be irreversible in reporting their data. They note that at 100 °C the equilibrium for the reaction corresponds to a conversion greater than 94%. If the equilibrium constant for the reaction is expressed as the ratio of the *t*-butanol concentration to the isobutylene concentration and corrected for the temperature change in going from 100 °C to 85 °C, a value of 16.6 may be considered appropriate for use.

The ion exchange particles may be regarded as isothermal, and the effective diffusivity of isobutylene within the particles may be taken as 2.0 × 10^{-5} cm^2/sec. The resin particles may be considered as spheres with radii equal to 0.0213 cm. The density of the swollen resin is assumed to be equal to 1.0 g/cm^3.

Rate measurements on these particles indicate that at 85 °C the rate is equal to 1.11 × 10^{-2} moles/ksec-g catalyst, when the conversion achieved is only 3.9%. From thermodynamic equilibrium data contained within the article, it is estimated that under these conditions, the isobutylene concentration in the reactor at the exterior surface of the resin may be taken as (0.961)(1.72 × 10^{-5}) moles/cm^3.

7. A well-insulated pilot scale packed bed reactor (5 m long by 5 cm diameter) is being used to carry out an irreversible reaction of the type

$$M + N \rightarrow R + S$$

All species are gases at the conditions of interest. The reaction is quite exothermic ($\Delta H = -120$ kJ/mole). Side reactions are unimportant.

There are two different catalyst manufacturers whose products are being considered for use in a commercial scale reactor facility. Both products are 3.2 mm diameter spherical pellets that are believed to be essentially the same in chemical composition.

The pilot plant results indicate that the deactivation characteristics of the two catalysts are somewhat different. The same gas pressure, feed composition, and molal feed rate are employed in all cases. If the inlet temperature is 550 °C, the behavior indicated in Figure 12P.2 is observed. If the feed temperature is reduced

to 475 °C, both catalysts show little loss of activity with time onstream and both catalysts give essentially the same yield. (It is, however, significantly lower than the yield obtained with the 550 °C feed temperature.) The deactivation behavior of both catalysts is very reproducible.

(a) How do you interpret these data?

(b) What modifications in the manufacturing or operating conditions do you suggest to obtain improved performance from catalyst A?

8. Barnett, Weaver, and Gilkeson [*AIChE J.*, 7 (211), 1961] have studied the dehydrogenation of cyclohexane to benzene over a platinum on alumina pelleted catalyst. Using a 4:1 feed ratio of hydrogen to cyclohexane and an operating pressure of 200 psig, these individuals studied the effects of particle size and poisoning on the observed reaction rate. The reaction follows first-order kinetics over the temperature range from 640 to 910 °F. Arrhenius plots of their data are shown in Figure 12P.3. For the

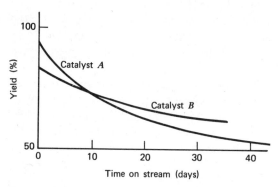

Figure 12P.2
Catalyst deactivation curves.

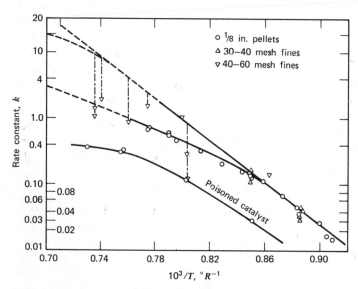

Figure 12P.3
Arrhenius plots for poisoned and unpoisoned catalysts. [From
AIChE J., 7 **(211), 1961. Used with permission.]**

poisoned catalyst, it is estimated that the fraction that is poisoned (α) is 0.78. This catalyst was used in runs 41 to 45 for which the following additional data are available.

Run	Thiele modulus for unpoisoned reaction	(Poisoned rate) (Unpoisoned rate)
45	0.77	0.22
44	1.78	0.26
41	4.24	0.214
42	4.40	0.210
43	6.95	0.153

What is your interpretation of these data? Do the data indicate whether homogeneous or pore-mouth poisoning takes place?

Catalyst Properties:

$$V_g = 0.48 \text{ cm}^3/\text{g}$$
$$S_g = 204 \text{ m}^2/\text{g}$$
$$\text{Average pore radius} = 47 \text{ Å}$$
$$\text{Pellet density } \rho_p = 1.332 \text{ g/cm}^3$$
$$\text{Skeletal density} = 3.25 \text{ g/cm}^3$$
$$\text{Pellet porosity } \varepsilon_p = 0.59$$
$$\text{Tortuosity factor} = 8$$

9. Before ethylene feedstocks produced by thermal cracking can be used chemically for most applications, it is necessary to remove the traces of acetylene present in such streams. This purification can be accomplished by selective hydrogenation of acetylene to ethylene. The process involves adding sufficient hydrogen to the feedstock so that the mole ratio of hydrogen to acetylene exceeds unity. Using a palladium-on-alumina catalyst under typical reaction conditions (25 atm, 50 to 200 °C), it is possible to achieve extremely high selectivity for the acetylene hydrogenation reaction. As long as acetylene is present, it is selectively adsorbed and hydrogenated. However, once it disappears, hydrogenation of ethylene takes place. The

competitive reactions may be written as

$$C_2H_2 + H_2 \rightarrow C_2H_4 \quad \text{(desired)}$$
$$C_2H_4 + H_2 \rightarrow C_2H_6 \quad \text{(undesired)}$$

The intrinsic rate expressions for these reactions are both first-order in hydrogen and zero-order in acetylene or ethylene. If there are diffusional limitations on the acetylene hydrogenation reaction, the acetylene concentration will go to zero at some point within the core of the catalyst pellet. Beyond this point within the central core of the catalyst, the undesired hydrogenation of ethylene takes place to the exclusion of the acetylene hydrogenation reaction.
(a) In the light of the above facts, what do the principles enunciated in this chapter have to say about the manner in which the reactor should be operated and the manner in which the catalyst should be fabricated?
(b) It is often observed that the catalysts used for this purpose in commercial installations do not achieve maximum *selectivity* until they have been on stream for several days. How do you explain this observation?

10. Cunningham, Carberry, and Smith [*AIChE J., 11* (636), 1965] have studied the catalytic hydrogenation of ethylene over a copper-magnesium oxide catalyst.

$$C_2H_4 + H_2 \rightarrow C_2H_6$$
$$\Delta H_{298} = -32,700 \text{ cal/g mole}$$

In order to minimize the complications arising from a change in the total moles within the catalyst particles, they restricted their studies to feeds containing less than 17% ethylene. They used continuous flow reactors operating at steady state to obtain the data reported on page 530. The pressure was one atmosphere. Two forms of catalyst were used in their studies.
1. Granular particles (100 to 150 Tyler mesh—equivalent diameter = 0.13 mm).

2. Spherical pellets (1.27 cm diameter) fabricated by compressing unreduced granular particles in a steel mold.

Pellets of three different densities were studied. They were obtained by varying the quantity of particles and the pressure used in the molding process. Pertinent physical property data for the various forms of catalyst used in these studies are summarized below.

Note. Engineering estimates of property values have been used, where necessary, to fill out the data.

11. It is instructive to consider the relative rates of mass transfer in fixed and fluidized bed reactors. The rapid rate in the fluidized bed is due not so much to the high mass transfer coefficients involved, but to the very large

Catalyst	Apparent density of particle or pellet (g/cm^3)	Equivalent diameter	S_g (m^2/g)	V_g (cm^3/g)	ε_{total}
Granular particles	2.14	0.13 mm	90	0.18	0.38
Pellet *A*	1.16	1.28 cm	90	0.46	0.53
Pellet *B*	0.92	1.27 cm	90	0.73	0.77
Pellet *C*	0.72	1.27 cm	90	1.17	0.84

ε_{total} refers to the ratio of the void volume within and between pellets to the total volume.

Reaction rate data were reported as a function of temperature and are shown in Figure 12P.4. Although the form of the intrinsic rate equation for ethylene hydrogenation for this specific catalyst is not known, one might anticipate an equation of the form

$$r = \frac{kK_{C_2H_4}P_{H_2}P_{C_2H_4}}{1 + K_{C_2H_4}P_{C_2H_4}}$$

on the basis of other studies of the catalytic hydrogenation of ethylene. What is your interpretation of the data shown in Figure 12P.4? If it is possible to back up your arguments using semiquantitative arguments, do so using *rough* numbers. *If multiple interpretations are possible, state them.* In the low temperature region the intrinsic activation energy based on the particle data is 17.8 kcal/g mole, but at high temperatures the apparent activation energy drops to 11.800 kcal/g mole.

exterior surface area per unit volume of reactor. If one assigns the weight of catalyst per unit volume of reactor the symbol ρ_B, the rate of mass transfer to the external surface per unit reactor volume is given by

$$r_{MT} = k_m a_m (C_B - C_{ES})\rho_B = k_m a_v (C_B - C_{ES})$$
$$(A)$$

where we have assumed equimolar counter-diffusion ($\delta = 0$) in order to simplify the analysis. The product $a_m \rho_B$ represents the external surface area of catalyst per unit volume of bed (a_v). For spherical pellets of radius R, this area is given by

$$a_m \rho_B = \left(\frac{4\pi R^2}{\rho_p \frac{4}{3}\pi R^3} \right) \rho_B = \frac{3}{R}\left(\frac{\rho_B}{\rho_p} \right)$$

where ρ_p is the apparent density of a catalyst pellet. The ratio of the catalyst density within the entire bed to that of a single pellet is equal to the ratio of the volume occupied by the

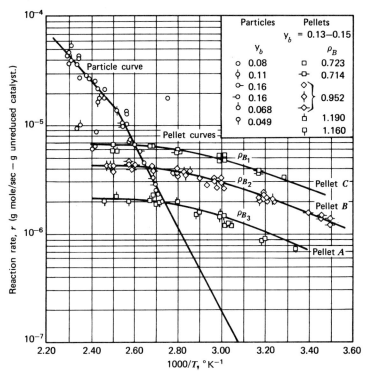

Figure 12P.4
Arrhenius plots for particles and pellets. (y_b = *ethylene mole fraction*)
[From AIChE J., 11 (636), 1965. Used with permission.]

pellets themselves to the entire bed volume. In terms of the external void fraction of the bed (ε_B), the volume fraction occupied by the pellets is equal to $1 - \varepsilon_B$. Hence

$$\frac{\rho_B}{\rho_p} = 1 - \varepsilon_B$$

Hence

$$a_m \rho_B = \frac{3}{R}(1 - \varepsilon_B) = \frac{6(1 - \varepsilon_B)}{D_p} \quad \text{(B)}$$

Combining equations A and B gives

$$r_{MT} = \frac{6k_m(1 - \varepsilon_B)}{D_p}(C_B - C_{ES}) \quad \text{(C)}$$

Assuming the same concentration driving force in both fixed and fluidized beds, use typical property values to determine the relative rates of mass transfer in these systems. Mass velocities for fixed and fluidized beds may be taken as 0.15 and 0.03 g/(cm²-sec), respectively. Bed void fractions may be taken as 0.30 and 0.80 for the fixed and fluidized beds, respectively. The corresponding catalyst sizes may be taken as 0.5 cm and 0.0063 cm (250 mesh). These numbers are chosen so as to favor fixed bed mass transfer. The reacting fluid may be regarded as a gas with a viscosity of 3.30×10^{-4} g/(cm-sec).

12.* The data on page 532 were obtained by Karpenko in an unpublished study described by

* Adapted with permission from C. N. Satterfield

Rylander in *Catalytic Hydrogenation Over Platinum Metals* (p. 39, Academic Press, New York, 1967). Nitrobenzene in ethanol was hydrogenated at room temperature and 1 atm over various amounts of 5% Pd on carbon. Four loading levels of catalyst were used. At each level, the reduction was carried out in two different types of batch reactor.

1. An equipoise shaker that gives very vigorous agitation.
2. A flask stirred by a rotating magnetic bar that gives relatively poor agitation.

Each reduction proceeded at nearly constant rate until the nitrobenzene was nearly exhausted.

13. Zajcew [*J. Am. Oil Chem. Soc.*, *37*(11), 1960] has studied the hydrogenation of fatty oils for shortening stock using a palladium catalyst. The experiments were carried out in a 1-gal hydrogenator provided with mechanical agitation, a gas dispersing system, and heating and cooling capability. Hydrogen gas is fed in at the reactor bottom and is rapidly consumed. Figure 12P.5 indicates the time necessary to achieve a specified conversion as a function of catalyst loading level. Note that semilog coordinates are employed. The table below reformulates these data in terms of the number of iodine units of reduction per minute per 1% of catalyst.

		Hydrogen uptake (cm^3/sec)			
Supported catalyst (mg)	Palladium (mg)	Total		Per mg Pd	
		Shaker	Stirrer	Shaker	Stirrer
52	2.6	0.042	0.042	0.016	0.016
105	5.2	0.72	0.52	0.138	0.100
210	10.5	1.97	0.92	0.187	0.088
420	21.0	2.57	0.92	0.122	0.044

What is your interpretation of the above data?

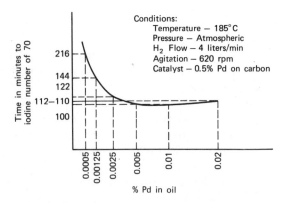

% Pd	Rate/% of catalyst
0.02	18
0.01	40
0.005	72
0.0025	130
0.00125	215
0.0005	354

What is your interpretation of these data?

Figure 12P.5

Hydrogenation rate as a function of catalyst concentration. [Adapted from M. Zajcew, *J. Am. Oil Chem. Soc.*, *37* (11), 1960. Used by permission of American Oil Chemists Society.]

14.* Price and Schiewetz [*Ind Eng. Chem. 49* (807), 1957] have studied the catalytic liquid phase hydrogenation of cyclohexene in a laboratory scale semibatch reactor. A supported platinum catalyst was suspended in a cyclohexene solution of the reactant by mechanical

* Adapted with permission from C. N. Satterfield

agitation of the solution. Hydrogen was bubbled through the solution continuously. The reactor is described in their words in the next paragraph.

"The reactor consisted of a 1-liter three-necked Morton flask. This flask has four equally spaced perpendicular indentations about its periphery and a concave bottom, which aid the action of the agitator. A four-bladed stirrer driven by a variable-speed motor extended into the flask through a packing gland in the center neck. Tubes were sealed into one side neck for hydrogen delivery, sample withdrawal, and a thermocouple well. An ice water condenser on the exit neck minimized solvent escape by entrainment or vaporization. Hydrogen was dispersed into the reacting solution by a 20 mm sintered-glass tube extending to the bottom of the flask. The entire assembly was immersed in a constant temperature bath maintained at $\pm 0.1\ °C$. However, because of the exothermic nature of the reaction, a temperature rise of 1 to 2 °C occurred in the reactor.*

At 25 °C the equilibrium constant for the reaction

$$\text{cyclohexene} + H_2 \longrightarrow \text{cyclohexane}$$

is 2×10^8. Consequently, the reaction may be regarded as irreversible. These investigators reported the data indicated on Figures 12P.6 to 12P.11. *Provide an interpretation of each figure individually* and then an overall analysis of the factors that govern the rate of this reaction. Where multiple interpretations of an individual figure are possible, indicate them. Notice that in Figure 12P.11, the apparent activation energies determined from the slopes of the solid lines are 4.80 and 12.8 kcal/g mole at 1300 and 2500 rpm, respectively.

* This material and the figures on pages 533–536 are reprinted, with permission, from R. H. Price and R. B. Schiewtz, *Ind. Eng. Chem.*, 49 (807), 1957. Copyright © by the American Chemical Society.

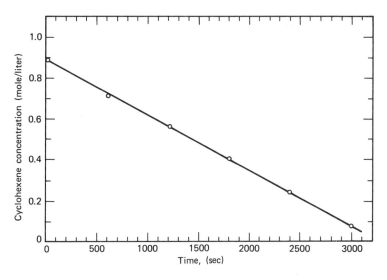

Figure 12P.6
Typical concentration-time relationship for experimental runs. Temperature, 26°C; pressure, 746 mm of mercury; hydrogen flow, 30.7 × 10^{-5} mole/sec; catalyst weight, 0.975 g; stirrer speed, 1100 rpm; slope, 13.7 × 10^{-5} mole/liter/sec.

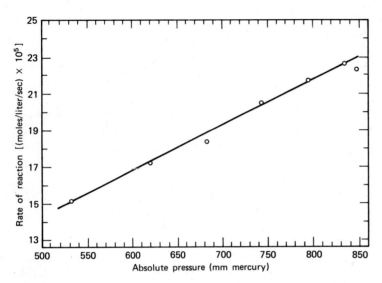

Figure 12P.7
Effect of pressure on rate of reaction. Temperature, 25–26 °C; hydrogen flow, 27.6×10^{-5} mole/sec; catalyst weight, 0.976 g; stirrer speed, 1500 rpm.

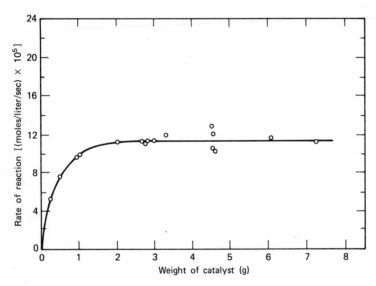

Figure 12P.8
Effect of catalyst weight on rate of reaction. Temperature, 25 °C; pressure, 748 mm of mercury; hydrogen flow, 31.3×10^{-5} mole/sec; stirrer speed, 1000 rpm.

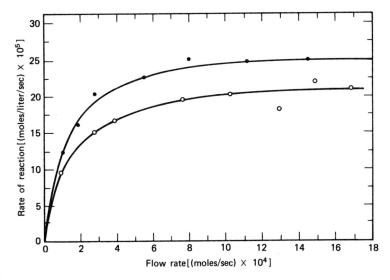

Figure 12P.9
Effect of hydrogen flow rate on rate of reaction. Temperature, 25–26 °C; pressure, ○ = 534 mm and ● = 741 mm of mercury; catalyst weight, 0.974 g; stirrer speed, 1500 rpm.

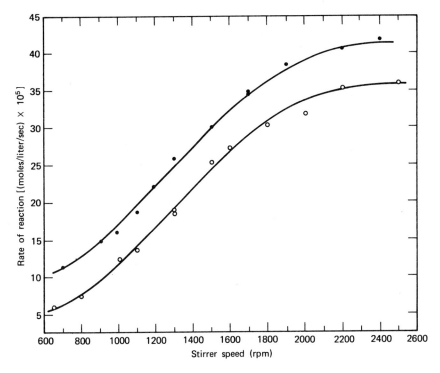

Figure 12P.10
Effect of stirring on rate of reaction. Temperature, 25–27 °C; pressure, 746 mm of mercury; hydrogen flow, ○ = 29.6 × 10⁻⁵ and ● = 81.5 × 10⁻⁵ mole/sec; catalyst weight, 0.976 g.

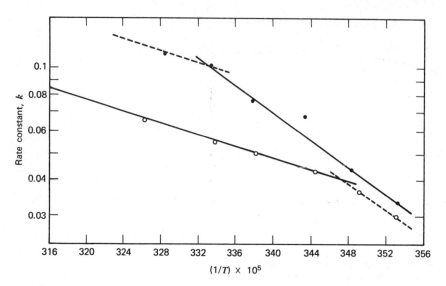

Figure 12P.11

Evaluation of apparent energies of activation. ○ = 1300 rpm and ● = 2500 rpm.

15. Vanadium pentoxide may be used as a catalyst for the oxidation of sulfur dioxide to sulfur trioxide.

$$SO_2 + \tfrac{1}{2}O_2 \rightarrow SO_3$$

In a series of laboratory scale experiments, streams of oxygen and sulfur dioxide were fed at different rates to a differential reactor containing 2.372 g of catalyst. The data below were recorded under essentially isothermal conditions

at 649 °F. The reaction rate expression pertinent to these experimental conditions is believed to be of the form:

$$r = \frac{kP_{SO_2}P_{O_2}^{1/2}}{(1 + K_{SO_2}P_{SO_2})^2}$$

The following data have been reported by Mathur and Thodos [*Chem. Eng. Sci., 21* (1191), 1966] and are sufficient for an initial rate analysis of this heterogeneous catalytic process.

Inlet pressure (atm)	Mass velocities (lb moles/hr-ft²)		Percent conversion of SO₂	Initial rate (lb moles SO₂ converted/ hr-lb catalyst)
	SO₂	O₂		
2.36	1.83	1.90	4.22	0.053
3.72	2.59	2.90	3.6	0.064
5.08	3.03	3.88	3.43	0.0717
5.76	3.24	4.26	3.10	0.0692
7.12	4.87	4.96	1.91	0.0642

(a) Determine the constants k and K_{SO_2}.

(b) If pure SO_2 and O_2 are both fed to a reactor at flow rates of 100 lb moles/hr, how many pounds of catalyst will be required for 20% conversion? The initial feed pressure is 1 atm. Assume that the aforementioned rate expression is valid over the range of variables concerned. The reactor may be assumed to operate isothermally at 649 °F. Set up the integral, but do not attempt to evaluate it. Make sure that all terms under the integral are expressed in terms of a single variable.

16. Wan [*Ind. Eng. Chem.*, *45* (234), 1953] has studied the partial oxidation of ethylene to ethylene oxide.

$$C_2H_4 + \tfrac{1}{2}O_2 \rightarrow C_2H_4O$$

The catalyst consists of silver supported on alumina and, while it is reasonably specific, appreciable amounts of CO_2 and H_2O are also formed. Over the range of interest, the yield of ethylene oxide is relatively constant so that for present purposes, we may regard the reaction stoichiometry as

$$C_2H_4 + 1.5O_2 \rightarrow$$
$$0.6C_2H_4O + 0.8CO_2 + 0.8H_2O$$

The rate of reaction may be expressed as

$$r = 1.17 \times 10^6 e^{-9713/T} P_{C_2H_4}^{0.341} P_{O_2}^{0.672}$$

where the partial pressures are expressed in atmospheres

T is expressed in degrees Kelvin

r is expressed in pound moles per pound of catalyst per hour

If 1/8 in. catalyst pellets are packed in 1 in. ID tubes, which in turn are immersed in a liquid bath that maintains the tube walls at 300 °F, consider the effects of varying the feed temperature and of diluting the feed with N_2 to moderate the thermal effects accompanying the reaction. Consider inlet temperatures from 350 to 480 °F and N_2/C_2H_4 ratios from 0 to 5.0.

Your analysis will be governed by the following constraints.

(a) If the temperature at any point in the reactor exceeds 550 °F, the conditions will be inappropriate in that explosions may occur in this regime.

(b) If the temperature decreases as one moves down the reactor, the reaction must be regarded as self-extinguishing.

(c) If the pressure drop exceeds 14 atm, the calculations must be terminated.

Determine the range of satisfactory performance and the resultant yields for various reactor lengths for the following operating specifications.

1. Inlet gas pressure = 15 atm.
2. Inlet C_2H_4/O_2 ratio (moles) = 4:1.
3. Superficial mass velocity = 8000 lb/hr-ft^2.
4. Bulk density of catalyst = 83.5 lb/ft^3.

External heat and mass transfer effects are to be neglected in your analysis, but you should estimate the potential magnitudes of these effects.

Not many problems dealing with the various fixed bed reactor models are set forth above, because I feel that the best means of demonstrating the features thereof is through case studies. I find that the principles involved are clarified by realistic examples involving the design of commercial scale reactors using machine computation where appropriate. Two brief examples that involve studies of specific process variables are considered in Chapter 13. One could employ studies of the effects of other process variables on the reactor performance as additional problems or else state the assignment in more general terms, as in the problem below. Space limitations preclude setting forth many such problems, but the book *Catalytic Processes* and *Proven Catalysts*, by C. L. Thomas, and the descriptions of new and current catalyst technology that appear frequently in *Hydrocarbon Processing* provide appropriate starting points for the acquisition of the necessary data for problem formulation. Old issues of *Industrial and Engineering Chemistry* and the more recent quarterlies evolving therefrom are also good sources of essential data on catalyst activity. The literature of various catalyst manufacturers provides information on the physical properties of general catalyst types when such information is not included with the activity data. As stated, the problems normally require literature searches, and the students usually find that information necessary to complete the design is often missing and must be estimated using good engineering judgment.

17. Oxidative Dehydrogenation of Butene to Butadiene

UW Chemical Corporation
Madison, Wisconsin
To: Student Team "Blue"
From: Big Red, Group Leader
Re: Butadiene production

A proposed expansion of the company's styrene-butadiene rubber production will require an additional 10,000 tons/year of butadiene as a raw material.

For many years, butadiene has been manufactured by dehydrogenating butene or butane over a catalyst at appropriate combinations of temperature and pressure. It is customary to dilute the butene feed with steam (10–20 moles H_2O/mole butene) to stabilize the temperature during the endothermic reaction and to help shift the equilibrium conversion in the desired direction by reducing the partial pressures of hydrogen and butadiene. The current processes suffer from two major disadvantages.

1. Catalyst on-stream periods are short, since coke builds up rapidly on the catalyst. Hence catalyst regeneration must take place at frequent intervals.

2. Equilibrium yields of butadiene are relatively low. For example, the yield is 35% at 930 °F and 71% at 1110 °F when using a hydrocarbon feed partial pressure of 0.1 atm.

In recent years, considerable research effort has been invested in trying to develop an oxidative process to accomplish the desired transformation. Several catalyst formulations with good selectivities for the desired reaction have been developed, and you are asked to determine the desirability of using one of these formulations in a fixed bed reactor configuration. Several advantages are claimed for the oxidative processes:

1. The constraint of thermodynamic equilibrium for the butene dehydrogenation reaction is effectively removed since hydrogen is converted to water by oxidation. Equilibrium yields then approach 100% over the complete temperature and partial pressure range of interest.

2. Selectivities are high, typically in excess of 90%.

3. Longer times on-stream are facilitated by the presence of oxygen in the feed. It inhibits carbon buildup on the catalyst, thereby permitting one to operate for months without regeneration.

4. Input energy requirements for the process are significantly reduced since the energy released by the exothermic oxidation reactions serves as a driving force for the endothermic dehydrogenation reaction.

For preliminary discussions of the proposed expansion program, it is desirable to determine the basic equipment requirements, although a detailed economic evaluation is not essential. An article by Sterrett and McIlvried [*Ind. Eng. Chem., Process Des. Develop., 13* (54), 1974] describes the use of a ferrite catalyst for the desired application. It contains kinetic data for this catalyst and several references to other oxidative dehydrogenation catalysts. Select a promising catalyst material and prepare a reactor design proposal using a fixed bed configuration. You are asked to consider factors such as:

1. The desirability of adding steam to the feed stream to moderate temperature excursions.

2. The optimum steam/butene ratio in the feed for operation as a single adiabatic reactor.

3. The desirability of using several adiabatic reactors in series with steam injection between stages.

4. Temperature profiles and the possibility of hot spots within a given reactor network configuration.

5. Volume requirements for different reactor network configurations.

6. Effect of catalyst pellet size on reactor volume and pumping cost requirements.

7. Interstage heat transfer requirements for different reactor network configurations.

8. Concentration profiles for a given reactor network.

9. Other factors you deem pertinent.

13 Illustrative Problems in Reactor Design

13.0 INTRODUCTION

This chapter contains a discussion of two intermediate level problems in chemical reactor design that indicate how the principles developed in previous chapters are applied in making preliminary design calculations for industrial scale units. The problems considered are the thermal cracking of propane in a tubular reactor and the production of phthalic anhydride in a fixed bed catalytic reactor. Space limitations preclude detailed case studies of these problems. In such studies one would systematically vary all relevant process parameters to arrive at an optimum reactor design. However, sufficient detail is provided within the illustrative problems to indicate the basic principles involved and to make it easy to extend the analysis to studies of other process variables. The conditions employed in these problems are not necessarily those used in current industrial practice, since the data are based on literature values that date back some years.

13.1 PYROLYSIS OF A PROPANE FEEDSTOCK

This problem indicates the considerations that enter into the design of a tubular reactor for an endothermic reaction. The necessity of supplying thermal energy to the reactor contents at an elevated temperature implies that the heat transfer considerations will be particularly important in determining the longitudinal temperature profile of the reacting fluid. This problem is based on an article by Fair and Rase (1).

13.1.1 Charge to Design Engineer and Pertinent Data

To: Kem Inguneer
From: Barney Boss

Re: Proposed Production of Ethylene by Pyrolysis of Propane

A proposed expansion of the corporation's polyethylene production capacity will require additional ethylene monomer as a feedstock. It is suggested that the ethylene be produced by the pyrolysis of a propane stream that is available at a rate of 7000 lb/hr.

The production of ethylene from light alkanes involves two steps: conversion of the charge by pyrolysis and separation of the ethylene from a complex product mixture. As normally practiced in the petroleum and petrochemicals industry, pyrolysis or thermal cracking is not a process that is readily characterized by simple stoichiometry. However, there is a wealth of data available on these homogeneous gas phase decomposition reactions that is adequate for preliminary design purposes. At the present stage of management's discussions of the expansion program, a detailed economic analysis is not essential, but it is necessary to determine the basic equipment requirements. You are to carry out the necessary engineering analysis and come up with a reactor design analysis that will meet corporate requirements.

The most common type of commercial pyrolysis equipment is the direct fired tubular heater in which the reacting material flows through several tubes connected in series. The tubes receive thermal energy by being immersed in an oil or gas furnace. The pyrolysis products are cooled rapidly after leaving the furnace and enter the separation train. Constraints on materials of construction limit the maximum temperature of the tubes to 1500 °F. Thus the effluent from the tubes should be restricted to temperatures of 1475 °F or less. You may presume that all reactor tubes and return bends are exposed to a thermal flux of 10,000 BTU/

(hr-ft^2 of external tube surface). You may also assume that the feed enters the reactor at 1100° F and 60 psia. In order to maintain flow through the downstream separation train the reactor effluent cannot be at a pressure lower than 25 psia. Propane conversion should be in the neighborhood of 85%. You are asked to prepare a preliminary reactor design for this system. Use the data tabulated below to determine length requirements and general suitability for 2, 4, and 6 in. nominal pipe sizes (schedule 40).

Data Summary. There is an extensive literature dealing with hydrocarbon pyrolysis reactions. The articles by Schutt (2) and Fair and Rase (1) were particularly useful in arriving at the numbers cited below.

Reaction Stoichiometry and Thermodynamic Data. The product distribution for a propane charge is shown in Figure 13.1. The reaction is not clean even at low conversions. At higher conversions secondary reactions of the primary products become more and more significant. However, it is reasonable to assume that the original reactants continue to decompose along the path of the original mechanism, and first-order kinetics is observed to quite high conversions. In the presence of established information on the product distribution as the conversion

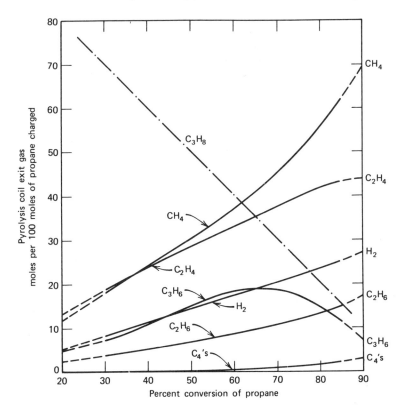

Figure 13.1
Product distribution for propane pyrolysis. [From Schutt, *Chemical Engineering Progress,* **50 (415), 1954. Used with permission.]**

increases, the establishment of rate expressions for the secondary and tertiary reactions is not essential from the standpoint of reactor design. For present purposes we may consider the reaction stoichiometry to be given by

$$C_3H_8 \rightarrow 0.300C_3H_6 + 0.065C_2H_6 + 0.6675C_2H_4$$
$$+ 0.635CH_4 + 0.300H_2 \qquad (13.1.1)$$

The volumetric expansion parameter δ may thus be taken as 0.9675. The product distribution will vary somewhat with temperature, but the stoichiometry indicated above is sufficient for preliminary design purposes. (We should also indicate that if one's primary goal is the production of ethylene, the obvious thing to do is to recycle the propylene and ethane and any unreacted propane after separation from the lighter components. In such cases the reactor feed would consist of a mixture of propane, propylene, and ethane, and the design analysis that we will present would have to be modified. For our purposes, however, the use of a mixed feed would involve significantly more computation without serving sufficient educational purpose.)

The following heat capacity data may be employed for the temperature range of interest. (Temperatures must be expressed in degrees Kelvin.) At 1100 °F the standard heat of reaction is equal to 21.960 kcal/g mole.

Compound	Heat capacity C_p (cal/g mole-°K)	
C_3H_8	$21.14 + 0.02056T$	(13.1.2)
C_3H_6	$17.88 + 0.01645T$	(13.1.3)
C_2H_6	$13.34 + 0.01589T$	(13.1.4)
C_2H_4	$12.29 + 0.01022T$	(13.1.5)
CH_4	$6.98 + 0.01012T$	(13.1.6)
H_2	$6.42 + 0.00082T$	(13.1.7)

Kinetic Data. The pyrolysis reaction obeys first-order kinetics with a rate constant equal to $3.98 \times 10^{12} e^{-59,100/RT} \text{ sec}^{-1}$, where T is ex-

pressed in degrees Kelvin and R in calories per gram mole degree Kelvin.

Viscosity Data. Our physical properties expert indicates that the following expression provides a good representation of the mixture viscosity data over the temperature, pressure, and composition range of interest.

$$\mu(\text{centipoise}) =$$
$$10^{-4}(2.37 + 0.708f - 0.411f^2)T^{2/3}$$
$$(13.1.8)$$

where

f is the fraction propane converted

T is expressed in degrees Kelvin

[*Note.* Kinetic theory predicts that μ should be proportional to $T^{1/2}$, but the $T^{2/3}$ dependence provides a better fit of the results calculated from methods outlined in Reid and Sherwood (3).]

13.1.2 Reactor Design Analysis

A preliminary design analysis for the proposed reactor may be carried out in terms of the plug flow model for a nonisothermal system. If the flow conditions lie in the turbulent regime, the assumption of linear velocity profiles will be reasonable and it will be appropriate to neglect radial variations in velocity, temperature, and composition. These assumptions are in essence the same as those employed in Section 10.4, and the discussion contained therein will be the basis for the development of our numerical results. However, we have added the additional complication of varying the pressure along the length of the reactor. This aspect of the problem is taken into account by breaking the reactor up into a number of segments, each of which is considered to operate at a constant pressure.

In order to characterize the flow regime and thereby justify the plug flow assumption, it is appropriate to calculate the Reynolds number

$$N_{Re} = \frac{DG}{\mu} \qquad (13.1.9)$$

where D is the tube diameter.

Now

$$G = \frac{7000}{\dfrac{\pi D^2}{4}} = \frac{8913}{D^2} \qquad (13.1.10)$$

where

D is measured in feet

G is measured in pounds mass per hour per square foot

Thus

$$N_{Re} = \frac{8913}{\mu D} \qquad (13.1.11)$$

where μ must be expressed in pounds mass per foot per hour.

At the reactor inlet (1100 °F or 866.49 °K)

$$\mu = 2.15 \times 10^{-2} \text{ cp} \approx 5.21 \times 10^{-2} \text{ lb}_m/\text{(ft-hr)}$$

Thus

$$N_{Re} = \frac{8913}{5.21 \times 10^{-2} D} = \frac{1.71 \times 10^5}{D} \qquad (13.1.12)$$

Since all of the tube diameters of interest are less than 1 ft, it is evident that in all three cases the flow will be highly turbulent and assumption of plug flow conditions will be quite appropriate.

For steady-state operation of a plug flow reactor the basic design equation (equation 8.2.9) can be written as

$$\frac{V_R}{\mathcal{V}_0} = C_{A0} \int_{f_{in}}^{f_{out}} \frac{df_A}{-r_A} \qquad (13.1.13)$$

where A refers to propane.

It is convenient to write the first-order reaction rate expression as

$$-r_A = \frac{kC_{A0}(1 - f_A)}{(1 + \delta_A f_A)} \left(\frac{T_0}{T}\right)\left(\frac{P}{P_0}\right) \qquad (13.1.14)$$

where C_{A0} is the propane concentration at zero fraction conversion and the reactor inlet temperature (T_0), and inlet pressure (P_0). (See equation 3.1.44.) Ideal gas behavior has been assumed, but this is quite appropriate for these nonpolar gases at the high temperatures and low pressures of interest.

Equations 13.1.13 and 13.1.14 may be combined to give

$$\frac{V_R}{\mathcal{V}_0} = \int_{f_{in}}^{f_{out}} \frac{(1 + \delta_A f_A)T P_0 \, df_A}{k(1 - f_A)T_0 P} \qquad (13.1.15)$$

or

$$\frac{\pi D^2 L}{4 \mathcal{V}_0} = \int_{f_{in}}^{f_{out}} \frac{(1 + \delta_A f_A)T P_0 \, df_A}{k(1 - f_A)T_0 P} \qquad (13.1.16)$$

In order to determine the required reactor volume one must relate the temperature (and thus k) and the local pressure P to the fraction conversion using an energy balance and conventional fluid flow equations.

For thermal cracking units, the majority of the required energy input is supplied by radiant heat transfer in a furnace. In such cases one finds it convenient to express the rate at which thermal energy is supplied to the reactor as the product of an energy flux and the external surface area of the tube. If we represent the flux by $\dot{q}$ and a segment of the reactor length by ΔL, then the total rate at which thermal energy is supplied $(\delta \dot{Q})$ is given by

$$\delta \dot{Q} = \dot{q} \pi D_0 \, \Delta L \qquad (13.1.17)$$

where D_0 is the outside diameter of the pipe. This expression may be substituted for the left side of equation 10.4.6 to give the following representation of an energy balance on a differential element of reactor length.

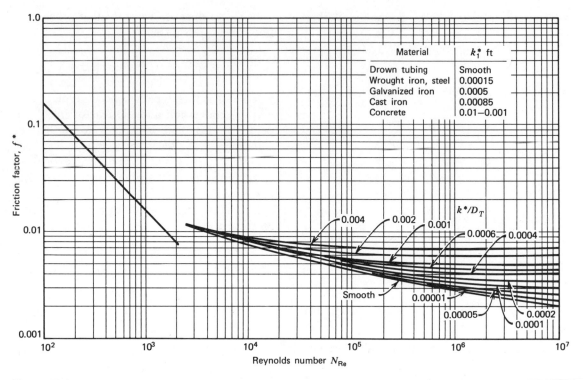

Figure 13.2
Friction-factor chart. (From *Unit Operations of Chemical Engineering* **by W. L. McCabe and J. C. Smith. Copyright © 1967. Used with permission of McGraw-Hill Book Company.)**

Table 13.1
Potential Range of Viscosities and Friction Factors

	Temperature 1100 °F (866 °K)		Temperature 1500 °F (1089 °K)	
	$f = 0$	$f = 0.8$	$f = 0$	$f = 0.8$
Viscosity (cp)	2.15×10^{-2}	2.43×10^{-2}	2.51×10^{-2}	2.83×10^{-2}
N_{Re} for 2-in. pipe	9.94×10^{5}	8.80×10^{5}	8.52×10^{5}	7.56×10^{5}
f^* for 2-in. pipe	0.0050	0.0050	0.0050	0.0051
N_{Re} for 4-in. pipe	5.11×10^{5}	4.52×10^{5}	4.37×10^{5}	3.88×10^{5}
f^* for 4-in. pipe	0.0043	0.0044	0.0044	0.0045
N_{Re} for 6-in. pipe	3.39×10^{5}	3.00×10^{5}	2.90×10^{5}	2.57×10^{5}
f^* for 6-in. pipe	0.0042	0.0044	0.0044	0.0045

Input by heat transfer = sensible heat change + heat of reaction

$$\dot{q}\pi D_0\,\Delta L = \sum\left(F_i\int_{T_0}^{T\text{ out of element}}C_{pi}\,dT\right) - \sum\left(F_i\int_{T_0}^{T\text{ into element}}C_{pi}\,dT\right) - \frac{F_{A0}\,\Delta H_{R\text{ at }T_0}(f_{A\text{ out}} - f_{A\text{ in}})}{\nu_A}$$

$$(13.1.18)$$

where the first summation involves the molal flow rates of the various species leaving the volume element and the second involves the molal flow rates of the various species entering the volume element. The datum for the enthalpy calculations has been taken as the reactor inlet temperature.

The pressure drop through a tubular reactor of length ΔL may be expressed in terms of the friction factor f^* as

$$\frac{\Delta P}{\rho} = \frac{2f^*\,\Delta L\langle u^2\rangle}{g_c D} \qquad (13.1.19)$$

where ρ is the mass density of the fluid and the g_c factor is introduced in order to make the units consistent.

The friction factor may be calculated from a knowledge of the fluid Reynolds number using the chart shown as Figure 13.2. Table 13.1 indicates the friction factors that correspond to zero and 80% conversion for the three pipe sizes of interest at two temperature levels. Examination

of the tabular entries indicates that the friction factor does not vary much with either temperature or fraction conversion and that an assumption of constant f^* for a given pipe size would be entirely appropriate. We will use $f^* = 0.0050$ for the 2-in. pipe and $f^* = 0.0044$ for the 4- and 6-in. pipes. Since the fluid density varies along the length of the reactor as the pressure and the temperature change, the local density must be used in equation 13.1.19 to evaluate the pressure drop over the length segment.

Equations 13.1.16, 13.1.18, and 13.1.19 must be solved simultaneously in order to determine the required reactor length for the various pipe sizes. We now wish to rewrite these equations in forms that are appropriate for solution using a finite difference approach. In the present case we will find it convenient to consider the reactor as consisting of a series of elements of length ΔL and employ these equations by incrementing the length. The following mole table may be prepared using stoichiometric principles.

Species	Effluent molal flow rate	$F_i C_{pi}$
C_3H_8	$(1 - f_A)F_{A0}$	$(21.14 + 0.02056T)(1 - f_A)F_{A0}$
C_3H_6	$0.300f_A F_{A0}$	$(5.36 + 0.00494T)f_A F_{A0}$
C_2H_6	$0.065f_A F_{A0}$	$(0.87 + 0.00103T)f_A F_{A0}$
C_2H_4	$0.6675f_A F_{A0}$	$(8.20 + 0.00682T)f_A F_{A0}$
CH_4	$0.635f_A F_{A0}$	$(4.43 + 0.00643T)f_A F_{A0}$
H_2	$0.300f_A F_{A0}$	$(1.93 + 0.00025T)f_A F_{A0}$
Total	$(1 + 0.9675f_A)F_{A0}$	$[(21.14 + 0.02056T) - f_A(0.35 + 0.00109T)]F_{A0}$

Now, if we let

$$f_{A \text{ out}} = f_{A \text{ in}} + \Delta f_A \tag{13.1.20}$$

the energy balance (equation 13.1.18) can be written as

$$
\dot{q} \pi D_0 \, \Delta L = F_{A0} \int_{T_0}^{T_{\text{out}}} (21.14 + 0.02056T) \, dT - F_{A0}(f_{A \text{ in}} + \Delta f_A) \int_{T_0}^{T_{\text{out}}} (0.35 + 0.00109T) \, dT
$$

$$
- \left[F_{A0} \int_{T_0}^{T_{\text{in}}} (21.14 + 0.02056T) \, dT - F_{A0}(f_{A \text{ in}}) \int_{T_0}^{T_{\text{in}}} (0.35 + 0.00109T) \, dT \right]
$$

$$
- F_{A0} \frac{\Delta H_{R \text{ at } T_0}}{v_A} \Delta f_A \tag{13.1.21}
$$

Since $\Delta H_{R \text{ at } T_0} = 21{,}960$ cal/g mole and the stoichiometric coefficient $v_A = -1$, the last equation can be written as

$$
\frac{\dot{q} \pi D_0 \, \Delta L}{F_{A0}} = \int_{T_{\text{in}}}^{T_{\text{out}}} (21.14 + 0.02056T) \, dT - f_{A \text{ in}} \int_{T_{\text{in}}}^{T_{\text{out}}} (0.35 + 0.00109T) \, dT
$$

$$
+ \Delta f_A \left[21{,}960 - \int_{T_0}^{T_{\text{out}}} (0.35 + 0.00109T) \, dT \right] \tag{13.1.22}
$$

where one may recognize the term in brackets as the heat of reaction at the temperature of the fluid leaving the length element. In this equation the units of T must be degrees Kelvin, and each term must have the units of calories per gram mole.

Now

$$
F_{A0} = \frac{7000(454)}{44.09} = 7.208 \times 10^4 \text{ g moles/hr}
$$

and

$$
\dot{q} = 10^4 \times 252 = 2.52 \times 10^6 \text{ cal/hr-ft}^2
$$

Thus, if D_0 is measured in inches and ΔL in feet,

$$
\frac{\dot{q} \pi D_0 \, \Delta L}{F_{A0}} = \frac{2.52 \times 10^6 \pi}{7.208 \times 10^4} \left(\frac{D_0}{12} \right) \Delta L = 9.153 D_0 \, \Delta L \, \frac{\text{cal}}{\text{g mole}}
$$

With $T_0 = 866.49$ °K (1100 °F), equation 13.1.22 becomes

$$
9.153 D_0 \, \Delta L = 21.14(T_{\text{out}} - T_{\text{in}}) + 0.01028(T_{\text{out}}^2 - T_{\text{in}}^2)
$$
$$
- f_{A \text{ in}}[0.35(T_{\text{out}} - T_{\text{in}}) + 5.45 \times 10^{-4}(T_{\text{out}}^2 - T_{\text{in}}^2)]
$$
$$
+ \Delta f_A[21{,}960 - 0.35(T_{\text{out}} - 866.49) - 5.45 \times 10^{-4}(T_{\text{out}}^2 - 866.49^2)] \tag{13.1.23}
$$

or

$$
9.153 D_0 \, \Delta L = 21.14(T_{\text{out}} - T_{\text{in}}) + 0.01028(T_{\text{out}}^2 - T_{\text{in}}^2)
$$
$$
- f_{A \text{ in}}[0.35(T_{\text{out}} - T_{\text{in}}) + 5.45 \times 10^{-4}(T_{\text{out}}^2 - T_{\text{in}}^2)]
$$
$$
+ \Delta f_A[22{,}672 - 0.35T_{\text{out}} - 5.45 \times 10^{-4}T_{\text{out}}^2] \tag{13.1.24}
$$

In order to provide a periodic check on the energy balance equation, one may use the form of equation 13.1.18, which results from integration between the reactor inlet and a distance L downstream where the fraction conversion is $f_{A\,out}$:

$$\dot{q}\pi D_0 L = \sum F_i \int_{T_0}^{T_{out}} C_{pi}\, dT - \frac{F_{A0}\,\Delta H_{R\,at\,T_0} f_{A\,out}}{v_A}$$

(13.1.25)

where the F_i are the molal flow rates at L.

Use of the mole table employed previously and of appropriate numerical values gives

$$9.153 D_0 L = 21.14T + 0.01028T^2 - 26{,}036 - 0.35 f_A T \\ - 5.45 \times 10^{-4} T^2 f_A + 22{,}672 f_A$$

(13.1.26)

This equation permits one to check the results after using equation 13.1.24 for several increments to make sure that round-off errors are not propagating to an excessive degree.

The volumetric flow rate of propane at the reactor inlet is given by the product of the molal flow rate, the molal volume of the gas at standard conditions, and the pressure and temperature correction factors implied by the ideal gas law.

$$\dot{V}_0 = F_{A0} V_{sc} \frac{T_0}{T_{sc}} \frac{P_{sc}}{P_0}$$

(13.1.27)

where the subscript sc denotes a standard condition property.

Hence

$$\dot{V}_0 = \frac{7000}{44.09}(359)\left(\frac{866.49}{273.16}\right)\left(\frac{14.7}{60}\right)$$

$$= 4.43 \times 10^4 \text{ ft}^3/\text{hr} \approx 12.30 \text{ ft}^3/\text{sec} \quad (13.1.28)$$

Combining equations 13.1.16 and 13.1.28 and inserting appropriate conversion factors to let us measure D in inches and L in feet gives

$$\frac{\pi D^2 L}{4(144)(12.30)} = \int_0^{f_A} \frac{(1 + 0.9675 f_A)T(14.7)\, df_A}{3.98 \times 10^{12} e^{-59{,}100/1.987T}(1 - f_A)(866.49)P}$$

(13.1.29)

where P is measured in pounds per square inch (absolute). Thus

$$D^2 L = \int_0^{f_A} \frac{(1 + 0.9675 f_A)T\, df_A}{1.040 \times 10^{11} e^{-59{,}100/1.987T}(1 - f_A)P}$$

(13.1.30)

Changes in the linear velocity arise from variations in the molal volume of the reacting fluid. Hence

$$u = u_0(1 + \delta_A f_A)\frac{T}{T_0}\frac{P_0}{P}$$

(13.1.31)

where u_0 is given by

$$u_0 = \frac{4\dot{V}_0}{\pi\left(\dfrac{D}{12}\right)^2} = \frac{4(12.30)}{\pi\left(\dfrac{D}{12}\right)^2}$$

$$= \frac{2.255 \times 10^3}{D^2}$$

(13.1.32)

where

D is measured in inches

u_0 is expressed in feet per second

Thus

$$u = \frac{2.255 \times 10^3(60)}{D^2(866.49)}(1 + \delta_A f_A)\frac{T}{P}$$

$$= 1.562 \times 10^2 \frac{(1 + \delta_A f_A)}{D^2}\frac{T}{P}$$

(13.1.33)

where

T is measured in degrees Kelvin

P is measured in pounds per square inch (absolute)

The mass density ρ may be determined from the ideal gas law.

$$\rho = \frac{PM}{RT} \qquad (13.1.34)$$

where

$$M = \sum y_i M_i \qquad (13.1.35)$$

The various mole fractions may be determined from the ratio of individual effluent flow rates to the total effluent flow rate. (The necessary data was tabulated previously for use in analysis of the heat capacity terms.) Thus

$$M = \frac{\begin{bmatrix} (1 - f_A)(44.09) + (0.300f_A)(42.08) + (0.065)f_A(30.07) \\ + 0.6675f_A(28.05) + 0.635f_A(16.04) + 0.300f_A(2.016) \end{bmatrix}}{(1 + 0.9675f_A)}$$

$$= \frac{44.09 + 2.12 \times 10^{-3}f_A}{1 + 0.9675f_A} \qquad (13.1.36)$$

or within the accuracy of the data.

$$M = \frac{44.09}{1 + 0.9675f_A} \qquad (13.1.37)$$

Hence

$$\rho = \frac{P(44.09)}{(1 + 0.9675f_A)(10.73)(1.8)T}$$

$$= \frac{2.283P}{(1 + 0.9675f_A)T} \qquad (13.1.38)$$

Combination of equations 13.1.19, 13.1.33, and 13.1.38 and conversion of ΔP to pounds per square inch gives

Inspection of this equation indicates that the pressure drop will be quite sensitive to the pipe diameter employed. To minimize the pressure drop, large tubes are desired, but the larger the pipe, the lower the heat transfer area per unit volume. The three specified pipe sizes have been chosen to illustrate the optimization problem. We are now prepared to carry out the calculations for a single length of each of these pipe sizes. For hand calculations we have the necessary relations in equations 13.1.24, 13.1.29, and 13.1.39. One manner in which we may proceed for each size diameter is as follows.

A. Choose a length increment ΔL.
B. Assume a gas temperature at a distance L from the reactor inlet and calculate an average gas temperature for the increment.
C. Assume an average gas pressure for the increment.
D. Determine the increment in the fraction conversion within the length element by using a finite difference form of equation 13.1.30.

$$\Delta P = \frac{2.283P2f^* \Delta L \left[1.562 \times 10^2 \dfrac{(1 + 0.9675f_A)}{D^2} \dfrac{T}{P} \right]^2}{(1 + 0.9675f_A)T(32.2)\left(\dfrac{D}{12}\right)(144)}$$

$$= 2.883 \times 10^2 \frac{f^*(1 + 0.9675f_A)}{D^5} \frac{T}{P} \Delta L \qquad (13.1.39)$$

$$\Delta f_A = 1.040 \times 10^{11} e^{-59,100/1.987T} \frac{(1 - f_A)P_{av}D^2(\Delta L)}{(1 + 0.9675f_A)T_{av}}$$ (13.1.40)

and iterating as necessary using an average value of f_A characteristic of the length element.

E. Check the assumed temperature using equation 13.1.24 to see if the same value of Δf_A is obtained as in part D.

F. Check the assumed pressure using equation 13.1.39.

The smaller the length element chosen, the more accurate the results will be, but more work will be involved. In the present case a length increment of 20 ft will be employed for each pipe size. Detailed calculations for the first few segments of a nominal 4-in. pipe size are carried out below. Table 13.2 summarizes the results of the calculations for the 4-in. pipe size. Figure 13.3 indicates the temperature and pressure profiles for this reactor.

First Increment.

A. Length = 20 ft.

B. Assume $T_1 = 885\,°K$
   $T_{av} = (866.49 + 885)/2 = 875.75\,°K$

C. Assume $P_{av} = 59.9$ psia.

D. Use $f_A = 0$ in equation 13.1.40 as a first estimate of an average value. Then, for a 4-in. pipe with $D = 4.026$ in.

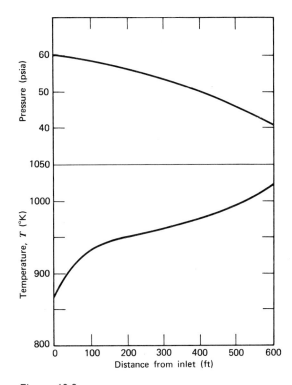

Figure 13.3
Temperature and pressure profiles for pyrolysis reactor.

$$\Delta f_{A1} = \frac{1.040 \times 10^{11} e^{-59,100/[(1.987)(875.75)]}(1 - 0.00)59.9(4.026)^2(20)}{[1 + 0.9675(0.00)]875.75} = 4.099 \times 10^{-3}$$

E. From equation 13.1.24, with $f_{A\,in} = 0.00$ and $D_0 = 4.500$.

$$\Delta f_{A1} = \frac{9.153(4.500)(20) - 21.14(885 - 866.49) - 0.01028[(885)^2 - (866.49)^2]}{22,672 - 0.35(885) - 5.45 \times 10^{-4}(885)^2}$$

$$= 4.522 \times 10^{-3}$$

Table 13.2
Summary of Stepwise Calculations for 4-in. Pipe Pyrolysis Reactor

Increment number	Distance from inlet (ft)	Temperature at inlet to element (°K)	ΔT (°K)	Average temperature (°K)	Average[a] pressure (psia)	ΔP^a (psia)	$\Delta f \times 10^3$	Effluent fraction conversion
1	20	866.49	18.75	875.87	59.82	0.352	4.1	0.004
2	40	885.24	16.57	893.53	59.47	0.363	7.7	0.012
3	60	901.81	13.63	908.62	59.10	0.376	12.9	0.025
4	80	915.44	10.35	920.62	58.72	0.389	18.7	0.043
5	100	925.79	7.42	929.50	58.32	0.403	24.0	0.067
6	120	933.21	5.27	935.85	57.91	0.418	28.0	0.095
7	140	938.48	3.91	940.44	57.48	0.435	30.4	0.126
8	160	942.39	3.17	943.98	57.04	0.452	31.8	0.158
9	180	945.56	2.78	946.95	56.58	0.469	32.6	0.190
10	200	948.34	2.59	949.64	56.10	0.488	32.9	0.223
11	220	950.94	2.53	952.21	55.60	0.506	33.0	0.256
12	240	953.47	2.52	954.73	55.09	0.526	33.1	0.289
13	260	955.99	2.55	957.27	54.55	0.546	33.0	0.322
14	280	958.54	2.60	959.84	53.99	0.566	32.9	0.355
15	300	961.14	2.66	962.47	53.42	0.588	32.8	0.388
16	320	963.82	2.73	965.19	52.82	0.610	32.7	0.421
17	340	966.55	2.82	967.96	52.20	0.633	32.5	0.453
18	360	969.37	2.92	970.83	51.55	0.657	32.3	0.485
19	380	972.29	3.03	973.81	50.89	0.682	32.1	0.518
20	400	975.32	3.17	976.91	50.19	0.708	31.9	0.549
21	420	978.49	3.31	980.15	49.47	0.736	31.6	0.581
22	440	981.80	3.45	983.53	48.72	0.764	31.4	0.613
23	460	985.25	3.64	987.07	47.94	0.794	31.0	0.644
24	480	988.89	3.84	990.81	47.13	0.826	30.7	0.674
25	500	992.73	4.07	994.77	46.28	0.860	30.2	0.704
26	520	996.80	4.33	998.97	45.40	0.896	29.7	0.734
27	540	1001.13	4.63	1003.45	44.49	0.933	29.2	0.763
28	560	1005.76	4.99	1008.26	43.54	0.974	28.5	0.792
29	580	1010.75	5.40	1013.45	42.54	1.017	27.7	0.820
30	600	1016.15	5.89	1019.05	41.50	1.064	26.8	0.846
31	620	1022.04	6.47	1025.28	40.41	1.115	25.7	0.872

[a] Where two columns appear to be inconsistent, round-off errors are responsible for the numerical discrepancies.

This result is not consistent with that obtained above, and a second iteration will be in order. Examination of equations 13.1.40 and 13.1.24 indicates that the assumed temperature will have to be increased to bring the two calculated values of Δf_{A1} closer together. Since $\Delta f_{A1} = f_{A1} - 0.000$, we may use the preliminary estimates of Δf_A to calculate average values of f_A in equation 13.1.40 ($f_{A,av} \simeq 2 \times 10^{-3}$). First, however, we must check the assumed pressure using equation 13.1.39 with $f^* = 0.0044$.

F.
$$\Delta P = \frac{(2.883 \times 10^2)(0.0044)[1 + 0.9675(2 \times 10^{-3})]875.75(20)}{(4.026)^5(59.9)} = 0.351 \text{ psia}$$

Thus,
$$P_{av} = 60 - \frac{0.351}{2} = 59.82.$$

For the second trial we may use our estimates of f_{A1} and P_{av} as input and increase our assumed value of T_1 to 885.24 K.

B.
$$T_{av} = (866.49 + 885.24)/2 = 875.865$$

D. From equation 13.1.40,

$$\Delta f_{A1} = \frac{1.040 \times 10^{11} e^{-59,100/[(1.987)(875.865)]}[1 - 2.0 \times 10^{-3}](59.82)(4.026)^2(20)}{[1 + 0.9675(2.0 \times 10^{-3})]875.865} = 4.104 \times 10^{-3}$$

E. From equation 13.1.24,

$$\Delta f_{A1} = \frac{(9.153)(4.500)(20) - 21.14(885.24 - 866.49) - 0.01028[(885.24)^2 - (866.49)^2]}{22,672 - 0.35(885.24) - 5.45 \times 10^{-4}(885.24)^2}$$
$$= 4.092 \times 10^{-3}$$

The two calculated values of Δf_{A1} are in sufficient agreement that we may presume that our assumed temperature is correct if our pressure drop calculations are also consistent with our assumed pressure.

F. From equation 13.1.39,

$$\Delta P = \frac{(2.883 \times 10^2)(0.0044)[1 + 0.9675(2.0 \times 10^{-3})](875.865)(20)}{(4.026)^5 59.82} = 0.352$$

Thus,
$$P_{out} = 60 - 0.352 = 59.65$$
$$P_{average} = 60 - 0.352/2 = 59.82 \text{ psia,}$$

which is a consistent result. The pressure drop is relatively insensitive to changes in assumed properties; this fact should be used in making the estimates of the pressure as one moves down the reactor.

Second Increment.
A. Length = 20 ft.
B. Assume $T_2 = 901.81 \,°K$

$$T_{av} = (885.24 + 901.81)/2 = 893.525 \,°K$$

C. Assume $P_{av} = 59.82 - 0.35 = 59.47$ psia.
D. Use $f_A = 0.008$ as a first estimate of the average fraction conversion for the second increment in

equation 13.1.40. Thus

$$\Delta f_{A2} = \frac{1.040 \times 10^{11} e^{-59,100/[(1.987)(893.525)]} (1 - 0.008)(59.47)(4.026)^2(20)}{[1 + 0.9675(0.008)]893.525} = 7.72 \times 10^{-3}$$

$$f_{A2\,\text{average}} = 4.10 \times 10^{-3} + \frac{7.72 \times 10^{-3}}{2} = 7.46 \times 10^{-3}$$

which is quite consistent with our first estimate considering the insensitivity of Δf_{A2} to small perturbations in the average fraction conversion for the increment.

E. From equation 13.1.24, with $f_{A\,\text{in}} = 4.10 \times 10^{-3}$,

$$\Delta f_{A2} = \frac{\begin{array}{c} 9.153(4.500)(20) - 21.14(901.81 - 885.24) - 0.01028[(901.81)^2 - (885.24)^2] \\ + (4.10 \times 10^{-3})\{0.35(901.81 - 885.24) + 5.45 \times 10^{-4}[(901.81)^2 - (885.24)^2]\} \end{array}}{22,672 - 0.35(901.81) - 5.45 \times 10^{-4}(901.81)^2}$$

or

$$\Delta f_{A2} = 7.72 \times 10^{-3}$$

so our assumed temperature is consistent thus far.

F. From equation 13.1.39,

$$\Delta P = \frac{2.883 \times 10^2(0.0044)[1 + 0.9675(7.46 \times 10^{-3})](893.525)(20)}{(4.026)^5(59.47)} = 0.363 \text{ psia}$$

or

$$P_{\text{av}} = 59.65 - \frac{0.363}{2} = 59.47 \text{ psia}$$

which is also consistent.

Therefore,

$$T_2 = 901.81 \text{ °K}$$
$$P_2 = 59.65 - 0.36 = 59.29 \text{ psia}$$
$$f_2 = 4.10 \times 10^{-3} + 7.72 \times 10^{-3}$$
$$= 11.82 \times 10^{-3}$$

One could iterate using these values as initial estimates of the properties, but the results would not change within the accuracy of the other parameters employed.

One may proceed downstream in similar fashion. The results of such calculations are summarized in Table 13.2. It is advisable to use an equation like 13.1.26 to check the calculations periodically for error and to ensure that the accumulation of round-off error is not excessive. No problems of this type were encountered. This design meets the specifications with regard to the maximum permissible temperature and the required outlet pressure.

One may now proceed in the same fashion to determine the temperature, pressure, and fraction conversion profiles for the 2-in. and 6-in. pipes. For the former case $D = 2.067$ in. and $D_0 = 2.375$ in.; for the latter $D = 6.065$ in. and $D_0 = 6.625$ in. The results for the 6-in. case are summarized in Table 13.3. For this case the pressure drop is minimal and the effluent temperature is below the stated limit. For the 2-in.

Table 13.3
Summary of Stepwise Calculations for 6-in. Pipe Pyrolysis Reactor

Increment number	Distance from inlet (ft)	Temperature at inlet to element (°K)	ΔT (°K)	Average temperature (°K)	Average pressure (psia)	ΔP psia	$\Delta f \times 10^3$	Effluent fraction conversion
1	20	866.49	25.09	879.04	59.98	0.046	10.44	0.010
2	40	891.58	18.29	900.73	59.93	0.047	22.26	0.033
3	60	909.87	11.12	915.43	59.88	0.050	35.13	0.068
4	80	920.99	6.38	924.18	59.83	0.052	43.75	0.112
5	100	927.37	4.26	929.50	59.78	0.054	47.64	0.159
6	120	931.63	3.47	933.37	59.72	0.057	49.10	0.208
7	140	935.10	3.26	936.73	59.66	0.059	49.49	0.258
8	160	938.36	3.25	939.99	59.60	0.062	49.51	0.307
9	180	941.61	3.33	943.28	59.54	0.065	49.39	0.357
10	200	944.94	3.44	946.66	59.47	0.067	49.19	0.406
11	220	948.38	3.59	950.18	59.41	0.070	48.93	0.455
12	240	951.97	3.77	953.86	59.33	0.073	48.61	0.503
13	260	955.74	3.99	957.74	59.26	0.075	48.21	0.552
14	280	959.73	4.25	961.86	59.18	0.078	47.74	0.599
15	300	963.98	4.57	966.27	59.10	0.081	47.15	0.647
16	320	968.55	4.97	971.04	59.02	0.084	46.43	0.693
17	340	973.52	5.47	976.26	58.93	0.087	45.53	0.738
18	360	978.99	6.09	982.04	58.85	0.090	44.38	0.783
19	380	985.08	6.90	988.53	58.75	0.092	42.87	0.826
20	400	991.98	7.98	995.97	58.66	0.095	40.88	0.867

Where two columns appear to be inconsistent, round-off errors are responsible for the numerical discrepancy.

reactor the pressure drop exceeds the stated limit in the first 60 ft of reactor length. The conversion achieved therein is less than 1%. The pressure drop problem could be circumvented by using many short tubes in parallel with much lower velocities within a given tube, but we will not consider this option. Only the 4-in. and 6-in. reactors will be considered further. Table 13.4 summarizes some key parameter values for these two reactors.

Both pipe sizes give reactors that are satisfactory from the standpoint of pressure drop and effluent temperature. Although the 4-in. reactor must be longer to achieve the desired conversion, it requires significantly less volume, and this is an important consideration in the design of a pyrolysis furnace. The 6-in. pipe has

Table 13.4
Summary of Design Parameters for 4 and 6-in. Pipe Pyrolysis Reactors

Parameter	4-in. reactor	6-in. reactor
Length (ft)	620	400
Internal volume (ft³)	54.8	80.3
Weight of pipes (lb)	6690	7588
Pressure drop (psi)	19.59	1.34
Effluent temperature (°K)	1028.51	999.96
Effluent temperature (°F)	1391.6	1340.2
Effluent conversion	0.872	0.867
Reactor space time (sec)	4.46	6.53

much less heat transfer surface per unit of reactor volume; this fact implies that a proportionately larger volume will be required. On the basis of these considerations, the 4-in. pipe size is to be preferred. It represents an engineering compromise between the high surface area/volume ratio characteristic of small pipe sizes and the low pressure drop associated with large pipe sizes.

The length of the reactor is comparable in magnitude to that of units that have been employed commercially, and it is smaller than many. If desired, the other process parameters could be changed, reducing the required length somewhat (e.g., one could design the furnace to give a somewhat higher heat flux, operate at a different inlet pressure, or use parallel combinations of reactors). The calculational technique used in generating the numbers in Tables 13.2 to 13.4 may be used to accomplish the necessary calculations. The equations are readily adaptable to machine computation and the latter route is preferable to hand calculations in seeking an overall optimum design. Figure 13.4 is a flow diagram for a program that could be used in machine computations to determine composition and temperature profiles from equations 13.1.40, 13.1.24, and 13.1.39. For machine computation one could use significantly smaller length increments than those employed in the numerical calculations of this section. The accuracy of the results could thereby be improved. Nonetheless, the numbers presented in Tables 13.2 and 13.3 provide good estimates of the required lengths of 4- and 6-in. pipe. One must recognize that the pressure drop that would be associated with U-bends in the piping has not been taken into account. Greater improvements in accuracy can be achieved by allowing for this factor in the hand calculations than can be achieved by going to shorter-length segments in the machine computations.

For a discussion of modern methods of olefin manufacture by pyrolysis, consult the paper by Ennis and co-workers (5).

13.2 PHTHALIC ANHYDRIDE PRODUCTION VIA THE CATALYTIC OXIDATION OF NAPHTHALENE IN A FIXED BED REACTOR

This problem treats some aspects of the design of a fixed bed reactor for a case where selectivity and safety considerations are very important.

13.2.1 Charge to Design Engineer:

To: Kem Inguneer
From: Barney Boss
Re: Design Specifications for Phthalic Anhydride Production in a Fixed Bed Reactor

A proposed expansion of the corporation's vinyl plastics operation will require a commitment by the company to produce its own plasticizer. Our long range planning group has suggested that 6 million pounds per year of new phthalic anhydride capacity would meet our internal needs and projected increases in demand from current customers.

Either naphthalene or ortho-xylene is an acceptable starting material for partial oxidation to phthalic anhydride, but current raw materials costs favor the former as a starting material. Both fixed and fluidized bed processes have been used on a commercial scale, but you are to focus your attention on the former. Figure 13.5 is a schematic flow diagram of the proposed process. Most research groups that have studied the catalytic oxidation of naphthalene over vanadium pentoxide agree that the principal reactions are

$$
\begin{array}{c}
\text{naphthoquinone} \\
\\
\text{naphthalene} \\
\\
\text{phthalic anhydride}
\end{array}
\quad \rightarrow \quad
\begin{array}{c}
\text{maleic anhydride} \\
\\
CO, CO_2, H_2O
\end{array}
\qquad (13.2.1)
$$

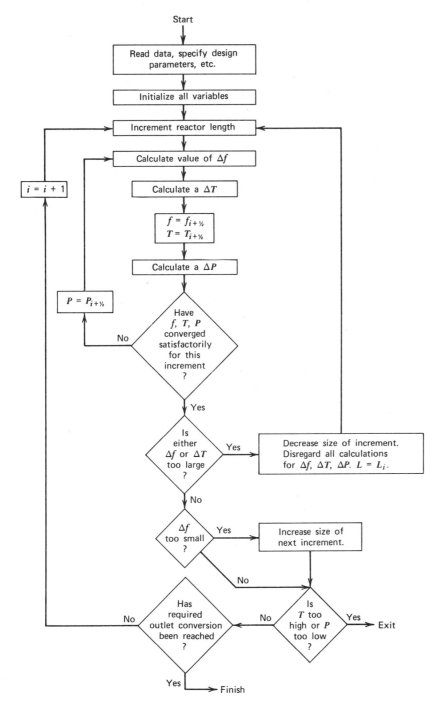

Figure 13.4
Generalized flow sheet for simulation of pyrolysis reactor by machine computation.

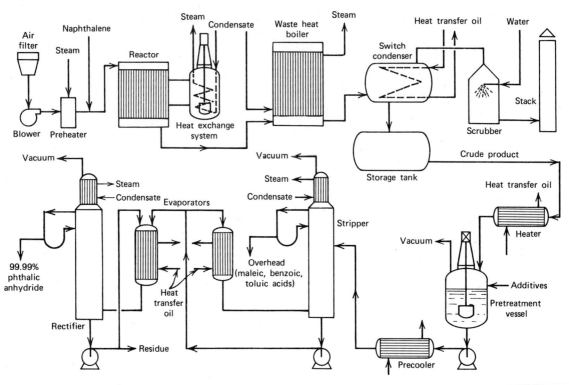

Figure 13.5
Schematic diagram of phthalic anhydride production process. (Adapted from P. Ellwood, *Chemical Engineering,* **June 2, 1969, pp. 81–82. Used with permission of** *Chemical Engineering.***)**

However, in order to design a reactor for phthalic anhydride production, the reaction rate constants for the temperature range of interest are such that the following simplified set of reactions has been suggested as appropriate (6, 7).

$$\text{naphthalene} \xrightarrow{k_1} \text{phthalic anhydride} \xrightarrow{k_2} CO_2, H_2O$$

$$(13.2.2)$$

The reactions are highly exothermic and very rapid. Consequently conventional practice in the design of fixed bed reactors for phthalic anhydride production has been based on the use of multitube reactors to ensure good heat transfer and good temperature control. These are required to ensure good selectivity. Often a thousand or more small diameter tubes may be

immersed in a liquid heat transfer medium to give the desired results. A thorough analysis will require a study of the effects of tube diameter on system performance but, for preliminary design calculations, 2-in. diameter tubes are to be employed. Cracking reactions begin to be significant at about 673 °K, and the system should not have any hot spots that exceed this temperature.

The necessity of avoiding potentially explosive conditions requires that the mole percent naphthalene in the feed be kept below 1%. It is suggested that preliminary design calculations be based on a feed composition of 0.75 mole percent naphthalene (remainder air).

Most commercial multitube reactors provide the necessary cooling capacity through the use of a molten salt that circulates around the reactor tubes and then to an external heat exchanger that can be used for the generation of high-pressure steam. It may be assumed that proper heat transfer equipment is available such that there is no substantial increment in the temperature of the heat transfer medium across the reactor.

It is known that high mass velocities are to be employed within the reactor tubes to minimize heat and mass transfer limitations on the catalytic reaction rates. It is also known that the effectiveness factors for the catalysts commonly employed often differ appreciably from unity.

Your assignment is to use the data and assumptions listed below to prepare a preliminary reactor design in which particular emphasis is placed on the variation of the maximum yield of phthalic anhydride with feed temperature.

The reactor inlet pressure may be taken as 25 psia. Consider feed temperatures of 630, 640, 650, and 660° K. In each case assume that the heat transfer medium maintains the tube walls throughout the reactor length at the feed temperature. Determine the required reactor length and temperature and composition profiles for each case. Obviously there will be a trade-off between the number of tubes employed and the required reactor length. For present purposes assume that 1000 schedule 40, 2-in. stainless steel tubes will be used to handle 6 million lb/yr of naphthalene feed. Use a one-dimensional model of the fixed bed reactor. The calculations should be terminated when one exceeds 99% conversion, when the pressure drops below 1 atm, or when the temperature at any point in the reactor exceeds 673 °K.

Data Summary and Permissable Assumptions. There are a number of groups that have attempted to simulate fixed bed performance for this reaction system (7–9). Many of the data below have been taken from these sources.

Reaction Stoichiometry and Thermodynamic Data. The stoichiometry of the reactions of interest may be represented as:

$$\text{(naphthalene)} + 4.5O_2 \xrightarrow{k_1} \text{(phthalic anhydride)} + 2CO_2 + 2H_2O \quad (13.2.3)$$

or

$$C_{10}H_8 + 4.5O_2 \xrightarrow{k_1} C_8H_4O_3 + 2CO_2 + 2H_2O \quad (13.2.3)$$

and

$$\text{(phthalic anhydride)} + 7.5O_2 \xrightarrow{k_2} 8CO_2 + 2H_2O \quad (13.2.4)$$

or

$$C_8H_4O_3 + 7.5O_2 \xrightarrow{k_2} 8CO_2 + 2H_2O \quad (13.2.4)$$

The second of these reactions may not go to completion, and the incomplete combustion may yield maleic anhydride, benzoic acid, CO, or other compounds as by-products.

There is considerable variation in the heat of reaction data employed in different articles in the literature that deals with this reaction. Cited values differ by more than an order of magnitude. If we utilize heat of combustion data for naphthalene and phthalic anhydride and correct for the fact that water will be a gas instead of a liquid at the conditions of interest, we find that for the first reaction (equation 13.2.3) the standard enthalpy change will be approximately -429 kcal/g mole; for the second reaction it will be approximately -760 kcal/g mole. These values will be used as appropriate for the temperature range of interest. Any variation of these parameters with temperature may be neglected.

For the case where the inlet naphthalene concentration is below the explosion limit and where one is successful in producing phthalic anhydride in commercial yields, the mole fraction oxygen in the reactant gases remains relatively constant. This point is readily seen by the construction of a mole table for a feed containing 0.75% naphthalene, of which 80% reacts to give phthalic anhydride and the remainder goes to CO_2.

Under these conditions the mole fraction oxygen will drop from 0.2084 to 0.1636, which represents a decrease of 21.5% or a variation of $\pm 12.1\%$. At lower conversions the difference will be even less substantial.

Because the feed is primarily air and because substantial amounts of N_2 and O_2 are present in the effluent stream, we will assume that the heat capacity of the reactant mixture may be taken as that of air at 630° K (0.255 cal/g-°K). The variations of the heat capacity with temperature and pressure may be neglected.

The Prandtl number of the fluid may be taken as 0.79.

Kinetic Data and Catalyst Data.   The simplified reactions indicated by equation 13.2.2 may be assumed on the basis of arguments set forth by DeMaria et al. (8). For the catalyst employed by this research group, pseudo first-order kinetics are followed by each of the two reactions, with

$$k_1 = 5.74 \times 10^{13}e^{-38,000/RT} \text{ sec}^{-1} \quad (13.2.5)$$

$$k_2 = 2.19 \times 10^{5}e^{-20,000/RT} \text{ sec}^{-1} \quad (13.2.6)$$

where T is expressed in degrees Kelvin and R in calories per gram mole-degree Kelvin. Both reactions may be regarded as irreversible.

Basis: 100 moles of feed

Species	Feed moles and/or mole percent	Effluent Moles		Effluent Mole percent
Naphthalene	0.75		0	0
Nitrogen	0.79(99.25) = 78.41		78.41	78.53
Phthalic anhydride	0	0.8(0.75) =	0.60	0.60
CO_2	0	2(0.8)(0.75) + 10(0.2)(0.75) =	2.70	2.70
H_2O	0	2(0.8)(0.75) + 4(0.2)(0.75) =	1.80	1.80
Oxygen	0.21(99.25) = 20.84	20.84 − $\frac{3}{2}$(0.60) − 2.70 − $\frac{1}{2}$(1.80) =	16.34	16.36
Total	100.0		99.85	99.99

The units on the rate constants reported by DeMaria et al. indicate that they are based on pseudo homogeneous rate expressions (i.e., the product of a catalyst bulk density and a reaction rate per unit mass of catalyst). It may be assumed that these relations pertain to the intrinsic reaction kinetics in the absence of any heat or mass transfer limitations.

The catalyst consists of V_2O_5 supported on silica gel with K_2SO_4 and other promoters also present. The physical property values tabulated below are typical of the low or intermediate surface area supports that one might expect to use in this application. DeMaria et al. (8) did not report data of these types for their catalyst.

> Pellet shape—spherical
>
> Pellet diameter = 0.318 cm
>
> $V_g = 0.26$ cm^3/g
>
> $S_g = 80$ m^2/g
>
> Bulk density = 0.84 g/cm^3
>
> Skeletal density = 2.2 g/cm^3
>
> Bed porosity $\varepsilon_B = 0.40$
>
> Pellet porosity $\varepsilon_p = 0.364$

One need not worry about the problem of catalyst deactivation for this reaction. Many process descriptions quote lifetimes of 3 years or more.

Transport Properties. Because the feed is primarily air and because substantial amounts of N_2 and O_2 are present in the effluent stream, we will assume that the fluid viscosity is that of air for purposes of pressure drop calculations. For the temperature range of interest, the fluid viscosity may be taken as equal to 320 micropoise. The pressure range of interest does not extend to levels where variations of viscosity with pressure need be considered. The effective diffusivities of naphthalene and phthalic anhydride in the catalyst pellet may be evaluated using the techniques developed in Section 12.2.

The overall heat transfer coefficient for thermal energy exchange between the tube wall and the reacting fluid may be taken as 1.0×10^{-3} cal/cm^2-sec-°K. The effective thermal conductivity of the catalyst pellets may be taken as equal to 6.5×10^{-4} cal/(sec-cm-°C).

13.2.2 Solution

The design of a fixed bed reactor for phthalic anhydride production in terms of a one-dimensional model involves only the integration of a set of ordinary differential equations. In the general case for the consecutive reactions of interest, the reaction selectivity will depend on several process variables. A thorough design analysis would require an investigation of the effects of variations in inlet temperature, coolant temperature, tube diameter, catalyst pellet size, mass velocity through the tubes as influenced by the number of tubes, system pressure, and the feed composition on the variation of composition and temperature with bed length. One should also investigate the sensitivity of the results to uncertainties in the wall heat transfer coefficient, reaction activation energies, etc. For this highly exothermic reaction, an investigation of a two-dimensional model of the reactor would also be appropriate. We will restrict our discussion to variations of the reactor inlet temperature for the case where the coolant is maintained at this temperature. Space limitations preclude our study of the influence of other variables, but the method discussed below can be used to determine the one-dimensional model predictions of reactor performance for variations in other process parameters.

One of the first things that should be done in the analysis is to determine if pressure variations along the length of a reasonable-size reactor will be significant for the specified operating conditions. This will require a knowledge of the superficial mass velocity through the tubes. This quantity may be calculated from the tube dimensions and the inlet flow rate and composition. The molal flow rate of naphthalene is equal to

$$6{,}000{,}000 \frac{\text{lb}}{\text{yr}} \times 454 \frac{\text{g}}{\text{lb}} \times \frac{\text{yr}}{365 \text{ day}} \times \frac{\text{day}}{24 \text{ hr}} \times \frac{\text{hr}}{3600 \text{ sec}} \times \frac{\text{g moles}}{128.16 \text{ g}} = 0.6740 \frac{\text{g mole}}{\text{sec}}$$

Per tube the molal flow rate of naphthalene is then 6.740×10^{-4} g moles/sec. The total inlet molal flow rate is equal to 0.6740/0.0075, or 89.86 g moles/sec. The average molecular weight of the feed is given by

The fluid density at inlet conditions may be calculated from the ideal gas law

$$\rho = \frac{PM}{RT} \qquad (13.2.9)$$

$$\bar{M} = \sum x_i M_i = (0.0075)(128.16) + (0.2084)(32.00) + 0.7841(28.00) = 29.58 \qquad (13.2.7)$$

The total mass flow rate is then $29.58(89.86) = 2658$ g/sec or 2.658 kg/sec. For 1000 tubes in parallel the available cross section for flow is given by

$$1000 \frac{\pi}{4} \left(\frac{2.067}{12}\right)^2 = 23.30 \text{ ft}^2 \approx 2.165 \text{ m}^2$$

The mass velocity is then 2.658/2.165 or 1.228 kg/m²-sec or 0.1228 g/cm²-sec.

The modified Reynolds number based on the catalyst pellet diameter of 1/8 in. or 0.318 cm is then

$$N_{\text{Re}} = \frac{D_p G}{\mu}$$

$$= \frac{(0.318 \text{ cm})(0.1228 \text{ g cm}^{-2}\text{-sec}^{-1})}{320 \times 10^{-6} \text{ g cm}^{-1}\text{-sec}^{-1}}$$

$$= 122 \qquad (13.2.8)$$

which indicates that the flow regime is somewhat beyond the point where it undergoes a transition from laminar to turbulent conditions ($N_{\text{Re}} = 40$).

If we consider the case where $T = 630\ °\text{K}$, then

$$\rho = \frac{\left(\frac{25}{14.696}\right)(29.58)}{(82.06)(630)} = 9.73 \times 10^{-4} \text{ g/cm}^3$$

$$(13.2.10)$$

Since $G = u\rho$, the fluid velocity at inlet conditions is then

$$u = \frac{G}{\rho} = \frac{0.1228}{9.73 \times 10^{-4}} = 126 \text{ cm/sec} \approx 1.26 \text{ m/sec}$$

$$(13.2.11)$$

The pressure drop through the packed bed reactor is governed by the Ergun equation discussed earlier as equation 12.7.4.

$$\frac{[\mathscr{P}_0 - \mathscr{P}_L]\rho}{G^2}\left(\frac{D_p^*}{L}\right)\left(\frac{\varepsilon_B^3}{1 - \varepsilon_B}\right) = \frac{150(1 - \varepsilon_B)}{(D_p^* G/\mu)} + 1.75$$

$$(13.2.12)$$

which can be written in differential form as

$$-\frac{d\mathscr{P}}{dL} = \frac{G^2}{D_p^*\rho}\frac{(1 - \varepsilon_B)}{\varepsilon_B^3}\left[\frac{150(1 - \varepsilon_B)}{(D_p^* G/\mu)} + 1.75\right]$$

$$(13.2.13)$$

The density may be eliminated using equation 13.2.9 to give

$$-\frac{d\mathcal{P}}{dL} = \frac{G^2 RT}{D_p^* PM} \frac{(1 - \varepsilon_B)}{\varepsilon_B^3} \left[\frac{150(1 - \varepsilon_B)}{(D_p^* G/\mu)} + 1.75 \right] \tag{13.2.14}$$

Substitution of numerical values gives

$$-\frac{d\mathcal{P}}{dL} = \frac{(0.1228)^2 (8.31 \times 10^7) T(1 - 0.4)}{(0.318) P (29.58)(0.4)^3} \left[\frac{150(1 - 0.4)}{122} + 1.75 \right] \tag{13.2.15}$$

$$= 3.107 \times 10^6 \frac{T}{P}$$

where

$\mathcal{P}$ and P are expressed in dynes per square centimeter

L is expressed in centimeters

T is expressed in degrees Kelvin.

The inlet pressure of 25 psia corresponds to 1.724×10^6 dynes/cm^2. Thus, when the inlet temperature is 630 °K, the pressure gradient at the start of the bed is equal to

$$(3.107 \times 10^6)(630)/(1.724 \times 10^6)$$

or 1135 dynes/cm^3. In more conventional engineering units this is equivalent to 0.502 psi/ft or 3.42×10^{-2} atm/ft. The magnitude of this parameter and the range of other process variables indicates that we will be restricted to relatively short reactors. This in turn implies that we may neglect hydrostatic head effects for this gaseous system and equate $\mathcal{P}$ with P. Hence,

$$-\frac{dP}{dL} = 3.107 \times 10^6 \frac{T}{P} \tag{13.2.16}$$

At this point it is instructive to consider the possible presence of intraparticle and external mass and heat transfer limitations using the methods developed in Chapter 12. In order to evaluate the catalyst effectiveness factor we first need to know the combined diffusivity for use in the single pore model of a catalyst particle. The Knudsen diffusivity is given by equation 12.2.4;

$$D_K = 9.7 \times 10^3 \bar{r} \sqrt{\frac{T}{M}} \tag{13.2.17}$$

where $\bar{r} = \dfrac{2V_g}{S_g}$ for the cylindrical pore model.

Thus

$$\bar{r} = \frac{2(0.26)}{80 \times 10^4} = 6.5 \times 10^{-7} \text{ cm} \approx 65 \text{ Å} \approx 6.5 \text{ nm} \tag{13.2.18}$$

and

$$D_K = 9.7 \times 10^3 (6.5 \times 10^{-7}) \sqrt{\frac{T}{M}} \tag{13.2.19}$$

At the reactor inlet where $T = 630°$ K,

$$D_K = (9.7 \times 10^3)(6.5 \times 10^{-7}) \sqrt{\frac{630}{128.16}}$$

$$= 1.40 \times 10^{-2} \text{ cm}^2/\text{sec} \tag{13.2.20}$$

For naphthalene in air the ordinary molecular diffusivity may be evaluated using a pseudo binary approach and the methods outlined in Reid and Sherwood (10). For an inlet temperature of 630 °K and an inlet pressure of 25 psia, the result is

$$D_{AB} = 0.0680 \text{ cm}^2/\text{sec} \tag{13.2.21}$$

The combined diffusivity is given by equation 12.2.8:

$$\mathcal{D}_c = \frac{1}{1/D_K + 1/D_{AB}} = \frac{D_{AB} D_K}{D_{AB} + D_K} \tag{13.2.22}$$

Thus

$$\mathscr{D}_c = \frac{0.0680(1.40 \times 10^{-2})}{0.0680 + 1.40 \times 10^{-2}} = 0.0116 \text{ cm}^2/\text{sec}$$

$$(13.2.23)$$

The average pore length is given by equation 12.3.3 as

$$\bar{L} = \frac{V_p}{S_x} = \frac{\dfrac{\pi D_p^3}{6}}{\pi D_p^2} = D_p/6 = 0.318/6 = 0.053 \text{ cm}$$

$$(13.2.24)$$

The reaction rate constants were reported on an effective volume basis. To convert to a unit surface area basis, the following relation must be employed.

$$k_v = k'' S_g \rho_B \qquad (13.2.25)$$

exothermic, and it is desirable to evaluate the energy generation function β of equation 12.3.108 to determine if the pellet conditions depart substantially from isothermality.

$$\beta = \frac{-\Delta H \mathscr{D}_{\text{eff}} C_0}{k_{\text{eff}} T_0} \qquad (13.2.30)$$

To employ this relation one needs to estimate a tortuosity factor in order to convert the combined diffusivity to an effective diffusivity by equation 12.2.9. If we assume a value of 3, then

$$\mathscr{D}_{\text{eff}} = \frac{\mathscr{D}_c \varepsilon_p}{\tau'} = \frac{(0.0116)(0.364)}{3}$$

$$= 1.41 \times 10^{-3} \text{ cm}^2/\text{sec} \quad (13.2.31)$$

and, for the first reaction in the sequence at the reactor inlet, where $C_0 = Y_0(\rho/M)$,

$$\beta = \frac{(+429{,}000)(1.41 \times 10^{-3})(0.0075)\left(\dfrac{9.73 \times 10^{-4}}{29.58}\right)}{(6.5 \times 10^{-4})(630)} = 3.64 \times 10^{-4} \quad (13.2.32)$$

Thus

$$k_1'' = \frac{5.74 \times 10^{13} e^{-38{,}000/(1.987T)}}{(80 \times 10^4)(0.84)}$$

$$= 8.54 \times 10^7 e^{-38{,}000/(1.987T)} \quad (13.2.26)$$

At an inlet temperature of 630 °K,

$$k_1'' = 5.60 \times 10^{-6} \text{ cm/sec} \quad (13.2.27)$$

From equation 12.3.11, at 630 °K,

$$h_T^2 = \frac{2(5.60 \times 10^{-6})(0.053)^2}{(6.5 \times 10^{-7})(0.0116)} \quad (13.2.28)$$

or

$$h_T = 2.04 \qquad (13.2.29)$$

From Figure 12.2 it is evident that the catalyst effectiveness factor for isothermal operation will be approximately 0.47. At higher temperatures the effectiveness factor will be smaller because the rate constant will increase more rapidly with temperature than will the combined diffusivity. However, the reactions in question are quite

This parameter is also equal to the ratio of the maximum possible temperature difference between the center of the pellet and the external surface to the external surface temperature. Thus

$$\Delta T_{\text{max}} = 630(3.64 \times 10^{-4}) = 0.23 \text{ °K}$$

$$(13.2.33)$$

The magnitude of this parameter indicates that we may neglect temperature variations within the pellet and use the isothermal effectiveness factor relation in our analysis.

$$\eta = \frac{\tanh h_T}{h_T} \qquad (13.2.34)$$

The astute reader will note that we have calculated β on the assumption that only the first reaction in the sequence is significant. One could modify the analysis by employing an effective ΔH that incorporated the energy release by the second reaction in the sequence. This would have the effect of increasing ΔT_{max} somewhat but, if phthalic anhydride is the dominant product species and yields are in excess of 70%, the assumption of an isothermal pellet is a good one. Furthermore, β will diminish as one moves through the reactor, since C_0 will decrease as one proceeds downstream. In any event we will presume throughout the remainder of our analysis that the pellets are isothermal.

Since our calculations indicate that intraparticle mass transfer limitations are significant, we must now consider the possibility that temperature and concentration differences will exist between the bulk fluid and the external surface of the catalyst. Appropriate mass and heat transfer coefficients must therefore be determined.

At the Reynolds number of interest, equation 12.4.10 indicates that

$$j_H = j_D = \frac{0.357}{\varepsilon_B N_{Re}^{0.359}} = \frac{0.357}{0.4(122)^{0.359}} = 0.159$$

$$(13.2.35)$$

The Schmidt and Prandtl numbers must be evaluated in order to be able to determine concentration and temperature differences between the bulk fluid and the external surface of the catalyst. The Schmidt number for naphthalene in the mixture may be evaluated using the ordinary molecular diffusivity employed earlier, the viscosity of the mixture, and the fluid density.

$$N_{Sc} = \frac{\mu}{\rho \mathcal{D}_{AB}} = \frac{320 \times 10^{-6}}{(9.73 \times 10^{-4})(0.068)} = 4.84$$

$$(13.2.36)$$

From the definition of the mass transfer factor (j_D),

$$k_{c,i} = \frac{G}{\rho} N_{Sc}^{-2/3} j_D \qquad (13.2.37)$$

Thus

$$k_{c,i} = \left(\frac{0.1228}{9.73 \times 10^{-4}}\right)(4.84)^{-2/3}(0.159)$$

$$= 7.02 \text{ cm/sec} \qquad (13.2.38)$$

From the definition of the j_H factor the heat transfer coefficient between the catalyst surface and the bulk fluid is given by

$$h_P = j_H C_p G N_{Pr}^{-2/3} \qquad (13.2.39)$$

or

$$h_P = (0.159(0.255)(0.1228)(0.79)^{-2/3}$$

$$= 5.83 \times 10^{-3} \text{ cal/(cm}^2\text{-sec-}^\circ\text{K)}$$

$$(13.2.40)$$

At steady state the rate of mass transfer must equal the reaction rate and, if one neglects the change in the number of moles on reaction, the drift factor may be taken as unity. Thus

$$k_{c,i}(C_{i,ES} - C_{i,B})a_m = v_i r_m \qquad (13.2.41)$$

where a_m is the gross external surface area of the catalyst pellets per unit mass. Now for spherical pellets with a pellet porosity of 0.364 and a skeletal density of 2.2.

$$a_m = \frac{\pi D_p^2}{\rho_{skeletal}(1 - \varepsilon_p)\dfrac{\pi D_p^3}{6}} = \frac{6}{\rho_{skeletal} D_p(1 - \varepsilon_p)}$$

$$(13.2.42)$$

Thus

$$a_m = \frac{6}{2.2(0.318)(1 - 0.364)} = 13.48 \text{ cm}^2/\text{g}$$

$$(13.2.43)$$

The reaction rate per unit mass at the reactor inlet is given by

$$r_m = \frac{\eta k_1 C_{N,ES}}{\rho_B} \qquad (13.2.44)$$

Thus, for naphthalene with $v_i = -1$,

$$k_{c,N}(C_{N,B} - C_{N,ES})a_m \rho_B = \eta k_1 C_{N,ES} \qquad (13.2.45)$$

or

$$\frac{C_{N,ES}}{C_{N,B}} = \frac{k_{c,N} a_m \rho_B}{\eta k_1 + k_{c,N} a_m \rho_B} \qquad (13.2.46)$$

At the reactor inlet

$$k_1 = 5.74 \times 10^{13} e^{-38,000/(1.987)(630)}$$

$$= 3.76 \text{ sec}^{-1}$$

Thus

$$\frac{C_{N,ES}}{C_{N,B}} = \frac{7.02(13.48)(0.84)}{(0.47)(3.76) + (7.02)(13.48)(0.84)}$$

$$= \frac{79.5}{1.77 + 79.5} = 0.978 \qquad (13.2.47)$$

This ratio is sufficiently large that we may call it unity for purposes of determining the required reactor size. The errors associated with assuming that for the cases of interest there will not be significant differences between bulk fluid and external surface concentrations will be small compared to other sources of error, particularly the errors inherent in the use of a one-dimensional model.

Now let us consider the possibility that there will be a significant temperature difference between the bulk fluid and the external surface of the catalyst pellet. Equation 12.5.6 indicates that the temperature and concentration gradients external to the particle are related as follows:

$$T_{ES} - T_B = \frac{k_c(\Delta H)(C_{N,B} - C_{N,ES})}{h_p v_N \left(1 - \frac{y_N}{v_N} \sum\limits_{j=1}^{c} v_j\right)}$$

$$(13.2.48)$$

where the lower term in parentheses is the drift factor. For present purposes it may be taken as unity.

Now the bulk fluid naphthalene concentration at the reactor inlet is given by

$$C_{N,B} = \frac{y_N \rho}{\overline{M}} = (0.0075)\frac{(9.73 \times 10^{-4})}{29.58}$$

$$= 2.47 \times 10^{-7} \text{ g moles/cm}^3$$

$$(13.2.49)$$

The concentration difference is then

$$C_{N,B} - C_{N,ES} = C_{N,B}\left(1 - \frac{C_{N,ES}}{C_{N,B}}\right)$$

$$= 2.47 \times 10^{-7}(1 - 0.978)$$

$$= 5.37 \times 10^{-9} \text{ g moles/cm}^3$$

$$(13.2.50)$$

The corresponding temperature difference is

$$T_{ES} - T_B \approx \frac{7.02(-429,000)(5.37 \times 10^{-9})}{5.83 \times 10^{-3}(-1)}$$

$$\approx 2.77 \, ^\circ K \qquad (13.2.51)$$

This estimate of the temperature difference might be expected to be somewhat on the low side in that it is based on the enthalpy change for just the first reaction in the sequence and does not include the effect of the energy released by further oxidation of phthalic anhydride to CO_2 and H_2O. On the other hand, it is evaluated at the reactor inlet where the reactant concentration is highest, so that in this sense it represents a quasimaximum value. The magnitude of the effect is such that it is debatable as to whether or not one should retain it for purposes of the design analysis. For the naphthalene oxidation reaction at approximately 630 °K a two degree temperature difference corresponds to a difference in rate constants of about 10%. For the phthalic anhydride reaction the effect gives only about a 5% difference. We have previously developed the relations by which such temperature gradients may be determined and used in a reactor design analysis. However, the inclusion of such effects clearly increases the labor involved in the design analysis; for purposes of preparing a preliminary estimate of the effects of variations in the fluid inlet temperature on temperature and conversion profiles, it is appropriate to neglect temperature differences between the bulk fluid and the catalyst surface. In a more refined analysis, perhaps with a two-dimensional model, the temperature difference should be taken into account.

In summary, our analysis indicates that intraparticle temperature gradients and external concentration gradients are clearly negligible while intraparticle concentration gradients are clearly significant. External temperature gradients do exist, but they are small.

If we consider a differential length element of our packed bed reactor (ΔX) and neglect the axial dispersion term, a material balance on naphthalene gives the following expression.

$$-\Delta F_N = \eta k_1 C_N \left(\frac{\pi D^2}{4}\right) \Delta X \tag{13.2.52}$$

or

$$F_{N0} \, \Delta f_N = \eta k_1 \left(\frac{\pi D^2}{4}\right) \Delta X C_{N0} \left(\frac{T_0}{T_{av}} \frac{P_{av}}{P_0}\right) (1 - f_N)_{av} \tag{13.2.53}$$

where we have recognized that changes in system pressure and temperature influence the reactant concentration through equation 3.1.44. We have, however, assumed that since the reactants are so dilute, the volume expansion parameter δ may be taken as approximately zero. Thus

$$\frac{\Delta f_N}{\Delta X} = \eta \frac{k_1 C_{N0}}{F_{N0}} \frac{T_0}{T_{av}} \frac{P_{av}}{P_0} \left(\frac{\pi D^2}{4}\right) (1 - f_N)_{av} \tag{13.2.54}$$

At any time the naphthalene concentration is

$$C_N = C_{N0} \left(\frac{T_0}{T}\right) \left(\frac{P}{P_0}\right) (1 - f_N) \tag{13.2.55}$$

Thus

$$\Delta C_N = \left(\frac{C_{N0} T_0}{P_0}\right) \left[\frac{P_{av}}{T_{av}} (-\Delta f_N) + (1 - f_N)_{av} \, \Delta \left(\frac{P}{T}\right)\right] \tag{13.2.56}$$

Equation 12.3.149 relates the changes in phthalic anhydride and naphthalene concentrations in the gas phase

$$\frac{\Delta C_{PA}}{\Delta C_N} = \frac{\mathscr{D}_{PA}}{\mathscr{D}_N} \left[\left(\frac{C_{PA}}{C_N} + \frac{k_1''/k_2''}{\frac{k_1'' \mathscr{D}_{PA}}{k_2'' \mathscr{D}_N} - 1}\right) \frac{\sqrt{\frac{k_2'' \mathscr{D}_N}{k_1'' \mathscr{D}_{PA}}} \tanh\left(h_T \sqrt{\frac{k_2'' \mathscr{D}_N}{k_1'' \mathscr{D}_{PA}}}\right)}{\tanh h_T} - \frac{k_1''/k_2''}{\left(\frac{k_1'' \mathscr{D}_{PA}}{k_2'' \mathscr{D}_N} - 1\right)}\right] \tag{13.2.57}$$

where C_{PA} and C_N are the local concentrations of phthalic anhydride and naphthalene, respectively. The diffusivity parameters take into account both knudsen diffusion and ordinary molecular diffusion. The inlet naphthalene concentration can be determined from the ideal gas law

$$C_{N0} - \frac{y_{N0} P_0}{R T_0} = \frac{0.0075 \left(\frac{25}{14.696}\right)}{82.06 T_0} = \frac{(1.555 \times 10^{-4})}{T_0} \text{ g moles/cm}^3 \tag{13.2.58}$$

and equations 13.2.56 and 13.2.57 then permit one to determine the naphthalene and phthalic anhydride concentrations at all points in the reactor by stepwise integration processes.

In order to use these equations, however, the local temperatures and pressures must be known. Equation 13.2.16 can be used to evaluate the latter parameter, while an energy balance on the reacting fluid gives the temperature. If the temperature difference between the bulk fluid and the external surface of the catalyst pellets is neglected for the reasons noted earlier, the energy balance on the reacting fluid can be written as:

Sensible heat change = energy gain by heat transfer through walls

+ energy effect accompanying reaction

$$G\left(\frac{\pi D^2}{4}\right) C_{pm}(T_{out} - T_{in}) = h_w \pi D(\Delta X)(T_W - T_{av}) - \rho_B \left(\frac{\pi D^2}{4}\right) \Delta X(r_{1m} \Delta H_1 + r_{2m} \Delta H_2)$$

(13.2.59)

where r_{1m} and r_{2m} are the rates of reactions 1 and 2 per unit mass of catalyst.

The rate of the first reaction is readily calculated from simple material balance considerations on our differential length element

$$F_N = (F_N + \Delta F_N) + r_{1m}\rho_B \frac{\pi D^2}{4} \Delta X$$

(13.2.60)

or

$$r_{1m} = \frac{-\Delta F_N}{\rho_B \left(\frac{\pi D^2}{4}\right) \Delta X} = \frac{F_{N0}(\Delta f_N)}{\rho_B \left(\frac{\pi D^2}{4}\right)(\Delta X)}$$

(13.2.61)

Now, at any point in the reactor,

$$\frac{dC_{PA}}{dC_N} = \frac{r_{1m} - r_{2m}}{-r_{1m}} = -1 + \frac{r_{2m}}{r_{1m}}$$

(13.2.62)

Thus,

$$r_{2m} = r_{1m}\left(\frac{dC_{PA}}{dC_N} + 1\right)$$

(13.2.63)

Thus,

$$r_{1m} \Delta H_1 + r_{2m} \Delta H_2 = r_{m1}\left[\Delta H_1 + \left(1 + \frac{dC_{PA}}{dC_N}\right)\Delta H_2\right]$$

$$= r_{m1}\left[(\Delta H_1 + \Delta H_2) + \Delta H_2 \frac{dC_{PA}}{dC_N}\right]$$

(13.2.64)

Combining equations 13.2.59, 13.2.61 and 13.2.64 gives

$$\frac{G\pi D^2}{4} C_{pm}(T_{out} - T_{in}) = h_w \pi D \Delta X(T_W - T_{av}) - F_{N0}(\Delta f_N)(\Delta H_1 + \Delta H_2) - F_{N0}(\Delta f_N)\frac{\Delta C_{PA}}{\Delta C_N} \Delta H_2$$

(13.2.65)

Equations 13.2.16, 13.2.54, 13.2.56, 13.2.57, and 13.2.65 constitute the set of relations that must be satisfied simultaneously within the volume element. One manner in which these equations may be employed in stepwise fashion to determine temperature and concentration profiles is as follows.

1. Calculate the average pressure characteristic of the length increment using the pressure drop equation 13.2.16 and property values characteristic of the previous length increment. This parameter is relatively insensitive to the slight changes in fluid temperature and pressure expected for normal operating conditions.
2. Assume a temperature characteristic of the fluid leaving the volume element.
3. Calculate an average temperature for the volume element.
4. Determine the effectiveness factor for the first reaction based on the average temperature.
5. Calculate the increment in the fraction naphthalene consumed using equation 13.2.54.
6. Calculate the changes in naphthalene and phthalic anhydride concentrations using equations 13.2.56 and 13.2.57.
7. Determine the temperature of the fluid leaving the volume element using the overall energy balance (13.2.65) and compare this result with the initial assumption.
8. Iterate as necessary until satisfactory agreement is obtained.

During the course of these calculations it is obviously necessary to employ property values characteristic of the volume element in question. Rate constants are extremely sensitive to temperature variations; diffusivities will vary as $T^{1/2}$ or $T^{3/2}$, depending on whether Knudsen or ordinary bulk diffusion dominates.

Numerical values may be substituted into the equations indicated in order to determine temperature and composition profiles along the length of the reactor. Obviously this is a task for machine computation, but it is a straightforward programming problem that can be handled by most students with some experience in this area. If differences in temperature between the bulk fluid and the external surface of the catalyst were considered, the programming task would be somewhat more difficult, but it still would be within the capabilities of many chemical engineering undergraduates.

The types of results that will be achieved are shown in Figures 13.6 and 13.7.

Figure 13.6 indicates that very large temperature gradients exist near the beginning of the bed and that the higher the inlet temperature, the greater the difference between this temperature and the maximum temperature achieved in the bed. Catalyst effectiveness factor profiles mirror the temperature profiles in an opposite

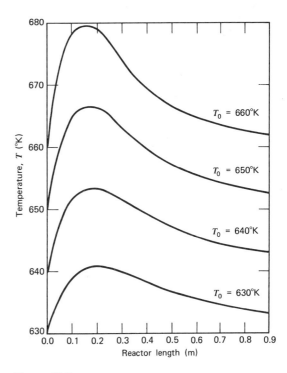

Figure 13.6

Temperature profiles for various inlet temperatures (wall temperature = inlet temperature).

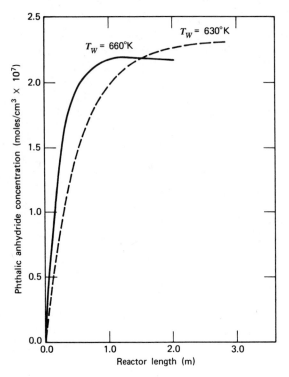

Figure 13.7
Phthalic anhydride concentration profile for two wall temperatures.

sense. The highest effectiveness factors are observed at the lowest temperatures and vice versa. For the conditions cited the effectiveness factors are all significantly less than unity, indicating the significance of intraparticle mass transfer limitations on observed conversion rates.

The yields predicted by the equations given above are considerably higher than would be expected in commercial reactors because of the simplifications we have made in the reaction kinetics. In industrial practice one expects yields to be around 0.85 lb phthalic anhydride/lb naphthalene fed. Typical reactor lengths for commercial scale facilities are about 5 m.

One could employ the techniques used in this example to investigate the effects of changes in tube diameter, catalyst pellet size, inlet pressure,

feed composition, number of tubes, etc., on the yields obtained and the optimum reactor length. One could also take into account temperature differences between the bulk fluid and the external surface of the catalyst or could employ a more sophisticated two-dimensional model of the reactor. What is most significant about the results of such calculations is not the absolute value of the numbers resulting from such analyses, but the steady-state response of the reactor to changes in process parameters. They permit one to study the relative influence on conversion and yield of variations in negotiable parameters such as feed temperature, pellet size, tube diameter, and Reynolds number, and thereby suggest to the reactor analyst measures that can be taken to enhance reactor performance.

Froment (9) and Carberry and White (7) have demonstrated that for the reactions in question the results obtained are sensitive to variations in the wall heat transfer coefficient, reaction activation energies, and feed composition. Consequently, one must regard the absolute numbers obtained from these analyses as suspect and must provide adequate safety factors in the design. The presence of both heat and mass transfer limitations on reaction rates introduces significant possibilities for error into the analysis, because the heat and mass transfer coefficients, effective diffusivities, and thermal conductivities are seldom known with a high degree of accuracy. Indeed, in systems such as this there is no substitute for experimental data (perhaps of the single tube type) to facilitate the design procedure.

I have found the above problem useful as the basis for a case study in a graduate course in chemical reactor design. The students (alone or in groups) carry out systematic studies of changes in process variables and in the physical configuration of the reactor. Their results serve as a framework for class discussions of the trade-offs one must make in attempting to develop a quasi-optimum reactor design.

LITERATURE CITATIONS

1. Fair, J. R., and Rase, H. F., *Chem. Eng. Prog.*, *50* (415), 1954.

2. Schutt, H. C., *Chem. Eng. Prog.*, *43* (103), 1947.

3. Reid, R. C., and Sherwood, T. K., *The Properties of Gases and Liquids*, Second Edition, Chapter 9, McGraw-Hill, New York, 1966.

4. McCabe, W. L., and Smith, J. C., *Unit Operations of Chemical Engineering*, p. 105, McGraw-Hill, New York, 1967.

5. Ennis, B. P., Boyd, H. B., and Orriss, R., *Chemtech*, *5* (693), 1975.

6. Westerterp, K. R., *Chem. Eng. Sci.*, *17* (423), 1962.

7. Carberry, J. J., and White, D., *Ind. Eng. Chem.*, *61*(7), p. 27, 1969.

8. DeMaria, R., Longfield, J. E., and Butler, G., *Ind. Eng. Chem.*, *53* (259), 1961.

9. Froment, G. F., *Ind. Eng. Chem.*, *59*(2), pp. 18–27, 1967.

10. Reid, R. C. and Sherwood, T. K., op. cit.

APPENDIX

 Thermochemical Data

The data in this appendix have been compiled from a number of sources. Nearly all of the critical property data are taken from Appendix A of *The Properties of Gases and Liquids*, Second Edition, by R. C. Reid and T. K. Sherwood, copyright © 1966, McGraw-Hill Book Company. They are used with the permission of McGraw-Hill Book Company. Most of the thermochemical data (ΔG_f^0, ΔH_f^0, and S^0) were obtained from the following sources.

1. Zwolinski, B. J., et al., "Selected Values of Properties of Hydrocarbons and Related Compounds," American Petroleum Institute Research Project 44, Thermodynamics Research Center, Texas A & M University,

College Station, Texas (Loose-leaf data sheets, extant, 1977).

2. Zwolinski, B. J., et al., "Selected Values of Properties of Chemical Compounds," Thermodynamics Research Center Data Project, Thermodynamics Research Center, Texas A & M University, College Station, Texas (Loose-Leaf data sheets, extant, 1977).

They are used with permission of the Thermodynamics Research Center. Other sources used include the *Handbook of Chemistry and Physics*, 57th edition, edited by R. C. Weast, CRC Press, Cleveland, 1974; and *Lange's Handbook of Chemistry*, 11th edition, edited by J. A. Dean, McGraw-Hill Book Company, New York, 1973.

Compound	State of aggregation	Molecular weight	$\Delta G_{f,25°C}^0$ kcal/g-mole	$\Delta H_{f,25°C}^0$ kcal/g-mole	$S_{25°C}^0$ Absolute entropy cal/g-mole-°K	T_c °K	P_c Atm
Elementary gases							
Hydrogen	Gas	2.016	0	0	31.211	33.3	12.80
Oxygen	Gas	32.000	0	0	49.003	154.8	50.1
Nitrogen	Gas	28.016	0	0	45.767	126.2	33.5
Fluorine	Gas	38.00	0	0	48.6	144.	55.
Chlorine	Gas	70.91	0	0	53.286	417.	76.1
Bromine	Liquid	159.83	0	0	36.4	584.	102.
Iodine	Crystal	253.82	0	0	27.9	785.	116.
Paraffins							
Methane	Gas	16.04	− 12.140	− 17.889	44.50	190.7	45.8
Ethane	Gas	30.07	− 7.860	− 20.236	54.85	305.4	48.2
Propane	Gas	44.09	− 5.614	− 24.820	64.51	369.9	42.0
n-Butane	Gas	58.12	− 4.10	− 30.15	74.12	425.2	37.5
Isobutane	Gas	58.12	− 5.00	− 32.15	70.42	408.1	36.0
n-Pentane	Gas	72.15	− 2.00	− 35.00	83.40	469.5	33.3
2-Methylbutane	Gas	72.15	− 3.50	− 36.92	82.12	460.4	32.9
Neopentane	Gas	72.15	− 3.64	− 39.67	73.23	433.8	31.6
n-Hexane	Gas	86.17	− 0.07	− 39.96	92.83	507.3	29.9
2-Methylpentane	Gas	86.17	− 1.20	− 41.66	90.95	496.5	30.0
3-Methylpentane	Gas	86.17	− 0.51	− 41.02	90.77	504.7	30.8

Compound	State of aggregation	Molecular weight	$\Delta G^0_{f,25\,C}$ kcal/g-mole	$\Delta H^0_{f,25\,C}$ kcal/g-mole	$S^0_{25\,C}$ Absolute entropy cal/g-mole-°K	T_c °K	P_c Atm
Olefins							
Ethylene	Gas	28.05	16.282	12.496	52.45	283.1	50.5
Propene	Gas	42.08	14.990	4.879	63.80	365.1	45.4
1-Butene	Gas	56.10	17.09	− 0.03	73.04	419.6	39.7
cis-2-Butene	Gas	56.10	15.74	− 1.67	71.90	434.6	40.5
trans-2-Butene	Gas	56.10	15.05	− 2.67	70.86	428.6	41.5
Isobutylene	Gas	56.10	13.88	− 4.04	70.17	419.7	39.5
1-Pentene	Gas	70.13	18.96	− 5.00	82.65	464.8	39.9
cis-2-Pentene	Gas	70.13	17.17	− 6.71	82.76	475.6	40.4
trans-2-Pentene	Gas	70.13	16.76	− 7.59	81.36	475.6	40.4
3-Methyl-1-butene	Gas	70.13	17.87	− 6.92	79.70	464.8	33.9
2-Methyl-2-butene	Gas	70.13	14.26	− 10.17	80.92	470.	34.
1-Hexene	Gas	84.16	20.94	− 9.96	91.93	504.0	31.1[a]
1-Octene	Gas	112.21	24.96	− 19.82	110.55	578.	25.5[a]
Cyclopentene	Gas	68.11	26.48	7.87	69.23	506.1	47.2[a]
Diolefins							
Propadiene	Gas	40.06	48.37	45.92	58.30	393.3	43.6[a]
1,3-Butadiene	Gas	54.09	36.43	26.75	66.62	425.	42.7
Acetylenes							
Acetylene	Gas	26.04	50.000	54.194	47.997	309.5	61.6
Propyne	Gas	40.06	46.313	44.319	59.30	401.	52.8
Ethylacetylene	Gas	54.09	48.30	39.48	69.51	463.7	37.5[a]
Dimethylacetylene	Gas	54.09	44.32	34.97	67.71	488.7	37.5[a]
Cycloparaffins							
Cyclopentane	Gas	70.13	9.29	− 18.41	70.00	511.8	44.6
Methylcyclopentane	Gas	84.16	8.55	− 25.50	81.24	532.7	37.4
Ethylcyclopentane	Gas	98.18	10.66	− 30.37	90.42	569.5	33.5
Cyclohexane	Gas	84.16	7.60	− 29.43	71.28	553.2	40.
Methylcyclohexane	Gas	98.18	6.52	− 36.99	82.06	572.1	34.3
Aromatics							
Benzene	Gas	78.11	30.989	19.820	64.34	562.1	48.6
Toluene	Gas	92.13	29.228	11.950	76.42	592.0	41.6
o-Xylene	Gas	106.16	29.177	4.540	84.31	631.6	35.7
m-Xylene	Gas	106.16	28.405	4.120	85.49	616.8	34.7
p-Xylene	Gas	106.16	28.952	4.290	84.23	618.8	33.9
Ethylbenzene	Gas	106.16	31.208	7.120	86.15	617.1	36.9
Alcohols							
Methyl alcohol	Gas	32.04	− 38.81	− 48.05	57.29	513.2	78.5
	Liquid		− 39.85	− 57.11	30.4		
Ethyl alcohol	Gas	46.07	− 40.13	− 56.03	67.54	516.3	63.0
	Liquid		− 41.65	− 66.20	38.53		

Compound	State of aggregation	Molecular weight	$\Delta G^0_{f, 25°C}$ kcal/g-mole	$\Delta H^0_{f, 25°C}$ kcal/g-mole	$S^0_{25°C}$ Absolute entropy cal/g-mole-°K	T_c °K	P_c Atm
n-Propyl alcohol	Gas	60.09	− 38.75	− 61.33	77.70	536.7	51.0
	Liquid		− 40.94	− 72.66	47.0		
Isopropyl alcohol	Gas	60.09	− 41.44	− 65.08	74.1	508.2	47.0
	Liquid		− 43.10	− 75.98	43.15		
n-Butyl alcohol	Gas	74.12	− 36.11	− 65.65	86.92	563.0	43.6
	Liquid		− 39.00	− 78.18	54.5		
sec-Butyl alcohol	Gas	74.12	− 42.33	− 70.00	85.8	536.0	41.4
	Liquid		− 40.12	− 81.86	53.4		
tert-Butyl alcohol	Gas	74.12	− 42.45	− 74.68	77.92	506.2	39.2
Isobutyl alcohol	Gas	74.12	− 40.12	− 70.00	85.8	547.7	42.4
	Liquid		− 42.33	− 81.06	53.4		
Phenol	Gas	94.11	− 7.61	− 22.98	75.43	694.3	60.5
	Liquid		− 12.17	− 39.44	34.9		
Ethers							
Dimethylether	Gas	46.07	− 26.96	− 43.99	63.74	400.1	52.6
Diethylether	Liquid	74.12	− 29.32	− 66.83	60.2	465.8	35.6
Isopropylether	Gas	102.17	− 28.5	− 76.24	91.2	500.1	28.4
	Liquid		− 30.09	− 84.00	70.4		
Ketones							
Acetone	Gas	58.08	− 36.50	− 51.79		509.1	47.
	Liquid		− 37.22	− 59.32			
Aldehydes							
Formaldehyde	Gas	30.03	− 27	− 28	52.26		
Acetaldehyde	Gas	44.05	− 30.81	− 39.72	59.8	461.	54.7[a]
Organic Acids							
Acetic acid	Liquid	60.05	− 93.8	−116.4	38.2	594.8	57.1
	Gas		− 91.24	−104.72			
Acetic anhydride	Gas	102.09	−119.29	−148.82		569.2	46.2
	Liquid		−121.75	−155.16			
Propionic acid	Gas	74.08	− 88.27	−108.75		612.7	53.0
	Liquid		− 91.65	−121.7			
Esters							
Methyl formate	Gas	60.05	− 71.37	− 83.7	29	487.2	59.2
	Liquid		− 71.53	− 90.60			
Ethyl acetate	Gas	88.10	− 74.93	−102.02		523.3	37.8
	Liquid		− 76.11	−110.72			
Ethyl propionate	Gas	102.13	− 77.37	−112.36		546.1	33.0
	Liquid		− 79.16	−122.16			
Nitrogen compounds							
Ammonia	Gas	17.03	− 3.94	− 11.02	45.97	405.6	112.5
Cyanogen	Gas	52.02	70.81	73.60	57.86	400.	60.0
Hydrogen cyanide	Gas	27.03	28.7	31.2	48.23	456.7	48.9
Methyl amine	Gas	31.06	+ 6.6	− 6.7	57.73	430.2	73.1

Compound	State of aggregation	Molecular weight	$\Delta G^0_{f,25°C}$ kcal/g-mole	$\Delta H^0_{f,25°C}$ kcal/g-mole	$S^0_{25°C}$ Absolute entropy cal/g-mole-°K	T_c °K	P_c Atm
Dimethyl amine	Gas	45.08	16.35	− 4.41	65.24	437.8	52.4
Ethyl amine	Gas	45.08	10.01	− 12.24		456.5	55.5
Propyl amine	Gas	59.11	14.38	− 16.45		497	46.8
Nitro methane	Liquid	61.04	− 2.26	− 21.28	41.1	588.	62.3
Sulfur compounds							
Sulfur	Crystal	32.06	0	0	7.62	1313.	116.
Carbonyl sulfide	Gas	60.07	− 40.85	− 33.83		378.	65.
Carbon disulfide	Gas	76.13	15.55	27.55	56.84	552.	78.
	Liquid		15.2	21.0	36.10		
Hydrogen sulfide	Gas	34.08	− 7.892	− 4.815	49.15	373.6	88.9
Dimethyl sulfide	Gas	62.13	3.7	− 6.9	68.28	503.1	54.6
Methyl mercaptan	Gas	48.10	− 2.23	− 5.34	60.90	470.	71.4
Inorganic halides							
Hydrogen fluoride	Gas	20.01	− 64.7	− 64.2	41.47	461.	64 ± 4
Hydrogen chloride	Gas	36.49	− 22.769	− 22.063	44.617	324.6	82.1
Hydrogen bromide	Gas	80.92	− 12.72	− 8.66	47.437	363.2	84.5
Hydrogen iodide	Gas	127.93	0.31	6.20	49.314	424.2	81.9
Organic halides							
Methyl chloride	Gas	50.49	− 14.0	− 19.6	55.97	416.3	65.9
Methyl bromide	Gas	94.95	− 6.2	− 8.5	58.74	467.2	83.4[a]
Methyl iodide	Gas	141.95	5.3	4.9	60.85	528.	72.7[a]
Methylene chloride	Gas	89.94	− 14.	− 21.	64.68	510.	60.
Chloroform	Gas	119.39	− 16.	− 24.	70.86	536.6	54.
	Liquid		− 17.1	− 31.5	48.5		
Carbon tetrachloride	Gas	153.84	− 15.3	− 25.50	73.95	556.4	45.0
	Liquid		− 16.4	− 33.3	51.25		
Ethyl chloride	Gas	64.52	− 12.7	− 25.1	65.90	460.4	52.0
1,2-Dichloroethane	Liquid	98.97	− 19.2	− 39.7	49.84	561.	
Oxides							
Carbon dioxide	Gas	44.01	− 94.2598	− 94.0518	51.061	304.2	72.9
Carbon monoxide	Gas	28.01	− 32.8077	− 26.4157	47.300	133.	34.5
Ethylene oxide	Gas	44.05	− 2.79	− 12.19	58.1	468.	71.0
Nitrous oxide	Gas	44.02	24.76	19.49	52.58	309.7	71.7
Nitric oxide	Gas	30.01	20.719	21.600	50.339	180.	64.
Nitrogen dioxide	Gas	46.01	12.390	8.091	57.47	431.	100.
Sulfur dioxide	Gas	64.06	− 71.79	− 70.96	59.40	430.7	77.7
Sulfur trioxide	Gas	80.06	− 88.52	− 94.45	61.24	491.4	81.4
Water	Gas	18.02	− 54.6351	− 57.7979	45.106	647.	218.3
	Liquid		− 56.6899	− 68.3174	16.716		

[a] Calculated by Lydersen's Critical Property Method. Reid and Sherwood, *The Properties of Gases and Liquids*, Second Edition, McGraw-Hill, 1966, Table 2-1, p. 9.

Generalized Fugacity Coefficients of Pure Gases and Liquids $(Z_c = 0.27)$

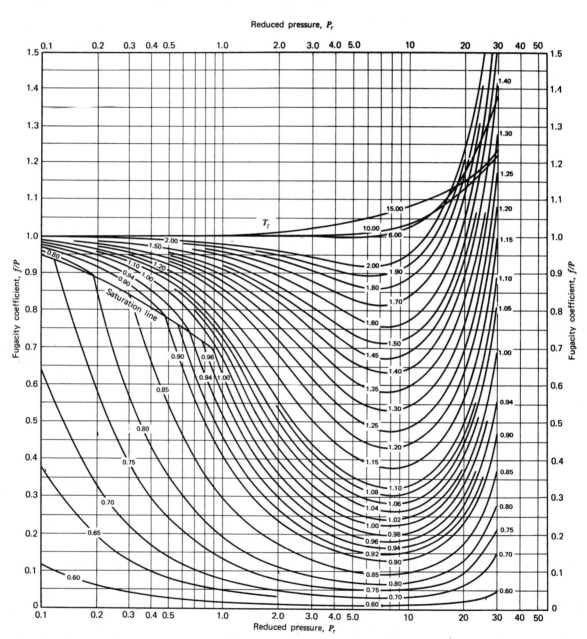

Reduced pressure, P_r

Fugacity coefficient, f/P

Reduced pressure, P_r

(Reprinted from *Chemical Process Principles Charts*, **Third Edition, by O. A. Hougen, K. M. Watson, and R. A. Ragatz. Copyright 1964. Reprinted by permission of John Wiley and Sons, Inc.)**

APPENDIX

C Nomenclature

Symbol	Meaning
a	activity
a	external (superficial) surface area
A	pre-exponential factor in rate constant
A_i	representative chemical species
C	concentration
C_{Ai}	concentration of rectant A at time t_i or in effluent from reactor i
C^*	reactant concentration at time zero in a CSTR operating under transient conditions
C_p	constant pressure heat capacity
D	diameter
D	dielectric constant
D_{AB}	bulk diffusivity
D_K	Knudsen diffusivity
$\mathscr{D}$	dispersion or diffusivity parameter
$\mathscr{D}_{Am}$	pseudo-binary diffusivity of species A in a multicomponent gas mixture
$\mathscr{D}_c$	combined diffusivity
E	energy or activation energy
E_c	relative kinetic energy directed along the line of centers in a collision (on a per mole basis)
$E_{\text{Diffusion}}$	activation energy for diffusion defined by equation 12.3.84
E_0	energy increase accompanying reaction at absolute zero
E_0	initial enzyme concentration
f	fraction conversion
f_i^0	fugacity of pure species i at pressure P
$\hat{f}_i$	fugacity of species i as it exists in the reaction mixture
f_M	friction factor
$(f/P)_i$	fugacity coefficient for species i
F	molal flow rate
F'_A	hypothetical molal flow rate of species A corresponding to a stream in which none of the A has reacted
$F(t)$	function used in characterization of fluid residence times (see Section 11.1)
$\mathscr{F}$	ratio of reaction rate for poisoned catalyst to that for unpoisoned catalyst
G	Gibbs free energy
G	mass velocity
h	specific enthalpy
h_c	convective heat transfer coefficient
h_i	Thiele modulus for ith order reaction
h_p	Thiele modulus for poisoned catalyst
h_T	Thiele modulus
j_D	Chilton-Colburn factor defined by equation 12.4.9
j_H	Chilton-Colburn factor defined by equation 12.5.1
J_i	Molar flux of species i relative to the molar average velocity
k	reaction rate constant
k	thermal conductivity
k_B	Boltzmann constant
k_c	mass transfer coefficient
k_f	thermal conductivity of bulk fluid
k_G	mass transfer coefficient defined by equation 12.4.2
k_{H^+}	rate constant for acid catalyzed reaction
k_m	mass transfer coefficient
k_0	rate constant in infinitely dilute solution or for uncatalyzed reaction
k_{OH^-}	rate constant for base catalyzed reaction
k'	pseudo rate constant for desorption process
K	equilibrium constant
K	Michaelis constant
K_a	equilibrium constant for reaction expressed in terms of activities
K_i	adsorption equilibrium constant
K_p	equilibrium constant expressed in terms of partial pressures
K_r	equilibrium constant for surface reaction expressed in terms of the fractional surface coverages θ_i

K_w	dissociation constant for water	Q_g	rate at which thermal energy is released by an exothermic chemical reaction
L	reactor length		
$\bar{L}$	average pore length		
$\mathscr{L}$	average chain length	Q_r	rate at which energy is removed from a system
ℓ	parameter involved in definition of termolecular collisions	$\dot{Q}$	heat transfer rate from surroundings to system
m	mass		
m	overall order of the reaction	r	radius
m_T	mass of tracer injected as a pulse input	r	reaction rate
$\dot{m}$	mass flow rate	$-r_A$	rate of disappearance of species A
M	molecular weight	$-r_{Ai}$	rate of disappearance of species A in CSTR i
n	number of stirred tank reactors in a cascade	r_i	rate of appearance of species i
n	number of permissible adsorption layers	r_v	reaction rate employed in pseudo homogeneous models of packed bed reactors (See Section 12.7.2.)
n	reaction order		
n_i	moles of species i	r_V	reaction rate in a constant volume system
n_i'	number density of molecules of species i		
		r_{xy}	interatomic separation distance between atoms x and y
n_p	number of pores per catalyst particle		
N	molar flux	$\bar{r}$	average pore radius
N	number of CSTR reactors in cascade	R	gas constant
N_i	number of moles of species i contained within a reactor	R	radius
		R	recycle ratio defined by equation 8.3.63
N_0	Avogadro's number		
N_{Pe}	Peclet number	S	number of squared terms contributing to the activation energy of a reaction
N_r	diffusive flux in radial direction		
N_{Re}	Reynolds number	S	selectivity
N_{Sc}	Schmidt number	S	space velocity
N_{Sh}	Sherwood number	S	surface area
P	pressure	S_g	specific surface area of catalyst
P_c	critical pressure	S_0	initial substrate concentration
P_0	saturation pressure or pressure at reactor inlet	S_x	gross geometric surface area of catalyst pellet
P_s	steric probability factor	t	time
$\mathscr{P}$	parameter defined by equation 12.7.3	t_f	time for a given fraction conversion
$\wp$	transfer function variable (equation 11.1.52)	t_i	time at which reactant concentration C_{Ai} is measured
q	charge on an electron	t_s	shutdown time in batch reactor
q	energy parameter defined by equation 4.1.37	$\bar{t}$	average residence time
		t^*	relaxation time defined by equation 5.1.51
q	heat transfer rate		
$\dot{q}$	heat flux	$\bar{t}^*$	ratio of reactor length to average linear velocity
Q	heat transferred from surroundings to system		
		T	temperature (absolute)
Q	partition function	T_c	temperature at center of catalyst pellet

	or critical temperature	α	dimensionless concentration variable defined in Section 5.4.2
T_m	temperature of heat source or sink		
u	linear velocity	α	area covered per adsorbed molecule
U	overall heat transfer coefficient	α	fraction of catalyst surface which is poisoned
v	volume of gas adsorbed		
v_m	volume of gas adsorbed in a monolayer	β	dimensionless concentration variable for competitive consecutive reactions (defined by equation 5.4.18)
$\bar{v}$	average molecular velocity		
$\vec{v}^{\neq}$	velocity with which activated complexes move from left to right across the transition state	β	energy generation function defined by equation 12.3.108
		β_i	order of the reaction with respect to species i
V	velocity of an enzymatic reaction		
V	volume	γ	activity coefficient
V_g	void volume per gram of catalyst	γ	Arrhenius number defined by equation 12.3.107
V_p	gross geometric volume of catalyst pellet	δ_A	volumetric expansion parameter defined by equation 3.1.41
V_R	reactor volume		
V'	volume of solid catalyst	δ^*	reaction progress variable for consecutive reactions (defined by equation 5.3.15)
$\mathscr{V}$	volumetric flow rate		
$\mathscr{V}_0$	volumetric flow rate at inlet of reactor network	$\delta^{\neq}$	characteristic length dimension of the transition state
w	weight fraction	Δ	Time separating rate measurements (see Sections 3.3.2.2 and 3.3.2.4)
W	weight of solid catalyst		
W	work done by system on surroundings	ΔE_{total}	change in total energy
$\dot{W}_s$	rate at which shaft work is done by system	ΔG	Gibbs free energy change
		ΔH	enthalpy change
x	distance from pore mouth	ΔS	entropy change
x	normalized pressure (P/P_0)	$\Delta \xi^*$	deviation from equilibrium conditions defined by equation 5.1.45
x_c	distance from pore mouth at which reactant concentration vanishes		
		ε_B	bed porosity
y	instantaneous fractional yield defined by equation 9.1.7	ε_c	relative kinetic energy directed along the line of centers in a collision
y	mole fraction	ε_p	porosity of pellet or particle
Y	yield	$\varepsilon_{\text{Total}}$	total porosity of packed bed
Y'_R	overall fractional yield of species R defined by equation 9.0.2	η	catalyst effectiveness factor
		θ_i	fraction of catalyst surface covered by species i
Z	distance from tubular reactor inlet		
Z_{AB}	bimolecular collision frequency for molecules A and B	θ_v	fraction of surface sites that are vacant
		κ	conductivity
Z_{ABC}	termolecular collision frequency	κ	ratio of rate constants for consecutive reactions defined by equation 5.3.13
Z_i	number (and sign) of charges on the ion i		
		λ	generalized physical property
		λ	mean free path

Greek Symbols

α	branching coefficient in chain reaction mechanism	μ	viscosity
		μ	ionic strength

μ_i	chemical potential of species i
μ_{AB}	reduced mass defined by equation 4.3.3
v	stoichiometric coefficient
ξ	extent of reaction
ξ^*	extent of reaction per unit volume in constant volume systems
$\prod_i$	this symbol indicates a product of the i terms that follow
ρ	density
ρ_B	bulk density of catalyst
ρ_i	Hammett reaction constant
σ	surface site
σ	emissivity of solid
σ_i	hard sphere diameter for molecule i
σ_i	Hammett substituent constant
σ_{AB}	effective hard sphere diameter for bimolecular collision (equation 4.3.2)
$\sigma_t^{\,2}$	variance of residence time distribution curve defined in Illustration 11.2
$\sum$	summation of the terms that follow
τ	reactor space time
τ_c	space time for a cascade of CSTR reactors
τ_i	space time for reactor i
τ_N	proportionality constant introduced in equation 7.4.14
τ_p	space time for a plug flow reactor
$\tau_{1/2}$	reaction half life
τ'	tortuosity factor
τ^*	dimensionless time variable for competitive consecutive reactions (defined by equation 5.4.18)
ϕ	parameter defined by equation 5.4.13
$\phi(C_i)$	concentration dependent portion of reaction rate expression (see equation 3.0.13)
ϕ_{Ln}	Thiele modulus defined by equation 12.3.74
ϕ_s	Thiele modulus defined by equation 12.3.56
Φ_m	mass flow rate
$\psi(C_i)$	function defined by equations 3.3.17 and 3.3.18

Subscripts

A	generalized chemical species A
B	bulk fluid property
e	equilibrium
eff	effective
E	effluent stream property or equilibrium property
ES	external surface of catalyst pellet
f	refers to reactions involving formation of a compound from its elements
f	refers to forward reaction
F	property evaluated at final or effluent conditions
g	gas
i	refers to species i, reaction i, or reactor i in a CSTR network
in	refers to inlet stream property
lim	refers to limiting reagent
L	property value at end of catalyst pore or in longitudinal direction
m	property value per unit mass of catalyst
max	maximum or extremum value
MT	mass transfer
n	property value leaving reactor n
out	refers to outlet stream property
p	pellet or particle value
PE	pseudo equilibrium
r	refers to reverse reaction
R	reaction or property value in radial direction
RLS	rate limiting step
S	solid
SS	steady state
sys	system
T	tube
v	property value per unit volume of catalyst bed
w	wall
z	property value in longitudinal direction
0	property value at time zero, reactor inlet, or at mouth of catalyst pore
∞	value at infinite time

Superscript

0   refers to variables associated with
           standard states of materials

$''$ per unit surface area

$'''$ per unit volume of solid catalyst

$\neq$ refers to variable associated with
           transition state

$\wedge$ average property value

Name Index

Subject Index

Subject index

Subject index